FORMULAS FROM GEOMETRY

Triangle

$h = a \sin \theta$

$\text{Area} = \dfrac{1}{2}bh$

(Law of Cosines)

$c^2 = a^2 + b^2 - 2ab \cos \theta$

Right Triangle

(Pythagorean Theorem)

$c^2 = a^2 + b^2$

Equilateral Triangle

$h = \dfrac{\sqrt{3}s}{2}$

$\text{Area} = \dfrac{\sqrt{3}s^2}{4}$

Parallelogram

$\text{Area} = bh$

Trapezoid

$\text{Area} = \dfrac{h}{2}(a + b)$

Circle

$\text{Area} = \pi r^2$

$\text{Circumference} = 2\pi r$

Sector of Circle

(θ in radians)

$\text{Area} = \dfrac{\theta r^2}{2}$

$s = r\theta$

Circular Ring

(p = average radius,
w = width of ring)

$\text{Area} = \pi(R^2 - r^2)$

$\quad\quad = 2\pi pw$

Sector of

(p = avera
w = widt
θ in radia
$\text{Area} = \theta pw$

Ellipse

$\text{Area} = \pi ab$

$\text{Circumference} \approx 2\pi \sqrt{\dfrac{a^2 + b^2}{2}}$

Cone

(A = area of base)

$\text{Volume} = \dfrac{Ah}{3}$

Right Circular Cone

$\text{Volume} = \dfrac{\pi r^2 h}{3}$

$\text{Lateral Surface Area} = \pi r \sqrt{r^2 + h^2}$

Frustum of Right Circular Cone

$\text{Volume} = \dfrac{\pi(r^2 + rR + R^2)h}{3}$

$\text{Lateral Surface Area} = \pi s(R + r)$

Right Circular Cylinder

$\text{Volume} = \pi r^2 h$

$\text{Lateral Surface Area} = 2\pi rh$

Sphere

$\text{Volume} = \dfrac{4}{3}\pi r^3$

$\text{Surface Area} = 4\pi r^2$

Wedge

(A = area of upper face,
B = area of base)

$A = B \sec \theta$

Calculus

CALCULUS

OF A SINGLE VARIABLE

Fourth Edition

ROLAND E. LARSON

ROBERT P. HOSTETLER

The Pennsylvania State University
The Behrend College

BRUCE H. EDWARDS

University of Florida

with the assistance of DAVID E. HEYD
The Pennsylvania State University
The Behrend College

D. C. HEATH AND COMPANY
Lexington, Massachusetts   Toronto

Chapter Opener Photo Credits

Chapter 1: Bettmann Archive.
Chapter 2: Ray Ellis/Photo Researchers.
Chapter 3: Dan Helms/duomo.
Chapter 4: Mark Antman/The Image Works.
Chapter 5: NASA/Johnson Space Center.
Chapter 6: Lionel Delevingne/Stock, Boston.
Chapter 7: Focus on Sports.
Chapter 8: NASA/Johnson Space Center.
Chapter 9: Focus on Sports.
Chapter 10: Kent and Donna Dannen/Photo Researchers.
Chapter 11: NASA/Johnson Space Center.

Acquisitions Editors: Mary Lu Walsh and Ann Marie Jones

Developmental Editor: Cathy Cantin

Production Editor: Laurie Johnson

Designer: Sally Steele

Production Coordinator: Michael O'Dea

Photo Researcher: Wendy Johnson

Composition: Jonathan Peck Typographers

Technical Art: Folium

Cover: Martucci Studio

Published simultaneously in Canada.

Printed in the United States of America.

International Standard Book Number: 0-669-24591-7

Library of Congress Catalog Card Number: 88-83340

10 9 8 7 6 5 4 3 2 1

Preface

Calculus of a Single Variable, Fourth Edition, is designed for use in a calculus course for students in engineering, economics, life sciences, mathematics, and the physical sciences. In writing the text, we were guided by two primary objectives that have crystallized over many years of teaching calculus. For the student, our objective was to write in a precise, readable manner and to clearly define and demonstrate the basic concepts of calculus. For the instructor, our objective was to design a comprehensive teaching instrument that uses proven pedagogical techniques, thus freeing the instructor to make the most efficient use of classroom time.

Changes in the Fourth Edition

In the Fourth Edition, all examples, theorems, definitions, and prose have been revised—or at least considered for revision. Many new examples, exercises, and applications have been added to the text. The major changes in the Fourth Edition, listed by chapter, are as follows:

- The basic organization of Chapter 1 is the same as in the Third Edition. Section 1.4 contains a new summary of Equations of Lines, and Section 1.5 contains a new summary of the graphs of basic functions.
- Chapter 2 begins with a new introduction to the tangent line problem. Section 2.1 has a new example dealing with the ε-δ definition applied to a quadratic function. Section 2.3 begins with a new summary of a strategy for finding limits.
- The "Four-Step Process" terminology has been eliminated from Section 3.1 in favor of simply applying the definition of the derivative. The introduction to the Chain Rule has been rewritten.
- Section 4.3 on the First Derivative Test has been reorganized. The section on business and economics applications has been moved to the end of the chapter.
- At the request of several users of the text, Chapter 5 has been reorganized. In an effort to get to the Fundamental Theorem of Calculus earlier in the

chapter, we moved the section on integration by substitution to Section 5.5. The material on sigma notation was condensed and now appears as the first part of Section 5.2. Numerical integration was moved from Chapter 8 to Section 5.6, and the introduction of the natural logarithmic function was postponed to Chapter 6.

- Chapter 6 has also been reorganized. It now begins with an introduction to the natural logarithmic function. There is a separate section on bases other than *e* and applications (Section 6.5) and a separate section on growth and decay (Section 6.7).
- In Chapter 7, we moved the material on arc length and surface area to Section 7.4, preceding the section on work.
- The section on numerical integration was moved from Chapter 8 to Chapter 5. In Section 8.2, the tabular method of repeated integration by parts was added to the text.
- At the request of many users of the Third Edition, the introduction to Taylor polynomials, which was in the first section of Chapter 9, has been moved to Section 9.7.
- In Chapter 10, the material on classifying conics was moved from Section 10.3 to Section 10.4.
- In Chapter 11, the section on area and arc length in polar coordinates now occurs before the section on polar equations for conics and Kepler's Laws.

Features

Order of Topics   The eleven chapters readily adapt to either semester or quarter systems. In each system both differentiation *and* integration can be introduced in the first course of the sequence. There is some flexibility in the order and depth in which the chapters can be covered. For instance, much of the precalculus material in Chapter 1 can be used as individual review. The ε-δ discussion of limits in Chapter 2 can be given minimal coverage. Sections 7.6 and 7.7 can be covered later in the course. Chapter 9 can be covered at any time after Chapter 8.

Definitions and Theorems   Special care has been taken to state the definitions and theorems simply, without sacrificing accuracy.

Proofs   We have chosen to include only those proofs that we have found to be both instructive and within the grasp of a beginning calculus student. Moreover, in presenting proofs, we have found that extensive detail often obscures rather than illuminates. For this reason, many of the proofs are presented in outline form, with an emphasis on the essence of the argument. (See the proof of the Product Rule in Section 3.4.) In some cases, we have included a more complete discussion of proofs in Appendix A. (See the proof of the Chain Rule in Section 3.5.)

Graphics   The Fourth Edition has over 1600 figures. Of these, over 700 are in the examples and exposition, over 480 are in the exercise sets, and over 430 are in the odd-numbered answers. The new art program in the Fourth Edition was computer generated for accuracy. Designed with additional colors used systematically for clarity, this improved art program will help students better visualize mathematical concepts, particularly in the presentation of complex, three-dimensional material. For example, axes and primary graphs are brown, while planes are blue and primary three-dimensional surfaces are red.

Enhanced Presentation As a new feature, the Fourth Edition has been designed with a functional use of four colors that strengthens the text as a pedagogical tool. Each color is used consistently to aid both reading and reference. For example, all theorems and definitions are highlighted by brown boxes, and equation side comments are given in red.

Exercises Over 650 new problems have been added, so the text now contains over 5000 exercises. The exercises are graded, progressing from skill-development problems to more challenging problems involving applications and proofs. Many exercise sets begin with a group of exercises that provides the graphs of functions involved. Review exercises are included at the end of each chapter.

Examples The text contains over 500 examples, each titled for easy reference. Many of the examples include red side comments that clarify the steps of the solution.

Computer/Calculator Exercises Another new feature in the Fourth Edition is the addition of exercises involving the use of a computer or graphics calculator. See, for instance, Exercises 69 and 70 in Section 1.3 or Exercises 79–82 in Section 1.6.

Numerical Methods With the increasing power and accessibility of computers, numerical techniques are becoming more widely used. This edition reflects this trend by introducing numerical integration earlier in the text, and by adding a tabular method of integration by parts. Calculators or computers are useful in these areas as well as with topics such as limits, Newton's Method, and Taylor polynomials.

Chapter Introductions As a new feature in the Fourth Edition, each chapter begins with a chapter overview and a special motivational application.

Applications We chose applications that offer variety and integrity, and that require a minimal knowledge of other fields. The Fourth Edition contains over 2000 different applications.

Summaries Many sections have summaries that identify core ideas and procedures—see Sections 3.5, 4.8, 6.5, and 9.7. In some instances, an entire section summarizes the preceding topics—see Sections 4.6 and 8.1.

Historical Notes Throughout the text, we include several short biographical notes about prominent mathematicians. These are designed to help students gain an appreciation of both the people involved in the development of calculus and the nature of the problems that calculus was designed to solve.

Remarks The text contains many special instructional notes to students in the form of "Remarks." These notes appear after definitions, theorems, or examples and are designed to give additional insight, help avoid common errors, or describe generalizations.

Supplements

For Students

- The *Study and Solutions Guide* by David E. Heyd contains detailed solutions to several representative problems from each exercise set. In the text, these exercises are identified in the exercise sets by blue numbers. The solutions to these exercises are given in greater detail than the examples in the text, with special care taken to show the algebra involved. In addition, the *Study and Solutions Guide* contains a review of algebra and a set of true-false questions for each chapter in the text.
- *The Algebra of Calculus* by Eric Braude provides additional calculus review. Keyed to the text, it offers a review of calculus topics, examples with solutions, related problems and their answers, and exercises with answers in the back of the book.

For Instructors

- A *Complete Solutions Guide* by Dianna L. Zook is available in three volumes. This guide contains brief solutions to every exercise in the text, including exercises requiring proofs.
- An *Instructor's Guide* by Ann R. Kraus contains, for each chapter in the text, sample tests, concept reviews in the form of true-false questions, and suggestions for classroom instruction.
- A package of color transparencies of figures from the text is available.

For Students and Instructors

- The *Calculus Applications in Engineering and Science* supplement contains solved examples and exercises (with answers to the odd-numbered ones) covering applications in engineering, physics, chemistry, biology, and other fields.

Technology

Software

A library of software products for this text is available for the IBM-PC, Macintosh, and Apple II. To make the most of the software products described below, instructors or students may choose to solve appropriate exercises from the text and those provided in a User Manual as well as analyze functions of the user's own selection. All of the following products can be used for classroom demonstration and have been class tested.

- *Math Lab Calculus*, The Math Lab (IBM-PC, Apple II)
  Offering both two-dimensional and three-dimensional computer graphics and numerical computations, the package consists of one disk and a User

Manual with 80 lab assignments. These assignments are presented in work-sheet format and are keyed to the text. Additionally, all computer/calculator exercises in the text can be solved using this software.

- *Math Utilities* (updated version 4.0), Bridge Software (IBM-PC)
  This is a program of three powerful graphing packages: CURVES, SURFS (Surfaces), and DIFFS (Differential Equations). Each package includes one disk and a User Manual. These enhanced graphing packages allow the user to save graphs and offer labelling and annotating capabilities for creating classroom handouts.
- *BestGrapher*, George Best (IBM-PC, Macintosh)
  This software graphs both two- and three-dimensional figures as well as offers some features that help teach concepts through simulation, such as rotation of a curve about an axis. A lab package that accompanies the software provides assignments in a worksheet format keyed to the text.
- *Computer Activities for Calculus*, Technology Training Associates (IBM-PC)
  This package includes three disks and a User Manual. Each easy-to-use disk has two units of tutorials for self-study and the Grapher, a graphing tool for two-dimensional figures that is designed to handle a wide variety of elementary functions.
- *TrueBASIC CALCULUS*, *TrueBASIC* (IBM-PC) and *TrueBASIC MULTI-VARIATE CALCULUS: MACFUNCTION* (Macintosh)
  Each package includes one disk and a User Manual. *TrueBASIC CALCULUS* performs numerical routines and two-dimensional graphing for topics in the first two semesters of calculus. *MACFUNCTION* is a tool for plotting and examining three-dimensional graphs of functions.
- *Workout* for Calculus, Fourth Edition, Joseph Mazur (IBM-PC)
  This tutorial software offers exercises with on-line hints and guidance for student self-study. Keyed on screen to the text, the software generates problems and then adjusts the difficulty of the exercises with sensitivity to the student's level of accomplishment.
- *Derive*, Soft Warehouse (IBM-PC)
  Special discounts on this computer algebra software are available upon adoption of the text.
- *Calculus*, Broderbund (Macintosh)
  Special discounts on this tutorial software are available upon adoption of the text.

Videotapes

- *Calculus Videotapes*, Video Tutorial Services, Inc.
  Interactive videotapes combine thorough coverage of major topics with state-of-the-art computer graphics. Keyed to the text, these videotapes will help students practice, learn, and review calculus. A *Study Guide* and an *Instructor's Guide* are also available.

Calculators

- *HP28S, HP48SX, Casio FX7000G, Casio FX7500G*
  Special discounts on these calculators are available upon adoption of the text.

Computerized Testing

- *HeathTest Plus for Calculus* (IBM-PC, Apple II, Macintosh)
  Instructors can produce chapter tests, mid-terms, and final exams easily and accurately. Instructors can also edit existing questions or add new ones as desired, or preview questions on screen and add them to a test with a single keystroke. The software supports graphics and offers both multiple-choice and open-ended questions. A User Manual and a printed test item file are available.
- *HeathTest for Calculus* (IBM-PC)
  This is an algorithm-based program that generates tests, quizzes, or worksheets in a multiple-choice format. A User Manual with printed test items accompanies the software.

Acknowledgments

We would like to thank the many people who have helped us at various stages of this project during the past seventeen years. Their encouragement, criticisms, and suggestions have been invaluable to us.

Fourth Edition Focus Group Homer F. Bechtell, University of New Hampshire; K. Elayn Gay, University of New Orleans; Hideaki Kaneko, Old Dominion University; Judith A. Palagallo, University of Akron; John Tweed, Old Dominion University

Fourth Edition Reviewers Keith Bergeron, United States Air Force Academy; Jorge Cossio, Miami-Dade Community College; Rosario Diprizio, Oakton Community College; Ali Hajjafar, University of Akron; Ransom Van B. Lynch, Phillips Exeter Academy; Bennet Manvel, Colorado State University; Duff A. Muir, United States Air Force Academy; Charlotte J. Newsom, Tidewater Community College; Terry J. Newton, United States Air Force Academy; Wayne J. Peeples, University of Texas; Jorge A. Perez, LaGuardia Community College; Barry J. Sarnacki, United States Air Force Academy; George W. Schultz, St. Petersburg Junior College; Frank Soler, De Anza College; Michael Steuer, Nassau Community College; John Tweed, Old Dominion University; Jay Wiestling, Palomar College; August J. Zarcone, College of Dupage; Li Fong, Johnson County Community College

Third Edition Reviewers Dennis Albér, Palm Beach Junior College; Garret J. Etgen, University of Houston; William R. Fuller, Purdue University; Timothy J. Kearns, Boston College; Norbert Lerner, State University of New York at Cortland; Robert L. Maynard, Tidewater Community College; Barbara L. Osofsky, Rutgers University; Jean E. Rubin, Purdue University; Lawrence A. Trivieri, Mohawk Valley Community College; J. Philip Smith, Southern Connecticut State University

Second Edition Reviewers Harry L. Baldwin, Jr., San Diego City College; Phillip A. Ferguson, Fresno City College; Thomas M. Green, Contra Costa College; Arnold J. Insel, Illinois State University; William J. Keane, Boston College; David C. Lantz, Colgate University; Richard E. Shermoen, Washburn University; Thomas W. Shilgalis, Illinois State University; Florence A. Warfel, University of Pittsburgh

First Edition Reviewers Paul W. Davis, Worcester Polytechnic Institute; Eric R. Immel, Georgia Institute of Technology; Frank T. Kocher, Jr., Pennsylvania State University; Joseph F.

Krebs, Boston College; Maurice L. Monahan, South Dakota State University; Robert A. Nowlan, Southern Connecticut State University; N. James Schoonmaker, University of Vermont; Bert K. Waits, Ohio State University

We would also like to thank users of the Third Edition who answered a questionnaire concerning changes they wanted in the new edition. They are George Anderson, Rhode Island College; Frank P. Battles, Massachusetts Maritime Academy; Derek I. Bloomfield, Orange County Community College; Karen J. Edwards, Paul Smith's College; Theodore Hanley, State University of New York at Utica/Rome; Peter Herron, Suffolk County Community College; Ann M. Joyce, Chestnut Hill College; Arthur Kaufman, College of Staten Island; Alan Levine, Franklin and Marshall College; James Magliano, Union County College; Frank Morgan, Castleton State College; Raymond Pluta, Castleton State College; M. Susan Richman, Pennsylvania State University; Carmen Vlad, Pace University; Christopher White, Castleton State College.

A special thanks to all the people at D. C. Heath and Company who worked with us in the development of the Fourth Edition, especially Ann Marie Jones and Mary Lu Walsh, Mathematics Acquisitions Editors; Cathy Cantin, Developmental Editor; Laurie Johnson, Production Editor; Sally Steele, Designer; Carolyn Johnson, Editorial Assistant; Mike O'Dea, Production Manager; and Wendy Johnson, Photo Researcher.

Several other people also worked on this project: David E. Heyd wrote the *Study and Solutions Guide* and solved the exercises, Dianna L. Zook wrote the *Complete Solutions Manual*, Ann R. Kraus wrote the *Instructor's Guide*, Timothy R. Larson prepared the art, Linda L. Kifer proofread the galleys, Linda M. Bollinger proofread the galleys and typed the supplements, Helen Medley solved the exercises and performed an accuracy check for the text, and Kathleen Evanoff, Randall Hammond, and Paula Sibeto solved the exercises and assisted with the production of the supplement package.

A special note of thanks goes to the over 500,000 students who have used earlier editions of the text.

On a personal level, we are grateful to our wives, Deanna Gilbert Larson, Eloise Hostetler, and Consuelo Edwards, for their love, patience, and support. Also, a special thanks goes to R. Scott O'Neil.

If you have suggestions for improving this text, please feel free to write to us. Over the past seventeen years we have received many useful comments from both instructors and students, and we value these very much.

Roland E. Larson, Robert P. Hostetler, Bruce H. Edwards

Calculus

To accommodate the different methods of teaching calculus, D. C. Heath also offers the two texts described below. Each has its own supplements package. The following is a brief discussion of how each book differs from CALCULUS OF A SINGLE VARIABLE, Fourth Edition.

Calculus with Analytic Geometry, Alternate Fourth Edition
Larson/Hostetler

This text, with late trigonometry, is also designed for a three-semester course. Additionally, the text offers a different treatment of the following topics: limits, applications of integration, exponential and logarithmic functions, and vectors.

Calculus with Analytic Geometry, Fourth Edition
Larson/Hostetler/Edwards

This text is designed for a three-semester course. All six trigonometric functions are reviewed in Chapter 1, then used throughout the text. Additionally, this text offers a different treatment of the following topics: limits, applications of integration, exponential and logarithmic functions, and vectors.

Contents

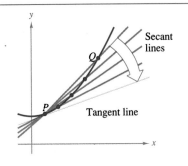

Differentiation   110

3.1   The Derivative and the Tangent Line Problem   111
3.2   Velocity, Acceleration, and Other Rates of Change   121
3.3   Differentiation Rules for Constant Multiples, Sums,
      Powers, Sines, and Cosines   129
3.4   Differentiation Rules for Products, Quotients, Secants,
      and Tangents   138
3.5   The Chain Rule   147
3.6   Implicit Differentiation   156
3.7   Related Rates   164
      Review Exercises   173

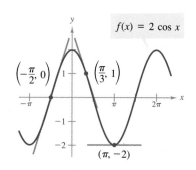

$f(x) = 2 \cos x$

$\left(-\frac{\pi}{2}, 0\right)$ $\left(\frac{\pi}{3}, 1\right)$

$(\pi, -2)$

4   Applications of Differentiation   176

4.1   Extrema on an Interval   177
4.2   Rolle's Theorem and the Mean Value Theorem   185
4.3   Increasing and Decreasing Functions and the First
       Derivative Test   190
4.4   Concavity and the Second Derivative Test   199
4.5   Limits at Infinity   206
4.6   A Summary of Curve Sketching   215
4.7   Optimization Problems   224
4.8   Newton's Method   233
4.9   Differentials   241
4.10   Business and Economics Applications   248
       Review Exercises   256

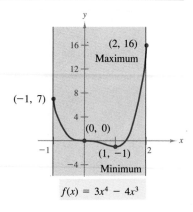

(2, 16)
Maximum
(−1, 7)
(0, 0)
(1, −1)
Minimum
$f(x) = 3x^4 - 4x^3$

5   Integration   258

5.1   Antiderivatives and Indefinite Integration   259
5.2   Area   269
5.3   Riemann Sums and the Definite Integral   281
5.4   The Fundamental Theorem of Calculus   291
5.5   Integration by Substitution   300
5.6   Numerical Integration   312
      Review Exercises   321

6 Logarithmic, Exponential, and Other Transcendental Functions   324

7 Applications of Integration   402

8 Integration Techniques, L'Hôpital's Rule, and Improper Integrals   464

9 Infinite Series 540

10 Conic Sections 626

Ellipse

Hyperbola

11 Plane Curves, Parametric Equations, and Polar Coordinates 660

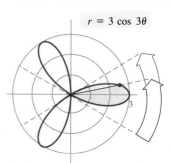

Calculus Multi-Media Library

**Coupons and information regarding software, videotapes, and graphing
calculators follow the index at the end of this book.**

What Is Calculus?

We begin to answer this question by saying that calculus is the reformulation of elementary mathematics through the use of a limit process. If limit processes are unfamiliar to you, then this answer is, at least for now, somewhat less than illuminating. From an elementary point of view, we may think of calculus as a "limit machine" that generates new formulas from old. Actually, the study of calculus involves three distinct stages of mathematics: *precalculus mathematics* (the length of a line segment, the area of a rectangle, and so forth), the *limit process*, and new *calculus* formulations (derivatives, integrals, and so forth).

Some students try to learn calculus as if it were simply a collection of new formulas. This is unfortunate. When students reduce calculus to the memorization of differentiation and integration formulas, they miss a great deal of understanding, self-confidence, and satisfaction.

On the following two pages we have listed some familiar precalculus concepts coupled with their more powerful calculus versions. Throughout this text, our goal is to show you how precalculus formulas and techniques are used as building blocks to produce the more general calculus formulas and techniques. Don't worry if you are unfamiliar with some of the "old formulas" listed on the following two pages—we will be reviewing all of them.

As you proceed through this text, we suggest that you come back to this discussion repeatedly. Try to keep track of where you are relative to the three stages involved in the study of calculus. For example, the first three chapters break down as follows: precalculus (Chapter 1), the limit process (Chapter 2), and new calculus formulas (Chapter 3). This cycle is repeated many times on a smaller scale throughout the text. We wish you well in your venture into calculus.

WITHOUT CALCULUS	WITH DIFFERENTIAL CALCULUS
value of $f(x)$ when $x = c$	limit of $f(x)$ as x approaches c
slope of a line	slope of a curve
secant line to a curve	tangent line to a curve
average rate of change between $t = a$ and $t = b$	instantaneous rate of change at $t = c$
curvature of a circle	curvature of a curve
height of a curve when $x = c$	maximum height of a curve on an interval
tangent plane to a sphere	tangent plane to a surface
direction of motion along a straight line	direction of motion along a curved line

WITHOUT CALCULUS	WITH INTEGRAL CALCULUS
area of a rectangle	area under a curve
work done by a constant force	work done by a variable force
center of a rectangle	centroid of a region
length of a line segment	length of an arc
surface area of a cylinder	surface area of a solid of revolution
mass of a solid of constant density	mass of a solid of variable density
volume of a rectangular solid	volume of a region under a surface
sum of a finite number of terms $\quad a_1 + a_2 + \cdots + a_n = S$	sum of an infinite number of terms $\quad a_1 + a_2 + a_3 \cdots = S$

Chapter 1 Application

The Ferris wheel was designed by the American mechanical engineer George Ferris (1859–1896). The first and largest Ferris wheel was built for the World's Columbian Exposition in Chicago in 1893, and later used at the World's Fair in St. Louis in 1904. It had a diameter of 250 feet, and each of its 36 cars could hold 60 passengers.

Height of a Ferris Wheel Car

A Ferris wheel with a radius of 50 feet rotates at a constant rate of 4 revolutions per minute. If the center of the Ferris wheel is considered to be the origin, then each car travels around the circle given by

$$x^2 + y^2 = 50^2$$

where x and y are measured in feet. The height of a car located at the point (x, y) is given by

$$h = 50 + y$$

where y is related to the angle θ by the equation

$$y = 50 \sin \theta$$

as shown in the accompanying figure. Since the wheel makes 4 revolutions per minute (with one revolution corresponding to 2π radians), it follows that

$$\theta = 4(2\pi)t = 8\pi t$$

where t is measured in minutes. Thus, as a function of time, the height of a car on the Ferris wheel is given by

$$h = 50 + 50 \sin 8\pi t.$$

Height above ground is $h = 50 + 50 \sin \theta$.

(x, y)

$y = 50 \sin \theta$

h

Ground

Chapter Overview

This first chapter contains a review of basic algebra, analytic geometry, and trigonometry. The more familiar you are with the material in this chapter, the more successful you will be in calculus.

Section 1.1 reviews the properties of the real numbers and the real number line. The next two sections review the fundamental concepts of plane analytic geometry, the Cartesian plane, and graphs of equations in two variables.

Section 1.4 discusses the slope of a line—this concept is critical in calculus. This section begins by showing how the slope of a line is related to the *average rate of change* of one variable with respect to another.

The concept of a **function** is also critical in calculus, and we review several fundamental ideas related to functions in Section 1.5. For instance, this section reviews the graphs of such basic functions as

$$f(x) = x \qquad f(x) = x^2$$
$$f(x) = x^3 \qquad f(x) = \sqrt{x}$$
$$f(x) = |x| \qquad f(x) = \frac{1}{x}.$$

Familiarity with the graphs of these functions will help you in later chapters.

Finally, Section 1.6 contains a brief review of trigonometry.

See Exercise 76, Section 1.6.

The Cartesian Plane and Functions

1.1 Real Numbers and the Real Line

Real numbers ▪ The real line ▪ Order and inequalities ▪ Absolute value ▪ Distance on the real line ▪ Intervals on the real line

In this first chapter we will lay the foundation for studying calculus. We assume that you have a good working knowledge of basic algebra. This is essential for the study of calculus.

The real line

The Real Line

FIGURE 1.1

To represent the set of real numbers we use a coordinate system called the **real line** or x-axis (Figure 1.1). The real number corresponding to a particular point on the real line is called the **coordinate** of the point. As Figure 1.1 shows, it is customary to identify those points whose coordinates are integers.

The point on the real line corresponding to zero is called the **origin** and is denoted by 0. The **positive direction** (to the right) is denoted by an arrowhead and indicates the direction of increasing values of x. Numbers to the right of the origin are **positive;** numbers to the left of the origin are **negative.** We use the term **nonnegative** to describe a number that is either positive or zero. Similarly, the term **nonpositive** is used to describe a number that is either negative or zero.

One-to-one correspondence between real numbers and points on the real line.

FIGURE 1.2

Each point on the real line corresponds to one and only one real number, and each real number corresponds to one and only one point on the real line. This type of relationship is called a **one-to-one correspondence.**

Each of the four points in Figure 1.2 corresponds to a real number that can be expressed as the ratio of two integers. $\left(\text{Note that } 4.5 = \frac{9}{2} \text{ and } -2.6 = -\frac{13}{5}.\right)$ We call such numbers **rational.** Rational numbers can be represented either by *terminating decimals* such as $\frac{2}{5} = 0.4$, or by *repeating decimals* such as $\frac{1}{3} = 0.333 \ldots = 0.\overline{3}$.

Real numbers that are not rational are called **irrational.** They cannot be represented as terminating or repeating decimals. To represent an irrational number, we usually resort to a decimal approximation. For example, $\sqrt{2} \approx 1.4142135623$, $\pi \approx 3.1415926535$, and $e \approx 2.7182818284$. (See Figure 1.3.)

$\sqrt{2} \quad e \ \pi$

Irrational Numbers

FIGURE 1.3

Order and inequalities

One important property of real numbers is that they are **ordered.**

DEFINITION OF ORDER ON THE REAL LINE	If a and b are real numbers, then a is **less than** b if $b - a$ is positive. We denote this order by the **inequality** $$a < b.$$ The symbol $a \leq b$ means that a is **less than or equal to** b. The statement b is **greater than** a is equivalent to saying a is less than b.

$a < b$ if and only if a lies to the left of b.

FIGURE 1.4

Geometrically, $a < b$ if and only if a lies to the *left* of b on the real line. (See Figure 1.4.) For example, $1 < 2$ because 1 lies to the left of 2 on the real line.

The following properties are often used to work with inequalities. Similar properties are obtained if $<$ is replaced by $\leq$ and $>$ is replaced by $\geq$.

THEOREM 1.1 PROPERTIES OF INEQUALITIES	1. If $a < b$ and $b < c$, then $a < c$. 2. If $a < b$ and $c < d$, then $a + c < b + d$. 3. If $a < b$ and k is any real number, then $a + k < b + k$. 4. If $a < b$ and $k > 0$, then $ak < bk$. 5. If $a < b$ and $k < 0$, then $ak > bk$.

REMARK   Note that we *reverse the inequality* when we multiply by a negative number. For example, if $x < 3$, then $-4x > -12$. This principle also applies to division by a negative number. Thus, if $-2x > 4$, then $x < -2$.

When three real numbers a, b, and c are ordered such that $a < b$ and $b < c$, we say that b is **between** a and c and we write $a < b < c$.

Occasionally it is convenient to use set notation to describe collections of real numbers. A **set** is a collection of elements. For example, the two major sets we have been discussing are the set of real numbers and the set of points on the real line. Often, we will restrict our interest to a **subset** of one of these two sets, in which case it is convenient to use **set notation** of the form

$$\{x: \text{condition on } x\}.$$

The set of all x such that a certain condition is true

For example, we can describe the set of positive real numbers as $\{x: 0 < x\}$. The **union** of two sets A and B is the set of elements that are members of A or B or both. This union is denoted by $A \cup B$. The **intersection** of two sets A and B is the set of elements that are members of A *and* B. This intersection is denoted by $A \cap B$. Two sets are called **disjoint** if they have no elements in common.

The most common sets we work with are subsets of the real line called **intervals.** For example, the **open** interval $(a, b) = \{x: a < x < b\}$ is the set of all real numbers greater than a and less than b, where a and b are called the **endpoints** of the interval. Note that the endpoints are not included in an open interval. Intervals that include their endpoints are called **closed** and are denoted by $[a, b] = \{x: a \le x \le b\}$. The nine basic types of intervals on the real line are shown in Table 1.1. The first four are called **bounded intervals** and the remaining five are called **unbounded intervals.**

TABLE 1.1   Intervals on the Real Line

	Interval notation	Set notation	Graph
Open interval	(a, b)	$\{x: a < x < b\}$	a b
Closed interval	$[a, b]$	$\{x: a \le x \le b\}$	a b
Half-open intervals	$[a, b)$	$\{x: a \le x < b\}$	a b
	$(a, b]$	$\{x: a < x \le b\}$	a b
Infinite intervals	$(-\infty, a]$	$\{x: x \le a\}$	a b
	$(-\infty, a)$	$\{x: x < a\}$	a b
	(b, ∞)	$\{x: b < x\}$	a b
	$[b, \infty)$	$\{x: b \le x\}$	a b
	$(-\infty, \infty)$	$\{x: x \text{ is a real number}\}$	a b

REMARK   We use the symbols ∞ and $-\infty$ to refer to positive and negative infinity. These symbols do not denote real numbers; they merely enable us to describe unbounded conditions more concisely. For instance, the interval $[b, \infty)$ is unbounded to the right since it includes *all* real numbers that are greater than or equal to b.

EXAMPLE 1   Intervals on the real line

Describe the intervals on the real line that correspond to the temperature ranges (in degrees Celsius) for water in the following two states.
(a) liquid   (b) gas

(a) Temperature range of water
(in degrees Celsius)

(b) Temperature range of steam
(in degrees Celsius)

FIGURE 1.5

SOLUTION

(a) Since water is in a liquid state at temperatures greater than $0°$ and less than $100°$, we have the interval

$$(0, 100) = \{x: 0 < x < 100\}$$

as shown in Figure 1.5(a).

(b) Since water is in a gaseous state (steam) at temperatures greater than or equal to $100°$, we have the interval

$$[100, \infty) = \{x: 100 \le x\}$$

as shown in Figure 1.5(b).   ▭

In calculus we are frequently asked to solve inequalities involving variable expressions such as $2x - 5 < 7$. We say that a is a **solution** of this inequality if the inequality is true when a is substituted for x. The set of all values of x that satisfy the inequality is called the **solution set** of the inequality.

EXAMPLE 2   Solving an inequality

Find the solution set of the inequality $2x - 5 < 7$.

SOLUTION

Using the properties in Theorem 1.1, we have

$$2x - 5 < 7$$
$$2x - 5 + 5 < 7 + 5 \qquad \text{Add 5 to both sides}$$
$$2x < 12$$
$$\frac{1}{2}(2x) < \frac{1}{2}(12) \qquad \text{Multiply both sides by } \frac{1}{2}$$
$$x < 6.$$

Thus, the interval representing the solution is $(-\infty, 6)$.   ▭

REMARK   In Example 2, all five inequalities listed as steps in the solution have the same solution set and are called **equivalent.**

Once you have solved an inequality, check some x-values in your solution interval to see whether they satisfy the original inequality. You also might check some values outside your solution interval to verify that they do not satisfy the inequality. For example, Figure 1.6 shows that when $x = 0$ or $x = 5$ the inequality is satisfied, but when $x = 7$ the inequality is not satisfied.

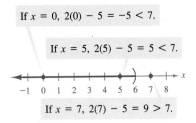

If $x = 0$, $2(0) - 5 = -5 < 7$.

If $x = 5$, $2(5) - 5 = 5 < 7$.

If $x = 7$, $2(7) - 5 = 9 > 7$.

FIGURE 1.6

EXAMPLE 3   Finding the intersection of two solution sets

Find the intersection of the solution sets of the inequalities

$$-3 \le 2 - 5x \qquad \text{and} \qquad 2 - 5x \le 12.$$

SOLUTION

We could solve both inequalities and then find the intersection of the resulting solution sets. However, since the expression $2 - 5x$ occurs on the left side of one inequality and the right side of the other, it is convenient to work with both inequalities at the same time.

$$-3 \leq \quad 2 - 5x \quad \leq 12$$
$$-3 - 2 \leq 2 - 5x - 2 \leq 12 - 2 \qquad \text{Subtract 2}$$
$$-5 \leq \quad -5x \quad \leq 10$$
$$\frac{-5}{-5} \geq \quad \frac{-5x}{-5} \quad \geq \frac{10}{-5} \qquad \text{Divide by } -5 \text{ and reverse the inequality}$$
$$1 \geq \quad x \quad \geq -2$$

FIGURE 1.7

Thus, the interval representing the solution is $[-2, 1]$, as shown in Figure 1.7.

The inequalities in Examples 2 and 3 involve first-degree polynomials. For inequalities involving polynomials of higher degree we use the fact that a polynomial can change signs *only* at its real zeros (a **zero** of a polynomial is a number at which the value of the polynomial is zero). Between two consecutive real zeros a polynomial must be entirely positive or entirely negative. This means that when the real zeros of a polynomial are put in order, they divide the real line into **test intervals** in which the polynomial has no sign changes. That is, if a polynomial has the factored form

$$(x - r_1)(x - r_2) \cdots (x - r_n), \qquad r_1 < r_2 < r_3 < \cdots < r_n$$

then the test intervals are

$$(-\infty, r_1), (r_1, r_2), \ldots, (r_{n-1}, r_n), \text{ and } (r_n, \infty).$$

For example, the polynomial

$$x^2 - x - 6 = (x - 3)(x + 2)$$

can change signs only at $x = -2$ and $x = 3$.

EXAMPLE 4   Solving an inequality involving a quadratic

Find the solution set of the inequality $x^2 < x + 6$.

SOLUTION

$$\begin{aligned} x^2 &< x + 6 & \text{Given} \\ x^2 - x - 6 &< 0 & \text{Polynomial form} \\ (x - 3)(x + 2) &< 0 & \text{Factor} \end{aligned}$$

Thus, the polynomial $x^2 - x - 6$ has $x = -2$ and $x = 3$ as its zeros, and we can solve the inequality by testing the sign of $x^2 - x - 6$ in each of the following open intervals.

$$(-\infty, -2), \qquad (-2, 3), \qquad (3, \infty)$$

Choose $x = -3$.
$(x - 3)(x + 2) > 0$

Choose $x = 4$.
$(x - 3)(x + 2) > 0$

Choose $x = 0$.
$(x - 3)(x + 2) < 0$

FIGURE 1.8

To test an interval, we choose an arbitrary number in the interval and compute the sign of each factor of $x^2 - x - 6$. For example, for any x in the open interval $(-\infty, -2)$, the factors $(x - 3)$ and $(x + 2)$ are both negative. Consequently, the product (of two negative numbers) is positive and the inequality is *not* satisfied in the interval $(-\infty, -2)$. We suggest that you use the testing format shown in Figure 1.8. Since the inequality $(x - 3)(x + 2) < 0$ is satisfied only for values of x in the center interval, we conclude that the solution set is the open interval $(-2, 3)$. ▭

Absolute value and distance

The **absolute value** of a real number a is denoted by $|a|$ and is defined as follows.

DEFINITION OF ABSOLUTE VALUE

If a is a real number, then the **absolute value** of a is

$$|a| = \begin{cases} a, & \text{if } a \geq 0 \\ -a, & \text{if } a < 0. \end{cases}$$

The absolute value of a number can never be negative. For example, let $a = -4$. Then, since $-4 < 0$, we have

$$|a| = |-4| = -(-4) = 4.$$

Remember that the symbol $-a$ does not necessarily mean that $-a$ is negative. Theorems 1.2 and 1.3 contain some useful properties of absolute value.

**THEOREM 1.2
OPERATIONS WITH
ABSOLUTE VALUE**

If a and b are real numbers and n is a positive integer, then the following properties are true.

1. $|ab| = |a||b|$ 2. $\left|\dfrac{a}{b}\right| = \dfrac{|a|}{|b|}, b \neq 0$

3. $|a| = \sqrt{a^2}$ 4. $|a^n| = |a|^n$

REMARK   You are asked to prove these properties in Exercises 67–71.

**THEOREM 1.3
INEQUALITIES AND
ABSOLUTE VALUE**

If a and b are real numbers and k is positive, then the following properties are true.

1. $-|a| \leq a \leq |a|$
2. $|a| \leq k$ if and only if $-k \leq a \leq k$.
3. $k \leq |a|$ if and only if $k \leq a$ or $a \leq -k$.
4. Triangle Inequality: $|a + b| \leq |a| + |b|$

Properties 2 and 3 are also true if $\leq$ is replaced by $<$.

PROOF   We give a proof of Property 4 and leave the proofs of the first three properties as exercises (see Exercises 72–74). Using Property 1, we have $-|a| \leq a \leq |a|$ and $-|b| \leq b \leq |b|$. Adding these two inequalities produces

$$-(|a| + |b|) \leq a + b \leq |a| + |b|.$$

Now, by Property 2 (using $k = |a| + |b|$), we can conclude that

$$|a + b| \leq |a| + |b|.$$

EXAMPLE 5   Solving an inequality involving an absolute value

Sketch the solution set of $|x - 3| \leq 2$.

SOLUTION

Using Property 2 of Theorem 1.3, we have

$$-2 \leq \quad x - 3 \quad \leq 2$$
$$-2 + 3 \leq x - 3 + 3 \leq 2 + 3$$
$$1 \leq \quad x \quad \leq 5.$$

Thus, the solution set is the closed interval $[1, 5]$, as shown in Figure 1.9.

FIGURE 1.9

EXAMPLE 6   A two-interval solution set

Find the solution set of $3 < |x + 2|$.

SOLUTION

Using Property 3 of Theorem 1.3, we have

$$3 < x + 2 \quad \text{or} \quad x + 2 < -3$$
$$1 < x \quad \text{or} \quad x < -5.$$

Thus, the solution set consists of the union of the disjoint intervals $(-\infty, -5)$ and $(1, \infty)$, as shown in Figure 1.10.

FIGURE 1.10

Examples 5 and 6 illustrate the general results shown in Figure 1.11. Note that if $d > 0$, the solution set for the inequality $|x - a| \leq d$ consists of a *single* interval, while the solution set for the inequality $|x - a| \geq d$ consists of *two* disjoint intervals.

FIGURE 1.11

The **distance between two points** a and b on the real line is given by

$$d = |a - b|.$$

The **directed distance from** a **to** b is $b - a$ and the **directed distance from** b **to** a is $a - b$, as shown in Figure 1.12.

FIGURE 1.12

EXAMPLE 7   Distance on the real line

(a) The distance between -3 and 4 is given by

$$|4 - (-3)| = |7| = 7 \quad \text{or} \quad |-3 - 4| = |-7| = 7.$$

FIGURE 1.13

(See Figure 1.13.)

(b) The directed distance from -3 to 4 is $4 - (-3) = 7$.

(c) The directed distance from 4 to -3 is $-3 - 4 = -7$.

To find the **midpoint** of an interval with endpoints a and b, we simply find the average value of a and b. That is,

$$\text{midpoint of interval } (a, b) = \frac{a + b}{2}.$$

To show that this is the midpoint, you need only show that $(a + b)/2$ is equidistant from a and b.

EXERCISES for Section 1.1

In Exercises 1–10, determine whether the real number is rational or irrational.

1. 0.7

2. -3678

3. $\dfrac{3\pi}{2}$

4. $3\sqrt{2} - 1$

5. $4.345\overline{1451}$

6. $\dfrac{22}{7}$

*7. $\sqrt[3]{64}$

8. 0.81778177

9. $4\dfrac{5}{8}$

10. $(\sqrt{2})^3$

*A blue number indicates that a detailed solution can be found in the *Study and Solutions Guide*.

In Exercises 11–14, express the repeating decimal as a ratio of integers using the following procedure. Let $x = 0.6363\ldots$. Then $100x = 63.6363\ldots$. Subtracting the first equation from the second produces $99x = 63$ or $x = \frac{63}{99} = \frac{7}{11}$.

11. $0.36\overline{36}$

12. $0.318\overline{18}$

13. $0.297\overline{297}$

14. $0.9900\overline{9900}$

15. Given $a < b$, determine which of the following are true.

(a) $a + 2 < b + 2$

(b) $5b < 5a$

(c) $5 - a > 5 - b$

(d) $\dfrac{1}{a} < \dfrac{1}{b}$

(e) $(a - b)(b - a) > 0$

(f) $a^2 < b^2$

16. For $A = \{x: 0 < x\}$, $B = \{x: -2 \le x \le 2\}$, and $C = \{x: x < 1\}$, find the indicated interval.
(a) $A \cup B$ (b) $A \cap B$
(c) $B \cap C$ (d) $A \cup C$
(e) $A \cap B \cap C$

In Exercises 17 and 18, complete the table by filling in the appropriate interval notation, set notation, and graph on the real line.

17.

Interval notation	Set notation	Graph
		(graph from -2 to -1 to 0)
$(-\infty, -4]$		
	$\{x: 3 \le x \le \frac{11}{2}\}$	
$(-1, 7)$		

18.

Interval notation	Set notation	Graph
		(graph: 98 99 100 101 102)
	$\{x: 10 < x\}$	
$(\sqrt{2}, 8]$		
	$\{x: \frac{1}{3} < x \le \frac{22}{7}\}$	

In Exercises 19–44, solve the inequality and graph the solution on the real line.

19. $x - 5 \ge 7$ **20.** $2x > 3$
21. $4x + 1 < 2x$ **22.** $2x + 7 < 3$
23. $2x - 1 \ge 0$ **24.** $3x + 1 \ge 2x + 2$
25. $-4 < 2x - 3 < 4$ **26.** $0 \le x + 3 < 5$
27. $\frac{3}{4}x > x + 1$ **28.** $-1 < -\frac{x}{3} < 1$
29. $\frac{x}{2} + \frac{x}{3} > 5$ **30.** $x > \frac{1}{x}$
31. $|x| < 1$ **32.** $\frac{x}{2} - \frac{x}{3} > 5$
33. $\left|\frac{x - 3}{2}\right| \ge 5$ **34.** $\left|\frac{x}{2}\right| > 3$
35. $|x - a| < b$ **36.** $|x + 2| < 5$

37. $|2x + 1| < 5$ **38.** $|3x + 1| \ge 4$
39. $\left|1 - \frac{2}{3}x\right| < 1$ **40.** $|9 - 2x| < 1$
41. $x^2 \le 3 - 2x$ **42.** $x^4 - x \le 0$
43. $x^2 + x - 1 \le 5$ **44.** $2x^2 + 1 < 9x - 3$

In Exercises 45–48, find the directed distance from a to b, the directed distance from b to a, and the distance between a and b.

45. (number line: $a = -1$, $b = 3$, from -2 to 4)

46. (number line: $a = -\frac{5}{2}$, $b = \frac{13}{4}$, from -3 to 4)

47. (a) $a = 126$, $b = 75$
(b) $a = -126$, $b = -75$
48. (a) $a = 9.34$, $b = -5.65$
(b) $a = \frac{16}{5}$, $b = \frac{112}{75}$

In Exercises 49–52, find the midpoint of the given interval.

49. (number line: $a = -1$, $b = 3$, from -2 to 4)

50. (number line: $a = -5$, $b = -\frac{3}{2}$, from -6 to 0)

51. (a) $[7, 21]$ (b) $[8.6, 11.4]$
52. (a) $[-6.85, 9.35]$ (b) $[-4.6, -1.3]$

In Exercises 53–58, use absolute values to define each interval (or pair of intervals) on the real line.

53. (number line: $a = -2$, $b = 2$, from -3 to 3)

54. (number line: $a = -3$, $b = 3$, from -4 to 4)

55. (number line: $a = 0$, $b = 4$, from -2 to 6)

56. (number line: $a = 20$, $b = 24$, from 18 to 26)

57. (a) All numbers that are at most 10 units from 12.
(b) All numbers that are at least 10 units from 12.
58. (a) y is at most 2 units from a.
(b) y is less than δ units from c.

59. The balance in an account after t years is given by

$$A = P + Prt$$

where P dollars is the initial investment and r is the simple interest rate (in decimal form). In order for an investment of $1000 to attain a balance that is greater than $1250 in two years, what should the interest rate be?

60. In the manufacture and sale of a certain product, the revenue for selling x units is

$$R = 115.95x$$

and the cost of producing x units is

$$C = 95x + 750.$$

In order for a profit to be realized, R must be greater than C. For what values of x will this product return a profit?

61. A utility company has a fleet of vans. The annual operating cost of each van is estimated to be

$$C = 0.32m + 2300$$

where C is measured in dollars and m is measured in miles. If the company wants the annual operating cost of each van to be less than $10,000, then m must be less than what value?

62. The heights, h, of two-thirds of the members of a certain population satisfy the inequality

$$\left|\frac{h - 68.5}{2.7}\right| \le 1$$

where h is measured in inches. Determine the interval on the real line in which these heights lie.

63. To determine if a coin is fair (has an equal probability of landing tails up or heads up), an experimenter tosses it 100 times and records the number of heads, x. Through statistical theory, the coin is declared unfair if

$$\left|\frac{x - 50}{5}\right| \ge 1.645.$$

For what values of x will the coin be declared unfair?

64. The estimated daily production, p, at a refinery is given by

$$|p - 2{,}250{,}000| < 125{,}000$$

where p is measured in barrels of oil. Determine the high and low production levels.

In Exercises 65 and 66, determine which of the two given real numbers is greater.

65. (a) π or $\dfrac{355}{113}$ (b) π or $\dfrac{22}{7}$

66. (a) $\dfrac{224}{151}$ or $\dfrac{144}{97}$ (b) $\dfrac{73}{81}$ or $\dfrac{6427}{7132}$

In Exercises 67–74, prove the given property.

67. $|ab| = |a||b|$
68. $|a - b| = |b - a|$ [Hint: Use Exercise 67 and the fact that $(a - b) = (-1)(b - a)$.]
69. $\left|\dfrac{a}{b}\right| = \dfrac{|a|}{|b|}, b \ne 0$
70. $|a| = \sqrt{a^2}$
71. $|a^n| = |a|^n, n = 1, 2, 3, \ldots$
72. $-|a| \le a \le |a|$
73. $|a| \le k$ if and only if $-k \le a \le k, k > 0$.
74. $k \le |a|$ if and only if $k \le a$ or $a \le -k, k > 0$.

1.2 The Cartesian Plane

The Cartesian plane ▪ The Distance Formula ▪ The Midpoint Formula ▪ Equations of circles ▪ Completing the square

Just as real numbers can be represented by points on the real line, we can represent ordered pairs of real numbers by points in a plane. An **ordered pair** (x, y) of real numbers has x as its *first* member and y as its *second* member. The model for representing ordered pairs is called the **rectangular coordinate system,** or the **Cartesian plane.** It is developed by considering two real lines intersecting at right angles (Figure 1.14).

The horizontal real line is usually called the **x-axis,** and the vertical real line is usually called the **y-axis.** Their point of intersection is called the **origin.** The two axes divide the plane into four parts called **quadrants.**

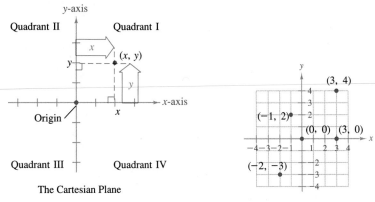

The Cartesian Plane

FIGURE 1.14

FIGURE 1.15

We identify each point in the plane by an ordered pair (x, y) of real numbers x and y, called **coordinates** of the point. The number x represents the directed distance from the y-axis to the point, and y represents the directed distance from the x-axis to the point (Figure 1.14). For the point (x, y), the first coordinate is called the x-coordinate or **abscissa,** and the second coordinate is called the y-coordinate or **ordinate.** For example, Figure 1.15 shows the location of the points $(-1, 2)$, $(3, 4)$, $(0, 0)$, $(3, 0)$, and $(-2, -3)$ in the Cartesian plane.

REMARK   Note that we use an ordered pair (a, b) to denote either a point in the plane *or* an open interval on the real line. As the nature of the problem clarifies whether a point in the plane or an open interval is being discussed, there should be no confusion.

The Distance and Midpoint Formulas

Section 1.1 defined the distance between two points x_1 and x_2 on the real line. We will now find the distance between two points in the plane. Recall from the Pythagorean Theorem that, for a right triangle with hypotenuse c and sides a and b, we have the relationship $a^2 + b^2 = c^2$. Conversely, if $a^2 + b^2 = c^2$, then the triangle is a right triangle (Figure 1.16).

Suppose we want to determine the distance d between the two points (x_1, y_1) and (x_2, y_2) in the plane. With these two points, a right triangle can be formed, as shown in Figure 1.17. The length of the vertical side of the triangle is $|y_2 - y_1|$. Similarly, the length of the horizontal side is $|x_2 - x_1|$. By the Pythagorean Theorem, it follows that

$$d^2 = |x_2 - x_1|^2 + |y_2 - y_1|^2$$
$$d = \sqrt{|x_2 - x_1|^2 + |y_2 - y_1|^2}.$$

Replacing $|x_2 - x_1|^2$ and $|y_2 - y_1|^2$ by the equivalent expressions $(x_2 - x_1)^2$ and $(y_2 - y_1)^2$, we obtain

$$d = \sqrt{(x_2 - x_1)^2 + (y_2 - y_1)^2}.$$

We choose the positive square root for d because the distance *between* two points is not a directed distance. We have therefore established the following theorem.

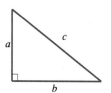

Pythagorean Theorem:
$a^2 + b^2 = c^2$

FIGURE 1.16

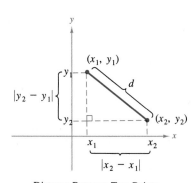

Distance Between Two Points

FIGURE 1.17

THEOREM 1.4 **DISTANCE FORMULA**	The distance d between the points (x_1, y_1) and (x_2, y_2) in the plane is given by $$d = \sqrt{(x_2 - x_1)^2 + (y_2 - y_1)^2}.$$

EXAMPLE 1 Finding the distance between two points

Find the distance between the points $(-2, 1)$ and $(3, 4)$.

SOLUTION

Applying the Distance Formula, we have
$$
\begin{aligned}
d &= \sqrt{[3 - (-2)]^2 + (4 - 1)^2} \\
&= \sqrt{(5)^2 + (3)^2} \\
&= \sqrt{25 + 9} \\
&= \sqrt{34} \approx 5.83.
\end{aligned}
$$

EXAMPLE 2 Verifying a right triangle

Plot the points $(2, 1)$, $(4, 0)$, and $(5, 7)$ and use the Distance Formula to show that the three points form the vertices of a right triangle.

SOLUTION

Figure 1.18 shows the triangle formed by the three points. Moreover, the three sides of the triangle have the following lengths.
$$
\begin{aligned}
d_1 &= \sqrt{(5 - 2)^2 + (7 - 1)^2} = \sqrt{9 + 36} = \sqrt{45} \\
d_2 &= \sqrt{(4 - 2)^2 + (0 - 1)^2} = \sqrt{4 + 1} = \sqrt{5} \\
d_3 &= \sqrt{(5 - 4)^2 + (7 - 0)^2} = \sqrt{1 + 49} = \sqrt{50}
\end{aligned}
$$
Since $d_1^2 + d_2^2 = 45 + 5 = 50 = d_3^2$, we can apply the Pythagorean Theorem to conclude that the triangle must be a right triangle.

FIGURE 1.18

The formula for the midpoint of a line segment in the plane is similar to that for an interval on the real line. The proof is left as an exercise (see Exercise 63).

THEOREM 1.5 **MIDPOINT FORMULA**	The midpoint of the line segment joining the points (x_1, y_1) and (x_2, y_2) is $$\left(\frac{x_1 + x_2}{2}, \frac{y_1 + y_2}{2}\right).$$

EXAMPLE 3   Finding the midpoint of a line segment

Find the midpoint of the line segment joining the points $(-5, -3)$ and $(9, 3)$.

SOLUTION

By the Midpoint Formula, the midpoint is

$$\left(\frac{-5 + 9}{2}, \frac{-3 + 3}{2}\right) = (2, 0).$$

(See Figure 1.19.)

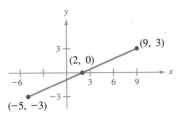

FIGURE 1.19

EXAMPLE 4   Finding points at a specified distance from a given point

Find x so that the distance between $(x, 3)$ and $(2, -1)$ is 5.

SOLUTION

Using the Distance Formula, we have

$$d = 5 = \sqrt{(x - 2)^2 + (3 + 1)^2}$$
$$25 = (x^2 - 4x + 4) + 16$$
$$0 = x^2 - 4x - 5$$
$$0 = (x - 5)(x + 1).$$

Therefore, $x = 5$ or $x = -1$, and we conclude that there are two solutions. That is, both of the points $(5, 3)$ and $(-1, 3)$ lie 5 units from the point $(2, -1)$, as shown in Figure 1.20.

FIGURE 1.20

Circles

One straightforward application of the Distance Formula is in developing an equation for a circle in the plane.

DEFINITION OF A CIRCLE IN THE PLANE

Let (h, k) be a point in the plane and let $r > 0$. The set of all points (x, y) such that r is the distance between (h, k) and (x, y) is called a **circle.** The point (h, k) is the **center** of the circle, and r is the **radius** (see Figure 1.21).

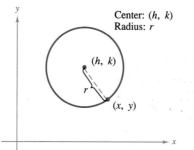

FIGURE 1.21

We can use the Distance Formula to write an equation for the circle with center (h, k) and radius r as

[distance between (h, k) and (x, y)] $= r$
$$\sqrt{(x - h)^2 + (y - k)^2} = r.$$

By squaring both sides of this equation, we obtain the **standard form of the equation of a circle,** as indicated in the following theorem.

THEOREM 1.6 STANDARD FORM OF THE EQUATION OF A CIRCLE

The point (x, y) lies on the circle of radius r and center (h, k) if and only if

$$(x - h)^2 + (y - k)^2 = r^2.$$

It follows from Theorem 1.6 that the standard form of the equation of a circle with center at the origin, $(h, k) = (0, 0)$, is

$$x^2 + y^2 = r^2.$$

If $r = 1$, then the graph of this equation is called the **unit circle.**

EXAMPLE 5 Finding the equation of a circle

The point $(3, 4)$ lies on a circle whose center is at $(-1, 2)$, as shown in Figure 1.22. Find an equation for the circle.

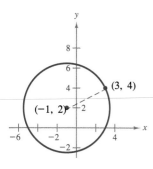

$$(x + 1)^2 + (y - 2)^2 = 20$$

FIGURE 1.22

SOLUTION

The radius of the circle is the distance between $(-1, 2)$ and $(3, 4)$. Thus,

$$r = \sqrt{[3 - (-1)]^2 + (4 - 2)^2} = \sqrt{16 + 4} = \sqrt{20}.$$

Therefore, the standard form of the equation of this circle is

$$[x - (-1)]^2 + (y - 2)^2 = (\sqrt{20})^2$$
$$(x + 1)^2 + (y - 2)^2 = 20.$$

By squaring and simplifying, the equation $(x - h)^2 + (y - k)^2 = r^2$ can be written in the following **general form of the equation of a circle.**

$$Ax^2 + Ay^2 + Cx + Dy + F = 0, \quad A \neq 0$$

To convert such an equation to the standard form

$$(x - h)^2 + (y - k)^2 = p$$

we use a process called **completing the square.** If $p > 0$, then the graph of the equation is a circle. If $p = 0$, then the graph is the single point (h, k). Finally, if $p < 0$, then the equation has no graph.

EXAMPLE 6   Completing the square

Sketch the graph of the circle whose general equation is

$$4x^2 + 4y^2 + 20x - 16y + 37 = 0.$$

SOLUTION

To complete the square, first divide by 4 so that the coefficients of x^2 and y^2 are both 1.

$$4x^2 + 4y^2 + 20x - 16y + 37 = 0 \qquad \text{General form}$$

$$x^2 + y^2 + 5x - 4y + \frac{37}{4} = 0 \qquad \text{Divide by 4}$$

$$(x^2 + 5x + \quad) + (y^2 - 4y + \quad) = -\frac{37}{4} \qquad \text{Group terms}$$

$$\left(x^2 + 5x + \frac{25}{4}\right) + (y^2 - 4y + 4) = -\frac{37}{4} + \frac{25}{4} + 4 \qquad \begin{array}{l}\text{Complete the}\\\text{square by}\\\text{adding } \frac{25}{4} \text{ and}\\\text{4 to both sides}\end{array}$$

$$\underbrace{\qquad}_{(\text{half})^2} \qquad \underbrace{\qquad}_{(\text{half})^2}$$

$$\left(x + \frac{5}{2}\right)^2 + (y - 2)^2 = 1 \qquad \text{Standard form}$$

Note that we complete the square by adding the square of half the coefficient of x *and* the square of half the coefficient of y to both sides of the equation. Therefore, the circle is centered at $\left(-\frac{5}{2}, 2\right)$, and its radius is 1, as shown in Figure 1.23.

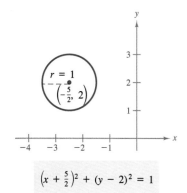

$$\left(x + \frac{5}{2}\right)^2 + (y - 2)^2 = 1$$

FIGURE 1.23

Pierre de Fermat

We have now introduced some fundamental concepts of *analytic geometry*. Because these concepts are in common use today, it is easy to overlook their revolutionary character. At the time analytic geometry was being developed by Pierre de Fermat (1601–1655) and René Descartes (1596–1650), the two major branches of mathematics—geometry and algebra—were largely independent of each other. Circles belonged to geometry and equations belonged to algebra. The coordination of the points on a circle and the solutions of an equation belongs to what is now called analytic geometry. It is important to become skilled in analytic geometry so that you may move easily between geometry and algebra. For instance, in Example 5, we are given a geometric description of a circle and are asked to find an algebraic equation for the circle. Thus, we are moving from geometry to algebra. Similarly, in Example 6 we are given an algebraic equation and asked to sketch a geometric picture. In this case, we are moving from algebra to geometry. These two examples illustrate the two most common problems in analytic geometry.

1. Given a graph, find its equation.

$$\boxed{\text{Geometry}} \rightarrow \boxed{\text{Algebra}}$$

2. Given an equation, find its graph.

$$\boxed{\text{Algebra}} \rightarrow \boxed{\text{Geometry}}$$

In the next two sections, we will look more closely at these two types of problems.

René Descartes

EXERCISES for Section 1.2

In Exercises 1–6, (a) plot the points, (b) find the distance between the points, and (c) find the midpoint of the line segment joining the points.

1. (2, 1), (4, 5)
2. (−3, 2), (3, −2)
3. $\left(\frac{1}{2}, 1\right), \left(-\frac{3}{2}, -5\right)$
4. $\left(\frac{2}{3}, -\frac{1}{3}\right), \left(\frac{5}{6}, 1\right)$
5. $(1, \sqrt{3}), (-1, 1)$
6. $(-2, 0), (0, \sqrt{2})$

In Exercises 7–10, show that the given points form the vertices of the indicated polygon. (A rhombus is a quadrilateral whose sides are all of the same length.)

Vertices	Figure
7. (4, 0), (2, 1), (−1, −5)	Right triangle
8. (1, −3), (3, 2), (−2, 4)	Isosceles triangle
9. (0, 0), (1, 2), (2, 1), (3, 3)	Rhombus
10. (0, 1), (3, 7), (4, 4), (1, −2)	Parallelogram

In Exercises 11–14, use the Distance Formula to determine whether the given points are collinear (lie on the same line).

11. (0, −4), (2, 0), (3, 2)
12. (0, 4), (7, −6), (−5, 11)
13. (−2, 1), (−1, 0), (2, −2)
14. (−1, 1), (3, 3), (5, 5)

In Exercises 15 and 16, find x so that the distance between the points is 5.

15. (0, 0), (x, −4)
16. (2, −1), (x, 2)

In Exercises 17 and 18, find y so that the distance between the points is 8.

17. (0, 0), (3, y)
18. (5, 1), (5, y)

In Exercises 19 and 20, find the relationship between x and y so that (x, y) is equidistant from the two given points.

19. $(4, -1)$, $(-2, 3)$ **20.** $\left(3, \dfrac{5}{2}\right)$, $(-7, -1)$

21. Use the Midpoint Formula to find the three points that divide the line segment joining (x_1, y_1) and (x_2, y_2) into four equal parts.
22. Use the result of Exercise 21 to find the points that divide the line segment joining the given points into four equal parts.
 (a) $(1, -2)$, $(4, -1)$.   (b) $(-2, -3)$, $(0, 0)$

In Exercises 23 and 24, complete the square for each expression.

23. (a) $x^2 + 5x$ (b) $x^2 + 8x + 7$
24. (a) $4x^2 - 4x - 39$ (b) $5x^2 + x$

In Exercises 25–30, match the given equation with its graph. [Graphs are labeled (a)–(f).]

25. $x^2 + y^2 = 1$
26. $(x - 1)^2 + (y - 3)^2 = 4$
27. $(x - 1)^2 + y^2 = 0$
28. $\left(x + \dfrac{1}{2}\right)^2 + \left(y - \dfrac{3}{4}\right)^2 = \dfrac{1}{4}$
29. $(x + 3)^2 + (y - 1)^2 = 16$
30. $x^2 + (y - 1)^2 = 1$

(a)   (b)

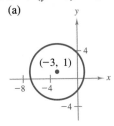

(-3, 1)

(1, 0)

(c)   (d)

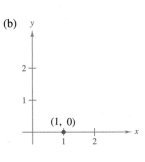

(1, 3)

(0, 0)

(e)   (f)

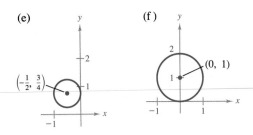

$\left(-\dfrac{1}{2}, \dfrac{3}{4}\right)$

(0, 1)

In Exercises 31–40, write the equation of the specified circle in general form.

31. Center: $(0, 0)$; radius: 3
32. Center: $(0, 0)$; radius: 5
33. Center: $(2, -1)$; radius: 4
34. Center: $(-4, 3)$; radius: $\dfrac{5}{8}$
35. Center: $(-1, 2)$; point on circle: $(0, 0)$
36. Center: $(3, -2)$; point on circle: $(-1, 1)$
37. Endpoints of diameter: $(2, 5)$, $(4, -1)$
38. Endpoints of diameter: $(1, 1)$, $(-1, -1)$
39. Points on circle: $(0, 0)$, $(0, 8)$, $(6, 0)$
40. Points on circle: $(1, -1)$, $(2, -2)$, $(0, -2)$

In Exercises 41–48, write the given equation (of a circle) in standard form and sketch its graph.

41. $x^2 + y^2 - 2x + 6y + 6 = 0$
42. $x^2 + y^2 - 2x + 6y - 15 = 0$
43. $x^2 + y^2 - 2x + 6y + 10 = 0$
44. $3x^2 + 3y^2 - 6y - 1 = 0$
45. $2x^2 + 2y^2 - 2x - 2y - 3 = 0$
46. $4x^2 + 4y^2 - 4x + 2y - 1 = 0$
47. $16x^2 + 16y^2 + 16x + 40y - 7 = 0$
48. $x^2 + y^2 - 4x + 2y + 3 = 0$

49. Find an equation for the path of a communications satellite in a circular orbit 22,000 miles above the earth. (Assume that the radius of the earth is 4000 miles.)
50. Find the equation of the circle passing through the points $(1, 2)$, $(-1, 2)$, and $(2, 1)$.
51. Find the equation of the circle passing through the points $(4, 3)$, $(-2, -5)$, and $(5, 2)$.
52. Find the equations of the circles passing through the points $(4, 1)$ and $(6, 3)$ and having radius $\sqrt{10}$.

In Exercises 53–56, sketch the set of all points satisfying the given inequality.

53. $x^2 + y^2 - 4x + 2y + 1 \le 0$
54. $x^2 + y^2 - 4x + 2y + 1 > 0$
55. $(x + 3)^2 + (y - 1)^2 < 9$
56. $(x - 1)^2 + \left(y - \dfrac{1}{2}\right)^2 > 1$

57. Prove that

$$\left(\frac{2x_1 + x_2}{3}, \frac{2y_1 + y_2}{3}\right)$$

is one of the points of trisection of the line segment joining (x_1, y_1) and (x_2, y_2). Also, find the midpoint of the line segment joining

$$\left(\frac{2x_1 + x_2}{3}, \frac{2y_1 + y_2}{3}\right)$$

and (x_2, y_2) to find the second point of trisection.

58. Use the results of Exercise 57 to find the points of trisection of the line segment joining the following points.
(a) $(1, -2)$ and $(4, 1)$ (b) $(-2, -3)$ and $(0, 0)$

59. Prove that the line segments joining the midpoints of the opposite sides of a quadrilateral bisect each other.

60. Prove that the midpoint of the hypotenuse of a right triangle is equidistant from each of the three vertices.

61. Prove that an angle inscribed in a semicircle is a right angle.

62. Prove that the perpendicular bisector of a chord of a circle passes through the center of the circle.

63. Prove the Midpoint Formula (Theorem 1.5).

1.3 Graphs of Equations

The graph of an equation ▪ Point-plotting method ▪ Intercepts of a graph ▪ Symmetry of a graph ▪ Points of intersection ▪ Mathematical models

Using a graph to show how two quantities are related is common. News magazines frequently show graphs that compare the gross national product or the unemployment rate to the time of year. Industries and businesses use graphs to report their monthly production and sales statistics. The value of such graphs is that they provide a geometric picture of the way one quantity changes with respect to another.

Frequently a relationship between two quantities is expressed as an equation. For instance, degrees on the Fahrenheit scale are related to degrees on the Celsius scale by the equation $F = \frac{9}{5}C + 32$. In this section we introduce a basic procedure for sketching the graph of such an equation.

TABLE 1.2

x	0	1	2	3	4
y	7	4	1	-2	-5

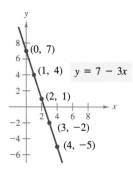

Solution points of $y = 7 - 3x$

FIGURE 1.24

The graph of an equation

Consider the equation $3x + y = 7$. If $x = 2$ and $y = 1$, the equation is satisfied and we call the point $(2, 1)$ a **solution point** of the equation. Of course, there are other solution points, such as $(1, 4)$ and $(0, 7)$. We can construct a **table of values** for x and y by choosing arbitrary values for x and determining the corresponding values for y. To determine the values for y, it is convenient to write the equation in the form

$$y = 7 - 3x.$$

Thus, $(0, 7)$, $(1, 4)$, $(2, 1)$, $(3, -2)$, and $(4, -5)$ are all solution points of the equation $3x + y = 7$, as shown in Table 1.2. Actually, there are infinitely many solution points of this equation, and the set of all such points is called the **graph** of the equation, as shown in Figure 1.24.

REMARK Even though we refer to the sketch shown in Figure 1.24 as the graph of $y = 7 - 3x$, it really represents only a *portion* of the graph. The entire graph would extend beyond the page.

DEFINITION OF THE GRAPH OF AN EQUATION IN TWO VARIABLES	The **graph of an equation** involving two variables x and y is the set of all points in the plane that are solution points of the equation.

EXAMPLE 1   The point-plotting method

Sketch the graph of the equation $y = x^2 - 2$.

SOLUTION

First, we make a table of values (Table 1.3) by choosing several convenient values of x and calculating the corresponding values of $y = x^2 - 2$. Next, we locate these points in the plane, as in Figure 1.25(a). Finally, we connect the points by a *smooth curve*, as shown in Figure 1.25(b). This particular graph is called a **parabola.** It is one of the conic sections we will study in Chapter 10.

TABLE 1.3

x	-2	-1	0	1	2	3
y	2	-1	-2	-1	2	7

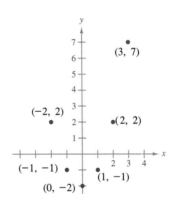

(a) Plot several points.

(b) Connect points with a smooth curve.

FIGURE 1.25

We call this method of sketching a graph the *point-plotting method*. It consists of three basic steps.

1. Make up a table of several solution points of the equation.
2. Plot these points in the plane.
3. Connect the points with a smooth curve.

In later chapters, we will discuss more sophisticated graphing techniques. In the meantime, when using the point-plotting method, we must plot a sufficient number of points to reveal the basic shape of the graph. With too few solution points, we can grossly misrepresent the graph of a given equation. For instance, how would you connect the four points shown in Figure 1.26? Without additional points or more information about the equation, any one of the three graphs shown in Figure 1.27 would be reasonable.

FIGURE 1.26

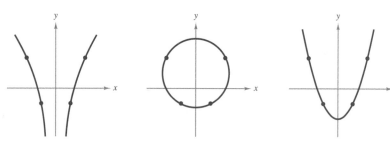

FIGURE 1.27

Two types of solution points that are especially useful are those having zero as either their *x*- or *y*-coordinate. Such points are called **intercepts,** because they are the points at which the graph intersects the *x*- or *y*-axis. Specifically, the point $(a, 0)$ is called an ***x*-intercept** of the graph of an equation if it is a solution point of the equation. Such points can be found by letting *y* be zero and solving the equation for *x*. Similarly, the point $(0, b)$ is called a ***y*-intercept** of the graph of an equation if it is a solution point of the equation. Such points can be found by letting *x* be zero and solving the equation for *y*.

REMARK   Some texts denote the *x*-intercept as the *x*-coordinate of the point $(a, 0)$ rather than the point itself. Unless it is necessary to make a distinction, we will use the term *intercept* to mean either the point or the coordinate.

It is possible for a graph to have no intercepts, or it might have several. For instance, consider the four graphs shown in Figure 1.28.

No *x*-intercept
One *y*-intercept

Three *x*-intercepts
One *y*-intercept

One *x*-intercept
Two *y*-intercepts

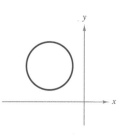

No intercepts

FIGURE 1.28

EXAMPLE 2 Finding *x*- and *y*-intercepts

Find the *x*- and *y*-intercepts for the graphs of the following equations.

(a) $y = x^3 - 4x$ (b) $y^2 - 3 = x$

SOLUTION

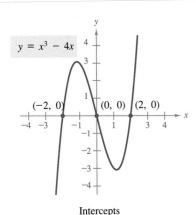

$y = x^3 - 4x$

(−2, 0)   (0, 0)   (2, 0)

Intercepts

FIGURE 1.29

(a) Let $y = 0$. Then, $0 = x(x^2 - 4)$ has solutions $x = 0$ and $x = \pm 2$.

    x-intercepts: (0, 0), (2, 0), (−2, 0)

  Let $x = 0$. Then $y = 0$.

    y-intercept: (0, 0)

  (See Figure 1.29.)

(b) Let $y = 0$. Then $-3 = x$.

    x-intercept: (−3, 0)

  Let $x = 0$. Then $y^2 - 3 = 0$ has solutions $y = \pm\sqrt{3}$.

    y-intercepts: $(0, \sqrt{3})$, $(0, -\sqrt{3})$

EXAMPLE 3 Sketching the graph of an equation

Sketch the graph of the equation $x^2 + 4y^2 = 16$.

SOLUTION

Sometimes it helps to rewrite an equation before calculating solution points. For example, if we rewrite the equation $x^2 + 4y^2 = 16$ as

$$x = \pm\sqrt{16 - 4y^2} = \pm 2\sqrt{4 - y^2}$$

then we can easily determine several solution points by choosing values for *y* and calculating the corresponding values for *x*. (Note that $x = \pm 2\sqrt{4 - y^2}$ is defined only when $|y| \le 2$.) By plotting these points and connecting them with a smooth curve, we create the graph shown in Figure 1.30. This particular graph is called an **ellipse.** It is one of the conic sections we will study in Chapter 10.

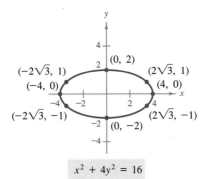

$(-2\sqrt{3}, 1)$
$(-4, 0)$
$(0, 2)$
$(2\sqrt{3}, 1)$
$(4, 0)$
$(-2\sqrt{3}, -1)$
$(2\sqrt{3}, -1)$
$(0, -2)$

$x^2 + 4y^2 = 16$

FIGURE 1.30

Symmetry of a graph

The graphs shown in Figures 1.25(b) and 1.30 are said to be **symmetric** with respect to the y-axis. This means that if the Cartesian plane were folded along the y-axis, the portion of the graph to the left of the y-axis would coincide with the portion to the right of the y-axis. Another way to describe this symmetry is to say that the graph is a reflection of itself with respect to the y-axis. Symmetry with respect to the x-axis can be described similarly.

Knowing that a graph has symmetry *before* attempting to sketch it is helpful because then we need only half as many solution points as we would otherwise. We define three basic types of symmetry, as shown in Figure 1.31.

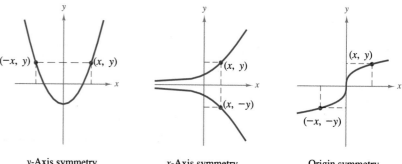

FIGURE 1.31 y-Axis symmetry x-Axis symmetry   Origin symmetry

---

DEFINITION OF SYMMETRY

A graph is said to be **symmetric with respect to the y-axis** if, whenever (x, y) is a point on the graph, $(-x, y)$ is also a point on the graph.

A graph is said to be **symmetric with respect to the x-axis** if, whenever (x, y) is a point on the graph, $(x, -y)$ is also a point on the graph.

A graph is said to be **symmetric with respect to the origin** if, whenever (x, y) is a point on the graph, $(-x, -y)$ is also a point on the graph.

REMARK   Note that a graph is symmetric with respect to the origin if a rotation of $180°$ (about the origin) leaves the graph unchanged.

Suppose we apply the definition of symmetry to the graph of the equation shown in Figure 1.25(b).

$$y = x^2 - 2 \qquad \text{Given equation}$$
$$y = (-x)^2 - 2 \qquad \text{Replace } x \text{ by } -x$$
$$y = x^2 - 2 \qquad \text{Equivalent equation}$$

Since substituting $-x$ for x produces an equivalent equation, it follows that if (x, y) is a solution point of the given equation, then $(-x, y)$ must also be a solution point. Therefore, the graph of $y = x^2 - 2$ is symmetric with respect to the y-axis.

A similar test can be made for symmetry with respect to the x-axis or the origin. These three tests are summarized as follows.

TESTS FOR SYMMETRY	1. The graph of an equation in x and y is symmetric with respect to the y-axis if replacing x by $-x$ yields an equivalent equation. 2. The graph of an equation in x and y is symmetric with respect to the x-axis if replacing y by $-y$ yields an equivalent equation. 3. The graph of an equation in x and y is symmetric with respect to the origin if replacing x by $-x$ *and* y by $-y$ yields an equivalent equation.

EXAMPLE 4   Testing for origin symmetry

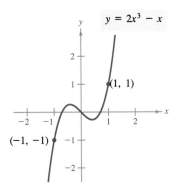

Origin Symmetry

FIGURE 1.32

Show that the graph of $y = 2x^3 - x$ is symmetric with respect to the origin.

SOLUTION

We apply the test for origin symmetry as follows.

$$y = 2x^3 - x \qquad \text{Given equation}$$
$$-y = 2(-x)^3 - (-x) \qquad \text{Replace } x \text{ by } -x \text{ and } y \text{ by } -y$$
$$-y = -2x^3 + x$$
$$y = 2x^3 - x \qquad \text{Equivalent equation}$$

Since the replacement produces an equivalent equation, we conclude that the graph of $y = 2x^3 - x$ is symmetric with respect to the origin, as shown in Figure 1.32. ▭

EXAMPLE 5   Using symmetry to sketch a graph

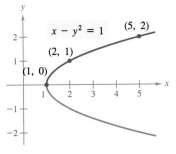

First, plot the points above the x-axis, then use symmetry to complete the graph.

FIGURE 1.33

Sketch the graph of $x - y^2 = 1$.

SOLUTION

The graph is symmetric with respect to the x-axis since replacing y by $-y$ yields

$$x - (-y)^2 = 1$$
$$x - y^2 = 1.$$

This means that the graph below the x-axis is a mirror image of the graph above the x-axis. Hence, we first sketch the graph above the x-axis and then reflect it to obtain the entire graph, as shown in Figure 1.33. ▭

Points of intersection

Since each point of a graph is a solution point of its corresponding equation, a **point of intersection** of two graphs is simply a solution point that satisfies both equations. Moreover, the points of intersection of two graphs can be found by solving the equations simultaneously.

EXAMPLE 6 Finding points of intersection

Find all points of intersection of the graphs of

$$x^2 - y = 3 \quad \text{and} \quad x - y = 1.$$

SOLUTION

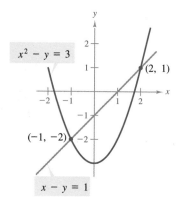

$x^2 - y = 3$

(2, 1)

(−1, −2)

$x - y = 1$

Two Points of Intersection

FIGURE 1.34

It is helpful to begin by making a sketch of each equation on the *same* coordinate plane, as shown in Figure 1.34. Having done this, it appears that the two graphs have two points of intersection. To find these two points, we proceed as follows.

$$y = x^2 - 3 \qquad \text{Solve first equation for } y$$
$$y = x - 1 \qquad \text{Solve second equation for } y$$
$$x^2 - 3 = x - 1 \qquad \text{Equate } y\text{-values}$$
$$x^2 - x - 2 = 0$$
$$(x - 2)(x + 1) = 0 \qquad \text{Solve for } x$$

The corresponding values of y are obtained by substituting $x = 2$ and $x = -1$ into either of the original equations. For instance, if we choose the equation $y = x - 1$, then the values of y are 1 and -2, respectively. Therefore, the two points of intersection are $(2, 1)$ and $(-1, -2)$.

Mathematical models

In applications we frequently use equations to form **mathematical models** of real-world phenomena. In developing a mathematical model to represent actual data, we strive for two (often conflicting) goals: accuracy and simplicity. That is, we want the model to be simple enough to be workable, yet accurate enough to produce meaningful results. Our next example describes a typical mathematical model.

EXAMPLE 7 A mathematical model

The median income (between 1955 and 1985) for married couples in the United States is given in Table 1.4. A mathematical model* for these data is given by

$$y = 0.033286t^2 - 0.130718t + 5.05716$$

where y represents the median income in thousands of dollars and t represents the year, with $t = 0$ corresponding to 1955. Using a graph, compare the data with the model and use the model to predict the median income for 1990.

---

*This model was developed using a procedure called the method of least squares. For a discussion of this method, see Section 14.9, Exercise 21.

TABLE 1.4

Year	1955	1960	1965	1970	1975	1980	1985
Income (in 1000s)	4.6	5.9	7.3	10.5	14.9	23.1	31.1

SOLUTION

Table 1.5 and Figure 1.35 compare the values given by the model with the actual values.

Model:
$y = 0.033286t^2 - 0.130718t + 5.05716$

FIGURE 1.35

TABLE 1.5

t	0	5	10	15	20	25	30
y		5.1	5.2	7.1	10.6	15.8	22.6	31.1
Actual (income)		4.6	5.9	7.3	10.5	14.9	23.1	31.1

To predict the median income for 1990, we let $t = 35$ and calculate y as follows.

$$y = 0.033286(35^2) - 0.130718(35) + 5.05716 \approx 41.3$$

Thus, we estimate that the 1990 median income will be $41,300.

EXERCISES for Section 1.3

In Exercises 1–6, match the given equation with its graph. [Graphs are labeled (a)–(f).]

1. $y = x - 2$ **2.** $y = -\dfrac{1}{2}x + 2$

3. $y = x^2 + 2x$ **4.** $y = \sqrt{9 - x^2}$

5. $y = 4 - x^2$ **6.** $y = x^3 - x$

(a)

(b)

(c)

(d)

(e)

(f)
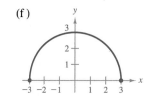

In Exercises 7–16, find the intercepts.

7. $y = 2x - 3$

8. $y = (x - 1)(x - 3)$

9. $y = x^2 + x - 2$

10. $y^2 = x^3 - 4x$

11. $y = x^2\sqrt{9 - x^2}$

12. $xy = 4$

13. $y = \dfrac{x - 1}{x - 2}$

14. $y = \dfrac{x^2 + 3x}{(3x + 1)^2}$

15. $x^2y - x^2 + 4y = 0$

16. $y = 2x - \sqrt{x^2 + 1}$

In Exercises 17–26, check for symmetry with respect to both axes and to the origin.

17. $y = x^2 - 2$

18. $y = x^4 - x^2 + 3$

19. $x^2y - x^2 + 4y = 0$

20. $x^2y - x^2 - 4y = 0$

21. $y^2 = x^3 - 4x$

22. $xy^2 = -10$

23. $y = x^3 + x$

24. $xy = 1$

25. $y = \dfrac{x}{x^2 + 1}$

26. $y = x^3 + x - 3$

In Exercises 27–46, sketch the graph of each equation. Identify the intercepts and test for symmetry.

27. $y = x$

28. $y = x - 2$

29. $y = x + 3$

30. $y = 2x - 3$

31. $y = -3x + 2$

32. $y = -\dfrac{1}{2}x + 2$

33. $y = \dfrac{1}{2}x - 4$

34. $y = x^2 + 3$

35. $y = 1 - x^2$

36. $y = 2x^2 + x$

37. $y = -2x^2 + x + 1$

38. $y = x^3 - 1$

39. $y = x^3 + 2$

40. $y = \sqrt{9 - x^2}$

41. $x^2 + 4y^2 = 4$

42. $9x^2 + y^2 = 9$

43. $y = (x + 2)^2$

44. $x = y^2 - 4$

45. $y = \dfrac{1}{x}$

46. $y = 2x^4$

In Exercises 47–56, find the points of intersection of the graphs of the equations; check your results.

47. $x + y = 2,\ 2x - y = 1$

48. $2x - 3y = 13,\ 5x + 3y = 1$

49. $x + y = 7,\ 3x - 2y = 11$

50. $x^2 + y^2 = 25,\ 2x + y = 10$

51. $x^2 + y^2 = 5,\ x - y = 1$

52. $x^2 + y = 4,\ 2x - y = 1$

53. $y = x^3,\ y = x$

54. $y = x^4 - 2x^2 + 1,\ y = 1 - x^2$

55. $y = x^3 - 2x^2 + x - 1,\ y = -x^2 + 3x - 1$

56. $x = 3 - y^2,\ y = x - 1$

In Exercises 57 and 58, find the sales necessary to break even ($R = C$) for the given cost C of x units and the given revenue R obtained by selling x units.

57. $C = 8650x + 250,000$
$R = 9950x$

58. $C = 5.5\sqrt{x} + 10,000$
$R = 3.29x$

In Exercises 59–62, determine whether the points lie on the graph of the given equation.

59. Equation: $2x - y - 3 = 0$
Points: $(1, 2), (1, -1), (4, 5)$

60. Equation: $x + y^2 = 4$
Points: $(1, -\sqrt{3}), \left(\dfrac{1}{2}, -1\right), \left(\dfrac{3}{2}, \dfrac{7}{2}\right)$

61. Equation: $x^2y - x^2 + 4y = 0$
Points: $\left(1, \dfrac{1}{5}\right), \left(2, \dfrac{1}{2}\right), (-1, -2)$

62. Equation: $x^2 - xy + 4y = 3$
Points: $(0, 2), \left(-2, -\dfrac{1}{6}\right), (3, -6)$

63. For what values of k does the graph of $y = kx^3$ pass through the given point?
(a) $(1, 4)$ (b) $(-2, 1)$
(c) $(0, 0)$ (d) $(-1, -1)$

64. For what values of k does the graph of $y^2 = 4kx$ pass through the given point?
(a) $(1, 1)$ (b) $(2, 4)$
(c) $(0, 0)$ (d) $(3, 3)$

65. The Consumer Price Index (CPI) for selected years is given in the following table.

Year	1970	1975	1980	1985	1987
CPI	116.3	161.2	246.8	322.2	333.9

A mathematical model for the CPI during this time period is

$$y = 0.1t^2 + 11.9t + 111.4$$

where y represents the CPI and t represents the year, with $t = 0$ corresponding to 1970.
(a) Use a graph to compare the CPI with the model.
(b) Use the model to predict the CPI for 1995.

66. From the model in Exercise 65, we obtain the model

$$V = \dfrac{1000}{t^2 + 119t + 1114}$$

where V represents the purchasing power of the dollar (in terms of constant 1967 dollars) and t represents the year, with $t = 0$ corresponding to 1970. Use the model to complete the following table.

t	0	5	10	15	20	25
V						

67. The farm population in the United States as a percentage of the total population for selected years is given in the following table.

Year	1950	1960	1970	1980	1985
Percentage	15.3	8.7	4.8	2.7	2.2

A mathematical model for these data is given by

$$y = \frac{1000}{11t + 27}$$

where y represents the percentage and t represents the year, with $t = 0$ corresponding to 1950.
(a) Use a graph to compare the actual percentage with that given by the model.
(b) Use the model to predict the farm percentage of the population in 1995.

68. The average number of acres per farm in the United States for selected years is given in the following table.

Year	1950	1960	1970	1980	1985
Number of acres	213	297	374	427	446

A mathematical model for these data is given by

$$y = -0.08t^2 + 9.69t + 211.79$$

where y represents the average acreage and t represents the year, with $t = 0$ corresponding to 1950.
(a) Use a graph to compare the actual number of acres per farm with that given by the model.
(b) Use the model to predict the average number of acres per farm in the United States in 1995.

In Exercises 69 and 70, use a computer or graphics calculator to sketch the graph of the equation and find its intercepts.

69. $y = \dfrac{1}{16}(x^5 - 6x^4 + 9x^3 + 32)$

70. $y = \dfrac{5}{x^2 + 1} - 1$

71. Prove that if a graph is symmetric with respect to the x-axis and to the y-axis, then it is symmetric with respect to the origin. Give an example to show that the converse is not true.

72. Prove that if a graph is symmetric with respect to one axis and the origin, then it is symmetric with respect to the other axis also.

1.4 Lines in the Plane

The slope of a line ▪ Equations of lines ▪ Sketching the graph of a line ▪ Parallel lines ▪ Perpendicular lines

In Chapter 3, you will see that one of the primary problems in calculus is measuring (instantaneous) rates of change. In this section, we discuss the noncalculus version of this problem—measuring an *average* rate of change. We begin with an example.

Consider an automobile that is traveling at a *constant* rate on a straight highway. At 2:00 P.M. the car has traveled 20 miles from a particular city, and at 4:00 P.M. the car has traveled 132 miles, as shown in Figure 1.36. How fast is the car traveling?

FIGURE 1.36

To measure the rate in *miles per hour*, we divide the distance traveled by the elapsed time.

$$\text{rate (mph)} = \frac{\text{distance (miles)}}{\text{time (hours)}} = \frac{132 - 20}{4 - 2} = \frac{112}{2} = 56 \text{ mph}$$

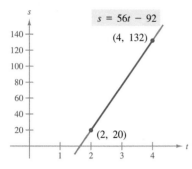

FIGURE 1.37

If we let s be the distance from the city in miles and t be the time in hours, then the distance is related to the time by the **linear equation**

$$s = 56t - 92, \qquad 2 \le t \le 4.$$

The graph of the equation $s = 56t - 92$ is a line (in this text, we use the term *line* to mean *straight line*), as shown in Figure 1.37. For every unit that t increases, the distance s increases 56 units. Mathematically, we say that this line has a *slope* of 56.

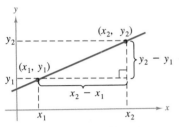

$\Delta y = y_2 - y_1 = \text{change in } y$
$\Delta x = x_2 - x_1 = \text{change in } x$

FIGURE 1.38

The slope of a line

By the **slope** of a (nonvertical) line, we mean the number of units a line rises (or falls) vertically for each unit of horizontal change from left to right. For instance, consider the two points (x_1, y_1) and (x_2, y_2) on the line in Figure 1.38. As we move from left to right along this line, a vertical change of $\Delta y = y_2 - y_1$ units corresponds to a horizontal change of $\Delta x = x_2 - x_1$ units. (Δ is the Greek uppercase letter *delta*, and the symbols Δy and Δx are read "delta y" and "delta x.") We use the ratio of Δy to Δx to define the slope of a line as follows.

DEFINITION OF THE SLOPE OF A LINE

The **slope** m of a nonvertical line passing through the points (x_1, y_1) and (x_2, y_2) is

$$m = \frac{\Delta y}{\Delta x} = \frac{y_2 - y_1}{x_2 - x_1}, \quad x_1 \ne x_2.$$

REMARK   Note that

$$\frac{y_2 - y_1}{x_2 - x_1} = \frac{-(y_1 - y_2)}{-(x_1 - x_2)} = \frac{y_1 - y_2}{x_1 - x_2}.$$

Hence, it does not matter in which order we subtract *as long as* we are consistent and both "subtracted coordinates" come from the same point.

EXAMPLE 1   The slope of a line passing through two points

(a) The slope of the line containing $(-2, 0)$ and $(3, 1)$ is

$$m = \frac{1 - 0}{3 - (-2)} = \frac{1}{3 + 2} = \frac{1}{5}.$$

(b) The slope of the line containing $(-1, 2)$ and $(2, 2)$ is

$$m = \frac{2 - 2}{2 - (-1)} = \frac{0}{3} = 0.$$

(c) The slope of the line containing $(0, 4)$ and $(1, -1)$ is

$$m = \frac{-1 - 4}{1 - 0} = \frac{-5}{1} = -5.$$

(d) We do not define the slope of the vertical line containing $(3, 4)$ and $(3, 1)$.

(See Figure 1.39.)

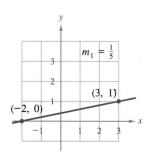

If m is positive, the line rises.

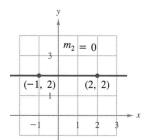

If m is zero, the line is horizontal.

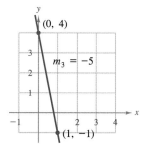

If m is negative, the line falls.

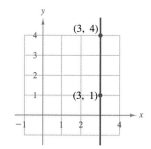

If the line is vertical, the slope is undefined.

FIGURE 1.39

REMARK   Note that we do not define the slope of a vertical line.

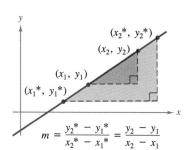

Any two points on a line can be used to determine its slope.

FIGURE 1.40

It is important to realize that *any* two points on a nonvertical line can be used to calculate its slope. This can be verified from the similar triangles shown in Figure 1.40. (See Exercise 77.) (Recall that the ratios of corresponding sides of similar triangles are equal.)

Equations of lines

If we know the slope of a line and one point on the line, how can we determine the equation of the line? Figure 1.40 leads us to the answer to this question. If (x_1, y_1) is a point lying on a line of slope m and (x, y) is any *other* point on the line, then

$$\frac{y - y_1}{x - x_1} = m.$$

This equation, involving the two variables x and y, can be rewritten in the form

$$y - y_1 = m(x - x_1)$$

which is called the **point-slope equation of a line.**

THEOREM 1.7 **POINT-SLOPE EQUATION** **OF A LINE**	The equation of the line with slope m passing through the point (x_1, y_1) is given by $$y - y_1 = m(x - x_1).$$

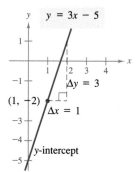

FIGURE 1.41

EXAMPLE 2   The point-slope equation of a line

Find an equation of the line that has a slope of 3 and passes through the point $(1, -2)$.

SOLUTION

$$y - y_1 = m(x - x_1) \qquad \text{Point-slope form}$$
$$y - (-2) = 3(x - 1)$$
$$y + 2 = 3x - 3$$
$$y = 3x - 5$$

(See Figure 1.41.)

EXAMPLE 3   An application: total U.S. sales

The total U.S. sales (including inventories) during the first two quarters of 1978 were 539.9 and 560.2 billion dollars, respectively. Assuming a *linear growth pattern*, estimate the total sales during the fourth quarter of 1978.

SOLUTION

Referring to Figure 1.42, we let $(1, 539.9)$ and $(2, 560.2)$ be two points on the line representing total U.S. sales. We let x represent the quarter and y represent the sales in billions of dollars. The slope of the line passing through these two points is

$$m = \frac{560.2 - 539.9}{2 - 1} = 20.3.$$

Thus, the equation of the line is

$$y - y_1 = m(x - x_1)$$
$$y - 539.9 = 20.3(x - 1)$$
$$y = 20.3(x - 1) + 539.9$$
$$y = 20.3x + 519.6.$$

Now, using this linear model, we estimate the fourth quarter sales ($x = 4$) to be

$$y = (20.3)(4) + 519.6 = 600.8 \text{ billion dollars.}$$

(In this particular case, the estimate proves to be quite good. The actual fourth quarter sales in 1978 were 600.5 billion dollars.)

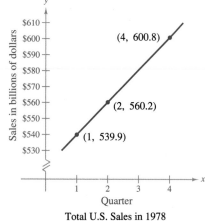

Total U.S. Sales in 1978

FIGURE 1.42

REMARK   The estimation method illustrated in Example 3 is called **linear extrapolation**. Note that the estimated point does not lie between the two given points. When the estimated point lies between the two given points, we call the procedure **linear interpolation**.

Sketching the graph of a line

In Section 1.2, we mentioned that many problems in analytic geometry can be classified in two basic categories: (1) Given a graph, what is its equation? and (2) Given an equation, what is its graph? The point-slope equation of a line fits in the first category. However, this form is *not* particularly useful for solving problems in the second category. The form that is best suited to sketching the graph of a line is called the **slope-intercept** form for the equation of a line.

THEOREM 1.8 **THE SLOPE-INTERCEPT** **EQUATION OF A LINE**	The graph of the equation $$y = mx + b$$ is a line having a *slope* of m and a *y-intercept* at $(0, b)$.

EXAMPLE 4   Sketching lines in the plane

Sketch the graphs of the following linear equations.

(a) $y = 2x + 1$ (b) $y = 2$ (c) $3y + x - 6 = 0$

SOLUTION

(a) Since $b = 1$, the y-intercept occurs at $(0, 1)$, and since the slope is $m = 2$, we know that this line rises 2 units for each unit it moves to the right. (See Figure 1.43(a).)

(b) Since $b = 2$, the y-intercept occurs at $(0, 2)$, and since the slope is $m = 0$, we know that the line is horizontal. That is, it doesn't rise or fall. (See Figure 1.43(b).)

(c) We begin by writing the equation in slope-intercept form.

$$3y + x - 6 = 0$$
$$3y = -x + 6$$
$$y = -\frac{1}{3}x + 2$$

Thus, the y-intercept occurs at $(0, 2)$ and the slope is $m = -\frac{1}{3}$. This means that the line falls 1 unit for every 3 units it moves to the right. (See Figure 1.43(c).)

(a) $m = 2$, line rises.

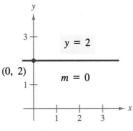

(b) $m = 0$, line is horizontal.

(c) $m = -\frac{1}{3}$, line falls.

FIGURE 1.43

Since the slope of a vertical line is not defined, its equation cannot be written in the slope-intercept form. However, the equation of *any* line can be written in the **general form**

$$Ax + By + C = 0$$

where A and B are not *both* zero. For instance, the vertical line given by $x = a$ can be represented by the general form $x - a = 0$. We summarize the five most common forms of equations of lines in the following list.

SUMMARY OF EQUATIONS OF LINES

1. General form: $Ax + By + C = 0$
2. Vertical line: $x = a$
3. Horizontal line: $y = b$
4. Point-slope form: $y - y_1 = m(x - x_1)$
5. Slope-intercept form: $y = mx + b$

Parallel and perpendicular lines

The slope of a line is a convenient tool for determining whether two lines are parallel or perpendicular. This is seen in the following two theorems.

THEOREM 1.9 PARALLEL LINES

Two distinct nonvertical lines are parallel if and only if their slopes are equal.

THEOREM 1.10 PERPENDICULAR LINES

Two nonvertical lines are perpendicular if and only if their slopes are related by the following equation.

$$m_1 = -\frac{1}{m_2}$$

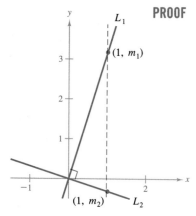

The slopes of perpendicular lines are negative reciprocals of each other.

FIGURE 1.44

PROOF We will prove only one direction of the theorem and leave the other direction as an exercise (see Exercise 78). Let us assume that we are given two nonvertical perpendicular lines L_1 and L_2 with slopes m_1 and m_2. For simplicity's sake let these two lines intersect at the origin, as shown in Figure 1.44. The vertical line $x = 1$ will intersect L_1 and L_2 at the respective points $(1, m_1)$ and $(1, m_2)$. Since the triangle formed by these two points and the origin is a right triangle, we can apply the Pythagorean Theorem and conclude that

$$\left(\begin{array}{c}\text{distance between}\\ (0, 0) \text{ and } (1, m_1)\end{array}\right)^2 + \left(\begin{array}{c}\text{distance between}\\ (0, 0) \text{ and } (1, m_2)\end{array}\right)^2 = \left(\begin{array}{c}\text{distance between}\\ (1, m_1) \text{ and } (1, m_2)\end{array}\right)^2.$$

Using the Distance Formula, we have

$$(\sqrt{1 + m_1^2})^2 + (\sqrt{1 + m_2^2})^2 = (\sqrt{0^2 + (m_1 - m_2)^2})^2$$
$$1 + m_1^2 + 1 + m_2^2 = (m_1 - m_2)^2$$
$$2 + m_1^2 + m_2^2 = m_1^2 - 2m_1m_2 + m_2^2$$
$$2 = -2m_1m_2$$
$$-\frac{1}{m_2} = m_1.$$

EXAMPLE 5   Finding parallel and perpendicular lines

Find an equation for the line that passes through the point $(2, -1)$ and is
(a) parallel to the line $2x - 3y = 5$
(b) perpendicular to the line $2x - 3y = 5$.

SOLUTION

Writing the equation $2x - 3y = 5$ in slope-intercept form, we have

$$y = \frac{2}{3}x - \frac{5}{3}.$$

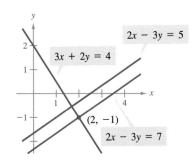

FIGURE 1.45

Therefore, the given line has a slope of $m = \frac{2}{3}$.

(a) The line through $(2, -1)$ that is parallel to the given line has an equation of the form

$$y - (-1) = \frac{2}{3}(x - 2)$$
$$3(y + 1) = 2(x - 2)$$
$$2x - 3y = 7.$$

(See Figure 1.45.) (Note the similarity to the original equation.)

(b) Using the negative reciprocal of the slope of the given line, we find the slope of a line perpendicular to the given line to be $-\frac{3}{2}$. Therefore, the line through the point $(2, -1)$ that is perpendicular to the given line has the equation

$$y - (-1) = -\frac{3}{2}(x - 2)$$
$$2(y + 1) = -3(x - 2)$$
$$3x + 2y = 4.$$

(See Figure 1.45.)

EXERCISES for Section 1.4

In Exercises 1–6, estimate the slope of the given line from its graph.

1.

2.

3.

4.

5.

6.
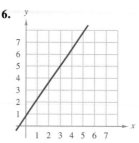

In Exercises 7–12, plot the given pair of points and find the slope of the line passing through them.

7. $(3, -4), (5, 2)$

8. $(-2, 1), (4, -3)$

9. $\left(\frac{1}{2}, 2\right), (6, 2)$

10. $(2, 1), (2, 5)$

11. $(1, 2), (-2, 4)$

12. $\left(\frac{7}{8}, \frac{3}{4}\right), \left(\frac{5}{4}, -\frac{1}{4}\right)$

In Exercises 13–16, use the given point on the line and the slope of the line to find three additional points that the line passes through. (The solution is not unique.)

Point	Slope	Point	Slope
13. $(2, 1)$	$m = 0$	**14.** $(-3, 4)$	m undefined
15. $(1, 7)$	$m = -3$	**16.** $(-2, -2)$	$m = 2$

In Exercises 17–20, find the slope and y-intercept (if possible) of the line specified by the given equation.

17. $x + 5y = 20$

18. $6x - 5y = 15$

19. $x = 4$

20. $y = -1$

In Exercises 21–26, find an equation for the line that passes through the given points, and sketch the graph of the line.

21. $(2, 1), (0, -3)$

22. $(-3, -4), (1, 4)$

23. $(0, 0), (-1, 3)$

24. $(-3, 6), (1, 2)$

25. $(1, -2), (3, -2)$

26. $\left(\frac{7}{8}, \frac{3}{4}\right), \left(\frac{5}{4}, -\frac{1}{4}\right)$

In Exercises 27–32, find an equation of the line that passes through the given point and has the indicated slope. Sketch the line.

Point	Slope	Point	Slope
27. $(0, 3)$	$m = \frac{3}{4}$	**28.** $(-1, 2)$	m undefined
29. $(0, 0)$	$m = \frac{2}{3}$	**30.** $(-2, 4)$	$m = -\frac{3}{5}$
31. $(0, 2)$	$m = 4$	**32.** $(0, 4)$	$m = 0$

33. Find an equation of the vertical line with x-intercept at 3.

34. Show that the line with intercepts $(a, 0)$ and $(0, b)$ has the following equation.

$$\frac{x}{a} + \frac{y}{b} = 1, \quad a \neq 0, b \neq 0$$

In Exercises 35–40, use the result of Exercise 34 to write an equation of the indicated line.

35. x-intercept: $(2, 0)$
y-intercept: $(0, 3)$

36. x-intercept: $(-3, 0)$
y-intercept: $(0, 4)$

37. x-intercept: $\left(-\frac{1}{6}, 0\right)$
y-intercept: $\left(0, -\frac{2}{3}\right)$

38. x-intercept: $\left(-\frac{2}{3}, 0\right)$
y-intercept: $(0, -2)$

39. Point on line: $(1, 2)$
x-intercept: $(a, 0)$
y-intercept: $(0, a)$
$(a \neq 0)$

40. Point on line: $(-3, 4)$
x-intercept: $(a, 0)$
y-intercept: $(0, a)$
$(a \neq 0)$

In Exercises 41–46, write an equation of the line through the given point (a) parallel to the given line and (b) perpendicular to the given line.

	Point	Line
41.	$(2, 1)$	$4x - 2y = 3$
42.	$(-3, 2)$	$x + y = 7$
43.	$\left(\frac{7}{8}, \frac{3}{4}\right)$	$5x + 3y = 0$
44.	$(-6, 4)$	$3x + 4y = 7$
45.	$(2, 5)$	$x = 4$
46.	$(-1, 0)$	$y = -3$

In Exercises 47–52, sketch the graph of the equation.

47. $y = -3$

48. $x = 4$

49. $2x - y - 3 = 0$

50. $x + 2y + 6 = 0$

51. $y = -2x + 1$

52. $y - 1 = 3(x + 4)$

In Exercises 53 and 54, find an equation of the line determined by the points of intersection of the graphs of the parabolas.

53. $y = x^2$
$y = 4x - x^2$

54. $y = x^2 - 4x + 3$
$y = -x^2 + 2x + 3$

In Exercises 55 and 56, determine whether the three given points are collinear (lie on the same straight line).

55. $(-2, 1)$, $(-1, 0)$, $(2, -2)$

56. $(0, 4)$, $(7, -6)$, $(-5, 11)$

In Exercises 57–60, refer to the triangle in the accompanying figure.

FIGURE FOR 57–60

57. Find the coordinates of the point of intersection of the perpendicular bisectors of the sides.

58. Find the coordinates of the point of intersection of the medians.

59. Find the coordinates of the point of intersection of the altitudes.

60. Show that the points of intersection of Exercises 57, 58, and 59 are collinear.

61. Find an equation of the line giving the relationship between the temperature in degrees Celsius C and degrees Fahrenheit F. Use the fact that water freezes at 0° Celsius (32° Fahrenheit) and boils at 100° Celsius (212° Fahrenheit).

62. Use the result of Exercise 61 to complete the following table.

C		$-10°$	$10°$			$177°$
F	0°			68°	90°	

63. A company reimburses its sales representatives $95 per day for lodging and meals plus 25¢ per mile driven. Write a linear equation giving the daily cost C to the company in terms of x, the number of miles driven.

64. A manufacturing company pays its assembly line workers $9.50 per hour *plus* an additional piecework rate of $0.75 per unit produced. Find a linear equation for the hourly wages W in terms of x, the number of units produced per hour.

65. A small business purchases a piece of equipment for $875. After 5 years the equipment will be obsolete and have no value. Write a linear equation giving the value y of the equipment during the 5 years it will be used. (Let t represent the time in years.)

66. A company constructs a warehouse for $825,000. It has an estimated useful life of 25 years, after which its value is expected to be $75,000. Use straight-line depreciation to write a linear equation giving the value y of the warehouse during its 25 years of useful life. (Let t represent the time in years.)

67. A real estate office handles an apartment complex with 50 units. When the rent is $380 per month, all 50 units are occupied. However, when the rent is $425, the average number of occupied units drops to 47. Assume that the relationship between the monthly rent p and the demand x is linear. (Note: Here we use the term *demand* to refer to the number of occupied units.)

(a) Write a linear equation giving the quantity demanded x in terms of the rent p.

(b) (Linear extrapolation) Use this equation to predict the number of units occupied if the rent is raised to $455.

(c) (Linear interpolation) Predict the number of units occupied if the rent is lowered to $395.

68. The number of subscribers to cable TV for the years 1980 and 1986 were 16 million and 37.5 million, respectively. Assume that the relationship between the year t and the number of subscribers y is linear.

(a) Write the equation giving the number of subscribers y in terms of t. (Let $t = 0$ represent 1980.)

(b) [Linear extrapolation] Use this equation to estimate the number of subscribers in 1990.

(c) [Linear interpolation] Estimate the number of subscribers in 1985.

(d) What information is given by the slope of the line in part (a)?

In Exercises 69–74, find the distance between the given point and line (or two lines) using the following formula for the distance between the point (x_1, y_1) and the line $Ax + By + C = 0$.

$$\frac{|Ax_1 + By_1 + C|}{\sqrt{A^2 + B^2}}$$

69. Point: $(0, 0)$; line: $4x + 3y = 10$
70. Point: $(2, 3)$; line: $4x + 3y = 10$
71. Point: $(-2, 1)$; line: $x - y - 2 = 0$
72. Point: $(6, 2)$; line: $x = -1$
73. Lines: $x + y = 1, x + y = 5$
74. Lines: $3x - 4y = 1, 3x - 4y = 10$

75. Prove that the diagonals of a rhombus intersect at right angles.

76. Prove that the figure formed by connecting consecutive midpoints of the sides of any quadrilateral is a parallelogram.
77. Prove that if the points (x_1, y_1) and (x_2, y_2) lie on the same line as (x_1^*, y_1^*) and (x_2^*, y_2^*), then

$$\frac{y_2^* - y_1^*}{x_2^* - x_1^*} = \frac{y_2 - y_1}{x_2 - x_1}.$$

Assume $x_1 \neq x_2$ and $x_2^* \neq x_1^*$.
78. Complete the proof of Theorem 1.10. That is, prove that if the slopes of two nonvertical lines are negative reciprocals of each other, then the lines are perpendicular.

1.5 Functions

Definition of function ▪ Function notation ▪ The graph of a function ▪ Transformations of graphs ▪ Classifications of functions ▪ Combinations of functions

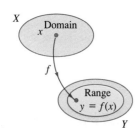

X

Domain
x

f

Range
$y = f(x)$

Y

Many common relationships involve two variables in such a way that the value of one of the variables depends on the value of the other. For example, the sales tax on an item depends on its selling price. The distance an object moves in a given time depends on its speed.

Consider the relationship between the area of a circle and its radius. This relationship can be expressed by the equation $A = \pi r^2$, where the value of A depends on the choice of r. We refer to A as the **dependent variable** and to r as the **independent variable.**

Of particular interest are relationships such that to every value of the independent variable there corresponds *one and only one* value of the dependent variable. We call this type of correspondence a **function.**

FIGURE 1.46

DEFINITION OF A FUNCTION

A **function** f from a set X into a set Y is a correspondence that assigns to each element x in X exactly one element y in Y. We call y the **image** of x under f and denote it by $f(x)$. The **domain** of f is the set X, and the **range** consists of all images of elements in X. (See Figure 1.46.)

If to each value in its range there corresponds exactly one value in its domain, the function is said to be **one-to-one.** Moreover, if the range of f consists of all of Y, then the function is said to be **onto.**

In the first twelve chapters of this text, we work with functions whose domains and ranges are sets of real numbers. We call such functions **real-valued functions of a real variable.**

Functions can be specified in a variety of ways. We will, however, concentrate primarily on functions that are given by equations involving the dependent and independent variables. To evaluate a function described by an equation, we generally isolate the dependent variable on the left side of the equation. For instance, the equation $x + 2y = 1$, written as

$$y = \frac{1 - x}{2}$$

describes y as a function of x, and we can denote this function as

$$f(x) = \frac{1 - x}{2}.$$

This function notation has the advantage of clearly identifying the dependent variable as $f(x)$ while at the same time telling us that x is the independent variable and that the function itself will be called "f." The symbol $f(x)$ is read "f of x." The $f(x)$ notation also allows us to be less wordy. Instead of asking, "What is the value of y that corresponds to $x = 3$?" we can ask, "What is $f(3)$?" In general, to denote the value of the dependent variable when $x = a$, we use the symbol $f(a)$. For example, the value of f when $x = 3$ is

$$f(3) = \frac{1 - (3)}{2} = \frac{-2}{2} = -1.$$

In an equation that defines a function, the role of the variable x is simply that of a placeholder. For instance, the function given by

$$f(x) = 2x^2 - 4x + 1$$

can be described by the form

$$f(\ ) = 2(\ )^2 - 4(\ ) + 1$$

where parentheses are used instead of x. Therefore, to evaluate $f(-2)$, we simply place -2 in each set of parentheses.

$$f(-2) = 2(-2)^2 - 4(-2) + 1 = 2(4) + 8 + 1 = 17$$

REMARK   Although we generally use f as a convenient function name and x as the independent variable, we also can use other symbols. For instance, the following equations all define the same function.

$$f(x) = x^2 - 4x + 7 \qquad f(t) = t^2 - 4t + 7 \qquad g(s) = s^2 - 4s + 7$$

EXAMPLE 1   Evaluating a function

For the function f defined by $f(x) = x^2 + 7$, evaluate the following.

(a) $f(3a)$ (b) $f(b - 1)$ (c) $\dfrac{f(x + \Delta x) - f(x)}{\Delta x}$, $\Delta x \neq 0$

SOLUTION

(a) $f(3a) = (3a)^2 + 7$ Replace x with $3a$
$$= 9a^2 + 7$$

(b) $f(b - 1) = (b - 1)^2 + 7$ Replace x with $b - 1$
$$= b^2 - 2b + 1 + 7$$
$$= b^2 - 2b + 8$$

(c) $\dfrac{f(x + \Delta x) - f(x)}{\Delta x} = \dfrac{[(x + \Delta x)^2 + 7] - [x^2 + 7]}{\Delta x}$

$$= \frac{x^2 + 2x\Delta x + (\Delta x)^2 + 7 - x^2 - 7}{\Delta x}$$

$$= \frac{2x\Delta x + (\Delta x)^2}{\Delta x}$$

$$= \frac{\Delta x(2x + \Delta x)}{\Delta x}$$

$$= 2x + \Delta x$$

REMARK   The ratio in Example 1(c) is called a difference quotient and has a special significance in calculus. We will say more about this in Chapter 3.

The domain of a function may be described explicitly, or it may be described *implicitly* by an equation used to define the function. (The implied domain is the set of all real numbers for which the equation is defined.) For example, the function given by

$$f(x) = \frac{1}{x^2 - 4}, \quad 4 \le x \le 5$$

has an explicitly defined domain given by $\{x: 4 \le x \le 5\}$. On the other hand, the function given by

$$g(x) = \frac{1}{x^2 - 4}$$

has an implied domain which is the set $\{x: x \ne \pm 2\}$. Another common type of implied domain is that used to avoid even roots of negative numbers. For example, the function given by

$$f(x) = \sqrt{x + 2}$$

has the implied domain $\{x: x \ge -2\}$.

EXAMPLE 2   Finding the domain and range of a function

Determine the domain and range for the function of x defined by
$$f(x) = \sqrt{x - 1}.$$

FIGURE 1.47

FIGURE 1.48

The Graph of a Function

FIGURE 1.49

SOLUTION

Since $\sqrt{x - 1}$ is not defined for $x - 1 < 0$ (that is, for $x < 1$), we must have $x \geq 1$. Therefore, the domain is the interval $[1, \infty)$.

To find the range, we observe that $f(x) = \sqrt{x - 1}$ is never negative. Moreover, as x takes on the various values in the domain, $f(x)$ takes on all nonnegative values and we find the range to be the interval $[0, \infty)$.

The graph of the function is shown in Figure 1.47.

EXAMPLE 3 A function defined by more than one equation

Determine the domain and range for the function of x given by

$$f(x) = \begin{cases} 1 - x, & \text{if } x < 1 \\ \sqrt{x - 1}, & \text{if } x \geq 1. \end{cases}$$

SOLUTION

Since f is defined for $x < 1$ and $x \geq 1$, the domain of the function is the entire set of real numbers.

On the portion of the domain for which $x \geq 1$, the function behaves as in Example 2. For $x < 1$, the value of $1 - x$ is positive, and therefore the range of the function is the interval $[0, \infty)$. (See Figure 1.48.)

REMARK Note that the function given in Example 2 is one-to-one whereas the function given in Example 3 is not one-to-one.

The graph of a function

As you study this section, remember that the graph of the function $y = f(x)$ consists of all points $(x, f(x))$ as shown in Figure 1.49, where

> x = the directed distance from the y-axis
> $f(x)$ = the directed distance from the x-axis.

Since, by the definition of a function, there is exactly one y-value for each x-value, it follows that a vertical line can intersect the graph of a function of x at most once. This observation provides us with a convenient visual test for functions. For example, in part (a) of Figure 1.50, we see that the graph does not define y as a function of x since a vertical line intersects the graph twice.

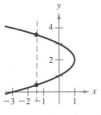

(a) Not a function of x

(b) A function of x

Vertical Line Test for Functions

(c) A function of x

FIGURE 1.50

Figure 1.51 shows the graphs of six basic functions. You need to know. these graphs well.

(a) Identity function

(b)

(c)

(d) Square root function

(e) Absolute value function

(f)

FIGURE 1.51

Function notation lends itself well to describing transformations of graphs in the plane. Some families of graphs all have the same basic shape. For example, consider the graph of $y = x^2$, as shown in Figure 1.52. Now compare this graph to those shown in Figure 1.53.

Original Graph

FIGURE 1.52

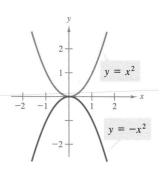

(a) Vertical shift upward

(b) Horizontal shift to the left

(c) Reflection

FIGURE 1.53

Each of the graphs in Figure 1.53 is a **transformation** of the graph of $y = x^2$. The three basic types of transformations illustrated by these graphs are (1) vertical shifts, (2) horizontal shifts, and (3) reflections.

BASIC TYPES OF TRANSFORMATIONS ($c > 0$)		
Original graph:	$y = f(x)$	
Horizontal shift c units to the **right:**	$y = f(x - c)$	
Horizontal shift c units to the **left:**	$y = f(x + c)$	
Vertical shift c units **downward:**	$y = f(x) - c$	
Vertical shift c units **upward:**	$y = f(x) + c$	
Reflection (about the x-axis):	$y = -f(x)$	

Classifications and combinations of functions

The modern notion of a function is derived from the efforts of many seventeenth- and eighteenth-century mathematicians. Of particular note was Leonhard Euler (1707–1783), to whom we are indebted for the function notation $y = f(x)$. By the end of the eighteenth century, mathematicians and scientists had concluded that most real-world phenomena can be represented by mathematical models taken from a basic collection of functions called **elementary functions.** Elementary functions are divided into three categories: (1) algebraic, (2) trigonometric, and (3) logarithmic and exponential. We will review the trigonometric functions in Section 1.6 and introduce the remaining elementary functions in Chapter 6.

The most common type of algebraic function is a **polynomial function**

$$f(x) = a_n x^n + a_{n-1} x^{n-1} + \cdots + a_2 x^2 + a_1 x + a_0, \quad a_n \neq 0$$

where the positive integer n is the **degree** of the polynomial function. The numbers a_i are called **coefficients,** with a_n the **leading coefficient** and a_0 the **constant term** of the polynomial function. It is common practice to use subscript notation for coefficients of general polynomial functions, but for polynomial functions of low degree we often use the following simpler forms.

Leonhard Euler

Zeroth degree:	$f(x) = a$	Constant function
First degree:	$f(x) = ax + b$	Linear function
Second degree:	$f(x) = ax^2 + bx + c$	Quadratic function
Third degree:	$f(x) = ax^3 + bx^2 + cx + d$	Cubic function

Although the graph of a polynomial function can have several turns, eventually the graph will rise or fall without bound as x moves to the right or left. Whether the graph of

$$f(x) = a_n x^n + a_{n-1} x^{n-1} + \cdots + a_2 x^2 + a_1 x + a_0$$

eventually rises or falls can be determined by the function's degree (odd or even) and by the leading coefficient a_n, as indicated in Figure 1.54. Note that the dashed portions of the graphs indicate that the **leading coefficient test** determines *only* the right and left behavior of the graph.

Graphs of polynomial functions of even degree   Graphs of polynomial functions of odd degree

FIGURE 1.54   Leading Coefficient Test for Polynomial Functions

Just as a rational number can be written as the quotient of two integers, a **rational function** can be written as the quotient of two polynomials. Specifically, a function f is **rational** if it has the form

$$f(x) = \frac{p(x)}{q(x)}, \quad q(x) \neq 0$$

where $p(x)$ and $q(x)$ are polynomials.

Polynomial functions and rational functions are two examples of a larger class of functions called **algebraic functions.** An algebraic function is one that can be expressed as a finite number of sums, differences, multiples, quotients, and radicals involving x^n. For example, the following functions are algebraic.

$$f(x) = \sqrt{x + 1} \quad \text{and} \quad g(x) = x + \frac{1}{\sqrt[3]{x + 1}}$$

Functions that are not algebraic are called **transcendental.** For instance, the trigonometric functions discussed in Section 1.6 are transcendental.

Two functions can be combined in various ways to create new functions. For example, if

$$f(x) = 2x - 3 \quad \text{and} \quad g(x) = x^2 + 1$$

we can form the following functions.

$$f(x) + g(x) = (2x - 3) + (x^2 + 1) = x^2 + 2x - 2 \qquad \text{Sum}$$
$$f(x) - g(x) = (2x - 3) - (x^2 + 1) = -x^2 + 2x - 4 \qquad \text{Difference}$$
$$f(x)g(x) = (2x - 3)(x^2 + 1) = 2x^3 - 3x^2 + 2x - 3 \qquad \text{Product}$$
$$\frac{f(x)}{g(x)} = \frac{2x - 3}{x^2 + 1} \qquad \text{Quotient}$$

We can combine two functions in yet another way to form what is called a **composite function.**

DEFINITION OF COMPOSITE FUNCTION	Let f and g be functions. The function given by $(f \circ g)(x) = f(g(x))$ is called the **composite** of f with g. The domain of $f \circ g$ is the set of all x in the domain of g such that $g(x)$ is in the domain of f.

It is important to realize that the composite of f with g may not be equal to the composite of g with f. This is illustrated in the following example.

EXAMPLE 4   Composition of functions

Given $f(x) = 2x - 3$ and $g(x) = x^2 + 1$, find $f \circ g$ and $g \circ f$.

SOLUTION

Since $f(x) = 2x - 3$, we have

$$(f \circ g)(x) = f(g(x)) = 2(g(x)) - 3 = 2(x^2 + 1) - 3 = 2x^2 - 1$$

and since $g(x) = x^2 + 1$, we have

$$(g \circ f)(x) = g(f(x)) = (f(x))^2 + 1 = (2x - 3)^2 + 1 = 4x^2 - 12x + 10.$$

Note that $(f \circ g)(x) \neq (g \circ f)(x)$.   ▭

In Section 1.3, we defined an x-intercept of a graph to be a point $(a, 0)$ at which the graph crosses the x-axis. If the graph represents a function f, then the number a is called a **zero** of f. In other words, the zeros of a function f are the solutions of the equation $f(x) = 0$. For example, the function $f(x) = x - 4$ has a zero at $x = 4$ because $f(4) = 0$.

In Section 1.3 we also discussed different types of symmetry. In the terminology of functions, we say that a function is **even** if its graph is symmetric with respect to the y-axis, and a function is **odd** if its graph is symmetric with respect to the origin. Thus, the symmetry tests in Section 1.3 yield the following test for even and odd functions.

THEOREM 1.11 TEST FOR EVEN AND ODD FUNCTIONS	The function $y = f(x)$ is **even** if $f(-x) = f(x)$. The function $y = f(x)$ is **odd** if $f(-x) = -f(x)$.

REMARK   Except for such trivial cases as the constant function $f(x) = 0$, the graph of a function cannot have symmetry with respect to the x-axis because it then would fail the vertical line test for the graph of a function.

EXAMPLE 5   Even and odd functions

Determine whether the following functions are even, odd, or neither. In each case find the zeros of the function.

(a) $f(x) = x^3 - x$ (b) $g(x) = x^2 + 1$

SOLUTION

(a) This function is odd since

$$f(-x) = (-x)^3 - (-x) = -x^3 + x = -(x^3 - x) = -f(x).$$

The zeros of f are found as follows.

$$x^3 - x = 0 \qquad \text{Let } f(x) = 0$$
$$x(x^2 - 1) = x(x - 1)(x + 1) = 0 \qquad \text{Factor}$$
$$x = 0, 1, -1 \qquad \text{Zeros of } f$$

(b) This function is even since

$$g(-x) = (-x)^2 + 1 = x^2 + 1 = g(x).$$

It has no zeros since $x^2 + 1$ is positive for all x. (See Figure 1.55.)

REMARK   Each of the functions in Example 5 is either even or odd. However, some functions such as

$$f(x) = x^2 + x + 1$$

are neither even nor odd.

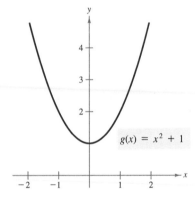

(a) Odd function

(b) Even function

FIGURE 1.55

EXERCISES for Section 1.5

1. Given $f(x) = 2x - 3$, find the following.
   (a) $f(0)$ (b) $f(-3)$
   (c) $f(b)$ (d) $f(x - 1)$
2. Given $f(x) = x^2 - 2x + 2$, find the following.
   (a) $f\left(\dfrac{1}{2}\right)$ (b) $f(-1)$
   (c) $f(c)$ (d) $f(x + \Delta x)$
3. Given $f(x) = \sqrt{x + 3}$, find the following.
   (a) $f(-2)$ (b) $f(6)$
   (c) $f(c)$ (d) $f(x + \Delta x)$

4. Given $f(x) = 1/\sqrt{x}$, find the following.
   (a) $f(2)$ (b) $f\left(\dfrac{1}{4}\right)$
   (c) $f(x + \Delta x)$ (d) $f(x + \Delta x) - f(x)$
5. Given $f(x) = |x|/x$, find the following.
   (a) $f(2)$ (b) $f(-2)$
   (c) $f(x^2)$ (d) $f(x - 1)$
6. Given $f(x) = |x| + 4$, find the following.
   (a) $f(2)$ (b) $f(-2)$
   (c) $f(x^2)$ (d) $f(x + \Delta x) - f(x)$

7. Given $f(x) = x^2 - x + 1$, find

$$\frac{f(2 + \Delta x) - f(2)}{\Delta x}.$$

8. Given $f(x) = 1/x$, find

$$\frac{f(1 + \Delta x) - f(1)}{\Delta x}.$$

9. Given $f(x) = x^3$, find

$$\frac{f(x + \Delta x) - f(x)}{\Delta x}.$$

10. Given $f(x) = 3x - 1$, find

$$\frac{f(x) - f(1)}{x - 1}.$$

11. Given $f(x) = 1/\sqrt{x - 1}$, find

$$\frac{f(x) - f(2)}{x - 2}.$$

12. Given $f(x) = x^3 - x$, find

$$\frac{f(x) - f(1)}{x - 1}.$$

In Exercises 13–22, find the domain and range of the given function, and sketch its graph.

13. $f(x) = 4 - x$

14. $f(x) = \frac{1}{3}x$

15. $f(x) = 4 - x^2$

16. $g(x) = \frac{4}{x}$

17. $h(x) = \sqrt{x - 1}$

18. $f(x) = \frac{1}{2}x^3 + 2$

19. $f(x) = \sqrt{9 - x^2}$

20. $h(x) = \sqrt{25 - x^2}$

21. $f(x) = |x - 2|$

22. $f(x) = \frac{|x|}{x}$

In Exercises 23–28, use the vertical line test to determine whether y is a function of x.

23. $y = x^2$

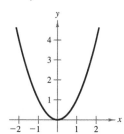

24. $y = x^3 - 1$

25. $x - y^2 = 0$

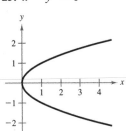

26. $x^2 + y^2 = 9$

27. $\sqrt{x^2 - 4} - y = 0$

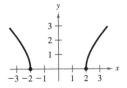

28. $x - xy + y + 1 = 0$

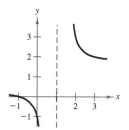

In Exercises 29–36, determine whether y is a function of x.

29. $x^2 + y^2 = 4$

30. $x = y^2$

31. $x^2 + y = 4$

32. $x + y^2 = 4$

33. $2x + 3y = 4$

34. $x^2 + y^2 - 4y = 0$

35. $y^2 = x^2 - 1$

36. $x^2y - x^2 + 4y = 0$

37. Use the graph of $f(x) = \sqrt{x}$ to sketch the graph of each of the following.
(a) $y = \sqrt{x} + 2$ (b) $y = -\sqrt{x}$
(c) $y = \sqrt{x} - 2$ (d) $y = \sqrt{x} + 3$
(e) $y = \sqrt{x} - 4$ (f) $y = 2\sqrt{x}$

38. Use the graph of $f(x) = 1/x$ to sketch the graph of each of the following.

(a) $y = \frac{1}{x} - 1$ (b) $y = \frac{1}{x + 1}$

(c) $y = \frac{1}{x - 1}$ (d) $y = -\frac{1}{x}$

(e) $y = \frac{4}{x}$ (f) $y = -\frac{1}{x} + 2$

39. Use the graph of $f(x) = x^2$ to determine a formula for the indicated function.
(a)   (b)

40. Use the graph of $f(x) = |x|$ to determine a formula for the indicated function.

(a)

(b)

41. Given $f(x) = \sqrt{x}$ and $g(x) = x^2 - 1$, find the following.

(a) $f(g(1))$ (b) $g(f(1))$
(c) $g(f(0))$ (d) $f(g(-4))$
(e) $f(g(x))$ (f) $g(f(x))$

42. Given $f(x) = 1/x$ and $g(x) = x^2 - 1$, find the following.

(a) $f(g(2))$ (b) $g(f(2))$

(c) $f\left(g\left(\dfrac{1}{\sqrt{2}}\right)\right)$ (d) $g\left(f\left(\dfrac{1}{\sqrt{2}}\right)\right)$

(e) $g(f(x))$ (f) $f(g(x))$

In Exercises 43–46, find the composite functions $(f \circ g)$ and $(g \circ f)$. What is the domain of each function? Are the two composite functions equal?

43. $f(x) = x^2,$ $g(x) = \sqrt{x}$
44. $f(x) = x^3,$ $g(x) = \sqrt[3]{x}$
45. $f(x) = x + 1,$ $g(x) = \dfrac{1}{x}$
46. $f(x) = x^2 - 1,$ $g(x) = x$

In Exercises 47–50, find the (real) zeros of the given function.

47. $f(x) = x^2 - 9$ **48.** $f(x) = x^3 - x$
49. $f(x) = \dfrac{3}{x-1} + \dfrac{4}{x-2}$ **50.** $f(x) = a + \dfrac{b}{x}$

In Exercises 51–54, determine whether the function is even, odd, or neither.

51. $f(x) = 4 - x^2$ **52.** $f(x) = \sqrt[3]{x}$
53. $f(x) = x(4 - x^2)$ **54.** $f(x) = 4x - x^2$

55. Show that the following function is odd.

$$f(x) = a_{2n+1}x^{2n+1} + \cdots + a_3x^3 + a_1x$$

56. Show that the following function is even.

$$f(x) = a_{2n}x^{2n} + a_{2n-2}x^{2n-2} + \cdots + a_2x^2 + a_0$$

57. Show that the product of two even (or two odd) functions is even.

58. Show that the product of an odd function and an even function is odd.

In Exercises 59–62, express the indicated values as functions of x.

59. R and r

60. R and r

61. h and p

62. h and p

63. A rectangle has a perimeter of 100 feet (see figure). Express the area A of the rectangle as a function of x.

64. A rancher has 200 feet of fencing to enclose two adjacent rectangular corrals (see figure). Express the area A of the enclosures as a function of x.

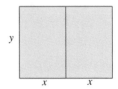

FIGURE FOR 63   FIGURE FOR 64

65. An open box is to be made from a square piece of material 12 inches on a side, by cutting equal squares from each corner and turning up the sides (see figure). Express the volume V as a function of x.

66. A rectangle is bounded by the x-axis and the semicircle $y = \sqrt{25 - x^2}$ (see figure). Write the area A of the rectangle as a function of x.

FIGURE FOR 65

$y = \sqrt{25 - x^2}$

FIGURE FOR 66

69. A man is in a boat 2 miles from the nearest point on the coast. He is to go to a point Q, 3 miles down the coast and 1 mile inland (see figure). He can row at 2 mph and walk at 4 mph. Express the total time T of the trip as a function of x.

70. The portion of the vertical line through the point $(x, 0)$ that lies between the x-axis and the graph of $y = \sqrt{x}$ is revolved about the x-axis. Express the area A of the resulting disk as a function of x (see figure).

67. A rectangular package with square cross sections has a combined length and girth (perimeter of a cross section) of 108 inches. Express the volume V as a function of x (see figure).

68. A closed box with a square base of side x has a surface area of 100 square feet (see figure). Express the volume V of the box as a function of x.

FIGURE FOR 69

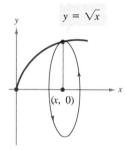

$y = \sqrt{x}$

FIGURE FOR 70

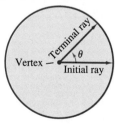

FIGURE FOR 67

FIGURE FOR 68

In Exercises 71 and 72, use a computer or graphics calculator to (a) sketch the graph of f, (b) find the zeros of f, and (c) determine the domain of f.

71. $f(x) = x\sqrt{9 - x^2}$

72. $f(x) = 2\left(\pi x^2 + \dfrac{6}{x}\right)$

1.6 Review of Trigonometric Functions

Angles and degree measure ▪ Radian measure ▪ The trigonometric functions ▪ Evaluation of trigonometric functions ▪ Solving trigonometric equations ▪ Graphs of trigonometric functions

Standard Position of an Angle

FIGURE 1.56

The concept of an angle is central to the study of trigonometry. As shown in Figure 1.56, an **angle** has three parts: an **initial ray,** a **terminal ray,** and a **vertex** (the point of intersection of the two rays). We say that an angle is in **standard position** if its initial ray coincides with the positive x-axis and its vertex is at the origin. We assume that you are familiar with the degree measure of an angle.* It is common practice to use θ (the Greek lowercase letter *theta*) to represent both an angle and its measure. We classify angles between $0°$ and $90°$ as **acute** and angles between $90°$ and $180°$ as **obtuse.** Positive angles are measured *counterclockwise*, beginning with the initial ray.

*For a more complete review of trigonometry, see *Algebra and Trigonometry*, 2nd edition, by Larson and Hostetler (Lexington, Mass., D. C. Heath and Company, 1989).

Negative angles are measured *clockwise*. For instance, Figure 1.57 shows an angle whose measure is $-45°$. We cannot assign a measure to an angle merely by knowing where its initial and terminal rays are located. To measure an angle, we must also know how the terminal ray was revolved. For example, Figure 1.57 shows that the angle measuring $-45°$ has the same terminal ray as the angle measuring $315°$. We call such angles **coterminal.**

An angle that is larger than $360°$ is one whose terminal ray has revolved more than one full revolution counterclockwise. Figure 1.58 shows an angle measuring more than $360°$. Similarly, we can generate angles whose measure is less than $-360°$ by revolving a terminal ray more than one full revolution clockwise.

Coterminal Angles

FIGURE 1.57

FIGURE 1.58

Radian measure

A second way to measure angles is by radian measure. To assign a radian measure to an angle θ, we consider θ to be the central angle of a circular sector of radius 1, as shown in Figure 1.59. The **radian measure** of θ is then defined to be the length of the arc of this sector. Recall that the total circumference of a circle is $2\pi r$. Thus, the circumference of a **unit circle** (that is, of radius 1) is simply 2π, and we may conclude that the radian measure of an angle measuring $360°$ is 2π. In other words, $360° = 2\pi$ radians.

Using radian measure, we have a simple formula for the length s of a circular arc of radius r, as shown in Figure 1.60.

$$\text{arclength} = s = r\theta \qquad \theta \text{ measured in radians}$$

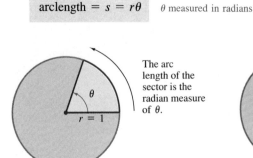

The arc length of the sector is the radian measure of θ.

Unit Circle

FIGURE 1.59

Circle of Radius r

FIGURE 1.60

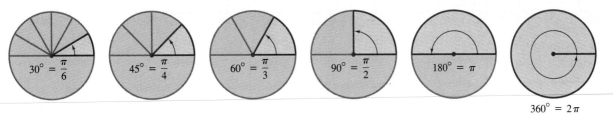

$$360° = 2\pi$$

FIGURE 1.61   Radian and Degree Measure for Several Common Angles

It is helpful to memorize the conversions of the common angles pictured in Figure 1.61. For other angles, you can use one of the following conversion rules.

CONVERSION RULES

$180° = \pi$ radians

Degrees ⟹ **Radians**   **Radians** ⟹ **Degrees**

$1° = \dfrac{\pi}{180}$ radians $1 \text{ radian} = \dfrac{180°}{\pi}$

EXAMPLE 1   Conversions between degrees and radians

(a) $40° = (40 \text{ deg})\left(\dfrac{\pi \text{ rad}}{180 \text{ deg}}\right) = \dfrac{2\pi}{9}$ radians

(b) $-270° = (-270 \text{ deg})\left(\dfrac{\pi \text{ rad}}{180 \text{ deg}}\right) = -\dfrac{3\pi}{2}$ radians

(c) $-\dfrac{\pi}{2}$ radians $= \left(-\dfrac{\pi}{2} \text{ rad}\right)\left(\dfrac{180 \text{ deg}}{\pi \text{ rad}}\right) = -90°$

(d) $\dfrac{9\pi}{2}$ radians $= \left(\dfrac{9\pi}{2} \text{ rad}\right)\left(\dfrac{180 \text{ deg}}{\pi \text{ rad}}\right) = 810°$

The trigonometric functions

There are two common approaches to the study of trigonometry. In one, the trigonometric functions are defined as ratios of two sides of a right triangle. In the other, these functions are defined in terms of a point on the terminal side of an angle in standard position. The first approach is generally used in surveying, navigation, and astronomy, where a typical problem involves a fixed triangle having three of its six parts (sides and angles) known and three to be determined. The second approach is normally used in physics, electronics, and biology, where the periodic nature of the trigonometric functions is emphasized. We define the six trigonometric functions, **sine, cosine, tangent, cotangent, secant,** and **cosecant** (abbreviated as sin, cos, etc.), from both viewpoints, as follows.

DEFINITION OF THE SIX TRIGONOMETRIC FUNCTIONS

Right triangle definitions, where $0 < \theta < \pi/2$. (Refer to Figure 1.62.)

$$\sin \theta = \frac{\text{opp.}}{\text{hyp.}} \qquad \csc \theta = \frac{\text{hyp.}}{\text{opp.}}$$

$$\cos \theta = \frac{\text{adj.}}{\text{hyp.}} \qquad \sec \theta = \frac{\text{hyp.}}{\text{adj.}}$$

$$\tan \theta = \frac{\text{opp.}}{\text{adj.}} \qquad \cot \theta = \frac{\text{adj.}}{\text{opp.}}$$

Circular function definitions, where θ is any angle. (Refer to Figure 1.63.)

$$\sin \theta = \frac{y}{r} \qquad \csc \theta = \frac{r}{y}$$

$$\cos \theta = \frac{x}{r} \qquad \sec \theta = \frac{r}{x}$$

$$\tan \theta = \frac{y}{x} \qquad \cot \theta = \frac{x}{y}$$

FIGURE 1.62

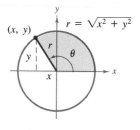

FIGURE 1.63

The following formulas are direct consequences of the definitions.

$$\csc \theta = \frac{1}{\sin \theta} \qquad \sec \theta = \frac{1}{\cos \theta} \qquad \cot \theta = \frac{1}{\tan \theta}$$

$$\tan \theta = \frac{\sin \theta}{\cos \theta} \qquad \cot \theta = \frac{\cos \theta}{\sin \theta}$$

Furthermore, since

$$\sin^2 \theta + \cos^2 \theta = \left(\frac{y}{r}\right)^2 + \left(\frac{x}{r}\right)^2 = \frac{x^2 + y^2}{r^2} = \frac{r^2}{r^2} = 1$$

we can readily obtain the Pythagorean Identity

$$\sin^2 \theta + \cos^2 \theta = 1.$$

Note that we use $\sin^2 \theta$ to mean $(\sin \theta)^2$. Additional trigonometric identities are listed next. (ϕ is the Greek letter *phi*.)

TRIGONOMETRIC IDENTITIES

Pythagorean identities:

$$\sin^2 \theta + \cos^2 \theta = 1$$
$$\tan^2 \theta + 1 = \sec^2 \theta$$
$$\cot^2 \theta + 1 = \csc^2 \theta$$

Sum or difference of two angles:

$$\sin (\theta \pm \phi) = \sin \theta \cos \phi \pm \cos \theta \sin \phi$$
$$\cos (\theta \pm \phi) = \cos \theta \cos \phi \mp \sin \theta \sin \phi$$
$$\tan (\theta \pm \phi) = \frac{\tan \theta \pm \tan \phi}{1 \mp \tan \theta \tan \phi}$$

Reduction formulas:

$$\sin (-\theta) = -\sin \theta$$
$$\cos (-\theta) = \cos \theta$$
$$\tan (-\theta) = -\tan \theta$$
$$\sin \theta = -\sin (\theta - \pi)$$
$$\cos \theta = -\cos (\theta - \pi)$$
$$\tan \theta = \tan (\theta - \pi)$$

Double angle formulas:

$\sin 2\theta = 2 \sin \theta \cos \theta$

$\cos 2\theta = 2 \cos^2 \theta - 1 = 1 - 2 \sin^2 \theta$
$\qquad\qquad = \cos^2 \theta - \sin^2 \theta$

Half angle formulas:

$\sin^2 \theta = \dfrac{1}{2}(1 - \cos 2\theta)$

$\cos^2 \theta = \dfrac{1}{2}(1 + \cos 2\theta)$

Law of Cosines:

$a^2 = b^2 + c^2 - 2bc \cos A$

Law of Cosines

REMARK   All angles in the remainder of this text are measured in radians unless stated otherwise. For example, when we write sin 3, we mean the sine of three radians, and when we write sin 3°, we mean the sine of three degrees.

Evaluation of trigonometric functions

There are two common methods of evaluating trigonometric functions: (1) decimal approximations with a calculator (or a table of trigonometric values) and (2) exact evaluations using trigonometric identities and formulas from geometry. We demonstrate the second method first.

EXAMPLE 2   Evaluating trigonometric functions

Evaluate the sine, cosine, and tangent of $\pi/3$.

SOLUTION

We begin by drawing the angle $\theta = \pi/3$ in the standard position, as shown in Figure 1.64. Then, since $60° = \pi/3$ radians, we obtain an equilateral triangle with sides of length 1 and θ as one of its angles. Since the altitude of this triangle bisects its base, we know that $x = \frac{1}{2}$. Now, using the Pythagorean Theorem, we have

$$y = \sqrt{r^2 - x^2} = \sqrt{1 - \left(\frac{1}{2}\right)^2} = \sqrt{\frac{3}{4}} = \frac{\sqrt{3}}{2}.$$

Thus,

$$\sin \frac{\pi}{3} = \frac{y}{r} = \frac{\sqrt{3}/2}{1} = \frac{\sqrt{3}}{2}$$

$$\cos \frac{\pi}{3} = \frac{x}{r} = \frac{1/2}{1} = \frac{1}{2}$$

$$\tan \frac{\pi}{3} = \frac{y}{x} = \frac{\sqrt{3}/2}{1/2} = \sqrt{3}.$$

FIGURE 1.64

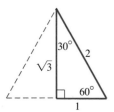

FIGURE 1.65

The degree and radian measures of several common angles are given in Table 1.6, along with the corresponding values of the sine, cosine, and tangent. (See Figure 1.65.)

TABLE 1.6   Common First Quadrant Angles

Degrees	0	30°	45°	60°	90°
Radians	0	$\dfrac{\pi}{6}$	$\dfrac{\pi}{4}$	$\dfrac{\pi}{3}$	$\dfrac{\pi}{2}$
$\sin\theta$	0	$\dfrac{1}{2}$	$\dfrac{\sqrt{2}}{2}$	$\dfrac{\sqrt{3}}{2}$	1
$\cos\theta$	1	$\dfrac{\sqrt{3}}{2}$	$\dfrac{\sqrt{2}}{2}$	$\dfrac{1}{2}$	0
$\tan\theta$	0	$\dfrac{\sqrt{3}}{3}$	1	$\sqrt{3}$	undefined

The quadrant signs of the various trigonometric functions are shown in Figure 1.66. To extend the use of Table 1.6 to angles in quadrants other than the first quadrant, we can use the concept of a **reference angle** (see Figure 1.67), with the appropriate quadrant sign.

FIGURE 1.66

FIGURE 1.67

For instance, the reference angle for $3\pi/4$ is $\pi/4$, and since the sine is positive in the second quadrant, we can write

$$\sin\frac{3\pi}{4} = +\sin\frac{\pi}{4} = \frac{\sqrt{2}}{2}.$$

Similarly, since the reference angle for 330° is 30°, and the tangent is negative in the fourth quadrant, we can write

$$\tan 330° = -\tan 30° = -\frac{\sqrt{3}}{3}.$$

EXAMPLE 3   Trigonometric identities and calculators

(a) Using the reduction formula $\sin(-\theta) = -\sin\theta$, we have

$$\sin\left(-\frac{\pi}{3}\right) = -\sin\frac{\pi}{3} = -\frac{\sqrt{3}}{2}.$$

(b) Using the reciprocal formula sec $\theta = 1/\cos \theta$, we have

$$\sec 60° = \frac{1}{\cos 60°} = \frac{1}{1/2} = 2.$$

(c) Using a calculator, we have

$$\cos (1.2) \approx 0.3624.$$

Remember that 1.2 is given in *radian* measure; consequently, your calculator must be set in radian mode.

Solving trigonometric equations

In Examples 2 and 3, we looked at techniques for evaluating trigonometric functions for given values of θ. In the next two examples, we look at the reverse problem. That is, if we are given the value of a trigonometric function, how can we solve for θ? For example, consider the equation

$$\sin \theta = 0.$$

We know $\theta = 0$ is one solution. But this is not the only solution. Any one of the following values of θ are also solutions.

$$\ldots, -3\pi, -2\pi, -\pi, 0, \pi, 2\pi, 3\pi, \ldots$$

We can write this infinite solution set as $\{n\pi \colon n$ is an integer$\}$.

EXAMPLE 4   Solving a trigonometric equation

Solve for θ in the following equation.

$$\sin \theta = -\frac{\sqrt{3}}{2}$$

SOLUTION

To solve the equation, we make two observations: the sine is negative in Quadrants III and IV *and* $\sin (\pi/3) = \sqrt{3}/2$. By combining these two observations, we conclude that we are seeking values of θ in the third and fourth quadrants that have a reference angle of $\pi/3$. In the interval $[0, 2\pi]$, the two angles fitting these criteria are

$$\theta = \pi + \frac{\pi}{3} = \frac{4\pi}{3} \quad \text{and} \quad \theta = 2\pi - \frac{\pi}{3} = \frac{5\pi}{3}.$$

Finally, we can add $2n\pi$ to either of these angles to obtain the solution set

$$\theta = \frac{4\pi}{3} + 2n\pi, \quad n \text{ is an integer}$$

or

$$\theta = \frac{5\pi}{3} + 2n\pi, \quad n \text{ is an integer}.$$

EXAMPLE 5   Solving a trigonometric equation

Solve the following equation for θ.

$$\cos 2\theta = 2 - 3 \sin \theta, \quad 0 \le \theta \le 2\pi$$

SOLUTION

Using the double angle identity $\cos 2\theta = 1 - 2 \sin^2 \theta$, we obtain the following polynomial (in $\sin \theta$).

$$1 - 2 \sin^2 \theta = 2 - 3 \sin \theta$$
$$0 = 2 \sin^2 \theta - 3 \sin \theta + 1$$
$$0 = (2 \sin \theta - 1)(\sin \theta - 1)$$

If $2 \sin \theta - 1 = 0$, we have $\sin \theta = 1/2$ and $\theta = \pi/6$ or $\theta = 5\pi/6$. If $\sin \theta - 1 = 0$, we have $\sin \theta = 1$ and $\theta = \pi/2$. Thus, for $0 \le \theta \le 2\pi$, there are three solutions to the given equation.

$$\theta = \frac{\pi}{6}, \frac{5\pi}{6}, \text{ or } \frac{\pi}{2}$$

Graphs of trigonometric functions

To sketch the graph of a trigonometric function in the xy-coordinate system, we usually use the variable x in place of θ. Moreover, when we write $y = \sin x$ or $y = \cos x$, we understand that x can have *any* real value and we evaluate the functions as if x were representing the radian measure of an angle.

REMARK   This is not the same use of x as that given in the definition of the six trigonometric functions. Generally, the context of a problem will distinguish clearly between these two uses of x.

One of the first things we notice about the graphs of all six trigonometric functions is that they are periodic. We call a function f **periodic** if there exists a nonzero number p such that

$$f(x + p) = f(x)$$

for all x in the domain of f. The smallest such positive value of p is called the **period** of f. Both the sine and cosine functions have a period of 2π, and by plotting several values in the interval $0 \le x \le 2\pi$, we obtain the graphs shown in Figure 1.68.

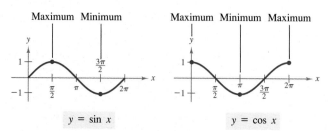

FIGURE 1.68 $y = \sin x$ $y = \cos x$

Domain: all $x \neq (2n - 1)\dfrac{\pi}{2}$

Range: $(-\infty, \infty)$

Period: π

Domain: all reals

Range: $[-1, 1]$

Period: 2π

$y = \sin x$

Domain: all reals

Range: $[-1, 1]$

Period: 2π

$y = \cos x$

$y = \tan x$

Domain: all $x \neq n\pi$

Range: $(-\infty, -1]$ and $[1, \infty)$

Period: 2π

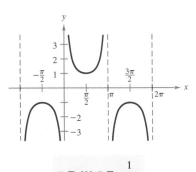

$y = \csc x = \dfrac{1}{\sin x}$

Domain: all $x \neq (2n - 1)\dfrac{\pi}{2}$

Range: $(-\infty, -1]$ and $[1, \infty)$

Period: 2π

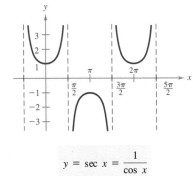

$y = \sec x = \dfrac{1}{\cos x}$

Domain: all $x \neq n\pi$

Range: $(-\infty, \infty)$

Period: π

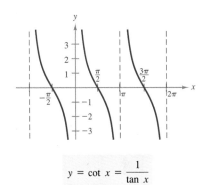

$y = \cot x = \dfrac{1}{\tan x}$

FIGURE 1.69   Graphs of the Six Trigonometric Functions

Note in Figure 1.68 that the maximum value of $\sin x$ is 1 and the minimum value is -1. Figure 1.69 shows the graphs of all six trigonometric functions. Familiarity with these six basic graphs will serve as a valuable aid in sketching the graphs of more complicated trigonometric functions.

The graph of the function $y = a \sin bx$ oscillates between $-a$ and a and hence has an **amplitude** of $|a|$. Furthermore, since $bx = 0$ when $x = 0$ and $bx = 2\pi$ when $x = 2\pi/b$, we may conclude that the function $y = a \sin bx$ has a period of $2\pi/|b|$. Table 1.7 summarizes the amplitudes and periods for some general types of trigonometric functions.

TABLE 1.7

Function	Period	Amplitude
$y = a \sin bx$ or $y = a \cos bx$	$\dfrac{2\pi}{|b|}$	$|a|$
$y = a \tan bx$ or $y = a \cot bx$	$\dfrac{\pi}{|b|}$	not applicable
$y = a \sec bx$ or $y = a \csc bx$	$\dfrac{2\pi}{|b|}$	not applicable

EXAMPLE 6   Sketching the graph of a trigonometric function

Sketch the graph of $f(x) = 3 \cos 2x$.

SOLUTION

The graph of $f(x) = 3 \cos 2x$ has the following characteristics.

amplitude: 3   period: $\dfrac{2\pi}{2} = \pi$

Using the basic shape of the graph of the cosine function, we sketch one period of the function on the interval $[0, \pi]$, using the following pattern.

maximum: $(0, 3)$ minimum: $\left(\dfrac{\pi}{2}, -3\right)$ maximum: $(\pi, 3)$

Then, by continuing this pattern, we sketch several cycles of the graph as shown in Figure 1.70.

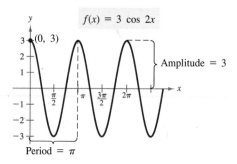

FIGURE 1.70

The discussion of horizontal shifts, vertical shifts, and reflections given in the previous section can be applied to the graphs of trigonometric functions. For instance, Figure 1.71 shows three different shifted graphs of sine functions.

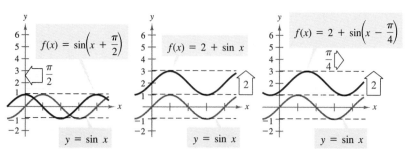

Horizontal shift to the left   Vertical shift upward   Horizontal and vertical shift

FIGURE 1.71

EXERCISES for Section 1.6

In Exercises 1 and 2, determine two coterminal angles (one positive and one negative) for the given angle. Give your answers in degrees.

1. (a)   (b)

2. (a)   (b)

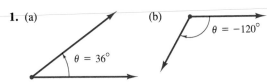

In Exercises 3 and 4, determine two coterminal angles (one positive and one negative) for the given angle. Give your answers in radians.

3. (a) $\theta = \dfrac{\pi}{9}$ (b) $\theta = \dfrac{4\pi}{3}$

4. (a) $\theta = -\dfrac{9\pi}{4}$ (b) $\theta = \dfrac{8\pi}{9}$

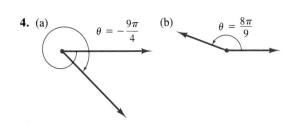

In Exercises 5 and 6, express the given angle in radian measure as a multiple of π.

5. (a) 30°   (b) 150°
(c) 315°   (d) 120°

6. (a) −20°   (b) −240°
(c) −270°   (d) 144°

In Exercises 7 and 8, express the given angle in degree measure.

7. (a) $\dfrac{3\pi}{2}$ (b) $\dfrac{7\pi}{6}$

(c) $-\dfrac{7\pi}{12}$ (d) $\dfrac{\pi}{9}$

8. (a) $\dfrac{7\pi}{3}$ (b) $-\dfrac{11\pi}{30}$

(c) $\dfrac{11\pi}{6}$ (d) $\dfrac{34\pi}{15}$

9. Let r represent the radius of a circle, θ the central angle (measured in radians), and s the length of the arc subtended by the angle. Use the relationship $\theta = s/r$ to complete the following table.

r	8 ft	15 in	85 cm		
s	12 ft			96 in	8642 mi
θ		1.6	$\dfrac{3\pi}{4}$	4	$\dfrac{2\pi}{3}$

10. The minute hand on a clock is $3\frac{1}{2}$ inches long (see figure). Through what distance does the tip of the minute hand move in 25 minutes?

11. A man bends his elbow through 75°. The distance from his elbow to the tip of his index finger is $18\frac{3}{4}$ inches (see figure).
(a) Find the radian measure of this angle.
(b) Find the distance the tip of the index finger moves.

FIGURE FOR 10   FIGURE FOR 11

12. A tractor tire 5 feet in diameter is partially filled with a liquid ballast for additional traction. To check the air pressure, the tractor operator rotates the tire until the valve stem is at the top so that the liquid will not enter the gauge. On a given occasion, the operator notes that the tire must be rotated 80° to have the stem in the proper position.
(a) Find the radian measure of this rotation.
(b) How far must the tractor be moved to get the valve stem in the proper position?

In Exercises 13 and 14, determine all six trigonometric functions for the given angle θ.

13. (a)

(b)

(8, −15)

14. (a)

(−12, −5)
(b)

(1, −1)

In Exercises 15 and 16, determine the quadrant in which θ lies.

15. $\sin \theta < 0$ and $\cos \theta < 0$
16. $\sin \theta > 0$ and $\cos \theta < 0$

In Exercises 17–22, find the indicated trigonometric function from the given one. (Assume $0 < \theta < \pi/2$.)

17. Given $\sin \theta = \dfrac{1}{2}$,
find $\csc \theta$.

18. Given $\sin \theta = \dfrac{1}{3}$,
find $\tan \theta$.

19. Given $\cos \theta = \dfrac{4}{5}$,
find $\cot \theta$.

20. Given $\sec \theta = \dfrac{13}{5}$,
find $\cot \theta$.

21. Given $\cot \theta = \dfrac{15}{8}$,
find $\sec \theta$.

22. Given $\tan \theta = \dfrac{1}{2}$,
find $\sin \theta$.

In Exercises 23–26, evaluate the sine, cosine, and tangent of the given angle *without* using a calculator.

23. (a) $60°$ (b) $\dfrac{2\pi}{3}$
    (c) $\dfrac{\pi}{4}$ (d) $\dfrac{5\pi}{4}$

24. (a) $-\dfrac{\pi}{6}$ (b) $150°$
    (c) $-\dfrac{\pi}{2}$ (d) $\dfrac{\pi}{2}$

25. (a) $225°$ (b) $-225°$
    (c) $300°$ (d) $330°$

26. (a) $750°$ (b) $510°$
    (c) $\dfrac{10\pi}{3}$ (d) $\dfrac{17\pi}{3}$

In Exercises 27–30, use a calculator to evaluate the given trigonometric function to four significant digits.

27. (a) $\sin 10°$ (b) $\csc 10°$
28. (a) $\sec 225°$ (b) $\sec 135°$
29. (a) $\tan \dfrac{\pi}{9}$ (b) $\tan \dfrac{10\pi}{9}$
30. (a) $\cot (1.35)$ (b) $\tan (1.35)$

In Exercises 31–34, find two values of θ corresponding to the given function. List the measure of θ in radians $(0 \le \theta < 2\pi)$. Do not use a calculator.

31. (a) $\cos \theta = \dfrac{\sqrt{2}}{2}$ (b) $\cos \theta = -\dfrac{\sqrt{2}}{2}$
32. (a) $\sec \theta = 2$ (b) $\sec \theta = -2$
33. (a) $\tan \theta = 1$ (b) $\cot \theta = -\sqrt{3}$
34. (a) $\sin \theta = \dfrac{\sqrt{3}}{2}$ (b) $\sin \theta = -\dfrac{\sqrt{3}}{2}$

In Exercises 35–42, solve the given equation for θ $(0 \le \theta < 2\pi)$. For some of the equations, you should use the trigonometric identities listed in this section.

35. $2 \sin^2 \theta = 1$ **36.** $\tan^2 \theta = 3$
37. $\tan^2 \theta - \tan \theta = 0$ **38.** $2 \cos^2 \theta - \cos \theta = 1$
39. $\sec \theta \csc \theta = 2 \csc \theta$ **40.** $\sin \theta = \cos \theta$
41. $\cos^2 \theta + \sin \theta = 1$ **42.** $\cos (\theta/2) - \cos \theta = 1$

In Exercises 43–46, solve for x, y, or r as indicated.

43. Solve for y.

44. Solve for x.

45. Solve for x.

46. Solve for r.

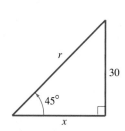

47. A 20-foot ladder leaning against the side of a house makes a 75° angle with the ground (see figure). How far up the side of the house does the ladder reach?

48. A biologist wants to know the width w of a river in order to set instruments properly to study the pollutants in the water. From point A, the biologist walks downstream 100 feet and sights point C to determine that $\theta = 50°$ (see figure). How wide is the river?

FIGURE FOR 47

FIGURE FOR 48

49. From a 150-foot observation tower on the coast, a Coast Guard officer sights a boat in difficulty. The angle of depression of the boat is 4° (see figure). How far is the boat from the shoreline?

FIGURE FOR 49

50. A ramp $17\frac{1}{2}$ feet long rises to a loading platform that is $3\frac{1}{2}$ feet off the ground (see figure). Find the angle that the ramp makes with the ground.

FIGURE FOR 50

In Exercises 51–56, determine the period and amplitude of the given function.

51. (a) $y = 2 \sin 2x$ (b) $y = \dfrac{1}{2} \sin \pi x$

52. (a) $y = \dfrac{3}{2} \cos \dfrac{x}{2}$ (b) $y = -2 \sin \dfrac{x}{3}$

53. $y = -2 \sin 10x$

54. $y = \dfrac{1}{2} \cos \dfrac{2x}{3}$

55. $y = 3 \sin 4\pi x$

56. $y = \dfrac{2}{3} \cos \dfrac{\pi x}{10}$

In Exercises 57–60, find the period of the given function.

57. $y = 5 \tan 2x$ **58.** $y = 7 \tan 2\pi x$
59. $y = \sec 5x$ **60.** $y = \csc 4x$

In Exercises 61–74, sketch the graph of the given function.

61. $y = \sin \dfrac{x}{2}$ **62.** $y = 2 \cos 2x$

63. $y = -2 \sin 6x$ **64.** $y = \cos 2\pi x$

65. $y = -\sin \dfrac{2\pi x}{3}$ **66.** $y = 2 \tan x$

67. $y = \csc \dfrac{x}{2}$ **68.** $y = \tan 2x$

69. $y = 2 \sec 2x$ **70.** $y = \csc 2\pi x$

71. $y = \sin (x + \pi)$ **72.** $y = \cos \left(x - \dfrac{\pi}{3} \right)$

73. $y = 1 + \cos \left(x - \dfrac{\pi}{2} \right)$ **74.** $y = 1 + \sin \left(x + \dfrac{\pi}{2} \right)$

75. For a person at rest, the rate of air intake v (in liters per second) during a respiratory cycle is

$$v = 0.85 \sin \dfrac{\pi t}{3}$$

where t is the time in seconds. Inhalation occurs when $v > 0$, and exhalation occurs when $v < 0$.
(a) Find the time for one full respiratory cycle.
(b) Find the number of cycles per minute.
(c) Sketch the graph of v as a function of t.
76. In the application at the beginning of this chapter, we developed the model

$$h = 50 + 50 \sin 8\pi t$$

for the height (in feet) of a Ferris wheel car, where t is measured in minutes. (The Ferris wheel has a radius of 50 feet.) This model yields a height of 50 feet when $t = 0$. Alter the model so that the height of the car is 0 feet when $t = 0$.
77. When tuning a piano, a technician strikes a tuning fork for the A above middle C, which creates a sound (a type of wave motion) that can be approximated by

$$y = 0.001 \sin 880\pi t$$

where y is measured in inches and t is the time in seconds.
(a) What is the period p of this function?
(b) What is the frequency f of this note ($f = 1/p$)?
(c) Sketch the graph of this function.

78. The function

$$P = 100 - 20 \cos \dfrac{5\pi t}{3}$$

approximates the blood pressure P in millimeters of mercury at time t in seconds for a person at rest.
(a) Find the period of the function.
(b) Find the number of heartbeats per minute.
(c) Sketch the graph of the pressure function.

In Exercises 79 and 80, use a computer or graphics calculator to sketch the graph of the given functions on the same coordinate axes where x is in the interval $[0, 2]$.

79. (a) $y = \dfrac{4}{\pi} \sin \pi x$

(b) $y = \dfrac{4}{\pi} \left(\sin \pi x + \dfrac{1}{3} \sin 3\pi x \right)$

80. (a) $y = \dfrac{1}{2} - \dfrac{4}{\pi^2} \cos \pi x$

(b) $y = \dfrac{1}{2} - \dfrac{4}{\pi^2} \left(\cos \pi x + \dfrac{1}{9} \cos 3\pi x \right)$

81. Use a computer or graphics calculator to sketch the functions $f(x) = x \sin x$ and $g(x) = x \cos x + \sin x$ on the same coordinate axes where x is in the interval $[0, \pi]$. The zero of $g(x)$ corresponds to what point on the graph of $f(x)$?
82. Sales S, in thousands of units, of a seasonal product is given by

$$S = 58.3 + 32.5 \cos \dfrac{\pi t}{6}$$

where t is the time in months (with $t = 1$ corresponding to January and $t = 12$ corresponding to December). Use a computer or graphics calculator to sketch the graph of S and determine the months when sales exceed 75,000 units.

REVIEW EXERCISES for Chapter 1

In Exercises 1–4, sketch the interval(s) defined by the given inequality.

1. $|x - 2| \leq 3$

2. $|3x - 2| \leq 0$

3. $4 < (x + 3)^2$

4. $\dfrac{1}{|x|} < 1$

In Exercises 5 and 6, find the midpoint of the given interval.

5. $\left[\dfrac{7}{8}, \dfrac{10}{4}\right]$

6. $\left[-1, \dfrac{3}{2}\right]$

7. Find the midpoints of the sides of the triangle whose vertices are $(1, 4)$ $(-3, 2)$, and $(5, 0)$.

8. Find the vertices of the triangle whose sides have midpoints $(0, 2)$, $(1, -1)$, and $(2, 1)$.

In Exercises 9–12, determine the radius and center of the given circle and sketch its graph.

9. $x^2 + y^2 + 6x - 2y + 1 = 0$

10. $4x^2 + 4y^2 - 4x + 8y = 11$

11. $x^2 + y^2 + 6x - 2y + 10 = 0$

12. $x^2 - 6x + y^2 + 8y = 0$

13. Determine the value of c so that the given circle has a radius of 2.

$$x^2 - 6x + y^2 + 8y = c$$

14. Find an equation in x and y such that the distance between (x, y) and $(-2, 0)$ is twice the distance between (x, y) and $(3, 1)$.

15. Find an equation for the circle whose center is $(1, 2)$ and whose radius is 3. Then determine whether the following points are inside, outside, or on the circle.
(a) $(1, 5)$ (b) $(0, 0)$
(c) $(-2, 1)$ (d) $(0, 4)$

16. Find an equation for the circle whose center is $(2, 1)$ and whose radius is 2. Then determine whether the following points are inside, outside, or on the circle.
(a) $(1, 1)$ (b) $(4, 2)$
(c) $(0, 1)$ (d) $(3, 1)$

In Exercises 17–20, sketch the graph of the given equation.

17. $y = \dfrac{-x + 3}{2}$

18. $y = 1 + \dfrac{1}{x}$

19. $y = 7 - 6x - x^2$

20. $y = 6x - x^2$

In Exercises 21 and 22, determine whether the given points lie on the same straight line.

21. $(-1, 3), (2, 9), (3, 1)$

22. $(2, 5), (4, 10), (6, 20)$

In Exercises 23–26, use the slope and y-intercept to sketch the graph of the given line.

23. $4x - 2y = 6$

24. $0.02x + 0.15y = 0.25$

25. $-\dfrac{1}{3}x + \dfrac{5}{6}y = 1$

26. $51x + 17y = 102$

27. Find equations of the lines passing through $(-2, 4)$ and having the following characteristics.
(a) Slope of $\frac{7}{16}$
(b) Parallel to the line $5x - 3y = 3$
(c) Passing through the origin
(d) Parallel to the y-axis

28. Find equations of the lines passing through $(1, 3)$ and having the following characteristics.
(a) Slope of $-\frac{2}{3}$
(b) Perpendicular to the line $x + y = 0$
(c) Passing through the point $(2, 4)$
(d) Parallel to the x-axis

29. The midpoint of a line segment is $(-1, 4)$. If one end of the line segment is $(2, 3)$, find the other end.

30. Find the point that is equidistant from $(0, 0)$, $(2, 3)$, and $(3, -2)$.

In Exercises 31 and 32, find the point(s) of intersection of the graphs of the given equations.

31. $3x - 4y = 8, \; x + y = 5$

32. $x - y + 1 = 0, \; y - x^2 = 7$

In Exercises 33–38, find a formula for the given function and find the domain.

33. The value v of a farm at \$850 per acre, with buildings, livestock, and equipment worth \$300,000, is a function of the number of acres a.

34. The value v of wheat at \$3.25 per bushel is a function of the number of bushels b.

35. The surface area s of a cube is a function of the length of an edge x.

36. The surface area s of a sphere is a function of the radius r.

37. The distance d traveled by a car at a speed of 45 miles per hour is a function of the time traveled t.

38. The area a of an equilateral triangle is a function of the length of one of its sides x.

39. The sum of two positive numbers is 500. Let one of the numbers be x, and express the product P of the two numbers as a function of x.

40. The product of two positive numbers is 120. Let one of the numbers be x, and express the sum of the two numbers as a function of x.

In Exercises 41–46, sketch the graph of the given equation and use the vertical line test to determine whether the equation expresses y as a function of x.

41. $x^2 - y = 0$
42. $x^2 + 4y^2 = 16$
43. $x - y^2 = 0$
44. $x^3 - y^2 + 1 = 0$
45. $y = x^2 - 2x$
46. $y = 36 - x^2$

47. Given $f(x) = 1 - x^2$ and $g(x) = 2x + 1$, find the following.
   (a) $f(x) + g(x)$
   (b) $f(x) - g(x)$
   (c) $f(x)g(x)$
   (d) $\dfrac{f(x)}{g(x)}$
   (e) $f(g(x))$
   (f) $g(f(x))$

48. Given $f(x) = 2x - 3$ and $g(x) = \sqrt{x + 1}$, find the following.
   (a) $f(x) + g(x)$
   (b) $f(x) - g(x)$
   (c) $f(x)g(x)$
   (d) $\dfrac{f(x)}{g(x)}$
   (e) $f(g(x))$
   (f) $g(f(x))$

49. Consider a plane flying at a constant rate on a direct route between two cities. The distance s (in miles) it has traveled in t hours is given by $s = 560t$.
   (a) Sketch the graph of this equation for $t \geq 0$.
   (b) What information is given by the slope of the line?

50. Find an equation of the line that bisects the acute angle formed by the lines $y = \sqrt{3}x$ and $y = 2$.

51. Sales representatives for a certain company are required to use their own cars for transportation. The cost to the company is $150 per day for lodging and meals, plus $0.30 per mile driven. Write a linear equation expressing the daily cost C to the company in terms of x, the number of miles driven.

52. A contractor purchases a piece of equipment for $36,500 that uses an average of $9.25 per hour for fuel and maintenance. The equipment operator is paid $13.50 per hour, and customers are charged $30 per hour.
   (a) Write an equation for the cost C of operating this equipment t hours.
   (b) Write an equation for the revenue R derived from t hours of use.
   (c) Find the break-even point for this equipment by finding the time at which $R = C$.

In Exercises 53 and 54, use a calculator to evaluate the given trigonometric function to four significant digits.

53. (a) tan 240°   (b) cot 210°
    (c) sin (−0.65)   (d) sec 3.1

54. (a) sin 5.63   (b) csc 2.62
    (c) csc 150°   (d) cos (−110°)

In Exercises 55–58, find two values of θ corresponding to the given function. List θ in degrees $(0 \leq \theta < 360°)$ and radians $(0 \leq \theta < 2\pi)$.

55. $\sin \theta = -\dfrac{1}{2}$
56. $\csc \theta = \sqrt{2}$
57. $\cos \theta = -\dfrac{\sqrt{3}}{2}$
58. $\tan \theta = -\dfrac{1}{\sqrt{3}}$

In Exercises 59–64, solve the given triangle for the indicated side and/or angle.

59.

60.

61.

62.

63.

64.

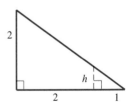

65. A 6-foot person standing 12 feet from a streetlight casts an 8-foot shadow (see figure). What is the height of the streetlight?

66. A guy wire is stretched from a broadcasting tower at a point 200 feet above the ground to an anchor 125 feet from the base of the tower (see figure). How long is the wire?

FIGURE FOR 65   FIGURE FOR 66

In Exercises 67–74, sketch a graph showing two periods for the given function.

67. $f(x) = 2 \sin \dfrac{2x}{3}$

68. $f(x) = \dfrac{1}{2} \cos \dfrac{x}{3}$

69. $f(x) = \cos \left(2x - \dfrac{\pi}{3} \right)$

70. $f(x) = -\sin \left(2x + \dfrac{\pi}{2} \right)$

71. $f(x) = \tan \dfrac{x}{2}$

72. $f(x) = \csc 2x$

73. $f(x) = \sec \left(x - \dfrac{\pi}{4} \right)$

74. $f(x) = \cot 3x$

75. The monthly sales S in thousands of units of a seasonal product are approximated by

$$S = 74.50 + 43.75 \sin \frac{\pi t}{6}$$

where t is the time in months with $t = 1$ corresponding to January. Sketch the graph of this sales function over one year.

76. A company that produces a seasonal product forecasts monthly sales over the next two years to be

$$S = 23.1 + 0.442t + 4.3 \sin \frac{\pi t}{6}$$

where S is measured in thousands of units and t is the time in months with $t = 1$ representing January 1989. Predict the sales for the following months.
(a) February 1989   (b) February 1990
(c) September 1989   (d) September 1990

Chapter 2 Application

On construction sites, otherwise harmless objects such as metal bolts can be life-threatening hazards. If a metal bolt is dropped from the hundredth floor of a skyscraper (approximately 1000 feet above ground), its velocity at ground level will be about 250 feet/second.

Velocity of a Free-Falling Object

A classic problem in calculus concerns the velocity of a free-falling object. For instance, consider an object that is dropped from a height of 25 feet above the earth's surface. We let $s(t)$ represent the height (in feet) of the object at time t (measured in seconds). Assuming that the only force acting on the object is that due to gravity, and neglecting air resistance, the height is given by the position function

$$s(t) = \frac{1}{2}gt^2 + v_0t + s_0$$

where $g = -32$ ft/sec^2 is the acceleration due to gravity, $v_0 = 0$ is the initial velocity, and $s_0 = 25$ is the initial height. Thus, for this object, the position function is

$$s(t) = -16t^2 + 25.$$

Suppose we want to determine the velocity of the object at time $t = 1$ second. When $t = 1$, the height is $s(1) = 9$ feet and its *average* velocity during the time interval $[0, 1]$ is $(9 - 25)/(1 - 0) = -16$ ft/sec. Similarly, the average velocity during the time interval $[0.5, 1]$ is

$$\frac{s(1) - s(0.5)}{1 - 0.5} = \frac{9 - 21}{0.5} = -24 \text{ ft/sec.}$$

In general, the average velocity during the time interval $[t, 1]$ is $[s(1) - s(t)]/(1 - t)$. To find the *instantaneous* velocity when $t = 1$, we let t approach 1 and evaluate the resulting limit.

$$v = \lim_{t \to 1} \frac{s(1) - s(t)}{1 - t} = \lim_{t \to 1} \frac{9 - (-16t^2 + 25)}{1 - t}$$

$$= \lim_{t \to 1} \frac{16(t^2 - 1)}{1 - t}$$

$$= \lim_{t \to 1} \frac{16(t + 1)(t - 1)}{-(t - 1)}$$

$$= \lim_{t \to 1} -16(t + 1)$$

$$= -32 \text{ ft/sec}$$

See Exercises 45 and 46, Section 2.3.

Chapter Overview

The concept of the **limit** of a function is the primary idea that distinguishes calculus from algebra and analytic geometry. Section 2.1 begins with a brief discussion of the way a limit will be used later (in Chapter 3) to solve the *tangent line problem*. The section then gives an informal description of the idea of the limit

$$\lim_{x \to a} f(x) = L.$$

This is followed by a theoretical definition—the so-called "ε-δ definition."

Section 2.2 discusses properties of limits. In this section it is important that you become familiar with several types of functions whose limits are easily found. For instance, the limit of $f(x) = x^2$ as x approaches 2 is simply $f(2) = 4$. Then, in Section 2.3, we use the properties discussed in Section 2.2 to find limits that are not so straightforward.

Section 2.4 introduces the notion of continuity. Informally, when we say that a function f is *continuous* on an interval (a, b), we mean that the graph of f has no holes, gaps, or jumps on the interval.

The last section in the chapter discusses infinite limits and vertical asymptotes. (Limits at infinity and horizontal asymptotes are discussed later in the text, in Section 4.5.)

Limits and Their Properties

2.1 An Introduction to Limits

The tangent line problem ▪ An introduction to limits ▪ Limits that fail to exist ▪ A formal definition of limit

The notion of a limit is fundamental to the study of calculus. Thus, it is important to acquire a good working knowledge of limits before moving on to other topics in calculus.

The tangent line problem

To give you some idea of the way we will be using limits, we begin this section with a brief introduction to a classic problem in calculus, called *the tangent line problem*. In this problem, we are given a function f and a point P on its graph and asked to find an equation of the tangent line to the graph at the point P, as shown in Figure 2.1. To answer this question, we must first define what we mean when we say a line is tangent to a curve at a point.

Except for cases involving a vertical tangent line, the problem of finding the **tangent line** at a point P is equivalent to finding the *slope* of the tangent line at P. We can approximate this slope by using a line through the point of tangency P and a second point on the curve, as shown in Figure 2.2(a). We call such a line a **secant line.** If $P = (c, f(c))$ is the point of tangency and $Q = (c + \Delta x, f(c + \Delta x))$ is a second point on the graph of f, then the slope of the secant line through these two points is given by

$$m_{\text{sec}} = \frac{f(c + \Delta x) - f(c)}{c + \Delta x - c} = \frac{f(c + \Delta x) - f(c)}{\Delta x}.$$

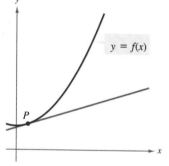

The tangent line to the graph of f at P

FIGURE 2.1

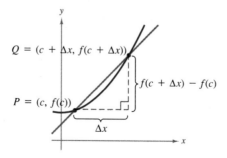

(a) The secant line through $(c, f(c))$ and $(c + \Delta x, f(c + \Delta x))$

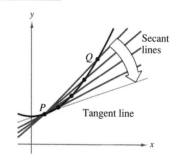

(b) As Q approaches P, the secant lines approach the tangent line.

FIGURE 2.2

The idea here is that by letting the point Q approach the point P, the slope of the secant line will approach the slope of the tangent line, as indicated in Figure 2.2(b) on page 65. When such a "limiting position" exists, we say that the slope of the tangent line is equal to the **limit** of the slope of the secant line. (Much more will be said about this important problem in Chapter 3.)

An introduction to limits

Suppose we are asked to sketch the graph of the function f given by

$$f(x) = \frac{x^3 - 1}{x - 1}, \quad x \neq 1.$$

For all values other than $x = 1$, we can use standard curve-sketching techniques. However, at $x = 1$, we are not sure what to expect. To get an idea of the behavior of the graph of f near $x = 1$, we use two sets of x-values—one set that approaches 1 from the left and one set that approaches 1 from the right, as shown in Table 2.1.

TABLE 2.1

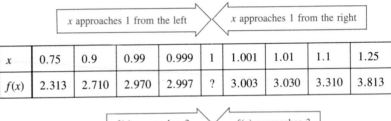

	x approaches 1 from the left					x approaches 1 from the right			
x	0.75	0.9	0.99	0.999	1	1.001	1.01	1.1	1.25
$f(x)$	2.313	2.710	2.970	2.997	?	3.003	3.030	3.310	3.813

$f(x)$ approaches 3 $f(x)$ approaches 3

When we plot these points, it appears that the graph of f is a parabola that has a gap at the point $(1, 3)$, as shown in Figure 2.3. Although x cannot equal 1, we can move arbitrarily close to 1, and as a result $f(x)$ moves arbitrarily close to 3. Using limit notation, we say that the *limit of* $f(x)$ *as* x *approaches 1 is 3* and write

$$\lim_{x \to 1} f(x) = 3.$$

This discussion leads to the following informal description of a limit. If $f(x)$ becomes arbitrarily close to a single number L as x approaches c from either side, then we say that the **limit** of $f(x)$, as x approaches c, is L and we write

$$\lim_{x \to c} f(x) = L.$$

For many functions, we can estimate the value of the limit L by using a calculator and evaluating the function at several points near c, as we did in Table 2.1.

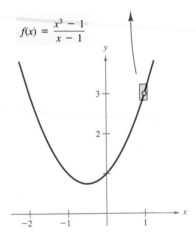

$$\lim_{x \to 1} f(x) = 3 \qquad (1, 3)$$

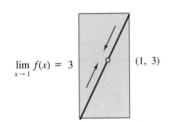

$$f(x) = \frac{x^3 - 1}{x - 1}$$

FIGURE 2.3

EXAMPLE 1 Using a calculator to estimate a limit

Evaluate $f(x) = x/(\sqrt{x + 1} - 1)$ at several points near $x = 0$ and use the result to estimate the limit

$$\lim_{x \to 0} \frac{x}{\sqrt{x + 1} - 1}.$$

SOLUTION

Table 2.2 lists the values of $f(x)$ for several x-values near 0.

TABLE 2.2

	x approaches 0 from the left				x approaches 0 from the right		
x	−0.01	−0.001	−0.0001	0	0.0001	0.001	0.01
$f(x)$	1.9950	1.9995	1.9999	?	2.0001	2.0005	2.0050
			$f(x)$ approaches 2		$f(x)$ approaches 2		

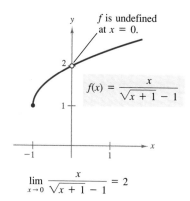

y

f is undefined at $x = 0$.

2

$f(x) = \dfrac{x}{\sqrt{x + 1} - 1}$

1

−1   1 x

$$\lim_{x \to 0} \frac{x}{\sqrt{x + 1} - 1} = 2$$

FIGURE 2.4

From the results shown in this table, we estimate the limit to be 2. The graph of f is shown in Figure 2.4.

In Example 1, note that the function is undefined at $x = 0$ and yet $f(x)$ appears to be approaching a limit as $x \to 0$. This often happens and it is important to realize that *the existence or nonexistence of $f(x)$ at $x = c$ has no bearing on the existence of the limit of $f(x)$ as x approaches c.* This is demonstrated further in the next example.

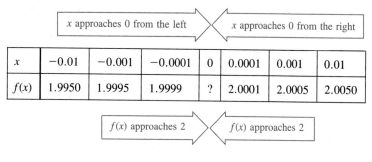

y

2

$f(x) = \begin{cases} 1, & x \neq 2 \\ 0, & x = 2 \end{cases}$

1

1   2   3 x

$$\lim_{x \to 2} f(x) = 1$$

FIGURE 2.5

EXAMPLE 2 Finding a limit

Find the limit of $f(x)$ as x approaches 2 where f is defined by

$$f(x) = \begin{cases} 1, & x \neq 2 \\ 0, & x = 2. \end{cases}$$

SOLUTION

Since $f(x) = 1$ for all x other than $x = 2$ and since the value of $f(2)$ is immaterial, we conclude that the limit is 1, as shown in Figure 2.5.

Limits that fail to exist

The next three examples examine some functions for which a limit does not exist.

EXAMPLE 3   Behavior that differs from the right and left

Show that the following limit does not exist.

$$\lim_{x \to 0} \frac{|x|}{x}$$

SOLUTION

We consider the function $f(x) = |x|/x$. From Figure 2.6, we see that for positive x, $|x|/x = 1$, and for negative x, $|x|/x = -1$. This means that no matter how close we get to zero, we will have both positive and negative x-values that yield $f(x) = 1$ and $f(x) = -1$. Specifically, if δ (the lowercase Greek letter *delta*) is a positive number, then for x-values satisfying the inequality $0 < |x| < \delta$, we can classify the values of $|x|/x$ as follows.

This implies that the limit does not exist.

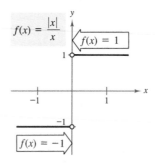

$$\lim_{x \to 0} f(x) \text{ does not exist.}$$

FIGURE 2.6

EXAMPLE 4   Unbounded behavior

Discuss the existence of the following limit.

$$\lim_{x \to 0} \frac{1}{x^2}$$

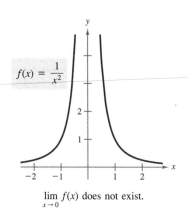

$f(x) = \dfrac{1}{x^2}$

$\lim\limits_{x \to 0} f(x)$ does not exist.

FIGURE 2.7

SOLUTION

We let $f(x) = 1/x^2$. From Figure 2.7, we see that as x approaches 0 from either the right or the left, $f(x)$ increases without bound. This means that by choosing x close enough to 0, we can force $f(x)$ to be as large as we want. For instance, $f(x)$ will be larger than 100 if we choose x that is within $\frac{1}{10}$ of 0. That is,

$$0 < |x| < \frac{1}{10} \implies f(x) = \frac{1}{x^2} > 100.$$

Similarly, we can force $f(x)$ to be larger than $1{,}000{,}000$, as follows.

$$0 < |x| < \frac{1}{1000} \implies f(x) = \frac{1}{x^2} > 1{,}000{,}000$$

Since $f(x)$ is not approaching a real number L as x approaches 0, we say that the limit does not exist.

EXAMPLE 5   Oscillating behavior

Discuss the existence of the limit

$$\lim_{x \to 0} \sin\left(\frac{1}{x}\right).$$

SOLUTION

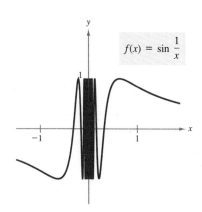

$f(x) = \sin\dfrac{1}{x}$

$\lim\limits_{x \to 0} f(x)$ does not exist.

FIGURE 2.8

We let $f(x) = \sin(1/x)$. In Figure 2.8, we see that as x approaches 0, $f(x)$ oscillates between -1 and 1. Therefore, the limit does not exist because no matter how small we choose δ, it is possible to choose x_1 and x_2 within δ units of 0 such that $\sin(1/x_1) = 1$ and $\sin(1/x_2) = -1$ as indicated in Table 2.3.

TABLE 2.3

x	$\dfrac{2}{\pi}$	$\dfrac{2}{3\pi}$	$\dfrac{2}{5\pi}$	$\dfrac{2}{7\pi}$	$\dfrac{2}{9\pi}$	$\dfrac{2}{11\pi}$	$x \to 0$
$\sin\left(\dfrac{1}{x}\right)$	1	-1	1	-1	1	-1	Limit does not exist.

Examples 3, 4, and 5 show three of the most common types of problems that lead to the nonexistence of a limit of $f(x)$ as x approaches c.

1. $f(x)$ approaches a different number from the right side of c than it approaches from the left side.
2. $f(x)$ increases or decreases without bound as x approaches c.
3. $f(x)$ oscillates between two fixed values as x approaches c.

There are many other interesting functions that have unusual limit behavior. An often cited one is the *Dirichlet function*

$$f(x) = \begin{cases} 0, & \text{if } x \text{ is rational} \\ 1, & \text{if } x \text{ is irrational.} \end{cases}$$

This function has *no limit* at any real number c.

In the early development of calculus, the definition of function was much more restricted than it is today, and "functions" such as the Dirichlet function would not have been considered. The modern definition of a function was given by the German mathematician Peter Gustav Dirichlet (1805–1859). Dirichlet made many contributions to mathematics and, together with Cauchy, Riemann, and Weierstrass, developed much of the rigor of modern calculus.

Peter Gustav Dirichlet

Augustin-Louis Cauchy

$$\lim_{x \to c} f(x) = L$$

FIGURE 2.9

A formal definition of limit

Let's take another look at our informal description of a limit. If $f(x)$ becomes arbitrarily close to a single number L as x approaches c from either side, then we say that the limit of $f(x)$, as x approaches c, is L and we write

$$\lim_{x \to c} f(x) = L.$$

At first glance, this description looks fairly technical—so why do we call it informal? The answer lies in the exact meaning of the two phrases

"$f(x)$ becomes arbitrarily close to L"

and

"x approaches c."

The first person to assign a mathematically rigorous meaning to these two phrases was Augustin-Louis Cauchy (1789–1857). His ε-δ definition of a limit is the standard used today.

In Figure 2.9 let ε represent a (small) positive number. Then the phrase "$f(x)$ becomes arbitrarily close to L" means that $f(x)$ lies in the interval $(L - \varepsilon, L + \varepsilon)$. In terms of absolute value, we write this as

$$|f(x) - L| < \varepsilon.$$

Similarly, the phrase "x approaches c" means that there exists a positive number δ such that x lies in either the interval $(c - \delta, c)$ or the interval $(c, c + \delta)$. This fact can be concisely expressed by the double inequality

$$0 < |x - c| < \delta.$$

The first inequality, $0 < |x - c|$, expresses the fact that $x \neq c$, while the second, $|x - c| < \delta$, says that x is within a distance δ of c.

This brings us to the following formal ε-δ definition of a **limit.**

DEFINITION OF LIMIT	Let f be a function defined on an open interval containing c (except possibly at c) and let L be a real number. The statement

$$\lim_{x \to c} f(x) = L$$

means that for each $\varepsilon > 0$ there exists a $\delta > 0$ such that

$$|f(x) - L| < \varepsilon \qquad \text{whenever} \qquad 0 < |x - c| < \delta.$$

REMARK   Throughout this text, when we write

$$\lim_{x \to c} f(x) = L$$

we imply two statements—the limit **exists** *and* the limit is L.

Some functions do not have a limit as $x \to c$, but those that do cannot have two different limits as $x \to c$. That is, *if the limit of a function exists, it is unique* (see Exercise 37).

The next two examples should help you develop a better understanding of the ε-δ definition of a limit.

EXAMPLE 6   Finding a δ for a given ε

Given the limit

$$\lim_{x \to 3} (2x - 5) = 1$$

find δ such that $|(2x - 5) - 1| < 0.01$ whenever $0 < |x - 3| < \delta$.

SOLUTION

In this problem, we are working with a given value of ε, namely $\varepsilon = 0.01$. To find an appropriate δ, we try to establish a connection between the two absolute values $|(2x - 5) - 1|$ and $|x - 3|$. By simplifying the first absolute value, we get

$$|(2x - 5) - 1| = |2x - 6| = 2|x - 3|.$$

Thus, the inequality $|(2x - 5) - 1| < 0.01$ is equivalent to $2|x - 3| < 0.01$, and we have

$$|x - 3| < \frac{0.01}{2} = 0.005.$$

Thus, we choose $\delta = 0.005$. This choice works because $0 < |x - 3| < 0.005$ implies that $|(2x - 5) - 1| = 2|x - 3| < 2(0.005) = 0.01$, as indicated in Figure 2.10.

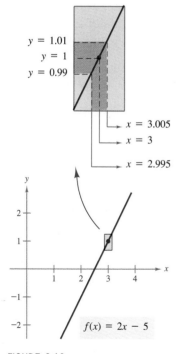

FIGURE 2.10

REMARK   Note in Example 6 that 0.005 is the *largest* value of δ that will guarantee $|(2x - 5) - 1| < 0.01$ whenever $0 < |x - 3| < \delta$. Any *smaller* positive value of δ would, of course, also work.

In Example 6, we found a δ-value for a *given* ε. This does not prove the existence of the limit. To do that, we must prove that we can find a δ for *any* ε as demonstrated in the next example.

EXAMPLE 7   Using the ε-δ definition of a limit

Use the ε-δ definition of a limit to prove that

$$\lim_{x \to 2} (3x - 2) = 4.$$

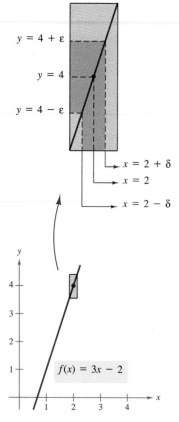

$y = 4 + \varepsilon$

$y = 4$

$y = 4 - \varepsilon$

$x = 2 + \delta$
$x = 2$
$x = 2 - \delta$

$f(x) = 3x - 2$

FIGURE 2.11

SOLUTION

We are required to show that for each $\varepsilon > 0$, there exists a $\delta > 0$ such that

$$|(3x - 2) - 4| < \varepsilon$$

whenever

$$0 < |x - 2| < \delta.$$

Since our choice for δ depends on ε, we try to establish a connection between the absolute values $|(3x - 2) - 4|$ and $|x - 2|$. By simplifying the first absolute value, we get

$$|(3x - 2) - 4| = |3x - 6| = 3|x - 2|.$$

Thus, the inequality $|(3x - 2) - 4| < \varepsilon$ requires $3|x - 2| < \varepsilon$ and we have

$$|x - 2| < \frac{\varepsilon}{3}.$$

Finally, we choose $\delta = \varepsilon/3$. This choice works because

$$0 < |x - 2| < \delta = \frac{\varepsilon}{3}$$

implies that

$$|(3x - 2) - 4| = 3|x - 2| < 3\left(\frac{\varepsilon}{3}\right) = \varepsilon$$

as shown in Figure 2.11.

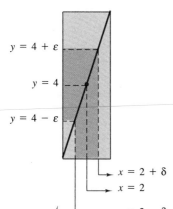

$y = 4 + \varepsilon$

$y = 4$

$y = 4 - \varepsilon$

$x = 2 + \delta$

$x = 2$

$x = 2 - \delta$

$f(x) = x^2$

FIGURE 2.12

EXAMPLE 8 Using the ε-δ definition of a limit

Use the ε-δ definition of a limit to prove that

$$\lim_{x \to 2} x^2 = 4.$$

SOLUTION

We must show that for each $\varepsilon > 0$, there exists a $\delta > 0$ such that

$$|x^2 - 4| < \varepsilon \qquad \text{whenever} \qquad 0 < |x - 2| < \delta.$$

We begin by writing $|x^2 - 4| = |x - 2||x + 2|$. For all x in the interval
$(1, 3)$, we know that $|x + 2| < 5$. Thus, letting δ be the minimum of $\varepsilon/5$
and 1, it follows that whenever $0 < |x - 2| < \delta$, we have

$$|x^2 - 4| = |x - 2||x + 2| < \left(\frac{\varepsilon}{5}\right)(5) = \varepsilon$$

as shown in Figure 2.12.

Throughout this chapter we will use the ε-δ definition of a limit primarily
to prove theorems about limits and to establish the existence or nonexistence
of particular types of limits. The ε-δ definition does not provide us with an
easy method of finding limits. We will develop other techniques for doing
that.

EXERCISES for Section 2.1

In Exercises 1–10, complete the table and use the result
to estimate the given limit.

1. $\lim_{x \to 2} \dfrac{x - 2}{x^2 - x - 2}$

x	1.9	1.99	1.999	2.001	2.01	2.1
$f(x)$						

2. $\lim_{x \to 2} \dfrac{x - 2}{x^2 - 4}$

x	1.9	1.99	1.999	2.001	2.01	2.1
$f(x)$						

3. $\lim_{x \to 0} \dfrac{\sqrt{x + 3} - \sqrt{3}}{x}$

x	−0.1	−0.01	−0.001	0.001	0.01	0.1
$f(x)$						

4. $\lim_{x \to -3} \dfrac{\sqrt{1 - x} - 2}{x + 3}$

x	−3.1	−3.01	−3.001
$f(x)$			

x	−2.999	−2.99	−2.9
$f(x)$			

*5. $\displaystyle\lim_{x\to3}\frac{[1/(x+1)]-(1/4)}{x-3}$

x	2.9	2.99	2.999	3.001	3.01	3.1
$f(x)$						

6. $\displaystyle\lim_{x\to4}\frac{[x/(x+1)]-(4/5)}{x-4}$

x	3.9	3.99	3.999	4.001	4.01	4.1
$f(x)$						

7. $\displaystyle\lim_{x\to\pi/4}\frac{\tan x-1}{x-(\pi/4)}$

x	0.70	0.75	0.784	0.786	0.80	0.85
$f(x)$						

8. $\displaystyle\lim_{x\to0}\frac{\sec x-1}{x}$

x	−0.1	−0.01	−0.001	0.001	0.01	0.1
$f(x)$						

9. $\displaystyle\lim_{x\to0}\frac{\sin x}{x}$

x	−0.1	−0.01	−0.001	0.001	0.01	0.1
$f(x)$						

10. $\displaystyle\lim_{x\to0}\frac{\cos x-1}{x}$

x	−0.1	−0.01	−0.001	0.001	0.01	0.1
$f(x)$						

*A blue number indicates that a detailed solution can be found in the *Student Solution Guide*.

In Exercises 11–20, use the given graph to find the limit (if it exists).

11. $\displaystyle\lim_{x\to3}(4-x)$

12. $\displaystyle\lim_{x\to1}(x^2+2)$

 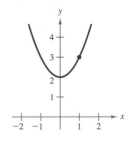

13. $\displaystyle\lim_{x\to2}f(x)$

$f(x)=\begin{cases}4-x,&x\neq2\\0,&x=2\end{cases}$

14. $\displaystyle\lim_{x\to1}f(x)$

$f(x)=\begin{cases}x^2+2,&x\neq1\\1,&x=1\end{cases}$

15. $\displaystyle\lim_{x\to5}\frac{|x-5|}{x-5}$

16. $\displaystyle\lim_{x\to3}\frac{1}{x-3}$

 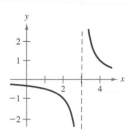

17. $\displaystyle\lim_{x\to\pi/2}\tan x$

18. $\displaystyle\lim_{x\to0}\sec x$

 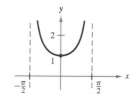

19. $\lim\limits_{x \to 0} \cos\left(\dfrac{1}{x}\right)$

20. $\lim\limits_{x \to 1} \sin \pi x$

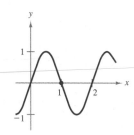

In Exercises 21–24, find the indicated limit L and then find $\delta > 0$ such that $|f(x) - L| < 0.01$ whenever $0 < |x - c| < \delta$.

21. $\lim\limits_{x \to 2} (3x + 2)$

22. $\lim\limits_{x \to 4} \left(4 - \dfrac{x}{2}\right)$

23. $\lim\limits_{x \to 2} (x^2 - 3)$

24. $\lim\limits_{x \to 5} (x^2 + 4)$

In Exercises 25–32, find the indicated limit L. Then for $\varepsilon > 0$, find $\delta > 0$ such that $|f(x) - L| < \varepsilon$ whenever $0 < |x - c| < \delta$.

25. $\lim\limits_{x \to 2} (x + 3)$

26. $\lim\limits_{x \to -3} (2x + 5)$

27. $\lim\limits_{x \to 6} 3$

28. $\lim\limits_{x \to 2} (-1)$

29. $\lim\limits_{x \to 0} \sqrt[3]{x}$

30. $\lim\limits_{x \to 3} |x - 3|$

31. $\lim\limits_{x \to 1} (x^2 + 1)$

32. $\lim\limits_{x \to -3} (x^2 + 3x)$

33. If $f(2) = 4$, can we conclude anything about

$$\lim\limits_{x \to 2} f(x)?$$

Give reasons for your answer.

34. If

$$\lim\limits_{x \to 2} f(x) = 4,$$

can we conclude anything about $f(2)$? Give reasons for your answer.

In Exercises 35 and 36, use a computer or graphics calculator to sketch the graph of the function f and find the specified limit (if it exists).

35. $f(x) = \dfrac{\sqrt{x + 5} - 3}{x - 4}, \qquad \lim\limits_{x \to 4} f(x)$

36. $f(x) = \dfrac{x - 3}{x^2 - 4x + 3}, \qquad \lim\limits_{x \to 3} f(x)$

37. Prove that if the limit of $f(x)$ as $x \to c$ exists, then the limit must be unique. [Hint: Let $\lim\limits_{x \to c} f(x) = L_1$ and $\lim\limits_{x \to c} f(x) = L_2$ and prove that $L_1 = L_2$.]

2.2 Properties of Limits

A strategy for finding limits ▪ Limits of algebraic functions ▪ Limits of trigonometric functions

We have pointed out already that the limit of $f(x)$ as $x \to c$ does not depend on the value of f at $x = c$. However, if it happens that the limit is precisely $f(c)$, then we say that the limit can be evaluated by **direct substitution.** That is,

$$\lim\limits_{x \to c} f(x) = f(c). \qquad \text{Substitute for } x$$

Such *well-behaved* functions are said to be **continuous at c** and we will examine this concept more closely in Section 2.4. An important application of direct substitution is shown in the following theorem.

THEOREM 2.1 **FUNCTIONS THAT AGREE** **AT ALL BUT ONE POINT**	Let c be a real number and $f(x) = g(x)$ for all $x \neq c$ in an open interval containing c. If the limit of $g(x)$ as $x \to c$ exists, then the limit of $f(x)$ also exists and $$\lim\limits_{x \to c} f(x) = \lim\limits_{x \to c} g(x).$$

PROOF   Assume that the limit of $g(x)$ as $x \to c$ is L. Then, by the definition of limit, for each $\varepsilon > 0$ there exists a $\delta > 0$ such that $f(x) = g(x)$ in the open intervals $(c - \delta, c)$ and $(c, c + \delta)$, and

$$|g(x) - L| < \varepsilon \qquad \text{whenever} \qquad 0 < |x - c| < \delta.$$

However, since $f(x) = g(x)$ for all x in the open interval other than $x = c$, it follows that

$$|f(x) - L| < \varepsilon \qquad \text{whenever} \qquad 0 < |x - c| < \delta.$$

Thus, we conclude that the limit of $f(x)$ as $x \to c$ is also L.

EXAMPLE 1   Two functions that agree at all but one point

Show that the functions $f(x) = \dfrac{x^3 - 1}{x - 1}$ and $g(x) = x^2 + x + 1$ have the same values for all x other than $x = 1$.

SOLUTION

By factoring the numerator of f, we have

$$f(x) = \frac{x^3 - 1}{x - 1} = \frac{(x - 1)(x^2 + x + 1)}{x - 1}.$$

Thus, if $x \neq 1$, we can cancel like factors to obtain

$$f(x) = \frac{\cancel{(x - 1)}(x^2 + x + 1)}{\cancel{x - 1}} = x^2 + x + 1 = g(x), \quad x \neq 1.$$

Now, in Figure 2.13, we see that for all points other than $x = 1$, the graphs of f and g are identical.   ◻

In Figure 2.13, note that there is no gap in the graph of the polynomial function g, and we will see in Theorem 2.4 that the limit of $g(x)$ as x approaches 1 is simply

$$\lim_{x \to 1} g(x) = \lim_{x \to 1} (x^2 + x + 1) = 1^2 + 1 + 1 = 3 = g(1).$$

Thus, by applying Theorem 2.1, we can conclude that the limit of $f(x)$ as $x \to 1$ is also 3.

This discussion identifies the following basic strategy for finding limits.

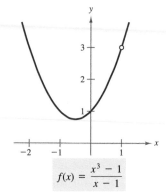

$f(x) = \dfrac{x^3 - 1}{x - 1}$

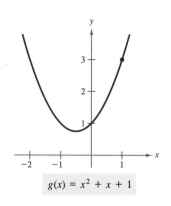

$g(x) = x^2 + x + 1$

FIGURE 2.13

A STRATEGY FOR FINDING LIMITS

1. Learn to recognize which limits can be evaluated by direct substitution.
2. If the limit of $f(x)$ as $x \to c$ *cannot* be evaluated by direct substitution, try to find a function g that agrees with f for all x other than $x = c$. (Choose g so that the limit of $g(x)$ *can* be evaluated by direct substitution.)
3. Apply Theorem 2.1 to conclude that

$$\lim_{x \to c} f(x) = \lim_{x \to c} g(x) = g(c).$$

We devote the remainder of this section to the study of limits that can be evaluated by direct substitution. In the next section, we will look at techniques for evaluating limits for which direct substitution fails.

Limits of algebraic functions

We begin with a look at the limits of some basic functions.

THEOREM 2.2
SOME BASIC LIMITS

If b and c are real numbers and n is a positive integer, then the following properties are true.

1. $\lim_{x \to c} b = b$ 2. $\lim_{x \to c} x = c$ 3. $\lim_{x \to c} x^n = c^n$

PROOF   We prove Property 2 and leave the proof of the other two properties as exercises (see Exercises 35 and 36). We need to show that for each $\varepsilon > 0$ there exists a $\delta > 0$ such that

$$|x - c| < \varepsilon \qquad \text{whenever} \qquad 0 < |x - c| < \delta.$$

Since the right-hand inequality is a stricter version of the left-hand one, we simply choose $\delta = \varepsilon$.

EXAMPLE 2   Evaluating a limit

Applying Theorem 2.2, we have the following.

(a) $\lim_{x \to 2} 3 = 3$ (b) $\lim_{x \to 2} x^2 = 2^2 = 4$ ▭

By combining Theorem 2.2 with the following theorem, we can find limits for a wide variety of algebraic functions.

THEOREM 2.3 **PROPERTIES OF LIMITS**	Let b and c be real numbers, and n a positive integer, and let f and g be functions whose limits exist as $x \to c$.

1. Scalar multiple: $\quad \lim\limits_{x \to c} [b(f(x))] = b \left[\lim\limits_{x \to c} f(x) \right]$

2. Sum or difference: $\quad \lim\limits_{x \to c} [f(x) \pm g(x)] = \lim\limits_{x \to c} f(x) \pm \lim\limits_{x \to c} g(x)$

3. Product: $\quad \lim\limits_{x \to c} [f(x)g(x)] = \left[\lim\limits_{x \to c} f(x) \right]\left[\lim\limits_{x \to c} g(x) \right]$

4. Quotient: $\quad \lim\limits_{x \to c} \dfrac{f(x)}{g(x)} = \dfrac{\lim\limits_{x \to c} f(x)}{\lim\limits_{x \to c} g(x)}$, provided $\lim\limits_{x \to c} g(x) \neq 0$

5. Power: $\quad \lim\limits_{x \to c} [f(x)]^n = \left[\lim\limits_{x \to c} f(x) \right]^n$

PROOF   The proof of Property 1 is left as an exercise (see Exercise 37). For proofs of Properties 3, 4, and 5, see Appendix A. We present a proof for the limit of a sum (Property 2). Assume that

$$\lim_{x \to c} f(x) = L$$

and

$$\lim_{x \to c} g(x) = K.$$

Choose $\varepsilon > 0$. Then, since $\varepsilon/2 > 0$, we know that there exist $\delta_1 > 0$ and $\delta_2 > 0$ such that $0 < |x - c| < \delta_1$ implies $|f(x) - L| < \varepsilon/2$, and $0 < |x - c| < \delta_2$ implies $|g(x) - K| < \varepsilon/2$. If δ is the smaller of δ_1 and δ_2, then $0 < |x - c| < \delta$ implies that

$$|f(x) - L| < \frac{\varepsilon}{2}$$

and

$$|g(x) - K| < \frac{\varepsilon}{2}.$$

Finally, we apply the Triangle Inequality to conclude that

$$|[f(x) + g(x)] - (L + K)| \le |f(x) - L| + |g(x) - K| < \frac{\varepsilon}{2} + \frac{\varepsilon}{2} = \varepsilon$$

which implies that

$$\lim_{x \to c} [f(x) + g(x)] = L + K = \lim_{x \to c} f(x) + \lim_{x \to c} g(x).$$

EXAMPLE 3   The limit of a polynomial

Find the following limit.

$$\lim_{x \to 2} (4x^2 + 3)$$

SOLUTION

Using the results of Example 2 and the properties in Theorem 2.3, we have

$$\lim_{x \to 2} (4x^2 + 3) = \lim_{x \to 2} 4x^2 + \lim_{x \to 2} 3 \qquad \text{Property 2}$$

$$= 4\left[\lim_{x \to 2} x^2\right] + \lim_{x \to 2} 3 \qquad \text{Property 1}$$

$$= 4(4) + 3 = 19.$$

In Example 3, note that the limit (as $x \to 2$) of the *polynomial function* $p(x) = 4x^2 + 3$ is simply the value of p at $x = 2$.

$$\lim_{x \to 2} p(x) = p(2) = 4(2^2) + 3 = 19$$

This *direct substitution* property is valid for all polynomial functions as stated in Theorem 2.4.

THEOREM 2.4
LIMIT OF A POLYNOMIAL
FUNCTION

If p is a polynomial function and c is a real number, then

$$\lim_{x \to c} p(x) = p(c).$$

PROOF   Let the polynomial function p be given by

$$p(x) = a_n x^n + \cdots + a_1 x + a_0.$$

Repeated applications of the sum and scalar multiple properties produce

$$\lim_{x \to c} p(x) = a_n\left[\lim_{x \to c} x^n\right] + \cdots + a_1\left[\lim_{x \to c} x\right] + \lim_{x \to c} a_0.$$

Finally, using Properties 1, 2, and 3 of Theorem 2.2, we obtain

$$\lim_{x \to c} p(x) = a_n c^n + \cdots + a_1 c + a_0 = p(c).$$

THEOREM 2.5
LIMIT OF A RATIONAL FUNCTION

If r is a rational function given by $r(x) = p(x)/q(x)$ and c is a real number such that $q(c) \neq 0$, then

$$\lim_{x \to c} r(x) = r(c) = \frac{p(c)}{q(c)}.$$

PROOF   By Theorem 2.4 we know that for the polynomial functions p and q, we have

$$\lim_{x \to c} p(x) = p(c) \qquad \text{and} \qquad \lim_{x \to c} q(x) = q(c).$$

Since $q(c) \neq 0$, we apply Property 4 of Theorem 2.3 to conclude that

$$\lim_{x \to c} r(x) = \lim_{x \to c} \frac{p(x)}{q(x)} = \frac{\lim_{x \to c} p(x)}{\lim_{x \to c} q(x)} = \frac{p(c)}{q(c)} = r(c).$$

EXAMPLE 4   The limit of a rational function

Find the following limit.

$$\lim_{x \to 1} \frac{x^2 + x + 2}{x + 1}$$

SOLUTION

Since the denominator is not zero when $x = 1$, we can apply Theorem 2.5 to obtain

$$\lim_{x \to 1} \frac{x^2 + x + 2}{x + 1} = \frac{1^2 + 1 + 2}{1 + 1} = \frac{4}{2} = 2.$$ ▭

Polynomial functions and rational functions constitute two of the three basic types of algebraic functions. The following theorem deals with the limit of the third type of algebraic function—one that involves radicals. The proof of this theorem is given in Appendix A.

THEOREM 2.6
LIMIT OF A FUNCTION
INVOLVING A RADICAL

If $c > 0$ and n is *any* positive integer, or if $c \leq 0$ and n is an *odd* positive integer, then

$$\lim_{x \to c} \sqrt[n]{x} = \sqrt[n]{c}.$$

The next theorem involves the limit of a composite function. Its proof is also given in Appendix A.

THEOREM 2.7
LIMIT OF A COMPOSITE
FUNCTION

If f and g are functions such that

$$\lim_{x \to c} g(x) = L \qquad \text{and} \qquad \lim_{x \to L} f(x) = f(L)$$

then

$$\lim_{x \to c} f(g(x)) = f(L).$$

To use Theorem 2.7 to find the limit of a composite function such as

$$\lim_{x \to 0} \sqrt{x^2 + 4}$$

we let $g(x) = x^2 + 4$, where $\lim_{x \to 0} (x^2 + 4) = g(0) = 4$, and $f(x) = \sqrt{x}$ where $\lim_{x \to 4} \sqrt{x} = 2$. Then it follows that

$$\lim_{x \to 0} f(g(x)) = \lim_{x \to 0} \sqrt{x^2 + 4} = 2.$$

This procedure is demonstrated further in Example 5.

EXAMPLE 5 The limit of a composite function

Find the following limit.

$$\lim_{x \to 3} \sqrt[3]{2x^2 - 10}$$

SOLUTION

Since

$$\lim_{x \to 3} (2x^2 - 10) = 2(3^2) - 10 = 8$$

we have

$$\lim_{x \to 3} \sqrt[3]{2x^2 - 10} = \sqrt[3]{8} = 2.$$ ▭

Limits of trigonometric functions

We have seen that the limits of many algebraic functions can be evaluated by direct substitution. The following theorem (which we present without proof) tell us that each of the six basic trigonometric functions also possesses this desirable quality.

THEOREM 2.8 **LIMITS OF TRIGONOMETRIC** **FUNCTIONS**	If c is a real number *in the domain of the given trigonometric function*, then the following properties are true.

1. $\lim\limits_{x \to c} \sin x = \sin c$ 　 2. $\lim\limits_{x \to c} \cos x = \cos c$

3. $\lim\limits_{x \to c} \tan x = \tan c$ 　 4. $\lim\limits_{x \to c} \cot x = \cot c$

5. $\lim\limits_{x \to c} \sec x = \sec c$ 　 6. $\lim\limits_{x \to c} \csc x = \csc c$

EXAMPLE 6 Limits involving trigonometric functions

(a) By Theorem 2.8, we have

$$\lim_{x \to 0} \sin x = \sin (0) = 0.$$

(b) By Theorem 2.8, and Property 3 of Theorem 2.3, we have

$$\lim_{x \to \pi} (x \cos x) = \left[\lim_{x \to \pi} x \right] \left[\lim_{x \to \pi} \cos x \right] = \pi \cos (\pi) = -\pi.$$ ▭

EXERCISES for Section 2.2

In Exercises 1–4, use the graph to visually determine the limit.

1. $f(x) = -2x + 1$

(a) $\lim_{x \to 0} f(x)$

(b) $\lim_{x \to -1} f(x)$

3. $g(x) = \sin x$

(a) $\lim_{x \to \pi/3} g(x)$

(b) $\lim_{x \to \pi/2} g(x)$

2. $f(x) = x^3 - 3x^2$

(a) $\lim_{x \to 1} f(x)$

(b) $\lim_{x \to 3} f(x)$

4. $h(x) = \tan x$

(a) $\lim_{x \to 0} h(x)$

(b) $\lim_{x \to 3\pi/4} h(x)$

In Exercises 5–28, find the indicated limit.

5. $\lim_{x \to 2} x^2$

6. $\lim_{x \to -3} (3x + 2)$

7. $\lim_{x \to 0} (2x - 1)$

8. $\lim_{x \to 1} (-x^2 + 1)$

9. $\lim_{x \to 2} (-x^2 + x - 2)$

10. $\lim_{x \to 1} (3x^3 - 2x^2 + 4)$

11. $\lim_{x \to 3} \sqrt{x + 1}$

12. $\lim_{x \to 4} \sqrt[3]{x + 4}$

13. $\lim_{x \to -4} (x + 3)^2$

14. $\lim_{x \to 0} (2x - 1)^3$

15. $\lim_{x \to 2} \dfrac{1}{x}$

16. $\lim_{x \to -3} \dfrac{2}{x + 2}$

17. $\lim_{x \to -1} \dfrac{x^2 + 1}{x}$

18. $\lim_{x \to 3} \dfrac{\sqrt{x + 1}}{x - 4}$

19. $\lim_{x \to \pi/2} \sin x$

20. $\lim_{x \to \pi} \tan x$

21. $\lim_{x \to 1} \cos \pi x$

22. $\lim_{x \to 1} \sin \dfrac{\pi x}{2}$

23. $\lim_{x \to 0} \sec 2x$

24. $\lim_{x \to \pi} \cos 3x$

25. $\lim_{x \to 5\pi/6} \sin x$

26. $\lim_{x \to 5\pi/3} \cos x$

27. $\lim_{x \to 3} \tan \left(\dfrac{\pi x}{4} \right)$

28. $\lim_{x \to 7} \sec \left(\dfrac{\pi x}{6} \right)$

29. If $\lim_{x \to c} f(x) = 2$ and $\lim_{x \to c} g(x) = 3$, find the following.

(a) $\lim_{x \to c} [5g(x)]$

(b) $\lim_{x \to c} [f(x) + g(x)]$

(c) $\lim_{x \to c} [f(x)g(x)]$

(d) $\lim_{x \to c} \dfrac{f(x)}{g(x)}$

30. If $\lim_{x \to c} f(x) = \frac{3}{2}$ and $\lim_{x \to c} g(x) = \frac{1}{2}$, find the following.

(a) $\lim_{x \to c} [4f(x)]$

(b) $\lim_{x \to c} [f(x) + g(x)]$

(c) $\lim_{x \to c} [f(x)g(x)]$

(d) $\lim_{x \to c} \dfrac{f(x)}{g(x)}$

31. If $\lim_{x \to c} f(x) = 4$, find the following.

(a) $\lim_{x \to c} [f(x)]^3$

(b) $\lim_{x \to c} \sqrt{f(x)}$

(c) $\lim_{x \to c} [3f(x)]$

(d) $\lim_{x \to c} [f(x)]^{3/2}$

32. If $\lim_{x \to c} f(x) = 27$, find the following.

(a) $\lim_{x \to c} \sqrt[3]{f(x)}$

(b) $\lim_{x \to c} \dfrac{f(x)}{18}$

(c) $\lim_{x \to c} [f(x)]^2$

(d) $\lim_{x \to c} [f(x)]^{2/3}$

33. Find two functions f and g such that

$$\lim_{x \to 0} f(x) \text{ and } \lim_{x \to 0} g(x)$$

do not exist, but

$$\lim_{x \to 0} [f(x) + g(x)]$$

does exist.

34. Prove that if

$$\lim_{x \to c} f(x) \text{ exists}$$

and

$$\lim_{x \to c} [f(x) + g(x)] \text{ does not exist,}$$

then

$$\lim_{x \to c} g(x) \text{ does not exist.}$$

35. Prove Property 1 of Theorem 2.2.

36. Prove Property 3 of Theorem 2.2. (You may use Property 3 of Theorem 2.3.)

37. Prove Property 1 of Theorem 2.3.

2.3 Techniques for Evaluating Limits

Cancellation technique ▪ Indeterminate forms ▪ Rationalization technique ▪ The Squeeze Theorem ▪ Limits of trigonometric functions

We begin this section by reviewing the strategy for finding the limit

$$\lim_{x \to c} f(x).$$

1. Learn to recognize which limits can be evaluated by direct substitution. (We did this in Section 2.2)
2. If the limit of $f(x)$ as $x \to c$ *cannot* be evaluated by direct substitution, try to find a function g that agrees with f for all x other than $x = c$. We choose g so that the limit of $g(x)$ can be evaluated by direct substitution.
3. Apply Theorem 2.1 to conclude that

$$\lim_{x \to c} f(x) = \lim_{x \to c} g(x) = g(c).$$

Two techniques for finding g are shown in Examples 1 and 2. The first technique involves cancelling common factors, and the second technique involves rationalizing the numerator of a fractional expression.

EXAMPLE 1 Cancellation technique

Find the following limit.

$$\lim_{x \to -3} \frac{x^2 + x - 6}{x + 3}$$

SOLUTION

Although we are taking the limit of a rational function, we *cannot* apply Theorem 2.5 because the limit of the denominator is zero.

$$\lim_{x \to -3} (x^2 + x - 6) = 0$$

$$\lim_{x \to -3} \frac{x^2 + x - 6}{x + 3}$$

Direct substitution fails

$$\lim_{x \to -3} (x + 3) = 0$$

However, since the limit of the numerator is also zero, we know that the numerator and denominator have a *common factor* of $(x + 3)$. Thus, for all $x \neq -3$, we can cancel this factor to obtain

$$\frac{x^2 + x - 6}{x + 3} = \frac{(x + 3)(x - 2)}{x + 3} = x - 2, \quad x \neq -3.$$

Finally, by Theorem 2.1, it follows that

$$\lim_{x \to -3} \frac{x^2 + x - 6}{x + 3} = \lim_{x \to -3} (x - 2) = -5.$$

This result is shown graphically in Figure 2.14. Note that the graph of the function f coincides with the graph of the function $g(x) = x - 2$, except that the graph of f has a hole at the point $(-3, -5)$.

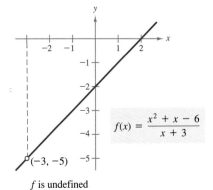

$$f(x) = \frac{x^2 + x - 6}{x + 3}$$

$(-3, -5)$

f is undefined when $x = -3$.

FIGURE 2.14

REMARK In Example 1, note that if $r(x) = p(x)/q(x)$ is a rational function such that $p(c) = 0$ and $q(c) = 0$, then the factor $(x - c)$ must be common to both $p(x)$ and $q(x)$.

In Example 1, direct substitution produced the meaningless fractional form $0/0$. We call such an expression an **indeterminate form** because we cannot (from the form alone) determine the limit. When you try to evaluate a limit and encounter this form, remember that you must rewrite the fraction so that the new denominator does not have zero as its limit. One way to do this is to *cancel like factors*, as shown in Example 1. A second way is to *rationalize the numerator*, as shown in Example 2.

EXAMPLE 2 Rationalization technique

Find the following limit.

$$\lim_{x \to 0} \frac{\sqrt{x + 1} - 1}{x}$$

SOLUTION

By direct substitution, we obtain the indeterminate form $0/0$.

$$\lim_{x \to 0} \frac{\sqrt{x + 1} - 1}{x} \qquad \begin{array}{l} \nearrow \lim_{x \to 0} (\sqrt{x + 1} - 1) = 0 \\ \\ \searrow \lim_{x \to 0} x = 0 \end{array}$$

In this case, we rewrite the fraction by rationalizing the numerator.

$$\frac{\sqrt{x + 1} - 1}{x} = \left(\frac{\sqrt{x + 1} - 1}{x}\right)\left(\frac{\sqrt{x + 1} + 1}{\sqrt{x + 1} + 1}\right)$$

$$= \frac{(x + 1) - 1}{x(\sqrt{x + 1} + 1)} = \frac{\cancel{x}}{\cancel{x}(\sqrt{x + 1} + 1)}$$

$$= \frac{1}{\sqrt{x + 1} + 1}, \quad x \neq 0$$

Therefore, by Theorem 2.1, we have

$$\lim_{x \to 0} \frac{\sqrt{x + 1} - 1}{x} = \lim_{x \to 0} \frac{1}{\sqrt{x + 1} + 1}$$

$$= \frac{1}{1 + 1} = \frac{1}{2}.$$

Table 2.4 reinforces our conclusion that this limit is $\frac{1}{2}$.

TABLE 2.4

	x approaches 0 from the left					x approaches 0 from the right			
x	−0.25	−0.1	−0.01	−0.001	0	0.001	0.01	0.1	0.25
$\dfrac{\sqrt{x+1}-1}{x}$	0.5359	0.5132	0.5013	0.5001	?	0.4999	0.4988	0.4881	0.4721

$f(x)$ approaches $\frac{1}{2}$ $f(x)$ approaches $\frac{1}{2}$

REMARK   The rationalization technique for evaluating limits is based on multiplication by a convenient form of 1. In Example 2, the convenient form is

$$1 = \frac{\sqrt{x+1}+1}{\sqrt{x+1}+1}.$$

EXAMPLE 3   Cancellation technique applied to trigonometric functions

Find the following limit.

$$\lim_{x \to 0} \frac{\tan x}{\sin x}$$

SOLUTION

Direct substitution produces the indeterminate form 0/0. However, using the fact that $\tan x = (\sin x)/(\cos x)$, we can rewrite the function as

$$\frac{\tan x}{\sin x} = \frac{(\sin x)/(\cos x)}{\sin x}$$

$$= \frac{\sin x}{(\cos x)(\sin x)} = \frac{1}{\cos x}, \quad \sin x \neq 0.$$

Thus, by Theorem 2.1, we have

$$\lim_{x \to 0} \frac{\tan x}{\sin x} = \lim_{x \to 0} \frac{1}{\cos x} = \frac{1}{1} = 1.$$

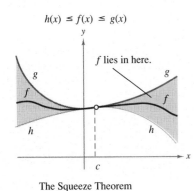

$h(x) \leq f(x) \leq g(x)$

f lies in here.

The Squeeze Theorem

FIGURE 2.15

When we encounter the indeterminate form 0/0 in a limit involving trigonometric functions, some ingenuity is usually required to determine the limit. The following theorem is helpful. It concerns the limiting behavior of a function that is squeezed between two other functions, each of which has the same limit at a given point, as shown in Figure 2.15.

THEOREM 2.9 THE SQUEEZE THEOREM	If $h(x) \leq f(x) \leq g(x)$ for all x in an open interval containing c, except possibly at c itself, and if $$\lim_{x \to c} h(x) = L = \lim_{x \to c} g(x)$$ then $$\lim_{x \to c} f(x) = L.$$

PROOF   For $\varepsilon > 0$ there exist δ_1 and δ_2 such that

$$|h(x) - L| < \varepsilon \qquad \text{whenever} \qquad 0 < |x - c| < \delta_1$$

and

$$|g(x) - L| < \varepsilon \qquad \text{whenever} \qquad 0 < |x - c| < \delta_2.$$

Let δ be the smaller of δ_1 and δ_2. Then, if $0 < |x - c| < \delta$, it follows that $|h(x) - L| < \varepsilon$ and $|g(x) - L| < \varepsilon$, which implies that

$$-\varepsilon < h(x) - L < \varepsilon \qquad \text{and} \qquad -\varepsilon < g(x) - L < \varepsilon$$
$$L - \varepsilon < h(x) \qquad \text{and} \qquad g(x) < L + \varepsilon.$$

Now, since $h(x) \leq f(x) \leq g(x)$, it follows that $L - \varepsilon < f(x) < L + \varepsilon$ and $|f(x) - L| < \varepsilon$. Therefore,

$$\lim_{x \to c} f(x) = L.$$

The usefulness of the Squeeze Theorem is seen in the proof of Theorem 2.10. We will use the two limits discussed in this theorem in Chapter 3.

THEOREM 2.10 TWO SPECIAL TRIGONOMETRIC LIMITS	1. $\displaystyle\lim_{x \to 0} \frac{\sin x}{x} = 1$ 2. $\displaystyle\lim_{x \to 0} \frac{1 - \cos x}{x} = 0$

PROOF   We prove the first limit and leave the proof of the second as an exercise (see Exercise 54). (Note that direct substitution produces the indeterminate form $0/0$ in both cases.) To avoid the confusion of two different uses of x, we present the proof using the variable θ, where θ is an acute positive angle (measured in radians). Figure 2.16 shows a circular section that is squeezed between two triangles.

 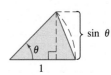

Area of triangle $\geq$ Area of sector $\geq$ Area of triangle

$$\frac{\tan\theta}{2} \geq \frac{\theta}{2} \geq \frac{\sin\theta}{2}$$

FIGURE 2.16

Multiplying each expression by $2/\sin \theta$ produces

$$\frac{1}{\cos \theta} \geq \frac{\theta}{\sin \theta} \geq 1$$

and taking reciprocals and reversing the inequalities yields

$$\cos \theta \leq \frac{\sin \theta}{\theta} \leq 1.$$

Since $\cos \theta = \cos (-\theta)$ and $(\sin \theta)/\theta = [\sin (-\theta)]/(-\theta)$, we can conclude that this inequality is valid for *all* nonzero θ in the open interval $(-\pi/2, \pi/2)$. Finally, since

$$\lim_{\theta \to 0} \cos \theta = 1$$

and

$$\lim_{\theta \to 0} 1 = 1$$

we can apply the Squeeze Theorem to conclude that

$$\lim_{\theta \to 0} \frac{\sin \theta}{\theta} = 1.$$

EXAMPLE 4   A limit involving a trigonometric function

Find the following limit.

$$\lim_{x \to 0} \frac{\tan x}{x}$$

SOLUTION

Direct substitution yields the indeterminate form $0/0$. To solve this problem, we write $\tan x = (\sin x)/(\cos x)$ and obtain

$$\lim_{x \to 0} \frac{\tan x}{x} = \lim_{x \to 0} \left(\frac{\sin x}{x}\right)\left(\frac{1}{\cos x}\right).$$

Now, since

$$\lim_{x \to 0} \frac{\sin x}{x} = 1$$

and

$$\lim_{x \to 0} \frac{1}{\cos x} = 1$$

we have

$$\lim_{x \to 0} \frac{\tan x}{x} = \left[\lim_{x \to 0} \frac{\sin x}{x}\right]\left[\lim_{x \to 0} \frac{1}{\cos x}\right] = (1)(1) = 1.$$

EXAMPLE 5 A limit involving a trigonometric function

Find the following limit.

$$\lim_{x \to 0} \frac{\sin 2x}{x}$$

SOLUTION

Direct substitution yields the indeterminate form $0/0$. To solve this problem, we rewrite the limit as

$$\lim_{x \to 0} \frac{\sin 2x}{x} = 2\left[\lim_{x \to 0} \frac{\sin 2x}{2x}\right].$$

Now by letting $y = 2x$ and observing that $x \to 0$ if and only if $y \to 0$, we can write

$$\lim_{x \to 0} \frac{\sin 2x}{x} = 2\left[\lim_{x \to 0} \frac{\sin 2x}{2x}\right]$$

$$= 2\left[\lim_{y \to 0} \frac{\sin y}{y}\right] = 2[1] = 2.$$

EXERCISES for Section 2.3

In Exercises 1–4, use the graph to determine the limit visually, if it exists.

1. $g(x) = \dfrac{-2x^2 + x}{x}$

2. $h(x) = \dfrac{x^2 - 3x}{x}$

3. $g(x) = \dfrac{x^3 - x}{x - 1}$

4. $f(x) = \dfrac{1}{x - 1}$

(a) $\lim_{x \to 0} g(x)$

(b) $\lim_{x \to -1} g(x)$

(a) $\lim_{x \to -2} h(x)$

(b) $\lim_{x \to 0} h(x)$

(a) $\lim_{x \to 1} g(x)$

(b) $\lim_{x \to -1} g(x)$

(a) $\lim_{x \to 1} f(x)$

(b) $\lim_{x \to 2} f(x)$

In Exercises 5–22, find the limit (if it exists).

5. $\lim\limits_{x \to -1} \dfrac{x^2 - 1}{x + 1}$

6. $\lim\limits_{x \to -1} \dfrac{2x^2 - x - 3}{x + 1}$

7. $\lim\limits_{x \to 3} \dfrac{x - 3}{x^2 - 9}$

8. $\lim\limits_{x \to -1} \dfrac{x^3 + 1}{x + 1}$

9. $\lim\limits_{x \to -2} \dfrac{x^3 + 8}{x + 2}$

10. $\lim\limits_{\Delta x \to 0} \dfrac{(x + \Delta x)^2 - x^2}{\Delta x}$

11. $\lim\limits_{\Delta x \to 0} \dfrac{2(x + \Delta x) - 2x}{\Delta x}$

12. $\lim\limits_{\Delta x \to 0} \dfrac{(x + \Delta x)^3 - x^3}{\Delta x}$

13. $\lim\limits_{\Delta x \to 0} \dfrac{(x + \Delta x)^2 - 2(x + \Delta x) + 1 - (x^2 - 2x + 1)}{\Delta x}$

14. $\lim\limits_{\Delta x \to 0} \dfrac{(1 + \Delta x)^3 - 1}{\Delta x}$

15. $\lim\limits_{x \to 5} \dfrac{x - 5}{x^2 - 25}$

16. $\lim\limits_{x \to 2} \dfrac{2 - x}{x^2 - 4}$

17. $\lim\limits_{x \to 1} \dfrac{x^2 + x - 2}{x^2 - 1}$

18. $\lim\limits_{x \to 0} \dfrac{\sqrt{2 + x} - \sqrt{2}}{x}$

19. $\lim\limits_{x \to 0} \dfrac{\sqrt{3 + x} - \sqrt{3}}{x}$

20. $\lim\limits_{x \to 0} \dfrac{[1/(x + 4)] - (1/4)}{x}$

21. $\lim\limits_{x \to 0} \dfrac{[1/(2 + x)] - (1/2)}{x}$

22. $\lim\limits_{x \to 3} \dfrac{\sqrt{x + 1} - 2}{x - 3}$

In Exercises 23–26, complete a table of values near $x = c$ to estimate the limit. Then, find the limit by analytic methods and compare the result to your estimated limit.

23. $\lim\limits_{x \to 0} \dfrac{\sqrt{x + 2} - \sqrt{2}}{x}$

24. $\lim\limits_{x \to 1} \dfrac{1 - x}{\sqrt{5 - x^2} - 2}$

25. $\lim\limits_{x \to 0} \dfrac{[1/(2 + x)] - (1/2)}{x}$

26. $\lim\limits_{x \to 2} \dfrac{x^5 - 32}{x - 2}$

In Exercises 27–40, determine the limit (if it exists).

27. $\lim\limits_{x \to 0} \dfrac{\sin x}{5x}$

28. $\lim\limits_{x \to 0} \dfrac{3(1 - \cos x)}{x}$

29. $\lim\limits_{\theta \to 0} \dfrac{\sec \theta - 1}{\theta \sec \theta}$

30. $\lim\limits_{\theta \to 0} \dfrac{\cos \theta \tan \theta}{\theta}$

31. $\lim\limits_{x \to 0} \dfrac{\sin^2 x}{x}$

32. $\lim\limits_{\phi \to \pi} \phi \sec \phi$

33. $\lim\limits_{x \to \pi/2} \dfrac{\cos x}{\cot x}$

34. $\lim\limits_{x \to \pi/4} \dfrac{1 - \tan x}{\sin x - \cos x}$

35. $\lim\limits_{t \to 0} \dfrac{\sin^2 t}{t^2}$ $\left[\text{Hint: Find } \lim\limits_{t \to 0} \left(\dfrac{\sin t}{t} \right)^2. \right]$

36. $\lim\limits_{t \to 0} \dfrac{\sin 3t}{t}$ $\left[\text{Hint: Find } \lim\limits_{t \to 0} 3 \left(\dfrac{\sin 3t}{3t} \right). \right]$

37. $\lim\limits_{x \to 0} \dfrac{\sin 2x}{\sin 3x}$

$\left[\text{Hint: Find } \lim\limits_{x \to 0} \left(\dfrac{2 \sin 2x}{2x} \right) \left(\dfrac{3x}{3 \sin 3x} \right). \right]$

38. $\lim\limits_{x \to 0} \dfrac{\tan^2 x}{x}$

39. $\lim\limits_{h \to 0} \dfrac{(1 - \cos h)^2}{h}$

40. $\lim\limits_{h \to 0} (1 + \cos 2h)$

In Exercises 41–44, use the Squeeze Theorem to find $\lim\limits_{x \to c} f(x)$.

41. $c = 0$

$4 - x^2 \le f(x) \le 4 + x^2$

42. $c = a$

$b - |x - a| \le f(x) \le b + |x - a|$

43. $c = 0, \quad f(x) = x \sin \dfrac{1}{x}$

$-|x| \le x \sin \dfrac{1}{x} \le |x|, \quad x \ne 0$

44. $c = 0, \quad f(x) = x^2 \cos \dfrac{1}{x}$

$-x^2 \le x^2 \cos \dfrac{1}{x} \le x^2, \quad x \ne 0$

In Exercises 45 and 46, use the position function $s(t) = -16t^2 + 25$ giving the height (in feet) of a free-falling object. As discussed in the Chapter 2 Application, the velocity at time $t = a$ seconds is given by

$$\lim\limits_{t \to a} \dfrac{s(a) - s(t)}{a - t}.$$

45. Find the velocity when $t = 0.5$ seconds.

46. Find the velocity when $t = 1.1$ seconds.

47. Prove each of the following, given that $\lim\limits_{x \to c} f(x) = 0$.

(a) $\lim\limits_{x \to c} |f(x)| = 0$

(b) $\lim\limits_{x \to c} f(x)g(x) = 0$

where $|g(x)| \le M$ for a fixed number M and all $x \ne c$. [Hint: Use the Squeeze Theorem.]

48. Prove that if $\lim\limits_{x \to c} |f(x)| = 0$, then $\lim\limits_{x \to c} f(x) = 0$. [Note: This is the converse of Exercise 47(a).]

49. Prove that if $\lim\limits_{x \to c} f(x) = L$ then $\lim\limits_{x \to c} |f(x)| = |L|$. [Hint: Use the inequality $\big| |f(x)| - |L| \big| \le |f(x) - L|$.]

50. Find a function f to show that the converse of Exercise 49 is not true.

[Hint: Find a function f such that $\lim\limits_{x \to c} |f(x)| = |L|$ but $\lim\limits_{x \to c} f(x)$ does not exist.]

In Exercises 51 and 52, use a computer or graphics calculator to sketch the graph of the function f and find the specified limit (if it exists).

51. $f(x) = \dfrac{\sin 5x}{\sin 2x}$, $\lim\limits_{x \to 0} f(x)$

52. $f(x) = \dfrac{x^2 - 5x + 6}{x^2 - 4x + 4}$, $\lim\limits_{x \to 2} f(x)$

53. Use a computer or graphics calculator to sketch the graph of

$$f(x) = \frac{\sin x}{x}, \quad x \neq 0.$$

Complete a table showing the values of $f(x)$ for $x = \pm 0.1$, $x = \pm 0.01$, and $x = \pm 0.001$.

54. Complete the proof of Theorem 2.10 by proving that

$$\lim_{x \to 0} \frac{1 - \cos x}{x} = 0.$$

2.4 Continuity and One-Sided Limits

Continuity at a point ▪ Continuity on an open interval ▪ One-sided limits ▪ Continuity on a closed interval ▪ Properties of continuity ▪ Intermediate Value Theorem

Three Points of Discontinuity

FIGURE 2.17

In mathematics the term *continuous* has much the same meaning as it does in our everyday usage. To say that a function is continuous at $x = c$ means that there is no interruption in the graph of f at c. That is, its graph is unbroken at c and there are no holes, jumps, or gaps. For example, Figure 2.17 identifies three values of x at which the graph of f is not continuous. At all other points of the interval (a, b), the graph of f is uninterrupted and we say it is **continuous** at such points. Thus, it appears that the continuity of a function at $x = c$ can be destroyed by any one of the following conditions.

1. The function is not defined at $x = c$.
2. The limit of $f(x)$ does not exist at $x = c$.
3. The limit of $f(x)$ exists at $x = c$, but it is not equal to $f(c)$.

This brings us to the following definition.

DEFINITION OF CONTINUITY	*Continuity at a Point:* A function f is called **continuous at c** if the following three conditions are met. 1. $f(c)$ is defined. 2. $\lim\limits_{x \to c} f(x)$ exists. 3. $\lim\limits_{x \to c} f(x) = f(c)$. *Continuity on an Open Interval:* A function is called **continuous on an open interval (a, b)** if it is continuous at each point in the interval. A function that is continuous on the entire real line $(-\infty, \infty)$ is called **everywhere continuous**.

FIGURE 2.18

A function f is said to be **discontinous at c** if f is defined on an open interval containing c (except possibly at c) and f is not continuous at c. Discontinuities fall into two categories: **removable** and **nonremovable.** A discontinuity at $x = c$ is called removable if f can be made continuous by appropriately defining (or redefining) f at $x = c$. For example, in Figure 2.18, the function f has two removable discontinuities and one nonremovable discontinuity.

In Section 2.2 we studied several types of functions that meet the three conditions for continuity. In Example 1 we use this knowledge of limits to examine two functions that are continuous and one that is not.

EXAMPLE 1   Testing for continuity

Determine whether the following functions are continuous on the given interval.

(a) $f(x) = \dfrac{1}{x}$, $(0, 1)$ (b) $f(x) = \dfrac{x^2 - 1}{x - 1}$, $(0, 2)$

(c) $f(x) = \sin x$, $(0, 2\pi)$

SOLUTION

The graphs of these three functions are shown in Figure 2.19.

(a) Since f is a rational function whose denominator is not zero in the interval $(0, 1)$, we can apply Theorem 2.5 to conclude that f is continuous on $(0, 1)$.

(b) Since f is undefined at $x = 1$, we conclude that it is discontinuous at $x = 1$ and continuous for all other values of x in the interval $(0, 2)$.

(c) Since the sine function is defined over the entire real line, we can apply Theorem 2.8 to conclude that f is continuous on $(0, 2\pi)$.

Continuous on $(0, 1)$

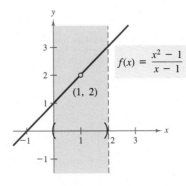

Discontinuous at $x = 1$

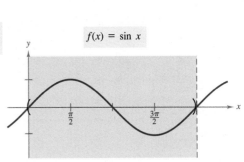

Continuous on $(0, 2\pi)$

FIGURE 2.19

REMARK   In part (b) of Example 1, the discontinuity at $x = 1$ is *removable*. Specifically, by defining $f(1)$ to be 2, we would obtain a function that is continuous on $(0, 2)$.

Each of the intervals in Example 1 is open. To discuss continuity on a closed interval, we first need to look at a different type of limit called a **one-sided limit**. For example, when we talk about the **limit from the right,** we mean that x approaches c from values greater than c. We denote this by

$$\lim_{x \to c^+} f(x) = L. \qquad \text{Limit from the right}$$

Similarly, the **limit from the left** means that x approaches c from values less than c. We denote this by

$$\lim_{x \to c^-} f(x) = L. \qquad \text{Limit from the left}$$

One-sided limits are useful in taking limits of functions involving radicals. For instance, if n is an even integer, then

$$\lim_{x \to 0^+} \sqrt[n]{x} = 0.$$

EXAMPLE 2   A one-sided limit

Find the limit of $f(x) = \sqrt{x - 1}$ as x approaches 1 from the right.

SOLUTION

As indicated in Figure 2.20, the limit as x approaches 1 from the right is given by

$$\lim_{x \to 1^+} \sqrt{x - 1} = 0. \qquad \blacksquare$$

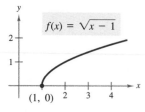

$f(x) = \sqrt{x - 1}$

FIGURE 2.20

Another use of one-sided limits is that of investigating the behavior of **step functions.** One common type of step function is the **greatest integer function** $[\![x]\!]$, defined by

$$[\![x]\!] = \text{greatest integer } n \text{ such that } n \le x.$$

EXAMPLE 3   The greatest integer function

Find the limit of $f(x) = [\![x]\!]$ as x approaches 0 from the left and from the right.

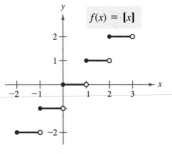

$f(x) = [\![x]\!]$

Greatest Integer Function

FIGURE 2.21

SOLUTION

The limit as x approaches 0 *from the left* is given by

$$\lim_{x \to 0^-} [\![x]\!] = -1$$

and the limit as x approaches 0 *from the right* is given by

$$\lim_{x \to 0^+} [\![x]\!] = 0.$$

(See Figure 2.21.)

In Figure 2.21 we see that the greatest integer function is discontinuous at $x = 0$ (and at every other integer). This discontinuity stems from the fact that the left and right limits at zero are different. In such cases we say that the (two-sided) limit *does not exist*. The following theorem makes this more explicit. The proof of this theorem follows directly from the definitions of limit and one-sided limits.

THEOREM 2.11
THE EXISTENCE OF A LIMIT

If f is a function and c and L are real numbers, then the limit of $f(x)$ as x approaches c is L if and only if

$$\lim_{x \to c^-} f(x) = L \quad \text{and} \quad \lim_{x \to c^+} f(x) = L.$$

Now that we have defined one-sided limits, we are ready to extend our definition of continuity to include closed intervals.

DEFINITION OF CONTINUITY
ON A CLOSED INTERVAL

A function f is **continuous on the closed interval** $[a, b]$ if it is continuous on the open interval (a, b) and

$$\lim_{x \to a^+} f(x) = f(a) \quad \text{and} \quad \lim_{x \to b^-} f(x) = f(b).$$

The function f is called **continuous from the right** at a and **continuous from the left** at b.

Similar definitions can be made to cover continuity on half-open intervals of the form $(a, b]$ and $[a, b)$, or on infinite intervals. For example, the function $f(x) = \sqrt{x}$ is continuous on the infinite interval $[0, \infty)$.

EXAMPLE 4   Continuity on a closed interval

Discuss the continuity of

$$g(x) = \begin{cases} 5 - x, & -1 \le x \le 2 \\ x^2 - 1, & 2 < x \le 3. \end{cases}$$

SOLUTION

From our work in the previous section, we know that the polynomial functions given by $5 - x$ and $x^2 - 1$ are continuous for all real x. Thus, to conclude that g is continuous on the entire interval $[-1, 3]$, we need only check the behavior of g when $x = 2$. By taking the one-sided limits when $x = 2$, we see that

$$\lim_{x \to 2^-} g(x) = \lim_{x \to 2^-} (5 - x) = 3 \qquad \text{Limit from the left}$$

and

$$\lim_{x \to 2^+} g(x) = \lim_{x \to 2^+} (x^2 - 1) = 3. \qquad \text{Limit from the right}$$

Since these two limits are equal, we can apply Theorem 2.11 to conclude that

$$\lim_{x \to 2} g(x) = g(2) = 3.$$

Thus, g is continuous at $x = 2$, and consequently it is continuous on the entire interval $[-1, 3]$. The graph of g is shown in Figure 2.22.   ▭

$$g(x) = \begin{cases} 5 - x, & -1 \le x \le 2 \\ x^2 - 1, & 2 < x \le 3 \end{cases}$$

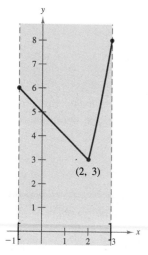

FIGURE 2.22

Properties of continuity

In Section 2.2, we looked at several properties of limits. Each property yields a corresponding property pertaining to the continuity of a function. The proof of Theorem 2.12 follows directly from Theorem 2.3.

THEOREM 2.12 **PROPERTIES OF CONTINUITY**	If b is a real number and f and g are continuous at $x = c$, then the following functions are also continuous at c. 1. Scalar multiple: bf 2. Sum and difference: $f \pm g$ 3. Product: fg 4. Quotient: $\dfrac{f}{g}$, if $g(c) \ne 0$

The following summary lists some common types of functions that are continuous at every point in their domains.

1. Polynomial functions: $p(x) = a_n x^n + a_{n-1} x^{n-1} + \cdots + a_1 x + a_0$

2. Rational functions: $r(x) = \dfrac{p(x)}{q(x)}$, $q(x) \ne 0$

3. Radical functions: $f(x) = \sqrt[n]{x}$

4. Trigonometric functions: $\sin x$, $\cos x$, $\tan x$, $\cot x$, $\sec x$, $\csc x$

By combining Theorem 2.12 with this summary, we can conclude that a wide variety of elementary functions are continuous at every point in their domains. For example, the following functions are continuous at every point in their domains.

$$f(x) = x + \sin x, \quad f(x) = 3 \tan x, \quad f(x) = \frac{x^2 + 1}{\cos x}$$

The next theorem, which is a consequence of Theorem 2.7, allows us to determine the continuity of *composite* functions such as

$$f(x) = \sin 3x, \quad f(x) = \sqrt{x^2 + 1}, \quad f(x) = \tan \frac{1}{x}.$$

THEOREM 2.13 **CONTINUITY OF A** **COMPOSITE FUNCTION**	If g is continuous at c and f is continuous at $g(c)$, then the composite function given by $(f \circ g)(x) = f(g(x))$ is continuous at c.

REMARK   Note that one consequence of the theorem is that if f and g satisfy the given conditions, we can determine the limit of $f(g(x))$ as x approaches c to be

$$\lim_{x \to c} f(g(x)) = f(g(c)).$$

EXAMPLE 5   Testing for continuity

Find the intervals for which the three functions shown in Figure 2.23 are continuous.

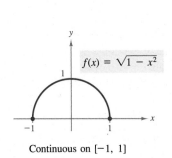

$f(x) = \sqrt{1 - x^2}$

Continuous on $[-1, 1]$

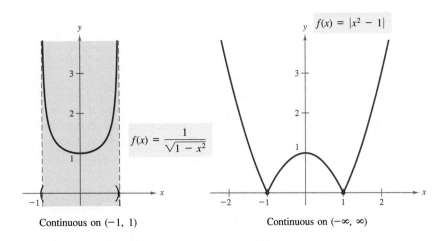

$f(x) = \dfrac{1}{\sqrt{1 - x^2}}$

Continuous on $(-1, 1)$

$f(x) = |x^2 - 1|$

Continuous on $(-\infty, \infty)$

FIGURE 2.23

SOLUTION

(a) The function $f(x) = \sqrt{1 - x^2}$ is continuous on the *closed* interval $[-1, 1]$.

(b) The function $f(x) = 1/\sqrt{1 - x^2}$ is continuous on the *open* interval $(-1, 1)$. (Note that f is *undefined* for all x such that $|x| \geq 1$.)

(c) At $x = \pm 1$, the limits from the right and left are zero. Thus, the function $f(x) = |x^2 - 1|$ is continuous on the entire real line. ▭

EXAMPLE 6 Testing for continuity

Find the intervals for which the three functions shown in Figure 2.24 are continuous.

$f(x) = \tan x$

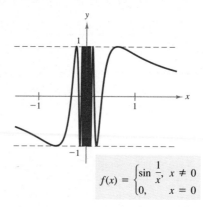

$f(x) = \begin{cases} \sin \dfrac{1}{x}, & x \neq 0 \\ 0, & x = 0 \end{cases}$

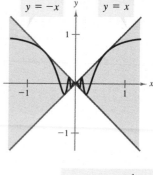

$f(x) = \begin{cases} x \sin \dfrac{1}{x}, & x \neq 0 \\ 0, & x = 0 \end{cases}$

FIGURE 2.24

SOLUTION

(a) The tangent function is undefined at $x = (\pi/2) + n\pi$. At all other points it is continuous. Thus, $f(x) = \tan x$ is continuous on the open intervals

$$\ldots, \left(-\frac{3\pi}{2}, -\frac{\pi}{2}\right), \left(-\frac{\pi}{2}, \frac{\pi}{2}\right), \left(\frac{\pi}{2}, \frac{3\pi}{2}\right), \ldots$$

(b) Since $y = 1/x$ is continuous except at $x = 0$ and the sine function is continuous everywhere, we need only check the continuity at $x = 0$. At this point, the limit does not exist (see Example 5, Section 2.1). Therefore, $f(x) = \sin (1/x)$ is continuous on the intervals $(-\infty, 0)$ and $(0, \infty)$.

(c) This function is similar to that in part (b) except that the oscillations are damped by the factor x. Using the Squeeze Theorem, we have

$$-|x| \leq x \sin \frac{1}{x} \leq |x|, \quad x \neq 0$$

and we conclude that the limit as $x \to 0$ is zero. Thus, f is continuous on the entire real line. ▭

Intermediate Value Theorem

We conclude this section with an important theorem concerning the behavior of functions that are continuous on a closed interval.

THEOREM 2.14 **INTERMEDIATE VALUE THEOREM**	If f is continuous on $[a, b]$ and k is any number between $f(a)$ and $f(b)$, then there is at least one number c in $[a, b]$ such that $f(c) = k$.

By referring to a text on advanced calculus, you will find that a proof of this theorem is based on a property of real numbers called *completeness*. The Intermediate Value Theorem states that for a continuous function f, if x takes on all values between a and b, then $f(x)$ must take on all values between $f(a)$ and $f(b)$. As a simple example of this theorem, consider a person's height. Suppose that a boy is 5 feet tall on his thirteenth birthday and 5 feet 7 inches tall on his fourteenth birthday. Then, for any height h between 5 feet and 5 feet 7 inches, there must have been a time t when his height was exactly h. This seems reasonable because we believe that normal human growth is continuous and a person's height could not abruptly change from one value to another.

The Intermediate Value Theorem guarantees the existence of *at least one* number c in the closed interval $[a, b]$. There may, of course, be more than one number c such that $f(c) = k$, as shown in Figure 2.25. A discontinuous function might not possess the intermediate value property. For example, the graph of the discontinuous function shown in Figure 2.26 jumps over the horizontal line given by $y = k$ and for this function there is no value of c in $[a, b]$ such that $f(c) = k$.

The Intermediate Value Theorem often can be used to locate the zeros of a function that is continuous on a closed interval. Specifically, if f is continuous on $[a, b]$ and $f(a)$ and $f(b)$ differ in sign, then the Intermediate Value Theorem guarantees the existence of at least one zero of f in the closed interval $[a, b]$.

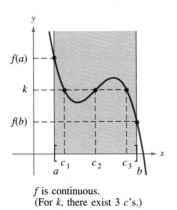

f is continuous.
(For k, there exist 3 c's.)

FIGURE 2.25

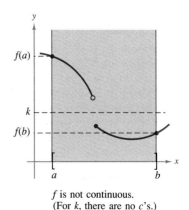

f is not continuous.
(For k, there are no c's.)

FIGURE 2.26

$f(x) = x^3 + 2x - 1$

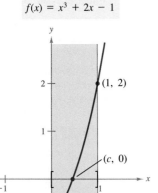

FIGURE 2.27

EXAMPLE 7 An application of the Intermediate Value Theorem

Use the Intermediate Value Theorem to show that the polynomial function $f(x) = x^3 + 2x - 1$ has a zero in the interval $[0, 1]$.

SOLUTION

Since

$$f(0) = 0^3 + 2(0) - 1 = -1, \quad f(0) < 0$$

and

$$f(1) = 1^3 + 2(1) - 1 = 2, \quad f(1) > 0$$

we can apply the Intermediate Value Theorem to conclude that there must be some c in $[0, 1]$ such that $f(c) = 0$, as shown in Figure 2.27.

REMARK Note that the Intermediate Value Theorem tells us that at least one c exists, but it does not give us a method for finding c. Such theorems are called **existence theorems.**

EXERCISES for Section 2.4

In Exercises 1–6, use the graph to determine the following visually.
(a) $\lim\limits_{x \to c^+} f(x)$ (b) $\lim\limits_{x \to c^-} f(x)$ (c) $\lim\limits_{x \to c} f(x)$

1.

2.

3.

4.

5.

6.
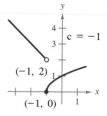

In Exercises 7–22, find the limit (if it exists).

7. $\lim\limits_{x \to 5^+} \dfrac{x - 5}{x^2 - 25}$

8. $\lim\limits_{x \to 2^+} \dfrac{2 - x}{x^2 - 4}$

9. $\lim\limits_{x \to 2^+} \dfrac{x}{\sqrt{x^2 - 4}}$

10. $\lim\limits_{x \to 4^-} \dfrac{\sqrt{x} - 2}{x - 4}$

11. $\lim\limits_{\Delta x \to 0^+} \dfrac{2(x + \Delta x) - 2x}{\Delta x}$

12. $\lim\limits_{\Delta x \to 0^+} \dfrac{(x + \Delta x)^2 + x + \Delta x - (x^2 + x)}{\Delta x}$

13. $\lim\limits_{\Delta x \to 0^-} \dfrac{\dfrac{1}{x + \Delta x} - \dfrac{1}{x}}{\Delta x}$

14. $\lim\limits_{x \to 1^-} \dfrac{x^2 - 2x + 1}{x - 1}$

15. $\lim\limits_{x \to 0} \dfrac{|x|}{x}$

16. $\lim\limits_{x \to 2} \dfrac{|x - 2|}{x - 2}$

17. $\lim\limits_{x \to 3} f(x)$, where $f(x) = \begin{cases} \dfrac{x + 2}{2}, & x \le 3 \\ \dfrac{12 - 2x}{3}, & x > 3 \end{cases}$

18. $\lim\limits_{x \to 2} f(x)$, where $f(x) = \begin{cases} x^2 - 4x + 6, & x < 2 \\ -x^2 + 4x - 2, & x \ge 2 \end{cases}$

19. $\lim\limits_{x \to 1} f(x)$, where $f(x) = \begin{cases} x^3 + 1, & x < 1 \\ x + 1, & x \ge 1 \end{cases}$

20. $\lim\limits_{x \to 1} f(x)$, where $f(x) = \begin{cases} x, & x \le 1 \\ 1 - x, & x > 1 \end{cases}$

21. $\lim\limits_{x \to \pi} \cot x$

22. $\lim\limits_{x \to \pi/2} \sec x$

In Exercises 23–28, find the discontinuities (if any).

23. $f(x) = -\dfrac{x^3}{2}$

24. $f(x) = \dfrac{x^2 - 1}{x}$

25. $f(x) = \dfrac{x^2 - 1}{x + 1}$

26. $f(x) = \dfrac{1}{x^2 - 4}$

27. $f(x) = \begin{cases} x, & x < 1 \\ 2, & x = 1 \\ 2x - 1, & 1 < x \end{cases}$

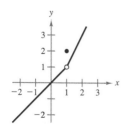

28. $f(x) = \dfrac{[\![x]\!]}{2} + x$

In Exercises 29–50, find the discontinuities (if any) for the given function. Which of the discontinuities are removable?

29. $f(x) = x^2 - 2x + 1$

30. $f(x) = \dfrac{1}{x^2 + 1}$

31. $f(x) = x + \sin x$

32. $f(x) = \cos \dfrac{\pi x}{2}$

33. $f(x) = \dfrac{1}{x - 1}$

34. $f(x) = \dfrac{x}{x^2 - 1}$

35. $f(x) = \dfrac{x}{x^2 + 1}$

36. $f(x) = \dfrac{x - 3}{x^2 - 9}$

37. $f(x) = \dfrac{x + 2}{x^2 - 3x - 10}$

38. $f(x) = \dfrac{x - 1}{x^2 + x - 2}$

39. $f(x) = \begin{cases} x, & x \le 1 \\ x^2, & x > 1 \end{cases}$

40. $f(x) = \begin{cases} -2x + 3, & x < 1 \\ x^2, & x \ge 1 \end{cases}$

41. $f(x) = \begin{cases} \dfrac{x}{2} + 1, & x \le 2 \\ 3 - x, & x > 2 \end{cases}$

42. $f(x) = \begin{cases} -2x, & x \le 2 \\ x^2 - 4x + 1, & x > 2 \end{cases}$

43. $f(x) = \dfrac{|x + 2|}{x + 2}$

44. $f(x) = \dfrac{|x - 3|}{x - 3}$

45. $f(x) = \begin{cases} \csc \dfrac{\pi x}{6}, & |x - 3| \le 2 \\ 2, & |x - 3| > 2 \end{cases}$

46. $f(x) = \begin{cases} \tan \dfrac{\pi x}{4}, & |x| < 1 \\ x, & |x| \ge 1 \end{cases}$

47. $f(x) = \csc 2x$

48. $f(x) = \tan \dfrac{\pi x}{2}$

49. $f(x) = [\![x - 1]\!]$

50. $f(x) = x - [\![x]\!]$

In Exercises 51–54, discuss the continuity of the composite function $h(x) = f(g(x))$.

51. $f(x) = x^2$, $g(x) = x - 1$

52. $f(x) = \dfrac{1}{\sqrt{x}}$, $g(x) = x - 1$

53. $f(x) = \dfrac{1}{(x - 1)}$, $g(x) = x^2 + 5$

54. $f(x) = \sin x$, $g(x) = x^2$

In Exercises 55–58, sketch the graph of the given function to determine any discontinuities.

55. $f(x) = \dfrac{x^2 - 16}{x - 4}$

56. $f(x) = \dfrac{x^3 - 8}{x - 2}$

57. $f(x) = [\![x]\!] - x$

58. $f(x) = \dfrac{\sin x}{x}$

In Exercises 59–64, find the interval(s) for which the function is continuous.

59. $f(x) = \dfrac{x^2}{x^2 - 36}$

60. $f(x) = x\sqrt{x + 3}$

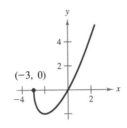

(-3, 0)

61. $f(x) = \dfrac{x}{x^2 + 1}$

62. $f(x) = \dfrac{x + 1}{\sqrt{x}}$

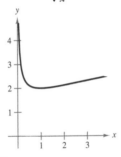

63. $f(x) = \csc \dfrac{x}{2}$

64. $f(x) = \cot x$

In Exercises 65 and 66, prove that the given function has a zero in the indicated interval.

65. $f(x) = x^2 - 4x + 3$, $[2, 4]$
66. $f(x) = x^3 + 3x - 2$, $[0, 1]$

In Exercises 67 and 68, use the Intermediate Value Theorem to approximate the zero of the given function in the interval $[0, 1]$. (a) Begin by locating the zero in a subinterval of length 0.1. (b) Refine your approximation by locating the zero in a subinterval of length 0.01.

67. $f(x) = x^3 + x - 1$ **68.** $f(x) = x^3 + 3x - 2$

In Exercises 69–72, verify the applicability of the Intermediate Value Theorem in the indicated interval and find the value of c guaranteed by the theorem.

69. $f(x) = x^2 + x - 1$, $[0, 5]$, $f(c) = 11$

70. $f(x) = x^2 - 6x + 8$, $[0, 3]$, $f(c) = 0$
71. $f(x) = x^3 - x^2 + x - 2$, $[0, 3]$, $f(c) = 4$
72. $f(x) = \dfrac{x^2 + x}{x - 1}$, $\left[\dfrac{5}{2}, 4\right]$, $f(c) = 6$

73. Determine the constant a so that the following function is continuous on the entire real line.

$$f(x) = \begin{cases} x^3, & x \le 2 \\ ax^2, & x > 2 \end{cases}$$

74. Determine the constants a and b so that the following function is continuous on the entire real line.

$$f(x) = \begin{cases} 2, & x \le -1 \\ ax + b, & -1 < x < 3 \\ -2, & x \ge 3 \end{cases}$$

75. A union contract guarantees a 9% increase for five years. For an initial annual salary of $28,500, the salary S is given by

$$S = 28{,}500(1.09)^{[\![t]\!]}$$

where $t = 0$ corresponds to 1990. Sketch a graph of this function and discuss its continuity.

76. A dial-direct long distance call between two cities costs $1.04 for the first two minutes and $0.36 for each additional minute or fraction thereof. Use the greatest integer function to write the cost C of a call in terms of the time t (in minutes). Sketch a graph of this function and discuss its continuity.

77. The number of units in inventory in a small company is given by

$$N(t) = 25\left(2\left[\!\left[\dfrac{t + 2}{2}\right]\!\right] - t\right)$$

where t is the time in months. Sketch the graph of this function and discuss its continuity. How often must this company replenish its inventory?

78. At 8:00 A.M. on Saturday a man begins running up the side of a mountain to his weekend campsite. On Sunday morning at 8:00 A.M. he runs back down the mountain. It takes him 20 minutes to run up, but only 10 minutes to run down. At some point on the way down he realizes that he passed the same place at exactly the same time on Saturday. Prove that he is correct. [Hint: Let $s(t)$ and $r(t)$ be the position functions for the run up and down, respectively, and apply the Intermediate Value Theorem to the function $f(t) = s(t) - r(t)$.]

In Exercises 79 and 80, use a computer or graphics calculator to sketch the graph of the function f and determine whether it is continuous on the entire real line.

79. $f(x) = \begin{cases} 2x - 4, & x \le 3 \\ x^2 - 2x, & x > 3 \end{cases}$

80. $f(x) = \begin{cases} \dfrac{\cos x - 1}{x}, & x < 0 \\ 5x, & x \geq 0 \end{cases}$

81. Prove that if f is continuous and has no zeros on $[a, b]$, then either

$f(x) > 0$ for all x in $[a, b]$

or

$f(x) < 0$ for all x in $[a, b]$.

82. The Dirichlet function f is defined as follows.

$$f(x) = \begin{cases} 0, & x \text{ rational} \\ 1, & x \text{ irrational} \end{cases}$$

Show that this function is discontinuous at every real number.

2.5 Infinite Limits

Infinite limits ▪ Vertical asymptotes

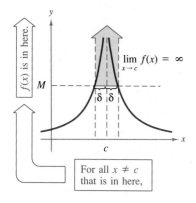

FIGURE 2.28

FIGURE 2.29

In this section we look at another important way in which a limit can fail to exist. We begin with an example. Let f be the function given by

$$f(x) = \frac{3}{x - 2}.$$

From Figure 2.28 and Table 2.5, we can see that $f(x)$ *decreases without bound* as x approaches 2 from the left, and $f(x)$ *increases without bound* as x approaches 2 from the right. Symbolically, we write

$$\lim_{x \to 2^-} \frac{3}{x - 2} = -\infty \qquad \text{and} \qquad \lim_{x \to 2^+} \frac{3}{x - 2} = \infty.$$

TABLE 2.5

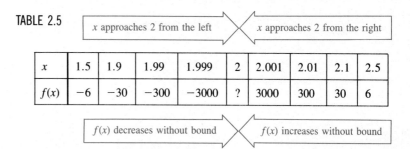

	x approaches 2 from the left					x approaches 2 from the right			
x	1.5	1.9	1.99	1.999	2	2.001	2.01	2.1	2.5
$f(x)$	-6	-30	-300	-3000	?	3000	300	30	6
	$f(x)$ decreases without bound					$f(x)$ increases without bound			

A limit in which $f(x)$ increases or decreases without bound as x approaches c is called an **infinite limit.**

DEFINITION OF INFINITE LIMITS

The statement

$$\lim_{x \to c} f(x) = \infty$$

means that for each $M > 0$ there exists a $\delta > 0$ such that $f(x) > M$ whenever $0 < |x - c| < \delta$ (see Figure 2.29). Similarly, the statement

$$\lim_{x \to c} f(x) = -\infty$$

means that for each $N < 0$ there exists a $\delta > 0$ such that $f(x) < N$ whenever $0 < |x - c| < \delta$.

REMARK   In the definition of infinite limits, we can define the **infinite limit from the left** by replacing $0 < |x - c| < \delta$ by $c - \delta < x < c$, and we can define the **infinite limit from the right** by replacing $0 < |x - c| < \delta$ by $c < x < c + \delta$.

Be sure to notice that the equal sign in the statement $\lim f(x) = \infty$ does not mean that the limit exists! On the contrary, it tells us how the limit *fails to exist* by denoting the unbounded behavior of $f(x)$ as x approaches c.

EXAMPLE 1   Determining infinite limits from a graph

Use Figure 2.30 to determine the limit of each function as $x \to 1$ from the left and from the right.

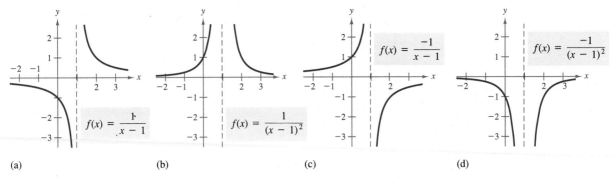

(a)   (b)   (c)   (d)

FIGURE 2.30

SOLUTION

(a) $\displaystyle\lim_{x \to 1^-} \frac{1}{x - 1} = -\infty$ and $\displaystyle\lim_{x \to 1^+} \frac{1}{x - 1} = \infty$

(b) $\displaystyle\lim_{x \to 1} \frac{1}{(x - 1)^2} = \infty$ Limit from both sides is ∞.

(c) $\displaystyle\lim_{x \to 1^-} \frac{-1}{x - 1} = \infty$ and $\displaystyle\lim_{x \to 1^+} \frac{-1}{x - 1} = -\infty$

(d) $\displaystyle\lim_{x \to 1} \frac{-1}{(x - 1)^2} = -\infty$ Limit from both sides is $-\infty$.

If it were possible to extend the graphs in Figure 2.30 toward infinity, you would see that each graph becomes arbitrarily close to the vertical line $x = 1$. We call this line a **vertical asymptote** of the graph of f.

DEFINITION OF VERTICAL ASYMPTOTE	If $f(x)$ approaches infinity (or negative infinity) as x approaches c from the right or left, then we call the line $x = c$ a **vertical asymptote** of the graph of f.

REMARK It should be clear that if a function f has a vertical asymptote at $x = c$, then f is *discontinuous* at c.

In Example 1, note that each of the functions is a *quotient* and that the vertical asymptote occurs at a number where the denominator is zero (and the numerator is not zero). The following theorem generalizes this observation. (A proof of the theorem is given in Appendix A.)

THEOREM 2.15
VERTICAL ASYMPTOTES

Let f and g be continuous on an open interval containing c. If $f(c) \neq 0$, $g(c) = 0$, and there exists an open interval containing c such that $g(x) \neq 0$ for all $x \neq c$ in the interval, then the graph of the function given by

$$F(x) = \frac{f(x)}{g(x)}$$

has a vertical asymptote at $x = c$.

$f(x) = \dfrac{1}{2(x + 1)}$

(a)

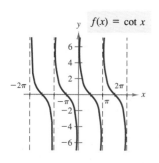

$f(x) = \dfrac{x^2 + 1}{x^2 - 1}$

$x = -1$ $x = 1$

(b)

EXAMPLE 2 Finding vertical asymptotes

Determine all vertical asymptotes for the graphs of the following functions.

(a) $f(x) = \dfrac{1}{2(x + 1)}$ (b) $f(x) = \dfrac{x^2 + 1}{x^2 - 1}$ (c) $f(x) = \cot x$

SOLUTION

(a) When $x = -1$, the denominator is zero and the numerator is not zero. Hence, by Theorem 2.15, we conclude that $x = -1$ is a vertical asymptote, as shown in Figure 2.31(a).

(b) By factoring the denominator as $(x^2 - 1) = (x - 1)(x + 1)$, we see that the denominator is zero at $x = -1$ and $x = 1$. Moreover, since the numerator is not zero at these two points, we apply Theorem 2.15 to conclude that the graph of f has the two vertical asymptotes shown in Figure 2.31(b).

(c) We can write the cotangent function in the form

$$f(x) = \cot x = \frac{\cos x}{\sin x}$$

and we know from Theorem 2.15 that vertical asymptotes occur at all values of x such that $\cos x \neq 0$ and $\sin x = 0$, as shown in Figure 2.31(c). Hence, the graph of this function has infinitely many vertical asymptotes. These occur when $x = n\pi$.

y $f(x) = \cot x$

(c)

FIGURE 2.31

Theorem 2.15 requires that the value of the numerator at $x = c$ be nonzero. If both the numerator and the denominator are zero at $x = c$, then we obtain the *indeterminate form* $0/0$, and we cannot determine the limit behavior at $x = c$ without further investigation. One case for which we can determine the limit is that in which the numerator and the denominator are polynomials. We do this by cancelling common factors, as shown in the next example.

EXAMPLE 3   A rational function with common factors

Determine all vertical asymptotes of the graph of

$$f(x) = \frac{x^2 + 2x - 8}{x^2 - 4}.$$

SOLUTION

Factoring both numerator and denominator produces

$$f(x) = \frac{x^2 + 2x - 8}{x^2 - 4} = \frac{(x + 4)(x - 2)}{(x + 2)(x - 2)} = \frac{x + 4}{x + 2}, \quad x \neq 2.$$

Now, at all x-values other than $x = 2$, the graph of f coincides with the graph of $g(x) = (x + 4)/(x + 2)$. Thus, we apply Theorem 2.15 to g to conclude that there is a vertical asymptote at $x = -2$, as shown in Figure 2.32. Note that $x = 2$ is *not* a vertical asymptote.

$f(x) = \dfrac{x^2 + 2x - 8}{x^2 - 4}$

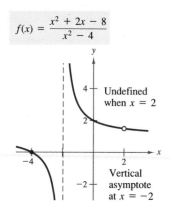

Undefined when $x = 2$

Vertical asymptote at $x = -2$

FIGURE 2.32

Since the graph of f in Example 3 has a vertical asymptote at $x = -2$, we know that the limit as $x \to -2$ from the right (or the left) is either ∞ or $-\infty$. But without looking at the graph, how can we determine the following limits?

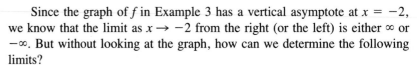

$$\lim_{x \to -2^-} \frac{x^2 + 2x - 8}{x^2 - 4} = -\infty \quad \text{and} \quad \lim_{x \to -2^+} \frac{x^2 + 2x - 8}{x^2 - 4} = \infty$$

It is cumbersome to answer this question analytically, and we have found the graphical method shown in the following example to be more efficient.

EXAMPLE 4   Determining infinite limits

Find the following limits.

$$\lim_{x \to 1^-} \frac{x^2 - 3x}{x - 1} \quad \text{and} \quad \lim_{x \to 1^+} \frac{x^2 - 3x}{x - 1}$$

SOLUTION

1. Factor both numerator and denominator (cancel any common factors).

$$f(x) = \frac{x^2 - 3x}{x - 1} = \frac{x(x - 3)}{x - 1}$$

2. The (remaining) factors of the numerator yield the x-intercepts of the function. In this case, the x-intercepts occur at $(0, 0)$ and $(3, 0)$.

3. The (remaining) factors of the denominator yield the vertical asymptotes of the function. In this case, a vertical asymptote occurs at $x = 1$.

4. Determine a few additional points on the graph, making certain that at least one point between each intercept and asymptote is chosen. Then plot the information obtained in the first four steps, as shown in Figure 2.33(a).

(a)

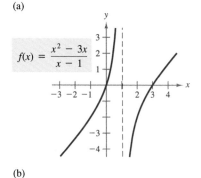

$$f(x) = \frac{x^2 - 3x}{x - 1}$$

(b)

FIGURE 2.33

x	-2	$\frac{1}{2}$	2	4
$f(x)$	$-\dfrac{10}{3}$	$\dfrac{5}{2}$	-2	$\dfrac{4}{3}$

5. Finally, complete the sketch as shown in Figure 2.33(b) and conclude that

$$\lim_{x \to 1^-} \frac{x^2 - 3x}{x - 1} = \infty \quad \text{and} \quad \lim_{x \to 1^+} \frac{x^2 - 3x}{x - 1} = -\infty.$$

We end this section with a theorem involving the limits of sums, products, and quotients of functions.

**THEOREM 2.16
PROPERTIES OF INFINITE LIMITS**

If c and L are real numbers and f and g are functions such that

$$\lim_{x \to c} f(x) = \infty \quad \text{and} \quad \lim_{x \to c} g(x) = L$$

then the following properties are true.

1. Sum or difference: $\quad \lim_{x \to c} [f(x) \pm g(x)] = \infty$

2. Product: $\quad \lim_{x \to c} [f(x)g(x)] = \infty, \quad L > 0$

$\qquad\qquad\qquad\qquad \lim_{x \to c} [f(x)g(x)] = -\infty, \quad L < 0$

3. Quotient: $\quad \lim_{x \to c} \dfrac{g(x)}{f(x)} = 0$

Similar properties hold for one-sided limits and for functions for which the limit of $f(x)$ as x approaches c is $-\infty$.

PROOF   We prove the first property and leave the proofs of the remaining properties as exercises (see Exercise 43). To show that the limit of $f(x) + g(x)$ is infinite, we choose $M > 0$ and need to find $\delta > 0$ such that $[f(x) + g(x)] > M$ whenever $0 < |x - c| < \delta$. For simplicity's sake, we assume L is positive and let $M_1 = M + 1$. Since the limit of $f(x)$ is infinite, there exists δ_1 such that $f(x) > M_1$ whenever $0 < |x - c| < \delta_1$. Also, since the limit of $g(x)$ is L, there exists δ_2 such that $|g(x) - L| < 1$ whenever $0 < |x - c| < \delta_2$. By letting δ be the smaller of δ_1 and δ_2, we conclude that $0 < |x - c| < \delta$ implies

$$f(x) > M + 1 \qquad \text{and} \qquad |g(x) - L| < 1.$$

The second of these two inequalities implies that $g(x) > L - 1$, and adding this to the first inequality we have

$$f(x) + g(x) > M + L > M.$$

Thus, we can conclude that

$$\lim_{x \to c} [f(x) + g(x)] = \infty.$$

The proof for $f(x) - g(x)$ is similar.

EXAMPLE 5   Determining limits

Find the following limits.

(a) $\lim_{x \to 0} \left(1 + \dfrac{1}{x^2}\right)$ (b) $\lim_{x \to 1^-} \dfrac{x^2 + 1}{1/(x - 1)}$ (c) $\lim_{x \to 0^+} 3 \cot x$

SOLUTION

(a) Since $\lim_{x \to 0} (1) = 1$ and $\lim_{x \to 0} (1/x^2) = \infty$, we can apply Property 1 of Theorem 2.16 to conclude that

$$\lim_{x \to 0} \left(1 + \frac{1}{x^2}\right) = \infty.$$

(b) Since $\lim_{x \to 1^-} (x^2 + 1) = 2$ and $\lim_{x \to 1^-} [1/(x - 1)] = -\infty$, we can apply Property 3 of Theorem 2.16 to conclude that

$$\lim_{x \to 1^-} \frac{x^2 + 1}{1/(x - 1)} = 0.$$

(c) Since $\lim_{x \to 0^+} (3) = 3$ and $\lim_{x \to 0^+} (\cot x) = \infty$, we can apply Property 2 of Theorem 2.16 to conclude that

$$\lim_{x \to 0^+} 3 \cot x = \infty.$$

EXERCISES for Section 2.5

In Exercises 1–4, determine whether $f(x)$ approaches ∞ or $-\infty$ as x approaches -2 from the left and from the right.

1. $f(x) = \dfrac{1}{(x + 2)^2}$

2. $f(x) = \dfrac{1}{x + 2}$

3. $f(x) = \tan \dfrac{\pi x}{4}$

4. $f(x) = \sec \dfrac{\pi x}{4}$

In Exercises 5–8, determine whether $f(x)$ approaches ∞ or $-\infty$ as x approaches -3 from the left and from the right.

5. $f(x) = \dfrac{1}{x^2 - 9}$

6. $f(x) = \dfrac{x}{x^2 - 9}$

7. $f(x) = \dfrac{x^2}{x^2 - 9}$

8. $f(x) = \sec \dfrac{\pi x}{6}$

In Exercises 9–18, find the vertical asymptotes (if any) of the given function.

9. $f(x) = \dfrac{1}{x^2}$

10. $f(x) = \dfrac{4}{(x - 2)^3}$

11. $f(x) = \dfrac{x^2 - 2}{x^2 - x - 2}$

12. $f(x) = \dfrac{2 + x}{1 - x}$

13. $f(x) = \dfrac{x^3}{x^2 - 1}$

14. $f(x) = \dfrac{-4x}{x^2 + 4}$

15. $f(x) = 1 - \dfrac{4}{x^2}$

16. $f(x) = \dfrac{-2}{(x - 2)^2}$

17. $f(x) = \dfrac{x}{x^2 + x - 2}$

18. $f(x) = \dfrac{1}{(x + 3)^4}$

In Exercises 19–22, determine whether the given function has a vertical asymptote or a removable discontinuity at $x = -1$.

19. $f(x) = \dfrac{x^2 - 1}{x + 1}$

20. $f(x) = \dfrac{x^2 - 6x - 7}{x + 1}$

21. $f(x) = \dfrac{x^2 + 1}{x + 1}$

22. $f(x) = \dfrac{\sin (x + 1)}{x + 1}$

In Exercises 23–34, find the indicated limit.

23. $\displaystyle \lim_{x \to 2^+} \dfrac{x - 3}{x - 2}$

24. $\displaystyle \lim_{x \to 1^+} \dfrac{2 + x}{1 - x}$

25. $\displaystyle \lim_{x \to 4^+} \dfrac{x^2}{x^2 - 16}$

26. $\displaystyle \lim_{x \to 4} \dfrac{x^2}{x^2 + 16}$

27. $\displaystyle \lim_{x \to 0^-} \left(1 + \dfrac{1}{x}\right)$

28. $\displaystyle \lim_{x \to 0^-} \left(x^2 - \dfrac{1}{x}\right)$

29. $\displaystyle \lim_{x \to 0^+} \dfrac{2}{\sin x}$

30. $\displaystyle \lim_{x \to (\pi/2)^+} \dfrac{-2}{\cos x}$

31. $\displaystyle \lim_{x \to 1} \dfrac{x^2 - x}{(x^2 + 1)(x - 1)}$

32. $\displaystyle \lim_{x \to 1} \dfrac{x^3 - 1}{x^2 + x + 1}$

33. $\displaystyle \lim_{x \to 1^+} \dfrac{x^2 + x + 1}{x^3 - 1}$

34. $\displaystyle \lim_{x \to 0^-} \dfrac{x^2 - 2x}{x^3}$

35. A 25-foot ladder is leaning against a house (see figure). If the base of the ladder is pulled away from the house at a rate of 2 feet per second, the top will move down the wall at a rate of

$$r = \dfrac{2x}{\sqrt{625 - x^2}} \text{ ft/sec.}$$

(a) Find the rate when x is 7 feet.

(b) Find the rate when x is 15 feet.

(c) Find the limit of r as $x \to 25^-$.

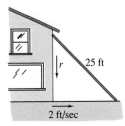

FIGURE FOR 35

36. A patrol car is parked 50 feet from a long warehouse. The revolving light on top of the car turns at a rate of $\frac{1}{2}$ revolution per second (see figure). The rate at which the light beam is moving along the wall is

$$r = 50\pi \sec^2 \theta \text{ ft/sec.}$$

(a) Find the rate when θ is $\pi/6$.
(b) Find the rate when θ is $\pi/3$.
(c) Find the limit of r as $\theta \to \pi/2^-$.

FIGURE FOR 36

37. The cost in millions of dollars for the federal government to seize x percent of a certain illegal drug as it enters the country is given by

$$C = \frac{528x}{100 - x}, \quad 0 \le x < 100.$$

(a) Find the cost of seizing 25 percent.
(b) Find the cost of seizing 50 percent.
(c) Find the cost of seizing 75 percent.
(d) Find the limit of C as $x \to 100^-$.

38. The cost in dollars of removing p percent of the air pollutants in the stack emission of a utility company that burns coal to generate electricity is

$$C = \frac{80,000p}{100 - p}, \quad 0 \le p < 100.$$

(a) Find the cost of removing 15 percent.
(b) Find the cost of removing 50 percent.
(c) Find the cost of removing 90 percent.
(d) Find the limit of C as $x \to 100^-$.

39. **Coulomb's Law** states that the force F of a point charge q_1 on a point charge q_2, when the charges are r units apart, is proportional to the product of the charges and inversely proportional to the square of the distance between them. If a point particle with charge $+1$ is placed on a straight line between two particles 5 units apart, each with a charge of -1, the net force on the particle with a positive charge is given by

$$F = -\frac{k}{x^2} + \frac{k}{(x - 5)^2}, \quad 0 < x < 5$$

where x is the distance shown in the figure. Sketch the graph of F.

FIGURE FOR 39

40. Find functions f and g such that

$$\lim_{x \to c} f(x) = \infty \quad \text{and} \quad \lim_{x \to c} g(x) = \infty$$

but

$$\lim_{x \to c} [f(x) - g(x)] \neq 0.$$

In Exercises 41 and 42, use a computer or graphics calculator to sketch the graph of the function and determine the one-sided limit.

41. $f(x) = \dfrac{1}{x^2 - 25}$, $\quad \lim\limits_{x \to 5^-} f(x)$

42. $f(x) = \sec\dfrac{\pi x}{6}$, $\quad \lim\limits_{x \to 3^+} f(x)$

43. Prove the remaining parts of Theorem 2.16.
44. Prove that if $\lim\limits_{x \to c} f(x) = \infty$, then

$$\lim_{x \to c} \frac{1}{f(x)} = 0.$$

REVIEW EXERCISES for Chapter 2

In Exercises 1–26, find the given limit (if it exists).

1. $\lim\limits_{x \to 2} (5x - 3)$

2. $\lim\limits_{x \to 2} (3x + 5)$

3. $\lim\limits_{x \to 2} (5x - 3)(3x + 5)$

4. $\lim\limits_{x \to 2} \dfrac{3x + 5}{5x - 3}$

5. $\lim\limits_{t \to 3} \dfrac{t^2 + 1}{t}$

6. $\lim\limits_{t \to 3} \dfrac{t^2 - 9}{t - 3}$

7. $\lim\limits_{t \to -2} \dfrac{t + 2}{t^2 - 4}$

8. $\lim\limits_{x \to 0} \dfrac{\sqrt{4 + x} - 2}{x}$

9. $\lim\limits_{x \to 0} \dfrac{[1/(x + 1)] - 1}{x}$

10. $\lim\limits_{s \to 0} \dfrac{(1/\sqrt{1 + s}) - 1}{s}$

11. $\lim\limits_{x \to -1} \dfrac{x^3 + 1}{x + 1}$

12. $\lim\limits_{x \to -2} \dfrac{x^2 - 4}{x^3 + 8}$

13. $\lim\limits_{x \to 0^+} \left(x - \dfrac{1}{x^3}\right)$

14. $\lim\limits_{x \to 2} \dfrac{1}{\sqrt[3]{x^2 - 4}}$

15. $\lim\limits_{x \to -2^-} \dfrac{2x^2 + x + 1}{x + 2}$

16. $\lim\limits_{x \to 1/2} \dfrac{2x - 1}{6x - 3}$

17. $\lim\limits_{x \to -1} \dfrac{x + 1}{x^3 + 1}$

18. $\lim\limits_{x \to -1} \dfrac{x + 1}{x^4 - 1}$

19. $\lim\limits_{x \to 1} \dfrac{x^2 - 2x + 1}{x + 1}$

20. $\lim\limits_{x \to -1^+} \dfrac{x^2 - 2x + 1}{x + 1}$

21. $\lim\limits_{x \to 0} \dfrac{\sin 4x}{5x}$

22. $\lim\limits_{x \to 0^+} \dfrac{\sec x}{x}$

23. $\lim\limits_{x \to 0^+} \dfrac{\csc 2x}{x}$

24. $\lim\limits_{x \to 0^-} \dfrac{\cos^2 x}{x}$

25. $\lim\limits_{\Delta x \to 0} \dfrac{\sin [(\pi/6)+ \Delta x] - (1/2)}{\Delta x}$

[Hint: $\sin (\theta + \phi) = \sin \theta \cos \phi + \cos \theta \sin \phi$]

26. $\lim\limits_{\Delta x \to 0} \dfrac{\cos (\pi + \Delta x) + 1}{\Delta x}$

[Hint: $\cos (\theta + \phi) = \cos \theta \cos \phi - \sin \theta \sin \phi$]

27. Estimate the limit

$$\lim_{x \to 1^+} \dfrac{\sqrt{2x + 1} - \sqrt{3}}{x - 1}$$

by completing the following table.

x	1.1	1.01	1.001	1.0001
$f(x)$				

28. Estimate the limit

$$\lim_{x \to 1^+} \dfrac{1 - \sqrt[3]{x}}{x - 1}$$

by completing the following table.

x	1.1	1.01	1.001	1.0001
$f(x)$				

29. Evaluate the limit in Exercise 27 by rationalizing the numerator.

30. Evaluate the limit in Exercise 28 by rationalizing the numerator.
[Hint: Use $a^3 - b^3 = (a - b)(a^2 + ab + b^2)$.]

In Exercises 31–36, determine whether the given limit statement is true or false.

31. $\lim\limits_{x \to 0} \dfrac{|x|}{x} = 1$

32. $\lim\limits_{x \to 0} x^3 = 0$

33. $\lim\limits_{x \to 2} f(x) = 3$, where $f(x) = \begin{cases} 3, & x \le 2 \\ 0, & x > 2 \end{cases}$

34. $\lim\limits_{x \to 3} f(x) = 1$, where

$$f(x) = \begin{cases} x - 2, & x \le 3 \\ -x^2 + 8x - 14, & x > 3 \end{cases}$$

35. $\lim\limits_{x \to 0^+} \sqrt{x} = 0$

36. $\lim\limits_{x \to 0} \sqrt[3]{x} = 0$

In Exercises 37–46, determine the intervals on which the given function is continuous.

37. $f(x) = [\![x + 3]\!]$

38. $f(x) = \dfrac{3x^2 - x - 2}{x - 1}$

39. $f(x) = \begin{cases} \dfrac{3x^2 - x - 2}{x - 1}, & x \ne 1 \\ 0, & x = 1 \end{cases}$

40. $f(x) = \begin{cases} 5 - x, & x \le 2 \\ 2x - 3, & x > 2 \end{cases}$

41. $f(x) = \dfrac{1}{(x - 2)^2}$

42. $f(x) = \sqrt{\dfrac{x + 2}{x}}$

43. $f(x) = \dfrac{3}{x + 1}$

44. $f(x) = \dfrac{x + 1}{2x + 2}$

45. $f(x) = \csc \dfrac{\pi x}{2}$

46. $f(x) = \tan 2x$

47. Determine the value of c so that the following function is continuous on the entire real line.

$$f(x) = \begin{cases} x + 3, & x \le 2 \\ cx + 6, & x > 2 \end{cases}$$

48. Determine the values of b and c so that the following function is continuous on the entire real line.

$$f(x) = \begin{cases} x + 1, & 1 < x < 3 \\ x^2 + bx + c, & |x - 2| \ge 1 \end{cases}$$

49. A sum of \$5000 is deposited in a savings plan that pays 12 percent compounded semiannually. The account's amount after t years is given by $A = 5000(1.06)^{[\![2t]\!]}$. Sketch a graph of this function and discuss its continuity.

50. A sum of \$1000 is deposited in a savings plan that pays 14 percent compounded annually. The amount in the account after t years is given by $A = 1000(1.14)^{[\![t]\!]}$. Sketch a graph of this function and discuss its continuity.

In Exercises 51–54, find the vertical asymptotes (if any) of the given function.

51. $g(x) = 1 + \dfrac{2}{x}$

52. $h(x) = \dfrac{4x}{4 - x^2}$

53. $f(x) = \dfrac{8}{(x - 10)^2}$

54. $f(x) = \csc \pi x$

55. The function f is defined as follows.

$$f(x) = \dfrac{\tan 2x}{x}, \quad x \ne 0.$$

(a) Find $\lim\limits_{x \to 0} \dfrac{\tan 2x}{x}$ (if it exists).

(b) Can the function f be defined so that it is continuous at $x = 0$?

Chapter 3 Application

High-diver Greg Louganis is shown in mid-air after diving from a 10-meter platform. Louganis is considered by many to be the greatest diver of all time. In 1976, at the age of 16, he won an Olympic silver medal for the United States in the platform competition. Later, in 1984 and again in 1988, he won both Olympic gold medals in diving—one in the platform competition and one in the springboard competition.

The Velocity of a High-Diver

A diver jumps from a 30-foot platform. The initial velocity of the diver is 5 feet per second (upward). When will the diver hit the water, and how fast will the diver be traveling at that time? To answer these questions, we use the classical model for the position of a free-falling object. This model assumes that air resistance is negligible and is given by

$$s(t) = \frac{1}{2}gt^2 + v_0 t + h_0$$

where $s(t)$ is the height of the diver in feet, t is the time in seconds, $g = -32$ ft/sec^2 is the acceleration due to gravity, v_0 is the initial velocity, and h_0 is the initial height. Using $v_0 = 5$ and $h_0 = 30$, the position function becomes

$$s(t) = -16t^2 + 5t + 30.$$

To find the time when the diver hits the water, we let $s(t) = 0$ and solve for t. The solution of this equation is $t \approx 1.53$ seconds. The velocity of the diver is given by the derivative of the position function

$$v(t) = s'(t) = -32t + 5$$

where $v(t)$ is measured in feet per second. Thus, when $t \approx 1.53$ seconds, the velocity is $v(1.53) \approx -44$ feet per second (or approximately 30 miles per hour).

Diver's Velocity at Water for Varying Platform Heights
(Ten meters is standard for Olympic competition.)

Height (in feet)	30	40	50	60
Velocity (in ft/sec)	−44.1	−50.8	−56.8	−62.2

Chapter Overview

Calculus has two fundamental operations—differentiation and integration. In this chapter we introduce the first of these.

Section 3.1 begins by taking a detailed look at the problem of finding the **tangent line** to the graph of a function at a specified point. The section also shows how the **derivative** of a function can be used to find the slope of the tangent line at a point.

Section 3.2 presents the second primary use of the derivative—that of finding the instantaneous rate of change of one variable with respect to another. For example, the velocity of an object moving along a straight line can be found by differentiating the position function of the object.

In the first two sections of the chapter, the emphasis is on the *use* of derivatives. In these two sections, the actual *calculation* of derivatives is done by finding a limit, and this can be quite tedious. In Sections 3.3–3.5, we streamline the differentiation process by presenting several "differentiation rules." It is important that you memorize each rule, so that you can find derivatives efficiently.

Section 3.6 introduces a procedure called **implicit differentiation.** This procedure can be used to find derivatives of functions that are implied by equations. For example, implicit differentiation can be applied to the circle $x^2 + y^2 = 1$ to find the slope of the tangent line at a given point on the circle.

The chapter closes by showing how differentiation can be used in applications involving rates of change (Section 3.7).

See Exercise 61, Section 3.3.

Differentiation

3.1 The Derivative and the Tangent Line Problem

Tangent line to a curve ▪ Vertical tangent lines ▪ The derivative of a function ▪ Differentiability and continuity

Isaac Newton

Calculus grew out of four major problems that European mathematicians were working on during the seventeenth century.

1. The tangent line problem (Sections 2.1 and 3.1)
2. The velocity and acceleration problem (Section 3.2)
3. The minimum and maximum problem (Section 4.1)
4. The area problem (Section 5.2)

Each problem involves the notion of a limit, and we could introduce calculus with any of the four problems.

Recall that we gave a brief introduction to the tangent line problem in Section 2.1. Although partial solutions to this problem were given by Pierre de Fermat (1601–1655), René Descartes (1596–1650), Christian Huygens (1629–1695), and Isaac Barrow (1630–1677), credit for the first general solution usually is given to Isaac Newton (1642–1727) and Gottfried Leibniz (1646–1716). Newton's work on the problem stemmed from his interest in optics and light refraction.

The tangent line problem

What do we mean when we say that a line is tangent to a curve at a point? For a circle, we can characterize the tangent line at point P as the line that is perpendicular to the radial line at point P, as shown in Figure 3.1. For a general curve, however, the problem is more difficult. For example, how would you define the tangent lines shown in Figure 3.2?

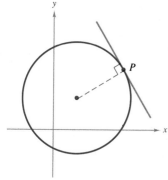

Tangent Line to a Circle

FIGURE 3.1

Tangent Line to a Curve at a Point

FIGURE 3.2

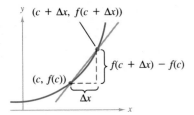

The secant line through $(c, f(c))$ and $(c + \Delta x, f(c + \Delta x))$

FIGURE 3.3

Essentially, the problem of finding the tangent line at a point P boils down to the problem of finding the *slope* of the tangent line at P. We can approximate this slope using a line through the point of tangency and a second point on the curve, as shown in Figure 3.3. We call such a line a **secant line***. If $(c, f(c))$ is the point of tangency and $(c + \Delta x, f(c + \Delta x))$ is a second point on the graph of f, then the slope of the secant line through these two points is given by

$$m_{\text{sec}} = \frac{f(c + \Delta x) - f(c)}{c + \Delta x - c}$$

$$= \frac{f(c + \Delta x) - f(c)}{\Delta x}. \qquad \text{Slope of secant line}$$

The right-hand side of this equation is called a **difference quotient.** The denominator Δx is called the **change in** x, and the numerator $\Delta y = f(c + \Delta x) - f(c)$ is called the **change in** y.

The beauty of this procedure is that we can obtain more and more accurate approximations to the slope of the tangent line by choosing a sequence of points closer and closer to the point of tangency, as shown in Figure 3.4.

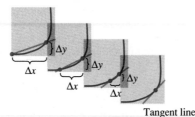

Tangent line   Tangent line

FIGURE 3.4

As $\Delta x \to 0$ from the left, the secant lines approach the tangent line.   As $\Delta x \to 0$ from the right, the secant lines approach the tangent line.

DEFINITION OF TANGENT LINE WITH SLOPE m

If f is defined on an open interval containing c and the limit

$$\lim_{\Delta x \to 0} \frac{\Delta y}{\Delta x} = \lim_{\Delta x \to 0} \frac{f(c + \Delta x) - f(c)}{\Delta x} = m$$

exists, then we call the line passing through $(c, f(c))$ with slope m the **tangent line** to the graph of f at the point $(c, f(c))$.

We often refer to the slope of the tangent line to the graph of f at the point $(c, f(c))$ as simply the **slope of the graph of f at $x = c$.**

*This use of the word *secant* comes from the Latin *secare*, meaning to cut, and not from a reference to the trigonometric function of the same name.

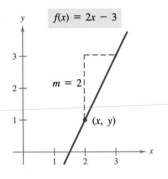

$f(x) = 2x - 3$

$m = 2$

(x, y)

FIGURE 3.5

EXAMPLE 1 The slope of the graph of a linear function

Find the slope of the graph of $f(x) = 2x - 3$ at the point $(2, 1)$.

SOLUTION

$$
\begin{aligned}
\lim_{\Delta x \to 0} \frac{f(2 + \Delta x) - f(2)}{\Delta x} &= \lim_{\Delta x \to 0} \frac{[2(2 + \Delta x) - 3] - [2(2) - 3]}{\Delta x} \\
&= \lim_{\Delta x \to 0} \frac{4 + 2\Delta x - 3 - 4 + 3}{\Delta x} \\
&= \lim_{\Delta x \to 0} \frac{2\Delta x}{\Delta x} = \lim_{\Delta x \to 0} 2 = 2
\end{aligned}
$$

Therefore, the slope of f at $(2, 1)$ is $m = 2$, as shown in Figure 3.5.

REMARK Note in Example 1 that the limit definition of the slope of f agrees with our usual definition of the slope of a line as discussed in Section 1.4.

We know that the graph of a linear function has the same slope at any point. This is not true of nonlinear functions, as can be seen in the following example.

EXAMPLE 2 Tangent lines to the graph of a nonlinear function

Find the slope of the tangent line to the graph of $f(x) = x^2 + 1$ at the points $(0, 1)$ and $(-1, 2)$, as shown in Figure 3.6.

SOLUTION

To solve this problem, we let $(x, f(x))$ represent an arbitrary point on the graph of f. Then the slope of the tangent line at $(x, f(x))$ is given by

$$
\begin{aligned}
\lim_{\Delta x \to 0} \frac{f(x + \Delta x) - f(x)}{\Delta x} &= \lim_{\Delta x \to 0} \frac{x^2 + 2x(\Delta x) + (\Delta x)^2 + 1 - x^2 - 1}{\Delta x} \\
&= \lim_{\Delta x \to 0} \frac{2x(\Delta x) + (\Delta x)^2}{\Delta x} \\
&= \lim_{\Delta x \to 0} (2x + \Delta x) = 2x.
\end{aligned}
$$

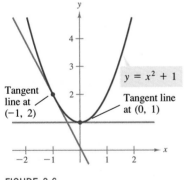

$y = x^2 + 1$

Tangent line at $(-1, 2)$

Tangent line at $(0, 1)$

FIGURE 3.6

Therefore, the slope at *any* point $(x, f(x))$ on the graph of f is given by $m = 2x$. At the point $(0, 1)$, the slope is $m = 2(0) = 0$, and at the point $(-1, 2)$, the slope is $m = 2(-1) = -2$.

REMARK In Example 2 note that x is held constant in the limit process (as $\Delta x \to 0$).

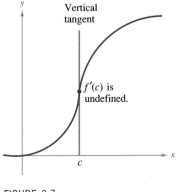

FIGURE 3.7

Our definition of a tangent line to a curve does not cover the possibility of a vertical tangent line. For vertical tangent lines, we give the following definition. If f is continuous at c and

$$\lim_{\Delta x \to 0} \left| \frac{f(c + \Delta x) - f(c)}{\Delta x} \right| = \infty$$

then the vertical line passing through $(c, f(c))$ is called a **vertical tangent line** to the graph of f. For example, the function shown in Figure 3.7 has a vertical tangent line at $x = c$. If the domain of f is a closed interval $[a, b]$, then we extend the definition of a vertical tangent line to include the endpoints by considering continuity and limits from the right (for $x = a$) and from the left (for $x = b$).

The derivative of a function

We have now arrived at a crucial point in the study of calculus. The limit used to define the slope of a tangent line is also used to define one of the two fundamental operations of calculus—**differentiation.**

DEFINITION OF THE DERIVATIVE OF A FUNCTION

The **derivative** of f at x is given by

$$f'(x) = \lim_{\Delta x \to 0} \frac{f(x + \Delta x) - f(x)}{\Delta x}$$

provided the limit exists.

The process of finding the derivative of a function is called **differentiation.** A function is called **differentiable** at x if its derivative exists at x and **differentiable on an open interval** (a, b) if it is differentiable at every point in the interval.

In addition to $f'(x)$, read "f prime of x," other notations are used to denote the derivative of $y = f(x)$. The most common are

$$f'(x), \quad \frac{dy}{dx}, \quad y', \quad \frac{d}{dx}[f(x)], \quad D_x[y]. \qquad \text{Notation for derivatives}$$

The notation dy/dx is read as "the derivative of y *with respect to x*." Using limit notation, we have

$$\frac{dy}{dx} = \lim_{\Delta x \to 0} \frac{\Delta y}{\Delta x} = \lim_{\Delta x \to 0} \frac{f(x + \Delta x) - f(x)}{\Delta x} = f'(x).$$

EXAMPLE 3   Finding the derivative by the limit process

Find the derivative of $f(x) = x^3 + 2x$.

SOLUTION

$$f'(x) = \lim_{\Delta x \to 0} \frac{f(x + \Delta x) - f(x)}{\Delta x}$$

$$= \lim_{\Delta x \to 0} \frac{(x + \Delta x)^3 + 2(x + \Delta x) - (x^3 + 2x)}{\Delta x}$$

$$= \lim_{\Delta x \to 0} \frac{3x^2\Delta x + 3x(\Delta x)^2 + (\Delta x)^3 + 2\Delta x}{\Delta x}$$

$$= \lim_{\Delta x \to 0} \frac{\Delta x[3x^2 + 3x\Delta x + (\Delta x)^2 + 2]}{\Delta x}$$

$$= \lim_{\Delta x \to 0} [3x^2 + 3x\Delta x + (\Delta x)^2 + 2]$$

$$= 3x^2 + 2$$

Remember that the derivative $f'(x)$ gives us a formula for finding the slope of the tangent line at the point $(x, f(x))$ on the graph of f. This is illustrated in the following example.

EXAMPLE 4 Using the derivative to find the slope at a point

Find $f'(x)$ for $f(x) = \sqrt{x}$, and use the result to find the slope of the graph of f at the points $(1, 1)$ and $(4, 2)$. Discuss the behavior of f at $(0, 0)$.

SOLUTION

We use the procedure for rationalizing numerators, as discussed in Section 2.3.

$$f'(x) = \lim_{\Delta x \to 0} \frac{f(x + \Delta x) - f(x)}{\Delta x}$$

$$= \lim_{\Delta x \to 0} \frac{\sqrt{x + \Delta x} - \sqrt{x}}{\Delta x}$$

$$= \lim_{\Delta x \to 0} \left(\frac{\sqrt{x + \Delta x} - \sqrt{x}}{\Delta x}\right)\left(\frac{\sqrt{x + \Delta x} + \sqrt{x}}{\sqrt{x + \Delta x} + \sqrt{x}}\right)$$

$$= \lim_{\Delta x \to 0} \frac{(x + \Delta x) - x}{\Delta x(\sqrt{x + \Delta x} + \sqrt{x})}$$

$$= \lim_{\Delta x \to 0} \frac{\Delta x}{\Delta x(\sqrt{x + \Delta x} + \sqrt{x})}$$

$$= \lim_{\Delta x \to 0} \frac{1}{\sqrt{x + \Delta x} + \sqrt{x}} = \frac{1}{2\sqrt{x}}$$

Therefore, at the point $(1, 1)$ the slope is $f'(1) = \frac{1}{2}$, and at the point $(4, 2)$ the slope is $f'(4) = \frac{1}{4}$, as shown in Figure 3.8. At the point $(0, 0)$ the slope is undefined, since substituting $x = 0$ in $f'(x)$ produces division by zero. Moreover, because the limit of $f'(x)$ as $x \to 0$ from the right is infinite, the graph of f has a vertical tangent line at $(0, 0)$.

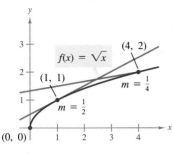

FIGURE 3.8

In many applications, it is convenient to use a variable other than x as the independent variable. Example 5 shows a function that uses t as the independent variable.

EXAMPLE 5 Finding the derivative of a function

Find the derivative with respect to t for the function $y = 2/t$.

SOLUTION

Considering $y = f(t)$, we obtain the following.

$$\frac{dy}{dt} = \lim_{\Delta t \to 0} \frac{f(t + \Delta t) - f(t)}{\Delta t}$$

$$= \lim_{\Delta t \to 0} \frac{\dfrac{2}{t + \Delta t} - \dfrac{2}{t}}{\Delta t} = \lim_{\Delta t \to 0} \frac{\dfrac{2t - 2(t + \Delta t)}{t(t + \Delta t)}}{\Delta t}$$

$$= \lim_{\Delta t \to 0} \frac{-2\Delta t}{\Delta t(t)(t + \Delta t)}$$

$$= \lim_{\Delta t \to 0} \frac{-2}{t(t + \Delta t)} = -\frac{2}{t^2}$$

Differentiability and continuity

There is a close relationship between differentiability and continuity. The following alternate limit form of the derivative is useful in investigating this relationship.

**THEOREM 3.1
ALTERNATE FORM
OF THE DERIVATIVE**

The derivative of f at c is given by

$$f'(c) = \lim_{x \to c} \frac{f(x) - f(c)}{x - c}$$

provided this limit exists.

PROOF The derivative of f at c is given by

$$f'(c) = \lim_{\Delta x \to 0} \frac{f(c + \Delta x) - f(c)}{\Delta x}.$$

In Figure 3.9, we can see that if $x = c + \Delta x$, then $x \to c$ as $\Delta x \to 0$. Thus, if we replace $c + \Delta x$ by x, we can write

$$f'(c) = \lim_{\Delta x \to 0} \frac{f(c + \Delta x) - f(c)}{\Delta x} = \lim_{x \to c} \frac{f(x) - f(c)}{x - c}.$$

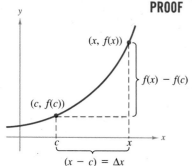

$(x - c) = \Delta x$

FIGURE 3.9

Note that the existence of the limit in this alternate form requires that the one-sided limits

$$\lim_{x \to c^-} \frac{f(x) - f(c)}{x - c}$$

and

$$\lim_{x \to c^+} \frac{f(x) - f(c)}{x - c}$$

exist and are equal. For convenience, we refer to these one-sided limits as the **derivatives from the left and from the right,** respectively. Keep in mind, however, that if these one-sided limits are not equal at c, then the derivative does not exist at c.

The next two examples will give you a sense of the relationship between differentiability and continuity.

EXAMPLE 6   A function whose one-sided derivatives are different

The function $f(x) = |x - 2|$ shown in Figure 3.10 is continuous at $x = 2$. However, the one-sided limits

$$\lim_{x \to 2^-} \frac{f(x) - f(2)}{x - 2} = \lim_{x \to 2^-} \frac{|x - 2| - 0}{x - 2} = -1 \qquad \text{Derivative from the left}$$

and

$$\lim_{x \to 2^+} \frac{f(x) - f(2)}{x - 2} = \lim_{x \to 2^+} \frac{|x - 2| - 0}{x - 2} = 1 \qquad \text{Derivative from the right}$$

are not equal. Therefore, f is not differentiable at $x = 2$ and the graph of f does not have a tangent line at the point $(2, 0)$. ▭

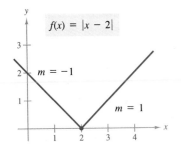

$f(x) = |x - 2|$

$m = -1$

$m = 1$

Not differentiable at $x = 2$, since the one-sided derivatives are not equal.

FIGURE 3.10

EXAMPLE 7   A function with a vertical tangent

The function $f(x) = x^{1/3}$ is continuous at $x = 0$, as shown in Figure 3.11. However, since the following limit is infinite,

$$\lim_{x \to 0} \frac{f(x) - f(0)}{x - 0} = \lim_{x \to 0} \frac{x^{1/3} - 0}{x}$$

$$= \lim_{x \to 0} \frac{1}{x^{2/3}} = \infty$$

we conclude that the tangent line is vertical at $x = 0$. Therefore, f is not differentiable at $x = 0$. ▭

$f(x) = x^{1/3}$

Not differentiable at $x = 0$, since f has a vertical tangent at $x = 0$.

FIGURE 3.11

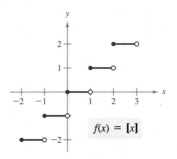

$f(x) = [x]$

Not differentiable at $x = 0$,
since f is not continuous at $x = 0$.

FIGURE 3.12

In Examples 6 and 7, we saw that a function is not differentiable at a point at which its graph has a sharp turn *or* a vertical tangent. Differentiability can also be destroyed by a discontinuity. For instance, the greatest integer function $f(x) = [x]$ is not continuous at $x = 0$ and hence is not differentiable at $x = 0$ (see Figure 3.12). Another way of saying this is that if a function is differentiable at a point, then it is continuous at the point. We formalize this result in the following theorem.

**THEOREM 3.2
DIFFERENTIABILITY
IMPLIES CONTINUITY**

If f is differentiable at $x = c$, then f is continuous at $x = c$.

PROOF

To prove that f is continuous at $x = c$, we will show that $f(x)$ approaches $f(c)$ as $x \to c$. To do this, we use the differentiability of f at $x = c$ and consider the following limit.

$$\lim_{x \to c} [f(x) - f(c)] = \lim_{x \to c} \left[(x - c)\left(\frac{f(x) - f(c)}{x - c} \right) \right]$$
$$= \left[\lim_{x \to c} (x - c) \right]\left[\lim_{x \to c} \frac{f(x) - f(c)}{x - c} \right]$$
$$= (0)[f'(c)] = 0$$

Since the difference $[f(x) - f(c)]$ approaches zero as $x \to c$, we conclude that

$$\lim_{x \to c} f(x) = f(c).$$

Therefore, f is continuous at $x = c$.

EXERCISES for Section 3.1

In Exercises 1 and 2, trace the curve on another piece of paper and sketch the tangent line at the point (x, y).

1.

2.

In Exercises 3 and 4, estimate the slope of the curve at the point (x, y).

3. (a)

(b)

4. (a)

(b)
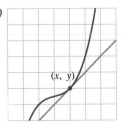

In Exercises 5–14, use the definition of the derivative to find $f'(x)$.

5. $f(x) = 3$ **6.** $f(x) = 3x + 2$
7. $f(x) = -5x$ **8.** $f(x) = 1 - x^2$
9. $f(x) = 2x^2 + x - 1$ **10.** $f(x) = \sqrt{x - 4}$
11. $f(x) = \dfrac{1}{x - 1}$ **12.** $f(x) = \dfrac{1}{x^2}$
13. $f(t) = t^3 - 12t$ **14.** $f(t) = t^3 + t^2$

In Exercises 15–20, find the equation of the tangent line to the graph of f at the indicated point. Then verify your answer by sketching both the graph of f and the tangent line.

Function	Point of Tangency
15. $f(x) = x^2 + 1$	$(2, 5)$
16. $f(x) = x^2 + 2x + 1$	$(-3, 4)$

Function	Point of Tangency
17. $f(x) = x^3$	$(2, 8)$
18. $f(x) = x^3$	$(-2, -8)$
19. $f(x) = \sqrt{x + 1}$	$(3, 2)$
20. $f(x) = \dfrac{1}{x + 1}$	$(0, 1)$

In Exercises 21–26, use the alternate form of the derivative (Theorem 3.1) to find the derivative at $x = c$ (if it exists).

21. $f(x) = x^2 - 1$, $c = 2$
22. $f(x) = x^3 + 2x$, $c = 1$
23. $f(x) = x^3 + 2x^2 + 1$, $c = -2$
24. $f(x) = \dfrac{1}{x}$, $c = 3$
25. $f(x) = (x - 1)^{2/3}$, $c = 1$
26. $f(x) = |x - 2|$, $c = 2$

In Exercises 27–36, find every point at which the function is differentiable.

27. $f(x) = |x + 3|$

28. $f(x) = |x^2 - 9|$

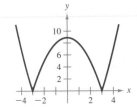

29. $f(x) = \dfrac{1}{x + 1}$

30. $f(x) = \dfrac{2x}{x - 1}$

31. $f(x) = (x - 3)^{2/3}$

32. $f(x) = x^{2/5}$

33. $f(x) = \sqrt{x - 1}$

34. $f(x) = \dfrac{x^2}{x^2 - 4}$

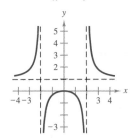

35. $f(x) = \begin{cases} 4 - x^2, & 0 < x \\ x^2 - 4, & x \le 0 \end{cases}$

36. $f(x) = \begin{cases} x^2 - 2x, & x > 1 \\ x^3 - 3x^2 + 3x, & x \le 1 \end{cases}$

FIGURE FOR 35

FIGURE FOR 36

In Exercises 37–40, find the derivatives from the left and from the right at $x = 1$ (if they exist). Is the function differentiable at $x = 1$?

37. $f(x) = \sqrt{1 - x^2}$

38. $f(x) = \begin{cases} x - 1, & x \le 1 \\ (x - 1)^2, & x > 1 \end{cases}$

39. $f(x) = \begin{cases} (x - 1)^3, & x \le 1 \\ (x - 1)^2, & x > 1 \end{cases}$

40. $f(x) = \begin{cases} x, & x \le 1 \\ x^2, & x > 1 \end{cases}$

In Exercises 41 and 42, find an equation of the line that is tangent to the graph of *f* and parallel to the given line.

Function	*Line*
41. $f(x) = x^3$	$3x - y + 1 = 0$
42. $f(x) = \dfrac{1}{\sqrt{x}}$	$x + 2y - 6 = 0$

In Exercises 43 and 44, find the equations of the two tangent lines to the graph of *f* that pass through the indicated point.

43. $f(x) = 4x - x^2$

44. $f(x) = x^2$

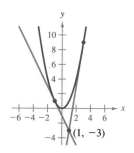

FIGURE FOR 43

FIGURE FOR 44

45. Assume $f'(c) = 3$. Find $f'(-c)$ for the following conditions.
 (a) *f* is an odd function.
 (b) *f* is an even function.

46. Sketch the graph of *f* and *f'* on the same set of axes for each of the following.
 (a) $f(x) = x^2$ (b) $f(x) = x^3$

In Exercises 47–50, determine whether the statement is true or false.

47. The slope of the graph of $y = x^2$ is different at every point on the curve.

48. If a function is continuous at a point, then it is differentiable at that point.

49. If a function is differentiable at a point, then it is continuous at that point.

50. If a function has derivatives from both the right and the left at a point, then it is differentiable at that point.

In Exercises 51 and 52, use a computer or graphics calculator to sketch the graph of *f* over the interval $[-2, 2]$ and complete the following table.

x	-2	-1.5	-1	-0.5	0	0.5	1	1.5	2
$f(x)$									
$f'(x)$									

51. $f(x) = \dfrac{1}{4}x^3$

52. $f(x) = \dfrac{4}{x}$

3.2 Velocity, Acceleration, and Other Rates of Change

Straight-line motion ▪ Average velocity ▪ Instantaneous velocity ▪ Acceleration ▪ Higher-order derivatives ▪ Other rates of change

We have seen how the derivative is used to determine slope. We now consider another use—to determine the rate of change in one variable with respect to another. Applications involving rates of change can be found in a wide variety of fields. A few examples are population growth rates, production rates, the rate of water flow, velocity, and acceleration.

Straight-line motion

A common use of a rate of change is to describe the motion of an object moving in a straight line. Such motion is called **rectilinear motion.** It is customary to use either a horizontal or vertical line with a designated origin to represent the line of motion. Movement to the right (or upward) is considered to be in the **positive direction,** and movement to the left (or down) is considered to be in the **negative direction.**

The function s that gives the position (relative to the origin) of an object as a function of time t is called a **position function.** If, over a period of time Δt, the object changes its position by the amount

$$\Delta s = s(t + \Delta t) - s(t) \qquad \text{Change in distance}$$

then, by the familiar formula

$$\text{rate} = \frac{\text{distance}}{\text{time}}$$

the **average rate of change** in distance with respect to time is given by

$$\frac{\text{change in distance}}{\text{change in time}} = \frac{\Delta s}{\Delta t}.$$

We call this average rate of change the **average velocity.**

DEFINITION OF AVERAGE VELOCITY

If $s(t)$ gives the position at time t of an object moving in a straight line, then the **average velocity** of the object over the interval $[t, t + \Delta t]$ is given by

$$\text{average velocity} = \frac{\Delta s}{\Delta t} = \frac{s(t + \Delta t) - s(t)}{\Delta t}.$$

EXAMPLE 1   Finding average velocities for a falling object

If a free-falling object is dropped from a height of 100 feet, its height s at time t is given by the position function $s = -16t^2 + 100$, where s is measured in feet and t is measured in seconds. Find the average rate of change of the height over the following intervals.

(a) $[1, 2]$ (b) $[1, 1.5]$ (c) $[1, 1.1]$

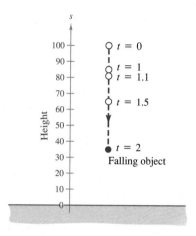

FIGURE 3.13

TABLE 3.1

s	84	80.64	64	36
t	1	1.1	1.5	2

SOLUTION

Using the position equation, we determine the heights at $t = 1$, 1.1, 1.5, and 2, as shown in Table 3.1 and Figure 3.13.

(a) For the interval [1, 2], the object falls from a height of 84 feet to a height of 36 feet. Thus, the average rate of change is

$$\frac{\Delta s}{\Delta t} = \frac{36 - 84}{2 - 1} = \frac{-48}{1} = -48 \text{ ft/sec.}$$

(b) For the interval [1, 1.5], the average rate of change is

$$\frac{\Delta s}{\Delta t} = \frac{64 - 84}{1.5 - 1} = \frac{-20}{0.5} = -40 \text{ ft/sec.}$$

(c) For the interval [1, 1.1], the average rate of change is

$$\frac{\Delta s}{\Delta t} = \frac{80.64 - 84}{1.1 - 1} = \frac{-3.36}{0.1} = -33.6 \text{ ft/sec.}$$

REMARK   Note that the average velocities in Example 1 are *negative*, indicating that the object is moving downward.

Suppose that in Example 1 we wanted to find the velocity at the instant, $t = 1$ second. We call this the **instantaneous velocity,** or simply the **velocity** of the object when $t = 1$. Just as we approximate the slope of the tangent line by calculating the slope of the secant line, we can approximate the velocity at $t = 1$ by calculating the average velocity over a small interval $[1, 1 + \Delta t]$, as shown in Table 3.2.

TABLE 3.2

Δt	1	0.5	0.1	0.01	0.001	0.0001
$\dfrac{\Delta s}{\Delta t}$	−48	−40	−33.6	−32.16	−32.016	−32.0016

From this table, it seems reasonable to conclude that the velocity when $t = 1$ is -32 feet per second. We will verify this conclusion after presenting the following definition.

DEFINITION OF (INSTANTANEOUS) VELOCITY

If $s = s(t)$ is the position function for an object moving along a straight line, then the **velocity** of the object at time t is given by

$$v(t) = \lim_{\Delta t \to 0} \frac{s(t + \Delta t) - s(t)}{\Delta t} = s'(t).$$

REMARK   Note that velocity is given by the *derivative* of the position function.

EXAMPLE 2   Using the derivative to find velocity

Find the velocity at $t = 1$ and $t = 2$ of a free-falling object whose position function is

$$s(t) = -16t^2 + 100$$

where s is measured in feet and t is measured in seconds.

SOLUTION

Using the limit definition of the derivative, we find the velocity function to be

$$
\begin{aligned}
v(t) = s'(t) &= \lim_{\Delta t \to 0} \frac{s(t + \Delta t) - s(t)}{\Delta t} \\
&= \lim_{\Delta t \to 0} \frac{-16(t + \Delta t)^2 + 100 - (-16t^2 + 100)}{\Delta t} \\
&= \lim_{\Delta t \to 0} \frac{-32t\Delta t - 16(\Delta t)^2}{\Delta t} \\
&= \lim_{\Delta t \to 0} (-32t - 16\Delta t) \\
&= -32t.
\end{aligned}
$$

Therefore, the velocity at $t = 1$ is $v(1) = -32$ ft/sec and the velocity at $t = 2$ is $v(2) = -64$ ft/sec.   ◻

Sometimes there is confusion about the terms "speed" and "velocity." We define **speed** as the absolute value of velocity; as such, it is always nonnegative. Thus, speed indicates only how fast an object is moving, whereas velocity indicates both the speed and the direction of motion relative to a given coordinate system.

Just as we can obtain the velocity function by differentiating the position function, we can obtain the **acceleration** function by differentiating the velocity function.

DEFINITION OF ACCELERATION	If s is the position function for an object moving along a straight line, then the **acceleration** of the object at time t is given by

$$a(t) = v'(t)$$

where $v(t)$ is the velocity at time t.

REMARK If the time t is expressed in seconds and the distance s is expressed in feet, then velocity will be expressed in feet per second (ft/sec) and acceleration will be given in feet per second per second (ft/sec^2).

EXAMPLE 3 Finding acceleration as the derivative of the velocity

Find the acceleration of a free-falling object whose position function is

$$s(t) = -16t^2 + 100.$$

SOLUTION

From Example 2, we know the velocity function for this object is

$$v(t) = -32t.$$

Thus, the acceleration is given by

$$a(t) = v'(t) = -32 \text{ ft/sec}^2.$$

The acceleration found in Example 3 is called the **acceleration due to gravity,** denoted by g, and its exact value depends on one's location on the earth. The *standard value* of g is -32.174 feet per second per second (or -9.81 meters per second per second).

In general, the position of a free-falling object (neglecting air resistance) under the influence of gravity can be represented by the equation

$$s(t) = \frac{1}{2}gt^2 + v_0t + s_0$$

where s_0 is the initial height of the object, v_0 is the initial velocity of the object, and g is the acceleration due to gravity. Considering the acceleration due to the earth's gravity to be $g = -32$ feet per second per second, we obtain the position function

$$s(t) = -16t^2 + v_0t + s_0.$$

Remember that for free-falling objects, we consider the velocity to be positive for upward motion and negative for downward motion.

EXAMPLE 4   Finding the acceleration of a moving object

Suppose that the velocity of an automobile starting from rest is given by

$$v = \frac{80t}{t + 5} \text{ ft/sec.}$$

Make a table to compare the velocity and acceleration of the automobile when $t = 0, 5, 10, \ldots, 60$ seconds.

SOLUTION

The position of the car is shown in Figure 3.14. The acceleration at time t is given by

$$a = \frac{dv}{dt} = \lim_{\Delta t \to 0} \frac{1}{\Delta t} \left[\frac{80(t + \Delta t)}{(t + \Delta t) + 5} - \frac{80t}{t + 5} \right]$$

$$= \lim_{\Delta t \to 0} \frac{80}{\Delta t} \left[\frac{5\Delta t}{(t + \Delta t + 5)(t + 5)} \right]$$

$$= \lim_{\Delta t \to 0} \frac{400}{(t + \Delta t + 5)(t + 5)} = \frac{400}{(t + 5)^2} \text{ ft/sec}^2.$$

Table 3.3 compares the velocity and acceleration of the automobile at five-second intervals during its first minute of travel.

FIGURE 3.14

TABLE 3.3

t	0	5	10	15	20	25	30	35	40	45	50	55	60
v	0	40	53.3	60.0	64.0	66.7	68.6	70.0	71.1	72.0	72.7	73.3	73.8
$\dfrac{dv}{dt}$	16	4	1.78	1.00	0.64	0.44	0.33	0.25	0.20	0.16	0.13	0.11	0.09

Note from Table 3.3 that the acceleration approaches zero as the velocity levels off. This observation should agree with your experience of riding in an accelerating car—you do not feel the velocity, but you feel the acceleration. In other words, you feel changes in velocity.

Higher-order derivatives

To derive the acceleration function from the position function, we need to differentiate the position function *twice*.

$$s(t) \qquad \text{Position function}$$

$$v(t) = s'(t) \qquad \text{Velocity function}$$

$$a(t) = v'(t) = s''(t) \qquad \text{Acceleration function}$$

We call $a(t)$ the **second derivative** of $s(t)$ and denote it by $s''(t)$.

The second derivative is an example of a **higher-order derivative.** We can define derivatives of any positive integer order. For instance, the **third derivative** is the derivative of the second derivative. We denote higher-order derivatives as follows.

First derivative: $\quad y', \quad f'(x), \quad \dfrac{dy}{dx}, \quad \dfrac{d}{dx}[f(x)], \quad D_x(y)$

Second derivative: $\quad y'', \quad f''(x), \quad \dfrac{d^2y}{dx^2}, \quad \dfrac{d^2}{dx^2}[f(x)], \quad D_x^2(y)$

Third derivative: $\quad y''', \quad f'''(x), \quad \dfrac{d^3y}{dx^3}, \quad \dfrac{d^3}{dx^3}[f(x)], \quad D_x^3(y)$

Fourth derivative: $\quad y^{(4)}, \quad f^{(4)}(x), \quad \dfrac{d^4y}{dx^4}, \quad \dfrac{d^4}{dx^4}[f(x)], \quad D_x^4(y)$

$$\vdots$$

nth derivative: $\quad y^{(n)}, \quad f^{(n)}(x), \quad \dfrac{d^ny}{dx^n}, \quad \dfrac{d^n}{dx^n}[f(x)], \quad D_x^n(y)$

Other rates of change

Velocity and acceleration are only two examples of rates of change. In general, we can use the derivative to measure the rate of change of any variable with respect to another [provided the two variables are related by a differentiable function $y = f(x)$]. When determining the rate of change of one variable with respect to another, we must be careful to distinguish between average and instantaneous rates of change. The distinction between these two rates of change is comparable to the distinction between the slope of the secant line through two points on a curve and the slope of the tangent line at one point on a curve.

REMARK   In future work with the derivative, we will use "rate of change" to mean "instantaneous rate of change."

EXAMPLE 5   Finding the average rate of change over an interval

The concentration of a drug in a patient's bloodstream is monitored over 10-minute intervals for two hours. Find the average rates of change (in milligrams per minute) over the time intervals [0, 10], [0, 20], and [100, 110] for the concentrations in Table 3.4.

TABLE 3.4

t (min)	0	10	20	30	40	50	60	70	80	90	100	110	120
C (mg)	0	2	17	37	55	73	89	103	111	113	113	103	68

Drug Concentration in Bloodstream

FIGURE 3.15

SOLUTION

For the interval [0, 10], the average rate of change is

$$\frac{\Delta C}{\Delta t} = \frac{2 - 0}{10 - 0} = \frac{2}{10} = 0.2 \text{ mg/min.}$$

For the interval [0, 20], the average rate of change is

$$\frac{\Delta C}{\Delta t} = \frac{17 - 0}{20 - 0} = \frac{17}{20} = 0.85 \text{ mg/min.}$$

For the interval [100, 110], the average rate of change is

$$\frac{\Delta C}{\Delta t} = \frac{103 - 113}{110 - 100} = \frac{-10}{10} = -1 \text{ mg/min.}$$

Note in Figure 3.15 that the average rate of change is positive when the concentration increases and negative when the concentration decreases. ▭

To conclude this section, we give a summary concerning the derivative and its interpretations.

INTERPRETATIONS OF THE DERIVATIVE

If the function given by $y = f(x)$ is differentiable at x, then its derivative

$$\frac{dy}{dx} = f'(x) = \lim_{\Delta x \to 0} \frac{f(x + \Delta x) - f(x)}{\Delta x}$$

denotes both

1. the *slope* of the graph of f at x and
2. the *instantaneous rate of change* in y with respect to x.

EXERCISES for Section 3.2

In Exercises 1–6, find the average rate of change of the given function over the indicated interval. Compare this average rate of change to the instantaneous rates of change at the endpoints of the interval.

Function	Interval
1. $f(t) = 2t + 7$	[1, 2]
2. $f(t) = 3t - 1$	$\left[0, \frac{1}{3}\right]$

Function	Interval
3. $f(x) = \dfrac{1}{x + 1}$	[0, 3]
4. $f(x) = \dfrac{-1}{x}$	[1, 2]
5. $f(t) = t^2 - 3$	[2, 2.1]
6. $f(x) = x^2 - 6x - 1$	[−1, 3]

7. The height s at time t of a silver dollar dropped from the World Trade Center is given by $s(t) = -16t^2 + 1350$, where s is measured in feet and t is measured in seconds [$s'(t) = -32t$].
   (a) Find the average velocity on the interval [1, 2].
   (b) Find the instantaneous velocity when $t = 1$ and $t = 2$.
   (c) How long will it take the dollar to hit the ground?
   (d) Find the velocity of the dollar when it hits the ground.

8. An automobile's velocity starting from rest is given by

   $$v = \frac{100t}{2t + 15}$$

   where v is measured in feet per second. Find the acceleration at the following times.
   (a) 5 seconds   (b) 10 seconds   (c) 20 seconds

In Exercises 9–14, use the following position and velocity functions for free-falling objects.

$$s(t) = -16t^2 + v_0 t + s_0$$
$$s'(t) = -32t + v_0$$

9. A projectile is shot upward from the surface of the earth with an initial velocity of 384 feet per second. What is its velocity after 5 seconds? After 10 seconds?

10. Repeat Exercise 9 for an initial velocity of 256 feet per second.

11. A pebble is dropped from a height of 600 feet. Find the pebble's velocity when it hits the ground.

12. A ball is thrown straight down from the top of a 220-foot building with an initial velocity of -22 feet per second. What is its velocity after 3 seconds? What is its velocity after falling 108 feet?

13. To estimate the height of a building, a stone is dropped from the top of the building into a pool of water at ground level. How high is the building if the splash is seen 6.8 seconds after the stone is dropped?

14. A ball is dropped from a height of 100 feet. One second later another ball is dropped from a height of 75 feet. Which ball hits the ground first?

15. A car is traveling at a rate of 66 feet per second (45 miles per hour) when the brakes are applied. The position function for the car is $s(t) = -8.25t^2 + 66t$, where s is measured in feet and t is measured in seconds. Use this function to complete the following table.

t	0	1	2	3	4
$s(t)$					
$v(t)$					
$a(t)$					

16. The position function for an object is given by $s(t) = 10t^2$, $0 \le t \le 10$, where s is measured in feet and t is measured in seconds. Use this function to complete the following table.

t	0	2	4	6	8	10
$s(t)$						
$v(t)$						
$a(t)$						

17. Find the average velocity of the car in Exercise 15 during each given time interval.

18. Find the average velocity of the object in Exercise 16 during each given time interval.

In Exercises 19–24, find the indicated derivative.

Given	Find
19. $f'(x) = x^2$	$f''(x)$
20. $f''(x) = x^3$	$f'''(x)$
21. $f''(x) = 2 - \dfrac{2}{x}$	$f'''(x)$
22. $f'''(x) = 2\sqrt{x} - 1$	$f^{(4)}(x)$
23. $f^{(4)}(x) = 2x + 1$	$f^{(6)}(x)$
24. $f(x) = 2x^2 - 2$	$f''(x)$

25. The annual inventory cost for a certain manufacturer is given by

    $$C = \frac{1,008,000}{Q} + 6.3Q$$

    where Q is the order size when the inventory is replenished. Find the change in annual cost when Q is increased from 350 to 351 and compare this with the rate of change

    $$\frac{dC}{dQ} = -\frac{1,008,000}{Q^2} + 6.3$$

    when $Q = 350$.

26. A car is driven 15,000 miles a year and gets x miles per gallon. Assume that the average fuel cost is $1.10 per gallon. Find the annual cost of fuel C as a function of x and use this function to complete the following table.

x	10	15	20	25	30	35	40
C							
$\dfrac{dC}{dx}$							

27. At 0° Celsius, the equation for heat loss H from the body in kilocalories per square meter per hour is

$$H(v) = 33(10\sqrt{v} - v + 10.45)$$

where v is the wind speed in meters per second. The rate of change of H is given by

$$\frac{dH}{dv} = 33\left(\frac{5}{\sqrt{v}} - 1\right).$$

Find the rate at which H is changing when (a) $v = 2$ and (b) $v = 5$.

28. When a guitar string is plucked, it vibrates with a frequency of

$$F = 200\sqrt{T}$$

where F is measured in vibrations per second and the tension T is measured in pounds. Find the rate of change of the frequency when (a) $T = 4$ and (b) $T = 9$.

29. Newton's Law of Gravitation states that the gravitational force between two (point) particles of masses m_1 and m_2, respectively, is

$$F = K\frac{m_1 m_2}{r^2}, \quad r > 0$$

where r is the distance between the particles. Find the rate of change of force between the particles with respect to r and explain why the rate of change is negative.

30. Newton's Law of Cooling states that the rate of change of the temperature of an object is proportional to the difference between the object's temperature T and the temperature T_a of the surrounding medium. Write an equation for this law.

31. The height of an object projected upward from an initial height of 100 feet above the ground is given by

$$s = -16t^2 + 27t + 100$$

where t is the time in seconds. Use a computer or graphics calculator to sketch the graph of the position function and find the time when the object reaches ground level. Find the velocity of the object when it reaches ground level.

3.3 Differentiation Rules for Constant Multiples, Sums, Powers, Sines, and Cosines

Constant Rule ▪ Power Rule ▪ Constant Multiple Rule ▪ Sum and Difference Rules ▪ Derivatives of sine and cosine ▪ Applications of the derivative

In Sections 3.1 and 3.2 we used the limit definition to find derivatives. In this and the next two sections we introduce several "differentiation rules" that allow us to find derivatives without the *direct* use of the limit definition.

THEOREM 3.3
CONSTANT RULE

The derivative of a constant is zero.

$$\frac{d}{dx}[c] = 0, \quad c \text{ is a real number}$$

---

y

$f(x) = c$

The slope of a horizontal line is zero.

The derivative of a constant function is zero.

x

PROOF Let $f(x) = c$. Then, by the limit definition of the derivative,

$$\frac{d}{dx}[c] = f'(x) = \lim_{\Delta x \to 0} \frac{f(x + \Delta x) - f(x)}{\Delta x} = \lim_{\Delta x \to 0} \frac{c - c}{\Delta x} = 0.$$

REMARK Note in Figure 3.16 that the Constant Rule is equivalent to saying that the slope of a horizontal line is zero.

FIGURE 3.16

EXAMPLE 1 The derivative of a constant function

	Function	*Derivative*
(a)	$y = 7$	$\dfrac{dy}{dx} = 0$
(b)	$f(x) = 0$	$f'(x) = 0$
(c)	$s'(t) = -3$	$s''(t) = 0$

Before proving the next rule, we review the procedure for expanding a binomial. Recall that

$$(x + \Delta x)^2 = x^2 + 2x\Delta x + (\Delta x)^2$$
$$(x + \Delta x)^3 = x^3 + 3x^2\Delta x + 3x(\Delta x)^2 + (\Delta x)^3$$

and the general binomial expansion for a positive integer n is

$$(x + \Delta x)^n = x^n + nx^{n-1}\Delta x + \underbrace{\frac{n(n-1)x^{n-2}}{2}(\Delta x)^2 + \cdots + (\Delta x)^n}.$$

$(\Delta x)^2$ is a factor of these terms

This binomial expansion is used in proving a special case of the following theorem.

THEOREM 3.4
POWER RULE

If n is a rational number, then

$$\frac{d}{dx}[x^n] = nx^{n-1}.$$

PROOF Here, we prove only the case in which n is a positive integer greater than 1. In Exercise 69, we ask you to prove the case for $n = 1$. Section 3.4 (Example 7) has a proof for the case in which n is a negative integer, and Section 3.5 (Example 5) has a proof for the case in which n is rational. (In Chapter 6, the Power Rule will be extended to cover irrational values of n.) If n is a positive integer greater than 1, then the binomial expansion produces

$$\frac{d}{dx}[x^n] = \lim_{\Delta x \to 0} \frac{(x + \Delta x)^n - x^n}{\Delta x}$$
$$= \lim_{\Delta x \to 0} \frac{x^n + nx^{n-1}(\Delta x) + [n(n-1)x^{n-2}/2](\Delta x)^2 + \cdots + (\Delta x)^n - x^n}{\Delta x}$$
$$= \lim_{\Delta x \to 0} \left[nx^{n-1} + \left(\frac{n(n-1)x^{n-2}}{2}\right)(\Delta x) + \cdots + (\Delta x)^{n-1} \right]$$
$$= nx^{n-1} + 0 + \cdots + 0 = nx^{n-1}.$$

In the Power Rule, the case in which $n = 1$ is worth memorizing as a separate differentiation rule. That is,

$$\frac{d}{dx}[x] = 1.$$

This rule is consistent with the fact that the slope of the line given by $y = x$ is 1.

EXAMPLE 2 Applying the Power Rule

Function	Derivative
(a) $f(x) = x^3$	$f'(x) = 3x^2$
(b) $y = \dfrac{1}{x^2}$	$\dfrac{dy}{dx} = \dfrac{d}{dx}[x^{-2}] = (-2)x^{-3} = -\dfrac{2}{x^3}$

In part (b) of Example 2, note that *before* differentiating we rewrote $1/x^2$ as x^{-2}. Rewriting is the first step in *many* differentiation problems.

Given:		Rewrite:		Differentiate:		Simplify:
$y = \dfrac{1}{x^2}$	$\longrightarrow$	$y = x^{-2}$	$\longrightarrow$	$\dfrac{dy}{dx} = (-2)x^{-3}$	$\longrightarrow$	$\dfrac{dy}{dx} = -\dfrac{2}{x^3}$

THEOREM 3.5
CONSTANT MULTIPLE RULE

If f is a differentiable function and c is a real number, then

$$\frac{d}{dx}[cf(x)] = cf'(x).$$

PROOF

$$\frac{d}{dx}[cf(x)] = \lim_{\Delta x \to 0} \frac{cf(x + \Delta x) - cf(x)}{\Delta x}$$

$$= \lim_{\Delta x \to 0} c\left[\frac{f(x + \Delta x) - f(x)}{\Delta x}\right]$$

$$= c\left[\lim_{\Delta x \to 0} \frac{f(x + \Delta x) - f(x)}{\Delta x}\right] = cf'(x)$$

Informally, the Constant Multiple Rule states that constants can be factored out of the differentiation process.

$$\frac{d}{dx}[cf(x)] = c\frac{d}{dx}[f(x)] = cf'(x)$$

This rule is often overlooked, especially when the constant appears in the denominator, as follows.

$$\frac{d}{dx}\left[\frac{f(x)}{c}\right] = \frac{d}{dx}\left[\left(\frac{1}{c}\right)f(x)\right] = \left(\frac{1}{c}\right)\frac{d}{dx}[f(x)] = \left(\frac{1}{c}\right)f'(x)$$

EXAMPLE 3   Applying the Constant Multiple Rule with the Power Rule

Function	*Derivative*

(a) $y = \dfrac{2}{x}$ $\qquad$ $\dfrac{dy}{dx} = \dfrac{d}{dx}[2x^{-1}] = 2\dfrac{d}{dx}[x^{-1}] = 2(-1)x^{-2} = -\dfrac{2}{x^2}$

(b) $f(t) = \dfrac{4t^2}{5}$ $\qquad$ $f'(t) = \dfrac{d}{dt}\left[\dfrac{4}{5}t^2\right] = \dfrac{4}{5}\dfrac{d}{dt}[t^2] = \dfrac{4}{5}(2t) = \dfrac{8}{5}t$ $\qquad$ ▭

REMARK   It is helpful to see that the Constant Multiple and Power Rules can be combined into one rule. The combination rule is $D_x[cx^n] = cnx^{n-1}$.

Although the two functions in the next example are very simple, errors are frequently made in differentiating functions involving a constant multiple of the first power of x. Keep in mind that $D_x[cx] = c$.

EXAMPLE 4   Applying the Constant Multiple Rule

(a) $\dfrac{d}{dx}\left[-\dfrac{3x}{2}\right] = -\dfrac{3}{2}$ $\qquad$ (b) $\dfrac{d}{dx}[3\pi x] = 3\pi$ $\qquad$ ▭

The Sum Rule given in the following theorem tells us that we can differentiate the sum of two functions separately and add the resulting derivatives.

THEOREM 3.6
SUM AND DIFFERENCE RULES

The derivative of the sum (or difference) of two differentiable functions is the sum (or difference) of their derivatives.

$$\dfrac{d}{dx}[f(x) + g(x)] = f'(x) + g'(x) \qquad \text{Sum Rule}$$

$$\dfrac{d}{dx}[f(x) - g(x)] = f'(x) - g'(x) \qquad \text{Difference Rule}$$

PROOF   A proof of the Sum Rule follows directly from Theorem 2.3 (the limit of a sum). The Difference Rule can be proved in a similar way.

$$\dfrac{d}{dx}[f(x) + g(x)] = \lim_{\Delta x \to 0} \dfrac{[f(x + \Delta x) + g(x + \Delta x)] - [f(x) + g(x)]}{\Delta x}$$

$$= \lim_{\Delta x \to 0} \dfrac{f(x + \Delta x) + g(x + \Delta x) - f(x) - g(x)}{\Delta x}$$

$$= \lim_{\Delta x \to 0} \left[\dfrac{f(x + \Delta x) - f(x)}{\Delta x} + \dfrac{g(x + \Delta x) - g(x)}{\Delta x}\right]$$

$$= \lim_{\Delta x \to 0} \dfrac{f(x + \Delta x) - f(x)}{\Delta x} + \lim_{\Delta x \to 0} \dfrac{g(x + \Delta x) - g(x)}{\Delta x}$$

$$= f'(x) + g'(x)$$

The Sum and Difference Rules can be extended to cover the derivative of any finite number of functions. For instance, if

$$F(x) = f(x) + g(x) - h(x) - k(x)$$

then

$$F'(x) = f'(x) + g'(x) - h'(x) - k'(x).$$

EXAMPLE 5 Applying the Sum and Difference Rules

Function	Derivative
(a) $f(x) = x^3 - 4x + 5$	$f'(x) = 3x^2 - 4$
(b) $g(x) = -\dfrac{x^4}{2} + 3x^3 - 2x$	$g'(x) = -2x^3 + 9x^2 - 2$

Parentheses can play an important role in the Power Rule and the Constant Multiple Rule, as shown in Example 6.

EXAMPLE 6 Using parentheses when differentiating

Given Function	Rewrite	Differentiate	Simplify
(a) $y = \dfrac{5}{2x^3}$	$y = \dfrac{5}{2}(x^{-3})$	$y' = \dfrac{5}{2}(-3x^{-4})$	$y' = -\dfrac{15}{2x^4}$
(b) $y = \dfrac{5}{(2x)^3}$	$y = \dfrac{5}{8}(x^{-3})$	$y' = \dfrac{5}{8}(-3x^{-4})$	$y' = -\dfrac{15}{8x^4}$
(c) $y = \dfrac{7}{3x^{-2}}$	$y = \dfrac{7}{3}(x^2)$	$y' = \dfrac{7}{3}(2x)$	$y' = \dfrac{14x}{3}$
(d) $y = \dfrac{7}{(3x)^{-2}}$	$y = 63(x^2)$	$y' = 63(2x)$	$y' = 126x$

When differentiating functions involving radicals, we rewrite the function in terms of rational exponents, as shown in the next example.

EXAMPLE 7 Differentiating a function involving a radical

Function	Derivative
(a) $y = 2\sqrt{x}$	$\dfrac{dy}{dx} = \dfrac{d}{dx}[2x^{1/2}] = 2\left(\dfrac{1}{2}x^{-1/2}\right) = x^{-1/2} = \dfrac{1}{\sqrt{x}}$
(b) $y = \dfrac{1}{2\sqrt[3]{x^2}}$	$\dfrac{dy}{dx} = \dfrac{d}{dx}\left[\dfrac{1}{2}x^{-2/3}\right] = \dfrac{1}{2}\left(-\dfrac{2}{3}\right)x^{-5/3} = -\dfrac{1}{3x^{5/3}}$

Derivatives of sine and cosine functions

In Section 2.3, we discussed the following limits.

$$\lim_{\Delta x \to 0} \frac{\sin \Delta x}{\Delta x} = 1 \quad \text{and} \quad \lim_{\Delta x \to 0} \frac{1 - \cos \Delta x}{\Delta x} = 0$$

These two limits are crucial in the proofs of the derivatives of the sine and cosine functions. (The derivatives of the other four trigonometric functions will be discussed in the next section.)

THEOREM 3.7 **DERIVATIVES OF** **SINE AND COSINE**	$\dfrac{d}{dx}[\sin x] = \cos x \qquad \dfrac{d}{dx}[\cos x] = -\sin x$

PROOF We prove the first of these two rules and leave the proof of the second as an exercise (see Exercise 70).

$$\frac{d}{dx}[\sin x] = \lim_{\Delta x \to 0} \frac{\sin (x + \Delta x) - \sin x}{\Delta x}$$

$$= \lim_{\Delta x \to 0} \frac{\sin x \cos \Delta x + \cos x \sin \Delta x - \sin x}{\Delta x}$$

$$= \lim_{\Delta x \to 0} \frac{\cos x \sin \Delta x - (\sin x)(1 - \cos \Delta x)}{\Delta x}$$

$$= \lim_{\Delta x \to 0} \left[(\cos x)\left(\frac{\sin \Delta x}{\Delta x}\right) - (\sin x)\left(\frac{1 - \cos \Delta x}{\Delta x}\right) \right]$$

$$= \cos x \left[\lim_{\Delta x \to 0} \frac{\sin \Delta x}{\Delta x} \right] - \sin x \left[\lim_{\Delta x \to 0} \frac{1 - \cos \Delta x}{\Delta x} \right]$$

$$= (\cos x)(1) - (\sin x)(0) = \cos x$$

This differentiation formula is shown graphically in Figure 3.17. Note that for each x the *slope* of the sine curve determines the *value* of the cosine curve.

$y = \sin x$

$y' = 0$

$y' = -1$ $y' = 1$

$y' = 1$

$y' = 0$

y increasing, y decreasing, y increasing,
y' positive y' negative y' positive

$y' = \cos x$

$$\frac{d}{dx}[\sin x] = \cos x$$

FIGURE 3.17

EXAMPLE 8 Derivatives involving sines and cosines

Function	*Derivative*
(a) $y = 3 \sin x$	$y' = 3 \cos x$
(b) $y = x + \cos x$	$y' = 1 - \sin x$
(c) $y = \dfrac{\sin x}{2} = \dfrac{1}{2} \sin x$	$y' = \dfrac{1}{2} \cos x = \dfrac{\cos x}{2}$

Applications of the derivative

The first two sections of this chapter included two important applications of the derivative—the slope of a curve and rate of change. We conclude this section with two examples of these applications.

EXAMPLE 9 Using the derivative to find the slope of a curve at a point

Find the slope of the graph of $f(x) = 2 \cos x$ at the following points.

(a) $\left(-\dfrac{\pi}{2}, 0\right)$ (b) $\left(\dfrac{\pi}{3}, 1\right)$ (c) $(\pi, -2)$

SOLUTION

The derivative of f is $f'(x) = -2 \sin x$. Therefore, the slopes at the indicated points are as follows.

(a) At $x = -\dfrac{\pi}{2}$, the slope is

$$f'\left(-\frac{\pi}{2}\right) = -2 \sin \left(-\frac{\pi}{2}\right) = -2(-1) = 2.$$

(b) At $x = \dfrac{\pi}{3}$, the slope is

$$f'\left(\frac{\pi}{3}\right) = -2 \sin \frac{\pi}{3} = -2\left(\frac{\sqrt{3}}{2}\right) = -\sqrt{3}.$$

(c) At $x = \pi$, the slope is

$$f'(\pi) = -2 \sin \pi = -2(0) = 0.$$

(See Figure 3.18.)

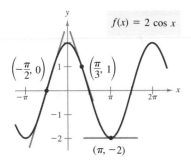

FIGURE 3.18

EXAMPLE 10 Using the derivative to find velocity

At time $t = 0$, a diver jumps from a diving board that is 32 feet above the water. The position of the diver is given by

$$s(t) = -16t^2 + 16t + 32$$

where s is measured in feet and t is measured in seconds. (See Figure 3.19.)
(a) When does the diver hit the water?
(b) What is the diver's velocity at impact?

SOLUTION

(a) To find the time at which the diver hits the water, we let $s = 0$ and solve for t.

$$-16t^2 + 16t + 32 = 0$$
$$-16(t^2 - t - 2) = 0$$
$$-16(t + 1)(t - 2) = 0$$
$$t = -1 \text{ or } 2$$

The solution $t = -1$ doesn't make sense, so we conclude that the diver hits the water at $t = 2$ seconds.

(b) The velocity at time t is given by the derivative

$$s'(t) = -32t + 16.$$

Therefore, the velocity at time $t = 2$ is

$$s'(2) = -32(2) + 16 = -48 \text{ ft/sec.}$$

REMARK   In Figure 3.19, note that the diver moves upward for the first half-second. This corresponds to the fact that the velocity is positive for $0 < t < \frac{1}{2}$.

FIGURE 3.19

EXERCISES for Section 3.3

In Exercises 1 and 2, find the slope of the tangent line to $y = x^n$ at the point $(1, 1)$.

1. (a) $y = x^{1/2}$ (b) $y = x^{3/2}$

(c) $y = x^2$ (d) $y = x^3$

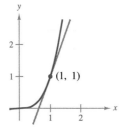

2. (a) $y = x^{-1/2}$ (b) $y = x^{-1}$

(c) $y = x^{-3/2}$ (d) $y = x^{-2}$

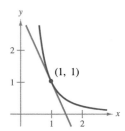

In Exercises 3–16, find the derivative of the given function.

3. $y = 3$
4. $f(x) = -2$
5. $f(x) = x + 1$
6. $g(x) = 3x - 1$
7. $g(x) = x^2 + 4$
8. $y = t^2 + 2t - 3$
9. $f(t) = -2t^2 + 3t - 6$
10. $y = x^3 - 9$
11. $s(t) = t^3 - 2t + 4$
12. $f(x) = 2x^3 - x^2 + 3x$
13. $y = x^2 - \dfrac{1}{2}\cos x$
14. $y = 5 + \sin x$
15. $y = \dfrac{1}{x} - 3\sin x$
16. $g(t) = \pi \cos t$

In Exercises 17–24, find the value of the derivative of the given function at the indicated point.

Function	Point
17. $f(x) = \dfrac{1}{x}$	$(1, 1)$
18. $f(t) = 3 - \dfrac{3}{5t}$	$\left(\dfrac{3}{5}, 2\right)$
19. $f(x) = -\dfrac{1}{2} + \dfrac{7}{5}x^3$	$\left(0, -\dfrac{1}{2}\right)$
20. $y = 3x\left(x^2 - \dfrac{2}{x}\right)$	$(2, 18)$
21. $y = (2x + 1)^2$	$(0, 1)$
22. $f(x) = 3(5 - x)^2$	$(5, 0)$
23. $f(\theta) = 4\sin\theta - \theta$	$(0, 0)$
24. $g(t) = 2 + 3\cos t$	$(\pi, -1)$

In Exercises 25–38, find $f'(x)$.

25. $f(x) = x^2 - \dfrac{4}{x}$

26. $f(x) = x^2 - 3x - 3x^{-2}$

27. $f(x) = x^3 - 3x - \dfrac{2}{x^4}$
28. $f(x) = \dfrac{2x^2 - 3x + 1}{x}$

29. $f(x) = \dfrac{x^3 - 3x^2 + 4}{x^2}$

30. $f(x) = (x^2 + 2x)(x + 1)$

31. $f(x) = x(x^2 + 1)$
32. $f(x) = x + \dfrac{1}{x^2}$

33. $f(x) = x^{4/5}$
34. $f(x) = x^{1/3} - 1$

35. $f(x) = \sqrt[3]{x} + \sqrt[5]{x}$
36. $f(x) = \dfrac{1}{\sqrt[3]{x^2}}$

37. $f(x) = 4\sqrt{x} + 3\cos x$
38. $f(x) = 2\sin x + 3\cos x$

In Exercises 39–44, complete the table, using Example 6 as a model.

Function	Rewrite	Derivative	Simplify
39. $y = \dfrac{1}{3x^3}$			
40. $y = \dfrac{2}{3x^2}$			
41. $y = \dfrac{1}{(3x)^3}$			
42. $y = \dfrac{\pi}{(3x)^2}$			
43. $y = \dfrac{\sqrt{x}}{x}$			
44. $y = \dfrac{4}{x^{-3}}$			

In Exercises 45 and 46, find an equation of the tangent line to the given function at the indicated point.

45. $y = x^4 - 3x^2 + 2$, $(1, 0)$
46. $y = x^3 + x$, $(-1, -2)$

In Exercises 47–52, determine the point(s) (if any) at which the given function has a horizontal tangent line.

47. $y = x^4 - 3x^2 + 2$
48. $y = x^3 + x$
49. $y = \dfrac{1}{x^2}$
50. $y = x^2 + 1$
51. $y = x + \sin x$, $0 \le x < 2\pi$
52. $y = \sqrt{3}x + 2\cos x$, $0 \le x < 2\pi$

53. Sketch the graphs of the two equations $y = x^2$ and $y = -x^2 + 6x - 5$, and sketch the two lines that are tangent to both graphs. Find the equations of these lines.

54. Show that the graphs of the two equations $y = x$ and $y = 1/x$ have tangent lines that are perpendicular to each other at their points of intersection.

55. The area of a square with sides of length s is given by $A = s^2$. Find the rate of change of the area with respect to s when $s = 4$.

56. The volume of a cube with sides of length s is given by $V = s^3$. Find the rate of change of the volume with respect to s when $s = 4$.

57. A company finds that charging p dollars per unit produces a monthly revenue

$$R = 12{,}000p - 1{,}000p^2, \quad 0 \le p \le 12.$$

(Note that the revenue is zero when $p = 12$ since no one is willing to pay that much.) Find the rate of change of R with respect to p when p has the following values.
(a) $p = 1$
(b) $p = 4$
(c) $p = 6$
(d) $p = 10$

58. Suppose that the profit P obtained in selling x units of a certain item each week is given by

$$P = 50\sqrt{x} - 0.5x - 500, \quad 0 \le x \le 8000.$$

Find the rate of change of P with respect to x when
(a) $x = 900$ (b) $x = 1600$
(c) $x = 2500$ (d) $x = 3600$
(Note that the eventual decline in profit occurs because the only way to sell larger quantities is to decrease the price per unit.)

59. Suppose that the effectiveness E of a painkilling drug t hours after entering the bloodstream is given by

$$E = \frac{1}{27}(9t + 3t^2 - t^3), \quad 0 \le t \le 4.5.$$

Find the rate of change of E with respect to t when
(a) $t = 1$ (b) $t = 2$
(c) $t = 3$ (d) $t = 4$

60. In a certain chemical reaction, the amount in grams Q of a substance produced in t hours is given by the equation

$$Q = 16t - 4t^2, \quad 0 < t \le 2.$$

Find the rate in grams per hour at which the substance is being produced when t has the following values.
(a) $t = \dfrac{1}{2}$ (b) $t = 1$
(c) $t = 2$

61. In the Chapter 3 Application, we developed the model

$$s(t) = -16t^2 + 5t + 30$$

for the height (in feet) of a diver at time t, where t is measured in seconds. Find the velocity of the diver when the diver hits the water.

62. In Exercise 61, the position equation corresponding to a diving platform of height h is

$$s(t) = -16t^2 + 5t + h.$$

What is the velocity of the diver when the diver hits the water for $h = 40$ ft, $h = 50$ ft, and $h = 60$ ft?

63. An astronaut standing on the moon throws a rock into the air. The height of the rock is given by

$$s = -\frac{27}{10}t^2 + 27t + 6$$

where s is measured in feet and t is measured in seconds. Find the acceleration of the rock and compare it with the acceleration due to gravity on the earth.

64. A ball is thrown upward from ground level, and its height at any time is given by

$$s(t) = -16t^2 + 48t$$

where s is measured in feet and t is measured in seconds.
(a) Find expressions for the velocity and acceleration of the ball.
(b) Find the time when the ball is at its highest point by finding the time when the velocity is zero.
(c) Find the height at the time given in part (b).

In Exercises 65 and 66, determine whether the given statement is true or false.

65. If $f(x) = g(x) + c$, then $f'(x) = g'(x)$.
66. If $y = \pi^2$, then $y' = 2\pi$.

67. Use a computer or graphics calculator to sketch the graph of f and f' over the interval $[0, 3]$ if

$$f(x) = 4.1x^3 - 12x^2 + 2.5x.$$

Determine the points (if any) at which f has horizontal tangents.

68. Use a computer or graphics calculator to find an equation of the tangent line to the graph of $f(x) = 2/x$, which passes through the point $(5, 0)$. Sketch the graph of f and the tangent line on the same set of coordinate axes. [Hint: To find the point of tangency on the graph of f, solve the equation

$$f'(x) = \frac{0 - y}{5 - x}.]$$

69. Prove that $D_x[x] = 1$.
70. Prove that $D_x[\cos x] = -\sin x$.

3.4 Differentiation Rules for Products, Quotients, Secants, and Tangents

Product Rule ▪ Quotient Rule ▪ Derivatives of tangent, secant, cotangent, and cosecant

In Section 3.3 we saw that the derivative of a sum or difference of two functions is simply the sum or difference of their derivatives. The rules for the derivative of a product or quotient of two functions are not as simple, and you may find the results surprising.

THEOREM 3.8
PRODUCT RULE

The derivative of the product of two differentiable functions is given by the first function times the derivative of the second plus the second function times the derivative of the first.

$$\frac{d}{dx}[f(x)g(x)] = f(x)g'(x) + g(x)f'(x)$$

PROOF   Some mathematical proofs, such as the proof of the Sum Rule in the previous section, are straightforward. Others involve clever steps that may appear unmotivated to a reader. This proof involves such a step—adding and subtracting the same quantity. Once you see this step (shown in color), you should be able to justify the limits shown in the other steps.

$$\frac{d}{dx}[f(x)g(x)] = \lim_{\Delta x \to 0} \frac{f(x + \Delta x)g(x + \Delta x) - f(x)g(x)}{\Delta x}$$

$$= \lim_{\Delta x \to 0} \frac{f(x + \Delta x)g(x + \Delta x) - f(x + \Delta x)g(x) + f(x + \Delta x)g(x) - f(x)g(x)}{\Delta x}$$

$$= \lim_{\Delta x \to 0} \left[f(x + \Delta x)\frac{g(x + \Delta x) - g(x)}{\Delta x} + g(x)\frac{f(x + \Delta x) - f(x)}{\Delta x} \right]$$

$$= \lim_{\Delta x \to 0} \left[f(x + \Delta x)\frac{g(x + \Delta x) - g(x)}{\Delta x} \right] + \lim_{\Delta x \to 0} \left[g(x)\frac{f(x + \Delta x) - f(x)}{\Delta x} \right]$$

$$= \lim_{\Delta x \to 0} f(x + \Delta x) \cdot \lim_{\Delta x \to 0} \frac{g(x + \Delta x) - g(x)}{\Delta x} + \lim_{\Delta x \to 0} g(x) \cdot \lim_{\Delta x \to 0} \frac{f(x + \Delta x) - f(x)}{\Delta x}$$

$$= f(x)g'(x) + g(x)f'(x)$$

The Product Rule can be extended to cover products involving more than two terms. For example, if f, g, and h are differentiable functions of x, then

$$\frac{d}{dx}[f(x)g(x)h(x)] = f'(x)g(x)h(x) + f(x)g'(x)h(x) + f(x)g(x)h'(x).$$

EXAMPLE 1   Differentiation with the Product Rule

Find the derivative of $f(x) = (3x - 2x^2)(5 + 4x)$.

SOLUTION

Using the Product Rule, we have

$$f'(x) = \overbrace{(3x - 2x^2)}^{\text{(first)}}\overbrace{\frac{d}{dx}[5 + 4x]}^{\left(\substack{\text{derivative} \\ \text{of second}}\right)} + \overbrace{(5 + 4x)}^{\text{(second)}}\overbrace{\frac{d}{dx}[3x - 2x^2]}^{\left(\substack{\text{derivative} \\ \text{of first}}\right)}$$

$$= (3x - 2x^2)(4) + (5 + 4x)(3 - 4x)$$

$$= (12x - 8x^2) + (15 - 8x - 16x^2)$$

$$= -24x^2 + 4x + 15.$$

REMARK   Note that the derivative of a product of two functions is not (in general) given by the product of the derivatives of the two functions. To see this, try comparing the product of the derivatives of $f(x) = 3x - 2x^2$ and $g(x) = 5 + 4x$ with the derivative found in Example 1.

In Example 1, we had the option of finding the derivative with or without the Product Rule. To find the derivative without the Product Rule, we can write

$$D_x[(3x - 2x^2)(5 + 4x)] = D_x[-8x^3 + 2x^2 + 15x]$$
$$= -24x^2 + 4x + 15.$$

In the next example, we must use the Product Rule to find the derivative.

EXAMPLE 2   Using the Product Rule

Find the derivative of $y = x \sin x$.

SOLUTION

$$\frac{d}{dx}[x \sin x] = x\frac{d}{dx}[\sin x] + \sin x\frac{d}{dx}[x]$$
$$= x \cos x + (\sin x)(1)$$
$$= x \cos x + \sin x$$

EXAMPLE 3   Using the Product Rule

Find the derivative of $y = 2x \cos x - 2 \sin x$.

SOLUTION

$$\frac{dy}{dx} = \overbrace{(2x)\left(\frac{d}{dx}[\cos x]\right) + (\cos x)\left(\frac{d}{dx}[2x]\right)}^{\text{Product Rule}} - \overbrace{2\frac{d}{dx}[\sin x]}^{\text{Constant Multiple Rule}}$$
$$= (2x)(-\sin x) + (\cos x)(2) - 2(\cos x)$$
$$= -2x \sin x$$

REMARK   In Example 3, notice that we use the Product Rule when both factors of the product are variable, and we use the Constant Multiple Rule when one of the factors is a constant.

**THEOREM 3.9
QUOTIENT RULE**

The derivative of the quotient of two differentiable functions is given by the denominator times the derivative of the numerator minus the numerator times the derivative of the denominator divided by the square of the denominator.

$$\frac{d}{dx}\left[\frac{f(x)}{g(x)}\right] = \frac{g(x)f'(x) - f(x)g'(x)}{[g(x)]^2}, \quad g(x) \neq 0$$

PROOF As in the proof of the Product Rule, a key step in this proof involves adding and subtracting the same quantity.

$$\frac{d}{dx}\left[\frac{f(x)}{g(x)}\right] = \lim_{\Delta x \to 0} \frac{\dfrac{f(x + \Delta x)}{g(x + \Delta x)} - \dfrac{f(x)}{g(x)}}{\Delta x}$$

$$= \lim_{\Delta x \to 0} \frac{g(x)f(x + \Delta x) - f(x)g(x + \Delta x)}{\Delta x g(x)g(x + \Delta x)}$$

$$= \lim_{\Delta x \to 0} \frac{g(x)f(x + \Delta x) - f(x)g(x) + f(x)g(x) - f(x)g(x + \Delta x)}{\Delta x g(x)g(x + \Delta x)}$$

$$= \frac{\lim\limits_{\Delta x \to 0} \dfrac{g(x)[f(x + \Delta x) - f(x)]}{\Delta x} - \lim\limits_{\Delta x \to 0} \dfrac{f(x)[g(x + \Delta x) - g(x)]}{\Delta x}}{\lim\limits_{\Delta x \to 0} [g(x)g(x + \Delta x)]}$$

$$= \frac{g(x)\left[\lim\limits_{\Delta x \to 0} \dfrac{f(x + \Delta x) - f(x)}{\Delta x}\right] - f(x)\left[\lim\limits_{\Delta x \to 0} \dfrac{g(x + \Delta x) - g(x)}{\Delta x}\right]}{\lim\limits_{\Delta x \to 0} [g(x)g(x + \Delta x)]}$$

$$= \frac{g(x)f'(x) - f(x)g'(x)}{[g(x)]^2}$$

EXAMPLE 4 Using the Quotient Rule

Differentiate

$$y = \frac{2x^2 - 4x + 3}{2 - 3x}.$$

SOLUTION

$$y' = \frac{(2 - 3x)\dfrac{d}{dx}[2x^2 - 4x + 3] - (2x^2 - 4x + 3)\dfrac{d}{dx}[2 - 3x]}{(2 - 3x)^2}$$

$$= \frac{(2 - 3x)(4x - 4) - (2x^2 - 4x + 3)(-3)}{(2 - 3x)^2}$$

$$= \frac{(-12x^2 + 20x - 8) - (-6x^2 + 12x - 9)}{(2 - 3x)^2}$$

$$= \frac{-6x^2 + 8x + 1}{(2 - 3x)^2}$$

REMARK   Note the use of parentheses in Example 4. A liberal use of parentheses is recommended for *all* types of differentiation problems. For instance, with the Quotient Rule, it is a good idea to enclose all factors and derivatives in parentheses and to pay special attention to the subtraction required in the numerator.

When we introduced differentiation rules in the previous section, we emphasized the need for rewriting *before* differentiating. The next example illustrates this point with the Quotient Rule.

EXAMPLE 5   Rewriting before differentiating

$$y = \frac{3 - (1/x)}{x + 5} \qquad \text{Given function}$$

$$= \frac{(3x - 1)/x}{x + 5} = \frac{3x - 1}{x(x + 5)} = \frac{3x - 1}{x^2 + 5x} \qquad \text{Rewrite}$$

$$\frac{dy}{dx} = \frac{(x^2 + 5x)(3) - (3x - 1)(2x + 5)}{(x^2 + 5x)^2} \qquad \text{Quotient Rule}$$

$$= \frac{(3x^2 + 15x) - (6x^2 + 13x - 5)}{(x^2 + 5x)^2}$$

$$= \frac{-3x^2 + 2x + 5}{(x^2 + 5x)^2} \qquad \text{Simplify} \qquad \blacksquare$$

Not every quotient needs to be differentiated by the Quotient Rule. For example, the quotients in the next example can each be considered as the product of a constant times a function of x. In such cases it is more convenient to use the Constant Multiple Rule than the Quotient Rule.

EXAMPLE 6   Differentiating quotients with the Constant Multiple Rule

	Given Function	Rewrite	Differentiate	Simplify
(a)	$y = \dfrac{x^2 + 3x}{6}$	$y = \dfrac{1}{6}(x^2 + 3x)$	$y' = \dfrac{1}{6}(2x + 3)$	$y' = \dfrac{2x + 3}{6}$
(b)	$y = \dfrac{5x^4}{8}$	$y = \dfrac{5}{8}x^4$	$y' = \dfrac{5}{8}(4x^3)$	$y' = \dfrac{5}{2}x^3$
(c)	$y = \dfrac{-3(3x - 2x^2)}{7x}$	$y = -\dfrac{3}{7}(3 - 2x)$	$y' = -\dfrac{3}{7}(-2)$	$y' = \dfrac{6}{7}$
(d)	$y = \dfrac{9}{5x^2}$	$y = \dfrac{9}{5}(x^{-2})$	$y' = \dfrac{9}{5}(-2x^{-3})$	$y' = -\dfrac{18}{5x^3}$

$\blacksquare$

In Section 3.3, we claimed that the Power Rule, $D_x[x^n] = nx^{n-1}$, is valid for any rational number n, but we proved only the case where n is a positive integer. In the next example, we prove the rule for the case where n is a negative integer.

EXAMPLE 7   Proof of the Power Rule for negative integers

Use the Quotient Rule to prove the Power Rule for the case when n is a negative integer.

SOLUTION

If n is a negative integer, then there exists a positive integer k such that $n = -k$. Thus, by the Quotient Rule, we have

$$\frac{d}{dx}[x^n] = \frac{d}{dx}\left[\frac{1}{x^k}\right] = \frac{x^k(0) - (1)(kx^{k-1})}{(x^k)^2}$$

$$= \frac{0 - kx^{k-1}}{x^{2k}} = -kx^{-k-1} = nx^{n-1}.$$

Derivatives of trigonometric functions

Knowing the derivatives of the sine and cosine functions, we can use the Quotient Rule to find the derivatives of the four remaining trigonometric functions.

**THEOREM 3.10
DERIVATIVES OF TRIGONOMETRIC
FUNCTIONS**

$$\frac{d}{dx}[\tan x] = \sec^2 x \qquad \frac{d}{dx}[\cot x] = -\csc^2 x$$

$$\frac{d}{dx}[\sec x] = \sec x \tan x \qquad \frac{d}{dx}[\csc x] = -\csc x \cot x$$

PROOF   We give a proof for the derivative of the tangent function and leave the proofs of the remaining three derivatives as exercises (see Exercise 70). Considering $\tan x = (\sin x)/(\cos x)$ and applying the Quotient Rule, we obtain

$$\frac{d}{dx}[\tan x] = \frac{(\cos x)(\cos x) - (\sin x)(-\sin x)}{\cos^2 x}$$

$$= \frac{\cos^2 x + \sin^2 x}{\cos^2 x} = \frac{1}{\cos^2 x} = \sec^2 x.$$

EXAMPLE 8   Differentiating trigonometric functions

Function	Derivative
(a) $y = x - \tan x$	$\dfrac{dy}{dx} = 1 - \sec^2 x$
(b) $y = x \sec x$	$y' = x(\sec x \tan x) + (\sec x)(1) = (\sec x)(1 + x \tan x)$

Because of the abundance of trigonometric identities, the derivative of a trigonometric function can take many forms. This presents a challenge when you are trying to match your answer to one given in the back of the text. To help, we will occasionally list more than one form of the answer to trigonometric problems. The next example illustrates the possible diversity of trigonometric forms.

EXAMPLE 9   Different forms of a derivative

Differentiate the function

$$y = \frac{1 - \cos x}{\sin x} = \csc x - \cot x$$

in two forms, and show that the derivatives are equal.

SOLUTION

For the first form, we can write

$$y = \frac{1 - \cos x}{\sin x}$$

$$y' = \frac{(\sin x)(\sin x) - (1 - \cos x)(\cos x)}{\sin^2 x}$$

$$= \frac{\sin^2 x + \cos^2 x - \cos x}{\sin^2 x} = \frac{1 - \cos x}{\sin^2 x}.$$

For the second form, we can write

$$y = \csc x - \cot x$$

$$y' = -\csc x \cot x + \csc^2 x.$$

To show that these two derivatives are equal, we write

$$\frac{1 - \cos x}{\sin^2 x} = \frac{1}{\sin^2 x} - \frac{\cos x}{\sin^2 x}$$

$$= \frac{1}{\sin^2 x} - \left(\frac{1}{\sin x}\right)\left(\frac{\cos x}{\sin x}\right)$$

$$= \csc^2 x - \csc x \cot x.$$

The summary in Table 3.5 shows that much of the work required to obtain the simplified form of a derivative occurs *after* differentiating. Note that two characteristics of a simplified form are the absence of negative exponents and the combining of like terms.

TABLE 3.5

	$f'(x)$ after differentiating	$f'(x)$ after simplifying
Example 1	$(3x - 2x^2)(4) + (5 + 4x)(3 - 4x)$	$-24x^2 + 4x + 15$
Example 3	$(2x)(-\sin x) + (\cos x)(2) - 2(\cos x)$	$-2x \sin x$
Example 4	$\dfrac{(2 - 3x)(4x - 4) - (2x^2 - 4x + 3)(-3)}{(2 - 3x)^2}$	$\dfrac{-6x^2 + 8x + 1}{(2 - 3x)^2}$
Example 5	$\dfrac{(x^2 + 5x)(3) - (3x - 1)(2x + 5)}{(x^2 + 5x)^2}$	$\dfrac{-3x^2 + 2x + 5}{(x^2 + 5x)^2}$
Example 9	$\dfrac{(\sin x)(\sin x) - (1 - \cos x)(\cos x)}{\sin^2 x}$	$\dfrac{1 - \cos x}{\sin^2 x}$

We conclude this section with a summary of the differentiation rules studied so far. To become skilled at differentiation, we recommend that you memorize each rule.

SUMMARY OF DIFFERENTIATION

General Differentiation Rules

Let u and v be differentiable functions of x.

Constant Multiple Rule: $\dfrac{d}{dx}[cu] = cu'$

Sum or Difference Rule: $\dfrac{d}{dx}[u \pm v] = u' \pm v'$

Product Rule: $\dfrac{d}{dx}[uv] = uv' + vu'$

Quotient Rule: $\dfrac{d}{dx}\left[\dfrac{u}{v}\right] = \dfrac{vu' - uv'}{v^2}$

Derivatives of Algebraic Functions

Constant Rule: $\dfrac{d}{dx}[c] = 0$

Power Rule: $\dfrac{d}{dx}[x^n] = nx^{n-1}$ $\dfrac{d}{dx}[x] = 1$

Derivatives of Trigonometric Functions

$\dfrac{d}{dx}[\sin x] = \cos x$ $\dfrac{d}{dx}[\tan x] = \sec^2 x$ $\dfrac{d}{dx}[\sec x] = \sec x \tan x$

$\dfrac{d}{dx}[\cos x] = -\sin x$ $\dfrac{d}{dx}[\cot x] = -\csc^2 x$ $\dfrac{d}{dx}[\csc x] = -\csc x \cot x$

REMARK   As an aid to memorization, note that the cofunctions (cosine, cotangent, and cosecant) require a negative sign as part of their derivatives.

EXERCISES for Section 3.4

In Exercises 1–8, find $f'(x)$ and $f'(c)$.

Function	Value of c
1. $f(x) = \frac{1}{3}(2x^3 - 4)$	$c = 0$
2. $f(x) = \frac{5 - 6x^2}{7}$	$c = 1$
3. $f(x) = 5x^{-2}(x + 3)$	$c = 1$
4. $f(x) = (x^2 - 2x + 1)(x^3 - 1)$	$c = 1$
5. $f(x) = (x^3 - 3x)(2x^2 + 3x + 5)$	$c = 0$
6. $f(x) = (x - 1)(x^2 - 3x + 2)$	$c = 0$
7. $f(x) = (x^5 - 3x)\left(\frac{1}{x^2}\right)$	$c = -1$
8. $f(x) = \frac{x + 1}{x - 1}$	$c = 2$

In Exercises 9–24, differentiate the given function.

9. $f(x) = \frac{3x - 2}{2x - 3}$

10. $f(x) = \frac{x^3 + 3x + 2}{x^2 - 1}$

11. $f(x) = \frac{3 - 2x - x^2}{x^2 - 1}$

12. $f(x) = x^4\left(1 - \frac{2}{x + 1}\right)$

13. $f(x) = \frac{x + 1}{\sqrt{x}}$

14. $f(x) = \sqrt[3]{x}(\sqrt{x} + 3)$

15. $h(t) = \frac{t + 1}{t^2 + 2t + 2}$

16. $h(x) = (x^2 - 1)^2$

17. $h(s) = (s^3 - 2)^2$

18. $f(x) = \left(\frac{x^2 - x - 3}{x^2 + 1}\right)(x^2 + x + 1)$

19. $g(x) = \left(\frac{x + 1}{x + 2}\right)(2x - 5)$

20. $f(x) = (x^2 - x)(x^2 + 1)(x^2 + x + 1)$

21. $f(x) = (3x^3 + 4x)(x - 5)(x + 1)$

22. $f(x) = \frac{x^2 + c^2}{x^2 - c^2}$, c is a constant

23. $f(x) = \frac{c^2 - x^2}{c^2 + x^2}$, c is a constant

24. $f(x) = \frac{x(x^2 - 1)}{x + 3}$

In Exercises 25–30, complete the table without using the Quotient Rule (see Example 6).

Function	Rewrite	Derivative	Simplify
25. $y = \frac{x^2 + 2x}{x}$			
26. $y = \frac{4x^{3/2}}{x}$			
27. $y = \frac{7}{3x^3}$			
28. $y = \frac{4}{5x^2}$			
29. $y = \frac{3x^2 - 5}{7}$			
30. $y = \frac{x^2 - 4}{x + 2}$			

In Exercises 31–50, find the derivative of the given trigonometric function.

31. $f(t) = t^2 \sin t$

32. $f(x) = \frac{\sin x}{x}$

33. $f(t) = \frac{\cos t}{t}$

34. $f(\theta) = (\theta + 1) \cos \theta$

35. $f(x) = -x + \tan x$

36. $y = x + \cot x$

37. $y = 5x \csc x$

38. $y = \frac{\sec x}{x}$

39. $y = -\csc x - \sin x$

40. $y = x \sin x + \cos x$

41. $y = x^2 \sin x + 2x \cos x$

42. $y = \frac{1 + \csc x}{1 - \csc x}$

43. $f(x) = \sin x \cos x$

44. $f(x) = \tan x \cot x$

45. $f(x) = x^2 \tan x$

46. $h(\theta) = 5 \sec \theta + \tan \theta$

47. $g(\theta) = \frac{\theta}{1 - \sin \theta}$

48. $f(\theta) = \frac{\sin \theta}{1 - \cos \theta}$

49. $h(t) = \frac{\sec t}{t}$

50. $f(x) = \sin x(\sin x + \cos x)$

In Exercises 51–56, find the second derivative of the given function.

51. $f(x) = 4x^{3/2}$

52. $f(x) = \frac{x^2 + 2x - 1}{x}$

53. $f(x) = \frac{x}{x - 1}$

54. $f(x) = x + \frac{32}{x^2}$

55. $f(x) = 3 \sin x$

56. $f(x) = \sec x$

In Exercises 57–62, find an equation of the tangent line to the graph of the given function at the indicated point.

Function	Point
57. $f(x) = \dfrac{x}{x-1}$	$(2, 2)$
58. $f(x) = (x-1)(x^2-2)$	$(0, 2)$
59. $f(x) = (x^3 - 3x + 1)(x + 2)$	$(1, -3)$
60. $f(x) = \dfrac{(x-1)}{(x+1)}$	$\left(2, \dfrac{1}{3}\right)$
61. $f(x) = \tan x$	$\left(\dfrac{\pi}{4}, 1\right)$
62. $f(x) = \sec x$	$\left(\dfrac{\pi}{3}, 2\right)$

In Exercises 63 and 64, determine the point(s) at which the graph of the function has a horizontal tangent.

63. $f(x) = \dfrac{x^2}{x-1}$

64. $f(x) = \dfrac{x^2}{x^2+1}$

65. The function

$$f(t) = \frac{t^2 - t + 1}{t^2 + 1}$$

measures the percentage of the normal level of oxygen in a pond, where t is the time in weeks after organic waste is dumped into the pond. Find the rate of change of f with respect to t when (a) $t = 0.5$, (b) $t = 2$, and (c) $t = 8$.

66. **Boyle's Law** states that if the temperature of a gas remains constant, its pressure is inversely proportional to its volume. Use the derivative to show that the rate of change of the pressure is inversely proportional to the square of the volume.

67. A population of 500 bacteria is introduced into a culture and grows in number according to the equation

$$P(t) = 500\left(1 + \frac{4t}{50 + t^2}\right)$$

where t is measured in hours. Find the rate at which the population is growing when $t = 2$.

68. Determine whether there exist any values of x in the interval $[0, 2\pi)$ such that the rate of change of $f(x) = \sec x$ and the rate of change of $g(x) = \csc x$ are equal.

69. Find $h'(c)$ where $h(x) = f(x)g(x)$ and f and g are differentiable functions such that $f'(c) = g'(c) = 0$.

70. Prove the following differentiation rules.

(a) $\dfrac{d}{dx}[\sec x] = \sec x \tan x$

(b) $\dfrac{d}{dx}[\csc x] = -\csc x \cot x$

(c) $\dfrac{d}{dx}[\cot x] = -\csc^2 x$

71. The velocity of a particle in straight-line motion is

$$v(t) = \frac{760t}{4t^2 + 25}, \quad 0 \le t \le 4.$$

Use a computer or graphics calculator to sketch the graph of $v(t)$ and $a(t)$. Determine the acceleration of the particle when $t = 2$.

3.5 The Chain Rule

The Chain Rule ▪ The General Power Rule ▪ Simplifying derivatives ▪ Trigonometric functions and the Chain Rule

We have yet to discuss one of the most powerful rules in differential calculus—the **Chain Rule.** This differentiation rule deals with composite functions and adds a surprising versatility to the rules we discussed in the two previous sections. For example, compare the following functions. Those on the left can be differentiated without the Chain Rule, and those on the right are best done with the Chain Rule.

Without the Chain Rule	With the Chain Rule
$y = x^2 + 1$	$y = \sqrt{x^2 + 1}$
$y = \sin x$	$y = \sin 2x$
$y = 3x + 2$	$y = (3x + 2)^5$
$y = x + \tan x$	$y = x + \tan x^2$

We begin with a physical example that should give you an intuitive idea about the nature of the Chain Rule. Basically, the rule says that if y changes dy/du times as fast as u, and u changes du/dx times as fast as x, then y changes $(dy/du)(du/dx)$ times as fast as x.

EXAMPLE 1   The derivative of a composite function

A set of gears is constructed as shown in Figure 3.20 so that the second and third gears are on the same axle. As the first axle revolves, it drives the second axle, which in turn drives the third axle. Let y, u, and x represent the number of revolutions per minute of the first, second, and third axles, respectively. Find dy/du, du/dx, and dy/dx, and show that

$$\frac{dy}{dx} = \left(\frac{dy}{du}\right)\left(\frac{du}{dx}\right).$$

SOLUTION

Since the circumference of the second gear is 3 times that of the first, the first axle must make 3 revolutions to turn the second axle once. Similarly, the second gear must make 2 revolutions to turn the third axle once, and we have

$$\frac{dy}{du} = 3 \quad \text{and} \quad \frac{du}{dx} = 2.$$

Combining these two results, we know that the first axle must make 6 revolutions to turn the third axle once. Thus, we can write

$$\frac{dy}{dx} = \left(\frac{dy}{du}\right)\left(\frac{du}{dx}\right) = (3)(2) = 6.$$

In other words, the rate of change of y with respect to x is the product of the rate of change of y with respect to u and the rate of change of u with respect to x.

Axle 1: y revolutions per minute
Axle 2: u revolutions per minute
Axle 3: x revolutions per minute

FIGURE 3.20

Example 1 illustrates a simple case of the Chain Rule. The general use is stated as follows.

THEOREM 3.11 **CHAIN RULE**	If $y = f(u)$ is a differentiable function of u and $u = g(x)$ is a differentiable function of x, then $y = f(g(x))$ is a differentiable function of x and $$\frac{dy}{dx} = \frac{dy}{du} \cdot \frac{du}{dx}$$ or, equivalently, $$\frac{d}{dx}[f(g(x))] = f'(g(x))g'(x).$$

PROOF Let $F(x) = f(g(x))$. Then, using the alternate form of the derivative (Theorem 3.1), we need to show that for $x = c$

$$F'(c) = f'(g(c))g'(c).$$

An important consideration in this proof is the behavior of g as x approaches c. A problem occurs if there are values of x, other than c, such that $g(x) = g(c)$. In Appendix A we show how to use the differentiability of f and g to overcome this problem. For now, we assume that $g(x) \neq g(c)$ for values of x other than c.

In the proof of the Product and Quotient Rules, we added and subtracted the same quantity to obtain the desired form. In this proof we use a similar technique—multiplying and dividing by the same (nonzero) quantity. Note that since g is differentiable, it is also continuous, and it follows that $g(x) \to g(c)$ as $x \to c$.

$$F'(c) = \lim_{x \to c} \frac{f(g(x)) - f(g(c))}{x - c}$$

$$= \lim_{x \to c} \left[\frac{f(g(x)) - f(g(c))}{g(x) - g(c)} \cdot \frac{g(x) - g(c)}{x - c} \right], \quad g(x) \neq g(c)$$

$$= \left[\lim_{x \to c} \frac{f(g(x)) - f(g(c))}{g(x) - g(c)} \right] \left[\lim_{x \to c} \frac{g(x) - g(c)}{x - c} \right]$$

$$= f'(g(c))g'(c)$$

When applying the Chain Rule, it is helpful to think of the composite function $f \circ g$ as having two parts—an *inside* and an *outside*, as follows.

$$y = f(\underbrace{g(x)}_{\substack{u = g(x) \\ \text{inside}}}) = f(\overbrace{u}^{\text{outside}})$$

The next example illustrates several composite functions.

EXAMPLE 2 Decomposition of a composite function

$y = f(g(x))$	$u = g(x)$	$y = f(u)$
(a) $y = \dfrac{1}{x + 1}$	$u = x + 1$	$y = \dfrac{1}{u}$
(b) $y = \sin 2x$	$u = 2x$	$y = \sin u$
(c) $y = \sqrt{3x^2 - x + 1}$	$u = 3x^2 - x + 1$	$y = \sqrt{u}$
(d) $y = \tan^2 x$	$u = \tan x$	$y = u^2$

EXAMPLE 3   Using the Chain Rule

Find dy/dx for $y = (x^2 + 1)^3$.

SOLUTION

To apply the Chain Rule, we identify the inside function u as follows.

$$y = \underbrace{(x^2 + 1)}_{u}{}^3 = u^3$$

Now, by the Chain Rule, we have

$$\frac{dy}{dx} = \underbrace{3(x^2 + 1)^2}_{\frac{dy}{du}}\underbrace{(2x)}_{\frac{du}{dx}} = 6x(x^2 + 1)^2.$$

The function in Example 3 is an instance of one of the most common types of composite functions, $y = [u(x)]^n$. The rule for differentiating such functions is called the **General Power Rule,** and it is a special case of the Chain Rule.

**THEOREM 3.12
GENERAL POWER RULE**

If $y = [u(x)]^n$, where u is a differentiable function of x and n is a rational number, then

$$\frac{dy}{dx} = n[u(x)]^{n-1}\frac{du}{dx}$$

or, equivalently,

$$\frac{d}{dx}[u^n] = nu^{n-1}u'.$$

PROOF   Since $y = u^n$, we apply the Chain Rule to obtain

$$\frac{dy}{dx} = \left(\frac{dy}{du}\right)\left(\frac{du}{dx}\right) = \frac{d}{du}[u^n]\frac{du}{dx}.$$

Now, by the (simple) Power Rule in Section 3.3, we have $D_u[u^n] = nu^{n-1}$, and it follows that

$$\frac{dy}{dx} = nu^{n-1}\frac{du}{dx}.$$

EXAMPLE 4   Applying the General Power Rule

Find the derivative of $f(x) = (3x - 2x^2)^3$.

SOLUTION

If we let $u = 3x - 2x^2$, then $f(x) = (3x - 2x^2)^3 = u^3$, and by the General Power Rule the derivative is

$$f'(x) = 3\overbrace{(3x - 2x^2)^2}^{u^{n-1}} \underbrace{\frac{d}{dx}[3x - 2x^2]}_{u'}$$

$$= 3(3x - 2x^2)^2(3 - 4x) = (9 - 12x)(3x - 2x^2)^2.$$

In Section 3.3 we claimed that the Power Rule is valid for *any* rational number n. At this point, however, we have given proofs for only two cases—positive integer powers (Section 3.3) and negative integer powers (Section 3.4). Using the Chain Rule, we can now prove the case for rational powers.

EXAMPLE 5 The Power Rule for rational exponents

Use the Chain Rule to show that for a rational number n,

$$\frac{d}{dx}[x^n] = nx^{n-1}.$$

SOLUTION

Let n be a rational number with denominator m. Then nm is an integer. We let $u = x^n$ and consider the equation

$$(x^n)^m = x^{nm}, \quad \text{where } m \text{ and } nm \text{ are integers.}$$

Using the Chain Rule on the left and the Power Rule (for integers) on the right, we differentiate both sides of this equation to obtain

$$\overbrace{m}^{m}\overbrace{(x^n)^{m-1}}^{u^{m-1}}\underbrace{\frac{d}{dx}[x^n]}_{u'} = nmx^{nm-1}$$

$$\frac{d}{dx}[x^n] = nx^{nm-1}x^{-nm+n} = nx^{n-1}.$$

EXAMPLE 6 Differentiating functions involving radicals

Find the derivative of $y = \sqrt[3]{(x^2 + 2)^2}$.

SOLUTION

We rewrite the function as $y = (x^2 + 2)^{2/3}$. Then, by the Power Rule (with $u = x^2 + 2$), we have

$$y' = \overbrace{\frac{2}{3}}^{n}\overbrace{(x^2 + 2)^{-1/3}}^{u^{n-1}}\overbrace{(2x)}^{u'} = \frac{4x}{3\sqrt[3]{x^2 + 2}}.$$

EXAMPLE 7 Differentiating quotients with constant numerators

Differentiate

$$g(t) = \frac{-7}{(2t-3)^2}.$$

SOLUTION

If we rewrite the function as $g(t) = -7(2t-3)^{-2}$, then the General Power Rule produces

$$g'(t) = \underbrace{(-7)}_{\substack{\text{Constant} \\ \text{Multiple Rule}}}\overbrace{(-2)}^{n}\overbrace{(2t-3)^{-3}}^{u^{n-1}}\overbrace{(2)}^{u'} = 28(2t-3)^{-3} = \frac{28}{(2t-3)^3}.$$

Simplifying derivatives

The next three examples illustrate some useful techniques for simplifying the "raw derivatives" of functions involving products, quotients, and composites.

EXAMPLE 8 Simplification by factoring out the least powers

Differentiate $f(x) = x^2\sqrt{1-x^2}$.

SOLUTION

$$f(x) = x^2(1-x^2)^{1/2} \qquad \text{Rewrite}$$

$$f'(x) = x^2\frac{d}{dx}[(1-x^2)^{1/2}] + (1-x^2)^{1/2}\frac{d}{dx}[x^2] \qquad \text{Product Rule}$$

$$= x^2\left[\frac{1}{2}(1-x^2)^{-1/2}(-2x)\right] + (1-x^2)^{1/2}(2x) \qquad \text{Power Rule}$$

$$= -x^3(1-x^2)^{-1/2} + 2x(1-x^2)^{1/2}$$

$$= x(1-x^2)^{-1/2}[-x^2(1) + 2(1-x^2)] \qquad \text{Factor}$$

$$= \frac{x(2-3x^2)}{\sqrt{1-x^2}} \qquad \text{Simplify}$$

EXAMPLE 9 Simplifying the derivative of a quotient

Differentiate

$$f(x) = \frac{x}{\sqrt[3]{x^2+4}}.$$

SOLUTION

$$f(x) = \frac{x}{(x^2 + 4)^{1/3}} \qquad \text{Rewrite}$$

$$f(x) = \frac{(x^2 + 4)^{1/3}(1) - x(1/3)(x^2 + 4)^{-2/3}(2x)}{(x^2 + 4)^{2/3}} \qquad \text{Quotient Rule}$$

$$= \frac{1}{3}(x^2 + 4)^{-2/3}\left[\frac{3(x^2 + 4) - (2x^2)(1)}{(x^2 + 4)^{2/3}}\right] \qquad \text{Factor}$$

$$= \frac{x^2 + 12}{3(x^2 + 4)^{4/3}} \qquad \text{Simplify} \qquad \blacksquare$$

EXAMPLE 10   Differentiating a quotient raised to a power

Differentiate

$$y = \left(\frac{3x - 1}{x^2 + 3}\right)^2.$$

SOLUTION

$$\frac{dy}{dx} = 2\overbrace{\left(\frac{3x - 1}{x^2 + 3}\right)}^{u^{n-1}}\overbrace{\frac{d}{dx}\left[\frac{3x - 1}{x^2 + 3}\right]}^{u'}$$

$$= \left[\frac{2(3x - 1)}{x^2 + 3}\right]\left[\frac{(x^2 + 3)(3) - (3x - 1)(2x)}{(x^2 + 3)^2}\right]$$

$$= \frac{2(3x - 1)(3x^2 + 9 - 6x^2 + 2x)}{(x^2 + 3)^3}$$

$$= \frac{2(3x - 1)(-3x^2 + 2x + 9)}{(x^2 + 3)^3}$$

Try finding y' using the Quotient Rule on $y = (3x - 1)^2/(x^2 + 3)^2$ and compare the results. $\blacksquare$

Trigonometric functions and the Chain Rule

We summarize the "Chain Rule versions" of the derivatives of the six trigonometric functions as follows.

DERIVATIVES OF TRIGONOMETRIC FUNCTIONS	$\dfrac{d}{dx}[\sin u] = \cos u \dfrac{du}{dx}$	$\dfrac{d}{dx}[\cos u] = -\sin u \dfrac{du}{dx}$
	$\dfrac{d}{dx}[\tan u] = \sec^2 u \dfrac{du}{dx}$	$\dfrac{d}{dx}[\cot u] = -\csc^2 u \dfrac{du}{dx}$
	$\dfrac{d}{dx}[\sec u] = \sec u \tan u \dfrac{du}{dx}$	$\dfrac{d}{dx}[\csc u] = -\csc u \cot u \dfrac{du}{dx}$

EXAMPLE 11 Applying the Chain Rule to trigonometric functions

Function	Derivative

(a) $y = \sin 2x$ $\quad y' = \overbrace{\cos 2x}^{\cos u}\overbrace{\dfrac{d}{dx}[2x]}^{u'} = (\cos 2x)(2) = 2\cos 2x$

where u is under $\sin 2x$

(b) $y = \cos(x - 1)$ $\quad y' = -\sin(x - 1)$

(c) $y = \tan 3x$ $\quad y' = 3\sec^2 3x$

Be sure that you understand the mathematical conventions regarding parentheses and trigonometric functions. For instance, in part (a) of Example 11, we write $\sin 2x$ to mean $\sin(2x)$. The next example shows the effect of different placements of parentheses.

EXAMPLE 12 Parentheses and trigonometric functions

Function	Derivative

(a) $y = \cos 3x^2 = \cos(3x^2)$ $\quad y' = (-\sin 3x^2)(6x) = -6x \sin 3x^2$

(b) $y = (\cos 3)x^2$ $\quad y' = (\cos 3)(2x) = 2x \cos 3$

(c) $y = \cos(3x)^2 = \cos(9x^2)$ $\quad y' = (-\sin 9x^2)(18x) = -18x \sin 9x^2$

(d) $y = \cos^2 3x = (\cos 3x)^2$ $\quad y' = (2\cos 3x)D_x[\cos 3x]$
$\qquad\qquad\qquad\qquad\qquad\quad = 2(\cos 3x)(-\sin 3x)(3)$
$\qquad\qquad\qquad\qquad\qquad\quad = -6\cos 3x \sin 3x$

EXAMPLE 13 Differentiating a composite function

Differentiate $f(t) = \sqrt{\sin 4t}$.

SOLUTION

By the Power Rule, we have

$$f(t) = (\sin 4t)^{1/2}$$

$$f'(t) = \frac{1}{2}(\sin 4t)^{-1/2}\frac{d}{dt}[\sin 4t]$$

$$= \frac{1}{2}(\sin 4t)^{-1/2}(4 \cos 4t) = \frac{2\cos 4t}{\sqrt{\sin 4t}}.$$

EXERCISES for Section 3.5

In Exercises 1–6, complete the table using Example 2 as a model.

$y = f(g(x))$	$u = g(x)$	$y = f(u)$

1. $y = (6x - 5)^4$

2. $y = \dfrac{1}{\sqrt{x+1}}$

3. $y = \sqrt{x^2 - 1}$

4. $y = \tan(\pi x + 1)$

5. $y = \csc^3 x$

6. $y = \cos \dfrac{3x}{2}$

In Exercises 7–44, find the derivative.

7. $y = (2x - 7)^3$

8. $y = (3x^2 + 1)^4$

9. $g(x) = 3(9x - 4)^4$

10. $f(x) = 2(x^2 - 1)^3$

11. $y = \dfrac{1}{x-2}$

12. $s(t) = \dfrac{1}{t^2 + 3t - 1}$

13. $f(t) = \left(\dfrac{1}{t-3}\right)^2$

14. $y = -\dfrac{4}{(t+2)^2}$

15. $f(x) = \dfrac{3}{x^3 - 4}$

16. $f(x) = \dfrac{1}{(x^2 - 3x)^2}$

17. $f(x) = x^2(x-2)^4$

18. $f(x) = x(3x - 9)^3$

19. $f(t) = \sqrt{1-t}$

20. $g(x) = \sqrt{3 - 2x}$

21. $s(t) = \sqrt{t^2 + 2t - 1}$

22. $y = \sqrt[3]{3x^3 + 4x}$

23. $y = \sqrt[3]{9x^2 + 4}$

24. $g(x) = \sqrt{x^2 - 2x + 1}$

25. $y = 2\sqrt{4 - x^2}$

26. $f(x) = -3\sqrt[4]{2 - 9x}$

27. $f(x) = (9 - x^2)^{2/3}$

28. $f(t) = (9t + 2)^{2/3}$

29. $y = \dfrac{1}{\sqrt{x+2}}$

30. $g(t) = \sqrt{\dfrac{1}{t^2 - 2}}$

31. $y = x\sqrt{1 - x^2}$

32. $y = x^2\sqrt{9 - x^2}$

33. $y = \dfrac{x}{\sqrt{x^2 + 1}}$

34. $y = \dfrac{x^2}{\sqrt{x^2 + 9}}$

35. $y = \dfrac{\sqrt{x+1}}{x^2 + 1}$

36. $f(x) = \dfrac{x+1}{2x - 3}$

37. $f(t) = \dfrac{3t + 2}{t - 1}$

38. $y = \sqrt{\dfrac{2x}{x+1}}$

39. $g(t) = \dfrac{3t^2}{\sqrt{t^2 + 2t - 1}}$

40. $f(x) = \sqrt{x}(2 - x)^2$

41. $y = \sqrt{\dfrac{x+1}{x}}$

42. $y = (t^2 - 9)\sqrt{t + 2}$

43. $s(t) = \dfrac{-2(2 - t)\sqrt{1 + t}}{3}$

44. $g(x) = \sqrt{x - 1} + \sqrt{x + 1}$

In Exercises 45–48, find an equation of the tangent line to the graph of f at the given point.

Function	*Point*
45. $f(x) = \sqrt{3x^2 - 2}$	$(3, 5)$
46. $f(x) = x\sqrt{x^2 + 5}$	$(2, 6)$
47. $f(x) = \sin 2x$	$(\pi, 0)$
48. $f(x) = \tan^2 x$	$\left(\dfrac{\pi}{4}, 1\right)$

In Exercises 49–52, find the second derivative of the given function.

49. $f(x) = 2(x^2 - 1)^3$

50. $f(x) = \dfrac{1}{x - 2}$

51. $f(x) = \sin x^2$

52. $f(t) = \dfrac{\sqrt{t^2 + 1}}{t}$

In Exercises 53 and 54, find the slope of the tangent line to the given sine function at the origin. Compare this value to the number of complete cycles in the interval $[0, 2\pi]$.

53. (a) $y = \sin x$

(b) $y = \sin 2x$

54. (a) $y = \sin 3x$

(b) $y = \sin \dfrac{x}{2}$

In Exercises 55–68, find dy/dx.

55. $y = \cos 3x$

56. $y = \sin \pi x$

57. $y = 3 \tan 4x$

58. $y = \sec(x^2)$

59. $y = \dfrac{1}{4}\sin^2 2x$

60. $y = 5\cos^2 \pi x$

61. $y = \dfrac{1}{4}\sin(2x)^2$

62. $y = 5\cos(\pi x)^2$

63. $y = \sec^3(2x)$

64. $y = \sqrt{\sin x}$

65. $y = \cot\left(\pi x - \dfrac{\pi}{2}\right)$

66. $y = \csc\left(\dfrac{x}{2} + \dfrac{\pi}{4}\right)$

67. $y = \dfrac{\cos x + 1}{x}$

68. $y = x^2 \tan \dfrac{1}{x}$

69. Let u be a differentiable function of x. Use the fact that $|u| = \sqrt{u^2}$ to prove that

$$\frac{d}{dx}[|u|] = u'\frac{u}{|u|}, \quad u \neq 0.$$

In Exercises 70–72, use the result of Exercise 69 to find the derivative of the given function.

70. $f(x) = |x^2 - 4|$
71. $f(x) = |x^3 + x|$
72. $f(x) = |\sin x|$

73. *(Doppler effect)*   The frequency F of a fire truck siren heard by a stationary observer is given by

$$F = \frac{132,400}{331 \pm v}$$

where $\pm v$ represents the velocity of the accelerating fire truck (see figure). Find the rate of change of F with respect to v when

(a) the fire truck is approaching at a velocity of 30 m/s [use $-v$], and then when

(b) the fire truck is moving away at a velocity of 30 m/s [use $+v$].

Doppler Effect

FIGURE FOR 73

74. The displacement from equilibrium of an object in harmonic motion on the end of a spring is given by

$$y = \frac{1}{3}\cos 12t - \frac{1}{4}\sin 12t$$

where y is measured in feet and t is time in seconds. Determine the position and velocity of the object when $t = \pi/8$.

75. A 6-inch pendulum moves according to the equation

$$\theta = 0.2 \cos 8t$$

where θ is the angular displacement from the vertical in radians and t is time in seconds. Determine the maximum angular displacement and the rate of change of θ when $t = 3$ seconds.

76. A buoy oscillates in simple harmonic motion ($y = A \cos \omega t$) as waves move past. At a given time, it is noted that the buoy moves a total of 3.5 feet from its low point to its high point, and that it returns to the high point every 10 seconds.

(a) Write an equation describing the motion of the buoy if at $t = 0$ it is at its high point.

(b) Determine the velocity of the buoy as a function of t.

77. The speed S of blood that is r centimeters from the center of an artery is given by $S = C(R^2 - r^2)$, where C is a constant, R is the radius of the artery, and S is measured in centimeters per second. Suppose a drug is administered and the artery begins dilating at the rate of dR/dt. At a constant distance r, find the rate at which S changes with respect to t for $C = 1.76 \times 10^5$, $R = 1.2 \times 10^{-2}$, and $dR/dt = 10^{-5}$.

78. For $f(x) = \sec^2 x$ and $g(x) = \tan^2 x$, show that $f'(x) = g'(x)$.

3.6 Implicit Differentiation

Implicit and explicit functions ▪ Implicit differentiation

So far, our equations involving two variables were generally expressed in the **explicit form** $y = f(x)$. That is, one of the two variables was explicitly given in terms of the other. For example,

$$y = 3x - 5, \qquad s = -16t^2 + 20t, \qquad u = 3w - w^2$$

all are written in explicit form, and we say that y, s, and u are functions of x, t, and w, respectively.

However, many functions are not given explicitly and are implied only by a given equation. For instance, the function $y = 1/x$ is defined **implicitly** by the equation $xy = 1$. Suppose that you were asked to find dy/dx in this equation. As it turns out, this is a relatively simple task, and you probably would begin by solving the equation for y.

Implicit form	*Explicit form*	*Derivative*
$xy = 1$	$y = \dfrac{1}{x} = x^{-1}$	$\dfrac{dy}{dx} = -x^{-2} = -\dfrac{1}{x^2}$

This procedure works well whenever we can easily solve for the function explicitly. However, we cannot use this procedure in cases in which we are unable to solve for y as a function of x. For instance, how would you find dy/dx in the equation

$$x^2 - 2y^3 + 4y = 2$$

where it is very difficult to express y as a function of x explicitly? To do this, we use a procedure called **implicit differentiation,** in which we assume y is a differentiable function of x.

To understand how to find dy/dx implicitly you must realize that the differentiation is taking place *with respect to x*. This means that when we differentiate terms involving x alone, we can differentiate as usual. *But* when we differentiate terms involving y, we must apply the Chain Rule because we are assuming that y is defined implicitly as a function of x. Study the next example carefully. Note in particular how the Chain Rule is used to introduce the dy/dx terms.

EXAMPLE 1 Applying the Chain Rule

Differentiate the following with respect to x.

(a) $3x^2$ (b) $2y^3$ (c) $x + 3y$ (d) xy^2

SOLUTION

(a) $\dfrac{d}{dx}[3x^2] = 6x$

(b) $\dfrac{d}{dx}[\overbrace{2y^3}^{u^n}] = \overbrace{2(3)y^2}^{nu^{n-1}}\overbrace{\dfrac{dy}{dx}}^{u'} = 6y^2\dfrac{dy}{dx}$ Chain Rule

(c) $\dfrac{d}{dx}[x + 3y] = 1 + 3\dfrac{dy}{dx}$

(d) $\dfrac{d}{dx}[xy^2] = x\dfrac{d}{dx}[y^2] + y^2\dfrac{d}{dx}[x]$ Product Rule

$\qquad\qquad = x\left(2y\dfrac{dy}{dx}\right) + y^2(1)$ Chain Rule

$\qquad\qquad = 2xy\dfrac{dy}{dx} + y^2$

For equations involving x and y, we suggest the following procedure for finding dy/dx implicitly.

IMPLICIT DIFFERENTIATION

Given an equation involving x and y, and assuming y is a differentiable function of x, we can find dy/dx as follows.

1. Differentiate both sides of the equation *with respect to x.*
2. Collect all terms involving dy/dx on the left side of the equation and move all other terms to the right side of the equation.
3. Factor dy/dx out of the left side of the equation.
4. Solve for dy/dx by dividing both sides of the equation by the left-hand factor that does not contain dy/dx.

EXAMPLE 2   Implicit differentiation

Find dy/dx given that $y^3 + y^2 - 5y - x^2 = -4$.

SOLUTION

1. Differentiate both sides of the equation with respect to x.

$$\frac{d}{dx}[y^3 + y^2 - 5y - x^2] = \frac{d}{dx}[-4]$$

$$\frac{d}{dx}[y^3] + \frac{d}{dx}[y^2] - \frac{d}{dx}[5y] - \frac{d}{dx}[x^2] = \frac{d}{dx}[-4]$$

$$3y^2\frac{dy}{dx} + 2y\frac{dy}{dx} - 5\frac{dy}{dx} - 2x = 0$$

2. Collect the dy/dx terms on the left side of the equation.

$$3y^2\frac{dy}{dx} + 2y\frac{dy}{dx} - 5\frac{dy}{dx} = 2x$$

3. Factor dy/dx out of the left side of the equation.

$$\frac{dy}{dx}(3y^2 + 2y - 5) = 2x$$

4. Solve for dy/dx by dividing by $(3y^2 + 2y - 5)$.

$$\frac{dy}{dx} = \frac{2x}{3y^2 + 2y - 5}$$

◻

REMARK   In Example 2, note that implicit differentiation can produce an expression for dy/dx that contains both x and y.

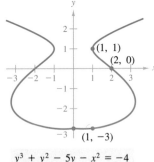

$y^3 + y^2 - 5y - x^2 = -4$

FIGURE 3.21

To see how we can use an *implicit derivative*, consider the graph of the equation $y^3 + y^2 - 5y - x^2 = -4$ as shown in Figure 3.21. The derivative found in Example 2 gives us a formula for the slope of the tangent line at a point on this graph. The slopes at several points on the graph are as follows.

Point on Graph	Slope of Graph
$(2, 0)$	$-\dfrac{4}{5}$
$(1, -3)$	$\dfrac{1}{8}$
$x = 0$	0
$(1, 1)$	undefined

When using implicit differentiation, keep in mind that it is meaningless to solve for dy/dx in an equation that has no solution points. (For example, $x^2 + y^2 = -4$ has no solution points.) If a segment of a graph can be represented by a differentiable function, then dy/dx will have meaning as the slope at each point on the segment. Recall that a function is not differentiable at (1) points with vertical tangents and (2) points of discontinuity. The following example illustrates this idea.

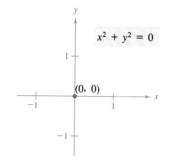

(a)

EXAMPLE 3   Representing a graph by differentiable functions

If possible, represent the graphs of the following equations by differentiable functions. (See Figure 3.22.)

(a) $x^2 + y^2 = 0$ (b) $x^2 + y^2 = 1$ (c) $x + y^2 = 1$

SOLUTION

(a) The graph of this equation is a single point. Therefore, it does not define y as a differentiable function of x.

(b) The graph of this equation is the unit circle, centered at $(0, 0)$. The upper semi-circle is given by the differentiable function

$$y = \sqrt{1 - x^2}, \quad -1 < x < 1$$

and the lower semi-circle is given by

$$y = -\sqrt{1 - x^2}, \quad -1 < x < 1.$$

At the point $(-1, 0)$ and $(1, 0)$, the slope of the graph is undefined.

(c) The upper half of this parabola is given by the differentiable function

$$y = \sqrt{1 - x}, \quad x < 1$$

and the lower half is given by

$$y = -\sqrt{1 - x}, \quad x < 1.$$

At the point $(1, 0)$, the slope of the graph is undefined.

(b)

(c)

FIGURE 3.22

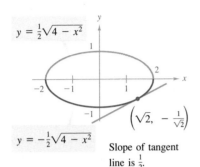

$y = \frac{1}{2}\sqrt{4 - x^2}$

$\left(\sqrt{2}, -\frac{1}{\sqrt{2}}\right)$

$y = -\frac{1}{2}\sqrt{4 - x^2}$

Slope of tangent line is $\frac{1}{2}$.

Ellipse: $x^2 + 4y^2 = 4$

FIGURE 3.23

EXAMPLE 4   Finding the slope of a curve implicitly

Determine the slope of the tangent line to the graph of $x^2 + 4y^2 = 4$ at the point $(\sqrt{2}, -1/\sqrt{2})$. (See Figure 3.23.)

SOLUTION

Implicit differentiation of the equation $x^2 + 4y^2 = 4$ with respect to x yields

$$2x + 8y\frac{dy}{dx} = 0$$

$$\frac{dy}{dx} = \frac{-2x}{8y} = \frac{-x}{4y}.$$

Therefore, at $(\sqrt{2}, -1/\sqrt{2})$, the slope is

$$\frac{dy}{dx} = \frac{-\sqrt{2}}{-4/\sqrt{2}} = \frac{1}{2}.$$

REMARK   To see the benefit of implicit differentiation, try doing Example 4 using the explicit function $y = -\frac{1}{2}\sqrt{4 - x^2}$. The graph of this function is the lower half of the ellipse.

EXAMPLE 5   Finding the slope of a curve implicitly

Determine the slope of the graph of $3(x^2 + y^2)^2 = 100xy$ at the point $(3, 1)$.

SOLUTION

$$\frac{d}{dx}[3(x^2 + y^2)^2] = \frac{d}{dx}[100xy]$$

$$3(2)(x^2 + y^2)\left(2x + 2y\frac{dy}{dx}\right) = 100\left[x\frac{dy}{dx} + y(1)\right]$$

$$12y(x^2 + y^2)\frac{dy}{dx} - 100x\frac{dy}{dx} = 100y - 12x(x^2 + y^2)$$

$$[12y(x^2 + y^2) - 100x]\frac{dy}{dx} = 100y - 12x(x^2 + y^2)$$

$$\frac{dy}{dx} = \frac{100y - 12x(x^2 + y^2)}{-100x + 12y(x^2 + y^2)} = \frac{25y - 3x(x^2 + y^2)}{-25x + 3y(x^2 + y^2)}$$

Now, at the point $(3, 1)$ the slope of the graph is

$$\frac{dy}{dx} = \frac{25(1) - 3(3)(3^2 + 1^2)}{-25(3) + 3(1)(3^2 + 1^2)} = \frac{25 - 90}{-75 + 30} = \frac{-65}{-45} = \frac{13}{9}$$

as shown in Figure 3.24. This graph is called a **lemniscate.**

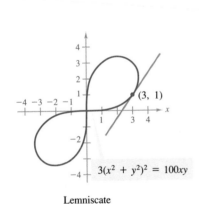

$3(x^2 + y^2)^2 = 100xy$

(3, 1)

Lemniscate

FIGURE 3.24

EXAMPLE 6   Implicit differentiation of trigonometric functions

Find dy/dx implicitly for the equation $\sin y = x$, and find the largest interval of the form $-a < y < a$ such that y is a differentiable function of x. (See Figure 3.25.)

SOLUTION

$$\frac{d}{dx}[\sin y] = \frac{d}{dx}[x]$$

$$\cos y \frac{dy}{dx} = 1$$

$$\frac{dy}{dx} = \frac{1}{\cos y}$$

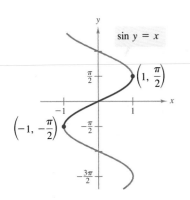

The largest interval about the origin for which y is a differentiable function of x is $-\pi/2 < y < \pi/2$. To see this, we note that $\cos y$ is positive for all y in this interval and zero at the endpoints. If we restrict y to the interval $-\pi/2 < y < \pi/2$, we should be able to write dy/dx explicitly as a function of x. To do this, we use the fact that $\sin y = x$, together with the Pythagorean Identity $\cos^2 y + \sin^2 y = 1$.

$$\cos y = \sqrt{1 - \sin^2 y} = \sqrt{1 - x^2}, \quad -\frac{\pi}{2} < y < \frac{\pi}{2}$$

Therefore,

$$\frac{dy}{dx} = \frac{1}{\sqrt{1 - x^2}}.$$

FIGURE 3.25

In implicit differentiation, the form of the derivative often can be simplified or changed to a single variable (see Example 6) by an appropriate use of the *original* equation. A similar technique can be used to find and simplify higher-order derivatives obtained implicitly.

EXAMPLE 7   Finding the second derivative implicitly

Given $x^2 + y^2 = 25$, find y''.

SOLUTION

Differentiating each term with respect to x, we obtain

$$2x + 2yy' = 0$$
$$2yy' = -2x$$
$$y' = \frac{-2x}{2y} = -\frac{x}{y}.$$

Differentiating a second time with respect to x yields

$$y'' = -\frac{(y)(1) - (x)(y')}{y^2}$$

$$= -\frac{y - (x)(-x/y)}{y^2} \qquad \text{Substitute } y' = -x/y$$

$$= -\frac{y^2 + x^2}{y^3} = -\frac{25}{y^3}. \qquad \text{Substitute } x^2 + y^2 = 25$$

EXAMPLE 8   Finding a tangent line to a graph

Find the tangent line to the **kappa curve** given by

$$x^2(x^2 + y^2) = y^2$$

at the point $(\sqrt{2}/2, \sqrt{2}/2)$ as shown in Figure 3.26.

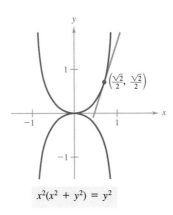

$$x^2(x^2 + y^2) = y^2$$

FIGURE 3.26

SOLUTION

By rewriting and differentiating implicitly, we have

$$x^4 + x^2y^2 - y^2 = 0$$
$$4x^3 + x^2(2yy') + 2xy^2 - 2yy' = 0$$
$$2y(x^2 - 1)y' = -2x(2x^2 + y^2)$$
$$y' = \frac{x(2x^2 + y^2)}{y(1 - x^2)}.$$

Now, at the point $(\sqrt{2}/2, \sqrt{2}/2)$, the slope is

$$m = \frac{(\sqrt{2}/2)[2(1/2) + (1/2)]}{(\sqrt{2}/2)[1 - (1/2)]} = \frac{3/2}{1/2} = 3$$

and the equation of the tangent line at this point is

$$y - \frac{\sqrt{2}}{2} = 3\left(x - \frac{\sqrt{2}}{2}\right)$$
$$y = 3x - \sqrt{2}.$$

Isaac Barrow

The graph in Example 8 is called the *kappa curve* because it resembles the Greek letter kappa. The general solution for the tangent line to this particular curve was first discovered by the English mathematician Isaac Barrow (1630–1677). Barrow and Newton were contemporaries and corresponded frequently regarding their work in the early development of calculus.

EXERCISES for Section 3.6

In Exercises 1–24, find dy/dx by implicit differentiation and evaluate the derivative at the indicated point.

Equation	Point
1. $x^2 + y^2 = 16$	$(3, \sqrt{7})$
2. $x^2 - y^2 = 16$	$(4, 0)$
3. $xy = 4$	$(-4, -1)$
4. $x^2 - y^3 = 0$	$(1, 1)$
5. $x^{1/2} + y^{1/2} = 9$	$(16, 25)$
6. $x^3 + y^3 = 8$	$(0, 2)$
7. $x^3 - xy + y^2 = 4$	$(0, -2)$
8. $x^2y + y^2x = -2$	$(2, -1)$
9. $y^2 = \dfrac{x^2 - 9}{x^2 + 9}$	$(3, 0)$
10. $(x + y)^3 = x^3 + y^3$	$(-1, 1)$
11. $x^3y^3 - y = x$	$(0, 0)$
12. $\sqrt{xy} = x - 2y$	$(4, 1)$
13. $x^{2/3} + y^{2/3} = 5$	$(8, 1)$
14. $x^3 + y^3 = 2xy$	$(1, 1)$
15. $x^3 - 2x^2y + 3xy^2 = 38$	$(2, 3)$
16. $2 \sin x \cos y = 1$	$\left(\dfrac{\pi}{4}, \dfrac{\pi}{4}\right)$
17. $\sin x + \cos 2y = 2$	$\left(\dfrac{\pi}{2}, 0\right)$
18. $(\sin \pi x + \cos \pi y)^2 = 2$	$\left(\dfrac{1}{4}, \dfrac{1}{4}\right)$
19. $\sin x = x(1 + \tan y)$	$\left(\pi, \dfrac{3\pi}{4}\right)$
20. $\cot y = x - y$	$\left(\dfrac{\pi}{2}, \dfrac{\pi}{2}\right)$
21. $\tan (x + y) = x$	$(0, 0)$
22. $x \cos y = 1$	$\left(2, \dfrac{\pi}{3}\right)$
23. $y = \sin (xy)$	$\left(\dfrac{\pi}{2}, 1\right)$
24. $x = \sec \dfrac{1}{y}$	$\left(\sqrt{2}, \dfrac{4}{\pi}\right)$

In Exercises 25–28, find the slope of the tangent line to the graph at the indicated point.

25. Witch of Agnesi:
$(x^2 + 4)y = 8$
Point: $(2, 1)$

26. Cissoid:
$(4 - x)y^2 = x^3$
Point: $(2, 2)$

27. Bifolium:
$(x^2 + y^2)^2 = 4x^2y$
Point: $(1, 1)$

28. Folium of Descartes:
$x^3 + y^3 - 6xy = 0$
Point: $\left(\dfrac{4}{3}, \dfrac{8}{3}\right)$

In Exercises 29–32, sketch the graph of the given equation. Then find dy/dx implicitly and explicitly and show that the two results are equivalent.

29. $x^2 + y^2 = 16$
30. $x^2 + y^2 - 4x + 6y + 9 = 0$
31. $9x^2 + 16y^2 = 144$
32. $4y^2 - x^2 = 4$

In Exercises 33–38, find d^2y/dx^2 in terms of x and y.

33. $x^2 + xy = 5$ **34.** $x^2y^2 - 2x = 3$
35. $x^2 - y^2 = 16$ **36.** $1 - xy = x - y$
37. $y^2 = x^3$ **38.** $y^2 = 4x$

In Exercises 39 and 40, find equations for the tangent line and normal line to the given circle at the indicated points. (The **normal line** at a point is perpendicular to the tangent line at the point.)

39. $x^2 + y^2 = 25$, $(4, 3)$ and $(-3, 4)$
40. $x^2 + y^2 = 9$, $(0, 3)$ and $(2, \sqrt{5})$

In Exercises 41 and 42, find the points at which the graph of the given equation has a vertical or horizontal tangent line.

41. $25x^2 + 16y^2 + 200x - 160y + 400 = 0$
42. $4x^2 + y^2 - 8x + 4y + 4 = 0$

In Exercises 43–46, sketch the intersecting graphs of the given equations and show that they are **orthogonal**. [Two graphs are orthogonal if at their point(s) of intersection, their tangent lines are perpendicular to each other.]

43. $2x^2 + y^2 = 6$ and $y^2 = 4x$
44. $y^2 = x^3$ and $2x^2 + 3y^2 = 5$
45. $x + y = 0$ and $x = \sin y$
46. $x^3 = 3(y - 1)$ and $x(3y - 29) = 3$

47. Two circles of radius 4 are tangent to the graph of $y^2 = 4x$ at the point $(1, 2)$. Find the equations for these two circles.

48. Show that the normal line (the line perpendicular to the tangent line) at any point on the circle $x^2 + y^2 = r^2$ passes through the origin.

49. Use Example 6 as a model to find dy/dx implicitly for the equation $\tan y = x$ and find the largest interval of the form $-a < y < a$ such that y is a differentiable function of x. Then express dy/dx as a function of x.

50. Show that

$$\frac{dy}{dx} = \frac{-(x - h)}{y - k} \text{ for the circle } (x - h)^2 + (y - k)^2 = r^2.$$

 In Exercises 51 and 52, use a computer to sketch the graph of the equation. Find the equation of the tangent line to the graph at the specified point and sketch its graph.

Equation	Point
51. $\sqrt{x} + \sqrt{y} = 3$	$(4, 1)$
52. $x^3 + y^3 - 6xy = 0$	$\left(\dfrac{4}{3}, \dfrac{8}{3}\right)$

3.7 Related Rates

Related rate problems

Volume is related to radius and height.

FIGURE 3.27

We have seen how the Chain Rule can be used to find dy/dx implicitly. Another important use of the Chain Rule is to find the rates of change of two or more related variables that are changing with respect to time.

For example, when water is drained out of a conical tank (see Figure 3.27), the volume V, the radius r, and the height h of the water level are all functions of time t. Knowing that these variables are related by the equation

$$V = \frac{\pi}{3}r^2h$$

we can differentiate implicitly with respect to t to obtain the **related rate** equation

$$\frac{dV}{dt} = \frac{\pi}{3}\left[r^2\frac{dh}{dt} + h\left(2r\frac{dr}{dt}\right)\right] = \frac{\pi}{3}\left[r^2\frac{dh}{dt} + 2rh\frac{dr}{dt}\right].$$

From this equation we see that the rate of change of V is related to the rates of change of both h and r. This is further demonstrated in Example 1.

EXAMPLE 1 Two rates that are related

Suppose x and y are both differentiable functions of t and are related by the equation $y = x^2 + 3$. Find dy/dt when $x = 1$, given that $dx/dt = 2$ when $x = 1$.

SOLUTION

Using the Chain Rule, we can differentiate both sides of the equation *with respect to t*.

$$y = x^2 + 3 \qquad \text{Given equation}$$

$$\frac{d}{dt}[y] = \frac{d}{dt}[x^2 + 3] \qquad \text{Differentiate with respect to } t$$

$$\frac{dy}{dt} = 2x\frac{dx}{dt} \qquad \text{Chain Rule}$$

Now, when $x = 1$ and $dx/dt = 2$, we have

$$\frac{dy}{dt} = 2(1)(2) = 4.$$

In Example 1, we are *given* the following mathematical model.

Given equation: $y = x^2 + 3$

Given rate: $\dfrac{dx}{dt} = 2$ when $x = 1$

Find: $\dfrac{dy}{dt}$ when $x = 1$

Now we look at an example in which we must *create* the model from a verbal description.

EXAMPLE 2   An application involving related rates

A pebble is dropped into a calm pond, causing ripples in the form of concentric circles, as shown in Figure 3.28. The radius r of the outer ripple is increasing at a constant rate of 1 foot per second. When this radius is 4 feet, at what rate is the total area A of the disturbed water increasing?

SOLUTION

The variables r and A are related by the equation for the area of a circle, $A = \pi r^2$. To solve this problem, we must remember that the rate of change of the radius r is given by its derivative, dr/dt. Thus, the problem can be summarized by the following model.

Given equation: $A = \pi r^2$

Given rate: $\dfrac{dr}{dt} = 1$ when $r = 4$

Find: $\dfrac{dA}{dt}$ when $r = 4$

With this information, we proceed as in Example 1.

$$\frac{d}{dt}[A] = \frac{d}{dt}[\pi r^2]$$

$$\frac{dA}{dt} = 2\pi r \frac{dr}{dt}$$

Thus, when $r = 4$, we have

$$\frac{dA}{dt} = 2\pi(4)(1) = 8\pi \text{ ft}^2/\text{sec.}$$

FIGURE 3.28

Following the procedure shown in Example 2, we suggest the following four solution steps for related rate problems.

PROCEDURE FOR SOLVING RELATED RATE PROBLEMS

1. Assign symbols to all *given* quantities and *quantities to be determined*. Make a sketch and label the quantities if feasible.
2. Write an equation involving the variables whose rates of change either are given or are to be determined.
3. Using the Chain Rule, implicitly differentiate both sides of the equation *with respect to time t.*
4. Substitute into the resulting equation all known values for the variables and their rates of change. Then solve for the required rate of change.

REMARK Be sure to complete Step 3 *before* starting Step 4. A common error is to introduce specific values for rates that are variable too early in the solution process.

Table 3.6 shows the mathematical models for some common rates of change which can be used in the first step of the solution to related rate problems.

TABLE 3.6

Verbal statement	*Mathematical model*
The velocity of a car after traveling one hour is 50 miles per hour.	x = distance traveled $\dfrac{dx}{dt} = 50$ when $t = 1$
Water is being pumped into a swimming pool at the rate of 10 cubic feet per minute.	V = volume of water in pool $\dfrac{dV}{dt} = 10$ ft³/min
A gear is revolving at the rate of 25 revolutions per minute (1 rev = 2π rad).	θ = angle of revolution $\dfrac{d\theta}{dt} = 25(2\pi)$ rad/min

EXAMPLE 3 An inflating balloon

Air is being pumped into a spherical balloon at the rate of 4.5 cubic inches per minute, as shown in Figure 3.29. Find the rate of change of the radius when the radius is 2 inches.

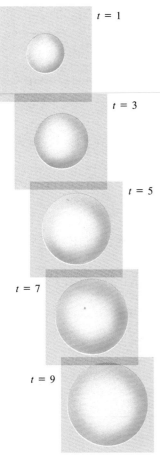

t = 1

t = 3

t = 5

t = 7

t = 9

Expanding Balloon

FIGURE 3.29

SOLUTION

1. We let V be the volume of the balloon and let r be its radius. Since the volume is increasing at the rate of 4.5 in³/min, we know that at time t the rate of change of the volume is $dV/dt = \frac{9}{2}$. Thus, the problem can be stated as follows.

Given: $\dfrac{dV}{dt} = \dfrac{9}{2}$ (constant rate)

Find: $\dfrac{dr}{dt}$ when $r = 2$

2. To find the rate of change of the radius, we must find an equation that relates the radius r to the volume V. This is given by the formula for the volume of a sphere.

$$V = \frac{4}{3}\pi r^3$$

3. Now we implicitly differentiate *with respect to t* to obtain

$$\frac{dV}{dt} = 4\pi r^2 \frac{dr}{dt}$$

$$\frac{dr}{dt} = \frac{1}{4\pi r^2}\left(\frac{dV}{dt}\right).$$

4. Finally, when $r = 2$, the rate of change of the radius is

$$\frac{dr}{dt} = \frac{1}{16\pi}\left(\frac{9}{2}\right) \approx 0.09 \text{ in/min.}$$

In Example 3, note that the volume is increasing at a *constant* rate but the radius is increasing at a *variable* rate. Thus, when we say that two rates are related, we do not mean that they are proportional. In this particular case, the radius is growing more and more slowly as t increases. This is illustrated in Figure 3.29 and Table 3.7.

TABLE 3.7

t	1	3	5	7	9	11
V	4.5	13.5	22.5	31.5	40.5	49.5
r	1.02	1.48	1.75	1.96	2.13	2.28
dr/dt	0.34	0.16	0.12	0.09	0.08	0.07

EXAMPLE 4 The velocity of an airplane tracked by radar

An airplane is flying at an elevation of 6 miles on a flight path that will take it directly over a radar tracking station. Let s represent the distance (measured in miles) between the radar station and the plane. If s is decreasing at a rate of 400 miles per hour when s is 10 miles, what is the velocity of the plane?

SOLUTION

1. We label the distances x and s in Figure 3.30.

   Given: $\dfrac{ds}{dt} = -400$ when $s = 10$

   Find: $\dfrac{dx}{dt}$ when $s = 10$

2. We use the Pythagorean Theorem to form an equation relating s and x.

   $$x^2 + 6^2 = s^2$$

3. Differentiating implicitly with respect to t, we have

   $$2x\frac{dx}{dt} = 2s\frac{ds}{dt}$$

   $$\frac{dx}{dt} = \frac{s}{x}\left(\frac{ds}{dt}\right).$$

4. To find dx/dt, we must first find x when $s = 10$.

   $$x = \sqrt{s^2 - 36} = \sqrt{100 - 36} = \sqrt{64} = 8$$

   Finally, when $s = 10$, we have

   $$\frac{dx}{dt} = \frac{10}{8}(-400) = -500 \text{ mph.}$$

Since the velocity is -500 miles per hour, the *speed* is 500 miles per hour.

FIGURE 3.30

EXAMPLE 5 A changing angle of elevation

A television camera at ground level is filming the lift-off of a space shuttle that is rising vertically according to the position equation $s = 50t^2$, where s is measured in feet and t is measured in seconds. The camera is 2000 feet from the launch pad. Find the rate of change in the angle of elevation of the camera 10 seconds after lift-off.

FIGURE 3.31

SOLUTION

1. We let θ be the angle of elevation, as shown in Figure 3.31. Then we can find the velocity of the rocket by differentiating s with respect to t to obtain $ds/dt = 100t$. Thus, we have the following model.

Given: $\dfrac{ds}{dt} = 100t = $ velocity

Find: $\dfrac{d\theta}{dt}$ when $t = 10$

2. Using Figure 3.31, we relate s and θ by the equation

$$\tan \theta = \frac{s}{2000}.$$

3. Implicit differentiation with respect to t yields

$$(\sec^2 \theta)\frac{d\theta}{dt} = \frac{1}{2000}\left(\frac{ds}{dt}\right)$$

$$\frac{d\theta}{dt} = \frac{\cos^2 \theta}{2000}\left(\frac{ds}{dt}\right).$$

4. Now, when $t = 10$, we know that $s = 50(10^2) = 5000$, and we have

$$\cos \theta = \frac{2000}{\sqrt{2000^2 + 5000^2}} = \frac{2}{\sqrt{29}}.$$

Finally, the rate of change of θ when $t = 10$ is

$$\frac{d\theta}{dt} = \left(\frac{2}{\sqrt{29}}\right)^2\left(\frac{1}{2000}\right)(100)(10) = \frac{2}{29} \text{ rad/sec.} \quad \blacksquare$$

EXAMPLE 6 The velocity of a piston

For the engine shown in Figure 3.32, a 7-inch connecting rod is fastened to a crank of radius 3 inches. The crankshaft rotates counterclockwise at a constant rate of 200 revolutions per minute. Find the velocity of the piston when $\theta = \pi/3$.

FIGURE 3.32

SOLUTION

1. We label the distances as shown in Figure 3.32. Furthermore, since a complete revolution corresponds to 2π radians, we can determine that $d\theta/dt = 200(2\pi) = 400\pi$ radians per minute.

$$\text{Given:} \quad \frac{d\theta}{dt} = 400\pi \quad \text{(constant rate)}$$

$$\text{Find:} \quad \frac{dx}{dt} \quad \text{when} \quad \theta = \frac{\pi}{3}$$

2. From the Law of Cosines (see Figure 3.33), we have

$$7^2 = 3^2 + x^2 - 2(3)(x) \cos \theta.$$

3. Implicit differentiation with respect to t yields the following.

$$0 = 2x\frac{dx}{dt} - 6\left[-x \sin \theta \frac{d\theta}{dt} + \cos \theta \frac{dx}{dt}\right]$$

$$(6 \cos \theta - 2x)\frac{dx}{dt} = 6x \sin \theta \frac{d\theta}{dt}$$

$$\frac{dx}{dt} = \frac{6x \sin \theta}{6 \cos \theta - 2x}\left(\frac{d\theta}{dt}\right)$$

4. When $\theta = \pi/3$, we can solve for x as follows.

$$7^2 = 3^2 + x^2 - 2(3)(x) \cos \frac{\pi}{3}$$

$$49 = 9 + x^2 - 6x\left(\frac{1}{2}\right)$$

$$0 = x^2 - 3x - 40$$

$$0 = (x - 8)(x + 5)$$

Choosing the positive solution, we have $x = 8$. Finally, when $\theta = \pi/3$ and $x = 8$, the velocity of the piston is

$$\frac{dx}{dt} = \frac{6(8)(\sqrt{3}/2)}{6(1/2) - 16}(400\pi) = \frac{9600\pi\sqrt{3}}{-13} \approx -4018 \text{ in/min.} \quad \blacksquare$$

REMARK   Note that the velocity in Example 6 is negative, because the piston is moving to the left.

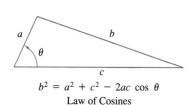

$$b^2 = a^2 + c^2 - 2ac \cos \theta$$
Law of Cosines

FIGURE 3.33

EXERCISES for Section 3.7

In Exercises 1–4, assume that x and y are both differentiable functions of t and find the indicated values of dy/dt and dx/dt.

Equation	*Find*	*Given*	*Equation*	*Find*	*Given*
1. $y = \sqrt{x}$	(a) $\dfrac{dy}{dt}$ when $x = 4$	$\dfrac{dx}{dt} = 3$	**2.** $y = x^2 - 3x$	(a) $\dfrac{dy}{dt}$ when $x = 3$	$\dfrac{dx}{dt} = 2$
	(b) $\dfrac{dx}{dt}$ when $x = 25$	$\dfrac{dy}{dt} = 2$		(b) $\dfrac{dx}{dt}$ when $x = 1$	$\dfrac{dy}{dt} = 5$

Equation	Find	Given
3. $xy = 4$	(a) $\dfrac{dy}{dt}$ when $x = 8$	$\dfrac{dx}{dt} = 10$
	(b) $\dfrac{dx}{dt}$ when $x = 1$	$\dfrac{dy}{dt} = -6$
4. $x^2 + y^2 = 25$	(a) $\dfrac{dy}{dt}$ when $x = 3$, $y = 4$	$\dfrac{dx}{dt} = 8$
	(b) $\dfrac{dx}{dt}$ when $x = 4$, $y = 3$	$\dfrac{dy}{dt} = -2$

5. The radius r of a circle is increasing at a rate of 2 inches per minute. Find the rate of change of the area when (a) $r = 6$ inches and (b) $r = 24$ inches.

6. The radius r of a sphere is increasing at a rate of 2 inches per minute. Find the rate of change of the volume when (a) $r = 6$ inches and (b) $r = 24$ inches.

7. Let A be the area of a circle of radius r that is changing with respect to time. If dr/dt is constant, is dA/dt constant? Explain why or why not.

8. Let V be the volume of a sphere of radius r that is changing with respect to time. If dr/dt is constant, is dV/dt constant? Explain why or why not.

9. A spherical balloon is inflated with gas at the rate of 20 cubic feet per minute. How fast is the radius of the balloon increasing at the instant the radius is (a) 1 foot and (b) 2 feet?

10. The formula for the volume of a cone is

$$V = \frac{1}{3}\pi r^2 h.$$

Find the rate of change of the volume if dr/dt is 2 inches per minute and $h = 3r$ when (a) $r = 6$ inches and (b) $r = 24$ inches.

11. At a sand and gravel plant, sand is falling off a conveyor and onto a conical pile at the rate of 10 cubic feet per minute. The diameter of the base of the cone is approximately three times the altitude. At what rate is the height of the pile changing when it is 15 feet high?

12. A conical tank (with vertex down) is 10 feet across the top and 12 feet deep. If water is flowing into the tank at the rate of 10 cubic feet per minute, find the rate of change of the depth of the water the instant it is 8 feet deep.

13. All edges of a cube are expanding at the rate of 3 centimeters per second. How fast is the volume changing when each edge is (a) 1 centimeter and (b) 10 centimeters?

14. The conditions are the same as in Exercise 13. Now measure how fast the *surface* area is changing when each edge is (a) 1 centimeter and (b) 10 centimeters.

15. A point is moving along the graph of $y = x^2$ so that dx/dt is 2 centimeters per minute. Find dy/dt when (a) $x = 0$ and (b) $x = 3$.

16. The conditions are the same as in Exercise 15, but now measure the rate of change of the distance between the point and the origin.

17. A point is moving along the graph of $y = 1/(1 + x^2)$ so that $dx/dt = 2$ centimeters per minute. Find dy/dt for the following values of x.
(a) $x = -2$ (b) $x = 0$
(c) $x = 2$ (d) $x = 10$

18. A point is moving along the graph of $y = \sin x$ so that $dx/dt = 2$ centimeters per minute. Find dy/dt for the following values of x.
(a) $x = \dfrac{\pi}{6}$ (b) $x = \dfrac{\pi}{3}$
(a) $x = \dfrac{\pi}{4}$ (d) $x = \dfrac{\pi}{2}$

19. A swimming pool is 40 feet long, 20 feet wide, 4 feet deep at the shallow end, and 9 feet deep at the deep end (see figure). Water is being pumped into the pool at 10 cubic feet per minute, and there is 4 feet of water at the deep end.
(a) What percentage of the pool is filled?
(b) At what rate is the water level rising?

20. A trough is 12 feet long and 3 feet across the top (see figure). Its ends are isosceles triangles with an altitude of 3 feet. If water is being pumped into the trough at 2 cubic feet per minute, how fast is the water level rising when it is 1 foot deep?

FIGURE FOR 19   FIGURE FOR 20

21. A ladder 25 feet long is leaning against the wall of a house (see figure). The base of the ladder is pulled away from the wall at a rate of 2 feet per second. How fast is the top moving down the wall when the base of the ladder is (a) 7 feet, (b) 15 feet, and (c) 24 feet from the wall?

FIGURE FOR 21

22. A construction worker pulls a 16-foot plank up the side of a building under construction by means of a rope tied to the end of the plank (see figure). Assume the opposite end of the plank follows a path perpendicular to the wall of the building and the worker pulls the rope at the rate of 0.5 feet per second. How fast is the end of the plank sliding along the ground when it is 8 feet from the wall of the building?

0.5 ft/sec

16 ft

r

FIGURE FOR 22

23. Consider the right triangle formed by the moving ladder, the side of the house, and the ground in Exercise 21. When the base is 7 feet from the wall, find the following rates.
(a) The rate at which the area of the triangle is changing.
(b) The rate at which the angle between the top of the ladder and the wall of the house is changing.

24. A boat is pulled in by means of a winch on the dock 12 feet above the deck of the boat (see figure). The winch pulls in rope at the rate of 4 feet per second. Determine the speed of the boat when there is 13 feet of rope out. What happens to the speed of the boat as it gets closer to the dock?

4 ft/sec

12 ft

13 ft

FIGURE FOR 24

25. An air traffic controller spots two planes at the same altitude converging on a point as they fly at right angles to each other (see figure). One plane is 150 miles from the point and is moving at 450 miles per hour. The

other plane is 200 miles from the point and has a speed of 600 miles per hour.
(a) At what rate is the distance between the planes decreasing?
(b) How much time does the traffic controller have to get one of the planes on a different flight path?

26. The point $(0, y)$ moves along the y-axis at a constant rate of R feet per second, while the point $(x, 0)$ moves along the x-axis at a constant rate of r feet per second. Find an expression for the rate of change of the distance between the two points.

27. A baseball diamond has the shape of a square with sides 90 feet long (see figure). A player 30 feet from third base is running at a speed of 28 feet per second. At what rate is the player's distance from home plate changing?

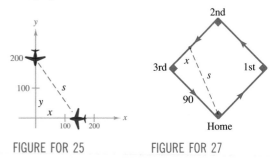

FIGURE FOR 25   FIGURE FOR 27

28. For the baseball diamond in Exercise 27, suppose the player is running from first to second at a speed of 28 feet per second. Find the rate at which the distance from home plate is changing when the player is 30 feet from second.

29. A man 6 feet tall walks at a rate of 5 feet per second away from a light that is 15 feet above the ground (see figure). When he is 10 feet from the base of the light,
(a) at what rate is the tip of his shadow moving?
(b) at what rate is the length of his shadow changing?

30. An airplane is flying at an altitude of 6 miles and passes directly over a radar antenna (see figure). When the plane is 10 miles away ($s = 10$), the radar detects that the distance s is changing at a rate of 240 miles per hour. What is the speed of the plane?

FIGURE FOR 29   FIGURE FOR 30

31. As a spherical raindrop falls, it reaches a layer of dry air and begins to evaporate at a rate that is proportional to its surface area ($S = 4\pi r^2$). Show that the radius decreases at a constant rate.

32. The combined electrical resistance R of R_1 and R_2, connected in parallel, is given by

$$\frac{1}{R} = \frac{1}{R_1} + \frac{1}{R_2}$$

where R, R_1, and R_2 are measured in ohms. R_1 and R_2 are increasing at rates of 1 and 1.5 ohms per second, respectively. At what rate is R changing when $R_1 = 50$ ohms and $R_2 = 75$ ohms?

33. When a certain polyatomic gas undergoes adiabatic expansion, its pressure p and volume v satisfy the equation

$$pv^{1.3} = k$$

where k is a constant. Find the relationship between the related rates dp/dt and dv/dt.

34. Cars on a certain roadway travel on a circular arc of radius r. In order not to rely on friction alone to overcome the centrifugal force, the road is banked at an angle of magnitude θ from the horizontal. The banking angle must satisfy the equation

$$rg \tan \theta = v^2$$

where v is the velocity of the cars and $g = 32$ ft/sec² is the acceleration due to gravity. Find the relationship between the related rates dv/dt and $d\theta/dt$.

35. A balloon rises at the rate of 10 feet per second from a point on the ground 100 feet from an observer. Find the rate of change of the angle of elevation of the balloon from the observer when the balloon is 100 feet above the ground.

36. A fish is reeled in at a rate of 1 foot per second from a bridge 15 feet above the water (see figure). At what rate is the angle between the line and the water changing when there is 25 feet of line out?

37. An airplane flies at an altitude of 5 miles toward a point directly over an observer (see figure). The speed of the plane is 600 miles per hour. Find the rate at which the angle of elevation θ is changing when the angle is (a) $\theta = 30°$, (b) $\theta = 60°$, and (c) $\theta = 75°$.

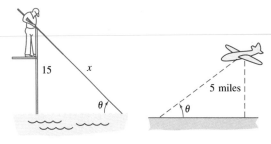

FIGURE FOR 36   FIGURE FOR 37

38. A patrol car is parked 50 feet from a long warehouse (see figure). The light on the car turns at a rate of 30 revolutions per minute. How fast is the light beam moving along the wall when the beam makes the following angles with the line perpendicular from the light to the wall?
(a) $\theta = 30°$ (b) $\theta = 60°$ (c) $\theta = 70°$

39. A wheel of radius 1 foot revolves at a rate of 10 revolutions per second. A dot is painted at a point P on the rim of the wheel (see figure). Find the rate of horizontal movement of the dot for the following angles.
(a) $\theta = 0°$ (b) $\theta = 30°$ (c) $\theta = 60°$

FIGURE FOR 38   FIGURE FOR 39

40. An airplane is flying in still air with an airspeed of 240 miles per hour. If it is climbing at an angle of 22°, find the rate at which it is gaining altitude.

REVIEW EXERCISES for Chapter 3

In Exercises 1–40, find the derivative of the given function.

1. $f(x) = x^3 - 3x^2$

2. $f(x) = \dfrac{2x^3 - 1}{x^2}$

3. $f(x) = x^{1/2} - x^{-1/2}$

4. $f(x) = \dfrac{x + 1}{x - 1}$

5. $g(t) = \dfrac{2}{3t^2}$

6. $h(x) = \dfrac{2}{(3x)^2}$

7. $f(x) = \sqrt{x^3 + 1}$

8. $f(x) = \sqrt[3]{x^2 - 1}$

9. $f(x) = (3x^2 + 7)(x^2 - 2x + 3)$

10. $f(x) = \left(x^2 + \dfrac{1}{x}\right)^5$

11. $f(s) = (s^2 - 1)^{5/2}(s^3 + 5)$

12. $h(\theta) = \dfrac{\theta}{(1 - \theta)^3}$

13. $f(x) = \dfrac{-2x^2}{x - 1}$

14. $f(x) = \dfrac{6x - 5}{x^2 + 1}$

15. $f(x) = \dfrac{x^2 + x - 1}{x^2 - 1}$

16. $f(t) = t^2(t - 1)^5$

17. $f(x) = -2(1 - 4x^2)^2$

18. $f(x) = [(x - 2)(x + 4)]^2$

19. $f(x) = \dfrac{1}{4 - 3x^2}$

20. $f(x) = \dfrac{9}{3x^2 - 2x}$

21. $g(x) = \dfrac{2x}{\sqrt{x + 1}}$

22. $g(x) = x\sqrt{x^2 + 1}$

23. $f(t) = \sqrt{t + 1}\sqrt[3]{t + 1}$

24. $y = \sqrt{3x}(x + 2)^3$

25. $y = \dfrac{1}{2} \csc 2x$

26. $y = \dfrac{1 + \sin x}{1 - \sin x}$

27. $y = \tan x - x$

28. $y = \dfrac{2}{3} \sin^{3/2} x - \dfrac{2}{7} \sin^{7/2} x$

29. $y = \dfrac{x}{2} - \dfrac{\sin 2x}{4}$

30. $y = \dfrac{\sec^7 x}{7} - \dfrac{\sec^5 x}{5}$

31. $y = 3 \cos (3x + 1)$

32. $y = 2 \csc^3 (\sqrt{x})$

33. $y = \tan \sqrt{1 - x}$

34. $y = 1 - \cos 2x + 2 \cos^2 x$

35. $y = \dfrac{\sin x}{x^2}$

36. $y = \csc 3x + \cot 3x$

37. $y = -x \tan x$

38. $y = \dfrac{\cos (x - 1)}{x - 1}$

39. $y = \dfrac{\sin 4x}{4} + x$

40. $y = x \cos x - \sin x$

In Exercises 41–50, find the second derivative of the given function.

41. $f(x) = \sqrt{x^2 + 9}$

42. $h(x) = x\sqrt{x^2 - 1}$

43. $f(t) = \dfrac{t}{(1 - t)^2}$

44. $h(x) = x^2 + \dfrac{3}{x}$

45. $g(x) = \dfrac{6x - 5}{x^2 + 1}$

46. $f(x) = (3x^2 + 7)(x^2 - 2x + 3)$

47. $f(x) = \cot x$

48. $y = \sin^2 x$

49. $y = \dfrac{\cos x}{x}$

50. $g(x) = x \tan x$

In Exercises 51–58, use implicit differentiation to find dy/dx.

51. $x^2 + 3xy + y^3 = 10$

52. $x^2 + 9y^2 - 4x + 3y - 7 = 0$

53. $y\sqrt{x} - x\sqrt{y} = 16$

54. $y^2 + x^2 - 6y - 2x - 5 = 0$

55. $y^2 - x^2 = 25$

56. $y^2 = (x - y)(x^2 + y)$

57. $x \sin y = y \cos x$

58. $\cos (x + y) = x$

In Exercises 59–64, find the equation of the tangent line and the normal line to the graph of the given equation at the indicated point.

Equation	Point
59. $y = (x + 3)^3$	$(-2, 1)$
60. $y = (x - 2)^2$	$(2, 0)$
61. $x^2 + y^2 = 20$	$(2, 4)$
62. $x^2 - y^2 = 16$	$(5, 3)$
63. $y = \sqrt[3]{(x - 2)^2}$	$(3, 1)$
64. $y = \dfrac{2x}{1 - x^2}$	$(0, 0)$

65. Find the points on the graph of

$$f(x) = \dfrac{1}{3}x^3 + x^2 - x - 1$$

at which the slope is (a) -1, (b) 2, and (c) 0.

66. Find the points on the graph of

$$f(x) = x^2 + 1$$

at which the slope is (a) -1, (b) 0, and (c) 1.

In Exercises 67–70, find the derivative of the given function by using the definition of the derivative.

67. $f(x) = \dfrac{1}{x^2}$

68. $f(x) = \dfrac{x + 1}{x - 1}$

69. $f(x) = \sqrt{x + 2}$

70. $f(x) = \dfrac{1}{\sqrt{x}}$

In Exercises 71 and 72, show that the function satisfies the specified equation.

Function	Equation
71. $y = 2 \sin x + 3 \cos x$	$y'' + y = 0$
72. $y = \dfrac{10 - \cos x}{x}$	$xy' + y = \sin x$

In Exercises 73 and 74, derive the equations for the velocity and acceleration of a particle having the given position function.

73. $s = t + \dfrac{1}{t + 1}$

74. $s = \dfrac{1}{t^2 + 2t + 1}$

75. Suppose that the temperature T of food placed in a freezer is given by the equation

$$T = \dfrac{700}{t^2 + 4t + 10}$$

where t is the time in hours. Find the rate of change of T with respect to t at the following times.
(a) $t = 1$ (b) $t = 3$ (c) $t = 5$ (d) $t = 10$

76. The emergent velocity v of a liquid flowing from a hole in the bottom of a tank is given by $v = \sqrt{2gh}$, where g is the acceleration due to gravity (32 ft/s^2) and h is the depth of the liquid in the tank. Find the rate of change of v with respect to h when (a) $h = 9$ and (b) $h = 4$.

77. What is the smallest initial velocity that is required to throw a stone to the top of a 49-foot silo?

78. A bomb is dropped from an airplane at an altitude of 14,400 feet. How long will it take to reach the ground? (Even though it will not be a vertical fall because of the motion of the plane, the time will be the same as that for a vertical fall.) The plane is moving at 600 miles per hour. How far will the bomb move horizontally after it is released from the plane?

79. A ball is thrown and follows a path described by $y = x - 0.02x^2$.
(a) Sketch a graph of the path.
(b) Find the total horizontal distance the ball was thrown.
(c) For what x-value does the ball reach its maximum height? (Use the symmetry of the path.)
(d) Find the equation that gives the instantaneous rate of change in the height of the ball with respect to the horizontal change and evaluate this equation at $x = 0, 10, 25, 30, 50$.
(e) What is the instantaneous rate of change of the height when the ball reaches its maximum height?

80. The path of a projectile thrown at an angle of 45° with level ground is given by

$$y = x - \frac{32}{v_0^2}(x^2)$$

where the initial velocity is v_0 feet per second.
(a) Sketch the path followed by the projectile.
(b) Find the x-coordinate of the point where the projectile strikes the ground. Use the symmetry of the path of the projectile to locate the x-coordinate of the point where the projectile reaches its maximum height.
(c) What is the instantaneous rate of change of the height when the projectile is at its maximum height?

81. The path of a projectile, thrown at an angle of 45° with the ground, is given by

$$y = x - \frac{32}{v_0^2}(x^2)$$

where the initial velocity is v_0 feet per second. Show that doubling the initial velocity of the projectile multiplies both the maximum height and the range by a factor of 4.

82. Use the equation given in Exercise 81 to find the maximum height and range of a projectile thrown with an initial velocity of 70 feet per second.

83. A point moves along the curve $y = \sqrt{x}$ in such a way that the y-value is increasing at the rate of 2 units per second. At what rate is x changing for the following values?
(a) $x = \dfrac{1}{2}$ (b) $x = 1$ (c) $x = 4$

84. The same conditions exist as in Exercise 83. Find the rate the distance between (x, y) and the origin is changing for the following.
(a) $x = \dfrac{1}{2}$ (b) $x = 1$ (c) $x = 4$

85. The cross section of a 5-foot trough is an isosceles trapezoid with a 2-foot lower base, a 3-foot upper base, and an altitude of 2 feet. Water is running into the trough at the rate of 1 cubic foot per minute. How fast is the water level rising when the water is 1 foot deep?

86. A rotating beacon is located 1 mile off a straight shoreline. If the beacon rotates at the rate of 3 revolutions per minute, how fast (in miles per hour) does the beam of light appear to be moving to a viewer who is $\frac{1}{2}$ mile down the shoreline?

87. The geometric mean of x and $x + n$ is

$$g = \sqrt{x(x + n)}$$

and the arithmetic mean is

$$a = \frac{x + (x + n)}{2}.$$

Show that $dg/dx = a/g$.

88. For $y = 1/x$, show that

$$y^{(n)} = \frac{(-1)^n n!}{x^{n+1}}.$$

[Note: $n! = n(n - 1)(n - 2) \cdots (3)(2)(1)$.]

89. Sketch the graph of $f(x) = 4 - |x - 2|$.
(a) Is f continuous at $x = 2$?
(b) Is f differentiable at $x = 2$? Why or why not?

90. Sketch the graph of

$$f(x) = \begin{cases} x^2 + 4x + 2, & x < -2 \\ 1 - 4x - x^2, & x \ge -2. \end{cases}$$

(a) Is f continuous at $x = -2$?
(b) Is f differentiable at $x = -2$? Why or why not?

Chapter 4 Application

Honey bees build their combs of pure wax, which can be produced only by the workers. Both sides of the combs are made up of geometrically perfect hexagonal cells and are constructed in a way that minimizes the amount of construction material. There are over 5 million honey bee colonies in the United States, producing over 250 million pounds of honey each year.

Construction of a Honeycomb

Honey bees use hexagonal cells to construct storage compartments. Each cell has a hexagonal base and three rhombic upper faces that meet the altitude of the cell at an angle θ. The *volume* of each cell is given by

$$V = \frac{3\sqrt{3}}{2}\, s^2 h$$

and is independent of the angle θ. On the other hand, the *surface area* of each cell, given by

$$S = 6hs + \frac{3s^2}{2}\left(\frac{\sqrt{3} - \cos\theta}{\sin\theta}\right)$$

does depend on the angle θ. The angle that minimizes the surface area (and thus the amount of wax needed to build the honeycomb), is given by solving the equation

$$\frac{dS}{d\theta} = \frac{3s^2}{2}\left(\frac{1 - \sqrt{3}\cos\theta}{\sin^2\theta}\right) = 0.$$

The solution of this equation is $\cos\theta = 1/\sqrt{3}$ or $\theta \approx 54.74°$.

Hexagonal cell used in the construction of a honeycomb

See Exercise 40, Section 4.1.

Chapter Overview

This chapter discusses several applications of the derivative of a function. These applications fall into three basic categories—curve sketching, optimization, and approximation techniques.

The chapter begins by showing how the derivative can be used to locate the minimum and maximum values of a function on a closed interval. Many of the results in the chapter depend on two important theorems called *Rolle's Theorem* and the *Mean Value Theorem*, and these are presented in Section 4.2. In Sections 4.3 through 4.5, we refine our curve-sketching techniques by showing how the first derivative can be used to determine whether a function is increasing or decreasing, and how the second derivative can be used to determine whether a function is concave upward or downward. We also present techniques for locating horizontal asymptotes of functions. Section 4.6 summarizes all of the curve-sketching techniques presented in the first four chapters of the text.

Section 4.7 is an important one because it shows how calculus can be used to solve optimization problems that occur in science and engineering. (Section 4.10 discusses optimization problems that occur in business and economics.)

Section 4.8 presents a numerical technique called *Newton's Method*. This method uses the derivative of a function to approximate the zeros of the function. Section 4.9 introduces the concept of the *differential* of a function. Finally, Section 4.10 discusses several applications of calculus in business and economics.

Applications of Differentiation

4.1 Extrema on an Interval

Extrema of a function ■ Relative extrema ■ Critical numbers

In calculus much effort is devoted to determining the behavior of a function f on an interval I. For instance, we are interested in several questions: Does f have a maximum value on I? Does it have a minimum value? Where is the function increasing? Where is it decreasing? In this chapter we will show how the derivative can be used to answer these questions. We will also show why these questions are important to applications.

We begin by looking at the minimum and maximum values of a function on an interval.

DEFINITION OF EXTREMA

Let f be defined on an interval I containing c.

1. $f(c)$ is the **minimum of f on I** if $f(c) \leq f(x)$ for all x in I.
2. $f(c)$ is the **maximum of f on I** if $f(c) \geq f(x)$ for all x in I.

The minimum and maximum of a function on an interval are called the **extreme values,** or **extrema,** of the function on the interval.

REMARK   The minimum and maximum of a function on an interval are sometimes called the **absolute minimum** and **absolute maximum** on the interval, respectively.

A function need not have a minimum or a maximum on an interval. For instance, Figure 4.1 shows three possibilities. By comparing the first and second graphs, we see that the function $f(x) = x^2 + 1$ has both a minimum and a maximum on the closed interval $[-1, 2]$, but does not have a maximum on the open interval $(-1, 2)$. Moreover, in the third graph, we see that a discontinuity (at $x = 0$) can affect the existence of an extremum on an interval. This suggests the following theorem, which identifies conditions that guarantee the existence of both a minimum and a maximum of a function on an interval. (A proof of this theorem is not within the scope of this text.)

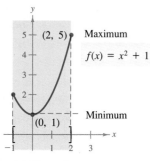

(a) f is continuous.
[−1, 2] is closed.

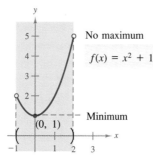

(b) f is continuous.
(−1, 2) is open.

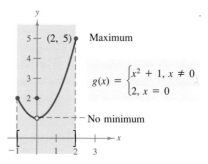

(c) g is discontinuous.
[−1, 2] is closed.

FIGURE 4.1

THEOREM 4.1 THE EXTREME VALUE THEOREM	If f is continuous on a closed interval $[a, b]$, then f has both a minimum and a maximum on the interval.

REMARK   Note that the Extreme Value Theorem (like the Intermediate Value Theorem) is an *existence theorem* because it tells of the existence of minimum and maximum values but does not show how to find these values.

From Figure 4.1 we can see that extrema can occur at interior points or endpoints of an interval. Extrema that occur at the endpoints are called **endpoint extrema.**

Relative extrema

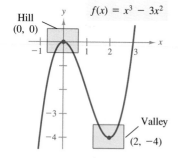

FIGURE 4.2

In Figure 4.2, the graph of $f(x) = x^3 - 3x^2$ has a **relative maximum** at the point (0, 0) and a **relative minimum** at the point (2, −4). Informally, we can think of a relative maximum occurring on a "hill" on the graph, and a relative minimum occurring in a "valley" on the graph. Such hills and valleys can occur in two ways. If the hill (or valley) is smooth and rounded, then the graph has a horizontal tangent line at the high point (or low point). On the other hand, if the hill (or valley) is sharp and peaked, then the graph represents a function that is not differentiable at the high point (or low point).

DEFINITION OF RELATIVE EXTREMA	1. If there is an open interval on which $f(c)$ is a maximum, then $f(c)$ is called a **relative maximum** of f. 2. If there is an open interval on which $f(c)$ is a minimum, then $f(c)$ is called a **relative minimum** of f.

REMARK   The plural of relative maximum is relative maxima, and the plural of relative minimum is relative minima.

$f(x) = \dfrac{9(x^2 - 3)}{x^3}$

(a)

(b)

(c)

FIGURE 4.3

In Example 1, we examine the derivative of a function at several *given* relative extrema. (We will say much more about *finding* the relative extrema of a function in Section 4.3.)

EXAMPLE 1 The value of the derivative at relative extrema

Find the value of the derivative at each of the relative extrema shown in Figure 4.3.

SOLUTION

(a) The derivative of this function is

$$f'(x) = \frac{x^3(18x) - (9)(x^2 - 3)(3x^2)}{(x^3)^2} = \frac{9(9 - x^2)}{x^4}.$$

Thus, at the point $(3, 2)$, the value of the derivative is $f'(3) = 0$.

(b) At $x = 0$, the derivative of $f(x) = |x|$ *does not exist* since the following one-sided limits differ.

$$\lim_{x \to 0^-} \frac{f(x) - f(0)}{x - 0} = \lim_{x \to 0^-} \frac{|x|}{x} = -1 \qquad \text{Limit from the left}$$

$$\lim_{x \to 0^+} \frac{f(x) - f(0)}{x - 0} = \lim_{x \to 0^+} \frac{|x|}{x} = +1 \qquad \text{Limit from the right}$$

(c) The derivative of this function is $f'(x) = \cos x$. Thus, at the point $(\pi/2, 1)$, the value of the derivative is $f'(\pi/2) = \cos(\pi/2) = 0$, and at the point $(3\pi/2, 1)$, the value of the derivative is $f'(3\pi/2) = \cos(3\pi/2) = 0$. ▭

Note in Example 1 that at a relative extremum the derivative is either zero or undefined. We call the x-values at these special points **critical numbers.**

DEFINITION OF CRITICAL NUMBER	If f is defined at c, then c is called a **critical number** of f if $f'(c) = 0$ or if f' is undefined at c.

Figure 4.4 illustrates the two types of critical numbers given in this definition.

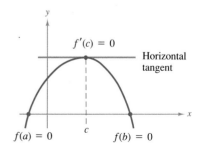

FIGURE 4.4

THEOREM 4.2 RELATIVE EXTREMA OCCUR ONLY AT CRITICAL NUMBERS	If f has a relative minimum or relative maximum at $x = c$, then c is a critical number of f.

PROOF

Case 1: If f is *not* differentiable at $x = c$, then by definition c is a critical number of f and we have nothing to prove.

Case 2: If f is differentiable at $x = c$, then $f'(c)$ must be either positive, negative, or zero. Suppose $f'(c)$ is positive, Then we have

$$f'(c) = \lim_{x \to c} \frac{f(x) - f(c)}{x - c} > 0.$$

But this implies that there exists an interval (a, b) containing c such that

$$\frac{f(x) - f(c)}{x - c} > 0, \quad \text{for all } x \neq c \text{ in } (a, b).$$

Since this quotient is positive, the signs of the denominator and numerator must agree; this produces the following inequalities for x-values in the interval (a, b).

Left of c: $x < c$ and $f(x) < f(c)$ $\Rightarrow$ $f(c)$ is not a relative minimum
Right of c: $x > c$ and $f(x) > f(c)$ $\Rightarrow$ $f(c)$ is not a relative maximum

Thus, our assumption that $f'(c) > 0$ contradicts the fact that $f(c)$ is a relative extremum. Assuming that $f'(c) < 0$ produces a similar contradiction, we are left with only one possibility, namely $f'(c) = 0$.

Theorem 4.2 tells us that the relative extrema of a function can occur *only* at the critical numbers of the function. By combining this observation with our knowledge of endpoint extrema, we obtain the following guidelines for finding extrema on a closed interval.

GUIDELINES FOR FINDING EXTREMA ON A CLOSED INTERVAL	To find the extrema of a continuous function f on a closed interval $[a, b]$, we suggest the following steps.

1. Find the critical numbers of f.
2. Evaluate f at each critical number in (a, b).
3. Evaluate f at each endpoint of $[a, b]$.
4. The least of these values is the minimum, and the greatest is the maximum.

EXAMPLE 2 Finding extrema on a closed interval

Find the extrema of $f(x) = 3x^4 - 4x^3$ on the interval $[-1, 2]$.

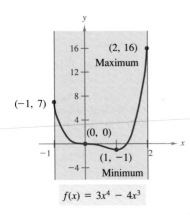

$f(x) = 3x^4 - 4x^3$

FIGURE 4.5

SOLUTION

To find the critical numbers, we differentiate to obtain

$$f'(x) = 12x^3 - 12x^2 = 0 \qquad \text{Set } f'(x) = 0$$
$$12x^2(x - 1) = 0 \qquad \text{Factor}$$
$$x = 0, 1. \qquad \text{Critical numbers}$$

Since f' is defined for all x, we conclude that these are the only critical numbers of f. By evaluating f at these two critical numbers and at the endpoints of $[-1, 2]$, we determine that the maximum is $f(2) = 16$ and the minimum is $f(1) = -1$, indicated in Table 4.1. The graph of f is shown in Figure 4.5.

TABLE 4.1

Left endpoint	Critical number	Critical number	Right endpoint
$f(-1) = 7$	$f(0) = 0$	$f(1) = -1$ Minimum	$f(2) = 16$ Maximum

REMARK Note in Figure 4.5 that the critical number $x = 0$ does not yield a relative minimum or a relative maximum. This tells us that the converse of Theorem 4.2 is not true. In other words, the critical numbers of a function need not produce relative extrema.

EXAMPLE 3 Finding extrema on a closed interval

Find the extrema of $f(x) = 2x - 3x^{2/3}$ on the interval $[-1, 3]$.

SOLUTION

Differentiating produces

$$f'(x) = 2 - \frac{2}{x^{1/3}} = 2\left(\frac{x^{1/3} - 1}{x^{1/3}}\right).$$

This gives us the following critical numbers.

$$x = 1 \qquad f'(1) = 0$$
$$x = 0 \qquad f' \text{ is undefined}$$

Finally, by evaluating f at these two points and at the endpoints of the interval, we conclude that the minimum is $f(-1) = -5$ and the maximum is $f(0) = 0$, as indicated in Table 4.2. The graph of f is shown in Figure 4.6.

$f(x) = 2x - 3x^{2/3}$

FIGURE 4.6

TABLE 4.2

Left endpoint	Critical number	Critical number	Right endpoint
$f(-1) = -5$ Minimum	$f(0) = 0$ Maximum	$f(1) = -1$	$f(3) = 6 - 3\sqrt[3]{9} \approx -0.24$

EXAMPLE 4 Finding extrema on a closed interval

Find the extrema of $f(x) = 2 \sin x - \cos 2x$ on the interval $[0, 2\pi]$.

SOLUTION

This function is differentiable for all real x, so we can find all critical numbers by setting $f'(x)$ equal to zero, as follows.

$$f'(x) = 2 \cos x + 2 \sin 2x = 0$$
$$2 \cos x + 4 \cos x \sin x = 0 \qquad \sin 2x = 2 \cos x \sin x$$
$$2(\cos x)(1 + 2 \sin x) = 0 \qquad \text{Factor}$$

Setting the two factors equal to zero and solving for x in the interval $[0, 2\pi]$ produces the following critical numbers.

$$\cos x = 0 \quad \Longrightarrow \quad x = \frac{\pi}{2}, \frac{3\pi}{2}$$

$$\sin x = -\frac{1}{2} \quad \Longrightarrow \quad x = \frac{7\pi}{6}, \frac{11\pi}{6}$$

Finally, by evaluating f at these four critical numbers and at the endpoints of the interval, we conclude that the maximum is $f(\pi/2) = 3$ and the minimum occurs at *two* points, $f(7\pi/6) = -3/2$ and $f(11\pi/6) = -3/2$, as indicated in Table 4.3. The graph is shown in Figure 4.7.

TABLE 4.3

Left endpoint	Critical number	Critical number	Critical number	Critical number	Right endpoint
$f(0) = -1$	$f\left(\dfrac{\pi}{2}\right) = 3$ Maximum	$f\left(\dfrac{7\pi}{6}\right) = -\dfrac{3}{2}$ Minimum	$f\left(\dfrac{3\pi}{2}\right) = -1$	$f\left(\dfrac{11\pi}{6}\right) = -\dfrac{3}{2}$ Minimum	$f(2\pi) = -1$

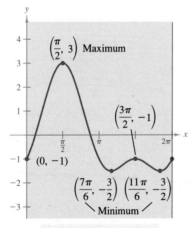

$$f(x) = 2 \sin x - \cos 2x$$

FIGURE 4.7

EXERCISES for Section 4.1

In Exercises 1–6, find the value of the derivative (if it exists) at the indicated extrema.

1. $f(x) = \dfrac{x^2}{x^2 + 4}$

2. $f(x) = \cos \dfrac{\pi x}{2}$

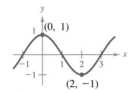

3. $f(x) = x + \dfrac{32}{x^2}$

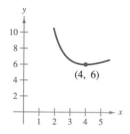

4. $f(x) = -3x\sqrt{x + 1}$

5. $f(x) = (x + 2)^{2/3}$

6. $f(x) = 4 - |x|$

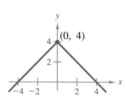

In Exercises 7–20, locate the absolute extrema of the given function on the indicated interval.

Function	Interval
7. $f(x) = 2(3 - x)$	$[-1, 2]$
8. $f(x) = \dfrac{2x + 5}{3}$	$[0, 5]$
9. $f(x) = -x^2 + 3x$	$[0, 3]$
10. $f(x) = x^2 + 2x - 4$	$[-1, 1]$
11. $f(x) = x^3 - 3x^2$	$[-1, 3]$
12. $f(x) = x^3 - 12x$	$[0, 4]$
13. $f(x) = 3x^{2/3} - 2x$	$[-1, 1]$
14. $g(x) = \sqrt[3]{x}$	$[-1, 1]$
15. $h(t) = 4 - |t - 4|$	$[1, 6]$
16. $g(t) = \dfrac{t^2}{t^2 + 3}$	$[-1, 1]$
17. $h(s) = \dfrac{1}{s - 2}$	$[0, 1]$
18. $h(t) = \dfrac{t}{t - 2}$	$[3, 5]$

Function	Interval
19. $f(x) = \cos \pi x$	$\left[0, \dfrac{1}{6}\right]$
20. $g(x) = \csc x$	$\left[\dfrac{\pi}{6}, \dfrac{\pi}{3}\right]$

21. Explain why the function $f(x) = \tan x$ has a maximum on $[0, \pi/4]$, but not on $[0, \pi]$.

22. Explain why the function $y = 1/(x + 1)$ has a minimum on $[0, 2]$, but not on $[-2, 0]$.

In Exercises 23–26, determine from the graph whether f possesses a minimum in the interval (a, b).

23. (a) (b)

24. (a) (b)

25. (a) (b)

26. (a) (b)

In Exercises 27 and 28, locate the absolute extrema of the function (if any exist) over the indicated interval.

27. $f(x) = 2x - 3$
  (a) $[0, 2]$
  (b) $[0, 2)$
  (c) $(0, 2]$
  (d) $(0, 2)$

28. $f(x) = 5 - x$
  (a) $[1, 4]$
  (b) $[1, 4)$
  (c) $(1, 4]$
  (d) $(1, 4)$

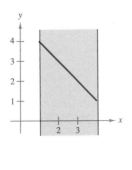

In Exercises 29–32, find the maximum value of $|f''(x)|$ on the indicated interval. (This value is used in the error estimate for the Trapezoidal Rule, as discussed in Section 5.6.)

Function	Interval
29. $f(x) = \dfrac{1}{x^2 + 1}$	$[0, 3]$
30. $f(x) = \dfrac{1}{x^2 + 1}$	$\left[\dfrac{1}{2}, 3\right]$
31. $f(x) = \sqrt{1 + x^3}$	$[0, 2]$
32. $f(x) = x^3(3x^2 - 10)$	$[0, 1]$

In Exercises 33–36, find the maximum value of $|f^{(4)}(x)|$ on the indicated interval. (This value is used in the error estimate for Simpson's Rule, as discussed in Section 5.6.)

Function	Interval
33. $f(x) = 15x^4 - \left(\dfrac{2x - 1}{2}\right)^6$	$[0, 1]$
34. $f(x) = x^5 - 5x^4 + 20x^3 + 600$	$\left[0, \dfrac{3}{2}\right]$
35. $f(x) = (x + 1)^{2/3}$	$[0, 2]$
36. $f(x) = \dfrac{1}{x^2 + 1}$	$[-1, 1]$

37. The formula for the power output P of a battery is given by

$$P = VI - RI^2$$

where V is the electromotive force in volts, R is the resistance, and I is the current. Find the current (measured in amperes) that corresponds to a maximum value

of P in a battery for which $V = 12$ volts and $R = 0.5$ ohms. (Assume that a 15-amp fuse bounds the output in the interval $0 \leq I \leq 15$.)

38. A retailer has determined that the cost C for ordering and storing x units of a certain product is

$$C = 2x + \frac{300,000}{x}, \quad 0 \leq x \leq 300.$$

Find the order size that will minimize cost if the delivery truck can bring a maximum of 300 units per order.

39. A lawn sprinkler is constructed in such a way that $d\theta/dt$ is constant, where θ ranges between 45° and 135° (see figure). The distance the water travels horizontally is

$$x = \frac{v^2 \sin 2\theta}{32}, \quad \frac{\pi}{4} \leq \theta \leq \frac{3\pi}{4}$$

where v is the speed of the water. Find dx/dt and explain why this model of lawn sprinkler does not water evenly. What part of the lawn receives the most water?

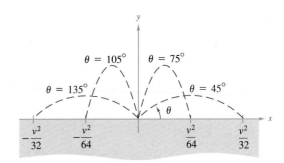

Water Sprinkler: $45° \leq \theta \leq 135°$

FIGURE FOR 39

40. In the Chapter 4 Application, we showed that the surface area of a cell in a honeycomb is given by

$$S = 6hs + \frac{3s^2}{2}\left(\frac{\sqrt{3} - \cos \theta}{\sin \theta}\right)$$

where h and s are positive constants and θ is the angle at which the upper faces meet the altitude of the cell. Find the angle θ ($\pi/6 \leq \theta \leq \pi/2$) that minimizes the surface area S.

In Exercises 41 and 42, use a computer or graphics calculator to sketch the function f and find its critical numbers over the specified interval.

Function	Interval
41. $f(x) = 3.2x^5 + 5x^3 - 3.5x$	$[0, 1]$
42. $f(x) = \dfrac{4}{3}x\sqrt{3 - x}$	$[0, 3]$

4.2 Rolle's Theorem and the Mean Value Theorem

Rolle's Theorem ▪ Mean Value Theorem ▪ Applications

The Extreme Value Theorem (Section 4.1) tells us that a continuous function on a closed interval $[a, b]$ must have both a minimum and maximum on the interval. However, both of these values can occur at the endpoints. We now present a theorem called **Rolle's Theorem,** named after the French mathematician Michel Rolle (1652–1719), that implies (under certain conditions) the existence of an extreme value in the interior of a closed interval.

THEOREM 4.3
ROLLE'S THEOREM

Let f be continuous on the closed interval $[a, b]$ and differentiable on the open interval (a, b). If

$$f(a) = f(b)$$

then there is at least one number c in (a, b) such that $f'(c) = 0$.

PROOF Let $f(a) = d = f(b)$.

Case 1: If $f(x) = d$ for all x in $[a, b]$, then f is constant on the interval, and by Theorem 3.3 $f'(x) = 0$ for all x in (a, b).

Case 2: If $f(x) > d$ for some x in (a, b), then, by the Extreme Value Theorem, we know that f has a maximum at some c in the interval. Moreover, since $f(c) > d$, this maximum does not occur at either endpoint. Therefore, f has a maximum in the *open* interval (a, b). This implies that $f(c)$ is a *relative maximum*, and by Theorem 4.2 we know c is a critical number of f. Finally, since f is differentiable at c, we can conclude that $f'(c) = 0$.

Case 3: If $f(x) < d$ for some x in (a, b), then we can use an argument similar to that in Case 2.

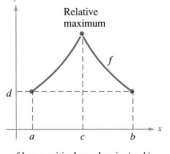

f has a critical number in (a, b).

FIGURE 4.8

If we drop the differentiability requirement from Rolle's Theorem, then f will still have a critical number in (a, b), but it need not yield a horizontal tangent. This is shown in Figure 4.8 and stated in the following corollary to Rolle's Theorem.

COROLLARY TO ROLLE'S THEOREM

Let f be continuous on the closed interval $[a, b]$. If $f(a) = f(b)$, then f has a critical number in the open interval (a, b).

EXAMPLE 1   An application of Rolle's Theorem

Find the two *x*-intercepts of $f(x) = x^2 - 3x + 2$ and show that $f'(x) = 0$ at some point between the two intercepts.

SOLUTION

Note that f is differentiable on the entire real line. Setting $f(x)$ equal to zero, we have

$$x^2 - 3x + 2 = 0$$
$$(x - 1)(x - 2) = 0.$$

Thus, $f(1) = f(2) = 0$, and from Rolle's Theorem we know that there exists c in the interval $(1, 2)$ such that $f'(c) = 0$. To find c we solve the equation

$$f'(x) = 2x - 3 = 0$$

and determine that $f'(x) = 0$ when $x = \frac{3}{2}$. This *x*-value lies in the open interval $(1, 2)$, as shown in Figure 4.9.

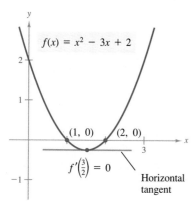

FIGURE 4.9

Rolle's Theorem states that if f satisfies the conditions of the theorem, then there must be *at least* one point between *a* and *b* at which the derivative is zero. There may of course be more than one such point, as illustrated in the next example.

EXAMPLE 2   An application of Rolle's Theorem

Let $f(x) = x^4 - 2x^2$. Find all *c* in the interval $(-2, 2)$ such that $f'(c) = 0$.

SOLUTION

Since $f(-2) = 8 = f(2)$ and f is differentiable, Rolle's Theorem guarantees the existence of at least one *c* in $(-2, 2)$ such that $f'(c) = 0$. Setting the derivative equal to zero produces

$$f'(x) = 4x^3 - 4x = 0$$
$$4x(x^2 - 1) = 0$$
$$x = 0, 1, -1.$$

Thus, in the interval $(-2, 2)$, the derivative is zero at each of these three *x*-values, as shown in Figure 4.10.

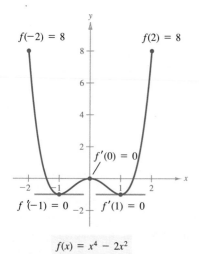

FIGURE 4.10

The Mean Value Theorem

Rolle's Theorem can be used to prove another well-known theorem in calculus—the **Mean Value Theorem.**

THEOREM 4.4
THE MEAN VALUE THEOREM

If f is continuous on the closed interval $[a, b]$ and differentiable on the open interval (a, b), then there exists a number c in (a, b) such that

$$f'(c) = \frac{f(b) - f(a)}{b - a}.$$

PROOF

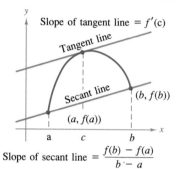

Slope of tangent line $= f'(c)$

Tangent line

Secant line

$(b, f(b))$

$(a, f(a))$

$a \quad c \quad b$

Slope of secant line $= \dfrac{f(b) - f(a)}{b - a}$

FIGURE 4.11

Refer to Figure 4.11. The equation of the secant line containing the points $(a, f(a))$ and $(b, f(b))$ is given by

$$y = \left[\frac{f(b) - f(a)}{b - a} \right](x - a) + f(a).$$

Let $g(x)$ be the difference between $f(x)$ and y. Then

$$g(x) = f(x) - y = f(x) - \left[\frac{f(b) - f(a)}{b - a} \right](x - a) - f(a).$$

Now, by evaluating g at a and b, we see that $g(a) = 0 = g(b)$. Furthermore, since f is differentiable, g is also differentiable, and we can apply Rolle's Theorem to the function g. Thus, there exists a point c in (a, b) such that $g'(c) = 0$. This means that

$$0 = g'(c) = f'(c) - \frac{f(b) - f(a)}{b - a}.$$

Therefore, there exists a point c in (a, b) such that

$$f'(c) = \frac{f(b) - f(a)}{b - a}.$$

REMARK   The "mean" in the Mean Value Theorem refers to the mean (or average) rate of change of f in the interval $[a, b]$.

Joseph-Louis Lagrange

Although the Mean Value Theorem can be used directly in problem solving, it is used more often to prove other theorems. In fact, some people consider this to be the most important theorem in calculus. It was proved by a famous French mathematician Joseph-Louis Lagrange (1736–1813), and it is closely related to the Fundamental Theorem of Calculus discussed in Chapter 5. For now, you can get an idea of the versatility of this theorem by looking at the results stated in Exercises 37–42 in this section.

The Mean Value Theorem has implications for both basic interpretations of the derivative. Geometrically, the theorem guarantees the existence of a tangent line that is parallel to the secant line through the points $(a, f(a))$ and $(b, f(b))$, as shown in Figure 4.11. Example 3 illustrates this geometrical interpretation of the Mean Value Theorem. In terms of rates of change, the Mean Value Theorem tells us that there must be a point in the open interval (a, b) at which the instantaneous rate of change is equal to the average rate of change over the interval $[a, b]$. This is illustrated in Example 4.

EXAMPLE 3 A tangent line application of the Mean Value Theorem

Given $f(x) = 5 - (4/x)$, find all c in the interval $(1, 4)$ such that

$$f'(c) = \frac{f(4) - f(1)}{4 - 1}.$$

SOLUTION

The slope of the secant line through $(1, f(1))$ and $(4, f(4))$ is

$$\frac{f(4) - f(1)}{4 - 1} = \frac{4 - 1}{4 - 1} = 1.$$

Since f satisfies the conditions of the Mean Value Theorem, there exists at least one c in $(1, 4)$ such that $f'(c) = 1$. Solving the equation $f'(x) = 1$ yields the following.

$$f'(x) = \frac{4}{x^2} = 1 \quad \Longrightarrow \quad x = \pm 2$$

Finally, in the interval $(1, 4)$ we choose $c = 2$, as shown in Figure 4.12.

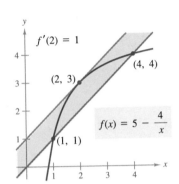

FIGURE 4.12

EXAMPLE 4 A rate of change application of the Mean Value Theorem

Two stationary patrol cars equipped with radar are 5 miles apart on a highway, as shown in Figure 4.13. As a truck passes the first patrol car, its speed is clocked at 55 miles per hour. Four minutes later, when the truck passes the second patrol car, its speed is clocked at 50 miles per hour. Prove that the truck must have exceeded the speed limit (of 55 miles per hour) at some time during the four minutes.

FIGURE 4.13

SOLUTION

We let $t = 0$ be the time (in hours) when the truck passes the first patrol car. Then, the time when the truck passes the second patrol car is

$$t = \frac{4}{60} = \frac{1}{15} \text{ hr.}$$

Now, if we let $s(t)$ represent the distance (in miles) traveled by the truck, we have $s(0) = 0$ and $s(\frac{1}{15}) = 5$. Therefore, the average velocity of the truck over the five-mile stretch of highway is given by

$$\text{average velocity} = \frac{s(1/15) - s(0)}{(1/15) - 0} = \frac{5}{1/15} = 75 \text{ mph.}$$

Assuming that the position function is differentiable, we can apply the Mean Value Theorem to conclude that the truck must have been traveling at a rate of 75 miles per hour sometime during the four minutes.

REMARK   A useful alternate form of the Mean Value Theorem is as follows: If f is continuous on $[a, b]$ and differentiable on (a, b), then there exists a number c in (a, b) such that $f(b) = f(a) + (b - a)f'(c)$.

When working the exercises for this section, keep in mind that polynomial functions, rational functions, and trigonometric functions are differentiable at all points in their domains.

EXERCISES for Section 4.2

In Exercises 1 and 2, state why Rolle's Theorem does not apply to the function even though there exist a and b such that $f(a) = f(b) = 0$.

1. $f(x) = 1 - |x - 1|$ **2.** $f(x) = \cot \dfrac{x}{2}$

In Exercises 3–20, determine whether Rolle's Theorem can be applied to f on the indicated interval. If Rolle's Theorem can be applied, find all values of c in the interval such that $f'(c) = 0$.

Function	Interval
3. $f(x) = x^2 - 2x$	$[0, 2]$
4. $f(x) = x^2 - 3x + 2$	$[1, 2]$
5. $f(x) = (x - 1)(x - 2)(x - 3)$	$[1, 3]$
6. $f(x) = (x - 3)(x + 1)^2$	$[-1, 3]$
7. $f(x) = |x| - 1$	$[-1, 1]$
8. $f(x) = 3 - |x - 3|$	$[0, 6]$
9. $f(x) = x^{2/3} - 1$	$[-8, 8]$
10. $f(x) = x - x^{1/3}$	$[0, 1]$
11. $f(x) = \dfrac{x^2 - 2x - 3}{x + 2}$	$[-1, 3]$
12. $f(x) = \dfrac{x^2 - 1}{x}$	$[-1, 1]$
13. $f(x) = \sin x$	$[0, 2\pi]$
14. $f(x) = \cos x$	$[0, 2\pi]$
15. $f(x) = \sin 2x$	$\left[\dfrac{\pi}{6}, \dfrac{\pi}{3}\right]$
16. $f(x) = 4x - \tan \pi x$	$\left[-\dfrac{1}{4}, \dfrac{1}{4}\right]$

Function	Interval
17. $f(x) = \dfrac{x}{2} - \sin \dfrac{\pi x}{6}$	$[-1, 0]$
18. $f(x) = \dfrac{6x}{\pi} - 4 \sin^2 x$	$\left[0, \dfrac{\pi}{6}\right]$
19. $f(x) = \tan x$	$[0, \pi]$
20. $f(x) = \sec x$	$\left[-\dfrac{\pi}{4}, \dfrac{\pi}{4}\right]$

In Exercises 21–30, apply the Mean Value Theorem to f on the indicated interval. In each case, find all values of c in the interval (a, b) such that

$$f'(c) = \frac{f(b) - f(a)}{b - a}.$$

Function	Interval
21. $f(x) = x^2$	$[-2, 1]$
22. $f(x) = x(x^2 - x - 2)$	$[-1, 1]$
23. $f(x) = x^{2/3}$	$[0, 1]$
24. $f(x) = \dfrac{x + 1}{x}$	$\left[\dfrac{1}{2}, 2\right]$
25. $f(x) = \dfrac{x}{x + 1}$	$\left[-\dfrac{1}{2}, 2\right]$
26. $f(x) = \sqrt{x - 2}$	$[2, 6]$
27. $f(x) = x^3$	$[0, 1]$
28. $f(x) = \sin x$	$[0, \pi]$
29. $f(x) = x - 2 \sin x$	$[-\pi, \pi]$
30. $f(x) = 2 \sin x + \sin 2x$	$[0, \pi]$

31. The height of a ball t seconds after it is thrown is given by

$$f(t) = -16t^2 + 48t + 32.$$

(a) Verify that $f(1) = f(2)$.
(b) According to Rolle's Theorem, what must be the velocity at some time in the interval $[1, 2]$?

32. The ordering and transportation cost C of components used in a manufacturing process is approximated by

$$C(x) = 10\left(\frac{1}{x} + \frac{x}{x+3}\right)$$

where C is measured in thousands of dollars and x is the order size in hundreds.

(a) Verify that $C(3) = C(6)$.

(b) According to Rolle's Theorem, the rate of change of cost must be zero for some order size in the interval [3, 6]. Find that order size.

33. The height of an object t seconds after it was dropped from a height of 500 feet is given by

$$s(t) = -16t^2 + 500.$$

(a) Find the average velocity of the object during the first 3 seconds.

(b) Use the Mean Value Theorem to verify that at some time during the first three seconds of fall the instantaneous velocity equals the average velocity. Find that time.

34. A company introduces a new product for which the number of units sold S is given by

$$S(t) = 200\left(5 - \frac{9}{2+t}\right)$$

where t is the time in months.

(a) Find the average rate of change of $S(t)$ during the first year.

(b) During what month does $S'(t)$ equal its average rate of change during the first year?

35. Given the function

$$f(x) = \frac{1}{x-4}$$

show that for the interval (2, 6) there exists no real number c such that

$$f'(c) = \frac{f(6) - f(2)}{6 - 2}.$$

State whether this contradicts the Mean Value Theorem and give the reason for your answer.

36. Prove the Corollary to Rolle's Theorem.

37. If $a > 0$ and n is any integer, prove that the polynomial function

$$p(x) = x^{2n+1} + ax + b$$

cannot have two real roots.

38. Let p be a *nonconstant* polynomial function.

(a) Prove that between any two consecutive zeros of p', there is at most one zero of p.

(b) If p has three distinct zeros in the interval $[a, b]$, prove that $p''(c) = 0$ for some real number c in (a, b).

39. Prove that if $f'(x) = 0$ for all x in an interval (a, b), then f is constant on the interval.

40. Let $p(x) = Ax^2 + Bx + C$. Prove that for any interval $[a, b]$, the value c guaranteed by the Mean Value Theorem is the midpoint of the interval.

41. Prove that if two functions f and g have the same derivatives on an interval, then they must differ only by a constant on the interval. [Hint: Let $h(x) = f(x) - g(x)$ and use the result of Exercise 39.]

42. Prove that if f is differentiable on $(-\infty, \infty)$ and $f'(x) < 1$ for all real numbers, then f has at most one fixed point. A **fixed point** of a function f is a real number c such that $f(c) = c$.

43. Use a computer or graphics calculator to sketch the graph of $f(x) = \sqrt{x}$ over the interval [1, 9].

(a) Find an equation of the secant line to the graph of f passing through the points $(1, f(1))$ and $(9, f(9))$. Sketch the graph of the secant line on the same axes as the graph of f.

(b) Find the value of c in the interval (1, 9) such that

$$f'(c) = \frac{f(9) - f(1)}{9 - 1}.$$

Find the equation of the tangent line to the graph of f at the point $(c, f(c))$ and sketch its graph on the same axes as the graph of f. Note that the secant line and tangent line are parallel.

4.3 Increasing and Decreasing Functions and the First Derivative Test

Increasing and decreasing functions ▪ The First Derivative Test ▪ Strictly monotonic functions

We now know that the derivative is useful in *locating* the relative extrema of a function. In this section we will show that the derivative also can be used to *classify* relative extrema as either relative minima or relative maxima. We begin by defining what is meant when we say a function increases (or decreases) on an interval.

DEFINITION OF INCREASING AND DECREASING FUNCTIONS

A function f is said to be **increasing** on an interval if for any two numbers x_1 and x_2 in the interval,

$$x_1 < x_2 \quad \text{implies} \quad f(x_1) < f(x_2).$$

A function f is said to be **decreasing** on an interval if for any two numbers x_1 and x_2 in the interval,

$$x_1 < x_2 \quad \text{implies} \quad f(x_1) > f(x_2).$$

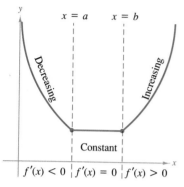

FIGURE 4.14

From this definition we can see that a function is increasing if its graph moves up as x moves to the right and a function is decreasing if its graph moves down as x moves to the right. For example, the function in Figure 4.14 is decreasing on the interval $(-\infty, a)$, is constant on the interval (a, b), and is increasing on the interval (b, ∞).

The derivative is useful in determining whether a function is increasing or decreasing on an interval. Specifically, as Figure 4.14 shows, a positive derivative implies that the graph slopes upward and the function is increasing. Similarly, a negative derivative implies that the function is decreasing. Finally, a zero derivative on an entire interval implies that the function is constant on the interval.

THEOREM 4.5 TEST FOR INCREASING OR DECREASING FUNCTIONS

Let f be a function that is differentiable on the interval (a, b).

1. If $f'(x) > 0$ for all x in (a, b), then f is increasing on (a, b).
2. If $f'(x) < 0$ for all x in (a, b), then f is decreasing on (a, b).
3. If $f'(x) = 0$ for all x in (a, b), then f is constant on (a, b).

PROOF

To prove the first case, we assume that $f'(x) > 0$ for all x in the interval (a, b) and let $x_1 < x_2$ be any two points in the interval. By the Mean Value Theorem, we know there exists a number c such that $x_1 < c < x_2$, and

$$f'(c) = \frac{f(x_2) - f(x_1)}{x_2 - x_1}.$$

Since $f'(c) > 0$ and $x_2 - x_1 > 0$, we know that $f(x_2) - f(x_1) > 0$, which implies that

$$f(x_1) < f(x_2).$$

Thus, f is increasing on the interval. The second case has a similar proof (see Exercise 59), and the third case was given as Exercise 39 in Section 4.2.

To apply Theorem 4.5, note that for a continuous function on an interval (a, b), $f'(x)$ can change sign only at its critical numbers. This suggests the following guidelines. These guidelines are also valid if the interval (a, b) is replaced by an interval of the form $(-\infty, b)$, (a, ∞), or $(-\infty, \infty)$.

GUIDELINES FOR FINDING INTERVALS ON WHICH A FUNCTION IS INCREASING OR DECREASING	Let f be continuous on the interval (a, b). To find the open intervals on which f is increasing or decreasing, we suggest the following steps. 1. Locate the critical numbers of f in (a, b), and use these numbers to determine test intervals. 2. Determine the sign of $f'(x)$ at one value in each of the test intervals. 3. Use Theorem 4.5 to decide whether f is increasing or decreasing on each interval.

REMARK These guidelines require that f be *continuous* on the interval (a, b). The guidelines must be modified for functions with points of discontinuity. (See Example 4.)

EXAMPLE 1 Determining intervals on which f is increasing or decreasing

Find the open intervals on which $f(x) = x^3 - \frac{3}{2}x^2$ is increasing or decreasing.

SOLUTION

Note that f is continuous on the entire real line. To determine the critical numbers of f, we set $f'(x)$ equal to zero.

$$f'(x) = 3x^2 - 3x = 0 \qquad \text{Let } f'(x) = 0$$
$$3(x)(x - 1) = 0 \qquad \text{Factor}$$
$$x = 0, 1 \qquad \text{Critical numbers}$$

Since there are no points for which f' is undefined, we conclude that $x = 0$ and $x = 1$ are the only critical numbers. Table 4.4 summarizes the testing of the three intervals determined by these critical numbers.

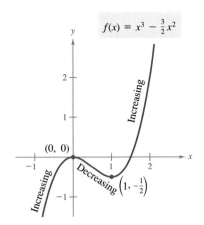

$f(x) = x^3 - \frac{3}{2}x^2$

(0, 0)

Increasing

Decreasing $\left(1, -\frac{1}{2}\right)$

Increasing

FIGURE 4.15

TABLE 4.4

Interval	$-\infty < x < 0$	$0 < x < 1$	$1 < x < \infty$
Test value	$x = -1$	$x = \dfrac{1}{2}$	$x = 2$
Sign of $f'(x)$	$f'(-1) = 6 > 0$	$f'\left(\dfrac{1}{2}\right) = -\dfrac{3}{4} < 0$	$f'(2) = 6 > 0$
Conclusion	increasing	decreasing	increasing

The graph of f is shown in Figure 4.15. Note that the test values in Table 4.4 were chosen for convenience; other points could have been used.

The First Derivative Test

Once we have determined the intervals on which a function is increasing or decreasing, we can locate the relative extrema of the function easily. For instance, in Figure 4.15 (Example 1), the function $f(x) = x^3 - \frac{3}{2}x^2$ has a relative maximum at the point $(0, 0)$ because f is increasing immediately to the left of $x = 0$ and decreasing immediately to the right of $x = 0$. Similarly, f has a relative minimum at the point $\left(1, -\frac{1}{2}\right)$ because f is decreasing immediately to the left of $x = 1$ and increasing immediately to the right of $x = 1$. The following theorem, called the First Derivative Test, makes this more explicit.

THEOREM 4.6
THE FIRST DERIVATIVE TEST

Let c be a critical number of a function f that is continuous on an open interval I containing c. If f is differentiable on the interval, except possibly at c, then $f(c)$ can be classified as follows.

1. If f' changes from negative to positive at c, then $f(c)$ is a **relative minimum** of f.
2. If f' changes from positive to negative at c, then $f(c)$ is a **relative maximum** of f.
3. If f' does not change signs at c, then $f(c)$ is neither a relative minimum nor a relative maximum.

Relative minimum

Relative maximum

Neither relative minimum nor maximum

PROOF   We prove the first case and leave the other two cases as exercises (see Exercise 60). Assume that f' changes from negative to positive at c. Then there exist a and b in I such that

$$f'(x) < 0 \text{ for all } x \text{ in } (a, c)$$

and

$$f'(x) > 0 \text{ for all } x \text{ in } (c, b).$$

By Theorem 4.5, f is decreasing on (a, c) and increasing on (c, b). Therefore, $f(c)$ is a minimum of f on the open interval (a, b) and, consequently, a relative minimum of f.

To apply the First Derivative Test, we suggest the same tabular format used in Example 1.

EXAMPLE 2 Applying the First Derivative Test

Find the relative extrema of the function given by $f(x) = \frac{1}{2}x - \sin x$ in the interval $(0, 2\pi)$.

SOLUTION

Note that f is continuous on the interval $(0, 2\pi)$. To determine the critical numbers of f in this interval, we set $f'(x)$ equal to zero.

$$f'(x) = \frac{1}{2} - \cos x = 0 \qquad \text{Let } f'(x) = 0$$

$$\cos x = \frac{1}{2}$$

$$x = \frac{\pi}{3}, \frac{5\pi}{3} \qquad \text{Critical numbers}$$

Since there are no points for which f' is undefined, we conclude that $x = \pi/3$ and $x = 5\pi/3$ are the only critical numbers. Table 4.5 summarizes the testing of the three intervals determined by these critical numbers.

TABLE 4.5

Interval	$0 < x < \dfrac{\pi}{3}$	$\dfrac{\pi}{3} < x < \dfrac{5\pi}{3}$	$\dfrac{5\pi}{3} < x < 2\pi$
Test value	$x = \dfrac{\pi}{4}$	$x = \pi$	$x = \dfrac{7\pi}{4}$
Sign of $f'(x)$	$f'\left(\dfrac{\pi}{4}\right) < 0$	$f'(\pi) > 0$	$f'\left(\dfrac{7\pi}{4}\right) < 0$
Conclusion	decreasing	increasing	decreasing

$$f(x) = \frac{x}{2} - \sin x$$

FIGURE 4.16

Finally, applying the First Derivative Test, we conclude that f has a relative minimum at $x = \pi/3$ and a relative maximum at $x = 5\pi/3$, as shown in Figure 4.16.

Note that in Examples 1 and 2 the given functions are differentiable on the entire real line. For such functions, the only critical numbers are those for which $f'(x) = 0$. In Example 3, we look at a function that has two types of critical numbers—those for which $f'(x) = 0$ and those for which f' is undefined.

EXAMPLE 3 Applying the First Derivative Test

Find the relative extrema of $f(x) = (x^2 - 4)^{2/3}$.

SOLUTION

We begin by noting that f is continuous on the entire real line. The derivative of f,

$$f'(x) = \frac{2}{3}(x^2 - 4)^{-1/3}(2x) = \frac{4x}{3(x^2 - 4)^{1/3}}$$

is zero when $x = 0$ and undefined when $x = \pm 2$. Thus, the critical numbers are $x = -2$, $x = 0$, and $x = 2$. Table 4.6 summarizes the testing of the four test intervals determined by these three critical numbers.

TABLE 4.6

Interval	$-\infty < x < -2$	$-2 < x < 0$	$0 < x < 2$	$2 < x < \infty$
Test value	$x = -3$	$x = -1$	$x = 1$	$x = 3$
Sign of $f'(x)$	$f'(-3) < 0$	$f'(-1) > 0$	$f'(1) < 0$	$f'(3) > 0$
Conclusion	decreasing	increasing	decreasing	increasing

Finally, applying the First Derivative Test, we conclude that f has a relative minimum at the point $(-2, 0)$, a relative maximum at the point $(0, 4^{2/3})$, and another relative minimum at the point $(2, 0)$, as shown in Figure 4.17.

$f(x) = (x^2 - 4)^{2/3}$

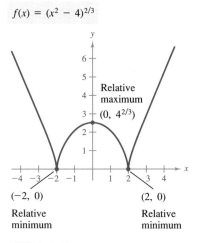

FIGURE 4.17

If a function has points of discontinuity, then these points should be used along with the critical numbers to determine the test intervals. This is demonstrated in Example 4.

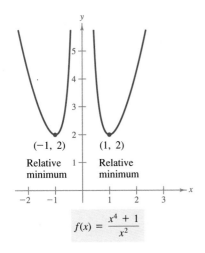

$$f(x) = \frac{x^4 + 1}{x^2}$$

FIGURE 4.18

EXAMPLE 4 The First Derivative Test and points of discontinuity

Find the relative extrema of $f(x) = (x^4 + 1)/x^2$.

SOLUTION

Using the Quotient Rule, we find the derivative of f to be

$$f'(x) = \frac{2(x^4 - 1)}{x^3} = \frac{2(x^2 + 1)(x - 1)(x + 1)}{x^3}.$$

Since $f'(x)$ is zero at $x = \pm 1$ and f is discontinuous at $x = 0$, we use these three x-values to determine the test intervals.

$x = -1$ and $x = 1$ Critical numbers

$x = 0$ Point of discontinuity

Table 4.7 summarizes the testing of the four intervals determined by these three x-values.

TABLE 4.7

Interval	$-\infty < x < -1$	$-1 < x < 0$	$0 < x < 1$	$1 < x < \infty$
Test value	$x = -2$	$x = -\dfrac{1}{2}$	$x = \dfrac{1}{2}$	$x = 2$
Sign of $f'(x)$	$f'(-2) < 0$	$f'\left(-\dfrac{1}{2}\right) > 0$	$f'\left(\dfrac{1}{2}\right) < 0$	$f'(2) > 0$
Conclusion	decreasing	increasing	decreasing	increasing

Finally, applying the First Derivative Test, we conclude that f has one relative minimum at the point $(-1, 2)$ and another at the point $(1, 2)$, as shown in Figure 4.18.

(a) Strictly monotonic function

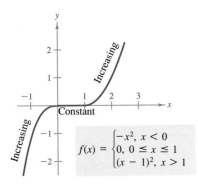

(b) Not a strictly monotonic function

FIGURE 4.19

Strictly monotonic functions

A function is called **strictly monotonic** on an interval if it is either increasing on the entire interval or decreasing on the entire interval. For instance, the function $f(x) = x^3$ is strictly monotonic on the entire real line because it is increasing on the entire real line, as shown in Figure 4.19(a). The function shown in Figure 4.19(b) is not *strictly* monotonic on the entire real line because it is constant on the interval [0, 1].

Strictly monotonic functions play a special role in calculus. (We will say more about this in Chapter 6.)

EXERCISES for Section 4.3

In Exercises 1–6, identify the open intervals on which the function is increasing or decreasing.

1. $f(x) = x^2 - 6x + 8$ **2.** $y = -(x + 1)^2$

3. $y = \dfrac{x^3}{4} - 3x$ **4.** $f(x) = x^4 - 2x^2$

 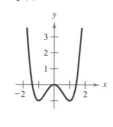

5. $f(x) = \dfrac{1}{x^2}$ **6.** $y = \dfrac{x^2}{x + 1}$

In Exercises 7–30, find the critical numbers of f (if any), find the open intervals on which f is increasing or decreasing, and locate all relative extrema.

7. $f(x) = -2x^2 + 4x + 3$ **8.** $f(x) = x^2 + 8x + 10$

9. $f(x) = x^2 - 6x$ **10.** $f(x) = (x - 1)^2(x + 2)$

11. $f(x) = 2x^3 + 3x^2 - 12x$

12. $f(x) = (x - 3)^3$

13. $f(x) = x^3 - 6x^2 + 15$ **14.** $f(x) = x^4 - 2x^3$

15. $f(x) = (x - 1)^{2/3}$ **16.** $f(x) = (x - 1)^{1/3}$

17. $f(x) = x^{1/3} + 1$ **18.** $f(x) = x^{2/3}(x - 5)$

19. $f(x) = x + \dfrac{1}{x}$ **20.** $f(x) = \dfrac{x}{x + 1}$

21. $f(x) = \dfrac{x^2}{x^2 - 9}$ **22.** $f(x) = \dfrac{x + 3}{x^2}$

23. $f(x) = \dfrac{x^5 - 5x}{5}$ **24.** $f(x) = x^4 - 32x + 4$

25. $f(x) = \dfrac{x^2 - 2x + 1}{x + 1}$ **26.** $f(x) = \dfrac{x^2 - 3x - 4}{x - 2}$

27. $f(x) = \dfrac{x}{2} + \cos x, \; 0 \le x < 2\pi$

28. $f(x) = \sin x \cos x, \; 0 \le x < 2\pi$

29. $f(x) = \sin^2 x + \sin x, \; 0 \le x \le 2\pi$

30. $f(x) = \dfrac{\cos x}{1 + \sin^2 x}, \; 0 \le x < 2\pi$

In Exercises 31–34, determine whether the given function is strictly monotonic on the indicated interval.

31. $f(x) = x^2$

(a) $(-\infty, \infty)$ (b) $(-\infty, 0)$ (c) $(0, \infty)$

32. $f(x) = x^3 - x$

(a) $(-1, 0)$ (b) $\left(-1, -\dfrac{1}{2}\right)$ (c) $(-1, 1)$

33. $f(x) = \sin x$

(a) $(0, 2\pi)$ (b) $(0, \pi)$ (c) $\left(-\dfrac{\pi}{2}, \dfrac{\pi}{2}\right)$

34. $f(x) = \cos x$

(a) $(0, 2\pi)$ (b) $(0, \pi)$ (c) $\left(-\dfrac{\pi}{2}, \dfrac{\pi}{2}\right)$

35. The height (in feet) of a ball at time t (in seconds) is given by the position function

$$s(t) = 96t - 16t^2.$$

Find the open interval on which the ball is moving up and the open interval on which it is moving down. What is the maximum height of the ball?

36. Repeat Exercise 35 using the position function

$$s(t) = -16t^2 + 64t.$$

37. Coughing forces the trachea (windpipe) to contract, which affects the velocity v of the air passing through the trachea. Suppose the velocity of the air during coughing is

$$v = k(R - r)r^2$$

where k is a constant, R is the normal radius of the trachea, and r is the radius during coughing. What radius will produce the maximum air velocity?

38. The concentration C of a certain chemical in the bloodstream t hours after injection into muscle tissue is given by

$$C = \dfrac{3t}{27 + t^3}.$$

When is the concentration greatest?

39. After a drug is administered to a patient, the drug concentration in the patient's bloodstream over a two-hour period is given by

$$C = 0.29483t + 0.04253t^2 - 0.00035t^3$$

where C is measured in milligrams and t is the time in minutes. Find the open interval on which C is increasing or decreasing.

40. A fast-food restaurant sells x hamburgers to make a profit P given by

$$P = 2.44x - \frac{x^2}{20,000} - 5000, \quad 0 \le x \le 35,000.$$

Find the open interval on which P is increasing or decreasing.

41. After birth, an infant normally will lose weight for a few days and then start gaining. A model for the average weight W of infants over the first two weeks following birth is

$$W = 0.033t^2 - 0.3974t + 7.3032.$$

Find the open intervals on which W is increasing or decreasing.

42. The electric power P in watts in a direct-current circuit with two resistors R_1 and R_2 connected in series is

$$P = \frac{vR_1R_2}{(R_1 + R_2)^2}$$

where v is the voltage. If v and R_1 are held constant, what resistance R_2 produces maximum power?

43. The resistance R of a certain type of resistor is given by

$$R = \sqrt{0.001T^4 - 4T + 100}$$

where R is measured in ohms and the temperature T is measured in degrees Celsius. What temperature produces a minimum resistance for this type of resistor?

44. Consider the functions $f(x) = x$ and $g(x) = \sin x$ on the interval $(0, \pi)$.
   (a) Prove that $f(x) > g(x)$. [Hint: Show that $h'(x) > 0$ where $h = f - g$.]
   (b) Sketch the graphs of f and g on the same set of axes.

45. A manufacturer of fertilizer finds that the national sales of fertilizer roughly follow the cyclical pattern

$$F = 100,000\left[1 + \sin\left(\frac{2\pi[t - 60]}{365}\right)\right]$$

where F is measured in pounds and t is measured in days. If $t = 1$ represents January 1, on which day of the year is the maximum amount of fertilizer sold?

46. The normal average daily temperature (in degrees Fahrenheit) for a certain city is given by

$$T = 45 - 23\cos\left[\frac{2\pi(t - 32)}{365}\right]$$

where t is the time in days with $t = 1$ corresponding to January 1. Find the expected date of (a) the warmest day and (b) the coldest day.

47. Find a, b, c, and d so that the function given by

$$f(x) = ax^3 + bx^2 + cx + d$$

has a relative minimum at $(0, 0)$ and a relative maximum at $(2, 2)$.

48. Find a, b, and c so that the function given by

$$f(x) = ax^2 + bx + c$$

has a relative maximum at $(5, 20)$ and passes through the point $(2, 10)$.

In Exercises 49–54, assume that f is differentiable for all x. The sign of f' is as follows.

$f'(x) > 0$ on $(-\infty, -4)$

$f'(x) < 0$ on $(-4, 6)$

$f'(x) > 0$ on $(6, \infty)$

In each exercise, supply the appropriate inequality for the indicated value of c.

Function	Sign of $g'(c)$
49. $g(x) = f(x) + 5$	$g'(0)$ ___ 0
50. $g(x) = 3f(x) - 3$	$g'(-5)$ ___ 0
51. $g(x) = -f(x)$	$g'(-6)$ ___ 0
52. $g(x) = -f(x)$	$g'(0)$ ___ 0
53. $g(x) = f(x - 10)$	$g'(0)$ ___ 0
54. $g(x) = f(x - 10)$	$g'(8)$ ___ 0

In Exercises 55–58, use a computer or graphics calculator (a) to sketch the graph of f and f' on the same coordinate axes over the specified interval, (b) to find the critical numbers of f, and (c) to find the interval(s) on which f' is positive and the interval(s) on which it is negative. Note the behavior of f in relation to the sign of f'.

Function	Interval
55. $f(x) = 2x\sqrt{9 - x^2}$	$[-3, 3]$
56. $f(x) = 10(5 - \sqrt{x^2 - 3x + 16})$	$[0, 5]$
57. $f(t) = t^2 \sin t$	$[0, 2\pi]$
58. $f(x) = \dfrac{x}{2} + \cos\dfrac{x}{2}$	$[0, 4\pi]$

59. Prove the second case of Theorem 4.5.

60. Prove the second and third cases of Theorem 4.6.

4.4 Concavity and the Second Derivative Test

Concavity ▪ Points of inflection ▪ The Second Derivative Test

We have already seen that locating the intervals in which a function f increases or decreases helps to determine its graph. In this section we show that, by locating the intervals in which f' increases or decreases, we can determine where the graph of f is *curving upward* or *curving downward*. We refer to this notion of curving upward or downward as **concavity.**

DEFINITION OF CONCAVITY

Let f be differentiable on an open interval. We say that the graph of f is **concave upward** if f' is increasing on the interval and **concave downward** if f' is decreasing on the interval.

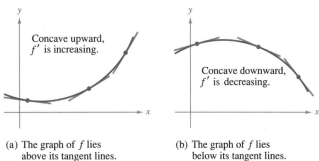

(a) The graph of f lies above its tangent lines.

(b) The graph of f lies below its tangent lines.

FIGURE 4.20

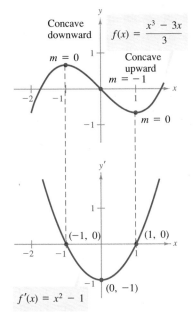

$f'(x) = x^2 - 1$

f' is decreasing. f' is increasing.

FIGURE 4.21

A comparable definition of concavity is given in the two statements that follow. (See Appendix A for a proof of the equivalence of the two definitions.)

1. Let f be differentiable at c. The graph of f is **concave upward** at $(c, f(c))$ if the graph of f lies *above* the tangent line at $(c, f(c))$ on some open interval containing c. (See Figure 4.20(a).)
2. Let f be differentiable at c. The graph of f is **concave downward** at $(c, f(c))$ if the graph of f lies *below* the tangent line at $(c, f(c))$ on some open interval containing c. (See Figure 4.20(b).)

To find the open intervals on which the graph of a function f is concave upward or downward, we need to find the intervals on which f' is increasing or decreasing. For instance, the graph of $f(x) = \frac{1}{3}x^3 - x$ is concave downward on the open interval $(-\infty, 0)$ because f' is decreasing there. (See Figure 4.21.) Similarly, the graph of f is concave upward on the interval $(0, \infty)$ because f' is increasing on $(0, \infty)$.

The following theorem shows how to use the *second* derivative of a function f to determine intervals on which the graph of f is concave upward or downward. A proof of this theorem follows directly from Theorem 4.5 and the definition of concavity.

THEOREM 4.7 TEST FOR CONCAVITY	Let f be a function whose second derivative exists on an open interval I.
	1. If $f''(x) > 0$ for all x in I, then the graph of f is concave upward.
	2. If $f''(x) < 0$ for all x in I, then the graph of f is concave downward.

REMARK A third conclusion to Theorem 4.7 is that if $f''(x) = 0$ for all x in I, then f is linear. Note, however, that we do not define concavity for straight lines. In other words, a straight line is neither concave upward nor concave downward.

We suggest the following guidelines for applying Theorem 4.7. First, locate the x-values at which $f''(x) = 0$ or f'' is undefined. Second, use these x-values to determine test intervals. Finally, test the sign of $f''(x)$ in each of the test intervals. We illustrate the procedure in Example 1.

EXAMPLE 1   Determining concavity

Determine the open intervals on which the graph of $f(x) = 6(x^2 + 3)^{-1}$ is concave upward or downward.

SOLUTION

We begin by observing that f is continuous on the entire real line. Next, we find the second derivative of f.

$$f'(x) = (-6)(2x)(x^2 + 3)^{-2} = \frac{-12x}{(x^2 + 3)^2}$$

$$f''(x) = \frac{(x^2 + 3)^2(-12) - (-12x)(2)(2x)(x^2 + 3)}{(x^2 + 3)^4} = \frac{36(x^2 - 1)}{(x^2 + 3)^3}$$

Since $f''(x) = 0$ when $x = \pm 1$ and f'' is defined on the entire real line, we test f'' in the intervals $(-\infty, -1)$, $(-1, 1)$, and $(1, \infty)$. The results are shown in Table 4.8 and Figure 4.22.

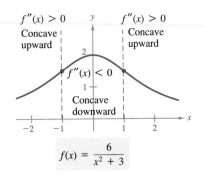

$f''(x) > 0$ Concave upward $f''(x) > 0$ Concave upward

$f''(x) < 0$ Concave downward

$$f(x) = \frac{6}{x^2 + 3}$$

FIGURE 4.22

TABLE 4.8

Interval	$-\infty < x < -1$	$-1 < x < 1$	$1 < x < \infty$
Test value	$x = -2$	$x = 0$	$x = 2$
Sign of $f''(x)$	$f''(-2) > 0$	$f''(0) < 0$	$f''(2) > 0$
Conclusion	Concave upward	Concave downward	Concave upward

The function given in Example 1 is continuous on the entire real line. If a function has one or more points of discontinuity, then these points should be used (along with the points at which $f''(x)$ is zero or undefined) to form the test intervals. This is illustrated in Example 2.

EXAMPLE 2   Determining concavity for a discontinuous function

Determine the open intervals in which the graph of

$$f(x) = \frac{x^2 + 1}{x^2 - 4}$$

is concave upward or downward.

SOLUTION

$$f'(x) = \frac{(x^2 - 4)(2x) - (x^2 + 1)(2x)}{(x^2 - 4)^2} = \frac{-10x}{(x^2 - 4)^2}$$

$$f''(x) = -10 \frac{(x^2 - 4)^2(1) - (x)(2)(2x)(x^2 - 4)}{(x^2 - 4)^4} = \frac{10(3x^2 + 4)}{(x^2 - 4)^3}$$

There are no points at which f'' is zero, but at $x = \pm 2$ the function f is discontinuous, so we test for concavity in the intervals $(-\infty, -2)$, $(-2, 2)$, and $(2, \infty)$, as shown in Table 4.9. The graph of f is shown in Figure 4.23.

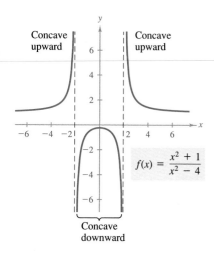

Concave upward   Concave upward

$$f(x) = \frac{x^2 + 1}{x^2 - 4}$$

Concave downward

FIGURE 4.23

TABLE 4.9

Interval	$-\infty < x < -2$	$-2 < x < 2$	$2 < x < \infty$
Test value	$x = -3$	$x = 0$	$x = 3$
Sign of $f''(x)$	$f''(-3) > 0$	$f''(0) < 0$	$f''(3) > 0$
Conclusion	Concave upward	Concave downward	Concave upward

Points of inflection

The graph in Figure 4.22 has two points at which the concavity changes. If the tangent line to the graph exists at such a point, the point is called a **point of inflection.** Three types of points of inflection are shown in Figure 4.24.

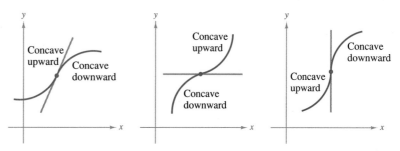

Concave upward   Concave downward

Concave upward   Concave downward

Concave downward   Concave upward

FIGURE 4.24   The graph *crosses* its tangent line at a point of inflection.

DEFINITION OF POINT OF INFLECTION

Let f be a function whose graph has a tangent line at $(c, f(c))$. The point $(c, f(c))$ is called a **point of inflection** if the concavity of f changes from upward to downward (or vice versa).

REMARK   Note in Figure 4.24 that the graph crosses its tangent line at a point of inflection.

Since a point of inflection occurs where the concavity of a graph changes, it must be true that the sign of f'' changes at such points. Thus, to locate possible points of inflection, we need only determine the values of x for which $f''(x) = 0$ or for which f'' is undefined. This is similar to the procedure for locating relative extrema of f.

THEOREM 4.8 POINTS OF INFLECTION	If $(c, f(c))$ is a point of inflection of the graph of f, then either $f''(c) = 0$ or f'' is undefined at $x = c$.

EXAMPLE 3   Finding points of inflection

Determine the points of inflection and discuss the concavity of the graph of $f(x) = x^4 - 4x^3$.

SOLUTION

Differentiating twice produces

$$f'(x) = 4x^3 - 12x^2$$
$$f'' = 12x^2 - 24x = 12x(x - 2).$$

Possible points of inflection occur at $x = 0$ and $x = 2$. By testing the intervals determined by these x-values, we conclude that they both yield points of inflection. A summary of this testing is shown in Table 4.10, and the graph of f is shown in Figure 4.25.

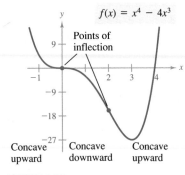

$f(x) = x^4 - 4x^3$

Points of inflection

Concave upward   Concave downward   Concave upward

FIGURE 4.25

TABLE 4.10

Interval	$-\infty < x < 0$	$0 < x < 2$	$2 < x < \infty$
Test value	$x = -1$	$x = 1$	$x = 3$
Sign of $f''(x)$	$f''(-1) > 0$	$f''(1) < 0$	$f''(3) > 0$
Conclusion	Concave upward	Concave downward	Concave upward

It is possible for the second derivative to be zero at a point that is *not* a point of inflection. For instance, the graph of $f(x) = x^4$ is shown in Figure 4.26. The second derivative is zero when $x = 0$, but the point $(0, 0)$ is not a point of inflection because the graph of f is concave upward to the left and to the right of $x = 0$.

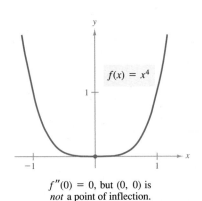

$f(x) = x^4$

$f''(0) = 0$, but $(0, 0)$ is *not* a point of inflection.

FIGURE 4.26

The Second Derivative Test

The second derivative sometimes can be used to perform a simple test for relative maxima and minima. The test is based on the fact that if f is a function such that $f'(c) = 0$ and there exists an open interval containing c on which the graph of f is concave upward, then $f(c)$ must be a relative minimum of f. Similarly, if f is a function such that $f'(c) = 0$ and there exists an open interval containing c on which the graph of f is concave downward, then $f(c)$ must be a relative maximum of f. (See Figure 4.27.)

FIGURE 4.27   Relative maximum   Relative minimum

THEOREM 4.9
SECOND DERIVATIVE TEST

Let f be a function such that $f'(c) = 0$ and the second derivative of f exists on an open interval containing c.

1. If $f''(c) > 0$, then $f(c)$ is a relative minimum.
2. If $f''(c) < 0$, then $f(c)$ is a relative maximum.
3. If $f''(c) = 0$, then the test fails.

PROOF This theorem follows from Theorem 4.7. We outline a proof of the first case. (The rest of the proof is left to you.) If $f''(c) > 0$, then f is concave upward in some interval I containing c. This implies that the graph of f lies above its tangent lines in I. Since $f'(c) = 0$, the tangent line must be horizontal at $(c, f(c))$. Thus, we can conclude that $f(c)$ is a minimum of f in the interval I, and consequently $f(c)$ must be a relative minimum of f.

REMARK   Be sure to understand that if $f''(c) = 0$, the Second Derivative Test does not apply. In such cases we can use the First Derivative Test.

EXAMPLE 4 Using the Second Derivative Test

Find the relative extrema for $f(x) = -3x^5 + 5x^3$.

SOLUTION

We begin by finding the critical numbers of f.

$$f'(x) = -15x^4 + 15x^2 = 15x^2(1 - x^2) = 0$$
$$x = -1, 0, 1 \qquad\qquad \text{Critical numbers}$$

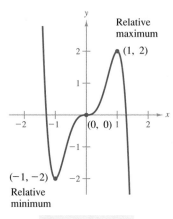

Relative maximum

(1, 2)

(0, 0)

(−1, −2)

Relative minimum

$f(x) = -3x^5 + 5x^3$

FIGURE 4.28

Using $f''(x) = 30(-2x^3 + x)$, we apply the Second Derivative Test as follows.

Point	Sign of f''	Conclusion
$(-1, -2)$	$f''(-1) = 30 > 0$	Relative minimum
$(1, 2)$	$f''(1) = -30 < 0$	Relative maximum
$(0, 0)$	$f''(0) = 0$	Test fails

Since the Second Derivative Test fails at $(0, 0)$, we use the First Derivative Test and observe that f increases to the left and right of $x = 0$. Thus, $(0, 0)$ is neither a relative minimum nor a relative maximum (even though the graph has a horizontal tangent line at this point). The graph of f is shown in Figure 4.28.

EXERCISES for Section 4.4

In Exercises 1–6, find the open intervals on which the graph of the given function is concave upward and those on which it is concave downward.

1. $y = x^2 - x - 2$

2. $f(x) = \dfrac{24}{x^2 + 12}$

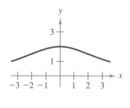

3. $y = -x^3 + 3x^2 - 2$

4. $f(x) = \dfrac{x^2 - 1}{2x + 1}$

5. $f(x) = \dfrac{x^2 + 1}{x^2 - 1}$

6. $y = \dfrac{-3x^5 + 40x^3 + 135x}{270}$

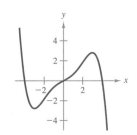

In Exercises 7–20, identify all relative extrema. Use the Second Derivative Test where applicable.

7. $f(x) = 6x - x^2$

8. $f(x) = x^2 + 3x - 8$

9. $f(x) = (x - 5)^2$

10. $f(x) = -(x - 5)^2$

11. $f(x) = x^3 - 3x^2 + 3$

12. $f(x) = 5 + 3x^2 - x^3$

13. $f(x) = x^4 - 4x^3 + 2$

14. $f(x) = x^3 - 9x^2 + 27x - 26$

15. $f(x) = x^{2/3} - 3$

16. $f(x) = \sqrt{x^2 + 1}$

17. $f(x) = x + \dfrac{4}{x}$

18. $f(x) = \dfrac{x}{x - 1}$

19. $f(x) = \cos x - x,\ 0 \le x \le 4\pi$

20. $f(x) = 2 \sin x + \cos 2x,\ 0 \le x \le 2\pi$

In Exercises 21–42, sketch the graph of the given function and identify all relative extrema and points of inflection.

21. $f(x) = x^3 - 12x$

22. $f(x) = x^3 + 1$

23. $f(x) = x^3 - 6x^2 + 12x - 8$

24. $f(x) = 2x^3 - 3x^2 - 12x + 8$

25. $f(x) = \dfrac{1}{4}x^4 - 2x^2$

26. $f(x) = 2x^4 - 8x + 3$

27. $f(x) = x(x - 4)^3$

28. $f(x) = x^3(x - 4)$

29. $f(x) = x^2 + \dfrac{1}{x^2}$

30. $f(x) = \dfrac{x^2}{x^2 - 1}$

31. $f(x) = x\sqrt{x + 3}$

32. $f(x) = x\sqrt{x + 1}$

33. $f(x) = \dfrac{x}{x^2 - 4}$

34. $f(x) = \dfrac{1}{x^2 - x - 2}$

35. $f(x) = \dfrac{x - 2}{x^2 - 4x + 3}$

36. $f(x) = \dfrac{x + 1}{x^2 + x + 1}$

37. $f(x) = \sin \dfrac{x}{2},\quad 0 \le x \le 4\pi$

38. $f(x) = 2 \csc \dfrac{3x}{2},\quad 0 \le x \le 2\pi$

39. $f(x) = \sec \left(x - \dfrac{\pi}{2} \right),\quad 0 \le x \le 4\pi$

40. $f(x) = \sin x + \cos x,\quad 0 \le x \le 2\pi$

41. $f(x) = 2 \sin x + \sin 2x,\quad 0 \le x \le 2\pi$

42. $f(x) = x - \sin x,\quad 0 \le x \le 4\pi$

43. Sketch the graph of a function f having the following characteristics.

(a) $f(2) = f(4) = 0$

$f'(x) < 0$ if $x < 3$

$f'(3)$ is undefined

$f'(x) > 0$ if $x > 3$

$f''(x) < 0,\ x \ne 3$

(b) $f(0) = f(2) = 0$

$f'(x) > 0$ if $x < 1$

$f'(1) = 0$

$f'(x) < 0$ if $x > 1$

$f''(x) < 0$

44. Sketch the graph of a function f having the following characteristics.

(a) $f(2) = f(4) = 0$

$f'(x) > 0$ if $x < 3$

$f'(3)$ is undefined

$f'(x) < 0$ if $x > 3$

$f''(x) > 0,\ x \ne 3$

(b) $f(0) = f(2) = 0$

$f'(x) < 0$ if $x < 1$

$f'(1) = 0$

$f'(x) > 0$ if $x > 1$

$f''(x) > 0$

In Exercises 45–48, trace the given graph of f. On the same set of axes sketch the graph of f' and f''.

45.

46.

47.

48.

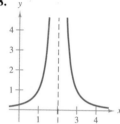

49. Find the inflection points (if any) and sketch the graph of $f(x) = (x - c)^n$ for $n = 1, 2, 3,$ and 4. What conclusion can be made about the value of n and the existence of an inflection point on the graph of f?

50. (a) Sketch the graph of $f(x) = \sqrt[3]{x}$ and identify the inflection point.

(b) Does $f''(x)$ exist at the inflection point?

51. Find a cubic polynomial function that has a relative maximum at $(3, 3)$, a relative minimum at $(5, 1)$, and a point of inflection at $(4, 2)$.

52. Find a cubic polynomial function that has a relative maximum at $(2, 4)$, a relative minimum at $(4, 2)$, and a point of inflection at $(3, 3)$.

53. Show that the point of inflection of

$$f(x) = x(x - 6)^2$$

lies midway between the relative extrema of f.

54. Prove that a cubic function with three distinct real zeros has a point of inflection whose x-coordinate is the average of the three zeros.

55. The deflection D of a particular beam of length L is given by

$$D = 2x^4 - 5Lx^3 + 3L^2x^2$$

where x is the distance from one end of the beam. Find the value of x that yields the maximum deflection.

56. The equation

$$E = \frac{kqx}{(x^2 + a^2)^{3/2}}$$

gives the electric field intensity on the axis of a uniformly charged ring, where q is the total charge on the ring, k is a constant, and a is the radius of the ring. At what value of x is E maximum?

57. A manufacturer has determined that the total cost C of operating a certain facility is given by

$$C = 0.5x^2 + 15x + 5000$$

where x is the number of units produced. At what level of production will the average cost per unit be minimum? (The average cost per unit is given by C/x.)

58. The total cost C for ordering and storing x units is

$$C = 2x + \frac{300{,}000}{x}.$$

What order size will produce a minimum cost?

59. For the engine shown in the accompanying figure, a connecting rod 7 inches long is fastened to a crank of radius 3 inches at point P. The crankshaft rotates counterclockwise at a constant rate of 200 revolutions per minute. The horizontal velocity (in/min) of point P is given by

$$v = -1200\pi \sin \theta$$

where θ is the central angle of the crankshaft. What values of θ produce a maximum velocity?

FIGURE FOR 59

60. The graph of $y = x \sin (1/x)$ is shown in the accompanying figure. Show that the graph is concave downward to the right of $x = 1/\pi$.

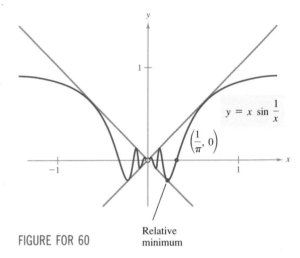

FIGURE FOR 60

Relative minimum

▦ In Exercises 61–64, use a computer or graphics calculator to (a) sketch the graphs of f, f', and f'' on the same coordinate axes over the specified interval, (b) find any critical numbers and the coordinates of the relative extrema of f, and (c) find any points of inflection of the graph of f.

Function	*Interval*
61. $f(x) = \dfrac{2x}{x^2 - x + 1}$	$[-2, 2]$
62. $f(x) = \dfrac{4x^2}{x^2 + 2}$	$[-3, 3]$
63. $f(x) = \sin x - \dfrac{1}{3} \sin 3x + \dfrac{1}{5} \sin 5x$	$[0, \pi]$
64. $f(x) = \sqrt{2x} \sin x$	$[0, 2\pi]$

4.5 Limits at Infinity

Limits at infinity ▪ Horizontal asymptotes

In this section we take a further look at the behavior of a function on *infinite* intervals. Consider the graph of

$$f(x) = \frac{3x^2}{x^2 + 1}$$

as shown in Figure 4.29. Table 4.11 suggests that the value of $f(x)$ approaches 3 as x increases without bound ($x \to \infty$). Similarly, $f(x)$ approaches 3 as $x \to -\infty$. We denote these **limits at infinity** by

$$\lim_{x \to -\infty} f(x) = 3 \quad \text{and} \quad \lim_{x \to \infty} f(x) = 3.$$

$f(x) = \dfrac{3x^2}{x^2 + 1}$

$f(x) \to 3$ as $x \to -\infty$

$f(x) \to 3$ as $x \to \infty$

FIGURE 4.29

TABLE 4.11

	← x decreases without bound					x increases without bound →			
x	$-\infty \leftarrow$	-100	-10	-1	0	1	10	100	$\rightarrow \infty$
$f(x)$	$3 \quad \leftarrow$	2.9997	2.97	1.5	0	1.5	2.97	2.9997	$\rightarrow 3$
		← $f(x)$ approaches 3				$f(x)$ approaches 3 →			

To say that a statement is true as x increases *without bound* means that for any (large) real number M, the statement is true for *all* x in the interval $\{x: x > M\}$. The following definition uses this concept.

DEFINITION OF LIMITS AT INFINITY

The statement

$$\lim_{x \to \infty} f(x) = L$$

means that for each $\varepsilon > 0$ there exists an $M > 0$ such that $|f(x) - L| < \varepsilon$ whenever $x > M$. Similarly, the statement

$$\lim_{x \to -\infty} f(x) = L$$

means that for each $\varepsilon > 0$ there exists an $N < 0$ such that $|f(x) - L| < \varepsilon$ whenever $x < N$.

REMARK   Remember that when we write

$$\lim_{x \to -\infty} f(x) = L \qquad \text{or} \qquad \lim_{x \to \infty} f(x) = L$$

we mean that the limit exists *and* the limit is equal to the number L.

The definition of a limit at infinity is illustrated graphically in Figure 4.30. In this figure, note that for a given positive number ε there exists a postive number M such that, *to the right of $x = M$, the graph of f will lie between the horizontal lines given by $y = L \pm \varepsilon$.*

In Figure 4.30, the graph of f approaches the line $y = L$ as x increases without bound. We call the line $y = L$ a **horizontal asymptote** of the graph of f.

FIGURE 4.30

DEFINITION OF HORIZONTAL ASYMPTOTE

If

$$\lim_{x \to -\infty} f(x) = L \qquad \text{or} \qquad \lim_{x \to \infty} f(x) = L$$

then the line $y = L$ is called a **horizontal asymptote** of the graph of f.

REMARK   From this definition, it follows that the graph of a *function* of x can have at most two horizontal asymptotes—one to the right and one to the left.

Limits at infinity have many of the same properties of limits discussed in Section 2.2. For example, if $\lim_{x \to \infty} f(x)$ and $\lim_{x \to \infty} g(x)$ both exist, then

$$\lim_{x \to \infty} [f(x) + g(x)] = \lim_{x \to \infty} f(x) + \lim_{x \to \infty} g(x)$$

and

$$\lim_{x \to \infty} [f(x)g(x)] = \left(\lim_{x \to \infty} f(x) \right) \left(\lim_{x \to \infty} g(x) \right).$$

Similar properties hold for limits at $-\infty$.

When evaluating limits at infinity, the following theorem is helpful. (A proof of this theorem is given in Appendix A.)

**THEOREM 4.10
LIMITS AT INFINITY**

If r is a positive rational number and c is any real number, then

$$\lim_{x \to \infty} \frac{c}{x^r} = 0.$$

Furthermore, if x^r is defined when $x < 0$, then $\lim_{x \to -\infty} (c/x^r) = 0.$

EXAMPLE 1 Evaluating a limit at infinity

Find the following limit.

$$\lim_{x \to \infty} \left(5 - \frac{2}{x^2} \right)$$

SOLUTION

Using Theorem 4.10, we obtain

$$\lim_{x \to \infty} \left(5 - \frac{2}{x^2} \right) = \lim_{x \to \infty} 5 - \lim_{x \to \infty} \frac{2}{x^2} = 5 - 0 = 5. \qquad \square$$

EXAMPLE 2 Evaluating a limit at infinity

Find the following limit.

$$\lim_{x \to \infty} \frac{2x - 1}{x + 1}$$

SOLUTION

Note that both the numerator and the denominator approach infinity as x approaches infinity.

$$\lim_{x \to \infty} \frac{2x - 1}{x + 1}$$

$$\lim_{x \to \infty} (2x - 1) \to \infty$$

$$\lim_{x \to \infty} (x + 1) \to \infty$$

To resolve this difficulty, we divide both the numerator and the denominator by x. After this division, the limit may be evaluated as follows.

$$\lim_{x \to \infty} \frac{2x - 1}{x + 1} = \lim_{x \to \infty} \frac{2 - (1/x)}{1 + (1/x)} = \frac{\lim_{x \to \infty} 2 - \lim_{x \to \infty} \frac{1}{x}}{\lim_{x \to \infty} 1 + \lim_{x \to \infty} \frac{1}{x}} = \frac{2 - 0}{1 + 0} = 2$$

Thus, the line $y = 2$ is a horizontal asymptote to the right. By taking the limit as $x \to -\infty$, we can see that $y = 2$ is also a horizontal asymptote to the left. The graph of this function is shown in Figure 4.31. ▭

$y = 2$ is horizontal asymptote.

FIGURE 4.31

REMARK   We can test the reasonableness of the limit found in Example 2 by evaluating $f(x)$ for a few large positive values of x. For instance, $f(100) \approx 1.9703$, $f(1000) \approx 1.9970$, and $f(10000) \approx 1.9997$.

In Example 2, our first attempt to evaluate the limit resulted in the **indeterminate form** ∞/∞. We were able to resolve the difficulty by rewriting the given expression in an equivalent form. Specifically, we divided both the numerator and the denominator by x. In general, we suggest dividing by the highest power of x in the *denominator*. This is illustrated in the following example.

EXAMPLE 3   A comparison of three rational functions

Find the following limits.

(a) $\lim\limits_{x \to \infty} \dfrac{2x + 5}{3x^2 + 1}$ (b) $\lim\limits_{x \to \infty} \dfrac{2x^2 + 5}{3x^2 + 1}$ (c) $\lim\limits_{x \to \infty} \dfrac{2x^3 + 5}{3x^2 + 1}$

SOLUTION

(a) We divide both the numerator and the denominator by x^2 and obtain

$$\lim_{x \to \infty} \frac{2x + 5}{3x^2 + 1} = \lim_{x \to \infty} \frac{(2/x) + (5/x^2)}{3 + (1/x^2)} = \frac{0 + 0}{3 + 0} = \frac{0}{3} = 0.$$

(b) In this case, we also divide by x^2 and obtain

$$\lim_{x \to \infty} \frac{2x^2 + 5}{3x^2 + 1} = \lim_{x \to \infty} \frac{2 + (5/x^2)}{3 + (1/x^2)}$$

$$= \frac{2 + 0}{3 + 0} = \frac{2}{3}.$$

(c) Again, we divide by x^2 and obtain

$$\lim_{x \to \infty} \frac{2x^3 + 5}{3x^2 + 1} = \lim_{x \to \infty} \frac{2x + (5/x^2)}{3 + (1/x^2)} \cdot$$

$$\lim_{x \to \infty} \left[2x + \frac{5}{x^2} \right] \to \infty$$

$$\lim_{x \to \infty} \left[3 + \frac{1}{x^2} \right] \to 3$$

In this case, we conclude that the limit *does not exist* because the numerator increases without bound while the modified denominator approaches 3.

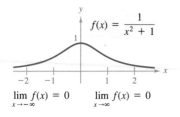

$$f(x) = \frac{1}{x^2 + 1}$$

$$\lim_{x \to -\infty} f(x) = 0 \qquad \lim_{x \to \infty} f(x) = 0$$

FIGURE 4.32

It is instructive to compare the three rational functions in Example 3. In part (a) the degree of the numerator is *less* than the degree of the denominator and the limit of the rational function is zero. In part (b) the degrees of the numerator and the denominator are *equal* and the limit is simply the ratio of the two leading coefficients 2 and 3. Finally, in part (c) the degree of the numerator is *greater* than that of the denominator and the limit does not exist. This seems reasonable when we realize that for large values of x the highest-powered term of the rational function is the most "influential" in determining the limit. For instance, the limit as x approaches infinity of the function given by

$$f(x) = \frac{1}{x^2 + 1}$$

is zero since the denominator overpowers the numerator as x increases or decreases without bound, as shown in Figure 4.32.

The function shown in Figure 4.32 is a special case of a type of curve studied by the Italian mathematician Maria Gaetana Agnesi (1717–1783). The general form of this function is $f(x) = a^3/(x^2 + a^2)$ and, through a mistranslation of the Italian word *vertéré*, the curve has come to be known as the witch of Agnesi. Agnesi's work with this curve first appeared in a comprehensive text on calculus that was published in 1748.

In Figure 4.32, we can see that the function $f(x) = 1/(x^2 + 1)$ approaches the same horizontal asymptote to the right and to the left. That is,

$$\lim_{x \to -\infty} f(x) = 0$$

$$= \lim_{x \to \infty} f(x).$$

This is always the case with rational functions. Functions that are not rational, however, may approach different horizontal asymptotes to the right and to the left, as shown in Example 4.

Maria Agnesi

EXAMPLE 4   A function with two horizontal asymptotes

Determine the following limits.

(a) $\displaystyle\lim_{x\to\infty} \frac{3x - 2}{\sqrt{2x^2 + 1}}$ (b) $\displaystyle\lim_{x\to-\infty} \frac{3x - 2}{\sqrt{2x^2 + 1}}$

SOLUTION

(a) For $x > 0$, we have $x = \sqrt{x^2}$. Thus, dividing both the numerator and the denominator by x produces

$$\frac{3x - 2}{\sqrt{2x^2 + 1}} = \frac{\dfrac{3x - 2}{x}}{\dfrac{\sqrt{2x^2 + 1}}{\sqrt{x^2}}} = \frac{3 - \dfrac{2}{x}}{\sqrt{\dfrac{2x^2 + 1}{x^2}}} = \frac{3 - \dfrac{2}{x}}{\sqrt{2 + \dfrac{1}{x^2}}}$$

and we can take the limit as follows.

$$\lim_{x\to\infty} \frac{3x - 2}{\sqrt{2x^2 + 1}} = \lim_{x\to\infty} \frac{3 - \dfrac{2}{x}}{\sqrt{2 + \dfrac{1}{x^2}}} = \frac{3 - 0}{\sqrt{2 + 0}} = \frac{3}{\sqrt{2}}$$

(b) For $x < 0$, we have $x = -\sqrt{x^2}$. Thus, dividing both the numerator and the denominator by x produces

$$\frac{3x - 2}{\sqrt{2x^2 + 1}} = \frac{\dfrac{3x - 2}{x}}{\dfrac{\sqrt{2x^2 + 1}}{-\sqrt{x^2}}} = \frac{3 - \dfrac{2}{x}}{-\sqrt{\dfrac{2x^2 + 1}{x^2}}} = \frac{3 - \dfrac{2}{x}}{-\sqrt{2 + \dfrac{1}{x^2}}}$$

and we can take the limit as follows.

$$\lim_{x\to-\infty} \frac{3x - 2}{\sqrt{2x^2 + 1}} = \lim_{x\to-\infty} \frac{3 - \dfrac{2}{x}}{-\sqrt{2 + \dfrac{1}{x^2}}} = \frac{3 - 0}{-\sqrt{2 + 0}} = -\frac{3}{\sqrt{2}}$$

The graph of $f(x) = (3x - 2)/\sqrt{2x^2 + 1}$ is shown in Figure 4.33.

$y = \dfrac{3x - 2}{\sqrt{2x^2 + 1}}$

$y = \dfrac{3}{\sqrt{2}}$, horizontal asymptote to the right

$y = \dfrac{-3}{\sqrt{2}}$, horizontal asymptote to the left

FIGURE 4.33

In Section 2.4 (Example 6), we used the Squeeze Theorem to evaluate limits involving trigonometric functions. This theorem is also valid for limits at infinity, and its use is demonstrated in the next example.

EXAMPLE 5   Limits involving trigonometric functions

Determine the following limits.

(a) $\displaystyle\lim_{x\to\infty} \sin x$ (b) $\displaystyle\lim_{x\to\infty} \frac{\sin x}{x}$

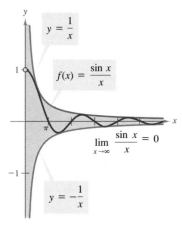

$y = \dfrac{1}{x}$

$f(x) = \dfrac{\sin x}{x}$

$\displaystyle\lim_{x\to\infty} \dfrac{\sin x}{x} = 0$

$y = -\dfrac{1}{x}$

FIGURE 4.34

SOLUTION

(a) As x approaches infinity, the sine function oscillates between 1 and -1, and we conclude that this limit does not exist.

(b) Since $-1 \le \sin x \le 1$, it follows that

$$-\frac{1}{x} \le \frac{\sin x}{x} \le \frac{1}{x}$$

where

$$\lim_{x\to\infty} \left(-\frac{1}{x}\right) = 0 \quad \text{and} \quad \lim_{x\to\infty} \left(\frac{1}{x}\right) = 0.$$

Therefore, by the Squeeze Theorem, we have

$$\lim_{x\to\infty} \frac{\sin x}{x} = 0$$

as indicated in Figure 4.34.

There are many examples of asymptotic behavior in the physical sciences. For instance, the following example describes the asymptotic recovery of oxygen in a pond.

EXAMPLE 6   An application involving oxygen levels

Suppose that $f(t)$ measures the level of oxygen in a pond, where $f(t) = 1$ is the normal (unpolluted) level and the time t is measured in weeks. When $t = 0$, organic waste is dumped into the pond, and as the waste material oxidizes, the level of oxygen in the pond is given by

$$f(t) = \frac{t^2 - t + 1}{t^2 + 1}.$$

What percentage of the normal level of oxygen exists in the pond after 1 week? After 2 weeks? After 10 weeks? What is the limit as t approaches infinity?

SOLUTION

When $t = 1$, 2, and 10, the levels of oxygen are as follows.

$$f(1) = \frac{1^2 - 1 + 1}{1^2 + 1} = \frac{1}{2} = 50\% \qquad \text{1 week}$$

$$f(2) = \frac{2^2 - 2 + 1}{2^2 + 1} = \frac{3}{5} = 60\% \qquad \text{2 weeks}$$

$$f(10) = \frac{10^2 - 10 + 1}{10^2 + 1} = \frac{91}{101} \approx 90.1\% \qquad \text{10 weeks}$$

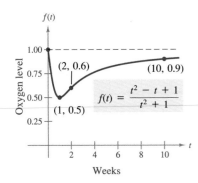

f(t)

(2, 0.6)

(10, 0.9)

(1, 0.5)

$f(t) = \dfrac{t^2 - t + 1}{t^2 + 1}$

Oxygen level

Weeks

FIGURE 4.35

To take the limit as t approaches infinity, we divide the numerator and the denominator by t^2 to obtain

$$\lim_{t \to \infty} \frac{t^2 - t + 1}{t^2 + 1} = \lim_{t \to \infty} \frac{1 - (1/t) + (1/t^2)}{1 + (1/t^2)} = \frac{1 - 0 + 0}{1 + 0} = 1 = 100\%.$$

(See Figure 4.35.)

EXERCISES for Section 4.5

In Exercises 1–8, match the given function to one of the graphs (a)–(h), using horizontal asymptotes as an aid.

1. $f(x) = \dfrac{3x^2}{x^2 + 2}$

2. $f(x) = \dfrac{2x}{\sqrt{x^2 + 2}}$

3. $f(x) = \dfrac{x}{x^2 + 2}$

4. $f(x) = 2 + \dfrac{x^2}{x^4 + 1}$

5. $f(x) = \dfrac{-6x}{\sqrt{4x^2 + 5}}$

6. $f(x) = 5 - \dfrac{1}{x^2 + 1}$

7. $f(x) = \dfrac{4 \sin x}{x^2 + 1}$

8. $f(x) = \dfrac{2x^2 - 3x + 5}{x^2 + 1}$

(a)

(b)

(c)

(d)

(e)

(f)

(g)

(h)

In Exercises 9–28, find the indicated limit.

9. $\displaystyle\lim_{x \to \infty} \frac{2x - 1}{3x + 2}$

10. $\displaystyle\lim_{x \to \infty} \frac{5x^3 + 1}{10x^3 - 3x^2 + 7}$

11. $\displaystyle\lim_{x \to \infty} \frac{x}{x^2 - 1}$

12. $\displaystyle\lim_{x \to \infty} \frac{2x^{10} - 1}{10x^{11} - 3}$

13. $\displaystyle\lim_{x \to -\infty} \frac{5x^2}{x + 3}$

14. $\displaystyle\lim_{x \to \infty} \frac{x^3 - 2x^2 + 3x + 1}{x^2 - 3x + 2}$

15. $\displaystyle\lim_{x \to \infty} \left(2x - \frac{1}{x^2}\right)$

16. $\displaystyle\lim_{x \to \infty} (x + 3)^{-2}$

17. $\displaystyle\lim_{x \to -\infty} \left(\frac{2x}{x - 1} + \frac{3x}{x + 1}\right)$

18. $\displaystyle\lim_{x \to \infty} \left(\frac{2x^2}{x - 1} + \frac{3x}{x + 1}\right)$

19. $\displaystyle\lim_{x \to -\infty} \frac{x}{\sqrt{x^2 - x}}$

20. $\displaystyle\lim_{x \to \infty} \frac{x}{\sqrt{x^2 + 1}}$

21. $\displaystyle\lim_{x \to \infty} \frac{2x + 1}{\sqrt{x^2 - x}}$

22. $\displaystyle\lim_{x \to -\infty} \frac{-3x + 1}{\sqrt{x^2 + x}}$

23. $\displaystyle\lim_{x \to \infty} \frac{x^2 - x}{\sqrt{x^4 + x}}$

24. $\displaystyle\lim_{x \to \infty} \frac{2x}{\sqrt{4x^2 + 1}}$

25. $\displaystyle\lim_{x \to \infty} \frac{\sin 2x}{x}$

26. $\displaystyle\lim_{x \to \infty} \frac{x - \cos x}{x}$

27. $\displaystyle\lim_{x \to \infty} \frac{1}{2x + \sin x}$

28. $\displaystyle\lim_{x \to \infty} \sin \frac{1}{x}$

In Exercises 29 and 30, find the indicated limit. [Hint: Let $x = 1/t$ and find the limit as $t \to 0^+$.]

29. $\displaystyle\lim_{x \to \infty} x \sin \frac{1}{x}$

30. $\displaystyle\lim_{x \to \infty} x \tan \frac{1}{x}$

In Exercises 31–34, find the indicated limit. [Hint: Treat the expression as a fraction whose denominator is 1, and rationalize the numerator.]

31. $\lim\limits_{x \to -\infty} (x + \sqrt{x^2 + 3})$ **32.** $\lim\limits_{x \to \infty} (2x - \sqrt{4x^2 + 1})$

33. $\lim\limits_{x \to \infty} (x - \sqrt{x^2 + x})$ **34.** $\lim\limits_{x \to -\infty} (3x + \sqrt{9x^2 - x})$

In Exercises 35–46, sketch the graph of the given equation. As a sketching aid, examine each equation for intercepts, symmetry, and asymptotes.

35. $y = \dfrac{2 + x}{1 - x}$ **36.** $y = \dfrac{x - 3}{x - 2}$

37. $y = \dfrac{x^2}{x^2 + 9}$ **38.** $y = \dfrac{x^2}{x^2 - 9}$

39. $xy^2 = 4$ **40.** $x^2 y = 4$

41. $y = \dfrac{2x}{1 - x}$ **42.** $y = \dfrac{2x}{1 - x^2}$

43. $y = 2 - \dfrac{3}{x^2}$ **44.** $y = 1 + \dfrac{1}{x}$

45. $y = \dfrac{x^3}{\sqrt{x^2 - 4}}$ **46.** $y = \dfrac{x}{\sqrt{x^2 - 4}}$

In Exercises 47–50, complete the given table and estimate the limit of $f(x)$ as x approaches infinity. Then find the limit analytically and compare your results.

47. $f(x) = \dfrac{x + 1}{x\sqrt{x}}$

x	10^0	10^1	10^2	10^3	10^4	10^5	10^6
$f(x)$							

48. $f(x) = x - \sqrt{x(x - 1)}$

x	10^0	10^1	10^2	10^3	10^4	10^5	10^6
$f(x)$							

49. $f(x) = x \sin \dfrac{1}{2x}$

x	10^0	10^1	10^2	10^3	10^4	10^5	10^6
$f(x)$							

50. $f(x) = x^2 - x\sqrt{x(x - 1)}$

x	10^0	10^1	10^2	10^3	10^4	10^5	10^6
$f(x)$							

51. A business has a cost of $C = 0.5x + 500$ for producing x units. The average cost per unit is given by $\overline{C} = C/x$. Find the limit of $\overline{C}$ as x approaches infinity.

52. According to the theory of relativity, the mass m of a particle depends on its velocity v. That is,

$$m = \frac{m_0}{\sqrt{1 - (v^2/c^2)}}$$

where m_0 is the mass when the particle is at rest and c is the speed of light. Find the limit of the mass as v approaches c.

53. The efficiency of an internal combustion engine is defined to be

$$\text{efficiency } (\%) = 100\left[1 - \frac{1}{(v_1/v_2)^c}\right]$$

where v_1/v_2 is the ratio of the uncompressed gas to the compressed gas and c is a constant dependent upon the engine design. Find the limit of the efficiency as the compression ratio approaches infinity.

54. Verify that each of the following functions has two horizontal asymptotes.

(a) $f(x) = \dfrac{|x|}{x + 1}$ (b) $f(x) = \dfrac{2x}{\sqrt{x^2 + 1}}$

In Exercises 55 and 56, use a computer or graphics calculator to sketch the graph of the function and from the sketch locate any horizontal or vertical asymptotes.

55. $f(x) = \dfrac{3x}{\sqrt{4x^2 + 1}}$ **56.** $f(x) = \dfrac{2 \sin 2x}{x}$

57. Prove that if

$$p(x) = a_n x^n + \cdots + a_1 x + a_0$$
$$q(x) = b_m x^m + \cdots + b_1 x + b_0$$

then

$$\lim_{x \to \infty} \frac{p(x)}{q(x)} = \begin{cases} 0, & n < m \\ \dfrac{a_n}{b_m}, & n = m \\ \pm\infty, & n > m. \end{cases}$$

4.6   A Summary of Curve Sketching

Summary of curve-sketching techniques

It would be difficult to overstate the importance of curve sketching in mathematics. Descartes' introduction of this concept contributed significantly to the rapid advances in calculus that began during the mid-seventeenth century. In the words of Lagrange, "As long as algebra and geometry traveled separate paths their advance was slow and their applications limited. But when these two sciences joined company, they drew from each other fresh vitality and thenceforth marched on at a rapid pace toward perfection."

Today, government, science, industry, business, education, and the social and health sciences all make widespread use of graphs to describe and predict relationships between variables. We have seen, however, that sketching a graph sometimes can require considerable ingenuity.

So far, we have discussed several concepts that are useful in sketching the graph of a function.

- Domain and Range   (Section 1.5)
- x-intercepts and y-intercepts (Section 1.3)
- Symmetry   (Section 1.3)
- Points of discontinuity   (Section 2.4)
- Vertical asymptotes   (Section 2.5)
- Horizontal asymptotes   (Section 4.5)
- Points of nondifferentiability   (Section 3.1)
- Relative extrema   (Section 4.3)
- Concavity   (Section 4.4)
- Points of inflection   (Section 4.4)

In this section we give several examples that incorporate these concepts into an effective procedure for sketching the graph of a function. The following list of suggestions for curve sketching should be helpful.

SUGGESTIONS FOR SKETCHING THE GRAPH OF A FUNCTION	1. Make a rough preliminary sketch that includes any easily determined intercepts and asymptotes. 2. Locate the x-values where $f'(x)$ and $f''(x)$ are either zero or undefined. 3. Test the behavior of f at and between each of these x-values. 4. Sharpen the accuracy of the final sketch by plotting the relative extrema, the points of inflection, and a few points between.

REMARK   Note in these guidelines the importance of *algebra* (as well as calculus) for solving the equations $f(x) = 0$, $f'(x) = 0$, and $f''(x) = 0$.

EXAMPLE 1   Sketching the graph of a rational function

Sketch the graph of the function given by

$$f(x) = \frac{2(x^2 - 9)}{x^2 - 4}.$$

FIGURE 4.36

$$f(x) = \frac{2(x^2 - 9)}{x^2 - 4}$$

FIGURE 4.37

SOLUTION

First derivative: $f'(x) = \dfrac{20x}{(x^2 - 4)^2}$

Second derivative: $f''(x) = -\dfrac{20(3x^2 + 4)}{(x^2 - 4)^3}$

x-intercepts: $(-3, 0)$, $(3, 0)$

y-intercept: $\left(0, \dfrac{9}{2}\right)$

Vertical asymptotes: $x = -2$, $x = 2$

Horizontal asymptote: $y = 2$

Critical number: $x = 0$

Possible points of inflection: None

Points of discontinuity: $x = -2$, $x = 2$

Symmetry: With respect to y-axis

Test intervals: $(-\infty, -2)$, $(-2, 0)$, $(0, 2)$, $(2, \infty)$

The intercepts and asymptotes can be used to give preliminary clues about the graph of f, as shown in Figure 4.36. Then, we complete the graph as shown in Figure 4.37 by using a few additional points and the conclusions summarized in Table 4.12.

TABLE 4.12

	$f(x)$	$f'(x)$	$f''(x)$	*Shape of graph*
$-\infty < x < -2$		$-$	$-$	decreasing, concave down
$x = -2$	undefined	undefined	undefined	vertical asymptote
$-2 < x < 0$		$-$	$+$	decreasing, concave up
$x = 0$	$\dfrac{9}{2}$	0	$+$	relative minimum
$0 < x < 2$		$+$	$+$	increasing, concave up
$x = 2$	undefined	undefined	undefined	vertical asymptote
$2 < x < \infty$		$+$	$-$	increasing, concave down

A computer software package that generates the graph of a function would be helpful in duplicating the graphs given in this section. If you have such a package, try making changes in each function to see how the changes affect the graph. For instance, how would the graph of

$$f(x) = \frac{2(x^2 - 4)}{x^2 - 9}$$

compare to the graph of the function given in Example 1?

EXAMPLE 2   Sketching the graph of a rational function

Sketch the graph of the function given by

$$f(x) = \frac{x^2 - 2x + 4}{x - 2}.$$

SOLUTION

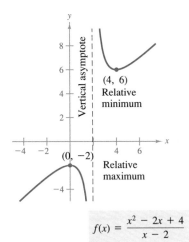

$$f(x) = \frac{x^2 - 2x + 4}{x - 2}$$

FIGURE 4.38

First derivative: $f'(x) = \dfrac{x(x - 4)}{(x - 2)^2}$

Second derivative: $f''(x) = \dfrac{8}{(x - 2)^3}$

x-intercepts: None

y-intercept: $(0, -2)$

Vertical asymptote: $x = 2$

Horizontal asymptotes: None

Critical numbers: $x = 0$, $x = 4$

Possible points of inflection: None

Points of discontinuity: $x = 2$

Test intervals: $(-\infty, 0)$, $(0, 2)$, $(2, 4)$, $(4, \infty)$

The analysis of the graph of f is shown in Table 4.13, and the graph is shown in Figure 4.38.

TABLE 4.13

	$f(x)$	$f'(x)$	$f''(x)$	*Shape of graph*
$-\infty < x < 0$		$+$	$-$	increasing, concave down
$x = 0$	-2	0	$-$	relative maximum
$0 < x < 2$		$-$	$-$	decreasing, concave down
$x = 2$	undefined	undefined	undefined	vertical asymptote
$2 < x < 4$		$-$	$+$	decreasing, concave up
$x = 4$	6	0	$+$	relative minimum
$4 < x < \infty$		$+$	$+$	increasing, concave up

Although the graph of the function in Example 2 has no horizontal asymptote, it does have a slant asymptote. The graph of a rational function (having no common factors) has a **slant asymptote** if the degree of the numerator exceeds the degree of the denominator by one. To find the slant asymptote, we use division to rewrite the rational function as the sum of a first-degree polynomial and another rational function.

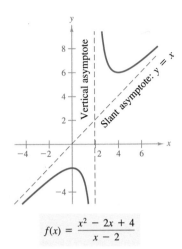

$$f(x) = \frac{x^2 - 2x + 4}{x - 2}$$

FIGURE 4.39

$$f(x) = \frac{-x^3 + x^2 + 4}{x^2}$$

FIGURE 4.40

$$f(x) = \frac{x^2 - 2x + 4}{x - 2} = x + \frac{4}{x - 2}$$

$y = x$ is a slant asymptote

In Figure 4.39 note that the graph of f approaches the slant asymptote $y = x$ as x approaches $-\infty$ or ∞.

EXAMPLE 3   Sketching a rational function with a slant asymptote

Sketch the graph of the function given by

$$f(x) = \frac{-x^3 + x^2 + 4}{x^2} = \frac{(-x + 2)(x^2 + x + 2)}{x^2}.$$

SOLUTION

First derivative: $f'(x) = -\dfrac{x^3 + 8}{x^3}$

Second derivative: $f''(x) = \dfrac{24}{x^4}$

x-intercept: $(2, 0)$

y-intercept: None

Vertical asymptote: $x = 0$

Slant asymptote: $y = -x + 1$ since $f(x) = -x + 1 + \dfrac{4}{x^2}$

Critical number: $x = -2$

Possible points of inflection: None

Points of discontinuity: $x = 0$

Test intervals: $(-\infty, -2), (-2, 0), (0, \infty)$

The analysis of the graph of f is shown in Table 4.14, and the graph is shown in Figure 4.40.

TABLE 4.14

	$f(x)$	$f'(x)$	$f''(x)$	*Shape of graph*
$-\infty < x < -2$		$-$	$+$	decreasing, concave up
$x = -2$	4	0	$+$	relative minimum
$-2 < x < 0$		$+$	$+$	increasing, concave up
$x = 0$	undefined	undefined	undefined	vertical asymptote
$0 < x < \infty$		$-$	$+$	decreasing, concave up

EXAMPLE 4   Sketching the graph of a function involving a radical

Sketch the graph of the function given by

$$f(x) = \frac{x}{\sqrt{x^2 + 2}}.$$

SOLUTION

First derivative: $f'(x) = \dfrac{2}{(x^2 + 2)^{3/2}}$

Second derivative: $f''(x) = -\dfrac{6x}{(x^2 + 2)^{5/2}}$

x-intercept: $(0, 0)$

y-intercept: $(0, 0)$

Vertical asymptotes: None

Horizontal asymptotes: $y = 1$ (to the right), $y = -1$ (to the left)

Critical numbers: None

Possible points of inflection: $x = 0$

Points of discontinuity: None

Symmetry: With respect to origin

Test intervals: $(-\infty, 0)$, $(0, \infty)$

The analysis of the graph of f is shown in Table 4.15, and the graph is shown in Figure 4.41.

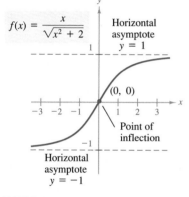

$f(x) = \dfrac{x}{\sqrt{x^2 + 2}}$

Horizontal asymptote $y = 1$

$(0, 0)$

Point of inflection

Horizontal asymptote $y = -1$

FIGURE 4.41

TABLE 4.15

	$f(x)$	$f'(x)$	$f''(x)$	*Shape of graph*
$-\infty < x < 0$		$+$	$+$	increasing, concave up
$x = 0$	0	$\dfrac{1}{\sqrt{2}}$	0	point of inflection
$0 < x < \infty$		$+$	$-$	increasing, concave down

REMARK   Although the function in Example 4 is defined over the entire real line, a square root often indicates a restricted domain. For example,

$$f(x) = \sqrt{x^2 - 4}$$

has $(-\infty, -2] \cup [2, \infty)$ as its domain.

EXAMPLE 5   Sketching the graph of a function involving cube roots

Sketch the graph of the function given by

$$f(x) = 2x^{5/3} - 5x^{4/3}.$$

SOLUTION

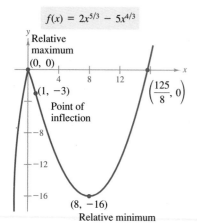

$f(x) = 2x^{5/3} - 5x^{4/3}$

Relative maximum (0, 0)

(1, −3)

Point of inflection

$\left(\dfrac{125}{8}, 0\right)$

−8

−12

−16

(8, −16)

Relative minimum

FIGURE 4.42

First derivative: $f'(x) = \dfrac{10}{3}x^{1/3}(x^{1/3} - 2)$

Second derivative: $f''(x) = \dfrac{20(x^{1/3} - 1)}{9x^{2/3}}$

x-intercepts: $(0, 0), \left(\dfrac{125}{8}, 0\right)$

y-intercept: $(0, 0)$

Vertical asymptotes: None

Horizontal asymptotes: None

Critical numbers: $x = 0, x = 8$

Possible points of inflection: $x = 0, x = 1$

Points of discontinuity: None

Test intervals: $(-\infty, 0), (0, 1), (1, 8), (8, \infty)$

The analysis of the graph of f is shown in Table 4.16, and the graph is shown in Figure 4.42.

TABLE 4.16

	$f(x)$	$f'(x)$	$f''(x)$	*Shape of graph*
$-\infty < x < 0$		+	−	increasing, concave down
$x = 0$	0	0	undefined	relative maximum
$0 < x < 1$		−	−	decreasing, concave down
$x = 1$	−3	−	0	point of inflection
$1 < x < 8$		−	+	decreasing, concave up
$x = 8$	−16	0	+	relative minimum
$8 < x < \infty$		+	+	increasing, concave up

REMARK   Although the function in Example 5 is differentiable over the entire real line, cube roots often indicate that there are points where the derivative is undefined. For example, $f(x) = x^{1/3}$ has a vertical tangent at $(0, 0)$.

EXAMPLE 6   Sketching the graph of a polynomial function

Sketch the graph of the function given by

$$f(x) = x^4 - 12x^3 + 48x^2 - 64x$$
$$= x(x - 4)^3.$$

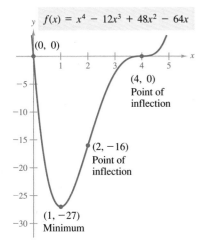

$f(x) = x^4 - 12x^3 + 48x^2 - 64x$

(0, 0)

(4, 0)
Point of
inflection

(2, −16)
Point of
inflection

(1, −27)
Minimum

FIGURE 4.43

SOLUTION

$$\text{First derivative: } f'(x) = 4(x - 1)(x - 4)^2$$
$$\text{Second derivative: } f''(x) = 12(x - 4)(x - 2)$$
$$x\text{-intercepts: } (0, 0), (4, 0)$$
$$y\text{-intercept: } (0, 0)$$
$$\text{Vertical asymptotes: None}$$
$$\text{Horizontal asymptotes: None}$$
$$\text{Critical numbers: } x = 1, x = 4$$
$$\text{Possible points of inflection: } x = 2, x = 4$$
$$\text{Points of discontinuity: None}$$
$$\text{Test intervals: } (-\infty, 1), (1, 2), (2, 4), (4, \infty)$$

The analysis of the graph of f is shown in Table 4.17, and the graph is shown in Figure 4.43.

TABLE 4.17

	$f(x)$	$f'(x)$	$f''(x)$	*Shape of graph*
$-\infty < x < 1$		−	+	decreasing, concave up
$x = 1$	−27	0	+	relative minimum
$1 < x < 2$		+	+	increasing, concave up
$x = 2$	−16	+	0	point of inflection
$2 < x < 4$		+	−	increasing, concave down
$x = 4$	0	0	0	point of inflection
$4 < x < \infty$		+	+	increasing, concave up

REMARK   The fourth degree polynomial function in Example 6 has one relative minimum and no relative maxima. In general, a polynomial function of degree n can have *at most* $n - 1$ relative extrema. Moreover, polynomial functions of even degree must have *at least* one relative extremum.

EXAMPLE 7 Sketching the graph of a trigonometric function

Sketch the graph of the function given by

$$f(x) = \frac{\cos x}{1 + \sin x}.$$

SOLUTION

This function has a period of 2π, and we begin by analyzing the graph in the interval $(-\pi/2, \, 3\pi/2)$.

$$\text{First derivative: } f'(x) = -\frac{1}{1 + \sin x}$$

$$\text{Second derivative: } f''(x) = \frac{\cos x}{(1 + \sin x)^2}$$

$$\text{Period: } 2\pi$$

$$x\text{-intercept: } \left(\frac{\pi}{2}, 0\right)$$

$$y\text{-intercept: } (0, \, 1)$$

$$\text{Vertical asymptotes: } x = -\frac{\pi}{2}, \, x = \frac{3\pi}{2}$$

$$\text{Horizontal asymptotes: None}$$

$$\text{Critical numbers: None}$$

$$\text{Possible points of inflection: } x = \frac{\pi}{2}$$

$$\text{Points of discontinuity: None in the interval } \left(-\frac{\pi}{2}, \frac{3\pi}{2}\right)$$

$$\text{Test intervals: } \left(-\frac{\pi}{2}, \frac{\pi}{2}\right), \left(\frac{\pi}{2}, \frac{3\pi}{2}\right)$$

The analysis of the graph of f on the interval $(-\pi/2, \, 3\pi/2)$ is shown in Table 4.18, and the graph is shown in Figure 4.44.

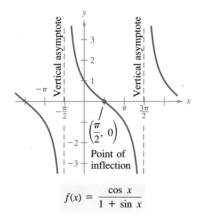

$$f(x) = \frac{\cos x}{1 + \sin x}$$

FIGURE 4.44

TABLE 4.18

	$f(x)$	$f'(x)$	$f''(x)$	*Shape of graph*
$x = -\dfrac{\pi}{2}$	undefined	undefined	undefined	vertical asymptote
$-\dfrac{\pi}{2} < x < \dfrac{\pi}{2}$		$-$	$+$	decreasing, concave up
$x = \dfrac{\pi}{2}$	0	$-\dfrac{1}{2}$	0	point of inflection
$\dfrac{\pi}{2} < x < \dfrac{3\pi}{2}$		$-$	$-$	decreasing, concave down
$x = \dfrac{3\pi}{2}$	undefined	undefined	undefined	vertical asymptote

REMARK   The work involved in sketching the graph of a trigonometric function can be lessened sometimes by the use of trigonometric identities. For instance, the function in Example 7 can be rewritten as

$$f(x) = \frac{\cos x}{1 + \sin x} = \cot\left(\frac{x}{2} + \frac{\pi}{4}\right).$$

In this form, we recognize the familiar cotangent graph shown in Figure 4.44.

EXERCISES for Section 4.6

In Exercises 1–30, sketch the graph of the given function. Choose a scale that allows all relative extrema and points of inflection to be identified on the sketch.

1. $y = x^3 - 3x^2 + 3$

2. $y = -\dfrac{1}{3}(x^3 - 3x + 2)$

3. $y = 2 - x - x^3$
4. $y = x^3 + 3x^2 + 3x + 2$
5. $f(x) = 3x^3 - 9x + 1$
6. $f(x) = (x + 1)(x - 2)(x - 5)$
7. $f(x) = -x^3 + 3x^2 + 9x - 2$

8. $f(x) = \dfrac{1}{3}(x - 1)^3 + 2$

9. $y = 3x^4 + 4x^3$
10. $y = 3x^4 - 6x^2$
11. $f(x) = x^4 - 4x^3 + 16x$
12. $f(x) = x^4 - 8x^3 + 18x^2 - 16x + 5$
13. $f(x) = x^4 - 4x^3 + 16x - 16$
14. $f(x) = x^5 + 1$
15. $y = x^5 - 5x$ **16.** $y = (x - 1)^5$
17. $y = |2x - 3|$ **18.** $y = |x^2 - 6x + 5|$

19. $y = \dfrac{x^2}{x^2 + 3}$ **20.** $y = \dfrac{x}{x^2 + 1}$

21. $y = x\sqrt{4 - x}$ **22.** $y = x\sqrt{4 - x^2}$

23. $y = 3x^{2/3} - 2x$ **24.** $y = 3x^{2/3} - x^2$

25. $f(x) = \dfrac{x}{\sqrt{x^2 + 7}}$ **26.** $f(x) = \dfrac{4x}{\sqrt{x^2 + 15}}$

27. $y = \sin x - \dfrac{1}{18}\sin 3x,\ 0 \le x \le 2\pi$

28. $y = \cos x - \dfrac{1}{2}\cos 2x,\ 0 \le x \le 2\pi$

29. $y = 2x - \tan x,\ -\dfrac{\pi}{2} < x < \dfrac{\pi}{2}$

30. $y = 2x + \cot x,\ 0 < x < \pi$

In Exercises 31–40, sketch the graph of the given function. In each case label the intercepts, relative extrema, points of inflection, and asymptotes.

31. $y = \dfrac{1}{x - 2} - 3$ **32.** $y = \dfrac{x^2 + 1}{x^2 - 2}$

33. $y = \dfrac{2x}{x^2 - 1}$ **34.** $y = \dfrac{x^2 - 6x + 12}{x - 4}$

35. $f(x) = \dfrac{x + 2}{x}$ **36.** $f(x) = x + \dfrac{32}{x^2}$

37. $f(x) = \dfrac{x^2 + 1}{x}$ **38.** $f(x) = \dfrac{x^3}{x^2 - 1}$

39. $y = \dfrac{x^3}{2x^2 - 8}$ **40.** $y = \dfrac{2x^2 - 5x + 5}{x - 2}$

In Exercises 41–46, determine the conditions on the coefficients of

$$f(x) = ax^3 + bx^2 + cx + d$$

such that the graph of f will resemble the given graph.

41.   **42.**

43.   **44.**

45.   **46.**

In Exercises 47–52, use the given graph (of f' or f'') to sketch a graph of the function f. [Hint: The solutions are not unique.]

47.   **48.**

49.   **50.**

51.   **52.**

4.7 Optimization Problems

Applied minimum and maximum problems

One of the most common applications of calculus involves the determination of minimum and maximum values. Consider how frequently we hear or read terms like greatest profit, least cost, least time, greatest voltage, optimum size, least area, greatest strength, or greatest distance. Before outlining a general method of solution for such problems, we present an example.

EXAMPLE 1   Finding the maximum volume

Open Box with Square Base:
$S = x^2 + 4xh = 108$

FIGURE 4.45

A manufacturer wants to design an open box having a square base and a surface area of 108 square inches. What dimensions will produce a box with maximum volume?

SOLUTION

Since the box (in Figure 4.45) has a square base, its volume is

$$V = x^2h. \qquad \text{Primary equation}$$

(We call this the primary equation because it gives a formula for the quantity we wish to optimize.) Furthermore, since the box is open at the top, its surface area is

$$S = \text{(area of base)} + \text{(area of four sides)}$$
$$S = x^2 + 4xh = 108. \qquad \text{Secondary equation}$$

Since V is to be maximized, we express it as a function of just one variable. To do this, we solve the equation $x^2 + 4xh = 108$ for h in terms of x to obtain

$$h = \frac{108 - x^2}{4x}.$$

Substituting in the equation for volume produces

$$V = x^2h = x^2\left(\frac{108 - x^2}{4x}\right) = 27x - \frac{x^3}{4}. \qquad \text{Function of one variable}$$

Before we try to find which x-value will yield a maximum value of V, we should determine the *feasible domain*. That is, what values of x make sense in this problem? We know that x should be nonnegative and that the area of the base ($A = x^2$) is at most 108. Thus, we have

$$0 \le x \le \sqrt{108}. \qquad \text{Feasible domain}$$

Now, to maximize V we find the critical numbers as follows.

$$\frac{dV}{dx} = 27 - \frac{3x^2}{4} = 0$$
$$3x^2 = 108 \implies x = \pm 6 \qquad \text{Critical numbers}$$

Evaluating V at the critical numbers in the domain and at the endpoints of the domain produces

$$V(0) = 0, \qquad V(6) = 108, \qquad \text{and} \qquad V(\sqrt{108}) = 0.$$

We conclude that V is maximum when $x = 6$ and the dimensions of the box are 6 in. $\times$ 6 in. $\times$ 3 in.   ▭

Some students have trouble with applied problems because they are too eager to use formulas. For instance, in Example 1, you should realize that there are infinitely many open boxes having 108 square inches of surface area. You might begin by asking yourself which basic shape would seem to yield a maximum volume. Should the box be tall, squat, or more cubical?

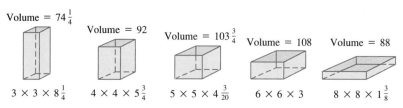

Volume $= 74\frac{1}{4}$ Volume $= 92$ Volume $= 103\frac{3}{4}$ Volume $= 108$ Volume $= 88$

$3 \times 3 \times 8\frac{1}{4}$ $4 \times 4 \times 5\frac{3}{4}$ $5 \times 5 \times 4\frac{3}{20}$ $6 \times 6 \times 3$ $8 \times 8 \times 1\frac{3}{8}$

FIGURE 4.46

Which size box has the maximum volume?

You might even try calculating a few volumes, as shown in Figure 4.46, to see if you can get a better feeling for what the optimum dimensions should be. Remember that you are not ready to begin solving a problem until you have clearly identified what the problem is.

Example 1 illustrates the following five-step procedure for solving applied minimum and maximum problems.

PROCEDURES FOR SOLVING APPLIED MINIMUM AND MAXIMUM PROBLEMS

1. Assign symbols to all *given* quantities and quantities *to be determined*. When feasible, make a sketch.
2. Write a **primary equation** for the quantity that is to be maximized (or minimized). (A review of several useful formulas from geometry is given inside the back cover of the text.)
3. Reduce the primary equation to one having a single independent variable. This may involve the use of **secondary equations** relating the independent variables of the primary equation.
4. Determine the domain of the primary equation. That is, determine the values for which the stated problem makes sense.
5. Determine the desired maximum or minimum value by the techniques discussed in Sections 4.1 through 4.4

REMARK   When performing step 5, recall that to determine the maximum or minimum value of a continuous function f on a closed interval, we compare the values of f at its critical numbers to the values of f at the endpoints of the interval.

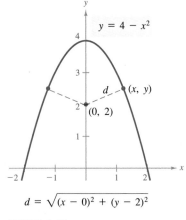

$d = \sqrt{(x - 0)^2 + (y - 2)^2}$

FIGURE 4.47

EXAMPLE 2   Finding the minimum distance

Find the points on the graph of $y = 4 - x^2$ that are closest to the point $(0, 2)$.

SOLUTION

Figure 4.47 indicates that there are two points at a minimum distance from the point $(0, 2)$. The distance between the point $(0, 2)$ and a point (x, y) on the graph of $y = 4 - x^2$ is given by

$$d = \sqrt{(x - 0)^2 + (y - 2)^2}. \qquad \text{Primary equation}$$

Using the secondary equation $y = 4 - x^2$, we rewrite the primary equation as

$$d = \sqrt{x^2 + (4 - x^2 - 2)^2} = \sqrt{x^4 - 3x^2 + 4}.$$

Since d is smallest when the expression inside the radical is smallest, we need only find the critical numbers of $f(x) = x^4 - 3x^2 + 4$. Note that the domain of f is the entire real line. Moreover, differentiation yields

$$f'(x) = 4x^3 - 6x = 2x(2x^2 - 3) = 0$$

$$x = 0, \ \sqrt{\frac{3}{2}}, \ -\sqrt{\frac{3}{2}}.$$

The First Derivative Test verifies that $x = 0$ yields a relative maximum, while both $x = \sqrt{3/2}$ and $x = -\sqrt{3/2}$ yield a minimum distance. Hence, the closest points are $(\sqrt{3/2}, 5/2)$ and $(-\sqrt{3/2}, 5/2)$.

EXAMPLE 3 Finding the minimum area

A rectangular page is to contain 24 square inches of print. The margins at the top and bottom of the page are each $1\frac{1}{2}$ inches. The margins on each side are 1 inch. What should the dimensions of the page be so that the least amount of paper is used?

SOLUTION

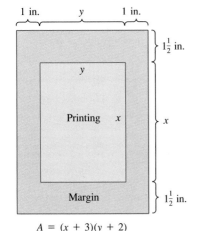

1 in. y 1 in.
$1\frac{1}{2}$ in.
y
Printing x
x
Margin
$1\frac{1}{2}$ in.

$A = (x + 3)(y + 2)$

FIGURE 4.48

Letting A be the area to be minimized, we have, from Figure 4.48,

$$A = (x + 3)(y + 2). \qquad \text{Primary equation}$$

The printed area inside the margins is given by

$$24 = xy. \qquad \text{Secondary equation}$$

Solving this equation for y produces

$$y = \frac{24}{x}$$

and the primary equation becomes

$$A = (x + 3)\left(\frac{24}{x} + 2\right) = 30 + 2x + \frac{72}{x}. \qquad \text{Function of one variable}$$

Since x must be positive, we are interested only in values of A when $x > 0$. Now, to find the minimum area, we differentiate with respect to x and obtain

$$\frac{dA}{dx} = 2 - \frac{72}{x^2} = 0 \quad \Longrightarrow \quad x^2 = 36.$$

Thus, the critical numbers are $x = \pm6$. Since -6 is outside the domain, we choose $x = 6$. The First Derivative Test confirms that A is a minimum at this point. Therefore, $y = \frac{24}{6} = 4$ and the dimensions of the page should be

$$x + 3 = 9 \text{ inches} \qquad \text{by} \qquad y + 2 = 6 \text{ inches}.$$

EXAMPLE 4 Finding the minimum length

Two posts, one 12 feet high and the other 28 feet high, stand 30 feet apart. They are to be stayed by two wires, attached to a single stake, running from ground level to the top of each post. Where should the stake be placed to use the least wire?

SOLUTION

Let W be the wire length to be minimized. Using Figure 4.49, we have

$$W = y + z. \qquad \text{Primary equation}$$

In this problem, rather than solving for y in terms of z (or vice versa), we solve for both y and z in terms of a third variable x, as shown in Figure 4.49. From the Pythagorean Theorem, we have

$$x^2 + 12^2 = y^2 \quad \Longrightarrow \quad y = \sqrt{x^2 + 144}$$
$$(30 - x)^2 + 28^2 = z^2 \quad \Longrightarrow \quad z = \sqrt{x^2 - 60x + 1684}.$$

Thus, W is given by

$$W = y + z = \sqrt{x^2 + 144} + \sqrt{x^2 - 60x + 1684}, \quad 0 \le x \le 30.$$

Differentiating W with respect to x yields

$$\frac{dW}{dx} = \frac{x}{\sqrt{x^2 + 144}} + \frac{x - 30}{\sqrt{x^2 - 60x + 1684}}.$$

By letting $dW/dx = 0$, we obtain

$$x^2(x^2 - 60x + 1684) = (30 - x)^2(x^2 + 144)$$

which simplifies to

$$320(x - 9)(2x + 45) = 0.$$

Since $x = -22.5$ is not in the interval $[0, 30]$ and

$$W(0) \approx 53.04, \qquad W(9) = 50, \qquad \text{and} \qquad W(30) \approx 60.31$$

we conclude that the wire should be staked at 9 feet from the 12-foot pole.

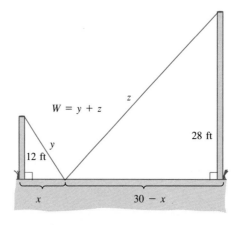

$$W = y + z$$

28 ft

12 ft

y

z

x

$30 - x$

FIGURE 4.49

In each of the first four examples, the extreme value occurs at a critical number. Although this happens often, remember that an extreme value can occur at an endpoint of an interval. The next example illustrates this.

EXAMPLE 5 An endpoint maximum

Four feet of wire is to be used to form a square and a circle. How much of the wire should be used for the square and how much should be used for the circle to enclose the maximum total area?

SOLUTION

4 feet

x

x Area: x^2

Perimeter: $4x$

?

r

Area: πr^2

Circumference: $2\pi r$

$4x + 2\pi r = 4$
$A = x^2 + \pi r^2$

FIGURE 4.50

From Figure 4.50, the total area is given by

$A =$ (area of square) + (area of circle)

$A = x^2 + \pi r^2.$ Primary equation

Since the total amount of wire is 4 feet, we have

$4 =$ (perimeter of square) + (circumference of circle)

$4 = 4x + 2\pi r.$

Thus, $r = 2(1 - x)/\pi$, and by substituting in the primary equation we have

$$A = x^2 + \pi \left[\frac{2(1-x)}{\pi}\right]^2 = x^2 + \frac{4(1-x)^2}{\pi} = \frac{1}{\pi}[(\pi + 4)x^2 - 8x + 4].$$

The feasible domain is $0 \le x \le 1$. Since $dA/dx = [2(\pi + 4)x - 8]/\pi$, the only critical number is $x = 4/(\pi + 4) \approx 0.56$. Therefore, since

$$A(0) \approx 1.273, \qquad A(0.56) \approx 0.56, \qquad \text{and} \qquad A(1) = 1$$

we conclude that the maximum area occurs when $x = 0$. That is, *all* the wire is used for the circle. ▭

Let's review the primary equations developed in the first five examples. As applications go, these five examples are fairly simple, and yet the resulting primary equations are quite complicated.

$$V = 27x - \frac{x^3}{4} \qquad\qquad W = \sqrt{x^2 + 144} + \sqrt{x^2 - 60x + 1684}$$

$$d = \sqrt{x^4 - 3x^2 + 4} \qquad A = \frac{1}{\pi}[(\pi + 4)x^2 - 8x + 4]$$

$$A = 30 + 2x + \frac{72}{x}$$

You must expect that real applications often involve equations that are at least as complicated as these five. Remember that one of the main goals of this course is to learn to use calculus to analyze equations that initially seem formidable.

For our last example, we look at a problem in which the primary equation is familiar enough that we do not have to use calculus to discover the extrema.

EXAMPLE 6 Finding the minimum and maximum rate of growth

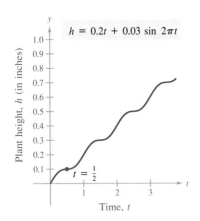

$$h = 0.2t + 0.03 \sin 2\pi t$$

Plant height, h (in inches)

$t = \frac{1}{2}$

Time, t

FIGURE 4.51

Plants do not grow at constant rates during a normal 24-hour period since their growth is affected by sunlight, as indicated in Figure 4.51. Suppose that the growth of a certain plant in a controlled environment is given by the model $h = 0.2t + 0.03 \sin 2\pi t$, where h is the height of the plant in inches and t is the time in days with $t = 0$ corresponding to midnight.
(a) During what time of day is the *rate of growth* a maximum?
(b) During what time of day is the *rate of growth* a minimum?

SOLUTION

Figure 4.51 shows the growth pattern for this plant. Since the rate of growth is given by dh/dt, we have

$$h = 0.2t + 0.03 \sin 2\pi t$$

$$\frac{dh}{dt} = 0.2 + (2\pi)(0.03) \cos 2\pi t. \qquad \text{Primary equation}$$

In this case the primary equation is a familiar one, and without taking its derivative we know that $\cos 2\pi t$ has a maximum of 1 and a minimum of -1.
(a) If $t = 0$, the rate is a maximum since

$$\cos [2\pi(0)] = 1.$$

(b) If $t = \frac{1}{2}$, the rate is a minimum since

$$\cos \left[2\pi \left(\frac{1}{2} \right) \right] = -1.$$

Thus, the maximum rate of growth occurs at midnight and the minimum at noon. ▭

EXERCISES for Section 4.7

In Exercises 1–6, find two positive numbers satisfying the given requirements.

1. The sum is 110 and the product is maximum.
2. The sum is S and the product is maximum.
3. The product is 192 and the sum is minimum.
4. The product is 192 and the sum of the first plus three times the second is minimum.
5. The second number is the reciprocal of the first and their sum is minimum.
6. The sum of the first and twice the second is 100 and the product is maximum.

In Exercises 7 and 8, find the length and width of a rectangle of maximum area for the given perimeter.

7. Perimeter: 100 feet   8. Perimeter: P units

In Exercises 9 and 10, find the length and width of a rectangle of minimum perimeter for the given area.

9. Area: 64 square feet
10. Area: A square feet

In Exercises 11 and 12, find the point on the graph of the function closest to the given point.

Function	Point
11. $f(x) = \sqrt{x}$	$(4, 0)$
12. $f(x) = x^2$	$\left(2, \frac{1}{2} \right)$

13. In an autocatalytic chemical reaction, the product formed is a catalyst for the reaction. If Q is the amount of the original substance and x is the amount of catalyst formed, then the rate of chemical reaction is given by

$$\frac{dQ}{dx} = kx(Q - x).$$

For what value of x will the rate of chemical reaction be greatest?

14. Show that among all positive numbers x and y with $x^2 + y^2 = r^2$, the sum $x + y$ is largest when $x = y$.

15. A dairy farmer plans to fence in a rectangular pasture adjacent to a river. The pasture must contain 180,000 square meters in order to provide enough grass for the herd. What dimensions would require the least amount of fencing if no fencing is needed along the river?

16. A rancher has 200 feet of fencing with which to enclose two adjacent rectangular corrals (see figure). What dimensions should be used so that the enclosed area will be a maximum?

17. An open box is to be made from a square piece of material, 12 inches on a side, by cutting equal squares from each corner and turning up the sides (see figure). Find the volume of the largest box that can be made in this manner.

FIGURE FOR 16

x $12 - 2x$ x

FIGURE FOR 17

18. (a) Solve Exercise 17 given that the square of material is s inches on a side.
(b) If the dimensions of the square piece of material are doubled, how does the volume change?

19. An open box is to be made from a rectangular piece of material by cutting equal squares from each corner and turning up the sides. Find the dimensions of the box of maximum volume if the material has dimensions of 2 feet by 3 feet.

20. A net enclosure for practicing golf is open at one end (see figure). Find the dimensions that require the least amount of netting if the volume of the enclosure is to be $83\frac{1}{3}$ cubic meters.

21. A Norman window is constructed by adjoining a semicircle to the top of an ordinary rectangular window (see figure). Find the dimensions of a Norman window of maximum area if the total perimeter is 16 feet.

FIGURE FOR 20

FIGURE FOR 21

22. A physical fitness room consists of a rectangular region with a semicircle on each end. If the perimeter of the room is to be a 200-meter running track, find the dimensions that will make the area of the rectangular region as large as possible.

23. A rectangle is bounded by the x- and y-axes and the graph of $y = (6 - x)/2$ (see figure). What length and width should the rectangle have so that its area is a maximum?

24. A right triangle is formed in the first quadrant by the x- and y-axes and a line through the point $(1, 2)$ (see figure). Find the vertices of the triangle so that the length of the hypotenuse is minimum.

FIGURE FOR 23

FIGURE FOR 24

25. A right triangle is formed in the first quadrant by the x- and y-axes and a line through the point $(2, 3)$. Find the vertices of the triangle so that its area is minimum.

26. Find the dimensions of the largest isosceles triangle that can be inscribed in a circle of radius 4 (see figure).

27. A rectangle is bounded by the x-axis and the semicircle $y = \sqrt{25 - x^2}$ (see figure). What length and width should the rectangle have so that its area is a maximum?

FIGURE FOR 26

FIGURE FOR 27

28. Find the dimensions of the largest rectangle that can be inscribed in a semicircle of radius r (see Exercise 27).

29. Find the dimensions of the trapezoid of greatest area that can be inscribed in a semicircle of radius r.

30. A page is to contain 30 square inches of print. The margins at the top and bottom of the page are 2 inches. The margins on each side are only 1 inch. Find the dimensions of the page so that the least paper is used.

31. A right circular cylinder is to be designed to hold 12 fluid ounces of a soft drink and to use a minimum of material in its construction. Find the required dimensions for the container. [1 fl oz $\approx$ 1.80469 in.^{3}]

32. Rework Exercise 31 if the volume of the cylinder is V_0 cubic units.

33. A rectangular package to be sent by a postal service can have a maximum combined length and girth (perimeter of a cross section) of 108 inches (see figure). Find the dimensions of the package of maximum volume that can be sent. (Assume the cross section is square.)

34. Rework Exercise 33 for a cylindrical package. (The cross sections are circular.)

35. Find the volume of the largest right circular cone that can be inscribed in a sphere of radius r (see figure).

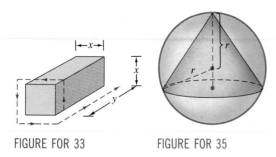

FIGURE FOR 33   FIGURE FOR 35

36. Find the volume of the largest right circular cylinder that can be inscribed in a sphere of radius r.

37. A solid is formed by adjoining two hemispheres to each end of a right circular cylinder. The total volume of the figure is 12 cubic inches. Find the radius of the cylinder that produces the minimum surface area.

38. An industrial tank of the shape described in Exercise 37 must have a volume of 3000 cubic feet. If the construction cost of the hemispherical ends is twice as much per square foot of surface area as the sides, find the dimensions that will minimize cost.

39. The combined perimeter of an equilateral triangle and a square is 10. Find the dimensions of the triangle and square that produce a minimum total area.

40. The combined perimeter of a circle and a square is 16. Find the dimensions of the circle and square that produce a minimum total area.

41. Ten feet of wire is to be used to form an isosceles right triangle and a circle. How much of the wire should be used for the circle if the total area enclosed is to be (a) minimum and (b) maximum?

42. Twenty feet of wire is to be used to form two figures. In each of the following cases, how much should be used for each figure so that the total enclosed area is a maximum?
    (a) equilateral triangle and square
    (b) square and regular pentagon
    (c) regular pentagon and hexagon
    (d) regular hexagon and circle
    What can you conclude from this pattern? [Hint: The area of a regular polygon with n sides of length x is $A = (n/4)(\cot \pi/n)x^2$.]

43. A wooden beam has a rectangular cross section of height h and width w (see figure). The strength S of the beam is directly proportional to the width and the square of the height. What are the dimensions of the strongest beam that can be cut from a round log of diameter 24 inches? [Hint: $S = kh^2w$, where k is the proportionality constant.]

44. The illumination from a light source is directly proportional to the strength of the source and inversely proportional to the square of the distance from the source. Two light sources of intensities I_1 and I_2 are d units apart. At what point on the line segment joining the two sources is the illumination least?

45. A man is in a boat 2 miles from the nearest point on the coast. He is to go to a point Q, 3 miles down the coast and 1 mile inland (see figure). If he can row at 2 miles per hour and walk at 4 miles per hour, toward what point on the coast should he row in order to reach point Q in the least time?

46. The conditions are the same as in Exercise 45 except that the man can row at 4 miles per hour (see figure). How does this change the solution?

47. The conditions are the same as in Exercise 45 except that the man can row at v_1 miles per hour and walk at v_2 miles per hour. If θ_1 and θ_2 are the magnitudes of the angles shown in the accompanying figure, show that the man will reach point Q in the least time when
$$\frac{\sin \theta_1}{v_1} = \frac{\sin \theta_2}{v_2}.$$

FIGURE FOR 45–47

FIGURE FOR 43

48. When light waves, traveling in a transparent medium, strike the surface of a second transparent medium, they begin to travel in a different direction. This change of direction is called refraction and is defined by **Snell's Law of Refraction,**

$$\frac{\sin \theta_1}{v_1} = \frac{\sin \theta_2}{v_2}$$

where θ_1 and θ_2 are the magnitudes of the angles shown in the accompanying figure and v_1 and v_2 are the velocities of light in the two mediums. Show that this problem is equivalent to that of Exercise 47, and that light waves traveling from P to Q follow the path of minimum time.

49. A component is designed to slide a block of steel of weight W across a table and into a chute (see figure). The motion of the block is resisted by a frictional force proportional to its apparent weight. (Let k be the constant of proportionality.) Find the minimum force F needed to slide the block, and find the corresponding value of θ. [Hint: $F \cos \theta$ is the force in the direction of motion, and $F \sin \theta$ is the amount of force tending to lift the block. Therefore, the apparent weight of the block is $W - F \sin \theta$.]

FIGURE FOR 48

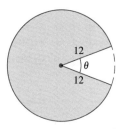

FIGURE FOR 49

50. The range R of a projectile fired with an initial velocity v_0 at an angle θ with the horizontal is given by

$$R = \frac{v_0{}^2 \sin 2\theta}{g}$$

where g is the acceleration due to gravity. Find the angle θ so that the range is maximum.

51. A sector with central angle θ is cut from a circle of radius 12 inches, and the resulting edges are brought together to form a cone (see figure). Find the magnitude of θ so that the volume of the cone is maximum.

52. The cross sections of an irrigation canal are isosceles trapezoids where the length of three of the sides is 8 feet (see figure). Determine the angle of elevation θ of the sides so that the area of the cross section is maximum.

FIGURE FOR 52

FIGURE FOR 51

In Exercises 53 and 54, use a computer or graphics calculator to sketch the graphs of the primary equation and its first derivative in the given applied extrema problem. From the graphs, find the required extrema.

53. Find the dimensions of the rectangle of maximum area that can be inscribed in a semicircle of radius 10.

54. Find the length of the longest pipe that can be carried level around a right-angle corner if the two intersecting corridors are of width 5 feet and 8 feet, respectively.

4.8 Newton's Method

Newton's Method ▪ Algebraic solutions of polynomial equations

In the first seven sections of this chapter, we frequently needed to find the zeros of a function. Until now our functions have been chosen carefully so that elementary algebraic techniques suffice for finding the zeros. For instance, the zeros of $f(x) = x^3 - 2x^2 - x + 2$ can be found easily by factoring. In applied problems we frequently encounter functions whose zeros are more difficult to find. For instance, the zeros of a function as simple as

$$f(x) = x^3 - x + 1$$

cannot be found by elementary algebraic methods. (More is said about this at the end of this section.)

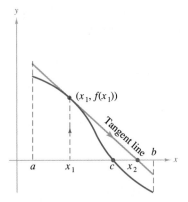

The *x*-intercept of the tangent line approximates the zero of *f*.

FIGURE 4.52

To introduce Newton's Method for approximating the real zeros of a function, we consider a function *f* that is continuous on the interval $[a, b]$ and differentiable on the interval (a, b). If $f(a)$ and $f(b)$ differ in sign, then by the Intermediate Value Theorem *f* must possess at least one zero in the interval (a, b). Suppose we estimate this zero to occur at $x = x_1$, as shown in Figure 4.52. Newton's method is based on the assumption that the graph of *f* and the tangent line at $(x_1, f(x_1))$ both cross the *x*-axis at *about* the same point. Since we can easily calculate the *x*-intercept for this tangent line, we use it as our second (and, we hope, better) estimate for the zero of *f*. The tangent line passes through the point $(x_1, f(x_1))$ with a slope of $f'(x_1)$. In point-slope form, the equation of the tangent line is therefore

$$y - f(x_1) = f'(x_1)(x - x_1)$$
$$y = f'(x_1)(x - x_1) + f(x_1).$$

Letting $y = 0$ and solving for *x*, we have

$$x = x_1 - \frac{f(x_1)}{f'(x_1)}.$$

Thus, from our initial guess we arrive at a new estimate

$$x_2 = x_1 - \frac{f(x_1)}{f'(x_1)}.$$

We may improve upon x_2 and calculate yet a third estimate

$$x_3 = x_2 - \frac{f(x_2)}{f'(x_2)}.$$

Repeated application of this process is called **Newton's Method.**

NEWTON'S METHOD FOR APPROXIMATING THE ZEROS OF A FUNCTION

Let $f(c) = 0$, where *f* is differentiable on an open interval containing *c*. Then, to approximate *c*, we use the following steps.

1. Make an initial estimate x_1 that is "close" to *c*. (A graph is helpful.)
2. Determine a new approximation

$$x_{n+1} = x_n - \frac{f(x_n)}{f'(x_n)}.$$

3. If $|x_n - x_{n+1}|$ is less than the desired accuracy, let x_{n+1} serve as the final approximation. Otherwise, return to step 2 and calculate a new approximation.

Each successive application of this procedure is called an **iteration.**

For many functions, just a few iterations of Newton's Method will produce approximations having very small errors. To demonstrate this, we begin with an example in which we know the exact zero.

EXAMPLE 1   Using Newton's Method

Calculate three iterations of Newton's Method to approximate a zero of $f(x) = x^2 - 2$. Use $x_1 = 1$ as the initial guess.

SOLUTION

Since $f(x) = x^2 - 2$, we have $f'(x) = 2x$ and the iterative process is given by the formula

$$x_{n+1} = x_n - \frac{f(x_n)}{f'(x_n)} = x_n - \frac{x_n^2 - 2}{2x_n}.$$

The calculations for three iterations are shown in Table 4.19.

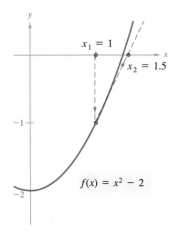

$x_1 = 1$

$x_2 = 1.5$

$f(x) = x^2 - 2$

FIGURE 4.53

TABLE 4.19

n	x_n	$f(x_n)$	$f'(x_n)$	$\dfrac{f(x_n)}{f'(x_n)}$	$x_n - \dfrac{f(x_n)}{f'(x_n)}$
1	1.000000	−1.000000	2.000000	−0.500000	1.500000
2	1.500000	0.250000	3.000000	0.083333	1.416667
3	1.416667	0.006945	2.833334	0.002451	1.414216
4	1.414216				

Of course, in this case we know that the two zeros of the function are $\pm\sqrt{2}$. To six decimal places, $\sqrt{2} = 1.414214$. Thus, after only three iterations of Newton's Method, we have obtained an approximation that is within 0.000002 of an actual root. The first iteration of this process is shown in Figure 4.53. ▭

REMARK   From the calculations in Example 1, you can see that Newton's Method has become more practical with the access to calculators and computers.

EXAMPLE 2   Using Newton's Method

Use Newton's Method to approximate the zeros of $f(x) = 2x^3 + x^2 - x + 1$. Continue the iterations until two successive approximations differ by less than 0.0001.

$f(x) = 2x^3 + x^2 - x + 1$

FIGURE 4.54

SOLUTION

We begin by sketching a graph of f and observing that it has only one zero, which occurs near $x = -1.2$, as shown in Figure 4.54. Now, we differentiate f to form the iterative formula

$$x_{n+1} = x_n - \frac{f(x_n)}{f'(x_n)} = x_n - \frac{2x_n^3 + x_n^2 - x_n + 1}{6x_n^2 + 2x_n - 1}.$$

The calculations are shown in Table 4.20.

TABLE 4.20

n	x_n	$f(x_n)$	$f'(x_n)$	$\dfrac{f(x_n)}{f'(x_n)}$	$x_n - \dfrac{f(x_n)}{f'(x_n)}$
1	-1.20000	0.18400	5.24000	0.03511	-1.23511
2	-1.23511	-0.00771	5.68276	-0.00136	-1.23375
3	-1.23375	0.00001	5.66533	0.00000	-1.23375
4	-1.23375				

Thus, we estimate the zero of f to be -1.23375, since two successive approximations differ by less than the required 0.0001. ▭

When, as in Examples 1 and 2, the approximations approach a limit, we say that the sequence

$$x_1, x_2, x_3, \ldots, x_n, \ldots$$

converges. Moreover, if the limit is c, then it can be shown that c must be a zero of f.

It is important to realize that Newton's Method does not always yield a convergent sequence. In such cases, we say that the method **fails**. One way this can happen is shown in Figure 4.55. Since Newton's Method involves division by $f'(x_n)$, it is clear that the method will fail if the derivative is zero for any x_n in the sequence. When you encounter this problem, you can usually overcome it by choosing a different value for x_1.

Another way Newton's Method can fail is illustrated in the next example.

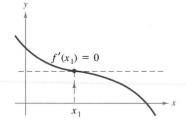

FIGURE 4.55

EXAMPLE 3 An example in which Newton's Method fails

Using $x_1 = 0.1$, show that Newton's Method fails to converge for the function $f(x) = x^{1/3}$.

SOLUTION

Since $f'(x) = \frac{1}{3}x^{-2/3}$, the iterative formula is

$$x_{n+1} = x_n - \frac{f(x_n)}{f'(x_n)} = x_n - \frac{x_n^{1/3}}{\frac{1}{3}x_n^{-2/3}} = x_n - 3x_n = -2x_n.$$

The calculations are given in Table 4.21. This table and Figure 4.56 indicate that x_n increases in magnitude as $n \to \infty$, and thus the limit of the sequence does not exist.

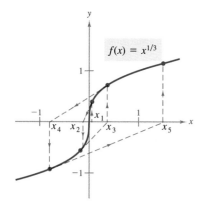

FIGURE 4.56

TABLE 4.21

n	x_n	$f(x_n)$	$f'(x_n)$	$\dfrac{f(x_n)}{f'(x_n)}$	$x_n - \dfrac{f(x_n)}{f'(x_n)}$
1	0.10000	0.46416	1.54720	0.30000	-0.20000
2	-0.20000	-0.58480	0.97467	-0.60000	0.40000
3	0.40000	0.73681	0.61401	1.20000	-0.80000
4	-0.80000	-0.92832	0.38680	-2.40000	1.60000

▭

REMARK   In Example 3, the initial estimate $x_1 = 0.1$ fails to produce a convergent sequence. Try showing that Newton's Method also fails for every other choice of x_1 (other than the actual zero).

It can be shown that a sufficient condition to produce convergence of Newton's Method to a zero of f is that

$$\left| \frac{f(x)f''(x)}{[f'(x)]^2} \right| < 1$$

on an open interval containing the zero. For instance, in Example 1 this test would yield $f(x) = x^2 - 2$, $f'(x) = 2x$, $f''(x) = 2$, and

$$\left| \frac{f(x)f''(x)}{[f'(x)]^2} \right| = \left| \frac{(x^2 - 2)(2)}{4x^2} \right| = \left| \frac{1}{2} - \frac{1}{x^2} \right|. \qquad \text{Example 1}$$

On the interval $(1, 3)$ this quantity is less than 1 and therefore the convergence of Newton's Method is guaranteed. On the other hand, in Example 3, we have $f(x) = x^{1/3}$, $f'(x) = \frac{1}{3}x^{-2/3}$, $f''(x) = -\frac{2}{9}x^{-5/3}$, and

$$\left| \frac{f(x)f''(x)}{[f'(x)]^2} \right| = \left| \frac{x^{1/3}(-2/9)(x^{-5/3})}{(1/9)(x^{-4/3})} \right| = 2 \qquad \text{Example 3}$$

which is not less than 1 for any value of x.

EXAMPLE 4   Using Newton's Method to find a point of intersection

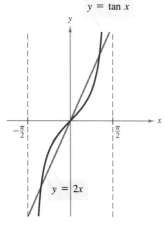

$y = \tan x$

$y = 2x$

FIGURE 4.57

Estimate the point of intersection of the graphs of

$$y = \tan x \qquad \text{and} \qquad y = 2x$$

as shown in Figure 4.57. Use Newton's Method and continue the iterations until two successive approximations differ by less than 0.0001.

SOLUTION

We let $\tan x = 2x$. Since this implies that $2x - \tan x = 0$, we need to find the zeros of the function given by

$$f(x) = 2x - \tan x.$$

Thus, the iterative formula for Newton's Method takes the form

$$x_{n+1} = x_n - \frac{f(x_n)}{f'(x_n)} = x_n - \frac{2x_n - \tan x_n}{2 - \sec^2 x_n}.$$

The calculations are shown in Table 4.22, beginning with an initial guess of $x_1 = 1.25$.

Niels Henrik Abel

TABLE 4.22

n	x_n	$f(x_n)$	$f'(x_n)$	$\dfrac{f(x_n)}{f'(x_n)}$	$x_n - \dfrac{f(x_n)}{f'(x_n)}$
1	1.25000	−0.50957	−8.05751	0.06324	1.18676
2	1.18676	−0.10110	−5.12374	0.01973	1.16703
3	1.16703	−0.00653	−4.47831	0.00146	1.16557
4	1.16557	−0.00003	−4.43436	0.00001	1.16556

Thus, we approximate the point of intersection to occur when $x = 1.16556$.

Algebraic solutions of polynomial equations

At the beginning of this section we mentioned that the zeros of the cubic function $f(x) = x^3 - x + 1$ cannot be found by *elementary* alebraic methods. This particular function happens to have only one real zero, and using more advanced algebraic techniques we can determine this value to be

$$x = -\sqrt[3]{\frac{3 - \sqrt{23/3}}{6}} - \sqrt[3]{\frac{3 + \sqrt{23/3}}{6}}.$$

Since the *exact* solution is written in terms of square roots and cube roots, we call it a **solution by radicals.**

The determination of radical solutions to a polynomial equation is one of the fundamental problems of algebra. The earliest result in this category is the well-known Quadratic Formula, and it dates back at least to Babylonian times. The general formula for the zeros of a cubic function was developed much later. In the sixteenth century an Italian mathematician, Jerome Cardan, published a method for finding radical solutions to cubic and quartic equations. Then, for 300 years the problem of finding a general quintic formula remained open. Finally, in the nineteenth century, the problem was answered independently by two young mathematicians. Niels Henrik Abel (1802–1829), a Norwegian mathematician, and Evariste Galois (1811–1832), a French mathematician, proved that it is not possible to solve a *general* fifth (or higher) degree polynomial equation by radicals. Of course, we can solve particular fifth-degree equations such as $x^5 - 1 = 0$, but Abel and Galois were able to show that no general *radical* solution exists.

Evariste Galois

EXERCISES for Section 4.8

In Exercises 1–4, complete one iteration of Newton's Method for the given function using the indicated initial guess.

Function	*Initial Guess*
1. $f(x) = x^2 - 3$	$x_1 = 1.7$
2. $f(x) = 3x^2 - 2$	$x_1 = 1$

Function	*Initial Guess*
3. $f(x) = \sin x$	$x_1 = 3$
4. $f(x) = \tan x$	$x_1 = 0.1$

In Exercises 5–12, approximate the indicated zero(s) of the function. Use Newton's Method and continue the process until two successive approximations differ by less than 0.001.

5. $f(x) = x^3 + x - 1$

6. $f(x) = x^5 + x - 1$

7. $f(x) = 3\sqrt{x-1} - x$

8. $f(x) = x^3 - 3.9x^2 + 4.79x - 1.881$

FIGURE FOR 7

FIGURE FOR 8

9. $f(x) = x^4 - 10x^2 - 11$ **10.** $f(x) = x^3 + 3$

11. $f(x) = x + \sin(x + 1)$ **12.** $f(x) = x^3 - \cos x$

In Exercises 13–16, apply Newton's Method to approximate the x-value of the indicated point(s) of intersection of the two graphs. Continue the process until two successive approximations differ by less than 0.001.

13. $f(x) = 2x + 1$;

$g(x) = \sqrt{x + 4}$

14. $f(x) = 3 - x$;

$g(x) = \dfrac{1}{x^2 + 1}$

15. $f(x) = x$;
$g(x) = \tan x$

16. $f(x) = x^2$;
$g(x) = \cos x$

In Exercises 17–20, apply Newton's Method, using the indicated initial guess, and explain why the method fails.

17. $y = 2x^3 - 6x^2 + 6x - 1$
$x_1 = 1$

18. $y = 4x^3 - 12x^2 + 12x - 3$
$x_1 = \dfrac{3}{2}$

FIGURE FOR 17

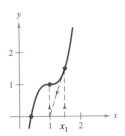

FIGURE FOR 18

19. $y = -x^3 + 3x^2 - x + 1$

$x_1 = 1$

20. $f(x) = 2 \sin x + \cos 2x$

$x_1 = \dfrac{3\pi}{2}$

FIGURE FOR 19   FIGURE FOR 20

In Exercises 21 and 22, use Newton's Method to obtain a general rule for approximating the required radical.

21. $\sqrt{a}$ [Hint: Consider $f(x) = x^2 - a$.]
22. $\sqrt[n]{a}$ [Hint: Consider $f(x) = x^n - a$.]

In Exercises 23–26, use the results of Exercises 21 and 22 to approximate the indicated radical to three decimal places.

23. $\sqrt{7}$ **24.** $\sqrt{5}$
25. $\sqrt[4]{6}$ **26.** $\sqrt[3]{15}$

27. Use Newton's Method to show that the equation

$$x_{n+1} = x_n(2 - ax_n)$$

can be used to approximate $1/a$ if x_1 is an initial guess of the reciprocal of a. Note that this method of approximating reciprocals uses only the operations of multiplication and subtraction. [Hint: Consider $f(x) = (1/x) - a$.]

28. Use the result of Exercise 27 to approximate the indicated reciprocals to three decimal places.

(a) $\dfrac{1}{3}$ (b) $\dfrac{1}{11}$

In Exercises 29 and 30, approximate π to three decimal places using Newton's Method and the given function.

29. $f(x) = 1 + \cos x$ **30.** $f(x) = \tan x$

In Exercises 31 and 32, approximate the critical number of f on the indicated interval.

31. $f(x) = x \cos x$, $0 \le x \le \pi$
32. $f(x) = x \sin x$, $0 \le x \le \pi$
33. Use the result of Exercise 31 to sketch the graph of $f(x) = x \cos x$ on the interval $[0, \pi]$.
34. Use the result of Exercise 32 to sketch the graph of $f(x) = x \sin x$ on the interval $[0, \pi]$.

In Exercises 35–38, we review some typical problems from the previous sections of this chapter. In each case, use Newton's Method to approximate the solution.

35. Find the point on the graph of $f(x) = 4 - x^2$ that is closest to the point $(1, 0)$.
36. Find the point on the graph of $f(x) = x^2$ that is closest to the point $(4, -3)$.
37. A woman is in a boat 2 miles from the nearest point on the coast (see figure). She is to go to a point Q, which is 3 miles down the coast and 1 mile inland. She can row at 3 miles per hour and walk at 4 miles per hour. Toward what point on the coast should she row in order to reach Q in the least time?

FIGURE FOR 37

38. The concentration C of a certain chemical in the bloodstream t hours after injection into muscle tissue is given by

$$C = \frac{3t^2 + t}{50 + t^3}$$

When is the concentration greatest?

In Exercises 39 and 40, use a computer or graphics calculator to approximate all the real zeros of the function by Newton's Method. Sketch the graph of the function in order to make the initial estimate of a zero.

39. $f(x) = \dfrac{1}{4}x^3 - 3x^2 + \dfrac{3}{4}x - 2$
40. $f(x) = \sqrt{4 - x^2} \sin (x - 2)$

4.9 Differentials

Differentials ▪ Error propagation ▪ Differential formulas

We previously defined the derivative as the limit ($\Delta x \to 0$) of the ratio $\Delta y / \Delta x$, and it seemed natural to retain the quotient symbolism for the limit itself. Thus, we denoted the derivative by

$$\frac{dy}{dx} = \lim_{\Delta x \to 0} \frac{\Delta y}{\Delta x}$$

even though we did not think of dy/dx as the quotient of the two separate quantities dy and dx. In this section we give separate meanings to dy and dx in such a way that their quotient, when $dx \neq 0$, is equal to the derivative of y with respect to x. We do this in the following manner.

DEFINITION OF DIFFERENTIALS	Let $y = f(x)$ represent a function that is differentiable in an open interval containing x. The **differential of x** (denoted by dx) is any nonzero real number. The **differential of y** (denoted by dy) is given by $$dy = f'(x) \, dx.$$

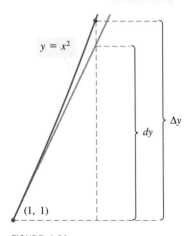

FIGURE 4.58

FIGURE 4.59

Note that in this definition dx can have any nonzero value. In most applications of differentials, however, we choose dx to be small and we denote this choice by $dx = \Delta x$.

One use of differentials is to approximate the change in $f(x)$ that corresponds to a change in x as shown in Figure 4.58. We denote this change by

$$\Delta y = f(x + \Delta x) - f(x).$$

The following example compares the values of dy and Δy.

EXAMPLE 1 Comparing Δy and dy

Let $y = x^2$. Find dy when $x = 1$ and $dx = 0.01$. Compare this value to Δy when $x = 1$ and $\Delta x = 0.01$.

SOLUTION

Since $y = f(x) = x^2$, we have $f'(x) = 2x$, and the differential dy is given by

$$dy = f'(x) \, dx = f'(1)(0.01) \qquad \text{Differential of } y$$
$$= 2(0.01) = 0.02.$$

Now, since $\Delta x = 0.01$, the change in y is

$$\Delta y = f(x + \Delta x) - f(x) = f(1.01) - f(1) \qquad \text{Change in } y$$
$$= (1.01)^2 - 1^2 = 0.0201.$$

Figure 4.59 shows the geometric comparison of dy and Δy. ▭

For the function in Example 1, we can see in Table 4.23 that *dy* approximates Δ*y* more and more closely as *dx* approaches zero.

TABLE 4.23

$dx = \Delta x$	dy	Δy	$\Delta y - dy$
0.1000	0.2000	0.21000000	0.01000000
0.0100	0.0200	0.02010000	0.00010000
0.0010	0.0020	0.00200100	0.00000100
0.0001	0.0002	0.00020001	0.00000001

REMARK   Note that for the sequence shown in Table 4.23, as Δ*x* decreases by tenths, Δ*y* − *dy* decreases by hundredths.

This type of approximation is called a **tangent line approximation** because we are using the tangent line at a point to approximate the graph of the function near the point. (Actually, this is not the first time we have encountered such an approximation. Recall from Section 4.8 that Newton's Method uses tangent lines to approximate an *x*-intercept of the graph of a function.)

The validity of using a tangent line to approximate a curve stems from the limit definition of the tangent line. That is, the existence of the limit

$$f'(x) = \lim_{\Delta x \to 0} \frac{f(x + \Delta x) - f(x)}{\Delta x}$$

implies that when Δ*x* is close to zero (within δ units of zero), then $f'(x)$ is close to the difference quotient (within ε units) and we have

$$\frac{f(x + \Delta x) - f(x)}{\Delta x} \approx f'(x), \quad \Delta x \neq 0$$

$$f(x + \Delta x) - f(x) \approx f'(x) \, \Delta x$$

$$\Delta y \approx f'(x) \, \Delta x.$$

Substituting *dx* for Δ*x* produces

$$\Delta y \approx f'(x) \, dx = dy, \, dx \neq 0.$$

We formalize this result in the following theorem. The proof follows directly from the definition of the derivative.

THEOREM 4.11
THE RELATIVE SIZE
OF *dy* AND Δ*y*

Let $y = f(x)$ be differentiable at *x* such that $f'(x) \neq 0$. If $dx = \Delta x$, then

$$\lim_{\Delta x \to 0} \frac{\Delta y}{dy} = 1$$

where $dy = f'(x) \, dx$ and $\Delta y = f(x + \Delta x) - f(x)$.

Error propagation

Physicists and engineers tend to make liberal use of the approximation of Δy by dy. One way this occurs in practice is in the estimation of errors propagated by physical measuring devices. For example, if we let x represent the measured value of a variable and let $x + \Delta x$ represent the exact value, then Δx is the *error in measurement*. Finally, if the measured value x is used to compute another value $f(x)$, then the difference between $f(x + \Delta x)$ and $f(x)$ is called the **propagated error.**

$$\underbrace{f(x + \overbrace{\Delta x}^{\substack{\text{measurement}\\\text{error}}})}_{\substack{\text{exact}\\\text{value}}} - \underbrace{f(x)}_{\substack{\text{measured}\\\text{value}}} = \underbrace{\Delta y}_{\substack{\text{propagated}\\\text{error}}}$$

EXAMPLE 2   Estimation of error

The radius of a ball bearing is measured to be 0.7 inch, as shown in Figure 4.60. If the measurement is correct to within 0.01 inch, estimate the propagated error in the volume V of the ball bearing.

SOLUTION

The formula for the volume of a sphere is $V = \frac{4}{3}\pi r^3$, where r is the radius of the sphere. Thus, we have

$$r = 0.7 \qquad \text{Measured radius}$$

and

$$-0.01 \le \Delta r \le 0.01. \qquad \text{Possible error}$$

To approximate the propagated error in the volume, we differentiate V to obtain $dV/dr = 4\pi r^2$ and write

$$\Delta V \approx dV = 4\pi r^2 \, dr = 4\pi(0.7)^2(\pm 0.01) \approx \pm 0.06158 \text{ in}^3. \qquad \blacksquare$$

Would you say that the propagated error in Example 2 is large or small? The answer is best given in *relative* terms by comparing dV to V. We obtain the ratio

$$\frac{dV}{V} = \frac{4\pi r^2 \, dr}{\frac{4}{3}\pi r^3} = \frac{3 \, dr}{r} \approx \frac{3}{0.7}(\pm 0.01) \approx \pm 0.0429$$

which is called the **relative error.** The corresponding **percentage error** is

$$\frac{dV}{V}(100) \approx 4.29\%.$$

FIGURE 4.60

We can use the definition of a differential to rewrite each of the derivative rules in **differential form.** For example, suppose u and v are differentiable functions of x. By the definition of differentials, we have $du = u'dx$ and $dv = v'dx$. Therefore, we can write the differential form of the Product Rule as follows.

$$d[uv] = \frac{d}{dx}[uv]\ dx \qquad \text{Differential of } uv$$

$$= [uv' + vu']\ dx \qquad \text{Product Rule}$$

$$= uv'\ dx + vu'\ dx$$

$$= u\ dv + v\ du$$

The following summary lists the differential forms corresponding to several differentiation rules.

GENERAL DIFFERENTIAL FORMULAS

Let u and v be differentiable functions of x.

Constant multiple: $d[cu] = c\ du$
Sum or difference: $d[u \pm v] = du \pm dv$
Product: $\qquad\quad d[uv] = u\ dv + v\ du$
Quotient: $\qquad\quad d\left[\dfrac{u}{v}\right] = \dfrac{v\ du - u\ dv}{v^2}$

In the next example, we compare the derivatives and differentials of several functions.

EXAMPLE 3 Finding differentials

Function	Derivative	Differential
(a) $y = x^2$	$\dfrac{dy}{dx} = 2x$	$dy = 2x\ dx$
(b) $y = 2\sin x$	$\dfrac{dy}{dx} = 2\cos x$	$dy = 2\cos x\ dx$
(c) $y = x\cos x$	$\dfrac{dy}{dx} = -x\sin x + \cos x$	$dy = (-x\sin x + \cos x)\ dx$
(d) $y = \dfrac{1}{x}$	$\dfrac{dy}{dx} = -\dfrac{1}{x^2}$	$dy = -\dfrac{dx}{x^2}$

Gottfried Wilhelm Leibniz

The notation in Example 3 is called the **Leibniz notation** for derivatives and differentials, named after the German mathematician Gottfried Wilhelm Leibniz (1646–1716). The beauty of this notation is that it provides us with an easy way to remember several important calculus formulas by making it seem as though the formulas were derived from algebraic manipulations of differentials. We will encounter several instances of this later in the text (in substitutions, inverse functions, parametric equations, and polar coordinates).

For now, we simply compare the *Chain Rule* in Leibniz notation to other possible notations.

$$\frac{dy}{dx} = \frac{dy}{du}\frac{du}{dx}$$ Leibniz notation

$$f'(g(x)) = f'(u)g'(x), \quad \text{where } u = g(x)$$ Function notation

$$D_x[y] = D_u[y]D_x[u]$$ Operator notation

The Leibniz form of the rule probably would appear to be true to a student in elementary algebra. Of course, it appears to be true for the *wrong reason*—that we have simply cancelled the *du*'s. Even so, the notation's many advantages overshadow the potential problem of drawing valid conclusions from invalid arguments.

EXAMPLE 4   Differential of a composite function

Use the Chain Rule to find dy for $y = \sin 3x$.

SOLUTION

$$y = f(x) = \sin 3x$$
$$f'(x) = 3\cos 3x$$
$$dy = f'(x)\,dx = 3\cos 3x\,dx$$ ▭

EXAMPLE 5   Differential of a composite function

Use the Chain Rule to find dy for $y = \sqrt{x^2 + 1}$.

SOLUTION

$$y = f(x) = (x^2 + 1)^{1/2}$$
$$f'(x) = \frac{1}{2}(2x)(x^2 + 1)^{-1/2} = \frac{x}{\sqrt{x^2 + 1}}$$

$$dy = f'(x)\,dx = \frac{x}{\sqrt{x^2 + 1}}\,dx$$ ▭

Differentials can be used to approximate function values. To do this for the function given by $y = f(x)$, we make use of the formula

$$f(x + \Delta x) \approx f(x) + dy = f(x) + f'(x)\,dx$$

which is derived from the approximation $\Delta y = f(x + \Delta x) - f(x) \approx dy$. The key to using this formula is choosing a value for x that makes the calculations easier. For instance, to approximate $\sqrt{101}$ we would let $x = 100$ and $\Delta x = 1$. This is illustrated in the next example.

EXAMPLE 6 Approximating function values

Use differentials to approximate the following function values.

(a) $\cos \dfrac{11\pi}{36}$ (b) $\sqrt{16.5}$

SOLUTION

(a) We use the function $f(x) = \cos x$ and choose the convenient value $x = \pi/3$. Then

$$dx = \frac{11\pi}{36} - \frac{\pi}{3} = -\frac{\pi}{36}$$

and we obtain

$$f(x + \Delta x) \approx f(x) + f'(x)\,dx = \cos x - \sin x\,dx$$

$$\cos \frac{11\pi}{36} \approx \cos \frac{\pi}{3} - \left(\sin \frac{\pi}{3}\right)\left(-\frac{\pi}{36}\right)$$

$$= \frac{1}{2} - \left(\frac{\sqrt{3}}{2}\right)\left(-\frac{\pi}{36}\right) \approx 0.5 + 0.0756 = 0.5756.$$

(b) We use the function $f(x) = \sqrt{x}$ and choose $x = 16$. Then $dx = 0.5$, and we obtain

$$f(x + \Delta x) \approx f(x) + f'(x)\,dx = \sqrt{x} + \frac{1}{2\sqrt{x}}\,dx$$

$$\sqrt{16.5} \approx \sqrt{16} + \frac{1}{2\sqrt{16}}(0.5) = 4 + \left(\frac{1}{8}\right)\left(\frac{1}{2}\right)$$

$$= 4 + \frac{1}{16} = 4.0625. \qquad \Box$$

REMARK   The use of differentials to approximate function values diminished with the availability of calculators and computers. Using a calculator, we obtain the values

$$\cos \frac{11\pi}{36} \approx 0.5736 \quad \text{and} \quad \sqrt{16.5} \approx 4.0620$$

which indicate the accuracy of the differential method in Example 6.

EXERCISES for Section 4.9

In Exercises 1–10, find the differential dy of the given function.

1. $y = 3x^2 - 4$

2. $y = 2x^{3/2}$

3. $y = \dfrac{x + 1}{2x - 1}$

4. $y = \sqrt{x^2 - 4}$

5. $y = x\sqrt{1 - x^2}$

6. $y = \sqrt{x} + \dfrac{1}{\sqrt{x}}$

7. $y = \dfrac{\sec^2 x}{x^2 + 1}$

8. $y = x \sin x$

9. $y = \dfrac{1}{3} \cos\left(\dfrac{6\pi x - 1}{2}\right)$

10. $y = x - \tan^2 x$

In Exercises 11–16, let $x = 2$ and use the given function and value of $\Delta x = dx$ to complete the table.

$dx = \Delta x$	dy	Δy	$\Delta y - dy$	$\dfrac{dy}{\Delta y}$
1.000				
0.500				
0.100				
0.010				
0.001				

11. $y = x^2$

12. $y = \dfrac{1}{x^2}$

13. $y = x^5$

14. $y = \sqrt{x}$

15. $y = \sin x$

16. $y = \csc x$

17. The area of a square of side x is given by $A(x) = x^2$.
  (a) Compute dA and ΔA in terms of x and Δx.
  (b) Use the accompanying figure to identify the region whose area is dA.
  (c) Use the accompanying figure to identify the region whose area is $\Delta A - dA$.

FIGURE FOR 17

18. The measurement of the side of a square is found to be 12 inches, with a possible error of $\frac{1}{64}$ inch. Use differentials to approximate the possible error in computing the area of the square.

19. The measurement of the radius of the end of a log is found to be 14 inches, with a possible error of $\frac{1}{4}$ inch. Use differentials to approximate the possible error in computing the area of the end of the log.

20. The measurement of the edge of a cube is found to be 12 inches, with a possible error of 0.03 inch. Use differentials to approximate the maximum possible error in computing the following.
  (a) the volume of the cube
  (b) the surface area of the cube

21. The measurement of a side of a square is found to be 15 centimeters.

(a) Approximate the percentage error in computing the area of the square if the possible error in measuring the side is 0.05 centimeters.
(b) Estimate the maximum allowable percentage error in measuring the side if the error in computing the area cannot exceed 2.5%.

22. The measurement of the circumference of a circle is found to be 56 centimeters.
  (a) Approximate the percentage error in computing the area of the circle if the possible error in measuring the circumference is 1.2 centimeters.
  (b) Estimate the maximum allowable percentage error in measuring the circumference if the error in computing the area cannot exceed 3%.

23. The radius of a sphere is claimed to be 6 inches, with a possible error of 0.02 inch. Use differentials to approximate the maximum possible error in calculating (a) the volume of the sphere and (b) the surface area of the sphere. (c) What is the relative error in parts (a) and (b)?

24. The profit P for a company is given by

$$P = (500x - x^2) - \left(\frac{1}{2}x^2 - 77x + 3000\right).$$

Approximate the change in profit as production changes from $x = 115$ to $x = 120$ units. Approximate the percentage change in profit when x changes from $x = 115$ to $x = 120$ units.

25. The period of a pendulum is given by

$$T = 2\pi\sqrt{\frac{L}{g}}$$

where L is the length of the pendulum in feet, g is the acceleration due to gravity, and T is the time in seconds. Suppose that the pendulum has been subjected to an increase in temperature so that the length increases by $\frac{1}{2}$ percent.
  (a) Find the approximate percentage change in the period.
  (b) Using the result of part (a), find the approximate error in this pendulum clock in one day.

26. Suppose a pendulum of length L is to be used to find the acceleration of gravity at a given point on the earth's surface. Use the formula

$$T = 2\pi\sqrt{\frac{L}{g}}$$

to approximate the percentage error in the value of g if you can identify the period T to within 0.1 percent of its true value.

27. A current of I amps passes through a resistor of R ohms. **Ohms Law** states that the voltage E applied to the resistor is given by

$$E = IR.$$

If the voltage is constant, show that the magnitude of the relative error in R caused by a change in I is equal in magnitude to the relative error in I.

28. The cost in dollars of removing $p\%$ of the air pollutants in the stack emission of a utility company that burns coal to generate electricity is

$$C = \frac{80,000p}{100 - p}, \quad 0 \le p < 100.$$

Use differentials to approximate the increase in cost if the government requires the utility company to remove 2% more of the pollutants and p is currently
(a) 40% (b) 75%

29. The measurement of one side of a right triangle is found to be 9.5 inches, and the angle opposite that side is $26°\ 45'$ with a possible error of $15'$.
(a) Approximate the percentage error in computing the length of the hypotenuse.
(b) Estimate the maximum allowable percentage error in measuring the angle if the error in computing the length of the hypotenuse cannot exceed 2%.

30. The measurement of one side of a right triangle is found to be 9.5 inches, and the angle opposite that side is $26°\ 45'$ with a possible error of $15'$. Approximate the percentage error in computing its area.

31. The range of a projectile is

$$R = \frac{v_0^2}{32}(\sin 2\theta)$$

where v_0 is the initial velocity in feet per second and θ is the angle of elevation. If $v_0 = 2200$ feet per second and θ is changed from $10°$ to $11°$, use differentials to approximate the change in the range.

32. A surveyor is standing 50 feet from the base of a large tree. He measures the angle of elevation to the top of the tree as $71.5°$. How accurately must the angle be measured if the percentage error in estimating the height of the tree is to be less than 6 percent?

33. Show that if $y = f(x)$ is a differentiable function, then

$$\Delta y - dy = \varepsilon \, \Delta x$$

where $\varepsilon \to 0$ as $\Delta x \to 0$.

4.10 Business and Economics Applications

Marginals ▪ Demand function

In Section 3.8 we mentioned that one of the most common ways to measure change is with respect to time. In this section we study some important rates of change in economics that are not measured with respect to time. For example, economists refer to **marginal profit, marginal revenue,** and **marginal cost** as the rates of change of the profit, revenue, and cost with respect to the number of units produced or sold.

We begin with a summary of some basic terms and formulas.

SUMMARY OF BUSINESS TERMS AND FORMULAS

Basic Terms — *Basic Formulas*

x is the number of units produced (or sold)
p is the price per unit
R is the total revenue from selling x units $R = xp$
C is the total cost of producing x units
$\overline{C}$ is the average cost per unit $\overline{C} = C/x$
P is the total profit from selling x units $P = R - C$

The **break-even point** is the number of units for which $R = C$.

Marginals

$\dfrac{dR}{dx}$ = marginal revenue $\approx$ the *extra* revenue for selling one additional unit

$\dfrac{dC}{dx}$ = marginal cost $\approx$ the *extra* cost of producing one additional unit

$\dfrac{dP}{dx}$ = marginal profit $\approx$ the *extra* profit for selling one additional unit

Marginal revenue

1 unit

Extra revenue for one unit

Revenue Function

FIGURE 4.61

REMARK In this summary, note that marginals can be used to approximate the *extra* revenue, cost, or profit associated with selling or producing one additional unit. This is illustrated graphically for marginal revenue in Figure 4.61.

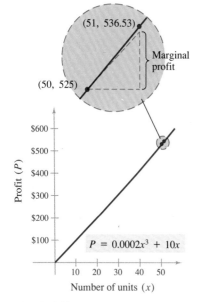

(51, 536.53)

Marginal profit

(50, 525)

$600
$500
$400
$300
$200
$100

Profit (P)

$P = 0.0002x^3 + 10x$

10 20 30 40 50

Number of units (x)

FIGURE 4.62

EXAMPLE 1 Using marginals as approximations

A manufacturer determines that the profit derived from selling x units of a certain item is given by $P = 0.0002x^3 + 10x$.
(a) Find the marginal profit for a production level of 50 units.
(b) Compare this to the actual gain in profit obtained by increasing the production from 50 to 51 units. (See Figure 4.62.)

SOLUTION

(a) The marginal profit is given by

$$\frac{dP}{dx} = 0.0006x^2 + 10.$$

When $x = 50$, the marginal profit is

$$\frac{dP}{dx} = (0.0006)(50)^2 + 10 = \$11.50. \qquad \text{Marginal profit}$$

(b) For $x = 50$ and 51, the actual profit is

$$P = (0.0002)(50)^3 + 10(50) = 25 + 500 = \$525.00$$
$$P = (0.0002)(51)^3 + 10(51) = 26.53 + 510 = \$536.53.$$

Thus, the additional profit obtained by increasing the production level from 50 to 51 units is

$$536.53 - 525.00 = \$11.53. \qquad \text{Extra profit for one unit} \quad \square$$

The profit function in Example 1 is unusual in that the profit continues to increase as long as the number of units sold increases. In practice, it is more common to encounter situations in which sales can be increased only by lowering the price per item, and such reductions in price ultimately will cause the profit to decline. The number of units x that consumers are willing to purchase at a given price per unit p is defined as the **demand function**

$$p = f(x). \qquad \text{Demand function}$$

EXAMPLE 2 Finding the demand function

A business sells 2000 items per month at a price of $10 each. It is predicted that monthly sales will increase by 250 items for each $0.25 reduction in price. Find the demand function corresponding to this prediction.

SOLUTION

From the given prediction, x increases 250 units each time p drops $0.25 from the original cost of $10. This is described by the equation

$$x = 2000 + 250\left(\frac{10 - p}{0.25}\right) = 12,000 - 1000p$$

or

$$p = 12 - \frac{x}{1000}. \qquad \text{Demand function} \quad \square$$

EXAMPLE 3 Finding the marginal revenue

A fast-food restaurant has determined that the monthly demand for their hamburgers is given by

$$p = \frac{60,000 - x}{20,000}.$$

Find the increase in revenue per hamburger for monthly sales of 20,000 hamburgers. In other words, find the marginal revenue when $x = 20,000$.

SOLUTION

Since the total revenue is given by $R = xp$, we have

$$R = xp = x\left(\frac{60,000 - x}{20,000}\right) = \frac{1}{20,000}(60,000x - x^2)$$

and the marginal revenue is

$$\frac{dR}{dx} = \frac{1}{20,000}(60,000 - 2x).$$

Finally, when $x = 20,000$, the marginal revenue is

$$\frac{dR}{dx} = \frac{1}{20,000}[60,000 - 2(20,000)] = \frac{20,000}{20,000} = \$1/\text{unit}.$$

REMARK Note that the demand function in Example 3 is typical in that a high demand corresponds to a low price, as seen in Figure 4.63 and Table 4.24.

As the price decreases, more hamburgers are sold.

$$p = \frac{60,000 - x}{20,000}$$

FIGURE 4.63

TABLE 4.24

p	$0.00	$0.50	$1.00	$1.50	$2.00	$2.50	$3.00
x	60,000	50,000	40,000	30,000	20,000	10,000	0

EXAMPLE 4   Finding the marginal profit

Suppose that in Example 3 the cost of producing x hamburgers is

$$C = 5000 + 0.56x.$$

Find the total profit and the marginal profit for (a) 20,000, (b) 24,400, and (c) 30,000 units.

SOLUTION

Since $P = R - C$, we use the revenue function in Example 3 to obtain

$$P = \frac{1}{20,000}(60,000x - x^2) - 5,000 - 0.56x$$

$$= 2.44x - \frac{x^2}{20,000} - 5,000.$$

Thus, the marginal profit is

$$\frac{dP}{dx} = 2.44 - \frac{x}{10,000}.$$

Table 4.25 shows the total profit and the marginal profit for the three indicated demands.

Maximum profit occurs
when $dP/dx = 0$.

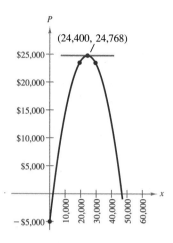

$$P = 2.44x - \frac{x^2}{20,000} - 5,000$$

FIGURE 4.64

TABLE 4.25

Demand	20,000	24,400	30,000
Profit	$23,800	$24,768	$23,200
Marginal Profit	$0.44	$0.00	−$0.56

REMARK In Example 4 we can see that when more than 24,400 hamburgers are sold, the marginal profit is negative. This means that increasing production beyond this point will *reduce* rather than increase profit. Figure 4.64 shows that the maximum profit corresponds to the point where the marginal profit is zero.

It is worth noting that many problems in economics are minimum and maximum problems. For such problems, the procedure used in Section 4.7 is an appropriate model to follow.

EXAMPLE 5 Finding the maximum profit

In marketing a certain item, a business has discovered that the demand for the item is represented by

$$p = \frac{50}{\sqrt{x}}.$$ Demand function

The cost of producing x items is given by $C = 0.5x + 500$. Find the price per unit that yields a maximum profit.

SOLUTION

From the given cost function, we have

$$P = R - C = xp - (0.5x + 500).$$ Primary equation

Substituting for p (from the demand function), we have

$$P = x\left(\frac{50}{\sqrt{x}}\right) - (0.5x + 500) = 50\sqrt{x} - 0.5x - 500.$$

Setting the marginal profit equal to zero, we obtain

$$\frac{dP}{dx} = \frac{25}{\sqrt{x}} - 0.5 = 0$$

$$\sqrt{x} = \frac{25}{0.5} = 50 \implies x = 2500.$$

Finally, we conclude that the maximum profit occurs when the price is

$$p = \frac{50}{\sqrt{2500}} = \frac{50}{50} = \$1.00.$$

FIGURE 4.65

REMARK   To find the maximum profit in Example 5, we differentiated the profit function, $P = R - C$, and set dP/dx equal to zero. From the equation

$$\frac{dP}{dx} = \frac{dR}{dx} - \frac{dC}{dx} = 0$$

it follows that the maximum profit occurs when the marginal revenue is equal to the marginal cost, as shown in Figure 4.65.

To study the effect of production levels on cost, economists use the average cost function $\overline{C}$ defined as

$$\overline{C} = \frac{C}{x}.$$ Average cost function

A use of this function is illustrated in the next example.

EXAMPLE 6   Minimizing the average cost

A company estimates that the cost (in dollars) of producing x units of a certain product is given by $C = 800 + 0.04x + 0.0002x^2$. Find the production level that minimizes the average cost per unit.

SOLUTION

Substituting from the given equation for C produces

$$\overline{C} = \frac{C}{x} = \frac{800 + 0.04x + 0.0002x^2}{x} = \frac{800}{x} + 0.04 + 0.0002x.$$

Setting the derivative $d\overline{C}/dx$ equal to zero yields

$$\frac{d\overline{C}}{dx} = -\frac{800}{x^2} + 0.0002 = 0$$

$$x^2 = \frac{800}{0.0002} = 4{,}000{,}000 \implies x = 2{,}000 \text{ units.}$$

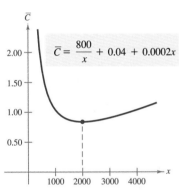

FIGURE 4.66

(See Figure 4.66.)

EXERCISES for Section 4.10

In Exercises 1–4, find the number of units x that produce a maximum revenue R.

1. $R = 900x - 0.1x^2$

2. $R = 600x^2 - 0.02x^3$

3. $R = \dfrac{1{,}000{,}000x}{0.02x^2 + 1800}$

4. $R = 30x^{2/3} - 2x$

In Exercises 5–8, find the number of units x that produce the minimum average cost per unit $\overline{C}$.

5. $C = 0.125x^2 + 20x + 5000$

6. $C = 0.001x^3 - 5x + 250$

7. $C = 3000x - x^2\sqrt{300 - x}$

8. $C = \dfrac{2x^3 - x^2 + 5000x}{x^2 + 2500}$

In Exercises 9–12, find the price per unit p that produces the maximum profit P.

Cost function	Demand function
9. $C = 100 + 30x$	$p = 90 - x$
10. $C = 2400x + 5200$	$p = 6000 - 0.4x^2$

Cost function	Demand function
11. $C = 4000 - 40x + 0.02x^2$	$p = 50 - \dfrac{x}{100}$
12. $C = 35x + 2\sqrt{x - 1}$	$p = 40 - \sqrt{x - 1}$

13. A manufacturer of lighting fixtures has daily production costs of

$$C = 800 - 10x + \frac{1}{4}x^2.$$

How many fixtures x should be produced each day to minimize costs?

14. The profit for a certain company is given by

$$P = 230 + 20s - \frac{1}{2}s^2$$

where s is the amount (in hundreds of dollars) spent on advertising. What amount of advertising gives the maximum profit?

15. A manufacturer of radios charges $90 per unit when the average production cost per unit is $60. To encourage large orders from distributors, the manufacturer will reduce the charge by $0.10 per unit for each unit ordered in excess of 100 (for example, there would be a charge of $88 per radio for an order size of 120). Find the largest order the manufacturer should allow so as to realize maximum profit.

16. A real estate office handles 50 apartment units. When the rent is $540 per month, all units are occupied. However, on the average, for each $30 increase in rent, one unit becomes vacant. Each occupied unit requires an average of $36 per month for service and repairs. What rent should be charged to realize the most profit?

17. A power station is on one side of a river that is $\frac{1}{2}$ mile wide, and a factory is 6 miles downstream on the other side. It costs $6 per foot to run power lines overland and $8 per foot to run them underwater. Find the most economical path for the transmission line from the power station to the factory.

18. An offshore oil well is 1 mile off the coast. The refinery is 2 miles down the coast. If laying pipe in the ocean is twice as expensive as on land, what path should the pipe follow in order to minimize the cost?

19. Assume that the amount of money deposited in a bank is proportional to the square of the interest rate the bank pays on this money. Furthermore, the bank can reinvest this money at 12 percent. Find the interest rate the bank should pay to maximize profit. (Use the simple interest formula.)

20. Prove that the average cost is minimum at the value of x where the average cost equals the marginal cost.

In Exercises 21 and 22, use the given cost function to find the value of x where the average cost is minimum. For that value of x, show that the marginal cost and average cost are equal (see Exercise 20).

21. $C = 2x^2 + 5x + 18$
22. $C = x^3 - 6x^2 + 13x$

23. A given commodity has a demand function given by $p = 100 - \frac{1}{2}x^2$ and a total cost function of $C = 4x + 375$.
(a) What price gives the maximum profit?
(b) What is the average cost per unit if production is set to give maximum profit?

24. When a wholesaler sold a certain product at $25 per unit, sales were 800 units each week. After a price increase of $5, the average number of units sold dropped to 775 per week. Assume that the demand function is linear and find the price that will maximize the total revenue.

25. The ordering and transportation cost C of the components used in manufacturing a certain product is given by

$$C = 100\left(\frac{200}{x^2} + \frac{x}{x + 30}\right), \quad 1 \le x$$

where C is measured in thousands of dollars and x is the order size in hundreds. Find the order size that minimizes the cost. [Hint: Use Newton's Method.]

26. A company estimates that the cost in dollars of producing x units of a certain product is given by the model

$$C = 800 + 0.4x + 0.02x^2 + 0.0001x^3.$$

Find the production level that minimizes the average cost per unit. [Hint: Use Newton's Method.]

27. The revenue R for a company selling x units is given by

$$R = 900x - 0.1x^2.$$

Use differentials to approximate the change in revenue if sales increase from $x = 3000$ to $x = 3100$ units.

28. The profit P for a company selling x units is given by

$$P = (500x - x^2) - \left(\frac{1}{2}x^2 - 77x + 3000\right).$$

Use differentials to approximate the change and percentage change in profit as production changes from $x = 175$ to $x = 180$ units.

29. Domestic energy consumption in the United States is seasonal. Suppose the consumption is approximated by the model

$$Q = 6.2 + \cos\left[\frac{\pi(2t - 1)}{12}\right]$$

where Q is the total consumption in quadrillion BTUs (quads) and t is the time in months with $0 \le t \le 1$ corresponding to January (see figure).

(a) On which day does this model predict the greatest consumption, and what is the monthly rate for that day?

(b) On which day does this model predict the least consumption, and what is the monthly rate for that day?

Energy consumption: 1985

FIGURE FOR 29

30. Sales of electricity in the United States have had both an increasing annual sales pattern and a seasonal monthly sales pattern. For the years 1983–1985, the sales pattern can be approximated by the model

$$S = 180.6 + 0.55t + 13.60 \cos\left(\frac{\pi t}{3}\right)$$

where S is the sales (per month) in billions of kilowatt hours and t is the time in months (see figure).

(a) Find the relative extrema of this function for 1983 and 1984.

(b) Use this model to predict the sales in August, 1985. (Use $t = 31.5$.)

Electricity sales

FIGURE FOR 30

31. The model for energy consumption in Exercise 29 does not account for the increase in energy consumption

resulting from air conditioning during the summer. A model (see figure) that does account for this additional use is given by

$$Q = 6.2 + \frac{2}{3} \sin\left[\frac{\pi(2t + 5)}{12}\right] - \frac{1}{3} \cos\left[\frac{\pi(2t + 5)}{6}\right].$$

(a) On which day does this model predict the greatest consumption, and what is the monthly rate for that day?

(b) On which days does this model predict the least consumption, and what is the monthly rate for these days?

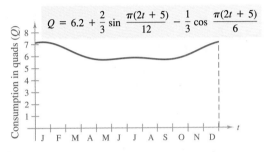

Energy consumption: 1985

FIGURE FOR 31

The relative responsiveness of consumers to a change in the price of an item is called the **price elasticity of demand**. If $p = f(x)$ is a differentiable demand function, then the price elasticity of demand is given by

$$\eta = \frac{p/x}{dp/dx}.$$

For a given price, if $|\eta| < 1$, the demand is **inelastic**, and if $|\eta| > 1$, the demand is **elastic**.

32. If the demand equation is given by $xp^m = a$, where a is a constant and $m > 1$, show that $\eta = -m$. (In other words, in terms of approximate price changes, a 1 percent increase in price results in an m percent decrease in quantity demanded.)

In Exercises 33–36, find η for the given demand function at the indicated x-value. Is the demand elastic, inelastic, or neither at the indicated x-value?

Demand Function	Quantity Demanded
33. $p = 400 - 3x$	$x = 20$
34. $p = 5 - 0.03x$	$x = 100$
35. $p = 400 - 0.5x^2$	$x = 20$
36. $p = \dfrac{500}{x + 2}$	$x = 23$

REVIEW EXERCISES for Chapter 4

In Exercises 1–26, make use of domain, range, symmetry, asymptotes, intercepts, relative extrema, or points of inflection to obtain an accurate graph of the given function.

1. $f(x) = 4x - x^2$ 　　　**2.** $f(x) = 4x^3 - x^4$

3. $f(x) = x\sqrt{16 - x^2}$ 　　**4.** $f(x) = x + \dfrac{4}{x^2}$

5. $f(x) = \dfrac{x + 1}{x - 1}$ 　　　**6.** $f(x) = x^2 + \dfrac{1}{x}$

7. $f(x) = x^3 + x + \dfrac{4}{x}$ 　**8.** $f(x) = x^3(x + 1)$

9. $f(x) = (x - 1)^3(x - 3)^2$

10. $f(x) = (x - 3)(x + 2)^3$

11. $f(x) = (5 - x)^3$ 　　**12.** $f(x) = (x^2 - 4)^2$

13. $f(x) = x^{1/3}(x + 3)^{2/3}$

14. $f(x) = (x - 2)^{1/3}(x + 1)^{2/3}$

15. $f(x) = x^3 + \dfrac{243}{x}$ 　**16.** $f(x) = \dfrac{2x}{1 + x^2}$

17. $f(x) = \dfrac{4}{1 + x^2}$ 　**18.** $f(x) = \dfrac{x^2}{1 + x^4}$

19. $f(x) = |x^2 - 9|$ 　　**20.** $f(x) = |9 - x^2|$

21. $f(x) = |x^3 - 3x^2 + 2x|$

22. $f(x) = |x - 1| + |x - 3|$

23. $f(x) = \dfrac{1}{|x - 1|}$ 　**24.** $f(x) = \dfrac{x - 1}{1 + 3x^2}$

25. $f(x) = x + \cos x, \quad 0 \le x \le 2\pi$

26. $f(x) = \dfrac{1}{\pi}(2 \sin \pi x - \sin 2\pi x), \quad -1 \le x \le 1$

27. Find the maximum and minimum points on the graph of

$$x^2 + 4y^2 - 2x - 16y + 13 = 0.$$

(a) without using calculus and (b) using calculus.

28. Consider the function $f(x) = x^n$ for positive integer values of n.

  (a) For what values of n does the function have a relative minimum at the origin?

  (b) For what values of n does the function have a point of inflection at the origin?

In Exercises 29–34, find the point(s) guaranteed by the Mean Value Theorem for the indicated interval.

Function	*Interval*
29. $f(x) = \dfrac{2x + 3}{3x + 2}$	$1 \le x \le 5$
30. $f(x) = \dfrac{1}{x}$	$1 \le x \le 4$
31. $f(x) = x^{2/3}$	$1 \le x \le 8$
32. $f(x) = |x^2 - 9|$	$0 \le x \le 2$
33. $f(x) = x - \cos x$	$-\dfrac{\pi}{2} \le x \le \dfrac{\pi}{2}$
34. $f(x) = \sqrt{x} - 2x$	$0 \le x \le 4$

35. Can the Mean Value Theorem be applied to the function $f(x) = 1/x^2$ on the interval $[-2, 1]$?

36. If $f(x) = 3 - |x - 4|$, verify that $f(1) = f(7)$ and yet $f'(x)$ is not equal to zero for any x in $[1, 7]$. Does this contradict Rolle's Theorem?

37. For the function

$$f(x) = Ax^2 + Bx + C$$

determine the value of c guaranteed by the Mean Value Theorem on the interval $[x_1, x_2]$.

38. Demonstrate the result of Exercise 37 for

$$f(x) = 2x^2 - 3x + 1$$

on the interval $[0, 4]$.

39. At noon ship A was 100 miles due east of ship B. Ship A is sailing west at 12 miles per hour, and ship B is sailing south at 10 miles per hour. At what time will the ships be nearest to each other, and what will this distance be?

40. The cost C of producing x units per day is

$$C = \frac{1}{4}x^2 + 62x + 125$$

and the price p per unit is

$$p = 75 - \frac{1}{3}x.$$

  (a) What daily output produces maximum profit?

  (b) What daily output produces minimum average cost?

  (c) Find the elasticity of demand.

41. Find the maximum profit if the demand equation is

$$p = 36 - 4x$$

and the total cost is

$$C = 2x^2 + 6.$$

42. Find the dimensions of the rectangle of maximum area, with sides parallel to the coordinate axes, that can be inscribed in the ellipse given by

$$\frac{x^2}{144} + \frac{y^2}{16} = 1.$$

43. A right triangle in the first quadrant has the coordinate axes as sides, and the hypotenuse passes through the point (1, 8). Find the vertices of the triangle so that the length of the hypotenuse is minimum.

44. The wall of a building is to be braced by a beam that must pass over a parallel fence 5 feet high and 4 feet from the building. Find the length of the shortest beam that can be used.

45. For groups of 80 or more, a charter bus company determines the rate per person according to the following formula.

Rate = $8.00 - $0.05(n - 80), \quad n \geq 80$

What number of passengers will give the bus company maximum revenue?

46. Show that the greatest area of any rectangle inscribed in a triangle is one-half that of the triangle.

47. Three sides of a trapezoid have the same length s. Of all such possible trapezoids, show that the one of maximum area has a fourth side of length $2s$.

48. The cost of fuel for running a locomotive is proportional to the $\frac{3}{2}$ power of the speed ($s^{3/2}$) and is $50 per hour for a speed of 25 miles per hour. Other fixed costs amount to an average of $100 per hour. Find the speed that will minimize the cost per mile.

49. Find the length of the longest pipe that can be carried level around a right-angle corner if the two intersecting corridors are of width 4 feet and 6 feet. (Do not use trigonometry.)

50. Rework Exercise 49, given corridors of width a feet and b feet. (Do not use trigonometry.)

51. A hallway of width 6 feet meets a hallway of width 9 feet at right angles. Find the length of the longest pipe that can be carried horizontally around this corner. [Hint: If L is the length of the pipe, show that

$$L = 6 \csc \theta + 9 \csc \left[\frac{\pi}{2} - \theta\right]$$

where θ is the angle between the pipe and the wall of the narrower hallway.]

52. Rework Exercise 51, given that one of the hallways is of width a feet and one is of width b feet. Show that the result is the same as in Exercise 50.

53. The height of an object attached to a spring is given by the harmonic equation

$$y = \frac{1}{3} \cos 12t - \frac{1}{4} \sin 12t$$

where y is measured in inches and t is measured in seconds.

(a) Calculate the height and velocity of the object when $t = \pi/8$ second.
(b) Show that the maximum displacement of the object is 5/12 inch.
(c) Find the period P of y. Find the frequency f (number of oscillations per second) if $f = 1/P$.

54. The general equation giving the height of an oscillating object attached to a spring is

$$y = A \sin \sqrt{\frac{k}{m}} t + B \cos \sqrt{\frac{k}{m}} t$$

where k is the spring constant and m is the mass of the object. Show that the maximum displacement of the object is $\sqrt{A^2 + B^2}$. Show that the object oscillates with a frequency of $(1/2\pi)\sqrt{k/m}$. How is the frequency changed if the stiffness k of the spring is increased? How is the frequency changed if the mass m of the object is increased?

55. The cost of inventory depends on ordering cost and storage cost, according to the following inventory model.

$$C = \left(\frac{Q}{x}\right)s + \left(\frac{x}{2}\right)r$$

Determine the order size that will minimize the cost, assuming that sales occur at a constant rate, Q is the number of units sold per year, r is the cost of storing one unit for one year, s is the cost of placing an order, and x is the number of units per order.

56. The demand and cost equations for a certain product are $p = 600 - 3x$ and $C = 0.3x^2 + 6x + 600$ where p is the price per unit, x is the number of units, and C is the cost of producing x units. If t is the excise tax per unit, the profit for producing x units is $P = xp - C - xt$. Find the maximum profit for (a) $t = 5$, (b) $t = 10$, and (c) $t = 20$.

57. Approximate, to three decimal places, the zero of $f(x) = x^3 - 3x - 1$ in the interval $[-1, 0]$.

58. Approximate, to three decimal places, the x-value of the points of intersection of the equations $y = x^4$ and $y = x + 3$.

59. Approximate, to three decimal places, the real root of

$$x^3 + 2x + 1 = 0.$$

60. The diameter of a sphere is measured to be 18 inches with a maximum possible error of 0.05 inch. Use the differential to approximate the possible error in the surface area and the volume of the sphere.

61. If a 1 percent error is made in measuring the edge of a cube, approximately what percentage error will be made in calculating the surface area and the volume of the cube?

62. A company finds that the demand for its commodity is given by

$$p = 75 - \frac{1}{4}x.$$

If x changes from 7 to 8, find the corresponding change in p. Compare the value of Δp and dp.

Chapter 5 Application

Voyager 1 and Voyager 2 were launched by the United States in 1977 to observe Jupiter and Saturn. Voyager 1 reached Jupiter in 1979 and Saturn in 1980. Voyager 2 also passed by Jupiter (1979) and Saturn (1981) and then continued on to observe the outer planets. In 1986, Voyager 2 transmitted stunning pictures of the rings of Uranus.

Escape Velocity from the Earth

The *escape velocity* from the earth is the minimum initial velocity that will allow an object, projected straight upward, to escape the earth's gravitational pull. If y is the distance from the center of the earth to the projected object, then the object will continue to move away from the earth provided that, for all t, the velocity of the object is positive. That is,

$$v = \frac{dy}{dt} > 0.$$

From Newton's Law of gravitation, we can conclude that the acceleration of the object is given by

$$\frac{dv}{dt} = -\frac{GM}{y^2}$$

where G is the gravitational constant and M is the mass of the earth. Using the Chain Rule, we have $dv/dt = (dv/dy)(dy/dt) = v(dv/dy)$. Thus, we can write

$$\int v\, dv = -GM \int \frac{1}{y^2} dy$$

$$\frac{1}{2}v^2 = \frac{GM}{y} + C.$$

Solving for the constant of integration C produces

$$v^2 = v_0{}^2 + 2GM\left(\frac{1}{y} - \frac{1}{R}\right)$$

where v_0 is the initial velocity of the object and R is the radius of the earth. Since $y > 0$, it follows that

$$v^2 > v_0{}^2 - \frac{2GM}{R}$$

which implies that v will be positive provided that

$$v_0 > \sqrt{\frac{2GM}{R}} \approx 24{,}995 \text{ mi/hr}.$$

See Exercise 54, Section 5.1.

Chapter Overview

In this chapter we introduce the second fundamental operation of calculus—**antidifferentiation** or **integration.** The chapter begins by looking at some basic rules for finding antiderivatives. For example, the family of antiderivatives of $f(x) = 2x$ is given by

$$F(x) = \int 2x\, dx = x^2 + C$$

where C is the constant of integration. Antidifferentiation has many uses. For instance, if the velocity function of an object is known, then antidifferentiation can be used to find its position function.

In Section 5.2, we introduce the problem of finding the area of a plane region. For instance, the area of the region bounded by the graphs of $f(x) = x^2$, $y = 0$, $x = 0$, and $x = 1$ is given by the **definite integral**

$$\text{area} = \int_0^1 x^2\, dx.$$

Section 5.3 discusses several properties of definite integrals, and in Section 5.4, we present the **Fundamental Theorem of Calculus.** This theorem shows how antiderivatives can be used to solve definite integrals. The chapter closes by looking at two methods of approximating definite integrals—the Trapezoidal Rule and Simpson's Rule.

Integration

5

5.1 Antiderivatives and Indefinite Integration

Antiderivatives ▪ Notation for antiderivatives ▪ Basic integration rules ▪ Initial conditions and particular solutions

Until now, our study of calculus has been concerned primarily with this problem: *given a function, find its derivative.* Now, however, we will study the inverse problem: *given the derivative of a function, find the original function.* For example, suppose you were asked to find a function F that has the following derivative.

$$F'(x) = 3x^2$$

From your knowledge of derivatives, you would probably say that

$$F(x) = x^3 \quad \text{because} \quad \frac{d}{dx}[x^3] = 3x^2$$

We call the function F an **antiderivative** of F'. For convenience, the phrase *$F(x)$ is an antiderivative of $f(x)$* is used synonymously with *F is an antiderivative of f.* For instance, we say that x^3 is an antiderivative of $3x^2$.

DEFINITION OF ANTIDERIVATIVE

A function F is called an **antiderivative** of the function f if for every x in the domain of f

$$F'(x) = f(x).$$

In this definition, we call F *an* antiderivative of f, rather than *the* antiderivative of f. To see why, consider that $F_1(x) = x^3$, $F_2(x) = x^3 - 5$, and $F_3(x) = x^3 + 97$ are all antiderivatives of $f(x) = 3x^2$. This suggests that for any constant C, the function given by $F(x) = x^3 + C$ is an antiderivative of f. This result is part of the following theorem.

THEOREM 5.1 REPRESENTATION OF ANTIDERIVATIVES	If F is an antiderivative of f on an interval I, then G is an antiderivative of f on the interval I if and only if G is of the form $$G(x) = F(x) + C, \quad \text{for all } x \text{ in } I$$ where C is a constant.

PROOF There are two directions in this theorem. The proof of one direction is straightforward. That is, if $F'(x) = f(x)$ and C is a constant, then

$$G'(x) = \frac{d}{dx}[F(x) + C] = F'(x) + 0 = f(x).$$

The proof of the other direction requires a little more work, but it can be accomplished with the Mean Value Theorem, by defining a function H such that

$$H(x) = G(x) - F(x).$$

If H is not constant on the interval I, then there must exist a and b ($a < b$) in the interval such that $H(a) \neq H(b)$. Moreover, because H is differentiable on $[a, b]$, we can apply the Mean Value Theorem to conclude that there exists some c in (a, b) such that

$$H'(c) = \frac{H(b) - H(a)}{b - a}.$$

Since $H(b) \neq H(a)$, it follows that $H'(c) \neq 0$. However, since $G'(c) = F'(c)$, we know that

$$H'(c) = G'(c) - F'(c) = 0$$

and we have a contradiction. Consequently, our assumption that $H(x)$ is not constant must be false, and we conclude that $H(x) = C$. Therefore, $G(x) - F(x) = C$ and we conclude that $G(x) = F(x) + C$.

The point of Theorem 5.1 is that we can represent the entire family of antiderivatives of a function by adding a constant to a *known* antiderivative. For example, knowing that $D_x[x^2] = 2x$, we can represent the family of *all* antiderivatives of $f(x) = 2x$ by $G(x) = x^2 + C$, where C is a constant. We call G the **general antiderivative** of f and $G(x) = x^2 + C$ the **general solution** of the equation $G'(x) = 2x$.

Notation for antiderivatives

If $y = F(x)$ is an antiderivative of f, then we say $F(x)$ is a solution of the equation

$$\frac{dy}{dx} = f(x).$$

When solving such an equation, it is convenient to write it in the equivalent differential form

$$dy = f(x)\ dx.$$

The operation of finding all solutions (the general antiderivative of f) of this equation is called **antidifferentiation** (or **integration**) and is denoted by an integral sign $\int$. The general solution of the equation $dy = f(x)\ dx$ is denoted by

Variable of integration

$$y = \int f(x)\ dx = F(x) + C.$$

Integrand

Constant of integration

We read $\int f(x)\ dx$ as the *antiderivative of f with respect to x*. Thus, the differential dx serves to identify x as the variable of integration. The term **indefinite integral** is a synonym for antiderivative.

DEFINITION OF INTEGRAL NOTATION FOR ANTIDERIVATIVES

The notation

$$\int f(x)\ dx = F(x) + C$$

where C is an arbitrary constant, means that F is an antiderivative of f. That is, $F'(x) = f(x)$ for all x in the domain of f.

The inverse nature of integration and differentiation can be verified by substituting $F'(x)$ for $f(x)$ in this definition to obtain

$$\int F'(x)\ dx = F(x) + C.$$ Integration is the inverse of differentiation

Moreover, if $\int f(x)\ dx = F(x) + C$, then

$$\frac{d}{dx}\left[\int f(x)\ dx\right] = f(x).$$ Differentiation is the inverse of integration

This allows us to obtain integration formulas directly from differentiation formulas, as shown in the following theorem.

THEOREM 5.2 BASIC INTEGRATION RULES	*Differentiation formula*	*Integration formula*
	$\dfrac{d}{dx}[C] = 0$	$\displaystyle\int 0 \, dx = C$
	$\dfrac{d}{dx}[kx] = k$	$\displaystyle\int k \, dx = kx + C$
	$\dfrac{d}{dx}[kf(x)] = kf'(x)$	$\displaystyle\int kf(x) \, dx = k \int f(x) \, dx$
	$\dfrac{d}{dx}[f(x) \pm g(x)] = f'(x) \pm g'(x)$	$\displaystyle\int [f(x) \pm g(x)] \, dx = \int f(x) \, dx \pm \int g(x) \, dx$
	$\dfrac{d}{dx}[x^n] = nx^{n-1}$	$\displaystyle\int x^n \, dx = \dfrac{x^{n+1}}{n+1} + C, \, n \neq -1$ Power Rule
	$\dfrac{d}{dx}[\sin x] = \cos x$	$\displaystyle\int \cos x \, dx = \sin x + C$
	$\dfrac{d}{dx}[\cos x] = -\sin x$	$\displaystyle\int \sin x \, dx = -\cos x + C$
	$\dfrac{d}{dx}[\tan x] = \sec^2 x$	$\displaystyle\int \sec^2 x \, dx = \tan x + C$
	$\dfrac{d}{dx}[\sec x] = \sec x \tan x$	$\displaystyle\int \sec x \tan x \, dx = \sec x + C$
	$\dfrac{d}{dx}[\cot x] = -\csc^2 x$	$\displaystyle\int \csc^2 x \, dx = -\cot x + C$
	$\dfrac{d}{dx}[\csc x] = -\csc x \cot x$	$\displaystyle\int \csc x \cot x \, dx = -\csc x + C$

REMARK Be sure you understand that the Power Rule for integration has the restriction that $n \neq -1$. The evaluation of $\int 1/x \, dx$ must wait until the introduction of the natural logarithm function in Chapter 6.

EXAMPLE 1 Applying the basic integration rules

$$\int 3x \, dx = 3 \int x \, dx \qquad \text{Constant Multiple Rule}$$

$$= 3 \int x^1 \, dx \qquad \text{Rewrite } (x = x^1)$$

$$= 3\left(\frac{x^2}{2}\right) + C \qquad \text{Power Rule } (n = 1)$$

$$= \frac{3}{2}x^2 + C \qquad \text{Simplify}$$

When evaluating indefinite integrals, a strict application of the basic integration rules tends to produce complicated constants of integration. For instance, in Example 1, we could have written

$$\int 3x \, dx = 3 \int x \, dx = 3\left[\frac{x^2}{2} + C\right] = \frac{3}{2}x^2 + 3C.$$

However, since C represents *any* constant, it is both cumbersome and unnecessary to write $3C$ as the constant of integration, and we choose the simpler form, $\frac{3}{2}x^2 + C$.

In Example 1, note that the general pattern of integration is similar to that of differentiation.

$$\boxed{\text{Given Integral}} \longrightarrow \boxed{\text{Rewrite}} \longrightarrow \boxed{\text{Integrate}} \longrightarrow \boxed{\text{Simplify}}$$

This pattern is followed in the next example.

EXAMPLE 2   Rewriting before integrating

	Given integral	*Rewrite*	*Integrate*	*Simplify*
(a)	$\displaystyle\int \frac{1}{x^3}\, dx$	$\displaystyle\int x^{-3}\, dx$	$\dfrac{x^{-2}}{-2} + C$	$-\dfrac{1}{2x^2} + C$
(b)	$\displaystyle\int \sqrt{x}\, dx$	$\displaystyle\int x^{1/2}\, dx$	$\dfrac{x^{3/2}}{3/2} + C$	$\dfrac{2}{3}x^{3/2} + C$
(c)	$\displaystyle\int 2\sin x\, dx$	$\displaystyle 2\int \sin x\, dx$	$2(-\cos x) + C$	$-2\cos x + C$

REMARK   Remember that we can check our answer to an antidifferentiation problem by differentiating. For instance, in Example 2(b), we can check that $\frac{2}{3}x^{3/2}$ is the correct antiderivative by differentiating to obtain $D_x\!\left[\frac{2}{3}x^{3/2}\right] = \left(\frac{2}{3}\right)\!\left(\frac{3}{2}\right)x^{1/2} = \sqrt{x}$.

The basic integration rules listed in Theorem 5.2 allow us to integrate *any* polynomial function. This is demonstrated in the next example.

EXAMPLE 3   Integrating polynomial functions

(a) $\displaystyle\int 1\, dx = x + C$

(b) $\displaystyle\int (x + 2)\, dx = \int x\, dx + \int 2\, dx = \frac{x^2}{2} + C_1 + 2x + C_2$

$$= \frac{x^2}{2} + 2x + C$$

We normally omit the third step by combining the two constants of integration, C_1 and C_2, to form a single constant, C.

(c) $\displaystyle\int (3x^4 - 5x^2 + x)\, dx = 3\left(\frac{x^5}{5}\right) - 5\left(\frac{x^3}{3}\right) + \frac{x^2}{2} + C$

$$= \frac{3}{5}x^5 - \frac{5}{3}x^3 + \frac{1}{2}x^2 + C$$

REMARK   The integral in Example 3(a) is usually simplified to the form

$$\int 1 \, dx = \int dx.$$

EXAMPLE 4   Rewriting before integrating

$$\int \frac{x+1}{\sqrt{x}} \, dx = \int \left(\frac{x}{\sqrt{x}} + \frac{1}{\sqrt{x}} \right) dx$$

$$= \int (x^{1/2} + x^{-1/2}) \, dx \qquad \text{Rewrite with fractional exponents}$$

$$= \frac{x^{3/2}}{3/2} + \frac{x^{1/2}}{1/2} + C \qquad \text{Integrate}$$

$$= \frac{2}{3}x^{3/2} + 2x^{1/2} + C \qquad \text{Simplify}$$

REMARK   When integrating quotients, do not integrate the numerator and denominator separately. This is no more valid in integration than it is in differentiation. For instance, in Example 4, be sure you understand that

$$\int \frac{x+1}{\sqrt{x}} \, dx \neq \frac{\int (x+1) \, dx}{\int \sqrt{x} \, dx}.$$

EXAMPLE 5   Rewriting before integrating

$$\int \frac{\sin x}{\cos^2 x} \, dx = \int \left(\frac{1}{\cos x} \right) \left(\frac{\sin x}{\cos x} \right) dx$$

$$= \int \sec x \tan x \, dx \qquad \text{Rewrite using trigonometric identity}$$

$$= \sec x + C \qquad \text{Integrate}$$

Initial conditions and particular solutions

We have mentioned already that the equation $y = \int f(x) \, dx$ has many solutions (each differing from the others by a constant). This means that the graphs of any two antiderivatives of f are vertical translations of each other. For example, Figure 5.1 shows the graphs of several antiderivatives of the form

$$y = \int (3x^2 - 1) \, dx = x^3 - x + C \qquad \text{General solution}$$

for various integer values of C. Each of these antiderivatives is a solution of the equation

$$\frac{dy}{dx} = 3x^2 - 1.$$

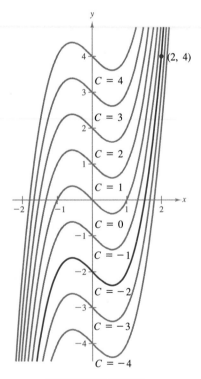

$F(x) = x^3 - x + C$

FIGURE 5.1

In many applications of integration, we are given enough information to determine a **particular solution.** To do this we need only know the value of $F(x)$ for one value of x. (This information is called an **initial condition.**) For example, in Figure 5.1, only one curve passes through the point (2, 4). To find this curve, we use the following information.

$$F(x) = x^3 - x + C \qquad \text{General solution}$$

$$F(2) = 4 \qquad \text{Initial condition}$$

By using the initial condition in the general solution, we determine that $F(2) = 8 - 2 + C = 4$, which implies that $C = -2$. Thus, we obtain

$$F(x) = x^3 - x - 2. \qquad \text{Particular solution}$$

EXAMPLE 6   Finding a particular solution

Find the general solution of the equation

$$F'(x) = \frac{1}{x^2}$$

and find the particular solution that satisfies the initial condition $F(1) = 0$.

SOLUTION

To find the general solution, we integrate to obtain

$$F(x) = \int \frac{1}{x^2}\, dx = \int x^{-2}\, dx$$

$$= \frac{x^{-1}}{-1} + C = -\frac{1}{x} + C.$$

Now, since $F(1) = 0$, we can solve for C as follows.

$$F(1) = -\frac{1}{1} + C = 0 \quad \Longrightarrow \quad C = 1$$

Thus, the particular solution, as shown in Figure 5.2, is

$$F(x) = -\frac{1}{x} + 1.$$

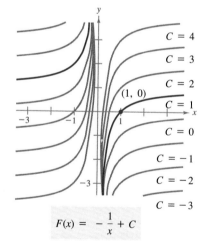

$$F(x) = -\frac{1}{x} + C$$

FIGURE 5.2

So far in this section we have been using x as the variable of integration. In applications, it is often convenient to use a different variable. For instance, in the following example involving *time*, we use t as the variable of integration.

FIGURE 5.3

EXAMPLE 7 An application involving gravity

A ball is thrown upward with an initial velocity of 64 feet per second from an initial height of 80 feet, as shown in Figure 5.3. (a) Find the position function giving the height s as a function of the time t. (b) When does the ball hit the ground?

SOLUTION

(a) We let $t = 0$ represent the initial time. Then the two given initial conditions can be written as follows.

$$s(0) = 80 \qquad \text{Initial height is 80 ft}$$

$$s'(0) = 64 \qquad \text{Initial velocity is 64 ft/sec}$$

Assuming the acceleration due to gravity to be -32 feet per second per second, we can write

$$s''(t) = -32$$

and, by integrating, we obtain

$$s'(t) = \int s''(t)\, dt = \int -32\, dt = -32t + C_1.$$

Using the initial velocity, we obtain $s'(0) = 64 = -32(0) + C_1$, which implies that $C_1 = 64$. Similarly, by integrating $s'(t)$, we have

$$s(t) = \int s'(t)\, dt = \int (-32t + 64)\, dt = -16t^2 + 64t + C_2.$$

Now, using the initial height, we obtain

$$s(0) = 80 = -16(0^2) + 64(0) + C_2$$

which implies that $C_2 = 80$. Therefore, the position function is

$$s(t) = -16t^2 + 64t + 80.$$

(b) Using the position function found in part (a), we can find the time that the ball hits the ground by solving the equation $s(t) = 0$.

$$s(t) = -16t^2 + 64t + 80 = 0$$
$$-16(t + 1)(t - 5) = 0$$
$$t = -1, 5$$

Since t must be positive, we conclude that the ball hits the ground 5 seconds after it was thrown. ▭

REMARK In Example 7, note that the position function has the form

$$s(t) = \frac{1}{2}gt^2 + v_0 t + s_0$$

where $g = -32$, v_0 is the initial velocity, and s_0 is the initial height, as presented earlier in Section 3.2.

Before you begin the exercise set for this section, be sure you realize that one of the most important steps in integration is *rewriting the integrand in a form that fits the basic integration rules.* To further illustrate this point, we list several additional examples in Table 5.1.

TABLE 5.1

Given	Rewrite	Integrate	Simplify
$\int \dfrac{2}{\sqrt{x}}\, dx$	$2\int x^{-1/2}\, dx$	$2\left(\dfrac{x^{1/2}}{1/2}\right) + C$	$4x^{1/2} + C$
$\int (t^2 + 1)^2\, dt$	$\int (t^4 + 2t^2 + 1)\, dt$	$\dfrac{t^5}{5} + 2\left(\dfrac{t^3}{3}\right) + t + C$	$\dfrac{1}{5}t^5 + \dfrac{2}{3}t^3 + t + C$
$\int \dfrac{x^3 + 3}{x^2}\, dx$	$\int (x + 3x^{-2})\, dx$	$\dfrac{x^2}{2} + 3\left(\dfrac{x^{-1}}{-1}\right) + C$	$\dfrac{1}{2}x^2 - \dfrac{3}{x} + C$
$\int \sqrt[3]{x}(x - 4)\, dx$	$\int (x^{4/3} - 4x^{1/3})\, dx$	$\dfrac{x^{7/3}}{7/3} - 4\left(\dfrac{x^{4/3}}{4/3}\right) + C$	$\dfrac{3}{7}x^{4/3}(x - 7) + C$

EXERCISES for Section 5.1

In Exercises 1–6, complete the table using Table 5.1 as a model.

Given	Rewrite	Integrate	Simplify

1. $\int \sqrt[3]{x}\, dx$

2. $\int \dfrac{1}{x^2}\, dx$

3. $\int \dfrac{1}{x\sqrt{x}}\, dx$

4. $\int x(x^2 + 3)\, dx$

5. $\int \dfrac{1}{2x^3}\, dx$

6. $\int \dfrac{1}{(2x)^3}\, dx$

In Exercises 7–34, evaluate the indefinite integral and check your result by differentiation.

7. $\int (x^3 + 2)\, dx$

8. $\int (x^2 - 2x + 3)\, dx$

9. $\int (x^{3/2} + 2x + 1)\, dx$

10. $\int \left(\sqrt{x} + \dfrac{1}{2\sqrt{x}}\right) dx$

11. $\int \sqrt[3]{x^2}\, dx$

12. $\int (\sqrt[4]{x^3} + 1)\, dx$

13. $\int \dfrac{1}{x^3}\, dx$

14. $\int \dfrac{1}{x^4}\, dx$

15. $\int \dfrac{1}{4x^2}\, dx$

16. $\int (2x + x^{-1/2})\, dx$

17. $\int \dfrac{x^2 + x + 1}{\sqrt{x}}\, dx$

18. $\int \dfrac{x^2 + 1}{x^2}\, dx$

19. $\int (x + 1)(3x - 2)\, dx$

20. $\int (2t^2 - 1)^2\, dt$

21. $\int \dfrac{t^2 + 2}{t^2}\, dt$

22. $\int (1 - 2y + 3y^2)\, dy$

23. $\int y^2\sqrt{y}\, dy$

24. $\int (1 + 3t)t^2\, dt$

25. $\int dx$

26. $\int 3\, dt$

27. $\int (2\sin x + 3\cos x)\, dx$

28. $\int (t^2 - \sin t)\, dt$

29. $\int (1 - \csc t \cot t)\, dt$

30. $\int (\theta^2 + \sec^2 \theta)\, d\theta$

31. $\int (\sec^2 \theta - \sin \theta)\, d\theta$

32. $\int \sec y(\tan y - \sec y)\, dy$

33. $\int (\tan^2 y + 1)\, dy$

34. $\int \dfrac{\sin x}{1 - \sin^2 x}\, dx$

In Exercises 35–38, find the equation of the curve, given the derivative and the indicated point on the curve.

35. $\dfrac{dy}{dx} = 2x - 1$

36. $\dfrac{dy}{dx} = 2(x - 1)$

(3, 2)

(1, 1)

37. $\dfrac{dy}{dx} = \cos x$

38. $\dfrac{dy}{dx} = -\dfrac{1}{x^2}$

(0, 4)

(1, 3)

In Exercises 39–42, find $y = f(x)$ satisfying the given conditions.

39. $f''(x) = 2, f'(2) = 5, f(2) = 10$
40. $f''(x) = x^2, f'(0) = 6, f(0) = 3$
41. $f''(x) = x^{-3/2}, f'(4) = 2, f(0) = 0$
42. $f''(x) = x^{-3/2}, f'(1) = 2, f(9) = -4$

In Exercises 43–47, use $a(t) = -32$ ft/s² as the acceleration due to gravity. (Neglect air resistance.)

43. An object is dropped from a balloon that is stationary at 1600 feet above the ground. Express its height above the ground as a function of t. How long does it take the object to reach the ground?

44. A ball is thrown vertically upward from the ground with an initial velocity of 60 feet per second. How high will the ball go?

45. With what initial velocity must an object be thrown upward (from ground level) to reach a maximum height of 550 feet (approximate height of the Washington Monument)?

46. Show that the height above the ground of an object thrown upward from a point s_0 feet above the ground with an initial velocity of v_0 feet per second is given by the function

$$f(t) = -16t^2 + v_0 t + s_0.$$

47. A balloon, rising vertically with a velocity of 16 feet per second, releases a sandbag at the instant when the balloon is 64 feet above the ground.
(a) How many seconds after its release will the bag strike the ground?
(b) With what velocity will it reach the ground?

48. Assume that a fully loaded plane starting from rest has a constant acceleration while moving down the runway. Find this acceleration if the plane requires, on the average, 0.7 miles of runway and a speed of 160 miles per hour before lifting off.

49. The maker of a certain automobile advertises that the automobile will accelerate from 15 to 50 miles per hour in high gear in 13 seconds. Assuming constant acceleration, compute
(a) the acceleration in feet per second per second
(b) the distance the car travels in the given time.

50. A car traveling at 45 miles per hour was brought to a stop, at constant deceleration, 132 feet from where the brakes were applied. How far had the car moved when its speed was reduced to (a) 30 miles per hour and (b) 15 miles per hour? (c) Draw the real number line from 0 to 132, and plot the points found in parts (a) and (b). What conclusions can you draw?

51. At the instant the traffic light turns green, an automobile that has been waiting at an intersection starts ahead with a constant acceleration of 6 feet per second per second. At the same instant, a truck traveling with a constant velocity of 30 feet per second overtakes and passes the car.
(a) How far beyond its starting point will the automobile overtake the truck?
(b) How fast will the automobile be traveling?

52. A ball is released from rest and rolls down an inclined plane with constant acceleration. In 4 seconds the ball travels 100 centimeters. What is its acceleration in centimeters per second per second?

53. Galileo Galilei (1564–1642) stated the following proposition concerning falling objects: The time in which any space is traversed by a body starting from rest and uniformly accelerated is equal to the time in which that same space would be traversed by the same body moving at a uniform speed whose value is the mean of the highest speed and the speed just before acceleration began. Use the techniques of this section to verify the statement.

54. In the Chapter 5 Application, we developed the equation

$$\int v\, dv = -GM \int \frac{1}{y^2}\, dy$$

where v is the velocity of an object projected from the earth, y is the distance from the center of earth, G is the gravitational constant, and M is the mass of the earth. Show that v and y are related by the equation

$$v^2 = v_0{}^2 + 2GM \left(\frac{1}{y} - \frac{1}{R} \right)$$

where v_0 is the initial velocity of the object and R is the radius of the earth.

55. If marginal cost is constant, show that the cost function is a straight line.

56. The marginal cost for production is

$$\frac{dC}{dx} = 2x - 12$$

and the fixed costs are \$50. Find the total cost function and the average cost function.

In Exercises 57 and 58, find the revenue and demand functions for the given marginal revenue.

57. $\dfrac{dR}{dx} = 100 - 5x$ **58.** $\dfrac{dR}{dx} = 10 - 6x - 2x^2$

5.2 Area

Sigma notation ▪ Area ▪ The area of a plane region ▪ Upper and lower sums

In the previous section, we introduced the concept of antidifferentiation. We now look at a new problem—that of finding the area of a region in the plane. At first glance, these two ideas seem unrelated, but we will discover in Section 5.4 that they are closely related by an important theorem called the Fundamental Theorem of Calculus.

One method for finding the area of a region in the plane involves the sum of many terms. Thus, we begin this section by introducing a concise notation for sums. This notation is called **sigma notation** because it uses the uppercase Greek letter sigma, written as Σ.

DEFINITION OF SIGMA NOTATION

The sum of n terms $a_1, a_2, a_3, \ldots, a_n$ is written as

$$\sum_{i=1}^{n} a_i = a_1 + a_2 + a_3 + \cdots + a_n$$

where i is called the **index of summation**, a_i is called the **ith term** of the sum, and the **upper and lower bounds of summation** are n and 1, respectively.

REMARK   The upper and lower bounds of summation must be constant *with respect to the index of summation*. However, the lower bound does not have to be 1. Any integer value less than (or equal to) the upper bound is legitimate.

EXAMPLE 1   Examples of sigma notation

(a) $\displaystyle\sum_{i=1}^{6} i = 1 + 2 + 3 + 4 + 5 + 6$

(b) $\displaystyle\sum_{j=3}^{7} j^2 = 3^2 + 4^2 + 5^2 + 6^2 + 7^2$

(c) $\displaystyle\sum_{k=1}^{n} \frac{1}{n}(k^2 + 1) = \frac{1}{n}(1^2 + 1) + \frac{1}{n}(2^2 + 1) + \cdots + \frac{1}{n}(n^2 + 1)$

(d) $\displaystyle\sum_{i=1}^{n} f(x_i)\Delta x = f(x_1)\Delta x + f(x_2)\Delta x + \cdots + f(x_n)\Delta x$ ▭

Although any variable can be used as the index of summation, we usually prefer i, j, and k. Note from Example 1 that the index of summation does not appear in the terms of the expanded sum.

To verify the following two properties of summation, we suggest that you write them in expanded form and apply the associative and commutative properties of addition and the distributive property of addition over multiplication.

THEOREM 5.3
SUMMATION PROPERTIES

1. $\displaystyle\sum_{i=1}^{n} ka_i = k \sum_{i=1}^{n} a_i, \quad k$ is a constant

2. $\displaystyle\sum_{i=1}^{n} [a_i \pm b_i] = \sum_{i=1}^{n} a_i \pm \sum_{i=1}^{n} b_i$

The next theorem gives some useful formulas for sums of powers. A proof of this theorem is discussed in Appendix A.

THEOREM 5.4
SUMMATION FORMULAS

1. $\displaystyle\sum_{i=1}^{n} c = cn$ 2. $\displaystyle\sum_{i=1}^{n} i = \frac{n(n + 1)}{2}$

3. $\displaystyle\sum_{i=1}^{n} i^2 = \frac{n(n + 1)(2n + 1)}{6}$ 4. $\displaystyle\sum_{i=1}^{n} i^3 = \frac{n^2(n + 1)^2}{4}$

The importance of Theorem 5.4 is that it provides us with a simple method of evaluating certain sums. For example,

$$\sum_{i=1}^{100} i = 1 + 2 + \cdots + 100 = \frac{100(101)}{2} = 5050.$$

This is illustrated further in the next example.

EXAMPLE 2 Evaluating a sum

Evaluate the sum

$$\sum_{i=1}^{n} \frac{i + 1}{n^2}$$

for $n = 10, 100, 1000,$ and $10,000$.

SOLUTION

Applying Theorems 5.3 and 5.4, we have

$$\sum_{i=1}^{n} \frac{i + 1}{n^2} = \frac{1}{n^2} \sum_{i=1}^{n} (i + 1) = \frac{1}{n^2} \left[\sum_{i=1}^{n} i + \sum_{i=1}^{n} 1 \right] = \frac{1}{n^2} \left[\frac{n(n + 1)}{2} + n \right]$$

$$= \frac{1}{n^2} \left[\frac{n^2 + 3n}{2} \right] = \frac{1}{2} + \frac{3}{2n} = \frac{n + 3}{2n}.$$

Now, with the formula

$$\sum_{i=1}^{n} \frac{i + 1}{n^2} = \frac{n + 3}{2n}$$

we can find the sum simply by substituting the appropriate values of n as shown in Table 5.2.

TABLE 5.2

n	10	100	1,000	10,000
$\sum_{i=1}^{n} \dfrac{i + 1}{n^2} = \dfrac{n + 3}{2n}$	0.65000	0.51500	0.50150	0.50015

In Table 5.2, note that the sum appears to approach a limit as n increases. Although our discussion of limits at infinity in Section 4.5 applied to a variable x, where x can be any real number, many of the same results hold true for limits involving the variable n, where n is restricted to positive integer values. Thus, to find the limit of $(n + 3)/2n$ as n approaches infinity, we can write

$$\lim_{n \to \infty} \frac{n + 3}{2n} = \frac{1}{2}.$$

This is further illustrated in Example 3.

EXAMPLE 3 Finding the limit of a sum

Let $s(n)$ be defined by

$$s(n) = \sum_{i=1}^{n} \left(2 + \frac{i}{n} \right)^2 \left(\frac{1}{n} \right)$$

and find the limit of $s(n)$ as $n \to \infty$.

SOLUTION

Using the summation formulas given in Theorem 5.4, we can write $s(n)$ as follows.

$$s(n) = \sum_{i=1}^{n} \left(2 + \frac{i}{n}\right)^2 \left(\frac{1}{n}\right)$$

$$= \sum_{i=1}^{n} \left(\frac{4n^2 + 4ni + i^2}{n^2}\right)\left(\frac{1}{n}\right)$$

$$= \frac{1}{n^3} \sum_{i=1}^{n} (4n^2 + 4ni + i^2)$$

$$= \frac{1}{n^3}\left[\sum_{i=1}^{n} 4n^2 + \sum_{i=1}^{n} 4ni + \sum_{i=1}^{n} i^2\right]$$

$$= \frac{1}{n^3}\left[4n^3 + 4n\left(\frac{n(n+1)}{2}\right) + \frac{n(n+1)(2n+1)}{6}\right]$$

$$= \frac{38n^3 + 15n^2 + n}{6n^3}$$

Thus,

$$\lim_{n \to \infty} s(n) = \lim_{n \to \infty} \frac{38n^3 + 15n^2 + n}{6n^3} = \frac{19}{3}.$$

FIGURE 5.4 Rectangle: $A = bh$

FIGURE 5.5 Triangle: $A = \frac{1}{2} bh$

Area

In Euclidean geometry, the simplest type of plane region is a rectangle. Although we often say that the *formula* for the area of a rectangle is $A = bh$, as shown in Figure 5.4, it is actually more proper to say that this is the *definition* of the **area of a rectangle.**

From this definition, we can develop formulas for the areas of many other plane regions. For example, to determine the area of a triangle, we can form a rectangle whose area is twice that of the triangle, as shown in Figure 5.5. Once we know how to find the area of a triangle, we can determine the area of any polygon by subdividing the polygon into triangular regions as shown in Figure 5.6.

FIGURE 5.6   Parallelogram   Hexagon   Polygon

When we move from polygons to more general plane regions, finding area becomes more difficult. The ancient Greeks were able to determine formulas for the area of some general regions (principally those bounded by conics) by the *exhaustion* method. The clearest description of this method was given by Archimedes (287–212 B.C.). Essentially, the method is a limiting process in which the area is squeezed between two polygons—one inscribed in the region and one circumscribed about the region. The process we use to determine the area of a plane region is similar to that used by Archimedes.

Archimedes

The area of a plane region

We introduce the general problem of finding the area of a region in the plane with an example.

EXAMPLE 4 Approximating the area of a plane region

Use the five rectangles in Figures 5.7(a) and 5.7(b) to find *two* approximations for the area of the region lying between the graph of $f(x) = -x^2 + 5$ and the x-axis between $x = 0$ and $x = 2$.

SOLUTION

(a) Since f is nonnegative, we can find the height of the five rectangles shown in Figure 5.7(a) by evaluating $f(x)$ at the right endpoint of each of the following intervals.

$$\left[0, \frac{2}{5}\right], \left[\frac{2}{5}, \frac{4}{5}\right], \left[\frac{4}{5}, \frac{6}{5}\right], \left[\frac{6}{5}, \frac{8}{5}\right], \left[\frac{8}{5}, \frac{10}{5}\right]$$

Evaluate $f(x)$ at the right endpoints of these intervals.

The width of each rectangle is $\frac{2}{5}$, and the right endpoints are given by

$$0 + \frac{2}{5}i, \quad i = 1, 2, \ldots, 5.$$

Thus, the sum of the areas of the five rectangles is

$$\sum_{i=1}^{5} \overbrace{f\left(\frac{2i}{5}\right)}^{\text{height}} \overbrace{\left(\frac{2}{5}\right)}^{\text{width}} = \sum_{i=1}^{5}\left[-\left(\frac{2i}{5}\right)^2 + 5\right]\left(\frac{2}{5}\right) = \frac{2}{5}\sum_{i=1}^{5}\frac{-4i^2 + 125}{25}$$

$$= \frac{2}{125}[121 + 109 + 89 + 61 + 25] = \frac{162}{25} = 6.48.$$

Each of the five rectangles lies inside the given region. Hence, we conclude that the area of the region is *greater* than 6.48.

(b) To find the sum of the areas of the five rectangles in Figure 5.7(b) we use the same basic procedure, except we evaluate f at the left endpoints of the intervals that are given by

$$0 + \frac{2}{5}(i - 1), \quad i = 1, 2, \ldots, 5$$

and obtain

$$\sum_{i=1}^{5} \overbrace{f\left(\frac{2i - 2}{5}\right)}^{\text{height}} \overbrace{\left(\frac{2}{5}\right)}^{\text{width}} = \sum_{i=1}^{5}\left[-\left(\frac{2i - 2}{5}\right)^2 + 5\right]\left(\frac{2}{5}\right) = \frac{202}{25} = 8.08.$$

From Figure 5.7(b), we see that the area of the given region is *less* than 8.08. By combining the results of parts (a) and (b), we conclude that

$$6.48 < (\text{area of region}) < 8.08.$$

(a)

(b)

FIGURE 5.7

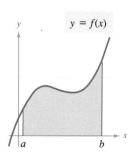

"Region under a Curve"

FIGURE 5.8

REMARK   By increasing the number of rectangles used in Example 4, we can obtain closer and closer approximations for the area of the region. For instance, using 25 rectangles of width $\frac{2}{25}$ each, we could conclude that

$$7.17 < \text{(area of region)} < 7.49.$$

We now generalize the procedure described in Example 4. To begin, we look at a plane region bounded above by the graph of a nonnegative, continuous function $y = f(x)$, as shown in Figure 5.8. The region is bounded below by the x-axis, and the left and right boundaries of the region are the vertical lines $x = a$ and $x = b$. We subdivide the interval $[a, b]$ into n subintervals, each of width $\Delta x = (b - a)/n$, as shown in Figure 5.9. The endpoints of the intervals are as follows.

$$\overbrace{a + 0(\Delta x)}^{a\,=\,x_0} < \overbrace{a + 1(\Delta x)}^{x_1} < \overbrace{a + 2(\Delta x)}^{x_2} < \cdots < \overbrace{a + n(\Delta x)}^{x_n\,=\,b}$$

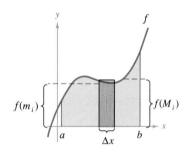

FIGURE 5.9

Since f is continuous, the Extreme Value Theorem guarantees the existence of a minimum and a maximum value of $f(x)$ in *each* subinterval.

$$f(m_i) = \text{minimum value of } f(x) \text{ in } i\text{th subinterval}$$
$$f(M_i) = \text{maximum value of } f(x) \text{ in } i\text{th subinterval}$$

Thus, we can define an **inscribed rectangle** lying *inside* the ith subregion and a **circumscribed rectangle** extending *outside* the ith subregion. We use the minimum value as the height of the inscribed rectangle and the maximum value as the height of the circumscribed rectangle, as shown in Figure 5.9. The areas of these two rectangles are related as follows.

$$\begin{pmatrix} \text{area of inscribed} \\ \text{rectangle} \end{pmatrix} = f(m_i)\Delta x \leq f(M_i)\Delta x = \begin{pmatrix} \text{area of circumscribed} \\ \text{rectangle} \end{pmatrix}$$

Area of inscribed rectangles is less than area of region.

Summing these areas produces

$$\textbf{lower sum} = s(n) = \sum_{i=1}^{n} f(m_i)\Delta x \qquad \text{Area of inscribed rectangles}$$

$$\textbf{upper sum} = S(n) = \sum_{i=1}^{n} f(M_i)\Delta x. \qquad \text{Area of circumscribed rectangles}$$

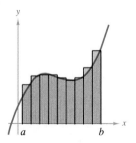

Area of circumscribed rectangles is greater than area of region.

FIGURE 5.10

From Figure 5.10, we can see that the lower sum $s(n)$ is less than or equal to the upper sum $S(n)$, and we have $s(n) \leq S(n)$. Moreover, the actual area of the region lies between these two sums. That is,

$$s(n) \leq \text{(area of region)} \leq S(n).$$

Inscribed rectangles

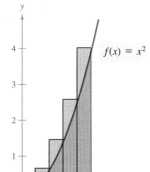

Circumscribed rectangles

FIGURE 5.11

EXAMPLE 5   Finding upper and lower sums for a region

Find the upper and lower sums for the region bounded by the graph of

$$f(x) = x^2$$

and the x-axis between $x = 0$ and $x = 2$.

SOLUTION

To begin, we partition the interval $[0, 2]$ into n subintervals, each of length

$$\Delta x = \frac{b - a}{n} = \frac{2 - 0}{n} = \frac{2}{n}.$$

Figure 5.11 shows the endpoints of the subintervals and several inscribed and circumscribed rectangles. Since f is increasing on the interval $[0, 2]$, the minimum value on each subinterval occurs at the left endpoint, and the maximum value occurs at the right endpoint.

Left endpoints	_Right endpoints_

$$m_i = 0 + (i - 1)\left(\frac{2}{n}\right) = \frac{2(i - 1)}{n} \qquad M_i = 0 + i\left(\frac{2}{n}\right) = \frac{2i}{n}$$

Using the left endpoints, the lower sum is

$$s(n) = \sum_{i=1}^{n} f(m_i)\Delta x = \sum_{i=1}^{n} f\left[\frac{2(i-1)}{n}\right]\left(\frac{2}{n}\right)$$

$$= \sum_{i=1}^{n} \left[\frac{2(i-1)}{n}\right]^2\left(\frac{2}{n}\right) = \sum_{i=1}^{n} \left(\frac{8}{n^3}\right)(i^2 - 2i + 1)$$

$$= \frac{8}{n^3}\left[\sum_{i=1}^{n} i^2 - 2\sum_{i=1}^{n} i + \sum_{i=1}^{n} 1\right]$$

$$= \frac{8}{n^3}\left(\frac{n(n + 1)(2n + 1)}{6} - 2\left[\frac{n(n + 1)}{2}\right] + n\right)$$

$$= \frac{4}{3n^3}(2n^3 - 3n^2 + n)$$

$$= \frac{8}{3} - \frac{4}{n} + \frac{4}{3n^2}.$$

Using the right endpoints, the upper sum is

$$S(n) = \sum_{i=1}^{n} f(M_i)\Delta x = \sum_{i=1}^{n} f\left(\frac{2i}{n}\right)\left(\frac{2}{n}\right)$$

$$= \sum_{i=1}^{n} \left(\frac{2^2 i^2}{n^2}\right)\left(\frac{2}{n}\right) = \sum_{i=1}^{n} \left(\frac{8}{n^3}\right)i^2$$

$$= \frac{8}{n^3}\left[\frac{n(n + 1)(2n + 1)}{6}\right]$$

$$= \frac{4}{3n^3}(2n^3 + 3n^2 + n)$$

$$= \frac{8}{3} + \frac{4}{n} + \frac{4}{3n^2}.$$

Although it is true that (for any value of n) the lower sum in Example 5 is less than the upper sum,

$$s(n) = \frac{8}{3} - \frac{4}{n} + \frac{4}{3n^2} < \frac{8}{3} + \frac{4}{n} + \frac{4}{3n^2} = S(n)$$

we can see that the difference between these two sums lessens as n increases. In fact, if we take the limit as $n \to \infty$, both the upper and the lower sums approach $\frac{8}{3}$.

$$\lim_{n\to\infty} s(n) = \lim_{n\to\infty} \left[\frac{8}{3} - \frac{4}{n} + \frac{4}{3n^2} \right] = \frac{8}{3} \qquad \text{Lower sum limit}$$

$$\lim_{n\to\infty} S(n) = \lim_{n\to\infty} \left[\frac{8}{3} + \frac{4}{n} + \frac{4}{3n^2} \right] = \frac{8}{3} \qquad \text{Upper sum limit}$$

The next theorem shows that the equivalence of the limit (as $n \to \infty$) of the upper and lower sums is not mere coincidence. It is true for all functions that satisfy the conditions stated in the following theorem. The proof of this theorem is best left to a course in advanced calculus.

**THEOREM 5.5
LIMIT OF THE LOWER
AND UPPER SUMS**

Let f be continuous and nonnegative on the interval $[a, b]$. The limits as $n \to \infty$ of both the lower and upper sums exist and are equal to each other. That is,

$$\lim_{n\to\infty} s(n) = \lim_{n\to\infty} \sum_{i=1}^{n} f(m_i)\Delta x = \lim_{n\to\infty} S(n) = \lim_{n\to\infty} \sum_{i=1}^{n} f(M_i)\Delta x$$

where $\Delta x = (b - a)/n$ and $f(m_i)$ and $f(M_i)$ are the minimum and maximum values of f on the ith subinterval.

From this theorem, we can deduce an important result. Since the same limit is attained for both the minimum value $f(m_i)$ and the maximum value $f(M_i)$, it follows from the Squeeze Theorem (Theorem 2.9) that the choice of x in the ith subinterval does not affect the limit. This means that we are free to choose an *arbitrary* x-value in the ith subinterval, and we do this in the following *definition of the area of a region in the plane*.

**DEFINITION OF AREA OF A
REGION IN THE PLANE**

Let f be continuous and nonnegative on the interval $[a, b]$. The **area** of the region bounded by the graph of f, the x-axis, and the vertical lines $x = a$ and $x = b$ is

$$\text{area} = \lim_{n\to\infty} \sum_{i=1}^{n} f(c_i)\Delta x, \qquad x_{i-1} \le c_i \le x_i$$

where $\Delta x = (b - a)/n$.

We can use this definition in two ways: (1) to *compute exact areas*, and (2) to *approximate areas*. Actually, with the techniques we have available now, there are only a few types of functions for which we can find exact areas. We are limited to the lower-degree polynomial functions covered by the summation formulas given in Theorem 5.4. In Section 5.4, however, we will extend this list greatly by a procedure described in an amazing theorem called the *Fundamental Theorem of Calculus*.

EXAMPLE 6 Finding area by the limit definition

Find the area of the region bounded by the graph of $f(x) = x^3$, the x-axis, and the vertical lines $x = 0$ and $x = 1$, as shown in Figure 5.12.

SOLUTION

We begin by noting that f is continuous and nonnegative on the interval $[0, 1]$. Next, we partition the interval $[0, 1]$ into n equal subintervals, each of width $\Delta x = 1/n$. According to the definition of area, we can choose any x-value in the ith subinterval, and for the sake of convenience we choose the right endpoint $c_i = i/n$.

$$\text{area} = \lim_{n \to \infty} \sum_{i=1}^{n} f(c_i)\Delta x = \lim_{n \to \infty} \sum_{i=1}^{n} \left(\frac{i}{n}\right)^3 \left(\frac{1}{n}\right)$$

$$= \lim_{n \to \infty} \frac{1}{n^4} \sum_{i=1}^{n} i^3$$

$$= \lim_{n \to \infty} \frac{n^2(n + 1)^2}{4n^4}$$

$$= \lim_{n \to \infty} \left(\frac{1}{4} + \frac{1}{2n} + \frac{1}{4n^2}\right) = \frac{1}{4}$$

FIGURE 5.12

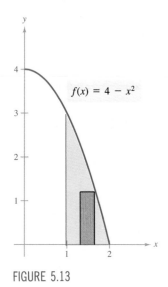

$f(x) = 4 - x^2$

FIGURE 5.13

EXAMPLE 7   Finding area by the limit definition

Find the area of the region bounded by the graph of $f(x) = 4 - x^2$, the x-axis, and the vertical lines $x = 1$ and $x = 2$, as shown in Figure 5.13.

SOLUTION

The function f is continuous and nonnegative on the interval $[1, 2]$, and so we begin by partitioning the interval into n equal subintervals, each of length $\Delta x = 1/n$. Choosing the right endpoint, $c_i = 1 + (i/n)$, of each subinterval, we obtain the following.

$$
\begin{aligned}
\text{area} &= \lim_{n \to \infty} \sum_{i=1}^{n} f(c_i)\Delta x \\
&= \lim_{n \to \infty} \sum_{i=1}^{n} \left[4 - \left(1 + \frac{i}{n} \right)^2 \right]\left(\frac{1}{n} \right) \\
&= \lim_{n \to \infty} \sum_{i=1}^{n} \left(3 - \frac{2i}{n} - \frac{i^2}{n^2} \right)\left(\frac{1}{n} \right) \\
&= \lim_{n \to \infty} \left[\frac{1}{n} \sum_{i=1}^{n} 3 - \frac{2}{n^2} \sum_{i=1}^{n} i - \frac{1}{n^3} \sum_{i=1}^{n} i^2 \right] \\
&= \lim_{n \to \infty} \left[3 - \left(1 + \frac{1}{n} \right) - \left(\frac{1}{3} + \frac{1}{2n} + \frac{1}{6n^2} \right) \right] \\
&= 3 - 1 - \frac{1}{3} = \frac{5}{3}
\end{aligned}
$$

In the next example we look at a region that is bounded by the y-axis (rather than the x-axis).

EXAMPLE 8   A region bounded by the y-axis

Find the area of the region bounded by the graph of $f(y) = y^2$ and the y-axis for $0 \le y \le 1$, as shown in Figure 5.14.

SOLUTION

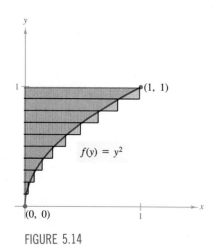

$f(y) = y^2$

(1, 1)

(0, 0)

FIGURE 5.14

When f is a continuous, nonnegative function of y, we still can use the same basic procedure illustrated in Example 6. We partition the interval $[0, 1]$ into n equal subintervals, each of width $\Delta y = 1/n$. Using the upper endpoints $c_i = i/n$, we obtain the following.

$$
\begin{aligned}
\text{area} &= \lim_{n \to \infty} \sum_{i=1}^{n} f(c_i)\Delta y = \lim_{n \to \infty} \sum_{i=1}^{n} \left(\frac{i}{n} \right)^2 \left(\frac{1}{n} \right) = \lim_{n \to \infty} \frac{1}{n^3} \sum_{i=1}^{n} i^2 \\
&= \lim_{n \to \infty} \frac{n(n + 1)(2n + 1)}{6n^3} \\
&= \lim_{n \to \infty} \left(\frac{1}{3} + \frac{1}{2n} + \frac{1}{6n^2} \right) = \frac{1}{3}
\end{aligned}
$$

EXERCISES for Section 5.2

In Exercises 1–8, find the given sum.

1. $\displaystyle\sum_{i=1}^{5} (2i + 1)$

2. $\displaystyle\sum_{i=1}^{6} 2i$

3. $\displaystyle\sum_{k=0}^{4} \frac{1}{k^2 + 1}$

4. $\displaystyle\sum_{j=3}^{5} \frac{1}{j}$

5. $\displaystyle\sum_{k=1}^{4} c$

6. $\displaystyle\sum_{n=1}^{10} \frac{3}{n + 1}$

7. $\displaystyle\sum_{i=1}^{4} [(i - 1)^2 + (i + 1)^3]$

8. $\displaystyle\sum_{k=2}^{5} (k + 1)(k - 3)$

In Exercises 9–18, use sigma notation to write the given sum.

9. $\dfrac{1}{3(1)} + \dfrac{1}{3(2)} + \dfrac{1}{3(3)} + \cdots + \dfrac{1}{3(9)}$

10. $\dfrac{5}{1 + 1} + \dfrac{5}{1 + 2} + \dfrac{5}{1 + 3} + \cdots + \dfrac{5}{1 + 15}$

11. $\left[2\left(\dfrac{1}{8}\right) + 3\right] + \left[2\left(\dfrac{2}{8}\right) + 3\right] + \cdots + \left[2\left(\dfrac{8}{8}\right) + 3\right]$

12. $\left[1 - \left(\dfrac{1}{4}\right)^2\right] + \left[1 - \left(\dfrac{2}{4}\right)^2\right] + \cdots + \left[1 - \left(\dfrac{4}{4}\right)^2\right]$

13. $\left[\left(\dfrac{1}{6}\right)^2 + 2\right]\left(\dfrac{1}{6}\right) + \cdots + \left[\left(\dfrac{6}{6}\right)^2 + 2\right]\left(\dfrac{1}{6}\right)$

14. $\left[\left(\dfrac{1}{n}\right)^2 + 2\right]\left(\dfrac{1}{n}\right) + \cdots + \left[\left(\dfrac{n}{n}\right)^2 + 2\right]\left(\dfrac{1}{n}\right)$

15. $\left[\left(\dfrac{2}{n}\right)^3 - \dfrac{2}{n}\right]\left(\dfrac{2}{n}\right) + \cdots + \left[\left(\dfrac{2n}{n}\right)^3 - \dfrac{2n}{n}\right]\left(\dfrac{2}{n}\right)$

16. $\left[1 - \left(\dfrac{2}{n} - 1\right)^2\right]\left(\dfrac{2}{n}\right) + \cdots + \left[1 - \left(\dfrac{2n}{n} - 1\right)^2\right]\left(\dfrac{2}{n}\right)$

17. $\left[2\left(1 + \dfrac{3}{n}\right)^2\right]\left(\dfrac{3}{n}\right) + \cdots + \left[2\left(1 + \dfrac{3n}{n}\right)^2\right]\left(\dfrac{3}{n}\right)$

18. $\left(\dfrac{1}{n}\right)\sqrt{1 - \left(\dfrac{0}{n}\right)^2} + \cdots + \left(\dfrac{1}{n}\right)\sqrt{1 - \left(\dfrac{n - 1}{n}\right)^2}$

In Exercises 19–24, use the properties of sigma notation and summation formulas to evaluate the given sum.

19. $\displaystyle\sum_{i=1}^{20} 2i$

20. $\displaystyle\sum_{i=1}^{10} i(i^2 + 1)$

21. $\displaystyle\sum_{i=1}^{20} (i - 1)^2$

22. $\displaystyle\sum_{i=1}^{15} (2i - 3)$

23. $\displaystyle\sum_{i=1}^{15} \frac{1}{n^3}(i - 1)^2$

24. $\displaystyle\sum_{i=1}^{10} (i^2 - 1)$

In Exercises 25–30, find the limit of $s(n)$ as $n \to \infty$.

25. $s(n) = \left(\dfrac{4}{3n^3}\right)(2n^3 + 3n^2 + n)$

26. $s(n) = \left(\dfrac{8}{3} + \dfrac{4}{n} + \dfrac{4}{3n^2}\right)$

27. $s(n) = \dfrac{81}{n^4}\left[\dfrac{n^2(n + 1)^2}{4}\right]$

28. $s(n) = \dfrac{64}{n^3}\left[\dfrac{n(n + 1)(2n + 1)}{6}\right]$

29. $s(n) = \dfrac{18}{n^2}\left[\dfrac{n(n + 1)}{2}\right]$

30. $s(n) = \dfrac{1}{n^2}\left[\dfrac{n(n + 1)}{2}\right]$

In Exercises 31–36, use the properties of sigma notation to find a formula for the given sum of n terms. Then use the formula to find the limit as $n \to \infty$.

31. $\displaystyle\lim_{n\to\infty} \sum_{i=1}^{n} \frac{1}{n^3}(i - 1)^2$

32. $\displaystyle\lim_{n\to\infty} \sum_{i=1}^{n} \left(1 + \frac{2i}{n}\right)^2\left(\frac{2}{n}\right)$

33. $\displaystyle\lim_{n\to\infty} \sum_{i=1}^{n} \frac{16i}{n^2}$

34. $\displaystyle\lim_{n\to\infty} \sum_{i=1}^{n} \left(\frac{2i}{n}\right)\left(\frac{2}{n}\right)$

35. $\displaystyle\lim_{n\to\infty} \sum_{i=1}^{n} \left(1 + \frac{2i}{n}\right)^3\left(\frac{2}{n}\right)$

36. $\displaystyle\lim_{n\to\infty} \sum_{i=1}^{n} \left(1 + \frac{i}{n}\right)\left(\frac{2}{n}\right)$

In Exercises 37–42, use the upper and lower sums to approximate the area of the given region using the indicated number of (equal) subintervals.

37. $y = \sqrt{x}$

38. $y = \sqrt{x} + 1$

39. $y = \dfrac{1}{x}$

40. $y = \dfrac{1}{x - 2}$

41. $y = \sqrt{1 - x^2}$

42. $y = \sqrt{x + 1}$

43. Consider the triangle of area 2 bounded by the graphs of $y = x$, $y = 0$, and $x = 2$.
   (a) Sketch the graph of the region.
   (b) Divide the interval $[0, 2]$ into n equal subintervals and show that the endpoints are

   $$0 < 1\left(\frac{2}{n}\right) < \cdots < (n - 1)\left(\frac{2}{n}\right) < n\left(\frac{2}{n}\right).$$

   (c) Show that $s(n) = \sum_{i=1}^{n} \left[(i - 1)\left(\frac{2}{n}\right)\right]\left(\frac{2}{n}\right)$.

   (d) Show that $S(n) = \sum_{i=1}^{n} \left[i\left(\frac{2}{n}\right)\right]\left(\frac{2}{n}\right)$.

   (e) Complete the following table.

n	5	10	50	100
$s(n)$				
$S(n)$				

   (f) Show that $\lim_{n \to \infty} s(n) = \lim_{n \to \infty} S(n) = 2$.

44. Consider the trapezoid of area 4 bounded by the graphs of $y = x$, $y = 0$, $x = 1$, and $x = 3$.
   (a) Sketch the graph of the region.
   (b) Divide the interval $[1, 3]$ into n equal subintervals and show that the endpoints are

   $$1 < 1 + 1\left(\frac{2}{n}\right) < \cdots < 1 + (n - 1)\left(\frac{2}{n}\right) < 1 + n\left(\frac{2}{n}\right).$$

   (c) Show that $s(n) = \sum_{i=1}^{n} \left[1 + (i - 1)\left(\frac{2}{n}\right)\right]\left(\frac{2}{n}\right)$.

   (d) Show that $S(n) = \sum_{i=1}^{n} \left[1 + i\left(\frac{2}{n}\right)\right]\left(\frac{2}{n}\right)$.

   (e) Complete the following table.

n	5	10	50	100
$s(n)$				
$S(n)$				

   (f) Show that $\lim_{n \to \infty} s(n) = \lim_{n \to \infty} S(n) = 4$.

In Exercises 45–54, use the limit process to find the area of the region between the graph of the function and the x-axis over the given interval. Sketch the region.

Function	Interval
45. $y = -2x + 3$	$[0, 1]$
46. $y = 3x - 4$	$[2, 5]$
47. $y = x^2 + 2$	$[0, 1]$
48. $y = 1 - x^2$	$[-1, 1]$
49. $y = 2x^2$	$[1, 3]$
50. $y = 2x^2 - x + 1$	$[0, 2]$
51. $y = 1 - x^3$	$[0, 1]$
52. $y = 2x - x^3$	$[0, 1]$
53. $y = x^2 - x^3$	$[-1, 1]$
54. $y = x^2 - x^3$	$[-1, 0]$

In Exercises 55 and 56, use the limit process to find the area of the region between the graph of the function and the y-axis over the given interval. Sketch the region.

Function	Interval
55. $f(y) = 3y$	$[0, 2]$
56. $f(y) = y^2$	$[0, 3]$

In Exercises 57–60, use the **Midpoint Rule**

$$\text{area} \approx \sum_{i=1}^{n} f\left(\frac{x_i + x_{i-1}}{2}\right)\Delta x$$

with $n = 4$ to approximate the area of the region bounded by the graph of the given function and the x-axis over the indicated interval.

Function	Interval
57. $f(x) = x^2 + 3$	$[0, 2]$
58. $f(x) = x^2 + 4x$	$[0, 4]$
59. $f(x) = \tan x$	$\left[0, \frac{\pi}{4}\right]$
60. $f(x) = \sin x$	$\left[0, \frac{\pi}{2}\right]$

61. The game commission in a certain state introduces 50 deer into some newly acquired state game land. It is believed that the size of the herd will increase according to the model

$$N = \frac{10(5 + 3t)}{1 + 0.04t}$$

where t is the time in years. Find the number in the herd when $t = 5$, 10, and 25 years. What is the limit of N as t approaches infinity?

62. (Learning Curve) Psychologists have developed models to predict performance as a function of the number of trials for a certain task. One such model is

$$P = \frac{b + \theta a(n-1)}{1 + \theta(n-1)}$$

where P is the percentage of correct responses after n trials and a, b, and θ are constants depending upon the actual learning situation. Find the limit of P as n approaches infinity.

In Exercises 63 and 64, use a computer or calculator to complete the following table of upper and lower sums for the given function over the given interval.

n	4	8	12	16	20
$s(n)$					
$S(n)$					

	Function	Interval
63.	$f(x) = \sqrt{x}$	$[0, 4]$
64.	$f(x) = \tan\left(\dfrac{\pi x}{5}\right)$	$[0, 2]$

5.3 Riemann Sums and the Definite Integral

Riemann sums ▪ The definite integral ▪ Properties of definite integrals

In the definition of area in Section 5.2, we required that the partitions have subintervals of *equal width*. This was done only for computational convenience. We begin this section with an example to show that it is not necessary to have subintervals of equal width.

EXAMPLE 1 A partition with subintervals of unequal widths

Consider the region bounded by the graph of $f(x) = \sqrt{x}$ and the x-axis for $0 \leq x \leq 1$, as shown in Figure 5.15 on page 282. Evaluate the limit

$$\lim_{n \to \infty} \sum_{i=1}^{n} f(c_i) \Delta x_i$$

where c_i is the right endpoint of the partition given by $x_i = i^2/n^2$ and Δx_i is the width of the ith interval.

FIGURE 5.15

FIGURE 5.16

Georg Friedrich Riemann

SOLUTION

The width of the ith interval is given by

$$\Delta x_i = \frac{i^2}{n^2} - \frac{(i-1)^2}{n^2} = \frac{i^2 - i^2 + 2i - 1}{n^2} = \frac{2i - 1}{n^2}.$$

Thus, we have

$$\lim_{n \to \infty} \sum_{i=1}^{n} f(c_i)\Delta x_i = \lim_{n \to \infty} \sum_{i=1}^{n} \sqrt{\frac{i^2}{n^2}} \left(\frac{2i - 1}{n^2}\right)$$

$$= \lim_{n \to \infty} \frac{1}{n^3} \sum_{i=1}^{n} (2i^2 - i)$$

$$= \lim_{n \to \infty} \frac{1}{n^3} \left[2\frac{n(n+1)(2n+1)}{6} - \frac{n(n+1)}{2}\right]$$

$$= \lim_{n \to \infty} \frac{4n^3 + 3n^2 - n}{6n^3} = \frac{2}{3}.$$

The two regions shown in Figures 5.15 and 5.16 can be united to form a square whose area is 1. Furthermore, we already know from Example 8 in Section 5.2 that the region in Figure 5.16 has an area of $\frac{1}{3}$. Thus, it is reasonable to conclude that the region in Figure 5.15 has an area of $\frac{2}{3}$. This agrees with the limit found in Example 1, even though we used a partition having subintervals of unequal widths. The reason we were able to use this particular partition is that as n increases, the *width of the largest subinterval approaches zero*. This is a key feature of the development of the definite integral.

Riemann sums

In the previous section we used the limit of a sum to define the area of a special type of region in the plane. Finding area by this means is only one of *many* applications involving the limit of a sum. We will soon see that a similar approach can be used to determine quantities as diverse as arc length, average value, centroids, volumes, work, and surface areas. The following development is named after Georg Friedrich Bernhard Riemann (1826–1866). Although the definite integral had been defined and used long before the time of Riemann, he generalized the concept to cover a broader category of functions.

In the following definition of a Riemann sum, note that the function f has no restrictions other than being defined on the interval $[a, b]$. (In the previous section, we assumed f to be nonnegative because we were dealing with the area under a curve.)

DEFINITION OF A RIEMANN SUM

Let f be defined on the closed interval $[a, b]$, and let Δ be an arbitrary partition of $[a, b]$,

$$a = x_0 < x_1 < x_2 < \cdots < x_{n-1} < x_n = b$$

where Δx_i is the width of the ith subinterval. If c_i is *any* point in the ith subinterval, then the sum

$$\sum_{i=1}^{n} f(c_i)\Delta x_i, \quad x_{i-1} \le c_i \le x_i$$

is called a **Riemann sum** of f for the partition Δ.

REMARK   The sums in Section 5.2 are examples of Riemann sums, but there are more general Riemann sums than those covered there.

For a given partition Δ, we call the width of the largest subinterval the **norm** of the partition, and we denote it by $\|\Delta\|$. If every subinterval is of equal width, then we call the partition **regular** and denote the norm by

$$\|\Delta\| = \Delta x = \frac{b-a}{n}. \qquad \text{Regular partition}$$

For a general partition, the norm is related to the number of subintervals of $[a, b]$ in the following way.

$$\frac{b-a}{\|\Delta\|} \le n \qquad \text{General partition}$$

Thus, we can see that the number of subintervals in a partition approaches infinity as the norm of the partition approaches zero.

$$\|\Delta\| \to 0 \quad \text{implies that} \quad n \to \infty$$

REMARK   The converse of this statement is not true. For example, let Δ_n be the partition of the interval $[0, 1]$ given by

$$0 < \frac{1}{2^n} < \frac{1}{2^{n-1}} < \cdots < \frac{1}{8} < \frac{1}{4} < \frac{1}{2} < 1.$$

For any positive value of n, the norm of the partition Δ_n is $\frac{1}{2}$. Thus, letting n approach infinity does not force $\|\Delta\|$ to approach zero. In a regular partition, the statements $\|\Delta\| \to 0$ and $n \to \infty$ are equivalent.

To define the definite integral, we use the following limit.

$$\lim_{\|\Delta\| \to 0} \sum_{i=1}^{n} f(c_i)\Delta x_i = L$$

To say that this limit exists means that for $\varepsilon > 0$ there exists a $\delta > 0$ such that for every partition with $\|\Delta\| < \delta$ it follows that

$$\left| L - \sum_{i=1}^{n} f(c_i)\Delta x_i \right| < \varepsilon.$$

(This must be true for any choice of c_i in the ith subinterval of Δ.)

DEFINITION OF THE DEFINITE INTEGRAL

If f is defined on the closed interval $[a, b]$ and the limit of a Riemann sum of f exists, then we say f is **integrable** on $[a, b]$ and we denote the limit by

$$\lim_{\|\Delta\| \to 0} \sum_{i=1}^{n} f(c_i)\Delta x_i = \int_{a}^{b} f(x) \, dx.$$

The limit is called the **definite integral** of f from a to b. The number a is the **lower limit** of integration, and the number b is the **upper limit** of integration.

It is not a coincidence that the notation for definite integrals is similar to that used for indefinite integrals. We will see why in the next section when we discuss the Fundamental Theorem of Calculus. For now it is important to see that definite integrals and indefinite integrals are different entities. A definite integral is a *number*, whereas an indefinite integral is a *family of functions*.

A sufficient condition for f to be integrable on $[a, b]$ is given in the following theorem.

THEOREM 5.6 CONTINUITY IMPLIES INTEGRABILITY

If a function f is continuous on the closed interval $[a, b]$, then f is integrable on $[a, b]$.

EXAMPLE 2 Evaluating a definite integral as a limit

Evaluate the definite integral

$$\int_{-2}^{1} 2x \, dx.$$

SOLUTION

Since $f(x) = 2x$ is continuous on the interval $[-2, 1]$, we know it is integrable. Moreover, the definition of integrability implies that any partition whose norm approaches zero can be used to determine the limit. For computational convenience, we define Δ by subdividing $[-2, 1]$ into n subintervals of equal width

$$\Delta x_i = \Delta x = \frac{b - a}{n} = \frac{3}{n}.$$

Choosing c_i as the right endpoint of each subinterval, we have

$$c_i = a + i(\Delta x) = -2 + \frac{3i}{n}.$$

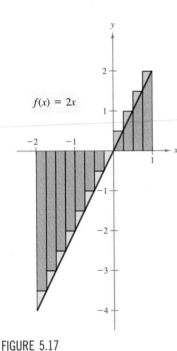

$f(x) = 2x$

FIGURE 5.17

Thus, the definite integral is given by

$$\int_{-2}^{1} 2x \, dx = \lim_{\|\Delta\| \to 0} \sum_{i=1}^{n} f(c_i)\Delta x_i = \lim_{n \to \infty} \sum_{i=1}^{n} f(c_i)\Delta x$$

$$= \lim_{n \to \infty} \sum_{i=1}^{n} 2\left(-2 + \frac{3i}{n}\right)\left(\frac{3}{n}\right)$$

$$= \lim_{n \to \infty} \frac{6}{n} \sum_{i=1}^{n} \left(-2 + \frac{3i}{n}\right)$$

$$= \lim_{n \to \infty} \frac{6}{n}\left[-2n + \frac{3}{n}\left(\frac{n(n+1)}{2}\right)\right]$$

$$= \lim_{n \to \infty} \left(-12 + 9 + \frac{9}{n}\right)$$

$$= -3.$$

Note that since the particular definite integral in Example 2 is negative, it *does not* represent the area of the region shown in Figure 5.17. Definite integrals can be positive, negative, or zero. For a definite integral to be interpreted as an area (as defined in Section 5.2), the function f must satisfy the conditions given in the following theorem.

THEOREM 5.7
THE DEFINITE INTEGRAL AS
THE AREA OF A REGION

If f is continuous and nonnegative on the closed interval $[a, b]$, then the area of the region bounded by the graph of f, the x-axis, and the vertical lines $x = a$ and $x = b$ is given by

$$\text{area} = \int_{a}^{b} f(x) \, dx.$$

PROOF   We can obtain a proof by using the definition of the area of a region with Theorem 5.6. From Theorem 5.6, we know that f is integrable on $[a, b]$. This means that the definite integral of f from a to b can be written as the limit of a Riemann sum. When such a limit exists, we can use any convenient partition so long as its norm approaches zero. Thus, we choose a regular partition and conclude that the limit is equal to the area of the given region.

As an example of Theorem 5.7, consider the region bounded by the graph of $f(x) = 4x - x^2$ and the x-axis, as shown in Figure 5.18. Since f is continuous and nonnegative on the closed interval $[0, 4]$, the area of the region is

$$\text{area} = \int_{0}^{4} (4x - x^2) \, dx.$$

From geometry we know several formulas for areas of plane regions. Since at this stage in our study it is not particularly easy to evaluate definite integrals using Riemann sums, it is a good idea to learn to recognize the definite integrals that correspond to areas of common geometric figures. We list some of these in the next example.

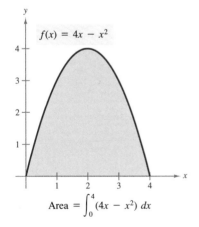

$f(x) = 4x - x^2$

$$\text{Area} = \int_{0}^{4} (4x - x^2) \, dx$$

FIGURE 5.18

(a) $\displaystyle\int_1^3 4\,dx$

Rectangle

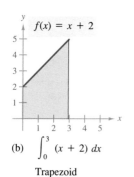

(b) $\displaystyle\int_0^3 (x + 2)\,dx$

Trapezoid

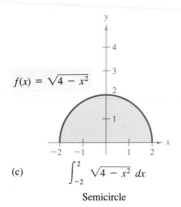

(c) $\displaystyle\int_{-2}^2 \sqrt{4 - x^2}\,dx$

Semicircle

FIGURE 5.19

EXAMPLE 3   Areas of common geometric figures

Sketch the region corresponding to each of the following definite integrals. Then evaluate each integral using a geometric formula.

(a) $\displaystyle\int_1^3 4\,dx$ (b) $\displaystyle\int_0^3 (x + 2)\,dx$ (c) $\displaystyle\int_{-2}^2 \sqrt{4 - x^2}\,dx$

SOLUTION

A sketch of each region is shown in Figure 5.19.

(a) This region is a rectangle of height 4 and width 2. Thus,

$$\int_1^3 4\,dx = 4(2) = 8.$$

(b) This region is a trapezoid with an altitude of 3 and parallel bases of lengths 2 and 5. The formula for the area of a trapezoid is $\frac{1}{2}h(b_1 + b_2)$, and so we have

$$\int_0^3 (x + 2)\,dx = \frac{1}{2}(3)(2 + 5) = \frac{21}{2}.$$

(c) This region is a semicircle of radius 2. Thus, the area is $\frac{1}{2}\pi r^2$, and we have

$$\int_{-2}^2 \sqrt{4 - x^2}\,dx = \frac{1}{2}\pi(2^2) = 2\pi.$$

Properties of definite integrals

In defining the definite integral of f on the interval $[a, b]$, we assumed that $a < b$. For the future, it will be convenient to extend this definition somewhat. Geometrically, the following two special definitions seem reasonable. For instance, it makes sense to define the area of a region of zero width and finite height to be zero.

DEFINITION OF TWO SPECIAL DEFINITE INTEGRALS	1. If f is defined at $x = a$, then $$\int_a^a f(x)\,dx = 0.$$ 2. If f is integrable on $[a, b]$, then $$\int_b^a f(x)\,dx = -\int_a^b f(x)\,dx.$$

EXAMPLE 4   Evaluating definite integrals

Evaluate the following definite integrals.

(a) $\displaystyle\int_{\pi}^{\pi} \sin x \, dx$ (b) $\displaystyle\int_{3}^{0} (x + 2) \, dx$

SOLUTION

(a) Since the sine is defined at $x = \pi$, and since the upper and lower limits of integration are equal, we have

$$\int_{\pi}^{\pi} \sin x \, dx = 0.$$

(b) This integral is the same as that given in Example 3(b) except that the upper and lower limits are interchanged. Since the integral in Example 3(b) has a value of $\frac{21}{2}$, we have

$$\int_{3}^{0} (x + 2) \, dx = -\int_{0}^{3} (x + 2) \, dx = -\frac{21}{2}.$$

It may be convenient to break a definite integral into two (or more) parts.

THEOREM 5.8
ADDITIVE INTERVAL PROPERTY

If f is integrable on the three closed intervals determined by a, b, and c, then

$$\int_{a}^{b} f(x) \, dx = \int_{a}^{c} f(x) \, dx + \int_{c}^{b} f(x) \, dx.$$

This theorem is valid for any integrable function and for any three numbers a, b, and c. However, rather than give a formal proof for the general case, it seems more instructive to give a geometric argument for the case in which $a < c < b$ and f is continuous and nonnegative. Figure 5.20 shows that the larger region can be divided at $x = c$ into two subregions whose intersection is a line segment. Since the line segment has zero area, it follows that the area of the larger region is equal to the sum of the areas of the two smaller regions.

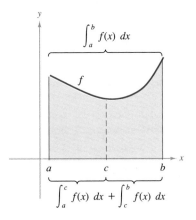

FIGURE 5.20

Since the definite integral is defined as the limit of a sum, it inherits the two properties of summation given in Theorem 5.3.

THEOREM 5.9 **PROPERTIES OF DEFINITE** **INTEGRALS**	If f and g are integrable on $[a, b]$ and k is a constant, then the following properties are true. 1. $\displaystyle\int_a^b kf(x)\,dx = k\int_a^b f(x)\,dx$ 2. $\displaystyle\int_a^b [f(x) \pm g(x)]\,dx = \int_a^b f(x)\,dx \pm \int_a^b g(x)\,dx$

REMARK   Property 2 in Theorem 5.9 can be extended to cover any finite number of functions. For example,

$$\int_a^b [f(x) + g(x) + h(x)]\,dx = \int_a^b f(x)\,dx + \int_a^b g(x)\,dx + \int_a^b h(x)\,dx.$$

EXAMPLE 5   Evaluation of a definite integral

Evaluate

$$\int_1^3 (-x^2 + 4x - 3)\,dx$$

using the following values:

$$\int_1^3 x^2\,dx = \frac{26}{3}, \qquad \int_1^3 x\,dx = 4, \qquad \text{and} \qquad \int_1^3 dx = 2.$$

SOLUTION

Using Theorem 5.9, we have

$$\int_1^3 (-x^2 + 4x - 3)\,dx = \int_1^3 (-x^2)\,dx + \int_1^3 4x\,dx + \int_1^3 (-3)\,dx$$

$$= -\int_1^3 x^2\,dx + 4\int_1^3 x\,dx - 3\int_1^3 dx$$

$$= -\left(\frac{26}{3}\right) + 4(4) - 3(2) = \frac{4}{3}.$$

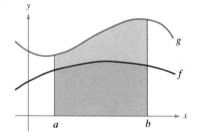

FIGURE 5.21

If f and g are continuous on the closed interval $[a, b]$ and $0 \le f(x) \le g(x)$ for $a \le x \le b$, then the following properties are true. First, the area of the region bounded by the graph of f and the x-axis (between a and b) must be nonnegative; second, this area must be less than or equal to the area of the region bounded by the graph of g and the x-axis (between a and b), as shown in Figure 5.21. These two results are generalized in the following theorem. (A proof of this theorem is given in Appendix A.)

THEOREM 5.10 PRESERVATION OF INEQUALITY	1. If f is integrable and nonnegative on the closed interval $[a, b]$, then $$0 \le \int_a^b f(x)\, dx.$$ 2. If f and g are integrable on the closed interval $[a, b]$ and $f(x) \le g(x)$ for every x in $[a, b]$, then $$\int_a^b f(x)\, dx \le \int_a^b g(x)\, dx.$$

EXERCISES for Section 5.3

In Exercises 1–10, set up a definite integral that yields the area of the given region. (Do not evaluate the integral.)

1. $f(x) = 3$

2. $f(x) = 4 - 2x$

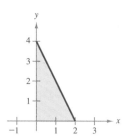

3. $f(x) = 4 - |x|$

4. $f(x) = x^2$

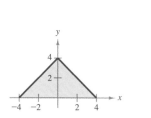

5. $f(x) = 4 - x^2$

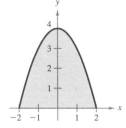

6. $f(y) = (y - 2)^2$

7. $g(y) = y^3$

8. $f(x) = \tan x$

9. $f(x) = \sin x$

10. $f(x) = \dfrac{1}{x^2 + 1}$

In Exercises 11–20, sketch the region whose area is indicated by the given definite integral. Then use a geometric formula to evaluate the integral ($a > 0, r > 0$).

11. $\displaystyle\int_0^3 4\, dx$

12. $\displaystyle\int_{-a}^a 4\, dx$

13. $\displaystyle\int_0^4 x\, dx$

14. $\displaystyle\int_0^4 \frac{x}{2}\, dx$

15. $\displaystyle\int_0^2 (2x + 5)\, dx$

16. $\displaystyle\int_0^5 (5 - x)\, dx$

17. $\displaystyle\int_{-1}^1 (1 - |x|)\, dx$

18. $\displaystyle\int_{-a}^a (a - |x|)\, dx$

19. $\displaystyle\int_{-3}^3 \sqrt{9 - x^2}\, dx$

20. $\displaystyle\int_{-r}^r \sqrt{r^2 - x^2}\, dx$

21. Given $\int_0^5 f(x)\,dx = 10$ and $\int_5^7 f(x)\,dx = 3$, find the following.

(a) $\int_0^7 f(x)\,dx$ (b) $\int_5^0 f(x)\,dx$

(c) $\int_5^5 f(x)\,dx$ (d) $\int_0^5 3f(x)\,dx$

22. Given $\int_0^3 f(x)\,dx = 4$ and $\int_3^6 f(x)\,dx = -1$, find the following.

(a) $\int_0^6 f(x)\,dx$ (b) $\int_6^3 f(x)\,dx$

(c) $\int_4^4 f(x)\,dx$ (d) $\int_3^6 -5f(x)\,dx$

23. Given $\int_2^6 f(x)\,dx = 10$ and $\int_2^6 g(x)\,dx = -2$, find the following.

(a) $\int_2^6 [f(x) + g(x)]\,dx$ (b) $\int_2^6 [g(x) - f(x)]\,dx$

(c) $\int_2^6 2g(x)\,dx$ (d) $\int_2^6 3f(x)\,dx$

24. Given $\int_{-1}^1 f(x)\,dx = 0$ and $\int_0^1 f(x)\,dx = 5$, find the following.

(a) $\int_{-1}^0 f(x)\,dx$ (b) $\int_0^1 f(x)\,dx - \int_{-1}^0 f(x)\,dx$

(c) $\int_{-1}^1 3f(x)\,dx$ (d) $\int_0^1 3f(x)\,dx$

In Exercises 25–30, evaluate the definite integral by the limit definition.

25. $\int_4^{10} 6\,dx$ **26.** $\int_{-2}^3 x\,dx$

27. $\int_{-1}^1 x^3\,dx$ **28.** $\int_0^1 x^3\,dx$

29. $\int_1^2 (x^2 + 1)\,dx$ **30.** $\int_1^2 4x^2\,dx$

In Exercises 31 and 32, use Example 1 as a model to evaluate the limit

$$\lim_{n \to \infty} \sum_{i=1}^n f(c_i)\Delta x_i$$

over the given region bounded by the graphs of the given equations.

31. $f(x) = \sqrt{x},\; y = 0,\; x = 0,\; x = 2$
    [Hint: Let $c_i = 2i^2/n^2$.]
32. $f(x) = \sqrt[3]{x},\; y = 0,\; x = 0,\; x = 1$
    [Hint: Let $c_i = i^3/n^3$.]

In Exercises 33–36, express the given limit as a definite integral on the interval $[a, b]$ where c_i is any point in the ith subinterval.

Limit	Interval
33. $\displaystyle \lim_{\|\Delta\| \to 0} \sum_{i=1}^n (3c_i + 10)\Delta x_i$	$[-1, 5]$
34. $\displaystyle \lim_{\|\Delta\| \to 0} \sum_{i=1}^n 6c_i(4 - c_i)^2\Delta x_i$	$[0, 4]$
35. $\displaystyle \lim_{\|\Delta\| \to 0} \sum_{i=1}^n \sqrt{c_i^2 + 4}\,\Delta x_i$	$[0, 3]$
36. $\displaystyle \lim_{\|\Delta\| \to 0} \sum_{i=1}^n \left(\frac{3}{c_i^2}\right)\Delta x_i$	$[1, 3]$

37. Prove that if f is a continuous function on a closed interval $[a, b]$, then

$$\left| \int_a^b f(x)\,dx \right| \le \int_a^b |f(x)|\,dx.$$

38. Determine whether the function

$$f(x) = \frac{1}{x - 4}$$

is integrable on the interval $[3, 5]$, and give a reason for your answer.

39. Determine whether the function

$$f(x) = \begin{cases} 1, & x \text{ is rational} \\ 0, & x \text{ is irrational} \end{cases}$$

is integrable on the interval $[0, 1]$ and give a reason for your answer.

40. Give an example of a function that is integrable on the interval $[-1, 1]$, but not continuous on $[-1, 1]$.

5.4 The Fundamental Theorem of Calculus

The Fundamental Theorem of Calculus ▪ The Mean Value Theorem for Integrals ▪ Average value of a function on an interval ▪ The Second Fundamental Theorem of Calculus

We have now looked at the two major branches of calculus: differential calculus (introduced with the tangent line problem) and integral calculus (introduced with the area problem). Initially, there seems no reason to assume that these two problems are related. But there is a very close connection. This connection was discovered independently by Isaac Newton and Gottfried Leibniz, and consequently these two men are usually credited with discovering calculus. The connection is stated in a theorem that is appropriately called the **Fundamental Theorem of Calculus.**

Roughly, the theorem tells us that differentiation and (definite) integration are inverse operations, in roughly the same sense as division and multiplication are inverse operations. To see how Newton and Leibniz might have anticipated this relationship, let's look at the approximations used in the development of the two operations. In Figure 5.22(a) we use the *quotient* $\Delta y / \Delta x$ (the slope of the secant line) to approximate the slope of the tangent line at (x, y). Similarly, in Figure 5.22(b) we use the *product* $\Delta y \Delta x$ (the area of a rectangle) to approximate the area under the curve. Thus, at least in the primitive approximation stage, the two operations appear to have an inverse relationship. The Fundamental Theorem of Calculus tells us that the limit processes (used to define the derivative and the definite integral) preserve this inverse relationship.

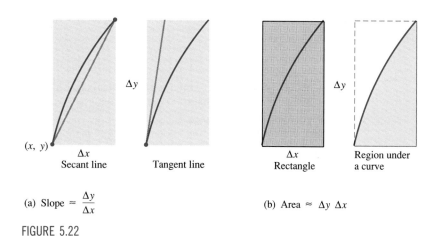

(a) Slope $\approx \dfrac{\Delta y}{\Delta x}$

(b) Area $\approx \Delta y \, \Delta x$

FIGURE 5.22

**THEOREM 5.11
THE FUNDAMENTAL
THEOREM OF CALCULUS**

If a function f is continuous on the closed interval $[a, b]$, then

$$\int_a^b f(x) \, dx = F(b) - F(a)$$

where F is any function that $F'(x) = f(x)$ for all x in $[a, b]$.

PROOF   The key to the proof is in writing the difference $F(b) - F(a)$ in a convenient form. Let Δ be the following partition of $[a, b]$.

$$a = x_0 < x_1 < x_2 < \cdots < x_{n-1} < x_n = b$$

By pairwise subtraction and addition of like terms, we can write

$$F(b) - F(a) = F(x_n) - F(x_{n-1}) + F(x_{n-1}) - \cdots - F(x_1) + F(x_1) - F(x_0)$$

$$= \sum_{i=1}^{n} [F(x_i) - F(x_{i-1})].$$

Now, by the Mean Value Theorem, we know that there exists a number c_i in the ith subinterval such that

$$F'(c_i) = \frac{F(x_i) - F(x_{i-1})}{x_i - x_{i-1}}.$$

Since $F'(c_i) = f(c_i)$, we let $\Delta x_i = x_i - x_{i-1}$ and write

$$F(b) - F(a) = \sum_{i=1}^{n} f(c_i)\Delta x_i.$$

This important equation tells us that by applying the Mean Value Theorem we can always find a collection of c_i's such that the *constant* $F(b) - F(a)$ is a Riemann sum of f on $[a, b]$. Taking the limit (as $\|\Delta\| \to 0$), we have

$$F(b) - F(a) = \int_a^b f(x)\ dx.$$

Three comments are in order regarding the Fundamental Theorem of Calculus. First, *provided we can find* an antiderivative of f, we now have a way to evaluate a definite integral without having to use the limit of a sum. Second, in applying this theorem it is helpful to use the notation

$$\int_a^b f(x)\ dx = \left[F(x) \right]_a^b = F(b) - F(a).$$

For instance, we write

$$\int_1^3 x^3\ dx = \left[\frac{x^4}{4} \right]_1^3$$

$$= \frac{3^4}{4} - \frac{1^4}{4} = \frac{81}{4} - \frac{1}{4} = 20.$$

Third, we observe that the constant of integration C can be dropped from the antiderivative, because

$$\int_a^b f(x)\ dx = \left[F(x) + C \right]_a^b$$

$$= [F(b) + C] - [F(a) + C] = F(b) - F(a).$$

EXAMPLE 1 Evaluating a definite integral

(a) $\displaystyle\int_1^2 (x^2 - 3)\, dx = \left[\frac{x^3}{3} - 3x\right]_1^2 = \left(\frac{8}{3} - 6\right) - \left(\frac{1}{3} - 3\right) = -\frac{2}{3}$

(b) $\displaystyle\int_1^4 3\sqrt{x}\, dx = 3\int_1^4 x^{1/2}\, dx = 3\left[\frac{x^{3/2}}{3/2}\right]_1^4 = 2(4)^{3/2} - 2(1)^{3/2} = 14$

(c) $\displaystyle\int_0^{\pi/4} \sec^2 x\, dx = \left[\tan x\right]_0^{\pi/4} = 1 - 0 = 1$ ▭

EXAMPLE 2 A definite integral involving absolute value

Evaluate

$$\int_0^2 |2x - 1|\, dx.$$

SOLUTION

From Figure 5.23 and the definition of absolute value, we note that

$$|2x - 1| = \begin{cases} -(2x - 1), & x < \dfrac{1}{2} \\ 2x - 1, & x \ge \dfrac{1}{2}. \end{cases}$$

Hence, we rewrite the integral in two parts, as follows.

$$\int_0^2 |2x - 1|\, dx = \int_0^{1/2} -(2x - 1)\, dx + \int_{1/2}^2 (2x - 1)\, dx$$

$$= \left[-x^2 + x\right]_0^{1/2} + \left[x^2 - x\right]_{1/2}^2$$

$$= \left(-\frac{1}{4} + \frac{1}{2}\right) - (0 + 0) + (4 - 2) - \left(\frac{1}{4} - \frac{1}{2}\right) = \frac{5}{2}$$ ▭

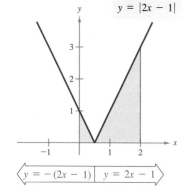

$y = |2x - 1|$

$\langle y = -(2x - 1) \mid y = 2x - 1 \rangle$

FIGURE 5.23

EXAMPLE 3 Using the Fundamental Theorem to find area

Find the area of the region bounded by the graph of $y = 2x^2 - 3x + 2$, the x-axis, and the vertical lines $x = 0$ and $x = 2$, as shown in Figure 5.24.

SOLUTION

$$\text{area} = \int_0^2 (2x^2 - 3x + 2)\, dx$$

$$= \left[\frac{2x^3}{3} - \frac{3x^2}{2} + 2x\right]_0^2$$

$$= \frac{16}{3} - 6 + 4 = \frac{10}{3}$$ ▭

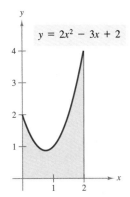

$y = 2x^2 - 3x + 2$

FIGURE 5.24

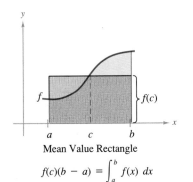

Mean Value Rectangle

$$f(c)(b - a) = \int_a^b f(x)\ dx$$

The Mean Value Theorem for Integrals

In Section 5.2, we observed that the area of a region under a curve is greater than the area of an inscribed rectangle and less than the area of a circumscribed rectangle. The Mean Value Theorem for Integrals states that somewhere "between" the inscribed and circumscribed rectangles there is a rectangle whose area is precisely equal to the area of the region under the curve, as shown in Figure 5.25.

FIGURE 5.25

**THEOREM 5.12
MEAN VALUE THEOREM
FOR INTEGRALS**

If f is continuous on the closed interval $[a, b]$, then there exists a number c in the open interval (a, b) such that

$$\int_a^b f(x)\ dx = f(c)(b - a).$$

PROOF

Case 1: If f is constant on the interval $[a, b]$, the result is trivial, since c can be any point in (a, b).

Case 2: If f is not constant on $[a, b]$, then by the Extreme Value Theorem we choose $f(m)$ and $f(M)$ to be the minimum and maximum values of f on $[a, b]$. Since $f(m) \leq f(x) \leq f(M)$ for all x in $[a, b]$, we conclude from Theorem 5.10 that

$$\int_a^b f(m)\ dx \leq \int_a^b f(x)\ dx \leq \int_a^b f(M)\ dx.$$

This inequality is depicted graphically in Figure 5.26.

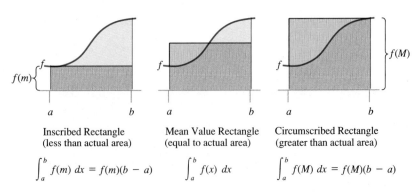

Inscribed Rectangle
(less than actual area)

$$\int_a^b f(m)\ dx = f(m)(b - a)$$

Mean Value Rectangle
(equal to actual area)

$$\int_a^b f(x)\ dx$$

Circumscribed Rectangle
(greater than actual area)

$$\int_a^b f(M)\ dx = f(M)(b - a)$$

FIGURE 5.26

Thus, we have

$$f(m)(b - a) \leq \int_a^b f(x)\ dx \leq f(M)(b - a)$$

$$f(m) \leq \frac{1}{b - a} \int_a^b f(x)\ dx \leq f(M).$$

Finally, by the Intermediate Value Theorem we conclude that there exists some c in (a, b) such that

$$f(c) = \frac{1}{b-a} \int_a^b f(x)\, dx$$

$$f(c)(b-a) = \int_a^b f(x)\, dx.$$

REMARK Note that the Mean Value Theorem for Integrals does not specify how to determine c. It merely guarantees the existence of at least one number c in the interval.

The value of $f(c)$, given in the Mean Value Theorem for Integrals, is called the **average value** of f on the interval $[a, b]$.

DEFINITION OF THE AVERAGE VALUE OF A FUNCTION ON AN INTERVAL

If f is continuous on $[a, b]$, then the **average value** of f on this interval is given by

$$\frac{1}{b-a} \int_a^b f(x)\, dx.$$

To see why we call this the average value of f, suppose that we partition $[a, b]$ into n subintervals of equal width $\Delta x = (b-a)/n$. If c_i is any point in the ith subinterval, then the arithmetic average (or mean) of the function values at the c_i's is given by

$$a_n = \frac{1}{n}[f(c_1) + f(c_2) + \cdots + f(c_n)]. \qquad \text{Average of } f(c_i), \ldots, f(c_n)$$

By multiplying and dividing by $(b-a)$, we can write the average as

$$a_n = \frac{1}{n} \sum_{i=1}^n f(c_i)\left(\frac{b-a}{b-a}\right) = \frac{1}{b-a} \sum_{i=1}^n f(c_i)\left(\frac{b-a}{n}\right)$$

$$= \frac{1}{b-a} \sum_{i=1}^n f(c_i)\Delta x.$$

Finally, taking the limit as $n \to \infty$ produces the average value of f on the interval $[a, b]$, as given in the above definition.

This development of the average value of a function on an interval is only one of many practical uses of definite integrals to represent summation processes. In Chapter 7 we will study other applications, such as volume, arc length, centers of mass, and work.

EXAMPLE 4 Finding the average value of a function

Find the average value of $f(x) = 3x^2 - 2x$ on the interval $[1, 4]$.

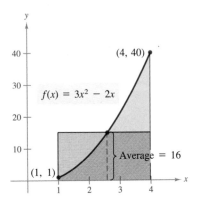

$f(x) = 3x^2 - 2x$

(1, 1)

(4, 40)

Average = 16

FIGURE 5.27

SOLUTION

The average value is given by

$$\frac{1}{b-a} \int_a^b f(x)\, dx = \frac{1}{3} \int_1^4 (3x^2 - 2x)\, dx = \frac{1}{3}\left[x^3 - x^2 \right]_1^4$$

$$= \frac{1}{3}[64 - 16 - (1 - 1)] = \frac{48}{3} = 16.$$

REMARK   Note in Figure 5.27 that the area of the region is equal to the area of the rectangle whose height is the average value.

The Second Fundamental Theorem of Calculus

When we defined the definite integral of f on the interval $[a, b]$, we used the constant b as the upper limit of integration and x as the variable of integration. We now look at a slightly different situation in which the variable x is used as the upper limit of integration. To avoid the confusion of using x in two different ways, we temporarily switch to using t as the variable of integration. (Remember that the definite integral is *not* a function of its variable of integration. Moreover, any variable can be used.)

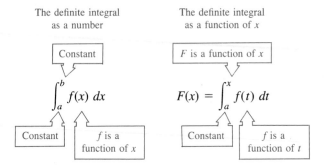

EXAMPLE 5   The definite integral as a function

Evaluate the function

$$F(x) = \int_0^x \cos t\, dt$$

at $x = 0$, $\pi/6$, $\pi/4$, $\pi/3$, and $\pi/2$.

SOLUTION

We could evaluate five different definite integrals, one for each of the given upper limits. However, it is much simpler to fix x (as a constant) temporarily and apply the Fundamental Theorem once, to obtain

$$\int_0^x \cos t\, dt = \left[\sin t \right]_0^x = \sin x - \sin 0 = \sin x.$$

Now, using $F(x) = \sin x$, we have the results shown in Figure 5.28.

FIGURE 5.28

$$F(x) = \int_0^x \cos t \, dt$$

In Example 5, note that the derivative of F is the original integrand (with only the variable changed). That is,

$$\frac{d}{dx}[F(x)] = \frac{d}{dx}[\sin x] = \frac{d}{dx}\left[\int_0^x \cos t \, dt\right] = \cos x.$$

We generalize this result in the following theorem, called the **Second Fundamental Theorem of Calculus.**

**THEOREM 5.13
THE SECOND FUNDAMENTAL
THEOREM OF CALCULUS**

If f is continuous on an open interval I containing a, then for every x in the interval,

$$\frac{d}{dx}\left[\int_a^x f(t) \, dt\right] = f(x).$$

PROOF   We define F as

$$F(x) = \int_a^x f(t) \, dt.$$

Then, by the definition of the derivative, we have

$$F'(x) = \lim_{\Delta x \to 0} \frac{F(x + \Delta x) - F(x)}{\Delta x}$$

$$= \lim_{\Delta x \to 0} \frac{1}{\Delta x}\left[\int_a^{x+\Delta x} f(t) \, dt - \int_a^x f(t) \, dt\right]$$

$$= \lim_{\Delta x \to 0} \frac{1}{\Delta x}\left[\int_a^{x+\Delta x} f(t) \, dt + \int_x^a f(t) \, dt\right]$$

$$= \lim_{\Delta x \to 0} \frac{1}{\Delta x}\left[\int_x^{x+\Delta x} f(t) \, dt\right].$$

Now, from the Mean Value Theorem for Integrals, we know there exists a number c in the interval $[x, x + \Delta x]$ such that the integral in the above expression is equal to $f(c)\Delta x$. Moreover, since $x \le c \le x + \Delta x$, it follows that $c \to x$ as $\Delta x \to 0$. Thus, we have

$$F'(x) = \lim_{\Delta x \to 0}\left[\frac{1}{\Delta x} f(c)\Delta x\right] = \lim_{\Delta x \to 0} f(c) = f(x).$$

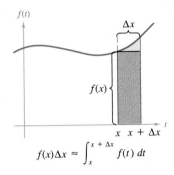

$$f(x)\Delta x \approx \int_x^{x+\Delta x} f(t)\,dt$$

FIGURE 5.29

REMARK   Using the area model for definite integrals, we can view the approximation

$$f(x)\Delta x \approx \int_x^{x+\Delta x} f(t)\,dt$$

as saying that the area of the rectangle of height $f(x)$ and width Δx is approximately equal to the area of the region lying between the graph of f and the x-axis on the interval $[x, x + \Delta x]$, as shown in Figure 5.29.

Note that the Second Fundamental Theorem of Calculus tells us that if a function is continuous, then we can be sure that it has an antiderivative. This antiderivative need not, however, be an elementary function. (Recall the discussion of elementary functions in Section 1.5.)

EXAMPLE 6   Applying the Second Fundamental Theorem of Calculus

Evaluate

$$\frac{d}{dx}\int_0^x \sqrt{t^2 + 1}\,dt.$$

SOLUTION

Note that $f(t) = \sqrt{t^2 + 1}$ is continuous on the entire real line. Thus, using the Second Fundamental Theorem of Calculus, we can write

$$\frac{d}{dx}\int_0^x \sqrt{t^2 + 1}\,dt = \sqrt{x^2 + 1}.$$

$\square$

EXERCISES for Section 5.4

In Exercises 1–30, evaluate the definite integral.

1. $\displaystyle\int_0^1 2x\,dx$

2. $\displaystyle\int_2^7 3\,dv$

3. $\displaystyle\int_{-1}^0 (x - 2)\,dx$

4. $\displaystyle\int_2^5 (-3v + 4)\,dv$

5. $\displaystyle\int_{-1}^1 (t^2 - 2)\,dt$

6. $\displaystyle\int_0^3 (3x^2 + x - 2)\,dx$

7. $\displaystyle\int_0^1 (2t - 1)^2\,dt$

8. $\displaystyle\int_{-1}^1 (t^3 - 9t)\,dt$

9. $\displaystyle\int_1^2 \left(\frac{3}{x^2} - 1\right)\,dx$

10. $\displaystyle\int_0^1 (3x^3 - 9x + 7)\,dx$

11. $\displaystyle\int_1^2 (5x^4 + 5)\,dx$

12. $\displaystyle\int_{-3}^3 v^{1/3}\,dv$

13. $\displaystyle\int_{-1}^1 (\sqrt[3]{t} - 2)\,dt$

14. $\displaystyle\int_{-2}^{-1} \sqrt{\frac{-2}{x}}\,dx$

15. $\displaystyle\int_1^4 \frac{u - 2}{\sqrt{u}}\,du$

16. $\displaystyle\int_{-2}^{-1} \left(u - \frac{1}{u^2}\right)\,du$

17. $\displaystyle\int_0^1 \frac{x - \sqrt{x}}{3}\,dx$

18. $\displaystyle\int_0^2 (2 - t)\sqrt{t}\,dt$

19. $\displaystyle\int_{-1}^0 (t^{1/3} - t^{2/3})\,dt$

20. $\displaystyle\int_{-8}^{-1} \frac{x - x^2}{2\sqrt[3]{x}}\,dx$

21. $\displaystyle\int_{-1}^1 |x|\,dx$

22. $\displaystyle\int_0^3 |2x - 3|\,dx$

23. $\displaystyle\int_0^4 |x^2 - 4x + 3|\,dx$

24. $\displaystyle\int_{-1}^1 |x^3|\,dx$

25. $\displaystyle\int_0^\pi (1 + \sin x)\,dx$

26. $\displaystyle\int_0^{\pi/4} \frac{1 - \sin^2 \theta}{\cos^2 \theta}\,d\theta$

27. $\displaystyle\int_{-\pi/6}^{\pi/6} \sec^2 x\,dx$

28. $\displaystyle\int_{\pi/4}^{\pi/2} (2 - \csc^2 x)\,dx$

29. $\displaystyle\int_{-\pi/3}^{\pi/3} 4 \sec \theta \tan \theta\,d\theta$

30. $\displaystyle\int_{-\pi/2}^{\pi/2} (2t + \cos t)\,dt$

In Exercises 31–38, determine the area of the indicated region.

31. $y = x - x^2$

32. $y = -x^2 + 2x + 3$

33. $y = 1 - x^4$

34. $y = \dfrac{1}{x^2}$

35. $y = \sqrt[3]{2x}$

36. $y = (3 - x)\sqrt{x}$

37. $y = \cos x$

38. $y = x + \sin x$

In Exercises 39–42, find the area of the region bounded by the graphs of the given equations.

39. $y = 3x^2 + 1$, $x = 0$, $x = 2$, $y = 0$
40. $y = 1 + \sqrt{x}$, $x = 0$, $x = 4$, $y = 0$
41. $y = x^3 + x$, $x = 2$, $y = 0$
42. $y = -x^2 + 3x$, $y = 0$

In Exercises 43–46, find the value of c guaranteed by the *Mean Value Theorem for Integrals* for the given function over the specified interval.

Function	Interval
43. $f(x) = x^3$	$[0, 2]$
44. $f(x) = \dfrac{9}{x^3}$	$[1, 3]$
45. $f(x) = 2 \sec^2 x$	$\left[-\dfrac{\pi}{4}, \dfrac{\pi}{4}\right]$
46. $f(x) = \cos x$	$\left[-\dfrac{\pi}{3}, \dfrac{\pi}{3}\right]$

In Exercises 47–50, sketch the graph of the given function over the specified interval. Find the average value of the function over the interval and all values of x where the function equals its average value.

Function	Interval
47. $f(x) = 4 - x^2$	$[-2, 2]$
48. $f(x) = \dfrac{x^2 + 1}{x^2}$	$\left[\dfrac{1}{2}, 2\right]$
49. $f(x) = \sin x$	$[0, \pi]$
50. $f(x) = \cos x$	$\left[0, \dfrac{\pi}{2}\right]$

In Exercises 51–56, (a) integrate to find F as a function of x and (b) demonstrate the Second Fundamental Theorem of Calculus by differentiating the result of part (a).

51. $F(x) = \displaystyle\int_0^x (t + 2)\, dt$

52. $F(x) = \displaystyle\int_0^x t(t^2 + 1)\, dt$

53. $F(x) = \displaystyle\int_8^x \sqrt[3]{t}\, dt$

54. $F(x) = \displaystyle\int_4^x \sqrt{t}\, dt$

55. $F(x) = \displaystyle\int_{\pi/4}^x \sec^2 t\, dt$

56. $F(x) = \displaystyle\int_{\pi/3}^x \sec t \tan t\, dt$

In Exercises 57–62, use the Second Fundamental Theorem of Calculus to find $F'(x)$.

57. $F(x) = \displaystyle\int_{-2}^x (t^2 - 2t + 5)\, dt$

58. $F(x) = \displaystyle\int_1^x \sqrt[4]{t}\, dt$

59. $F(x) = \displaystyle\int_{-1}^x \sqrt{t^4 + 1}\, dt$

60. $F(x) = \displaystyle\int_0^x \tan^4 t\, dt$

61. $F(x) = \displaystyle\int_0^x t \cos t\, dt$

62. $F(x) = \displaystyle\int_1^x \frac{t^2}{t^2 + 1}\, dt$

63. The volume V in liters of air in the lungs during a 5-second respiratory cycle is approximated by the model

$$V = 0.1729t + 0.1522t^2 - 0.0374t^3$$

where t is the time in seconds. Approximate the average volume of air in the lungs during one cycle.

64. The velocity v of the flow of blood at a distance r from the central axis of an artery of radius R is given by

$$v = k(R^2 - r^2)$$

where k is the constant of proportionality. Find the average rate of flow of blood along a radius of the artery. (Use zero and R as the limits of integration.)

65. The air temperature during a period of 12 hours is given by the model

$$T = 53 + 5t - 0.3t^2, \quad 0 \le t \le 12$$

where t is measured in hours and T in degrees Fahrenheit (see figure). Find the average temperature during (a) the first 6 hours of the period and (b) the entire period.

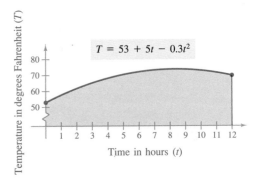

FIGURE FOR 65

66. (Buffon's Needle Experiment) A horizontal plane is ruled with parallel lines 2 inches apart. If a 2-inch needle is tossed randomly onto the plane, it can be shown that the probability that the needle will touch a line is given by

$$P = \frac{2}{\pi} \int_0^{\pi/2} \sin\theta\, d\theta$$

where θ is the acute angle between the needle and any one of the parallel lines. Find this probability.

5.5 Integration by Substitution

u-substitution ▪ Pattern recognition ▪ Change of variables ▪ The General Power Rule for Integration ▪ Change of variables for definite integrals ▪ Integration of even and odd functions

In this section we demonstrate techniques for integrating composite functions. We will split the discussion into two parts—*pattern recognition* and *change of variables*. Both techniques involve a **u-substitution.** With pattern recognition we perform the substitution mentally, and with change of variables we write the substitution steps.

The role of substitution in integration is comparable to the role of the Chain Rule in differentiation. Recall that for differentiable functions given by $y = F(u)$ and $u = g(x)$, the Chain Rule states that

$$\frac{d}{dx}[F(g(x))] = F'(g(x))g'(x).$$

From our definition of an antiderivative, it follows that

$$\int F'(g(x))g'(x)\, dx = F(g(x)) + C = F(u) + C.$$

We state these results in the following theorem.

THEOREM 5.14 ANTIDIFFERENTIATION OF A COMPOSITE FUNCTION	Let f and g be functions such that $f \circ g$ and g' are continuous on an interval I. If F is an antiderivative of f on I, then $$\int f(g(x))g'(x)\, dx = F(g(x)) + C.$$

Pattern recognition

There are several techniques for applying substitution, each differing slightly from the other. However, you should remember that the goal is the same with every technique—*we are trying to find an antiderivative of the integrand.*

Note that the statement of Theorem 5.14 doesn't tell how to distinguish between $f(g(x))$ and $g'(x)$ in the integrand. As you become more experienced at integration, your skill in doing this will increase. Of course, part of the key is familiarity with derivatives.

Examples 1 and 2 show how to apply the theorem *directly*, by recognizing the presence of $f(g(x))$ and $g'(x)$. Note that the composite function in the integrand has an *outside function f* and an *inside function g*. Moreover, the derivative $g'(x)$ is present as a factor of the integrand.

$$\int \underset{\substack{\uparrow \\ \text{Inside}}}{f(g(x))} \underset{\substack{\uparrow \\ \text{Derivative} \\ \text{of inside}}}{g'(x)} dx = F(g(x)) + C$$

Outside

EXAMPLE 1   Recognizing the $f(g(x))g'(x)$ pattern

Evaluate $\int (x^2 + 1)^2(2x)\, dx$.

SOLUTION

Letting $g(x) = x^2 + 1$, we have $g'(x) = 2x$ and $f(g(x)) = [g(x)]^2$, and we recognize that the integrand follows the $f(g(x))g'(x)$ pattern. Moreover, by the Power Rule, we know that $F(g(x)) = \frac{1}{3}[g(x)]^3$ is an antiderivative of f. Thus, by Theorem 5.14, we have

$$\int \underbrace{(x^2 + 1)^2}_{[g(x)]^2}\underbrace{(2x)}_{g'(x)}\, dx = \underbrace{F(x^2 + 1)}_{F(g(x))} + C = \underbrace{\frac{(x^2 + 1)^3}{3}}_{\frac{1}{3}[g(x)]^3} + C.$$

Try using the Chain Rule to check that the derivative of $\frac{1}{3}(x^2 + 1)^3 + C$ is the integrand of the original integral.

EXAMPLE 2 Recognizing the $f(g(x))g'(x)$ pattern

Evaluate $\int 5 \cos 5x \, dx$.

SOLUTION

Letting $g(x) = 5x$, we have $g'(x) = 5$, and we recognize that the integrand follows the $f(g(x))g'(x)$ pattern. Moreover, by the Sine Rule, we know that $F(g(x)) = \sin g(x)$ is an antiderivative of f. Thus, by Theorem 5.14, we have

$$\int \underbrace{(\cos 5x)}_{\cos\,[g(x)]}\underbrace{(5)}_{g'(x)} \, dx = \underbrace{F(5x)}_{F(g(x))} + C = \underbrace{\sin 5x}_{\sin\,[g(x)]} + C.$$

You can check this answer by differentiating $\sin 5x + C$ to obtain the original integrand. ▭

   Both of the integrands in Examples 1 and 2 fit the $f(g(x))g'(x)$ pattern exactly—we only had to recognize the pattern. We can extend this technique considerably with the Constant Multiple Rule

$$\int kf(x) \, dx = k \int f(x) \, dx.$$

Many integrands contain the essential part (the variable part) of $g'(x)$, but are missing a constant multiple. In such cases we can multiply and divide by the necessary constant multiple, as shown in the following example.

EXAMPLE 3 Multiplying and dividing by a constant

Evaluate $\int x(x^2 + 1)^2 \, dx$.

SOLUTION

This is similar to the integral given in Example 1, except that there is a missing factor of 2. Recognizing that $2x$ is the derivative of $x^2 + 1$, we let $g(x) = x^2 + 1$ and supply the $2x$ as follows.

$$\int x(x^2 + 1)^2 \, dx = \int (x^2 + 1)^2 \left(\frac{1}{2}\right)(2x) \, dx \qquad \text{Multiply and divide by 2}$$

$$= \frac{1}{2} \int (x^2 + 1)^2 (2x) \, dx \qquad \text{Constant Multiple Rule}$$

$$= \frac{1}{2} \int [g(x)]^2 g'(x) \, dx$$

$$= \frac{1}{2} \frac{[g(x)]^3}{3} + C \qquad \text{Integrate}$$

$$= \frac{(x^2 + 1)^3}{6} + C \qquad\qquad ▭$$

REMARK   Be sure you see that the *Constant* Multiple Rule applies only to constants. You cannot multiply and divide by a variable and then move the variable outside the integral sign. For instance,

$$\int (x^2 + 1)^2 \, dx \neq \frac{1}{2x} \int (x^2 + 1)^2 (2x) \, dx.$$

After all, if it were legitimate to move variable quantities outside the integral sign, you could move the entire integrand out and simplify the whole process! But the result would be incorrect.

Change of variables

The integration technique used in Examples 1 through 3 depends on the ability to recognize (or create) integrands of the form $f(g(x))g'(x)$. With a formal **change of variables,** we completely rewrite the integral in terms of u and du (or any other convenient variable). Although this procedure involves more written steps, it is useful for complicated integrands. The change of variable technique uses the Leibniz notation for the differential. That is, if $u = g(x)$, we write $du = g'(x) \, dx$, and the integral in Theorem 5.14 takes the form

$$\int f(g(x))g'(x) \, dx = \int f(u) \, du = F(u) + C.$$

We illustrate the procedure in the next several examples.

EXAMPLE 4   Change of variable

Evaluate $\int \sqrt{2x - 1} \, dx$.

SOLUTION

First, we let u be the inner function, $u = 2x - 1$. Then, we obtain

$$du = 2 \, dx. \qquad \text{Solve for } du$$

Now, since $\sqrt{2x - 1} = \sqrt{u}$ and $dx = du/2$, we substitute to obtain

$$\int \sqrt{2x - 1} \, dx = \int \sqrt{u} \left(\frac{du}{2}\right) \qquad \text{Integral in terms of } u$$

$$= \frac{1}{2} \int u^{1/2} \, du$$

$$= \frac{1}{2}\left(\frac{u^{3/2}}{3/2}\right) + C \qquad \text{Antiderivative in terms of } u$$

$$= \frac{1}{3}u^{3/2} + C.$$

Back-substitution of $u = 2x - 1$ yields

$$\int \sqrt{2x - 1} \, dx = \frac{1}{3}(2x - 1)^{3/2} + C. \qquad \text{Antiderivative in terms of } x$$

EXAMPLE 5   Change of variable

Evaluate $\int x\sqrt{2x-1}\ dx$.

SOLUTION

As in the previous example, we let $u = 2x - 1$ and obtain $dx = du/2$. Since the integrand contains a factor of x, we must also solve for x in terms of u as follows.

$$u = 2x - 1 \quad\Longrightarrow\quad x = \frac{u+1}{2} \qquad \text{Solve for } x \text{ in terms of } u$$

Thus, the integral becomes

$$\int x\sqrt{2x-1}\ dx = \int \left(\frac{u+1}{2}\right) u^{1/2} \left(\frac{du}{2}\right) = \frac{1}{4}\int (u^{3/2} + u^{1/2})\ du$$

$$= \frac{1}{4}\left[\frac{u^{5/2}}{5/2} + \frac{u^{3/2}}{3/2}\right] + C.$$

Back-substitution of $u = 2x - 1$ yields

$$\int x\sqrt{2x-1}\ dx = \frac{1}{10}(2x-1)^{5/2} + \frac{1}{6}(2x-1)^{3/2} + C. \qquad \blacksquare$$

To complete the change of variable in Example 5, we solved for x in terms of u. Sometimes this is very difficult. Fortunately it is not always necessary, as shown in the next example.

EXAMPLE 6   Change of variable

Evaluate $\int \sin^2 3x \cos 3x\ dx$.

SOLUTION

Since $\sin^2 3x = (\sin 3x)^2$, we let $u = \sin 3x$. Then

$$du = (\cos 3x)(3)\ dx.$$

Now, since $\cos 3x\ dx$ is part of the given integral, we write

$$\frac{du}{3} = \cos 3x\ dx.$$

Substituting u and $du/3$ in the given integral yields

$$\int \sin^2 3x \cos 3x\ dx = \int u^2\ \frac{du}{3} = \frac{1}{3}\int u^2\ du = \frac{1}{3}\left(\frac{u^3}{3}\right) + C.$$

Back-substitution of $u = \sin 3x$ yields

$$\int \sin^2 3x \cos 3x\ dx = \frac{1}{9}\sin^3 3x + C. \qquad \blacksquare$$

We summarize the steps used for integration by substitution in the following guidelines.

<table>
<tr><td>**GUIDELINES FOR INTEGRATION BY SUBSTITUTION**</td><td>1. Choose a substitution $u = g(x)$. Usually, it is best to choose the *inner* part of a composite function, such as a quantity raised to a power.
2. Compute $du = g'(x)\, dx$.
3. Rewrite the integral in terms of the variable u.
4. Evaluate the resulting integral in terms of u.
5. Replace u by $g(x)$ to obtain an antiderivative in terms of x.</td></tr>
</table>

The General Power Rule for Integration

One of the most common u-substitutions involves quantities in the integrand that are raised to a power. Because of the importance of this type of substitution, we give it a special name—the **General Power Rule.** A proof of this rule follows directly from the (simple) Power Rule for integration, together with Theorem 5.14.

<table>
<tr><td>**THEOREM 5.15**
THE GENERAL POWER RULE FOR INTEGRATION</td><td>If g is a differentiable function of x, then
$$\int [g(x)]^n g'(x)\, dx = \frac{[g(x)]^{n+1}}{n+1} + C, \quad n \ne -1.$$
Equivalently, if $u = g(x)$, then
$$\int u^n\, du = \frac{u^{n+1}}{n+1} + C, \quad n \ne -1.$$</td></tr>
</table>

Study the following example carefully to see the variety of integrals that can be evaluated with the General Power Rule.

EXAMPLE 7 Substitution and the General Power Rule

(a) $\displaystyle \int 3(3x-1)^4\, dx = \int \overbrace{(3x-1)^4}^{u^4}\overbrace{(3)\, dx}^{du} = \overbrace{\frac{(3x-1)^5}{5}}^{u^5/5} + C$

(b) $\displaystyle \int (2x+1)(x^2+x)\, dx = \int \overbrace{(x^2+x)^1}^{u^1}\overbrace{(2x+1)\, dx}^{du} = \overbrace{\frac{(x^2+x)^2}{2}}^{u^2/2} + C$

(c) $\displaystyle \int 3x^2\sqrt{x^3-2}\, dx = \int \overbrace{(x^3-2)^{1/2}}^{u^{1/2}}\overbrace{(3x^2)\, dx}^{du} = \overbrace{\frac{(x^3-2)^{3/2}}{3/2}}^{u^{3/2}/(3/2)} + C$

(d) $\displaystyle\int \frac{-4x}{(1-2x^2)^2}\, dx = \int \overbrace{(1-2x^2)^{-2}}^{u^{-2}}\overbrace{(-4x)\, dx}^{du} = \overbrace{\frac{(1-2x^2)^{-1}}{-1}}^{u^{-1}/(-1)} + C$

(e) $\displaystyle\int \cos^2 x \sin x\, dx = -\int \overbrace{(\cos x)^2}^{u^2}\overbrace{(-\sin x)\, dx}^{du} = \overbrace{-\frac{(\cos x)^3}{3}}^{u^3/3} + C$ ☐

Some integrals whose integrand involves a quantity raised to a power cannot be evaluated by the General Power Rule. Consider the two integrals

$$\int x(x^2+1)^2\, dx \qquad \text{and} \qquad \int (x^2+1)^2\, dx.$$

Can both be evaluated by the General Power Rule? We evaluated the first integral in Example 3 by letting $u = x^2 + 1$, from which we obtained $du = 2x\, dx$. However, in the second integral the substitution $u = x^2 + 1$ fails, since the integrand lacks the critical factor x needed for du. Fortunately, *for this particular integral,* we can expand the integrand into the polynomial form

$$(x^2 + 1)^2 = x^4 + 2x^2 + 1$$

and use the Power Rule to integrate each term.

EXAMPLE 8 Integration by *u*-substitution

Evaluate $\int x^2 \sin x^3\, dx$.

SOLUTION

Let $u = x^3$. Then we have

$$u = x^3 \ \Longrightarrow\ du = 3x^2\, dx \ \Longrightarrow\ \frac{du}{3} = x^2\, dx.$$

Thus,

$$\int x^2 \sin x^3\, dx = \int (\sin x^3)(x^2\, dx) = \int \sin u \left(\frac{du}{3}\right)$$

$$= \frac{1}{3}\int \sin u\, du = -\frac{1}{3}\cos u + C$$

$$= -\frac{1}{3}\cos x^3 + C.$$ ☐

Change of variables for definite integrals

When a definite integral involves a *u*-substitution, it is often convenient to determine the limits of integration for the variable *u* rather than to convert the antiderivative back to the variable *x* and evaluate at the original limits.

This change of variables is stated explicitly in the next theorem. The proof follows from Theorem 5.14 combined with the Fundamental Theorem of Calculus.

THEOREM 5.16 **CHANGE OF VARIABLES FOR** **DEFINITE INTEGRALS**	If the function $u = g(x)$ has a continuous derivative on the closed interval $[a, b]$ and f has an antiderivative over the range of g, then $$\int_a^b f(g(x))g'(x)\, dx = \int_{g(a)}^{g(b)} f(u)\, du.$$

EXAMPLE 9 Change of variables

Evaluate

$$\int_0^1 x(x^2 + 1)^3\, dx.$$

SOLUTION

To evaluate this integral, we let $u = x^2 + 1$. Then we have

$$u = x^2 + 1 \quad \Longrightarrow \quad du = 2x\, dx.$$

Before substituting, we determine the new upper and lower limits of integration.

Lower limit	Upper limit
When $x = 0$, $u = 0^2 + 1 = 1$.	When $x = 1$, $u = 1^2 + 1 = 2$.

Now, we substitute to obtain

$$\int_0^1 x(x^2 + 1)^3\, dx = \frac{1}{2}\int_0^1 (x^2 + 1)^3(2x)\, dx = \frac{1}{2}\int_1^2 u^3\, du$$

$$= \frac{1}{2}\left[\frac{u^4}{4}\right]_1^2 = \frac{1}{2}\left(4 - \frac{1}{4}\right) = \frac{15}{8}.$$

EXAMPLE 10 Change of variables

Evaluate

$$A = \int_1^5 \frac{x}{\sqrt{2x - 1}}\, dx.$$

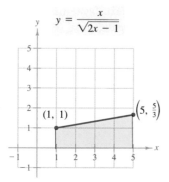

Region before substitution

FIGURE 5.30

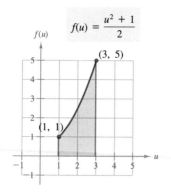

Region after substitution

FIGURE 5.31

SOLUTION

To evaluate this integral, we let $u = \sqrt{2x - 1}$. Then

$$u^2 = 2x - 1 \implies x = \frac{u^2 + 1}{2} \implies dx = u\,du.$$

Before substituting, we determine the new upper and lower limits of integration.

Lower limit	*Upper limit*
When $x = 1$, $u = \sqrt{2 - 1} = 1$.	When $x = 5$, $u = \sqrt{10 - 1} = 3$.

Now, we substitute to obtain

$$\int_1^5 \frac{x}{\sqrt{2x - 1}}\,dx = \int_1^3 \frac{1}{2}\left(\frac{u^2 + 1}{u}\right)u\,du = \frac{1}{2}\int_1^3 (u^2 + 1)\,du$$

$$= \frac{1}{2}\left[\frac{u^3}{3} + u\right]_1^3 = \frac{1}{2}\left(9 + 3 - \frac{1}{3} - 1\right) = \frac{16}{3}.$$

REMARK   Geometrically, we can interpret the equation

$$\int_1^5 \frac{x}{\sqrt{2x - 1}}\,dx = \int_1^3 \frac{u^2 + 1}{2}\,du$$

to mean that the two *different* regions shown in Figures 5.30 and 5.31 have the *same* area.

When evaluating definite integrals by substitution, do not be surprised if the upper limit of integration of the u-variable form is smaller than the lower limit. If this happens, don't rearrange the limits. Simply evaluate as usual. For example, after substituting $u = \sqrt{1 - x}$ in the integral

$$\int_0^1 x^2(1 - x)^{1/2}\,dx$$

we have $u = \sqrt{1 - 1} = 0$ when $x = 1$, and $u = \sqrt{1 - 0} = 1$ when $x = 0$. Thus, the correct u-variable form of this integral is

$$-2\int_1^0 (1 - u^2)^2 u^2\,du.$$

Integration of even and odd functions

Even by making a change of variables, integration often is not a simple task. Occasionally, you will be able to simplify the evaluation of a definite integral (over an interval that is symmetric to the origin) by recognizing the integrand to be an even or odd function. This result is described in the next theorem.

**THEOREM 5.17
INTEGRATION OF EVEN
AND ODD FUNCTIONS**

Let f be integrable on the closed interval $[-a, a]$.

1. If f is an *even* function, then

$$\int_{-a}^{a} f(x)\, dx = 2 \int_{0}^{a} f(x)\, dx.$$

2. If f is an *odd* function, then

$$\int_{-a}^{a} f(x)\, dx = 0.$$

PROOF   We prove Property 1 and leave the proof of Property 2 to you. Since f is even, we know that $f(x) = f(-x)$. Now we use Theorem 5.14 with the substitution $u = -x$ to obtain

$$\int_{-a}^{0} f(x)\, dx = \int_{a}^{0} f(-u)(-du) = -\int_{a}^{0} f(u)\, du = \int_{0}^{a} f(u)\, du = \int_{0}^{a} f(x)\, dx.$$

Finally, using Theorem 5.8, we have

$$\int_{-a}^{a} f(x)\, dx = \int_{-a}^{0} f(x)\, dx + \int_{0}^{a} f(x)\, dx$$

$$= \int_{0}^{a} f(x)\, dx + \int_{0}^{a} f(x)\, dx = 2 \int_{0}^{a} f(x)\, dx.$$

EXAMPLE 11   Integration of an odd function

Evaluate

$$\int_{-\pi/2}^{\pi/2} (\sin^3 x \cos x + \sin x \cos x)\, dx.$$

SOLUTION

By letting $f(x) = \sin^3 x \cos x + \sin x \cos x$, we have

$$f(-x) = \sin^3(-x) \cos(-x) + \sin(-x) \cos(-x)$$
$$= -\sin^3 x \cos x - \sin x \cos x = -f(x).$$

Thus, f is an odd function, and since $[-\pi/2, \pi/2]$ is symmetric about the origin, we can apply Theorem 5.17 to conclude that

$$\int_{-\pi/2}^{\pi/2} (\sin^3 x \cos x + \sin x \cos x)\, dx = 0.$$

REMARK   From Figure 5.32, we see that the two regions on either side of the y-axis have the same area. However, since one lies below the x-axis and one lies above, integration produces a cancellation effect. (We will say more about finding the area of a region below the x-axis in Section 7.1.)

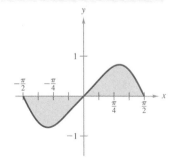

$f(x) = \sin^3 x \cos x + \sin x \cos x$

FIGURE 5.32

EXERCISES for Section 5.5

In Exercises 1–6, complete the table by identifying u and du for the given integral.

$$\int f(g(x))g'(x)\, dx \qquad u = g(x) \qquad du = g'(x)\, dx$$

1. $\displaystyle \int (5x^2 + 1)^2(10x)\, dx$

2. $\displaystyle \int x^2\sqrt{x^3 + 1}\, dx$

3. $\displaystyle \int \frac{x}{\sqrt{x^2 + 1}}\, dx$

4. $\displaystyle \int \sec 2x \tan 2x\, dx$

5. $\displaystyle \int \tan^2 x \sec^2 x\, dx$

6. $\displaystyle \int \frac{\cos x}{\sin^2 x}\, dx$

In Exercises 7–30, evaluate the indefinite integral and check the result by differentiation.

7. $\displaystyle \int (1 + 2x)^4(2)\, dx$

8. $\displaystyle \int (x^2 - 1)^3(2x)\, dx$

9. $\displaystyle \int \sqrt{9 - x^2}(-2x)\, dx$

10. $\displaystyle \int (1 - 2x^2)^3(-4x)\, dx$

11. $\displaystyle \int x^2(x^3 - 1)^4\, dx$

12. $\displaystyle \int x(4x^2 + 3)^3\, dx$

13. $\displaystyle \int 5x\sqrt[3]{1 - x^2}\, dx$

14. $\displaystyle \int u^3\sqrt{u^4 + 2}\, du$

15. $\displaystyle \int \frac{x^2}{(1 + x^3)^2}\, dx$

16. $\displaystyle \int \frac{x^2}{(16 - x^3)^2}\, dx$

17. $\displaystyle \int \frac{4x}{\sqrt{16 - x^2}}\, dx$

18. $\displaystyle \int \frac{10x^2}{\sqrt{1 + x^3}}\, dx$

19. $\displaystyle \int \frac{x + 1}{(x^2 + 2x - 3)^2}\, dx$

20. $\displaystyle \int \frac{x - 4}{\sqrt{x^2 - 8x + 1}}\, dx$

21. $\displaystyle \int \left(1 + \frac{1}{t}\right)^3\left(\frac{1}{t^2}\right) dt$

22. $\displaystyle \int \frac{1}{(3x)^2}\, dx$

23. $\displaystyle \int \frac{1}{\sqrt{2x}}\, dx$

24. $\displaystyle \int \frac{1}{2\sqrt{x}}\, dx$

25. $\displaystyle \int \frac{x^2 + 3x + 7}{\sqrt{x}}\, dx$

26. $\displaystyle \int \frac{t + 2t^2}{\sqrt{t}}\, dt$

27. $\displaystyle \int t^2\left(t - \frac{2}{t}\right) dt$

28. $\displaystyle \int \left(\frac{t^3}{3} + \frac{1}{4t^2}\right) dt$

29. $\displaystyle \int (9 - y)\sqrt{y}\, dy$

30. $\displaystyle \int 2\pi y(8 - y^{3/2})\, dy$

In Exercises 31–42, evaluate the indefinite integral.

31. $\displaystyle \int \sin 2x\, dx$

32. $\displaystyle \int x \sin x^2\, dx$

33. $\displaystyle \int x \cos x^2\, dx$

34. $\displaystyle \int \cos 6x\, dx$

35. $\displaystyle \int \sin 2x \cos 2x\, dx$

36. $\displaystyle \int \sec (1 - x) \tan (1 - x)\, dx$

37. $\displaystyle \int \tan^4 x \sec^2 x\, dx$

38. $\displaystyle \int \sqrt{\cot x}\, \csc^2 x\, dx$

39. $\displaystyle \int \frac{\csc^2 x}{\cot^3 x}\, dx$

40. $\displaystyle \int \frac{\sin x}{\cos^2 x}\, dx$

41. $\displaystyle \int \cot^2 x\, dx$

42. $\displaystyle \int \csc^2 \left(\frac{x}{2}\right) dx$

In Exercises 43–52, evaluate the indefinite integral by the method shown in Example 5.

43. $\displaystyle \int x\sqrt{x + 2}\, dx$

44. $\displaystyle \int x\sqrt{2x + 1}\, dx$

45. $\displaystyle \int x^2\sqrt{1 - x}\, dx$

46. $\displaystyle \int x^3\sqrt{x + 2}\, dx$

47. $\displaystyle \int \frac{x^2 - 1}{\sqrt{2x - 1}}\, dx$

48. $\displaystyle \int \frac{2x - 1}{\sqrt{x + 3}}\, dx$

49. $\displaystyle \int \frac{-x}{(x + 1) - \sqrt{x + 1}}\, dx$

50. $\displaystyle \int t\sqrt[3]{t - 4}\, dt$

51. $\displaystyle \int \frac{x}{\sqrt{2x + 1}}\, dx$

52. $\displaystyle \int (x + 1)\sqrt{2 - x}\, dx$

In Exercises 53–70, evaluate the definite integral.

53. $\displaystyle \int_{-1}^{1} x(x^2 + 1)^3\, dx$

54. $\displaystyle \int_{0}^{1} x\sqrt{1 - x^2}\, dx$

55. $\displaystyle \int_{0}^{4} \frac{1}{\sqrt{2x + 1}}\, dx$

56. $\displaystyle \int_{0}^{2} \frac{x}{\sqrt{1 + 2x^2}}\, dx$

57. $\displaystyle \int_{1}^{9} \frac{1}{\sqrt{x}(1 + \sqrt{x})^2}\, dx$

58. $\displaystyle \int_{0}^{2} x\sqrt[3]{4 + x^2}\, dx$

59. $\displaystyle \int_{1}^{2} (x - 1)\sqrt{2 - x}\, dx$

60. $\displaystyle \int_{0}^{4} \frac{x}{\sqrt{2x + 1}}\, dx$

61. $\displaystyle \int_{3}^{7} x\sqrt{x - 3}\, dx$

62. $\displaystyle \int_{0}^{1} \frac{1}{\sqrt{x} + \sqrt{x + 1}}\, dx$

63. $\displaystyle \int_{0}^{7} x\sqrt[3]{x + 1}\, dx$

64. $\displaystyle \int_{-2}^{6} x^2\sqrt[3]{x + 2}\, dx$

65. $\displaystyle\int_1^5 x^2\sqrt{x-1}\,dx$

66. $\displaystyle\int_0^{\pi/2} \sin 2x\,dx$

67. $\displaystyle\int_0^{\pi/2} \cos\left(\frac{2x}{3}\right) dx$

68. $\displaystyle\int_{\pi/3}^{\pi/2} (x+\cos x)\,dx$

69. $\displaystyle\int_{\pi/2}^{2\pi/3} \sec^2\left(\frac{x}{2}\right) dx$

70. $\displaystyle\int_{\pi/12}^{\pi/4} \csc 2x \cot 2x\,dx$

In Exercises 71 and 72, perform the integration in two ways and explain the difference in appearance of the answers.

71. $\displaystyle\int (2x-1)^2\,dx$

72. $\displaystyle\int \sin x \cos x\,dx$

In Exercises 73 and 74, find the area of the region.

73. $y = 2\sin x + \sin 2x$

74. $y = \sin x + \cos 2x$

75. Use the fact that

$$\int_0^2 x^2\,dx = \frac{8}{3}$$

to evaluate the following definite integrals without using the Fundamental Theorem of Calculus.

(a) $\displaystyle\int_{-2}^0 x^2\,dx$

(b) $\displaystyle\int_{-2}^2 x^2\,dx$

(c) $\displaystyle\int_0^2 -x^2\,dx$

(d) $\displaystyle\int_{-2}^0 3x^2\,dx$

76. Use the symmetry of the graphs of the sine and cosine functions as an aid in evaluating the following integrals.

(a) $\displaystyle\int_{-\pi/4}^{\pi/4} \sin x\,dx$

(b) $\displaystyle\int_{-\pi/4}^{\pi/4} \cos x\,dx$

(c) $\displaystyle\int_{-\pi/2}^{\pi/2} \cos x\,dx$

(d) $\displaystyle\int_{-\pi/2}^{\pi/2} \sin x \cos x\,dx$

77. Find the equation of the function f whose graph passes through the point $\left(0, \frac{7}{3}\right)$ and whose derivative is $f'(x) = x\sqrt{1-x^2}$.

78. A lumber company is seeking a model that yields the average weight loss W per log as a function of the number of days of drying time t. The model is to be reliable up to 100 days after the log is cut. Based on the weight loss during the first 30 days, it was determined that

$$\frac{dW}{dt} = \frac{12}{\sqrt{16t+9}}.$$

(a) Find W as a function of t. Note that no weight loss occurs until the tree is cut.

(b) Find the total weight loss after 100 days.

79. The marginal cost for a certain commodity has been determined to be

$$\frac{dC}{dx} = \frac{12}{\sqrt[3]{12x+1}}.$$

(a) Find the cost function if $C = 100$ when $x = 13$.

(b) Graph the marginal cost function and the cost function on the same set of axes.

80. The minimum stockpile level of gasoline in the United States can be approximated by the model

$$Q = 217 + 13\cos\frac{\pi(t-3)}{6}$$

where Q is measured in millions of barrels of gasoline and t is the time in months, with $t = 1$ corresponding to January. Find the average minimum level given by this model during the following periods.

(a) the first quarter $(0 \le t \le 3)$

(b) the second quarter $(3 \le t \le 6)$

(c) the entire year $(0 \le t \le 12)$

81. The sales of a seasonal product are given by the model

$$S = 74.50 + 43.75\sin\frac{\pi t}{6}$$

where S is measured in thousands of units and t is the time in months, with $t = 1$ corresponding to January. Find the average sales for the following periods.

(a) the first quarter $(0 \le t \le 3)$

(b) the second quarter $(3 \le t \le 6)$

(c) the entire year $(0 \le t \le 12)$

82. The oscillating current in an electrical circuit is given by

$$I = 2\sin(60\pi t) + \cos(120\pi t)$$

where I is measured in amperes and t is measured in seconds. Find the average current for the following time intervals.

(a) $0 \le t \le \dfrac{1}{60}$

(b) $0 \le t \le \dfrac{1}{240}$

(c) $0 \le t \le \dfrac{1}{30}$

5.6 Numerical Integration

The Trapezoidal Rule ▪ Simpson's Rule

Occasionally, we encounter functions for which we cannot find antiderivatives. Of course, that may be due to a lack of cleverness on our part. On the other hand, some elementary functions simply do not possess antiderivatives that are elementary functions. For example, there are no elementary functions that have any of the following functions as their derivative.

$$\sqrt[3]{x}\sqrt{1-x} \qquad \sqrt{x}\cos x \qquad \frac{\cos x}{x}$$

$$\sqrt{1-x^3} \qquad \sqrt{1+\sin^2 x} \qquad \sin x^2$$

If we wish to evaluate a definite integral involving a function whose antiderivative we cannot find, then the Fundamental Theorem of Calculus cannot be applied, and we must resort to an approximation technique. We describe two such techniques in this section.

The Trapezoidal Rule

One way to approximate a definite integral is by the use of n trapezoids, as shown in Figure 5.33. In the development of this method, we assume that f is continuous and positive on the interval $[a, b]$, and thus the definite integral $\int_a^b f(x)\,dx$ represents the area of the region bounded by the graph of f and the x-axis, from $x = a$ to $x = b$.

First, we partition the interval $[a, b]$ into n equal subintervals, each of width $\Delta x = (b - a)/n$, such that

$$a = x_0 < x_1 < x_2 < \cdots < x_n = b.$$

We then form trapezoids for each subinterval, as shown in Figure 5.34. The areas of these trapezoids are as follows.

$$\text{area of first trapezoid} = \left[\frac{f(x_0) + f(x_1)}{2}\right]\left(\frac{b-a}{n}\right)$$

$$\text{area of second trapezoid} = \left[\frac{f(x_1) + f(x_2)}{2}\right]\left(\frac{b-a}{n}\right)$$

$$\vdots$$

$$\text{area of } n\text{th trapezoid} = \left[\frac{f(x_{n-1}) + f(x_n)}{2}\right]\left(\frac{b-a}{n}\right)$$

Finally, the sum of the areas of the n trapezoids is

$$\text{area} = \left(\frac{b-a}{n}\right)\left[\frac{f(x_0) + f(x_1)}{2} + \frac{f(x_1) + f(x_2)}{2} + \cdots + \frac{f(x_{n-1}) + f(x_n)}{2}\right]$$

$$= \left(\frac{b-a}{2n}\right)[f(x_0) + \underbrace{f(x_1) + f(x_1)}_{2f(x_1)} + \underbrace{f(x_2) + \cdots}_{2f(x_2)} + \underbrace{f(x_{n-1})}_{2f(x_{n-1})} + f(x_n)]$$

$$= \left(\frac{b-a}{2n}\right)[f(x_0) + 2f(x_1) + 2f(x_2) + \cdots + 2f(x_{n-1}) + f(x_n)].$$

FIGURE 5.33   BLACK

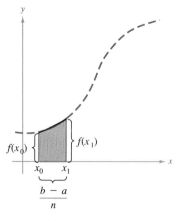

FIGURE 5.34

Letting $\Delta x = (b - a)/n$, we can take the limit as $n \to \infty$, to obtain

$$\lim_{n \to \infty} \left(\frac{b - a}{2n}\right) [f(x_0) + 2f(x_1) + \cdots + 2f(x_{n-1}) + f(x_n)]$$

$$= \lim_{n \to \infty} \left[\frac{[f(a) - f(b)]\Delta x}{2} + \sum_{i=1}^{n} f(x_i)\Delta x\right]$$

$$= \lim_{n \to \infty} \frac{[f(a) - f(b)](b - a)}{2n} + \lim_{n \to \infty} \sum_{i=1}^{n} f(x_i)\Delta x = 0 + \int_a^b f(x)\, dx.$$

This brings us to the following theorem, which we call the Trapezoidal Rule.

THEOREM 5.18
THE TRAPEZOIDAL RULE

Let f be continuous on $[a, b]$. The Trapezoidal Rule for approximating $\int_a^b f(x)\, dx$ is given by

$$\int_a^b f(x)\, dx \approx \frac{b - a}{2n} [f(x_0) + 2f(x_1) + 2f(x_2) + \cdots + 2f(x_{n-1}) + f(x_n)].$$

Moreover, as $n \to \infty$, the right-hand side approaches $\int_a^b f(x)\, dx$.

EXAMPLE 1 Approximation with the Trapezoidal Rule

Use the Trapezoidal Rule to approximate

$$\int_0^{\pi} \sin x\, dx.$$

Compare the results for $n = 4$ and $n = 8$, as shown in Figure 5.35.

SOLUTION

When $n = 4$, we have $\Delta x = \pi/4$, and by the Trapezoidal Rule we have

$$\int_0^{\pi} \sin x\, dx \approx \frac{\pi}{8}\left(\sin 0 + 2\sin\frac{\pi}{4} + 2\sin\frac{\pi}{2} + 2\sin\frac{3\pi}{4} + \sin \pi\right)$$

$$= \frac{\pi}{8}(0 + \sqrt{2} + 2 + \sqrt{2} + 0) = \frac{\pi(1 + \sqrt{2})}{4} \approx 1.896.$$

When $n = 8$, we have $\Delta x = \pi/8$, and

$$\int_0^{\pi} \sin x\, dx \approx \frac{\pi}{16}\left(\sin 0 + 2\sin\frac{\pi}{8} + 2\sin\frac{\pi}{4} + 2\sin\frac{3\pi}{8} + 2\sin\frac{\pi}{2}\right.$$

$$\left. + 2\sin\frac{5\pi}{8} + 2\sin\frac{3\pi}{4} + 2\sin\frac{7\pi}{8} + \sin \pi\right)$$

$$= \frac{\pi}{16}\left(2 + 2\sqrt{2} + 4\sin\frac{\pi}{8} + 4\sin\frac{3\pi}{8}\right) \approx 1.974.$$

For this particular integral, we could have found an antiderivative and determined that the exact area of the region is 2. ▭

Four subintervals

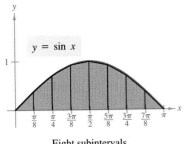

Eight subintervals

FIGURE 5.35

It is interesting to compare the Trapezoidal Rule for approximating definite integrals to the Midpoint Rule given in Section 5.2 (Exercises 57–60). For the Trapezoidal Rule we average the functional values at the endpoints of the subintervals, but for the Midpoint Rule we take the functional value of the subinterval midpoints.

Midpoint Rule

$$\int_a^b f(x)\,dx \approx \sum_{i=1}^n f\left(\frac{x_i + x_{i-1}}{2}\right)\Delta x$$

Trapezoidal Rule

$$\int_a^b f(x)\,dx \approx \sum_{i=1}^n \left(\frac{f(x_i) + f(x_{i-1})}{2}\right)\Delta x$$

There are two important points that should be made concerning the Trapezoidal Rule (or the Midpoint Rule). First, the approximation tends to become more accurate as n increases. For instance, in Example 1, if $n = 16$, the Trapezoidal Rule yields an approximation of 1.994. Second, although we could have used the Fundamental Theorem to evaluate the integral in Example 1, this theorem cannot be used to evaluate an integral such as

$$\int_0^\pi \sin x^2\,dx$$

(even though it looks simple) because $\sin x^2$ has no elementary antiderivative. Yet, the Trapezoidal Rule can be applied readily to this integral.

Simpson's Rule

One way to view the trapezoidal approximation of a definite integral is to say that on each subinterval we approximate f by a *first*-degree polynomial. In Simpson's Rule, named after the English mathematician Thomas Simpson (1710–1761), we take this procedure one step further and approximate f by *second*-degree polynomials.

Before presenting Simpson's Rule, we give a theorem for evaluating integrals of polynomials of degree 2 (or less).

THEOREM 5.19
INTEGRAL OF
$p(x) = Ax^2 + Bx + C$

If $p(x) = Ax^2 + Bx + C$, then

$$\int_a^b p(x)\,dx = \left(\frac{b - a}{6}\right)\left[p(a) + 4p\left(\frac{a + b}{2}\right) + p(b)\right].$$

PROOF

$$\int_a^b p(x)\,dx = \int_a^b (Ax^2 + Bx + C)\,dx = \left[\frac{Ax^3}{3} + \frac{Bx^2}{2} + Cx\right]_a^b$$

$$= \frac{A(b^3 - a^3)}{3} + \frac{B(b^2 - a^2)}{2} + C(b - a)$$

$$= \left(\frac{b - a}{6}\right)[2A(a^2 + ab + b^2) + 3B(b + a) + 6C]$$

Now, by expanding and collecting terms, the expression inside the brackets becomes

$$\underbrace{(Aa^2 + Ba + C)}_{p(a)} + \underbrace{4\left\{A\left(\frac{b+a}{2}\right)^2 + B\left(\frac{b+a}{2}\right) + C\right\}}_{4p\left(\frac{a+b}{2}\right)} + \underbrace{(Ab^2 + Bb + C)}_{p(b)}$$

and we have

$$\int_a^b p(x)\, dx = \left(\frac{b-a}{6}\right)\left[p(a) + 4p\left(\frac{a+b}{2}\right) + p(b)\right].$$

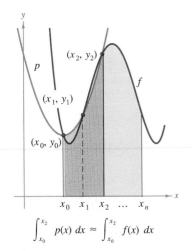

$$\int_{x_0}^{x_2} p(x)\, dx \approx \int_{x_0}^{x_2} f(x)\, dx$$

FIGURE 5.36

To develop Simpson's Rule for approximating a definite integral, we again partition the interval $[a,\, b]$ into n equal subintervals, each of width $\Delta x = (b - a)/n$. This time, however, we require n to be even, and group the subintervals into pairs such that

$$a = \underbrace{x_0 < x_1 < \underbrace{x_2 < x_3 < \underbrace{x_4}} < \cdots < \underbrace{x_{n-2} < x_{n-1} < x_n}} = b.$$
$$\quad \underset{[x_0,\, x_2]}{} \qquad \underset{[x_2,\, x_4]}{} \qquad\qquad\qquad \underset{[x_{n-2},\, x_n]}{}$$

Then on each (double) subinterval $[x_{i-2},\, x_i]$ we approximate f by a polynomial p of degree less than or equal to 2. For example, on the subinterval $[x_0,\, x_2]$, we choose the polynomial of least degree passing through the points $(x_0,\, y_0)$, $(x_1,\, y_1)$, and $(x_2,\, y_2)$, as shown in Figure 5.36. Now, using p as an approximation for f on this subinterval, we have

$$\int_{x_0}^{x_2} f(x)\, dx \approx \int_{x_0}^{x_2} p(x)\, dx = \frac{x_2 - x_0}{6}\left[p(x_0) + 4p\left(\frac{x_2 + x_0}{2}\right) + p(x_2)\right]$$

$$= \frac{2[(b-a)/n]}{6}[p(x_0) + 4p(x_1) + p(x_2)]$$

$$= \frac{b-a}{3n}[f(x_0) + 4f(x_1) + f(x_2)].$$

Repeating this procedure on the entire interval $[a,\, b]$ produces the following theorem.

THEOREM 5.20
SIMPSON'S RULE (n is even)

Let f be continuous on $[a,\, b]$. Simpson's Rule for approximating $\int_a^b f(x)\, dx$ is given by

$$\int_a^b f(x)\, dx \approx \frac{b-a}{3n}[f(x_0) + 4f(x_1) + 2f(x_2) + 4f(x_3) + \cdots + 4f(x_{n-1}) + f(x_n)].$$

Moreover, as $n \to \infty$, the right-hand side approaches $\int_a^b f(x)\, dx$.

REMARK   Note that the coefficients in Simpson's Rule have the following pattern.

$$1 \quad 4 \quad 2 \quad 4 \quad 2 \quad 4 \quad \cdots \quad 4 \quad 2 \quad 4 \quad 1$$

In Example 1 we used the Trapezoidal Rule to estimate $\int_0^\pi \sin x \, dx$. In the next example we see how well Simpson's Rule works for the same integral.

EXAMPLE 2 Approximation with Simpson's Rule

Use Simpson's Rule to approximate

$$\int_0^\pi \sin x \, dx.$$

Compare the results for $n = 4$ and $n = 8$.

SOLUTION

When $n = 4$, we have

$$\int_0^\pi \sin x \, dx \approx \frac{\pi}{12}\left(\sin 0 + 4 \sin \frac{\pi}{4} + 2 \sin \frac{\pi}{2} + 4 \sin \frac{3\pi}{4} + \sin \pi\right)$$

$$= \frac{\pi}{12}(4\sqrt{2} + 2) = \pi\left(\frac{2\sqrt{2} + 1}{6}\right) \approx 2.005.$$

When $n = 8$, we have

$$\int_0^\pi \sin x \, dx \approx \frac{\pi}{24}\left(\sin 0 + 4 \sin \frac{\pi}{8} + 2 \sin \frac{\pi}{4} + 4 \sin \frac{3\pi}{8} + 2 \sin \frac{\pi}{2}\right.$$

$$\left. + 4 \sin \frac{5\pi}{8} + 2 \sin \frac{3\pi}{4} + 4 \sin \frac{7\pi}{8} + \sin \pi\right)$$

$$= \frac{\pi}{24}\left(2 + 2\sqrt{2} + 8 \sin \frac{\pi}{8} + 8 \sin \frac{3\pi}{8}\right) \approx 2.0003.$$

In Examples 1 and 2 we were able to calculate the exact values of the integrals and compare those values to our approximations to see how close they were. In practice, of course, we would not bother with an approximation if it were possible to evaluate the integral exactly. However, if we must use an approximation technique, it is important to know how accurate we can expect the approximation to be. The following theorem, which we list without proof, gives the formulas for estimating the error involved in the use of Simpson's Rule and the Trapezoidal Rule.

THEOREM 5.21
ERROR IN THE TRAPEZOIDAL AND SIMPSON'S RULES

If f has a continuous second derivative on $[a, b]$, then the error E in approximating $\int_a^b f(x) \, dx$ by the Trapezoidal Rule is

$$E \le \frac{(b - a)^3}{12n^2}[\max |f''(x)|], \quad a \le x \le b. \qquad \text{Trapezoidal Rule}$$

Moreover, if f has a continuous fourth derivative on $[a, b]$, then the error E in approximating $\int_a^b f(x) \, dx$ by Simpson's Rule is

$$E \le \frac{(b - a)^5}{180n^4}[\max |f^{(4)}(x)|], \quad a \le x \le b. \qquad \text{Simpson's Rule}$$

Theorem 5.21 states that the errors generated by the Trapezoidal Rule and Simpson's Rule have upper bounds dependent on the extreme values of $f''(x)$ and $f^{(4)}(x)$, respectively, in the interval $[a, b]$. Furthermore, it is evident from this theorem that these errors can be made arbitrarily small by *increasing* n, provided that f'' and $f^{(4)}$ are continuous and therefore bounded in $[a, b]$. The next example shows how to determine a value of n that will bound the error within a predetermined **tolerance.**

EXAMPLE 3   The approximate error in the Trapezoidal Rule

Use the Trapezoidal Rule to estimate the value of

$$\int_0^1 \sqrt{1 + x^2} \, dx.$$

Determine n so that the approximation error is less than 0.01.

SOLUTION

Letting $f(x) = \sqrt{1 + x^2}$, we have

$$f'(x) = x(1 + x^2)^{-1/2}$$

$$f''(x) = x\left(-\frac{1}{2}\right)(2x)(1 + x^2)^{-3/2} + (1 + x^2)^{-1/2}$$

$$= (1 + x^2)^{-3/2}$$

which implies that the maximum value of $|f''(x)|$ on the interval $[0, 1]$ is $|f''(0)| = 1$. Thus, by Theorem 5.21, we can write

$$E \le \frac{(b - a)^3}{12n^2}|f''(0)| \le \frac{1}{12n^2}(1) = \frac{1}{12n^2}.$$

To obtain an error E that is less than 0.01, we must choose n so that $(1/12n^2) \le 1/100$. Thus,

$$100 \le 12n^2 \quad \Longrightarrow \quad 2.89 \approx \sqrt{\frac{100}{12}} \le n.$$

Therefore, we choose $n = 3$ (since n must be greater than or equal to 2.89) and apply the Trapezoidal Rule, as shown in Figure 5.37, to obtain

$$\int_0^1 \sqrt{1 + x^2} \, dx$$

$$\approx \frac{1}{6}\left[\sqrt{1 + 0^2} + 2\sqrt{1 + (1/3)^2} + 2\sqrt{1 + (2/3)^2} + \sqrt{1 + 1^2}\right]$$

$$\approx 1.154.$$

Thus, with an error no larger than 0.01, we know that

$$1.144 \le \int_0^1 \sqrt{1 + x^2} \, dx \le 1.164.$$

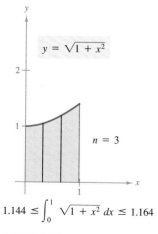

$y = \sqrt{1 + x^2}$

$n = 3$

$$1.144 \le \int_0^1 \sqrt{1 + x^2} \, dx \le 1.164$$

FIGURE 5.37

EXAMPLE 4 The approximate error in Simpson's Rule

Use Simpson's Rule to estimate the value of

$$\int_0^1 \cos x^2 \, dx.$$

Determine n so that the approximation error is less than 0.001.

SOLUTION

According to Theorem 5.21, the error in Simpson's Rule involves the fourth derivative. Hence, by successive differentiation, we have

$$
\begin{aligned}
f(x) &= \cos x^2 \\
f'(x) &= -2x \sin x^2 \\
f''(x) &= -2(2x^2 \cos x^2 + \sin x^2) \\
f'''(x) &= 4(2x^3 \sin x^2 - 3x \cos x^2) \\
f^{(4)}(x) &= 4(4x^4 \cos x^2 + 12x^2 \sin x^2 - 3 \cos x^2) \\
&= 4[(4x^4 - 3)(\cos x^2) + 12x^2 \sin x^2].
\end{aligned}
$$

Since, in the interval $[0, 1]$, $|f^{(4)}(x)|$ is greatest when $x = 1$, we know that

$$|f^{(4)}(x)| \le |f^{(4)}(1)| = 4[\cos 1 + 12 \sin 1] \approx 42.6 < 45.$$

Hence, we have

$$E \le \frac{(b-a)^5}{180n^4} |f^{(4)}(1)| = \frac{1}{180n^4} |f^{(4)}(1)| < \frac{45}{180n^4} = \frac{1}{4n^4}.$$

Now, choosing n so that $(1/4n^4) < 1/1000$, we have

$$\frac{1000}{4} < n^4 \implies 3.98 < n.$$

Therefore, we choose $n = 4$ as shown in Figure 5.38 and obtain

$$\int_0^1 \cos x^2 \, dx \approx \frac{1}{12}\left(\cos 0 + 4 \cos \frac{1}{16} + 2 \cos \frac{1}{4} + 4 \cos \frac{9}{16} + \cos 1\right)$$

$$\approx 0.9045$$

and we conclude that

$$0.9035 \le \int_0^1 \cos x^2 \, dx \le 0.9055.$$

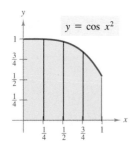

FIGURE 5.38

REMARK You may wonder why we introduced the Trapezoidal Rule, since for a fixed n Simpson's Rule usually gives a more accurate approximation. The main reason is that its error can be estimated more easily than the error involved in Simpson's Rule. For instance, if

$$f(x) = \sqrt{x} \sin (x + 1)$$

then to estimate the error in Simpson's Rule we would need to find the fourth derivative of f—a huge task! Therefore, we may prefer to use the Trapezoidal Rule, even if we have to use a larger n to obtain the desired accuracy.

EXERCISES for Section 5.6

In Exercises 1–10, use the Trapezoidal Rule and Simpson's Rule to approximate the value of the definite integral for the indicated value of n. Round the answer to four decimal places and compare the results with the exact value of the definite integral.

1. $\int_0^2 x^2 \, dx$

2. $\int_0^1 \left(\frac{x^2}{2} + 1 \right) dx$

3. $\int_0^2 x^3 \, dx$

4. $\int_1^2 \frac{1}{x^2} \, dx$

5. $\int_0^2 x^3 \, dx, \ n = 8$

6. $\int_0^8 \sqrt[3]{x} \, dx, \ n = 8$

7. $\int_4^9 \sqrt{x} \, dx, \ n = 8$

8. $\int_1^3 (4 - x^2) \, dx, \ n = 4$

9. $\int_1^2 \frac{1}{(x + 1)^2} \, dx, \ n = 4$

10. $\int_0^2 x\sqrt{x^2 + 1} \, dx, \ n = 4$

In Exercises 11–20, approximate each integral using (a) the Trapezoidal Rule and (b) Simpson's Rule.

11. $\int_0^{\sqrt{\pi/2}} \cos x^2 \, dx, \ n = 4$

12. $\int_0^{\sqrt{\pi/4}} \tan x^2 \, dx, \ n = 4$

13. $\int_0^2 \sqrt{1 + x^3} \, dx, \ n = 2$

14. $\int_0^2 \frac{1}{\sqrt{1 + x^3}} \, dx, \ n = 4$

15. $\int_0^1 \sqrt{x}\sqrt{1 - x} \, dx, \ n = 4$

16. $\int_0^{\pi} f(x) \, dx, \ f(x) = \begin{cases} \dfrac{\sin x}{x}, & x > 0 \\ 1, & x = 0 \end{cases}, \ n = 4$

17. $\int_1^{1.1} \sin x^2 \, dx, \ n = 2$

18. $\int_{\pi/2}^{\pi} \sqrt{x} \sin x \, dx, \ n = 4$

19. $\int_0^{\pi/4} x \tan x \, dx, \ n = 4$

20. $\int_0^{\pi/2} \sqrt{1 + \cos^2 x} \, dx, \ n = 2$

In Exercises 21–24, find the maximum possible error in approximating the given integral by (a) the Trapezoidal Rule and (b) Simpson's Rule.

21. $\int_0^2 x^3 \, dx, \ n = 4$

22. $\int_0^1 \frac{1}{x + 1} dx, \ n = 4$

23. $\int_0^1 \tan x^2 \, dx, \ n = 4$

24. $\int_0^1 \sin x^2 \, dx, \ n = 2$

In Exercises 25–28, find n so that the error in the approximation of the definite integral is less than 0.00001 using (a) the Trapezoidal Rule and (b) Simpson's Rule.

25. $\int_1^3 \frac{1}{x} \, dx$

26. $\int_0^1 \frac{1}{1 + x} \, dx$

27. $\int_0^2 \sqrt{1 + x} \, dx$

28. $\int_0^2 (x + 1)^{2/3} \, dx$

29. Approximate to two decimal places the area of the region bounded by the graphs of $y = \sqrt{x} \cos x$, $y = 0$, $x = 0$, and $x = \pi/2$. (For Simpson's Rule, $n = 14$.)

30. The **elliptic integral**

$$8\sqrt{3} \int_0^{\pi/2} \sqrt{1 - \frac{2}{3} \sin^2 \theta} \, d\theta$$

gives the circumference of an ellipse whose major axis has length $4\sqrt{3}$ and minor axis has length 4. Approximate to three decimal places the circumference of the ellipse. (For Simpson's Rule, use $n = 8$.)

31. Use Simpson's Rule with $n = 6$ to approximate π correct to five decimal places using the equation

$$\pi = \int_0^1 \frac{4}{1 + x^2} \, dx.$$

[In Section 6.8, we will be able to evaluate this integral using the inverse tangent function.]

32. Prove that Simpson's Rule is exact when approximating the integral of a cubic polynomial function, and demonstrate the result for

$$\int_0^1 x^3 \, dx, \ n = 2.$$

In Exercises 33 and 34, use the Trapezoidal Rule to estimate the number of square feet of land in a given lot where x and y are measured in feet, as shown in the accompanying figures. In each case the land is bounded by a stream and two straight roads that meet at right angles.

33.

x	y
0	125
100	125
200	120
300	112
400	90
500	90
600	95
700	88
800	75
900	35
1000	0

34.

x	y
0	75
10	81
20	84
30	76
40	67
50	68
60	69
70	72
80	68
90	56
100	42
110	23
120	0

35. To estimate the surface area of a pond, a surveyor takes several measurements, as shown in the accompanying figure. Use (a) the Trapezoidal Rule and (b) Simpson's Rule to estimate the surface area of this pond. [The measurements are given in feet.]

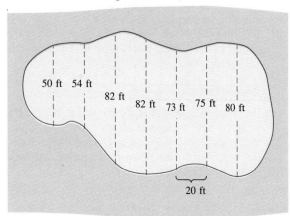

FIGURE FOR 35

36. The following table lists several physical measurements gathered in an experiment. Assume the function $y = f(x)$ is continuous and approximate the integral $\int_0^2 f(x)\, dx$ using (a) the Trapezoidal Rule and (b) Simpson's Rule.

x	0.00	0.25	0.50	0.75	1.00
y	4.32	4.36	4.58	5.79	6.14

x	1.25	1.50	1.75	2.00
y	7.25	7.64	8.08	8.14

In Exercises 37 and 38, use a computer or calculator to complete the following table of numerical approximations to the given integral.

n	Trapezoidal Rule	Simpson's Rule
4		
8		
12		
16		
20		

37. $\displaystyle\int_0^4 \sqrt{2 + 3x^2}\, dx$ **38.** $\displaystyle\int_0^4 \sin \sqrt{x}\, dx$

39. Use a computer or calculator and Simpson's Rule (with $n = 10$) to approximate t to three decimal places in the integral equation

$$\int_0^t \sin \sqrt{x}\, dx = 2.$$

REVIEW EXERCISES for Chapter 5

In Exercises 1–18, find the indefinite integral.

1. $\int \dfrac{2}{3\sqrt[3]{x}}\, dx$

2. $\int \dfrac{2}{\sqrt[3]{3x}}\, dx$

3. $\int (2x^2 + x - 1)\, dx$

4. $\int \dfrac{x^3 - 2x^2 + 1}{x^2}\, dx$

5. $\int \dfrac{(1 + x)^2}{\sqrt{x}}\, dx$

6. $\int x^2\sqrt{x^3 + 3}\, dx$

7. $\int \dfrac{x^2}{\sqrt{x^3 + 3}}\, dx$

8. $\int \dfrac{x^2 + 2x}{(x + 1)^2}\, dx$

9. $\int (x^2 + 1)^3\, dx$

10. $\int \sqrt{2 - 5x}\, dx$

11. $\int \sin^3 x \cos x\, dx$

12. $\int \dfrac{\cos x}{\sqrt{\sin x}}\, dx$

13. $\int \tan^n x \sec^2 x\, dx,\ n \neq -1$

14. $\int \dfrac{\sin \theta}{\sqrt{1 - \cos \theta}}\, d\theta$

15. $\int \sec 2x \tan 2x\, dx$

16. $\int x \sin 3x^2\, dx$

17. $\int \dfrac{x^3 + 1}{x^2}\, dx$

18. $\int \left(x + \dfrac{1}{x}\right)^2\, dx$

19. Write in sigma notation the sum of the following.
(a) the first ten positive odd integers
(b) the cubes of the first n positive integers
(c) $6 + 10 + 14 + 18 + \cdots + 42$

20. Evaluate the following sums for $x_1 = 2$, $x_2 = -1$, $x_3 = 5$, $x_4 = 3$, and $x_5 = 7$.
(a) $\dfrac{1}{5}\sum_{i=1}^{5} x_i$
(b) $\sum_{i=1}^{5} \dfrac{1}{x_i}$
(c) $\sum_{i=1}^{5} (2x_i - x_i^2)$
(d) $\sum_{i=2}^{5} (x_i - x_{i-1})$

In Exercises 21–30, use the Fundamental Theorem of Calculus to evaluate the definite integral.

21. $\displaystyle\int_0^4 (2 + x)\, dx$

22. $\displaystyle\int_{-1}^1 (t^2 + 2)\, dt$

23. $\displaystyle\int_{-1}^1 (4t^3 - 2t)\, dt$

24. $\displaystyle\int_3^6 \dfrac{x}{3\sqrt{x^2 - 8}}\, dx$

25. $\displaystyle\int_0^3 \dfrac{1}{\sqrt{1 + x}}\, dx$

26. $\displaystyle\int_0^1 x^2(x^3 + 1)^3\, dx$

27. $\displaystyle\int_4^9 x\sqrt{x}\, dx$

28. $\displaystyle\int_1^2 \left(\dfrac{1}{x^2} - \dfrac{1}{x^3}\right)\, dx$

29. $2\pi \displaystyle\int_0^1 (y + 1)\sqrt{1 - y}\, dy$

30. $2\pi \displaystyle\int_{-1}^0 x^2\sqrt{x + 1}\, dx$

31. Find the function f whose derivative is $f'(x) = -2x$ and whose graph passes through the point $(-1, 1)$.

32. A function f has a second derivative $f''(x) = 6(x - 1)$. Find the function if its graph passes through the point $(2, 1)$ and at that point is tangent to the line given by $3x - y - 5 = 0$.

33. An airplane taking off from a runway travels 3600 feet before lifting off. If it starts from rest, moves with constant acceleration, and makes the run in 30 seconds, with what speed does it lift off?

34. The speed of a car traveling in a straight line is reduced from 45 to 30 miles per hour in a distance of 264 feet. Find the distance in which the car can be brought to rest from 30 miles per hour, assuming the same constant acceleration.

35. A ball is thrown vertically upward from ground level with an initial velocity of 96 feet per second.
(a) How long will it take to rise to its maximum height?
(b) What is the maximum height?
(c) When is the velocity of the ball one-half the initial velocity?
(d) What is the height of the ball when its velocity is one-half the initial velocity?

36. Repeat Exercise 35 for an initial velocity of 128 feet per second.

37. Consider the region bounded by $y = mx$, $y = 0$, $x = 0$, and $x = b$.
(a) Find the upper and lower sum to approximate the area of the region when $\Delta x = b/4$.
(b) Find the upper and lower sum to approximate the area of the region when $\Delta x = b/n$.
(c) Find the area of the region by letting n approach infinity in both sums of part (b). Show that in each case you obtain the formula for the area of a triangle.
(d) Find the area of the region by using the Fundamental Theorem of Calculus.

38. (a) Find the area of the region bounded by the graphs of $y = x^3$, $y = 0$, $x = 1$, and $x = 3$ by the limit definition.
(b) Find the area of the given region by using the Fundamental Theorem of Calculus.

In Exercises 39–44, sketch the graph of the region whose area is given by the integral and find the area.

39. $\displaystyle\int_1^3 (2x - 1)\, dx$

40. $\displaystyle\int_0^2 (x + 4)\, dx$

41. $\displaystyle\int_3^4 (x^2 - 9)\, dx$

42. $\displaystyle\int_{-1}^2 (-x^2 + x + 2)\, dx$

43. $\displaystyle\int_0^1 (x - x^3)\, dx$

44. $\displaystyle\int_0^1 \sqrt{x}(1 - x)\, dx$

In Exercises 45–48, find the average value of the function over the given interval. Find the values of x where the function assumes its mean value and sketch the graph of the function.

Function	Interval
45. $f(x) = \dfrac{1}{\sqrt{x-1}}$	$[5, 10]$
46. $f(x) = x^3$	$[0, 2]$
47. $f(x) = x$	$[0, 4]$
48. $f(x) = x^2 - \dfrac{1}{x^2}$	$[1, 2]$

In Exercises 49 and 50, use Simpson's Rule (with $n = 4$) to approximate the definite integral.

49. $\displaystyle\int_1^2 \frac{1}{1 + x^3}\, dx$

50. $\displaystyle\int_0^1 \frac{x^{3/2}}{3 - x^2}\, dx$

51. For a person at rest, the rate of air intake v, in liters per second, during a respiratory cycle is

$$v = 0.85 \sin \frac{\pi t}{3}$$

where t is the time in seconds. Find the volume in liters of air inhaled during one cycle by integrating this function over the interval $[0, 3]$.

52. After exercising a few minutes, a person has a respiratory cycle for which the rate of air intake is

$$v = 1.75 \sin \frac{\pi t}{2}.$$

How much does the person's lung capacity increase as a result of exercising? (Compare your answer to that found in Exercise 51.)

53. The temperature in degrees Fahrenheit is given by

$$T = 72 + 12 \sin \frac{\pi(t - 8)}{12}$$

where t is the time in hours, with $t = 0$ representing midnight. Suppose the hourly cost of cooling a house is \$0.10 per degree.

(a) Find the cost C of cooling this house if its thermostat is set at 72° by evaluating the integral

$$C = 0.1 \int_8^{20} \left[72 + 12 \sin \frac{\pi(t - 8)}{12} - 72 \right] dt.$$

(See figure.)

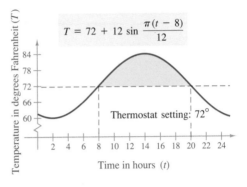

FIGURE FOR 53(a)

(b) Find the savings in resetting the thermostat to 78° by evaluating the integral

$$C = 0.1 \int_{10}^{18} \left[72 + 12 \sin \frac{\pi(t - 8)}{12} - 78 \right] dt.$$

(See figure.)

FIGURE FOR 53(b)

54. A manufacturer of fertilizer finds that national sales of fertilizer roughly follow the seasonal pattern

$$F = 100{,}000 \left[1 + \sin \frac{2\pi(t - 60)}{365} \right]$$

where F is measured in pounds and t is the time in days, with $t = 1$ representing January 1. The manufacturer wants to set up a schedule to produce a uniform amount each day. What should this amount be?

55. Suppose that gasoline is increasing in price according to the equation

$$p = 1 + 0.1t + 0.02t^2$$

where p is the dollar price per gallon and $t = 0$ represents the year 1983. If an automobile is driven 15,000 miles a year and gets M miles per gallon, then the annual fuel cost is

$$C = \frac{15,000}{M} \int_t^{t+1} p \; dt.$$

Find the annual fuel cost for the years (a) 1985 and (b) 1990.

In Exercises 56 and 57, the function

$$f(x) = kx^n(1 - x)^m, \qquad 0 \le x \le 1$$

where $n, m > 0$ and k is a constant, can be used to represent various probability distributions. If k is chosen so that

$$\int_0^1 f(x) \; dx = 1$$

the probability that x will fall between a and b is given by

$$P_{a,b} = \int_a^b f(x) \; dx.$$

56. The probability of recall in a certain experiment is found to be

$$P_{a,b} = \int_a^b \frac{15}{4} x \sqrt{1 - x} \; dx$$

where x represents the percentage of recall (see figure).

(a) For a randomly chosen individual, what is the probability that he or she will recall between 50% and 75% of the material?

(b) What is the median percentage recall? That is, for what value of b is it true that the probability from 0 to b is 0.5?

FIGURE FOR 56

FIGURE FOR 57

57. The probability of finding between a and b percentage of iron in ore samples taken from a certain region is given by

$$P_{a,b} = \int_a^b \frac{1155}{32} x^3 (1 - x)^{3/2} \; dx.$$

(See figure.) What is the probability that a sample will contain between

(a) 0% and 25%? (b) 50% and 100%?

Chapter 6 Application

The Gateway Arch, St. Louis, was designed by Eero Saarinen and completed in 1965. It is over 60 stories high, covered with over 900 tons of quarter-inch stainless steel, and has a total weight of over 16,000 tons. It is the tallest national monument of its kind: 75 feet taller than the Washington Monument and 175 feet taller than the Statue of Liberty.

The Catenary

When a cable is suspended between two points, it forms a curve called a catenary. The equation for a catenary

$$y = a \cosh \frac{x}{a} = a\left(\frac{e^{x/a} + e^{-x/a}}{2}\right)$$

involves the hyperbolic cosine. Telephone cables and clotheslines take this shape. Such a free-hanging cable has all of its internal forces in equilibrium. Because of this, a catenary (turned upside down) makes a perfect arch—all of its thrust passes through the legs to the base of the arch.

The Gateway Arch in St. Louis, Missouri is one example of an arch built in the form of a catenary. Its equation is

$$y \approx 757.71 - 127.71 \cosh \frac{x}{127.71}.$$

The height of the Gateway Arch is the same as the distance between its two legs—630 feet. To see this, note that $y \approx 630$ when $x = 0$, and $y \approx 0$ when $x = \pm 315$.

Chapter Overview

At this point in the text, we have discussed two types of elementary functions—algebraic functions and trigonometric functions. This chapter concludes our introduction of elementary functions. As each new type is introduced, we will develop its properties, derivative, and antiderivative.

The first two sections develop the **natural logarithmic function**

$$f(x) = \ln x$$

as the antiderivative of $1/x$. The following section discusses inverse functions, and this material is then used in Section 6.3 to develop properties of the **natural exponential function**

$$f(x) = e^x.$$

Section 6.5 introduces exponential and logarithmic functions to bases other than e. Section 6.5 also discusses compound interest and several other applications involving exponential functions. Then, in Section 6.6, we show how the exponential function can be used to solve problems involving growth and decay.

Sections 6.7 and 6.8 introduce the **inverse trigonometric functions,** and the last section in the chapter gives a brief introduction to the **hyperbolic functions.**

See Exercise 73, Section 6.9.

324

6

Logarithmic, Exponential, and Other Transcendental Functions

6.1 The Natural Logarithmic Function and Differentiation

The natural logarithmic function ▪ The number e ▪ The derivative of the natural logarithmic function ▪ Logarithmic differentiation

Recall that the Power Rule

$$\int x^n \, dx = \frac{x^{n+1}}{n+1} + C, \quad n \neq -1$$

has an important disclaimer—it doesn't apply when $n = -1$. Consequently, we have not yet found an antiderivative for the function $f(x) = 1/x$. But now, by using the Second Fundamental Theorem of Calculus, we can *define* such a function. This function is one we have not encountered previously in the text. It is neither algebraic nor trigonometric, but falls into a new class of functions called *logarithmic functions*. We call this particular function the **natural logarithmic function,** and you will see that it is the same logarithmic function that you studied in algebra. Logarithms were invented by the Scottish mathematician John Napier (1550–1617). Although he did not introduce the *natural* logarithmic function, it is sometimes called the *Naperian* logarithm.

John Napier

DEFINITION OF NATURAL LOGARITHMIC FUNCTION

The **natural logarithmic function** is

$$\ln x = \int_1^x \frac{1}{t} \, dt, \quad 0 < x.$$

325

REMARK Note that the function $f(t) = 1/t$ is continuous at every real number except $t = 0$. Thus, we can apply the Second Fundamental Theorem of Calculus to define an antiderivative on the interval $(-\infty, 0)$ or $(0, \infty)$. By convention, we choose the domain of the natural logarithmic function to be the set of positive real numbers.

We can study the properties of the natural logarithmic function from two points of view. As the *definite integral* of $f(t) = 1/t$ from 1 to x, we can tell that $\ln x$ is positive for $x > 1$ and negative for $0 < x < 1$, as shown in Figure 6.1. Moreover, we can see that $\ln (1) = 0$, since the upper and lower limits of integration are equal when $x = 1$.

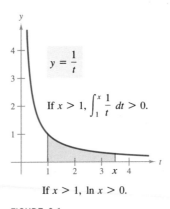

If $x > 1$, $\ln x > 0$.

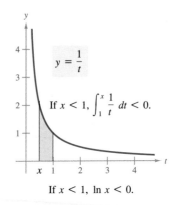

If $x < 1$, $\ln x < 0$.

FIGURE 6.1

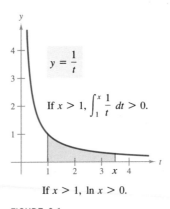

FIGURE 6.2

To picture the graph of $y = \ln x$, we can think of the natural logarithmic function as an *antiderivative* given by the equation

$$\frac{dy}{dx} = \frac{1}{x}.$$

Figure 6.2 is a computer-generated graph showing small line segments of slope $1/x$. The particular solution we want is the curve that passes through the point $(1, 0)$.

We list some basic properties of the natural logarithmic function in the following theorem.

THEOREM 6.1
PROPERTIES OF THE NATURAL LOGARITHMIC FUNCTION

The natural logarithmic function has the following properties.

1. The domain is $(0, \infty)$ and the range is $(-\infty, \infty)$.
2. The function is continuous, increasing, and one-to-one.
3. The graph is concave downward.

PROOF

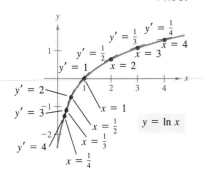

FIGURE 6.3

The domain of $f(x) = \ln x$ is $(0, \infty)$ by definition. Moreover, the function is continuous since it is differentiable. It is increasing since its derivative $f'(x) = 1/x$ is positive for $x > 0$, as shown in Figure 6.3. It is concave downward since $f''(x) = -1/x^2$ is negative. We leave the proof that f is one-to-one as an exercise (see Exercise 80). Since f is continuous, we can see by verifying the following limits that its range is the entire real line.

$$\lim_{x \to 0^+} \ln x = -\infty \qquad \text{and} \qquad \lim_{x \to \infty} \ln x = \infty$$

Verification of these two limits is given in Appendix A.

By using the definition of the natural logarithmic function, we are able to prove some important properties involving operations with natural logarithms. If you are already familiar with logarithms, you will recognize that these properties are characteristic of all logarithms.

**THEOREM 6.2
LOGARITHMIC PROPERTIES**

If a and b are positive numbers and n is rational, then the following properties are true.

1. $\ln (1) = 0$
2. $\ln (ab) = \ln a + \ln b$
3. $\ln (a^n) = n \ln a$
4. $\ln \left(\dfrac{a}{b}\right) = \ln a - \ln b$

PROOF

We have already discussed the first property. The proof of the second property follows from the fact that two antiderivatives of the same function differ at most by a constant. From the Second Fundamental Theorem of Calculus and the definition of the natural logarithmic function, we know that

$$\frac{d}{dx} [\ln x] = \frac{1}{x}.$$

Thus, we consider the two derivatives

$$\frac{d}{dx} [\ln (ax)] = \frac{a}{ax} = \frac{1}{x} \qquad \text{and} \qquad \frac{d}{dx} [\ln a + \ln x] = 0 + \frac{1}{x} = \frac{1}{x}.$$

Since $\ln (ax)$ and $(\ln a + \ln x)$ are both antiderivatives of $1/x$, they must differ at most by a constant

$$\ln (ax) = \ln a + \ln x + C.$$

By letting $x = 1$, we see that $C = 0$. The third property can be proved similarly by comparing the derivatives of $\ln (x^n)$ and $n \ln x$. Finally, using the second and third properties, we prove the fourth property:

$$\ln \left(\frac{a}{b}\right) = \ln [a(b^{-1})] = \ln a + \ln (b^{-1}) = \ln a - \ln b.$$

EXAMPLE 1 Expanding logarithmic expressions

(a) $\ln \dfrac{10}{9} = \ln 10 - \ln 9$ Property 4

(b) $\ln \sqrt{3x + 2} = \ln (3x + 2)^{1/2} = \dfrac{1}{2} \ln (3x + 2)$ Property 3

(c) $\ln \dfrac{6x}{5} = \ln (6x) - \ln 5 = \ln 6 + \ln x - \ln 5$ Properties 2 and 4

(d) $\ln \dfrac{(x^2 + 3)^2}{x\sqrt[3]{x^2 + 1}} = \ln (x^2 + 3)^2 - \ln (x\sqrt[3]{x^2 + 1})$

$$= 2 \ln (x^2 + 3) - [\ln x + \ln (x^2 + 1)^{1/3}]$$
$$= 2 \ln (x^2 + 3) - \ln x - \ln (x^2 + 1)^{1/3}$$
$$= 2 \ln (x^2 + 3) - \ln x - \frac{1}{3} \ln (x^2 + 1)$$

The number *e*

It is likely that you have previously studied logarithms in an algebra course. There, without the benefit of calculus, logarithms would have been defined in terms of a **base** number. For example, common logarithms have a base of 10 because $\log_{10} 10 = 1$. (We will say more about this in Section 6.5.)

To define the **base for the natural logarithm,** we use the fact that the natural logarithmic function is continuous, one-to-one, and has a range of $(-\infty, \infty)$. Hence, there must be a unique real number x such that $\ln x = 1$, as shown in Figure 6.4. We denote this number by the letter e. It can be shown that e is irrational and has the following decimal approximation.

$$e \approx 2.71828182846$$

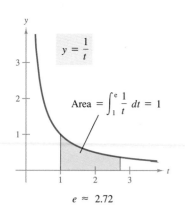

$y = \dfrac{1}{t}$

$\text{Area} = \displaystyle\int_1^e \frac{1}{t}\, dt = 1$

$e \approx 2.72$

FIGURE 6.4

DEFINITION OF *e*

The letter e denotes the positive real number such that

$$\ln e = \int_1^e \frac{1}{t}\, dt = 1.$$

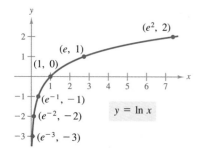

$(e^2, 2)$

$(e, 1)$

$(1, 0)$

$(e^{-1}, -1)$

$(e^{-2}, -2)$

$y = \ln x$

$(e^{-3}, -3)$

FIGURE 6.5

Once we know that $\ln e = 1$, we can use logarithmic properties to evaluate the natural logarithm of several other numbers. For example, by using the property

$$\ln (e^n) = n \ln e = n(1) = n$$

we can evaluate $\ln (e^n)$ for various powers of n, as shown in Table 6.1 and Figure 6.5.

TABLE 6.1

x	$\frac{1}{e^3} \approx 0.050$	$\frac{1}{e^2} \approx 0.135$	$\frac{1}{e} \approx 0.368$	$e^0 = 1$	$e \approx 2.718$	$e^2 \approx 7.389$
$\ln x$	-3	-2	-1	0	1	2

REMARK   The logarithms given in Table 6.1 are convenient because the x-values are integer powers of e. Of course, for most logarithms you would use a calculator. For example, $\ln (2) \approx 0.693$ and $\ln (0.1) \approx -2.303$.

The natural logarithmic function and differentiation

The derivative of the natural logarithmic function is given in the following theorem.

**THEOREM 6.3
DERIVATIVE OF THE NATURAL
LOGARITHMIC FUNCTION**

Let u be a differentiable function of x.

1. $\dfrac{d}{dx} [\ln x] = \dfrac{1}{x}, \quad x > 0$ 2. $\dfrac{d}{dx} [\ln u] = \dfrac{1}{u} \dfrac{du}{dx} = \dfrac{u'}{u}, \quad u > 0$

PROOF   The first part of Theorem 6.3 follows from the definition of the natural logarithmic function as an antiderivative and the Second Fundamental Theorem of Calculus. The second part of the theorem is simply the Chain Rule version of the first part.

EXAMPLE 2   Differentiation of logarithmic functions

(a) $\dfrac{d}{dx} [\ln (2x)] = \dfrac{u'}{u} = \dfrac{2}{2x} = \dfrac{1}{x}$ $\qquad\qquad u = 2x$

(b) $\dfrac{d}{dx} [\ln (x^2 + 1)] = \dfrac{u'}{u} = \dfrac{2x}{x^2 + 1}$ $\qquad\qquad u = x^2 + 1$

(c) $\dfrac{d}{dx} [x \ln x] = x\left(\dfrac{d}{dx} [\ln x]\right) + (\ln x)\left(\dfrac{d}{dx} [x]\right)$ $\qquad$ Product Rule

$\qquad\qquad\quad = x\left(\dfrac{1}{x}\right) + (\ln x)(1) = 1 + \ln x$ ▭

Napier used logarithmic properties to simplify *calculations* involving products, quotients, and powers. Of course, given the availability of calculators, we have little need for this particular application of logarithms. However, we do find a great value in using logarithmic properties to simplify *differentiation* involving products, quotients, and powers. The following examples illustrate this point.

EXAMPLE 3   Logarithmic properties as an aid to differentiation

Differentiate $f(x) = \ln \sqrt{x + 1}$.

SOLUTION

Since

$$f(x) = \ln \sqrt{x + 1} = \ln (x + 1)^{1/2} = \frac{1}{2} \ln (x + 1)$$

we have

$$f'(x) = \frac{1}{2}\left(\frac{1}{x + 1}\right) = \frac{1}{2(x + 1)}.$$

EXAMPLE 4   Logarithmic properties as an aid to differentiation

Differentiate

$$f(x) = \ln \frac{x(x^2 + 1)^2}{\sqrt{2x^3 - 1}}.$$

SOLUTION

Since

$$f(x) = \ln \frac{x(x^2 + 1)^2}{\sqrt{2x^3 - 1}} = \ln x + 2 \ln (x^2 + 1) - \frac{1}{2} \ln (2x^3 - 1)$$

we have

$$f'(x) = \frac{1}{x} + 2\left(\frac{2x}{x^2 + 1}\right) - \frac{1}{2}\left(\frac{6x^2}{2x^3 - 1}\right)$$

$$= \frac{1}{x} + \frac{4x}{x^2 + 1} - \frac{3x^2}{2x^3 - 1}.$$

REMARK   In Examples 3 and 4 be sure to see the benefit in applying logarithmic properties *before* differentiating. Consider, for instance, the difficulty of direct differentiation of the function given in Example 4.

On occasion, it is convenient to use logarithms as an aid in differentiating *non*logarithmic functions. We call this procedure **logarithmic differentiation,** and we illustrate its use in the next example.

EXAMPLE 5   Logarithmic differentiation

Find the derivative of

$$y = \frac{(x-2)^2}{\sqrt{x^2+1}}.$$

SOLUTION

We begin by taking the natural logarithm of both sides of the equation. Then we apply logarithmic properties and differentiate implicitly. Finally, we solve for y'.

$$\ln y = \ln \frac{(x-2)^2}{\sqrt{x^2+1}} \qquad \text{Take ln of both sides}$$

$$\ln y = 2 \ln (x-2) - \frac{1}{2} \ln (x^2+1) \qquad \text{Logarithmic properties}$$

$$\frac{y'}{y} = 2\left(\frac{1}{x-2}\right) - \frac{1}{2}\left(\frac{2x}{x^2+1}\right) \qquad \text{Differentiate}$$

$$= \frac{2}{x-2} - \frac{x}{x^2+1}$$

$$y' = y\left(\frac{2}{x-2} - \frac{x}{x^2+1}\right) \qquad \text{Solve for } y'$$

$$= \frac{(x-2)^2}{\sqrt{x^2+1}}\left[\frac{x^2+2x+2}{(x-2)(x^2+1)}\right] \qquad \text{Substitute for } y$$

$$= \frac{(x-2)(x^2+2x+2)}{(x^2+1)^{3/2}} \qquad \text{Simplify}$$

Since the natural logarithm is undefined for negative numbers, we often encounter expressions of the form $\ln |u|$. The following theorem tells us that we can differentiate functions of the form $y = \ln |u|$ as if the absolute value sign were not present.

THEOREM 6.4 **DERIVATIVE INVOLVING** **ABSOLUTE VALUE**	If u is a differentiable function of x such that $u \neq 0$, then $$\frac{d}{dx}[\ln |u|] = \frac{u'}{u}.$$

PROOF   If $u > 0$, then $|u| = u$, and the result follows from Theorem 6.3. If $u < 0$, then $|u| = -u$, and we have

$$\frac{d}{dx}[\ln |u|] = \frac{d}{dx}[\ln (-u)] = \frac{-u'}{-u} = \frac{u'}{u}.$$

EXAMPLE 6 Derivative involving absolute value

Find the derivative of $f(x) = \ln |\cos x|$.

SOLUTION

Using Theorem 6.4, we let $u = \cos x$ and write

$$\frac{d}{dx}[\ln |\cos x|] = \frac{u'}{u} = \frac{-\sin x}{\cos x} = -\tan x.$$

EXAMPLE 7 Finding relative extrema

Locate the relative extrema of $y = \ln (x^2 + 2x + 3)$.

SOLUTION

Differentiating y, we obtain

$$\frac{dy}{dx} = \frac{2x + 2}{x^2 + 2x + 3}.$$

Now, since $dy/dx = 0$ when $x = -1$, we apply the First Derivative Test and conclude that the point $(-1, \ln 2)$ is a relative minimum. Since there are no other critical points, we conclude that this is the only relative extremum. (See Figure 6.6.)

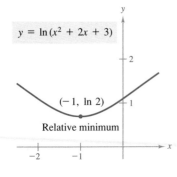

FIGURE 6.6

EXERCISES for Section 6.1

In Exercises 1–6, sketch the graph of the function.

1. $f(x) = 3 \ln x$ **2.** $f(x) = -2 \ln x$
3. $f(x) = \ln 2x$ **4.** $f(x) = \ln |x|$
5. $f(x) = \ln (x - 1)$ **6.** $g(x) = 2 + \ln x$

In Exercises 7 and 8, use the properties of logarithms to approximate the indicated logarithm given that $\ln 2 \approx 0.6931$ and $\ln 3 \approx 1.0986$.

7. (a) $\ln 6$ (b) $\ln \dfrac{2}{3}$
    (c) $\ln 81$ (d) $\ln \sqrt{3}$
8. (a) $\ln 0.25$ (b) $\ln 24$
    (c) $\ln \sqrt[3]{12}$ (d) $\ln \dfrac{1}{72}$

In Exercises 9–18, use the properties of logarithms to write each expression as a sum, difference, or multiple of logarithms.

9. $\ln \dfrac{2}{3}$

10. $\ln (xyz)$

11. $\ln \dfrac{xy}{z}$

12. $\ln \sqrt{a - 1}$

13. $\ln \sqrt{2^3}$

14. $\ln \dfrac{1}{5}$

15. $\ln \left(\dfrac{x^2 - 1}{x^3}\right)^3$

16. $\ln 3e^2$

17. $\ln z(z - 1)^2$

18. $\ln \dfrac{1}{e}$

In Exercises 19–24, write each expression as a logarithm of a single quantity.

19. $\ln (x - 2) - \ln (x + 2)$
20. $3 \ln x + 2 \ln y - 4 \ln z$
21. $\dfrac{1}{3}[2 \ln (x + 3) + \ln x - \ln (x^2 - 1)]$
22. $2[\ln x - \ln (x + 1) - \ln (x - 1)]$
23. $2 \ln 3 - \dfrac{1}{2} \ln (x^2 + 1)$

24. $\frac{3}{2}[\ln (x^2 + 1) - \ln (x + 1) - \ln (x - 1)]$

In Exercises 25–28, find the slope of the tangent line to the given logarithmic function at the point (1, 0).

25. $y = \ln x^3$

26. $y = \ln x^{3/2}$

27. $y = \ln x^2$

28. $y = \ln x^{1/2}$

In Exercises 29–54, find dy/dx.

29. $y = \ln x^2$

30. $y = \ln (x^2 + 3)$

31. $y = \ln \sqrt{x^4 - 4x}$

32. $y = \ln (1 - x)^{3/2}$

33. $y = (\ln x)^4$

34. $y = x \ln x$

35. $y = \ln (x\sqrt{x^2 - 1})$

36. $y = \ln \left(\frac{x}{x + 1}\right)$

37. $y = \ln \left(\frac{x}{x^2 + 1}\right)$

38. $y = \frac{\ln x}{x}$

39. $y = \frac{\ln x}{x^2}$

40. $y = \ln (\ln x)$

41. $y = \ln (\ln x^2)$

42. $y = \ln \sqrt{\frac{x - 1}{x + 1}}$

43. $y = \ln \sqrt{\frac{x + 1}{x - 1}}$

44. $y = \ln \sqrt{x^2 - 4}$

45. $y = \ln \left(\frac{\sqrt{4 + x^2}}{x}\right)$

46. $y = \ln (x + \sqrt{4 + x^2})$

47. $y = \frac{-\sqrt{x^2 + 1}}{x} + \ln (x + \sqrt{x^2 + 1})$

48. $y = \frac{-\sqrt{x^2 + 4}}{2x^2} - \frac{1}{4} \ln \left(\frac{2 + \sqrt{x^2 + 4}}{x}\right)$

49. $y = \ln |\sin x|$

50. $y = \ln |\sec x|$

51. $y = \ln \left|\frac{\cos x}{\cos x - 1}\right|$

52. $y = \ln |\sec x + \tan x|$

53. $y = \ln \left|\frac{-1 + \sin x}{2 + \sin x}\right|$

54. $y = \ln \sqrt{1 + \sin^2 x}$

In Exercises 55–60, find dy/dx using logarithmic differentiation.

55. $y = x\sqrt{x^2 - 1}$

56. $y = \sqrt{(x - 1)(x - 2)(x - 3)}$

57. $y = \frac{x^2\sqrt{3x - 2}}{(x - 1)^2}$

58. $y = \sqrt[3]{\frac{x^2 + 1}{x^2 - 1}}$

59. $y = \frac{x(x - 1)^{3/2}}{\sqrt{x + 1}}$

60. $y = \frac{(x + 1)(x + 2)}{(x - 1)(x - 2)}$

In Exercises 61 and 62, show that the given function is a solution to the differential equation.

Function	Differential equation
61. $y = 2 \ln x + 3$	$xy'' + y' = 0$
62. $y = x \ln x - 4x$	$x + y - xy' = 0$

In Exercises 63 and 64, find dy/dx by using implicit differentiation.

63. $x^2 - 3 \ln y + y^2 = 10$

64. $\ln xy + 5x = 30$

In Exercises 65 and 66, find an equation of the tangent line to the graph of the equation at the given point.

Equation	Point
65. $y = 3x^2 - \ln x$	(1, 3)
66. $x^2 + \ln (x + 1) + y^2 = 4$	(0, 2)

In Exercises 67–72, find any relative extrema and inflection points, and sketch the graph of the function.

67. $y = \frac{x^2}{2} - \ln x$

68. $y = x - \ln x$

69. $y = x \ln x$

70. $y = \frac{\ln x}{x}$

71. $y = \frac{x}{\ln x}$

72. $y = x^2 \ln x$

In Exercises 73 and 74, use Newton's Method to approximate, to three decimal places, the x-coordinate of the point of intersection of the graphs of the two equations.

73. $y = \ln x,$ $y = -x$

74. $y = \ln x,$ $y = 3 - x$

75. Apply the Mean Value Theorem to the function $f(x) = \ln x$ on the closed interval $[1, e]$. Find the value of c in the open interval $(1, e)$ such that

$$f'(c) = \frac{f(e) - f(1)}{e - 1}.$$

76. Use Simpson's Rule with $n = 4$ to show that

$$\int_1^{2.7} \frac{1}{t}\, dt < 1 < \int_1^{2.8} \frac{1}{t}\, dt.$$

Then use this given inequality to show that

$$\ln 2.7 < \ln e < \ln 2.8$$

and therefore $2.7 < e < 2.8$.

77. Show that

$$f(x) = \frac{\ln x^n}{x}$$

is a decreasing function for $x > e$ and $n > 0$.

78. A person walking along a dock drags a boat by a 10-foot rope. The boat travels along a path known as a *tractrix* (see figure). The equation of this path is

$$y = 10 \ln\left(\frac{10 + \sqrt{100 - x^2}}{x}\right) - \sqrt{100 - x^2}.$$

What is the slope of this path at the following x-values?

(a) $x = 10$ (b) $x = 5$

Tractrix

FIGURE FOR 78

79. There are 25 prime numbers less than 100. The **Prime Number Theorem** states that if $p(x)$ is the number of primes less than x, then the ratio of $p(x)$ to $x/(\ln x)$ approaches 1 as x approaches infinity. Compute $x/(\ln x)$ for $x = 1000$, $x = 1,000,000$, and $x = 1,000,000,000$. Then compute the ratio of $p(x)$ to $x/(\ln x)$ given that $p(1000) = 168$, $p(10^6) = 78,498$, and $p(10^9) = 50,847,478$.

80. Prove that the natural logarithmic function is one-to-one.

In Exercises 81 and 82, show that $f = g$ by using a computer to sketch the graph of f and g on the same coordinate axes. (Assume $x > 0$.)

81. $f(x) = \ln \dfrac{x^2}{4}$

$g(x) = 2 \ln x - \ln 4$

82. $f(x) = \ln \sqrt{x(x^2 + 1)}$

$g(x) = \dfrac{1}{2}[\ln x + \ln (x^2 + 1)]$

In Exercises 83 and 84, use a computer or calculator and Simpson's Rule with $n = 10$ to approximate the integral

$$\int_1^x \frac{1}{t}\, dt$$

and compare the result with $\ln x$ for the specified value of x.

83. $x = 3$ **84.** $x = 8.7$

6.2 The Natural Logarithmic Function and Integration

Log Rule for integration ▪ Integrals of trigonometric functions ▪ The six basic trigonometric integrals

In the previous section we defined the natural logarithmic function as an antiderivative of the function $y = 1/x$. With this definition, we are able to integrate several important functions that were not covered by previous integration rules. Specifically, the differentiation rules

$$\frac{d}{dx}\left[\ln |x|\right] = \frac{1}{x} \quad \text{and} \quad \frac{d}{dx}\left[\ln |u|\right] = \frac{u'}{u}$$

produce the following integration formulas.

THEOREM 6.5
LOG RULE FOR INTEGRATION

Let u be a differentiable function of x.

1. $\displaystyle\int \frac{1}{x}\, dx = \ln |x| + C$ 2. $\displaystyle\int \frac{1}{u}\, du = \ln |u| + C$

The second formula can also be written

$$\int \frac{u'}{u} \, dx = \ln |u| + C.$$

EXAMPLE 1   Using the Log Rule for integration

$$\int \frac{2}{x} \, dx = 2 \int \frac{1}{x} \, dx = 2 \ln |x| + C = \ln (x^2) + C \qquad \square$$

REMARK   In Example 1, the absolute value is unnecessary in the final form of the solution, since x^2 cannot be negative.

EXAMPLE 2   Using the Log Rule with a change of variable

Evaluate

$$\int \frac{1}{2x - 1} \, dx.$$

SOLUTION

Let $u = 2x - 1$, then $du = 2 \, dx$. After multiplying and dividing by 2, we have

$$\int \frac{1}{2x - 1} \, dx = \frac{1}{2} \int \left(\frac{1}{2x - 1} \right) 2 \, dx$$

$$= \frac{1}{2} \int \frac{1}{u} \, du$$

$$= \frac{1}{2} \ln |u| + C$$

$$= \frac{1}{2} \ln |2x - 1| + C. \qquad \square$$

In Example 3 we use the alternate form of the Log Rule, $\int u'/u \, dx = \ln |u| + C$. This form of the Log Rule is convenient, especially for simpler integrals. To apply this rule, look for quotients in which the numerator is the derivative of the denominator.

EXAMPLE 3   Finding area with the Log Rule

Find the area of the region bounded by the graph of $y = x/(x^2 + 1)$, the x-axis, and the line $x = 3$.

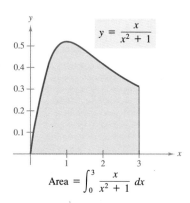

$$y = \frac{x}{x^2 + 1}$$

$$\text{Area} = \int_0^3 \frac{x}{x^2 + 1}\, dx$$

FIGURE 6.7

SOLUTION

From Figure 6.7 we see that the area of the region is given by the definite integral

$$\int_0^3 \frac{x}{x^2 + 1}\, dx.$$

Letting $u = x^2 + 1$, we have $u' = 2x$. Thus, to apply the Log Rule, we multiply and divide by 2 and write

$$\int_0^3 \frac{x}{x^2 + 1}\, dx = \frac{1}{2} \int_0^3 \frac{2x}{x^2 + 1}\, dx \qquad \int \frac{u'}{u}\, dx = \ln |u| + C$$

$$= \frac{1}{2} \left[\ln (x^2 + 1) \right]_0^3$$

$$= \frac{1}{2} (\ln 10 - \ln 1) = \frac{1}{2} \ln 10 \approx 1.151. \qquad \blacksquare$$

EXAMPLE 4 Recognizing quotient forms of the Log Rule

(a) $\displaystyle \int \frac{3x^2 + 1}{x^3 + x}\, dx = \ln |x^3 + x| + C \qquad u = x^3 + x$

(b) $\displaystyle \int \frac{\sec^2 x}{\tan x}\, dx = \ln |\tan x| + C \qquad u = \tan x$

(c) $\displaystyle \int \frac{x + 1}{x^2 + 2x}\, dx = \frac{1}{2} \int \frac{2x + 2}{x^2 + 2x}\, dx \qquad u = x^2 + 2x$

$$= \frac{1}{2} \ln |x^2 + 2x| + C \qquad \blacksquare$$

Integrals to which the Log Rule can be applied often appear in disguised form. For instance, if a rational function has a *numerator of degree greater than or equal to that of the denominator*, division may reveal a form to which we can apply the Log Rule. The next example illustrates this.

EXAMPLE 5 Using long division before integrating

Evaluate

$$\int \frac{x^2 + x + 1}{x^2 + 1}\, dx.$$

SOLUTION

First, by dividing, we have

$$\frac{x^2 + x + 1}{x^2 + 1} = 1 + \frac{x}{x^2 + 1}.$$

Now, we integrate to obtain

$$\int \frac{x^2 + x + 1}{x^2 + 1}\, dx = \int \left(1 + \frac{x}{x^2 + 1}\right) dx$$

$$= \int dx + \frac{1}{2} \int \frac{2x}{x^2 + 1}\, dx$$

$$= x + \frac{1}{2} \ln (x^2 + 1) + C.$$

REMARK   Try differentiating the antiderivative found in Example 5 to check that you obtain the original integrand.

The next example gives another instance in which the use of the Log Rule is disguised. In this case, a change of variables helps us recognize the Log Rule.

EXAMPLE 6   Change of variables with the Log Rule

Evaluate

$$\int \frac{2x}{(x + 1)^2}\, dx.$$

SOLUTION

Let $u = x + 1$, then $du = dx$ and $x = u - 1$.

$$\int \frac{2x}{(x + 1)^2}\, dx = \int \frac{2(u - 1)}{u^2}\, du$$

$$= 2 \int \left[\frac{u}{u^2} - \frac{1}{u^2}\right] du$$

$$= 2 \int \frac{du}{u} - 2 \int u^{-2}\, du$$

$$= 2 \ln |u| - 2\left(\frac{u^{-1}}{-1}\right) + C$$

$$= 2 \ln |u| + \frac{2}{u} + C$$

$$= 2 \ln |x + 1| + \frac{2}{x + 1} + C \qquad \text{Substitute for } u$$

As you study the two different methods shown in Examples 5 and 6, be aware that both methods involve rewriting a disguised integrand so that it fits one or more of the basic integration formulas. In the remaining sections of Chapter 6 and in Chapter 8 we will devote much time to integration techniques. To master these techniques you must recognize the "form-fitting" nature of integration. In this sense integration is not nearly as straightforward as differentiation. Differentiation takes the form "Here is the question; what is the answer?" Integration is more like "Here is the answer; what is the question?" We suggest the following guidelines for integration.

GUIDELINES FOR INTEGRATION	1. Memorize a basic list of integration formulas. (Including those given in this section, we now have twelve formulas: the Power Rule, the Log Rule, and ten trigonometric rules. By the end of Section 6.8 we will have expanded this list to nineteen basic formulas.)
	2. Find an integration formula that resembles all or part of the integrand, and by trial and error find a choice of u that will make the integrand conform to the formula.
	3. If you cannot find a u-substitution that works, try altering the integrand. You might try a trigonometric identity, division, addition, and subtraction of the same quantity. Be creative!

EXAMPLE 7   *u-substitution and the Log Rule*

Evaluate

$$\int \frac{1}{x \ln x} \, dx.$$

SOLUTION

Since the integrand is a quotient whose denominator is raised to the first power, we try to use the Log Rule. There are three basic choices for u. The choices $u = x$ and $u = x \ln x$ fail to fit the u'/u form of the Log Rule. However, the third choice does fit. Letting $u = \ln x$, we have $u' = 1/x$ and write

$$\int \frac{1}{x \ln x} \, dx = \int \frac{1/x}{\ln x} \, dx = \int \frac{u'}{u} \, dx = \ln |u| + C$$
$$= \ln |\ln x| + C.$$

Integrals of trigonometric functions

In Section 5.1 we looked at six trigonometric integration formulas—the six that correspond directly to differentiation rules. We now complete our set of basic trigonometric integration formulas by making use of the Log Rule.

EXAMPLE 8   *Using a trigonometric identity*

Evaluate $\int \tan x \, dx$.

SOLUTION

This integral seems not to fit any formulas on our basic list. However, by using a trigonometric identity, we obtain the following.

$$\int \tan x \, dx = \int \frac{\sin x}{\cos x} \, dx$$

Now, knowing that $D_x [\cos x] = -\sin x$, we consider $u = \cos x$ and write

$$\int \tan x \, dx = -\int \frac{-\sin x}{\cos x} \, dx = -\int \frac{u'}{u} \, dx = -\ln |u| + C$$
$$= -\ln |\cos x| + C.$$

In Example 8, we used a trigonometric identity to derive an integration formula for the tangent function. In the next example we take a rather unusual step (multiplying and dividing by the same quantity) to derive an integration formula for the secant function.

EXAMPLE 9   Derivation of the secant formula

Evaluate $\int \sec x \, dx$.

SOLUTION

Consider the following procedure.

$$\int \sec x \, dx = \int \sec x \left(\frac{\sec x + \tan x}{\sec x + \tan x} \right) dx = \int \frac{\sec^2 x + \sec x \tan x}{\sec x + \tan x} \, dx$$

Now, letting u be the denominator of this quotient, we have

$$u = \sec x + \tan x \quad \Longrightarrow \quad u' = \sec x \tan x + \sec^2 x.$$

Therefore, we conclude that

$$\int \sec x \, dx = \int \frac{\sec^2 x + \sec x \tan x}{\sec x + \tan x} \, dx = \int \frac{u'}{u} \, dx = \ln |u| + C$$
$$= \ln |\sec x + \tan x| + C.$$

With Examples 8 and 9 we now have integration formulas for $\sin x$, $\cos x$, $\tan x$, and $\sec x$. We omit the derivations for $\csc x$ and $\cot x$ and summarize all six trigonometric formulas in the following theorem.

THEOREM 6.6 **INTEGRALS OF THE SIX BASIC** **TRIGONOMETRIC FUNCTIONS**	$\int \sin u \, du = -\cos u + C$	$\int \cos u \, du = \sin u + C$
	$\int \tan u \, du = -\ln |\cos u| + C$	$\int \cot u \, du = \ln |\sin u| + C$
	$\int \sec u \, du = \ln |\sec u + \tan u| + C$	$\int \csc u \, du = -\ln |\csc u + \cot u| + C$

REMARK   As you memorize these formulas, note that the three formulas on the right follow the pattern set by the three on the left.

EXAMPLE 10   Integrating trigonometric functions

Evaluate

$$\int_0^{\pi/4} \sqrt{1 + \tan^2 x} \, dx.$$

SOLUTION

Since $1 + \tan^2 x = \sec^2 x$, we have

$$\int_0^{\pi/4} \sqrt{1 + \tan^2 x} \, dx = \int_0^{\pi/4} \sqrt{\sec^2 x} \, dx = \int_0^{\pi/4} \sec x \, dx$$

$$= \left[\ln |\sec x + \tan x| \right]_0^{\pi/4}$$

$$= \ln (\sqrt{2} + 1) - \ln 1 \approx 0.8814.$$

EXAMPLE 11   An application involving electric current

The electromotive force E of a particular electrical circuit is given by

$$E = 3 \sin 2t$$

where E is measured in volts and t is measured in seconds. Find the average value of E as t ranges from 0 to 0.5 second.

SOLUTION

The average value of E is given by

$$\frac{1}{b - a} \int_a^b f(t) \, dt = \frac{1}{0.5 - 0} \int_0^{0.5} 3 \sin 2t \, dt = 6 \int_0^{0.5} \sin 2t \, dt.$$

Let $u = 2t$, then $du = 2 \, dt$. Thus, we have

$$\text{average value} = 6 \left(\frac{1}{2} \right) \int_0^{0.5} (\sin 2t)(2) \, dt$$

$$= \left[3(-\cos 2t) \right]_0^{0.5}$$

$$= 3 \left[-\cos (1) + 1 \right] \approx 1.379 \text{ volts.}$$

EXERCISES for Section 6.2

In Exercises 1–40, evaluate the given integral.

1. $\displaystyle\int \frac{1}{x + 1} \, dx$

2. $\displaystyle\int \frac{1}{x - 5} \, dx$

3. $\displaystyle\int \frac{1}{3 - 2x} \, dx$

4. $\displaystyle\int \frac{1}{6x + 1} \, dx$

5. $\displaystyle\int \frac{x}{x^2 + 1} \, dx$

6. $\displaystyle\int \frac{x^2}{3 - x^3} \, dx$

7. $\displaystyle\int \frac{x^2 - 4}{x} \, dx$

8. $\displaystyle\int \frac{x + 5}{x} \, dx$

9. $\displaystyle\int_1^e \frac{\ln x}{2x} \, dx$

10. $\displaystyle\int_e^{e^2} \frac{1}{x \ln x} \, dx$

11. $\displaystyle\int_1^e \frac{(1 + \ln x)^2}{x}\,dx$ **12.** $\displaystyle\int_0^1 \frac{x - 1}{x + 1}\,dx$

13. $\displaystyle\int_0^2 \frac{x^2 - 2}{x + 1}\,dx$ **14.** $\displaystyle\int \frac{1}{(x + 1)^2}\,dx$

15. $\displaystyle\int \frac{1}{\sqrt{x} + 1}\,dx$ **16.** $\displaystyle\int \frac{x + 3}{x^2 + 6x + 7}\,dx$

17. $\displaystyle\int \frac{x^2 + 2x + 3}{x^3 + 3x^2 + 9x}\,dx$ **18.** $\displaystyle\int \frac{(\ln x)^2}{x}\,dx$

19. $\displaystyle\int \frac{1}{x^{2/3}(1 + x^{1/3})}\,dx$ **20.** $\displaystyle\int \frac{1}{x \ln (x^2)}\,dx$

21. $\displaystyle\int \frac{1}{1 + \sqrt{x}}\,dx$ **22.** $\displaystyle\int \frac{1 - \sqrt{x}}{1 + \sqrt{x}}\,dx$

23. $\displaystyle\int \frac{\sqrt{x}}{\sqrt{x} - 3}\,dx$ **24.** $\displaystyle\int_0^2 \frac{1}{1 + \sqrt{2x}}\,dx$

25. $\displaystyle\int \frac{\sqrt{x}}{1 - x\sqrt{x}}\,dx$ **26.** $\displaystyle\int \frac{2x}{(x - 1)^2}\,dx$

27. $\displaystyle\int \frac{x(x - 2)}{(x - 1)^3}\,dx$ **28.** $\displaystyle\int \tan 5x\,dx$

29. $\displaystyle\int \csc 2x\,dx$ **30.** $\displaystyle\int \sec \frac{x}{2}\,dx$

31. $\displaystyle\int \cos (1 - x)\,dx$ **32.** $\displaystyle\int \frac{\tan^2 2x}{\sec 2x}\,dx$

33. $\displaystyle\int \frac{\sec x \tan x}{\sec x - 1}\,dx$ **34.** $\displaystyle\int \frac{\sin x}{1 + \cos x}\,dx$

35. $\displaystyle\int \frac{\cos t}{1 + \sin t}\,dt$ **36.** $\displaystyle\int (\sec t + \tan t)\,dt$

37. $\displaystyle\int (\csc x - \sin x)\,dx$ **38.** $\displaystyle\int \frac{\sin^2 x - \cos^2 x}{\cos x}\,dx$

39. $\displaystyle\int \frac{1 - \cos \theta}{\theta - \sin \theta}\,d\theta$ **40.** $\displaystyle\int (\csc 2\theta - \cot 2\theta)^2\,d\theta$

In Exercises 41–44, show the equivalence of the given pair of formulas.

41. $\displaystyle\int \tan x\,dx = -\ln |\cos x| + C$

$\displaystyle\int \tan x\,dx = \ln |\sec x| + C$

42. $\displaystyle\int \cot x\,dx = \ln |\sin x| + C$

$\displaystyle\int \cot x\,dx = -\ln |\csc x| + C$

43. $\displaystyle\int \sec x\,dx = \ln |\sec x + \tan x| + C$

$\displaystyle\int \sec x\,dx = -\ln |\sec x - \tan x| + C$

44. $\displaystyle\int \csc x\,dx = -\ln |\csc x + \cot x| + C$

$\displaystyle\int \csc x\,dx = \ln |\csc x - \cot x| + C$

In Exercises 45 and 46, find the area of the indicated region.

45. $y = \dfrac{x^2 + 4}{x}$ **46.** $y = \dfrac{x + 5}{x}$

 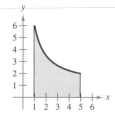

47. A population of bacteria is changing at the rate of

$$\frac{dP}{dt} = \frac{3000}{1 + 0.25t}$$

where t is the time in days. Assuming that the initial population (when $t = 0$) is 1000, write an equation that gives the population at any time t, and then find the population when $t = 3$ days.

48. Find the time required for an object to cool from $300°$ to $250°$ if that time is given by

$$t = \frac{10}{\ln 2} \int_{250}^{300} \frac{1}{T - 100}\,dT.$$

49. Find the average value of the function

$$f(x) = \frac{1}{x}$$

on the interval $1 \le x \le 5$.

50. The demand equation for a product is given by

$$p = \frac{90{,}000}{400 + 3x}.$$

Find the *average* price p on the interval $40 \le x \le 50$.

6.3 Inverse Functions

Inverse functions ▪ Existence of an inverse function ▪ Derivative of an inverse function

When we introduced the notion of a composite function in Section 1.5, we noted that composition is not commutative. That is, it is not necessarily true that $f(g(x))$ and $g(f(x))$ are equal. We now look at a special case for which composition is commutative—when f and g are inverses of each other.

DEFINITION OF INVERSE FUNCTION

A function g is the **inverse** of the function f if

$$f(g(x)) = x \quad \text{for each } x \text{ in the domain of } g$$

and

$$g(f(x)) = x \quad \text{for each } x \text{ in the domain of } f.$$

We denote g by f^{-1} (read "f inverse").

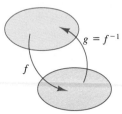

Domain of f = Range of g
Domain of g = Range of f

FIGURE 6.8

REMARK   Although the notation used to denote an inverse function resembles *exponential notation,* it is a different use of -1 as a superscript. That is, in general, $f^{-1}(x) \neq 1/f(x)$.

Here are some important observations about this definition.

1. If g is the inverse of f, then f is also the inverse of g.
2. The domain of f^{-1} is equal to the range of f (and vice versa), as indicated in Figure 6.8.
3. A function need not possess an inverse, but if it does, the inverse is unique. (See Exercise 57.)

To understand the concept of an inverse function, it is helpful to think of f^{-1} as undoing what has been done by f. For example, subtraction can be used to undo addition, and division can be used to undo multiplication. Use the definition of an inverse function to check the following inverses.

1. $f(x) = x + c$ and $f^{-1}(x) = x - c$

2. $f(x) = cx$ and $f^{-1}(x) = \dfrac{x}{c}, \quad c \neq 0$

EXAMPLE 1   Verifying inverse functions

Show that the following functions are inverses of each other.

$$f(x) = 2x^3 - 1 \quad \text{and} \quad g(x) = \sqrt[3]{\frac{x+1}{2}}$$

SOLUTION

First, note that both composite functions exist, since the domain and range of both f and g consist of the set of all real numbers. The composite of f with g is given by

$$f(g(x)) = 2\left(\sqrt[3]{\frac{x+1}{2}}\right)^3 - 1 = 2\left(\frac{x+1}{2}\right) - 1 = x + 1 - 1 = x.$$

The composite of g with f is given by

$$g(f(x)) = \sqrt[3]{\frac{(2x^3 - 1) + 1}{2}} = \sqrt[3]{\frac{2x^3}{2}} = \sqrt[3]{x^3} = x.$$

Since $f(g(x)) = g(f(x)) = x$, we conclude that f and g are inverses of each other. (See Figure 6.9). ▭

In Figure 6.9 the graphs of f and f^{-1} appear to be mirror images of each other with respect to the line $y = x$. We say that the graph of f^{-1} is a **reflection** of the graph of f in the line $y = x$. This idea is generalized in the following theorem.

$$g(x) = \sqrt[3]{\frac{x+1}{2}}$$

$y = x$

$f(x) = 2x^3 - 1$

FIGURE 6.9

THEOREM 6.7
REFLECTIVE PROPERTY
OF INVERSE FUNCTIONS

The graph of f contains the point (a, b) if and only if the graph of f^{-1} contains the point (b, a).

PROOF   If (a, b) is on the graph of f, then $f(a) = b$ and we have

$$f^{-1}(b) = f^{-1}(f(a)) = a.$$

Thus, (b, a) is on the graph of f^{-1}, as shown in Figure 6.10. A similar argument will prove the theorem in the other direction.

$y = x$

$y = f(x)$

(a, b)

(b, a)

$y = f^{-1}(x)$

The graph of f^{-1} is a reflection of the graph of f in the line $y = x$.

FIGURE 6.10

Not every function has an inverse, and Theorem 6.7 suggests a graphical test for those that do. It is called the **horizontal line test** for an inverse function, and it follows directly from the vertical line test for functions together with the reflective property of the graphs of f and f^{-1}. The test states that a function f has an inverse if and only if every horizontal line intersects the graph of f at most once. The following theorem formally states why the horizontal line test is valid. (Recall from Section 4.3 that a function is *strictly monotonic* if it is either increasing on its entire domain or decreasing on its entire domain.)

**THEOREM 6.8
THE EXISTENCE OF AN
INVERSE FUNCTION**

1. A function possesses an inverse if and only if it is one-to-one.
2. If f is strictly monotonic on its entire domain, then it is one-to-one and, hence, possesses an inverse.

PROOF   We leave the proof of the first part as an exercise (see Exercise 59). To prove the second part, recall from Section 1.5 that f is one-to-one if for x_1, x_2 in its domain

$$f(x_1) = f(x_2) \implies x_1 = x_2.$$

The *contrapositive* of this implication is logically equivalent and it states that

$$x_1 \neq x_2 \implies f(x_1) \neq f(x_2).$$

Now, choose x_1 and x_2 in the given interval. If $x_1 \neq x_2$, and since f is strictly monotonic, it follows that either $f(x_1) < f(x_2)$ or $f(x_1) > f(x_2)$. In either case, $f(x_1) \neq f(x_2)$. Thus, f is one-to-one on the interval.

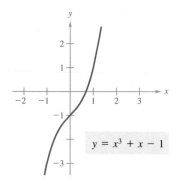

FIGURE 6.11

EXAMPLE 2   The existence of an inverse

Determine which of the following functions has an inverse.

(a) $f(x) = x^3 + x - 1$ (b) $f(x) = x^3 - x + 1$

SOLUTION

(a) From the graph of f given in Figure 6.11 it appears that f is increasing over its entire domain. To verify this, we note that the derivative, $f'(x) = 3x^2 + 1$, is positive for all real values of x. Therefore, f is strictly monotonic and it must have an inverse.
(b) From the graph given in Figure 6.12 we can see that the function does not pass the horizontal line test. In other words, it is not one-to-one. For instance, f has the same value when $x = -1$, 0, and 1.

$$f(-1) = f(1) = f(0) = 1 \qquad \text{Not one-to-one}$$

Therefore, by Theorem 6.8, f does not have an inverse.

FIGURE 6.12

REMARK   Often it is easier to prove that a function has an inverse than to find the inverse. For instance, it would be difficult algebraically to determine the inverse of the function in Example 2(a).

GUIDELINES FOR FINDING THE INVERSE OF A FUNCTION

1. Use Theorem 6.8 to determine whether the function given by $y = f(x)$ has an inverse.
2. Solve for x as a function of y: $x = g(y) = f^{-1}(y)$.
3. Define the domain of f^{-1} to be the range of f.
4. Verify that $f(f^{-1}(x)) = x$ and $f^{-1}(f(x)) = x$.

To avoid the confusion that could arise from using y as the independent variable for f^{-1}, it is customary to write f^{-1} as a *function of x* simply by interchanging the variables x and y after solving for x. This is illustrated in the next example.

EXAMPLE 3   Finding the inverse of a function

Find the inverse of the function given by $f(x) = \sqrt{2x - 3}$.

SOLUTION

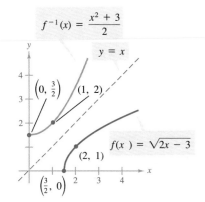

$$f^{-1}(x) = \frac{x^2 + 3}{2}$$

$f(x) = \sqrt{2x - 3}$

FIGURE 6.13

By Theorem 6.8, this function has an inverse because it is increasing on its entire domain, as shown in Figure 6.13. To find an equation for this inverse, we let $y = f(x)$ and solve for x in terms of y.

$$\sqrt{2x - 3} = y \qquad \text{Let } y = f(x)$$
$$2x - 3 = y^2$$
$$x = \frac{y^2 + 3}{2} \qquad \text{Solve for } x$$
$$f^{-1}(y) = \frac{y^2 + 3}{2}$$

Since the range of f is $[0, \infty)$, we define this interval to be the domain of f^{-1}. Finally, using x as the independent variable, we have

$$f^{-1}(x) = \frac{x^2 + 3}{2}, \quad 0 \le x. \qquad \text{Determine domain} \qquad \blacksquare$$

REMARK   Remember that any letter can be used to represent the independent variable. Thus,

$$f^{-1}(y) = \frac{y^2 + 3}{2}, \quad f^{-1}(x) = \frac{x^2 + 3}{2}, \quad \text{and} \quad f^{-1}(s) = \frac{s^2 + 3}{2}$$

all represent the same function.

Theorem 6.8 is useful in the following type of problem. Suppose you are given a function that is *not* one-to-one on its domain. By restricting the domain to an interval on which the function is strictly monotonic, you can conclude that the new function *is* one-to-one on the restricted domain. The next example illustrates this procedure.

EXAMPLE 4   Finding an interval on which a function is one-to-one

Show that the sine function, $f(x) = \sin x$, is not one-to-one on the entire real line. Then show that $[-\pi/2, \pi/2]$ is the largest interval, centered at the origin, for which f is strictly monotonic.

SOLUTION

FIGURE 6.14

It is clear that f is not one-to-one, since many different x-values yield the same y-value. For instance, $\sin(0) = 0 = \sin(\pi)$. Moreover, f is increasing on the open interval $(-\pi/2, \pi/2)$, since its derivative

$$f'(x) = \cos x$$

is positive there. Finally, since the left and right endpoints correspond to relative extrema of the sine, we can conclude that f is increasing on the closed interval $[-\pi/2, \pi/2]$ *and* that in any larger interval the function would not be strictly monotonic. (See Figure 6.14.)   ▭

The next two theorems discuss the derivative of an inverse function. The reasonableness of Theorem 6.9 follows from the reflective property of inverse functions as shown in Figure 6.10, and a formal proof of the theorem is given in Appendix A.

THEOREM 6.9 **CONTINUITY AND** **DIFFERENTIABILITY OF** **INVERSE FUNCTIONS**	Let f be a function that possesses an inverse. 1. If f is continuous on its domain, then f^{-1} is continuous on its domain. 2. If f is increasing on its domain, then f^{-1} is increasing on its domain. 3. If f is decreasing on its domain, then f^{-1} is decreasing on its domain. 4. If f is differentiable at c and $f'(c) \neq 0$, then f^{-1} is differentiable at $f(c)$.

THEOREM 6.10 **THE DERIVATIVE OF AN** **INVERSE FUNCTION**	If f is differentiable on its domain and possesses an inverse function g, then the derivative of g is given by $$g'(x) = \frac{1}{f'(g(x))}, \quad f'(g(x)) \neq 0.$$

PROOF   From Theorem 6.9 we know that g is differentiable. Using the Chain Rule we differentiate both sides of the equation $x = f(g(x))$ to obtain

$$1 = f'(g(x))\frac{d}{dx}[g(x)].$$

Since $f'(g(x)) \neq 0$, we can divide by this quantity to obtain

$$\frac{d}{dx}[g(x)] = \frac{1}{f'(g(x))}.$$

Geometrically, Theorem 6.10 tells us that the graphs of inverse functions have reciprocal slopes at the points (a, b) and (b, a), as illustrated in the next example. The equation

$$\frac{dy}{dx} = \frac{1}{dx/dy}$$

where $y = g(x)$, provides an easy way to remember this reciprocal relationship.

EXAMPLE 5   Graphs of inverse functions have reciprocal slopes

Let $f(x) = x^2$ (for $x \geq 0$) and let $f^{-1}(x) = \sqrt{x}$. Show that the slopes of the graphs of f and f^{-1} are reciprocals at the following points.

(a) $(2, 4)$ and $(4, 2)$ (b) $(3, 9)$ and $(9, 3)$

SOLUTION

The derivatives of f and f^{-1} are given by

$$f'(x) = 2x \quad \text{and} \quad (f^{-1})'(x) = \frac{1}{2\sqrt{x}}.$$

(a) At $(2, 4)$ the slope of the graph of f is $f'(2) = 2(2) = 4$. At $(4, 2)$, the slope of the graph of f^{-1} is

$$(f^{-1})'(4) = \frac{1}{2\sqrt{4}} = \frac{1}{2(2)} = \frac{1}{4}.$$

(b) At $(3, 9)$ the slope of the graph of f is $f'(3) = 2(3) = 6$. At $(9, 3)$, the slope of the graph of f^{-1} is

$$(f^{-1})'(9) = \frac{1}{2\sqrt{9}} = \frac{1}{2(3)} = \frac{1}{6}.$$

Thus, in both cases, the slopes are reciprocals, as shown in Figure 6.15.

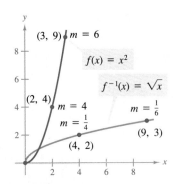

FIGURE 6.15

REMARK   In Example 5, note that f^{-1} is not differentiable at $(0, 0)$. This is consistent with Theorem 6.10 because the derivative of f at $(0, 0)$ is zero.

EXERCISES for Section 6.3

In Exercises 1–8, (a) show that f and g are inverse functions by showing that $f(g(x)) = x$ and $g(f(x)) = x$, and (b) graph f and g on the same set of coordinate axes.

1. $f(x) = x^3$ $\qquad g(x) = \sqrt[3]{x}$
2. $f(x) = x^{-1}$ $\qquad g(x) = x^{-1}$
3. $f(x) = 5x + 1$ $\qquad g(x) = \dfrac{1}{5}(x - 1)$
4. $f(x) = 3 - 4x$ $\qquad g(x) = \dfrac{1}{4}(3 - x)$
5. $f(x) = \sqrt{x - 4}$ $\qquad g(x) = x^2 + 4, \ x \ge 0$
6. $f(x) = 9 - x^2, \ x \ge 0$ $\qquad g(x) = \sqrt{9 - x}$
7. $f(x) = 1 - x^3$ $\qquad g(x) = \sqrt[3]{1 - x}$
8. $f(x) = x^{-2}, \ x > 0$ $\qquad g(x) = x^{-1/2}, \ x > 0$

In Exercises 9–22, find the inverse of f. Then graph both f and f^{-1}.

9. $f(x) = 2x - 3$ $\qquad$ **10.** $f(x) = 3x$
11. $f(x) = x^5$ $\qquad$ **12.** $f(x) = x^3 + 1$
13. $f(x) = \sqrt{x}$ $\qquad$ **14.** $f(x) = x^2, \ x \ge 0$
15. $f(x) = \sqrt{4 - x^2}, \ 0 \le x$
16. $f(x) = \sqrt{x^2 - 4}, \ x \ge 2$
17. $f(x) = \sqrt[3]{x - 1}$ $\qquad$ **18.** $f(x) = 3\sqrt[5]{2x - 1}$
19. $f(x) = x^{2/3}, \ x \ge 0$ $\qquad$ **20.** $f(x) = x^{3/5}$
21. $f(x) = \dfrac{x}{\sqrt{x^2 + 7}}$ $\qquad$ **22.** $f(x) = \dfrac{x + 2}{x}$

In Exercises 23 and 24, use the graph of the function f to complete the table and sketch the graph of f^{-1}.

23.

x	1	2	3	4
$f^{-1}(x)$				

24.

x	0	2	4
$f^{-1}(x)$			

In Exercises 25–28, use the functions

$$f(x) = \frac{1}{8}x - 3 \quad \text{and} \quad g(x) = x^3$$

to find the indicated value.

25. $(f^{-1} \circ g^{-1})(1)$ $\qquad$ **26.** $(g^{-1} \circ f^{-1})(-3)$
27. $(f^{-1} \circ f^{-1})(6)$ $\qquad$ **28.** $(g^{-1} \circ g^{-1})(-4)$

In Exercises 29–36, use the horizontal line test to determine whether the function is one-to-one on its entire domain and therefore has an inverse.

29. $f(x) = \dfrac{3}{4}x + 6$ $\qquad$ **30.** $f(x) = 5x - 3$

31. $f(\theta) = \sin \theta$ $\qquad$ **32.** $f(x) = \dfrac{x^2}{x^2 + 4}$

33. $h(s) = \dfrac{1}{s - 2} - 3$ $\qquad$ **34.** $g(t) = \dfrac{t}{\sqrt{t^2 + 1}}$

35. $f(x) = \ln x$

36. $f(x) = 3x\sqrt{x+1}$

In Exercises 37–42, use the derivative to determine whether the given function is strictly monotonic on its entire domain and therefore has an inverse.

37. $f(x) = (x + a)^3 + b$

38. $f(x) = \cos \dfrac{3x}{2}$

39. $f(x) = \dfrac{x^4}{4} - 2x^2$

40. $f(x) = x^3 - 6x^2 + 12x$

41. $f(x) = 2 - x - x^3$

42. $f(x) = \ln(x - 3)$

In Exercises 43–48, show that f is strictly monotonic on the given interval and therefore has an inverse on that interval.

Function	Interval
43. $f(x) = (x - 4)^2$	$[4, \infty)$
44. $f(x) = \lvert x + 2 \rvert$	$[-2, \infty)$
45. $f(x) = \dfrac{4}{x^2}$	$(0, \infty)$
46. $f(x) = \tan x$	$\left(-\dfrac{\pi}{2}, \dfrac{\pi}{2}\right)$
47. $f(x) = \cos x$	$[0, \pi]$
48. $f(x) = \sec x$	$\left[0, \dfrac{\pi}{2}\right) \cup \left(\dfrac{\pi}{2}, \pi\right]$

In Exercises 49–52, show that the slopes of the graphs of f and f^{-1} are reciprocals at the given points.

Functions	Point
49. $f(x) = x^3$	$\left(\dfrac{1}{2}, \dfrac{1}{8}\right)$
$f^{-1}(x) = \sqrt[3]{x}$	$\left(\dfrac{1}{8}, \dfrac{1}{2}\right)$
50. $f(x) = 3 - 4x$	$(1, -1)$
$f^{-1}(x) = \dfrac{3 - x}{4}$	$(-1, 1)$
51. $f(x) = \sqrt{x - 4}$	$(5, 1)$
$f^{-1}(x) = x^2 + 4$	$(1, 5)$
52. $f(x) = \dfrac{1}{1 + x^2}$	$\left(1, \dfrac{1}{2}\right)$
$f^{-1}(x) = \sqrt{\dfrac{1 - x}{x}}$	$\left(\dfrac{1}{2}, 1\right)$

In Exercises 53 and 54, the derivative of the function has the same sign for all x in its domain, but the function is not one-to-one. Explain why.

53. $f(x) = \tan x$

54. $f(x) = \dfrac{x}{x^2 - 4}$

In Exercises 55 and 56, find the inverse function of f over the specified interval. Use a computer or graphics calculator to sketch the graph of f and f^{-1} on the same coordinate axes and observe that the graph of f^{-1} is a reflection of the graph of f in the line $y = x$.

Function	Interval
55. $f(x) = \dfrac{x}{x^2 - 4}$	$(-2, 2)$
56. $f(x) = 2 - \dfrac{3}{x^2}$	$(0, 10)$

57. Prove that if a function has an inverse, then the inverse is unique.

58. Prove that if f has an inverse, then $(f^{-1})^{-1} = f$.

59. Prove that a function has an inverse if and only if it is one-to-one.

60. Prove that if f and g are one-to-one functions, then $(f \circ g)^{-1}(x) = (g^{-1} \circ f^{-1})(x)$.

6.4 Exponential Functions: Differentiation and Integration

The natural exponential function ▪ Differentiation ▪ Integration

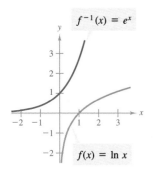

FIGURE 6.16

In Section 6.1 we saw that the function $f(x) = \ln x$ is increasing on its entire domain, and hence it has an inverse f^{-1}. Moreover, from the domain and range of f, we can conclude that the domain of f^{-1} is the set of all reals, and the range is the set of positive reals, as shown in Figure 6.16. Thus, for any real number x,

$$f(f^{-1}(x)) = \ln (f^{-1}(x)) = x. \qquad \text{\footnotesize x is any real number}$$

We also know that if x happens to be rational, then

$$\ln (e^x) = x \ln e = x(1) = x. \qquad \text{\footnotesize x is a rational number}$$

Since the natural logarithmic function is one-to-one, we conclude that $f^{-1}(x)$ and e^x agree for *rational* values of x. Because of this agreement, we extend the definition of e^x to cover *all* real numbers.

DEFINITION OF THE NATURAL EXPONENTIAL FUNCTION

The inverse of the natural logarithmic function $f(x) = \ln x$ is called the **natural exponential function** and is denoted by

$$f^{-1}(x) = e^x.$$

That is,

$$y = e^x \quad \text{if and only if} \quad x = \ln y.$$

We summarize the inverse relationship between the natural logarithmic function and the natural exponential function as follows.

$$\ln (e^x) = x \qquad \text{and} \qquad e^{\ln x} = x \qquad \text{\footnotesize Inverse relationships}$$

These two properties are useful in solving equations involving exponential and logarithmic functions, as demonstrated in the following example.

EXAMPLE 1 Solving exponential and logarithmic equations

Solve for x in the following equations.

(a) $7 = e^{x+1}$ (b) $\ln (2x - 3) = 5$

SOLUTION

(a) We can convert from exponential form to logarithmic form by *taking the natural log of both sides* of the exponential equation.

$$7 = e^{x+1} \quad \Longrightarrow \quad \ln 7 = \ln (e^{x+1}) \quad \Longrightarrow \quad \ln 7 = x + 1$$

Thus, $x = -1 + \ln 7 \approx 0.946$.

(b) To convert from logarithmic form to exponential form, we *apply the exponential function to both sides* of the logarithmic equation.

$$\ln (2x - 3) = 5 \quad \Longrightarrow \quad e^{\ln (2x-3)} = e^5 \quad \Longrightarrow \quad 2x - 3 = e^5$$

Thus, $x = \frac{1}{2}(e^5 + 3) \approx 75.707$.

The familiar rules for operating with rational exponents can be extended to the natural exponential function, as indicated in the following theorem.

**THEOREM 6.11
OPERATIONS WITH EXPONENTIAL
FUNCTIONS**

Let a and b be any real numbers. Then the following properties are true.

1. $e^a e^b = e^{a+b}$ 2. $\dfrac{e^a}{e^b} = e^{a-b}$ 3. $(e^a)^b = e^{ab}$

PROOF

FIGURE 6.17

We prove Property 1 and leave the proofs of the other two properties as exercises (see Exercises 87 and 88).

$$\ln (e^a e^b) = \ln (e^a) + \ln (e^b) = a + b = \ln (e^{a+b})$$

Thus, since the natural log function is one-to-one, we conclude that

$$e^a e^b = e^{a+b}.$$

In Section 6.3, we saw that an inverse function f^{-1} shares many properties with f. Thus, the natural exponential function inherits the following properties from the natural logarithmic function. (See Figure 6.17.)

**PROPERTIES OF THE NATURAL
EXPONENTIAL FUNCTION**

1. The domain of $f(x) = e^x$ is $(-\infty, \infty)$, and the range is $(0, \infty)$.
2. The function $f(x) = e^x$ is continuous, increasing, and one-to-one on its entire domain.
3. The graph of $f(x) = e^x$ is concave upward on its entire domain.
4. $\lim\limits_{x \to -\infty} e^x = 0$ and $\lim\limits_{x \to \infty} e^x = \infty$

Derivative of the natural exponential function

One of the most intriguing (and useful) characteristics of the natural exponential function is that *it is its own derivative*. In other words, it is a solution to the equation $y' = y$. This result is stated in the next theorem.

**THEOREM 6.12
THE DERIVATIVE OF THE NATURAL
EXPONENTIAL FUNCTION**

Let u be a differentiable function of x.

1. $\dfrac{d}{dx}[e^x] = e^x$ 2. $\dfrac{d}{dx}[e^u] = e^u \dfrac{du}{dx}$

PROOF Let $f(x) = \ln x$ and $g(x) = e^x$. Then $f'(x) = 1/x$, and by Theorem 6.10 we have

$$\frac{d}{dx}[e^x] = \frac{d}{dx}[g(x)] = \frac{1}{f'(g(x))} = \frac{1}{f'(e^x)} = \frac{1}{(1/e^x)} = e^x.$$

The derivative of e^u follows from the Chain Rule.

REMARK We can interpret this theorem geometrically by saying that the slope of the graph of $f(x) = e^x$ at any point (x, e^x) is equal to the y-coordinate of the point.

EXAMPLE 2 Differentiating exponential functions

(a) $\dfrac{d}{dx}[e^{2x-1}] = \dfrac{du}{dx}e^u = 2e^{2x-1}$ $u = 2x - 1$

(b) $\dfrac{d}{dx}[e^{-3/x}] = \dfrac{du}{dx}e^u = \left(\dfrac{3}{x^2}\right)e^{-3/x} = \dfrac{3e^{-3/x}}{x^2}$ $u = -\dfrac{3}{x}$ ▭

EXAMPLE 3 Locating relative extrema

Find the relative extrema of $f(x) = xe^x$.

SOLUTION

$f(x) = xe^x$

$(-1, -e^{-1})$
Relative minimum

FIGURE 6.18

The derivative of f is given by

$$f'(x) = x(e^x) + e^x(1) \qquad \text{Product Rule}$$
$$= e^x(x + 1).$$

Now, since e^x is never zero, the derivative is zero only when $x = -1$. Moreover, by the First Derivative Test, we can determine that this corresponds to a relative minimum, as shown in Figure 6.18. Since the derivative $f'(x) = e^x(x + 1)$ is defined for all x, there are no other critical points.   ▭

EXAMPLE 4 The normal probability density function

Show that the *normal probability density function*

$$f(x) = \frac{1}{\sqrt{2\pi}}e^{-x^2/2}$$

has points of inflection when $x = \pm 1$.

SOLUTION

To locate possible points of inflection, we find the x-values for which the second derivative is zero.

$$f'(x) = \frac{1}{\sqrt{2\pi}}(-x)e^{-x^2/2}$$

$$f''(x) = \frac{1}{\sqrt{2\pi}}[(-x)(-x)e^{-x^2/2} + (-1)e^{-x^2/2}] \qquad \text{Product Rule}$$

$$= \frac{1}{\sqrt{2\pi}}(e^{-x^2/2})(x^2 - 1)$$

Therefore, $f''(x) = 0$ when $x = \pm 1$, and we can apply the techniques of Chapter 4 to conclude that these values yield the two points of inflection shown in Figure 6.19.

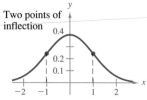

Two points of inflection

Bell-Shaped Curve Given By Normal Probability Density Function

FIGURE 6.19

REMARK   The general form of a normal probability density function is given by

$$f(x) = \frac{1}{\sigma\sqrt{2\pi}}e^{-x^2/2\sigma^2}$$

where σ is the standard deviation (σ is the lowercase Greek letter sigma). By following the procedure of Example 4, we can show that the bell-shaped curve of this function has points of inflection when $x = \pm\sigma$.

Integration of the natural exponential function

Each differentiation formula in Theorem 6.12 has a corresponding integration formula.

THEOREM 6.13 INTEGRATION RULES FOR EXPONENTIAL FUNCTIONS	Let u be a differentiable function of x. 1. $\int e^x \, dx = e^x + C$ 2. $\int e^u \, du = e^u + C$

EXAMPLE 5   Integrating exponential functions

Evaluate $\int e^{3x+1} \, dx$.

SOLUTION

Considering $u = 3x + 1$, we have $du = 3 \, dx$. Then

$$\int e^{3x+1} \, dx = \frac{1}{3} \int e^{3x+1}(3) \, dx = \frac{1}{3} \int e^u \, du$$

$$= \frac{1}{3} e^u + C = \frac{e^{3x+1}}{3} + C.$$

REMARK   In Example 5 we introduced the missing *constant* factor 3 to create *du* = 3 *dx*. However, remember that you cannot introduce a missing *variable* factor in the integrand. For instance,

$$\int e^{-x^2} \, dx \neq \frac{1}{x} \int e^{-x^2} (x \, dx).$$

EXAMPLE 6   Integrating exponential functions

Evaluate $\int 5xe^{-x^2} \, dx$.

SOLUTION

If we let $u = -x^2$, then $du = -2x \, dx$ or $x \, dx = -du/2$. Thus, we have

$$\int 5xe^{-x^2} \, dx = \int 5e^{-x^2}(x \, dx) = \int 5e^{u}\left(-\frac{du}{2}\right)$$

$$= -\frac{5}{2} \int e^{u} \, du$$

$$= -\frac{5}{2} e^{u} + C$$

$$= -\frac{5}{2} e^{-x^2} + C. \qquad \blacksquare$$

EXAMPLE 7   Integrating exponential functions

(a) $\displaystyle\int \underbrace{\frac{e^{1/x}}{x^2}}_{} \, dx = -\int \underbrace{e^{1/x}}_{e^u} \underbrace{\left(-\frac{1}{x^2}\right) dx}_{du} = -e^{1/x} + C \qquad\qquad u = \frac{1}{x}$

(b) $\displaystyle\int \sin x \, e^{\cos x} \, dx = -\int \underbrace{e^{\cos x}}_{e^u} \underbrace{(-\sin x \, dx)}_{du} = -e^{\cos x} + C \qquad u = \cos x$

$\blacksquare$

EXAMPLE 8   Finding area bounded by exponential functions

(a) $\displaystyle\int_0^1 e^{-x} \, dx = \left[-e^{-x}\right]_0^1 = -e^{-1} - (-1) = 1 - \frac{1}{e} \approx 0.632$

(b) $\displaystyle\int_0^1 \frac{e^x}{1 + e^x} \, dx = \left[\ln (1 + e^x)\right]_0^1 = \ln (1 + e) - \ln 2 \approx 0.620$

(c) $\displaystyle\int_{-1}^0 [e^x \cos (e^x)] \, dx = \left[\sin (e^x)\right]_{-1}^0 = [\sin 1 - \sin (e^{-1})] \approx 0.482$

(See Figure 6.20.)

$$y = e^{-x}$$

$$y = \frac{e^x}{1 + e^x}$$

$$y = e^x \cos(e^x)$$

FIGURE 6.20   (a)   (b)   (c)

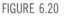

EXERCISES for Section 6.4

In Exercises 1–4, write the logarithmic equation as an exponential equation and vice versa.

1. (a) $e^0 = 1$ (b) $e^2 = 7.389 \ldots$
2. (a) $e^{-2} = 0.1353 \ldots$ (b) $e^{-1} = 0.3679 \ldots$
3. (a) $\ln 2 = 0.6931 \ldots$ (b) $\ln 8.4 = 2.128 \ldots$
4. (a) $\ln 0.5 = -0.6931 \ldots$
    (b) $\ln 1 = 0$

In Exercises 5–8, solve for x.

5. (a) $e^{\ln x} = 4$ (b) $\ln e^{2x} = 3$
6. (a) $e^{\ln 2x} = 12$ (b) $\ln e^{-x} = 0$
7. (a) $\ln x = 2$ (b) $e^x = 4$
8. (a) $\ln x^2 = 10$ (b) $e^{-4x} = 5$

In Exercises 9–12, sketch the graph of the given function.

9. $y = e^{-x}$ **10.** $y = \frac{1}{2}e^x$
11. $y = e^{-x^2}$ **12.** $y = e^{-x/2}$

In Exercises 13–16, show that the given functions are inverses of each other by sketching their graphs on the same coordinate system.

13. $f(x) = e^{2x}$, $g(x) = \ln \sqrt{x}$
14. $f(x) = e^{x/3}$, $g(x) = \ln x^3$
15. $f(x) = e^x - 1$, $g(x) = \ln (x + 1)$
16. $f(x) = e^{x-1}$, $g(x) = 1 + \ln x$

In Exercises 17 and 18, compare the given number to the natural number e.

17. (a) $\dfrac{271,801}{99,990}$ (b) $\dfrac{299}{110}$

18. (a) $1 + 1 + \dfrac{1}{2} + \dfrac{1}{6} + \dfrac{1}{24}$

    (b) $1 + 1 + \dfrac{1}{2} + \dfrac{1}{6} + \dfrac{1}{24} + \dfrac{1}{120} + \dfrac{1}{720} + \dfrac{1}{5040}$

In Exercises 19 and 20, find the slope of the tangent line to the given exponential function at the point $(0, 1)$.

19. (a) $y = e^{3x}$ (b) $y = e^{-3x}$

20. (a) $y = e^{2x}$ (b) $y = e^{-2x}$

In Exercises 21–42, find dy/dx.

21. $y = e^{2x}$ **22.** $y = e^{1-x}$
23. $y = e^{-2x+x^2}$ **24.** $y = e^{-x^2}$
25. $y = e^{\sqrt{x}}$ **26.** $y = x^2 e^{-x}$
27. $y = (e^{-x} + e^x)^3$ **28.** $y = e^{-1/x^2}$
29. $y = \ln (e^{x^2})$ **30.** $y = \ln \left(\dfrac{1 + e^x}{1 - e^x} \right)$
31. $y = \ln (1 + e^{2x})$ **32.** $y = \dfrac{2}{e^x + e^{-x}}$
33. $y = \ln \dfrac{e^x + e^{-x}}{2}$ **34.** $y = xe^x - e^x$
35. $y = x^2 e^x - 2xe^x + 2e^x$
36. $y = \dfrac{e^x - e^{-x}}{2}$
37. $y = e^{-x} \ln x$ **38.** $y = e^3 \ln x$
39. $y = e^x(\sin x + \cos x)$ **40.** $y = e^{\tan x}$
41. $y = \tan^2 (e^x)$ **42.** $y = \ln e^x$

In Exercises 43 and 44, use implicit differentiation to find dy/dx.

43. $xe^y - 10x + 3y = 0$ **44.** $e^{xy} + x^2 - y^2 = 10$

In Exercises 45 and 46, show that the function $y = f(x)$ is a solution of the given differential equation.

45. $y = e^x(\cos \sqrt{2}x + \sin \sqrt{2}x)$
$y'' - 2y' + 3y = 0$

46. $y = e^x(3 \cos 2x - 4 \sin 2x)$
$y'' - 2y' + 5y = 0$

In Exercises 47–52, find the extrema and the points of inflection (if any exist), and sketch the graph of the function.

47. $f(x) = \dfrac{1}{\sqrt{2\pi}}e^{-(x^2/2)}$ **48.** $f(x) = \dfrac{e^x - e^{-x}}{2}$

49. $f(x) = \dfrac{e^x + e^{-x}}{2}$ **50.** $f(x) = xe^{-x}$

51. $f(x) = x^2e^{-x}$
52. $f(x) = -2 + e^{3x}(4 - 2x)$

53. Find an equation of the line normal to the graph of $y = e^{-x}$ at $(0, 1)$.

54. Find the point on the graph of $y = e^{-x}$ where the normal line to the curve will pass through the origin.

55. Find the area of the largest rectangle that can be inscribed under the curve $y = e^{-x^2}$ in the first and second quadrants.

56. Find, to three decimal places, the value of x such that $e^{-x} = x$. [Use Newton's Method.]

In Exercises 57–76, evaluate the integral.

57. $\displaystyle\int_0^1 e^{-2x}\, dx$ **58.** $\displaystyle\int_1^2 e^{1-x}\, dx$

59. $\displaystyle\int_0^2 (x^2 - 1)e^{x^3-3x+1}\, dx$ **60.** $\displaystyle\int x^2e^{x^3}\, dx$

61. $\displaystyle\int \dfrac{e^{-x}}{1 + e^{-x}}\, dx$ **62.** $\displaystyle\int \dfrac{e^{2x}}{1 + e^{2x}}\, dx$

63. $\displaystyle\int xe^{ax^2}\, dx$ **64.** $\displaystyle\int_0^{\sqrt{2}} xe^{-(x^2/2)}\, dx$

65. $\displaystyle\int_1^3 \dfrac{e^{3/x}}{x^2}\, dx$ **66.** $\displaystyle\int (e^x - e^{-x})^2\, dx$

67. $\displaystyle\int e^x\sqrt{1 - e^x}\, dx$ **68.** $\displaystyle\int \dfrac{e^x - e^{-x}}{e^x + e^{-x}}\, dx$

69. $\displaystyle\int \dfrac{e^x + e^{-x}}{e^x - e^{-x}}\, dx$ **70.** $\displaystyle\int \dfrac{2e^x - 2e^{-x}}{(e^x + e^{-x})^2}\, dx$

71. $\displaystyle\int \dfrac{5 - e^x}{e^{2x}}\, dx$ **72.** $\displaystyle\int \dfrac{e^{2x} + 2e^x + 1}{e^x}\, dx$

73. $\displaystyle\int e^{\sin \pi x} \cos \pi x\, dx$ **74.** $\displaystyle\int e^{\tan 2x} \sec^2 2x\, dx$

75. $\displaystyle\int e^{-x} \tan (e^{-x})\, dx$ **76.** $\displaystyle\int \ln (e^{2x-1})\, dx$

In Exercises 77–80, find the area of the region bounded by the graphs of the equations.

77. $y = e^x$, $y = 0$, $x = 0$, $x = 5$
78. $y = e^{-x}$, $y = 0$, $x = a$, $x = b$
79. $y = xe^{-(x^2/2)}$, $y = 0$, $x = 0$, $x = \sqrt{2}$
80. $y = e^{-2x} + 2$, $y = 0$, $x = 0$, $x = 2$

81. Given $e^x \geq 1$ for $x \geq 0$, it follows that

$$\int_0^x e^t\, dt \geq \int_0^x 1\, dt.$$

Perform this integration to derive the inequality $e^x \geq 1 + x$ for $x \geq 0$.

82. Integrate each term of the following inequalities in a manner similar to that of Exercise 81 to obtain each succeeding inequality for $x \geq 0$. Then evaluate both sides of each inequality when $x = 1$.

(a) $e^x \geq 1 + x$

(b) $e^x \geq 1 + x + \dfrac{x^2}{2}$

(c) $e^x \geq 1 + x + \dfrac{x^2}{2} + \dfrac{x^3}{6}$

(d) $e^x \geq 1 + x + \dfrac{x^2}{2} + \dfrac{x^3}{6} + \dfrac{x^4}{24}$

In Exercises 83 and 84, find a function f that satisfies the given conditions.

83. $f''(x) = \dfrac{1}{2}(e^x + e^{-x})$,   **84.** $f''(x) = \sin x + e^{2x}$,
$f(0) = 1, f'(0) = 0$ $f(0) = \dfrac{1}{4}, f'(0) = \dfrac{1}{2}$

85. Given the function

$$f(x) = \dfrac{2}{1 + e^{1/x}}$$

use a computer or graphics calculator to (a) sketch the graph of f, (b) find any horizontal asymptotes, and (c) find $\lim_{x \to 0} f(x)$ (if it exists).

86. The displacement from equilibrium of a mass oscillating on the end of a spring suspended from the ceiling is given by $y = 1.56e^{-0.22t} \cos 4.9t$ where y is the displacement in feet and t is the time in seconds. Use a computer or graphics calculator to sketch the graph of the displacement function on the interval $[0, 10]$ and find the time past which the displacement does not exceed 3 inches from equilibrium.

87. Prove that $\dfrac{e^a}{e^b} = e^{a-b}$.

88. Prove that $(e^a)^b = e^{ab}$.

6.5 Bases Other Than *e* and Applications

Bases other than *e* ▪ Differentiation and integration ▪ Power Rule for real exponents ▪ Applications

The **base** of the natural exponential function is *e*. Using the third property in Theorem 6.11, we now assign a meaning to the exponential function $y = a^x$.

DEFINITION OF EXPONENTIAL FUNCTION TO BASE *a*

If *a* is a positive real number $(a \neq 1)$ and *x* is any real number, then $f(x) = a^x$ is given by

$$f(x) = a^x = (e^{\ln a})^x = e^{(\ln a)x}.$$

We call *f* the **exponential function to the base *a***.

REMARK   These functions obey the usual laws of exponents. For example, $a^x a^y = a^{x+y}$.

We can give a similar definition for logarithmic functions to other bases.

DEFINITION OF LOGARITHMIC FUNCTION TO BASE *a*

If *a* is a positive real number $(a \neq 1)$ and *x* is any positive real number, then $f(x) = \log_a x$ is given by

$$f(x) = \log_a x = \frac{1}{\ln a} \ln x.$$

We call *f* the **logarithmic function to the base *a***.

Logarithmic functions to the base *a* have properties similar to those of the natural logarithmic function given in Theorem 6.2. For instance, the following properties are valid.

$$\log_a 1 = 0$$
$$\log_a xy = \log_a x + \log_a y$$
$$\log_a \frac{x}{y} = \log_a x - \log_a y$$
$$\log_a x^n = n \log_a x$$

From the definition of the exponential and logarithmic functions to a base *a*, it follows that $f(x) = a^x$ and $g(x) = \log_a x$ are inverse functions of each other. Thus,

$$a^{\log_a x} = x \quad \text{and} \quad \log_a a^x = x.$$

Moreover,

$$y = a^x \quad \text{if and only if} \quad x = \log_a y.$$

The logarithmic function to the base 10 is called the **common logarithmic function.** Thus, for common logarithms, we have $y = 10^x$ if and only if $x = \log_{10} y$.

EXAMPLE 1   Bases other than *e*

Solve for x in the following equations.

(a) $3^x = \dfrac{1}{27}$ (b) $\log_2 x = -4$

SOLUTION

(a) Applying the logarithmic function to the base 3 to both sides produces

$$\log_3 3^x = \log_3 \frac{1}{27} = \log_3 3^{-3}$$

$$x = -3.$$

(b) We solve for x as follows.

$$\log_2 x = -4$$

$$2^{\log_2 x} = 2^{-4}$$

$$x = \frac{1}{2^4} = \frac{1}{16}$$

Differentiation and integration

To differentiate exponential and logarithmic functions to other bases, you have three options: (1) use the definition of a^x and $\log_a x$ and differentiate using the rules for the natural exponential and logarithmic functions, (2) use logarithmic differentiation, or (3) use the following differentiation rules for bases other than *e*.

THEOREM 6.14
DERIVATIVES FOR BASES
OTHER THAN *e*

Let a be a positive real number ($a \neq 1$) and let u be a differentiable function of x.

1. $\dfrac{d}{dx}[a^x] = (\ln a)a^x$ 2. $\dfrac{d}{dx}[a^u] = (\ln a)a^u \dfrac{du}{dx}$

3. $\dfrac{d}{dx}[\log_a x] = \dfrac{1}{(\ln a)x}$ 4. $\dfrac{d}{dx}[\log_a u] = \dfrac{1}{(\ln a)u}\dfrac{du}{dx}$

PROOF   By definition, $a^x = e^{(\ln a)x}$. Therefore, we can prove the first property by letting $u = (\ln a)x$ and differentiating with base e to obtain

$$\frac{d}{dx}[a^x] = \frac{d}{dx}[e^{(\ln a)x}] = e^u \frac{du}{dx} = e^{(\ln a)x}(\ln a) = (\ln a)a^x.$$

To prove the third property, we write

$$\frac{d}{dx}[\log_a x] = \frac{d}{dx}\left[\frac{1}{\ln a} \ln x\right] = \frac{1}{\ln a}\left(\frac{1}{x}\right) = \frac{1}{(\ln a)x}.$$

The second and fourth properties can be proved using the Chain Rule.

REMARK   Note that these differentiation rules are similar to those for the natural exponential function and natural logarithmic function. In fact, they differ only by the constant factors $\ln a$ and $1/\ln a$, respectively. This points out one reason why, for calculus, e is the most convenient base.

EXAMPLE 2   Differentiating functions to other bases

Find the derivatives of the following.

(a) $y = 2^x$ (b) $y = 2^{\sqrt{x}}$ (c) $y = \log_{10} \cos x$

SOLUTION

(a) $y' = \dfrac{d}{dx}[2^x] = (\ln 2)2^x$

(b) Let $u = \sqrt{x}$; then $u' = 1/(2\sqrt{x})$ and we have

$$y' = \frac{d}{dx}[2^{\sqrt{x}}] = (\ln 2)2^{\sqrt{x}}\left(\frac{1}{2\sqrt{x}}\right) = \frac{\ln 2}{2\sqrt{x}}2^{\sqrt{x}}.$$

(c) Let $u = \cos x$; then $u' = -\sin x$ and we have

$$y' = \frac{d}{dx}[\log_{10} \cos x] = \frac{-\sin x}{(\ln 10) \cos x} = -\frac{1}{\ln 10} \tan x. \qquad \blacksquare$$

Occasionally, an integrand involves an exponential function to a base other than e. When this occurs, there are two options: (1) convert to base e using the formula $a^x = e^{(\ln a)x}$ and then integrate, or (2) integrate directly, using the integration formula

$$\int a^x \, dx = \left(\frac{1}{\ln a}\right)a^x + C$$

which follows from Theorem 6.14.

EXAMPLE 3   Integrating an exponential function to another base

Evaluate $\int 2^x \, dx$.

SOLUTION

$$\int 2^x \, dx = \frac{1}{\ln 2} 2^x + C$$

When we introduced the Power Rule, $D_x[x^n] = nx^{n-1}$, in Chapter 3, we required the exponent n to be a rational number. We now extend the rule to cover any real value of n. Try to prove this theorem using logarithmic differentiation.

**THEOREM 6.15
THE POWER RULE FOR
REAL EXPONENTS**

Let n be any real number and let u be a differentiable function of x.

1. $\dfrac{d}{dx}[x^n] = nx^{n-1}$ 2. $\dfrac{d}{dx}[u^n] = nu^{n-1}\dfrac{du}{dx}$

In the next example, we differentiate four different types of functions. Each function uses a different differentiation formula, depending on whether the base and exponent are constants or variables.

EXAMPLE 4   Comparing variables and constants

(a) $\dfrac{d}{dx}[e^e] = 0$ Constant Rule

(b) $\dfrac{d}{dx}[e^x] = e^x$ Exponential Rule

(c) $\dfrac{d}{dx}[x^e] = ex^{e-1}$ Power Rule

(d) $y = x^x$ Logarithmic differentiation
$\ln y = x \ln x$
$\dfrac{y'}{y} = x\left(\dfrac{1}{x}\right) + (\ln x)(1) = 1 + \ln x$
$y' = y(1 + \ln x) = x^x(1 + \ln x)$

Applications of exponential functions

We now look at a business problem in which the limit definition of e plays an important role. Suppose P dollars are deposited in a savings account at an annual interest rate r. If accumulated interest is deposited into the account, what is the balance in the account at the end of one year? Of course, the answer depends on the number of times the interest is compounded according to the formula

$$A = P\left(1 + \frac{r}{n}\right)^n$$

where n is the number of compoundings per year. For instance, the results for a deposit of $1000 at 8 percent interest compounded n times a year are as shown in Table 6.2.

TABLE 6.2

n	1	2	4	12	365
A	$1080.00	$1081.60	$1082.43	$1083.00	$1083.28

It may surprise you to discover that as n increases, the balance A approaches a limit. To develop this limit, we use the following theorem.

THEOREM 6.16 **A LIMIT INVOLVING** e	$e = \lim_{x \to \infty} \left(1 + \frac{1}{x}\right)^x = \lim_{x \to \infty} \left(\frac{x+1}{x}\right)^x$

In Section 8.7, Example 5, we will give a formal proof of this theorem. For now, we can compute the value of $[(x + 1)/x]^x$ for several values of x, as shown in Table 6.3.

TABLE 6.3

x	10	100	1,000	10,000	100,000	1,000,000
$\left(\dfrac{x+1}{x}\right)^x$	2.59374	2.70481	2.71692	2.71815	2.71827	2.71828

Now, using Theorem 6.16, we let $x = n/r$. Then $x \to \infty$ as $n \to \infty$, and we have

$$A = \lim_{n \to \infty} P\left(1 + \frac{r}{n}\right)^n = P \lim_{n \to \infty} \left[\left(1 + \frac{1}{n/r}\right)^{n/r}\right]^r$$
$$= P\left[\lim_{x \to \infty} \left(1 + \frac{1}{x}\right)^x\right]^r$$
$$= Pe^r.$$

We call this limit the balance after one year of **continuous compounding.** Thus, for a deposit of $1000 at 8 percent interest, compounded continuously, the balance at the end of one year would be

$$A = 1000e^{0.08} \approx \$1083.29.$$

These results are summarized as follows.

**SUMMARY OF COMPOUND
INTEREST FORMULAS**

Let P = amount of deposit, t = number of years, A = balance after t years, and r = annual interest rate (decimal form).

1. Compounded n times per year: $A = P\left(1 + \dfrac{r}{n}\right)^{nt}$

2. Compounded continuously: $A = Pe^{rt}$

EXAMPLE 5 Comparing continuous compounding to quarterly compounding

A deposit of \$2500 is made in an account that pays an annual interest rate of 10 percent. Find the balance in the account at the end of 5 years if the interest is compounded (a) quarterly and (b) continuously.

SOLUTION

(a) The balance after quarterly compounding is

$$A = P\left(1 + \frac{r}{n}\right)^{nt} = 2500\left(1 + \frac{0.1}{4}\right)^{4(5)}$$
$$= 2500(1.025)^{20} = \$4096.54.$$

(b) The balance after continuous compounding is

$$A = Pe^{rt} = 2500[e^{0.1(5)}]$$
$$= 2500e^{0.5} = \$4121.80.$$
◻

EXAMPLE 6 An application from biology

A bacterial culture is growing according to the *logistics growth function*

$$y = \frac{1.25}{1 + 0.25e^{-0.4t}}, \quad 0 \le t$$

where y is the weight of the culture in grams and t is the time in hours. Find the weight of the culture after (a) 0 hours, (b) 1 hour, and (c) 10 hours. (d) What is the limit as t approaches infinity?

SOLUTION

(a) When $t = 0$,

$$y = \frac{1.25}{1 + 0.25e^0} = \frac{1.25}{1.25} = 1 \text{ g.}$$

(b) When $t = 1$,

$$y = \frac{1.25}{1 + 0.25e^{-0.4}} \approx 1.071 \text{ g.}$$

(c) When $t = 10$,

$$y = \frac{1.25}{1 + 0.25e^{-4}} \approx 1.244 \text{ g.}$$

(d) Finally, taking the limit as t approaches infinity, we have

$$\lim_{t \to \infty} \frac{1.25}{1 + 0.25e^{-0.4t}} = \frac{1.25}{1 + 0} = 1.25 \text{ g.}$$

(See Figure 6.21.)

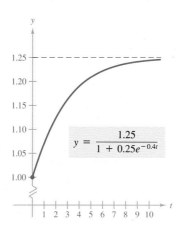

FIGURE 6.21

EXERCISES for Section 6.5

In Exercises 1–4, write the logarithmic equation as an exponential equation and vice versa.

1. (a) $2^3 = 8$ (b) $3^{-1} = \frac{1}{3}$

2. (a) $27^{2/3} = 9$ (b) $16^{3/4} = 8$

3. (a) $\log_{10} 0.01 = -2$ (b) $\log_{0.5} 8 = -3$

4. (a) $\log_3 \frac{1}{9} = -2$ (b) $49^{1/2} = 7$

In Exercises 5–10, solve for x (or b).

5. (a) $\log_{10} 1000 = x$ (b) $\log_{10} 0.1 = x$

6. (a) $\log_4 \frac{1}{64} = x$ (b) $\log_5 25 = x$

7. (a) $\log_3 x = -1$ (b) $\log_2 x = -4$

8. (a) $\log_b 27 = 3$ (b) $\log_b 125 = 3$

9. (a) $x^2 - x = \log_5 25$ (b) $3x + 5 = \log_2 64$

10. (a) $\log_3 x + \log_3 (x - 2) = 1$
   (b) $\log_{10} (x + 3) - \log_{10} x = 1$

In Exercises 11–14, sketch the graph of the given function.

11. $y = 3^x$ **12.** $y = 3^{x-1}$

13. $y = \left(\frac{1}{3}\right)^x$ **14.** $y = 2^{x^2}$

In Exercises 15 and 16, show that the given functions are inverses of each other by sketching their graphs on the same coordinate system.

15. $f(x) = 4^x$, $g(x) = \log_4 x$
16. $f(x) = 3^x$, $g(x) = \log_3 x$

In Exercises 17–28, find dy/dx.

17. $y = 4^x$ **18.** $y = 2^{-x}$

19. $y = 5^{x-2}$ **20.** $y = x(7^{-3x})$

21. $y = x^2 2^x$ **22.** $y = 2^{x^2} 3^{-x}$

23. $y = \log_3 x$ **24.** $y = \log_{10} 2x$

25. $y = \log_2 \frac{x^2}{x - 1}$ **26.** $y = \log_3 \frac{x\sqrt{x - 1}}{2}$

27. $y = \log_5 \sqrt{x^2 - 1}$ **28.** $y = \log_{10} \frac{x^2 - 1}{x}$

In Exercises 29–32, use logarithmic differentiation to find dy/dx.

29. $y = x^{2/x}$ **30.** $y = x^{x-1}$

31. $y = (x - 2)^{x+1}$ **32.** $y = (1 + x)^{1/x}$

In Exercises 33–38, evaluate the given integral.

33. $\displaystyle\int 3^x\, dx$ **34.** $\displaystyle\int 4^{-x}\, dx$

35. $\displaystyle\int_{-1}^{2} 2^x\, dx$ **36.** $\displaystyle\int_{-2}^{0} (3^3 - 5^2)\, dx$

37. $\displaystyle\int x5^{-x^2}\, dx$ **38.** $\displaystyle\int (3 - x)7^{(3-x)^2}\, dx$

39. Find the area of the region bounded by the graphs of $y = 3^x$, $y = 0$, $x = 0$, and $x = 3$.

40. Show that
$$\log_{10} 2 = \frac{\ln 2}{\ln 10}.$$

41. Find the amount of time necessary for P dollars to double if it is compounded continuously at $7\frac{1}{2}$ percent interest. Find the time necessary for it to triple.

42. Find the amount of time necessary for P dollars to double if it is compounded continuously at 9 percent interest. Find the time necessary for it to triple.

43. Complete the accompanying table to demonstrate that e can also be defined as
$$\lim_{x \to 0^+} (1 + x)^{1/x}.$$

x	1	10^{-1}	10^{-2}	10^{-4}	10^{-6}
$(1 + x)^{1/x}$					

44. Complete the accompanying table to find the time t necessary for P dollars to double if interest is compounded continuously at the rate r.

r	2%	4%	6%	8%	10%	12%
t						

45. If \$1000 is invested at $7\frac{1}{2}$ percent interest, find the amount after 10 years if interest is compounded
(a) annually   (b) semiannually
(c) quarterly   (d) monthly
(e) daily   (f) continuously

46. If \$2500 is invested at 12 percent interest, find the amount after 20 years if interest is compounded
(a) annually   (b) semiannually
(c) quarterly   (d) monthly
(e) daily   (f) continuously

47. The yield V (in millions of cubic feet per acre) for a stand of timber at age t is given by
$$V = 6.7e^{(-48.1)/t}$$
where t is measured in years.
(a) Find the limiting volume of wood per acre as t approaches infinity.
(b) Find the rate at which the yield is changing when $t = 20$ and $t = 60$ years.

48. The average typing speed (in the number of words per minute) after t weeks of lessons is given by
$$N = \frac{157}{1 + 5.4e^{-0.12t}}.$$
(a) Find the limiting number of words per minute as t approaches infinity.
(b) Find the rate at which typing speed is changing when $t = 5$ and $t = 25$ weeks.

49. In a group project in learning theory, a mathematical model for the proportion P of correct responses after n trials was found to be
$$P = \frac{0.83}{1 + e^{-0.2n}}.$$
(a) Find the limiting proportion of correct responses as n approaches infinity.
(b) Find the rate at which P is changing after $n = 3$ and $n = 10$ trials.

50. A lake is stocked with 500 fish, and their population increases according to the logistics curve
$$p(t) = \frac{10,000}{1 + 19e^{-t/5}}$$
where t is measured in months. At what rate is the fish population changing at the end of 1 month and at the end of 10 months? After how many months is the population increasing most rapidly? (See figure.)

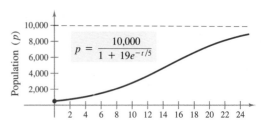

FIGURE FOR 50   Time in months (t)

51. Consider a deposit of \$100 placed in an account for 20 years at $r\%$ compounded continuously. Use a computer or graphics calculator to sketch the graphs on the same coordinate axes of the amount in the account if (a) $r = 6\%$, (b) $r = 9\%$, and (c) $r = 12\%$.

52. Repeat Exercise 51 for an account that earns interest for 50 years.

6.6 . Growth and Decay

Growth and decay ■ Applications

One practical application of exponential functions involves mathematical models in which the rate of change of a variable y is proportional to the value of y. If y is a function of time t, then we can write

| Rate of change of y | is | proportional to y |

$$\frac{dy}{dt} = ky.$$

The general solution to this **differential equation** is given in the following theorem.

**THEOREM 6.17
LAW OF EXPONENTIAL
GROWTH AND DECAY**

If y is a differentiable function of t such that $y > 0$ and $dy/dt = ky$, for some constant k, then

$$y = Ce^{kt}.$$

C is called the **initial value** of y, and k is called the **constant of proportionality**. **Exponential growth** is indicated by $k > 0$, and **exponential decay** by $k < 0$.

PROOF To solve the differentiable equation $y' = ky$, we use a technique called **separation of variables.**

$$y' = ky$$

$$\frac{y'}{y} = k \qquad \text{Divide by } y \text{ since } y \neq 0$$

$$\int \frac{y'}{y}\, dt = \int k\, dt \qquad \text{Integrate with respect to } t$$

$$\int \frac{1}{y}\, dy = \int k\, dt \qquad dy = y'\, dt$$

$$\ln y = kt + C_1$$

$$y = e^{kt+C_1} = e^{C_1}e^{kt} \qquad \text{Solve for } y$$

$$y = Ce^{kt} \qquad \text{Let } C = e^{C_1}$$

Example 1 involves radioactive decay, which is measured in terms of **half-life**—the number of years required for half of the atoms in a sample of radioactive material to decay. For example, the half-lives of some common radioactive isotopes are as follows.

Uranium (U^{238})	4,510,000,000 years
Plutonium (Pu^{230})	24,360 years
Carbon (C^{14})	5,730 years
Radium (Ra^{226})	1,620 years
Einsteinium (Es^{254})	276 days
Nobelium (No^{257})	23 seconds

Radioactive Half-life of 1620 Years

FIGURE 6.22

EXAMPLE 1 Radioactive decay

A sample contains 1 gram of radium. How much radium will remain after 1000 years? [Use a half-life of 1620 years, as shown in Figure 6.22.]

SOLUTION

Let y represent the mass (in grams) of radium in the sample. Since the rate of decay is proportional to y, we apply the Law of Exponential Decay to conclude that y is of the form $y = Ce^{kt}$, where t is measured in years. We are given the following values for the function y: $y = 1$ when $t = 0$, and $y = \frac{1}{2}$ when $t = 1620$. Applying the first of these two conditions, we have

$$y = 1 = Ce^{k(0)}. \qquad\qquad t = 0$$

Thus, $C = 1$. Now, by applying the second condition, we obtain

$$y = \frac{1}{2} = e^{k(1620)}. \qquad\qquad t = 1620$$

To solve for k, we take the natural log of both sides to obtain $\ln \frac{1}{2} = 1620k$. Therefore, $k = \left(\ln \frac{1}{2}\right)/1620 \approx -0.0004279$, and the equation for y as a function of time is

$$y = Ce^{kt} \approx e^{-0.0004279t}.$$

Finally, when $t = 1000$ years, the amount of radium remaining is

$$y \approx e^{-0.0004279(1000)} \approx 0.652 \text{ g.} \qquad t = 1000$$

REMARK   Example 1 demonstrates that finding exponential decay (or growth) functions basically involves solving for the constants C and k. Furthermore, we can see in Example 1 that it is easy to solve for C when we are given the value at $t = 0$.

In the next example, we demonstrate a procedure for solving for C and k in the growth equation $y = Ce^{kt}$ when we do not know the value of y at $t = 0$.

EXAMPLE 2 Population growth

Suppose that an experimental population of fruit flies increases according to the law of exponential growth. If there are 100 flies after the second day of the experiment and 300 flies after the fourth day, how many flies were in the original population?

SOLUTION

Let $y = Ce^{kt}$ be the number of flies at time t, where t is measured in days. Since $y = 100$ when $t = 2$ and $y = 300$ when $t = 4$, we have

$$100 = Ce^{2k} \qquad \text{and} \qquad 300 = Ce^{4k}.$$

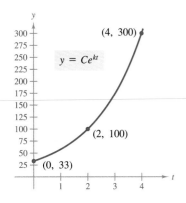

FIGURE 6.23

From the first equation, we have $C = 100e^{-2k}$. Substituting this value into the second equation, we have $300 = 100e^{-2k}e^{4k} = 100e^{2k}$. Now we can solve for k as follows.

$$\frac{300}{100} = e^{2k} \quad \Longrightarrow \quad \ln 3 = 2k \quad \Longrightarrow \quad k = \frac{\ln 3}{2} \approx 0.5493$$

Therefore, the equation for exponential growth is $y = Ce^{0.5493t}$. To solve for C, we can reapply the condition $y = 100$ when $t = 2$ and obtain

$$100 = Ce^{0.5493(2)} \quad \Longrightarrow \quad C = 100e^{-1.0986} \approx 33.$$

Thus, the original population (when $t = 0$) consisted of approximately $y = C = 33$ flies, as indicated in Figure 6.23.

EXAMPLE 3   Continuously compounded interest

Money is deposited in an account for which the interest is compounded continuously. If the balance doubles in 6 years, what is the annual percentage rate?

SOLUTION

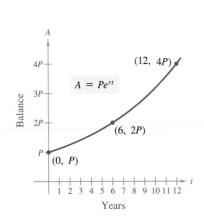

FIGURE 6.24

The formula for the balance A using continuously compounded interest is given by the exponential growth function

$$A = Pe^{rt}$$

where P is the original deposit, r is the annual rate (in decimal form), and t is the time in years. When $t = 6$, we know that $A = 2P$, as shown in Figure 6.24. Thus, we have

$$2P = Pe^{6r} \quad \Longrightarrow \quad 2 = e^{6r} \quad \Longrightarrow \quad \ln 2 = 6r$$

and we conclude that the annual percentage rate is

$$r = \frac{1}{6} \ln 2 \approx 0.1155 = 11.55\%.$$

EXAMPLE 4   Declining sales

Four months after it stopped advertising, a manufacturing company noticed that its sales had dropped from 100,000 units per month to 80,000 units per month. If the sales follow an exponential pattern of decline, what will they be after another two months?

SOLUTION

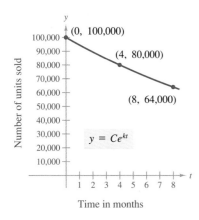

FIGURE 6.25

We use the exponential model $y = Ce^{kt}$, where t is measured in months, as shown in Figure 6.25. From the initial condition ($t = 0$), we know that $C = 100,000$. Moreover, since $y = 80,000$ when $t = 4$, we have

$$80,000 = 100,000e^{4k} \quad \Longrightarrow \quad 0.8 = e^{4k} \quad \Longrightarrow \quad \ln(0.8) = 4k.$$

Thus, $k = [\ln (0.8)]/4 \approx -0.0558$. Finally, after two more months ($t = 6$), we can expect the monthly sales rate to be

$$y = 100{,}000e^{-0.0558(6)} \approx 71{,}500 \text{ units.} \qquad \blacksquare$$

In each of the four examples describing exponential growth or decay, we did not actually have to solve the differential equation $y' = ky$. (We did that once in the proof of Theorem 6.17.) In the last example in this section, we demonstrate a problem whose solution involves the separation of variables technique used in proving the Law of Exponential Growth. The example involves **Newton's Law of Cooling,** which states that the rate of change in the temperature of an object is proportional to the difference between the object's temperature and the temperature of the surrounding medium.

EXAMPLE 5   Newton's Law of Cooling

Let y represent the temperature (in degrees Fahrenheit) of an object in a room whose temperature is kept at a constant 60°. If the object cools from 100° to 90° in 10 minutes, how much longer will it take for its temperature to decrease to 80°?

SOLUTION

From Newton's Law of Cooling, we know that the rate of change in y is proportional to the difference between y and 60. This gives us

$$\frac{dy}{dt} = k(y - 60), \quad 80 \le y \le 100$$

$$\left(\frac{1}{y - 60} \right) \frac{dy}{dt} = k$$

$$\int \frac{1}{y - 60}\, dy = \int k\, dt \qquad \text{Separation of variables}$$

$$\ln |y - 60| = kt + C.$$

Since $y > 60$, we can delete the absolute value signs. Now, using $y = 100$ when $t = 0$, we have $C = \ln 40$, which implies that

$$kt = \ln (y - 60) - \ln 40 = \ln \left(\frac{y - 60}{40} \right).$$

Furthermore, since $y = 90$ when $t = 10$, we have $k = \frac{1}{10} \ln \frac{3}{4}$, and thus

$$t = \left(\frac{10}{\ln (3/4)} \right) \ln \left(\frac{y - 60}{40} \right).$$

Finally, when $y = 80$,

$$t = \frac{10 \ln (1/2)}{\ln (3/4)} \approx 24.09 \text{ min.}$$

Therefore, it will require approximately 14.09 *more* minutes for the object to cool to a temperature of 80°. $\qquad \blacksquare$

EXERCISES for Section 6.6

In Exercises 1–4, find the exponential function $y = Ce^{kt}$ that passes through the two given points.

1.

2.

3.

4.

In Exercises 5–10, complete the table for a savings account in which interest is compounded continuously.

	Initial investment	Annual rate	Time to double	Amount after 10 yrs	Amount after 25 yrs
5.	$1,000	12%			
6.	$20,000	$10\frac{1}{2}\%$			
7.	$750		$7\frac{3}{4}$ years		
8.	$10,000		5 years		
9.	$500			$1,292.85	
10.	$2,000				$6,008.33

11. The management at a certain factory has found that a worker can produce at most 30 units in a day. The learning curve for the number of units N produced per day after a new employee has worked t days is given by

$$N = 30(1 - e^{kt}).$$

After 20 days on the job, a particular worker produced 19 units.
(a) Find the learning curve for this worker.
(b) How many days should pass before this worker is producing 25 units per day?

12. If in Exercise 11 the management requires a new employee to produce at least 20 units per day after 30 days on the job, find
(a) the learning curve that describes this minimum requirement,

(b) the number of days before a minimal achiever is producing 25 units per day.

13. The sales S (in thousands of units) of a new product after it is on the market t years is given by

$$S = Ce^{k/t}.$$

(a) Find S as a function of t if 5000 units have been sold after 1 year and the saturation point for the market is 30,000 (that is, $\lim_{t \to \infty} S = 30$).
(b) How many units have been sold after 5 years?
(c) Sketch a graph of this sales function.

14. The number of sales S (in thousands of units) of a new product after it has been on the market t years is given by

$$S = 30(1 - e^{kt}).$$

(a) Find S as a function of t if 5000 units have been sold after 1 year.
(b) How many units will saturate this market?
(c) How many units will have been sold after 5 years?
(d) Sketch a graph of this sales function.

15. A certain type of bacteria increases continuously at a rate proportional to the number present. If there are 100 present at a given time and 300 present 5 hours later, how many will there be 10 hours after the initial given time?

16. Given the conditions of Exercise 15, how long would it take for the number of bacteria to double?

17. In 1970 the population of a town was 2500, and in 1980 it was 3350. Assuming the population increases continuously at a constant rate proportional to the existing population, estimate the population in the year 2000.

18. Given the conditions of Exercise 17, how many years would be necessary for the population to double?

In Exercises 19–24, complete the table for the given radioactive isotope.

	Isotope	Half-life (in yrs)	Initial quantity	Amount after 1,000 yrs	Amount after 10,000 yrs
19.	Ra^{226}	1,620	10 grams		
20.	Ra^{226}	1,620		1.5 grams	
21.	C^{14}	5,730			2 grams
22.	C^{14}	5,730	3 grams		
23.	Pu^{230}	24,360		2.1 grams	
24.	Pu^{230}	24,360			0.4 gram

25. Radioactive radium has a half-life of approximately 1620 years. What percentage of a given amount remains after 100 years?

26. If radioactive material decays continuously at a rate proportional to the amount present, find the half-life of the material if after 1 year 99.57 percent of an initial amount still remains.

27. C^{14} dating assumes that the carbon dioxide on earth today has the same radioactive content as it did centuries ago. If this is true, then the amount of C^{14} absorbed by a tree that grew several centuries ago should be the same as the amount of C^{14} absorbed by a tree growing today. A piece of ancient charcoal contains only 15% as much of the radioactive carbon as a piece of modern charcoal. How long ago was the tree burned to make the ancient charcoal? [The half-life of C^{14} is 5730 years.]

28. On the Richter Scale, the magnitude R of an earthquake of intensity I is given by

$$R = \frac{\ln I - \ln I_0}{\ln 10}$$

where I_0 is the minimum intensity used for comparison. Assume $I_0 = 1$.
   (a) Find the intensity of the 1906 San Francisco earthquake when $R = 8.3$.
   (b) Find the factor by which the intensity is increased if the Richter Scale measurement is doubled.
   (c) Find dR/dI.

29. Using Newton's Law of Cooling (see Example 5), determine the reading on a thermometer 5 minutes after it is taken from a room at 72° Fahrenheit to the outdoors where the temperature is 20°, if the reading dropped to 48° after 1 minute.

30. An object in a room at 70° cools from 350° to 150° in 45 minutes. Using Newton's Law of Cooling, find the time necessary for the object to cool to 80°.

31. Using Newton's Law of Cooling, determine the outdoor temperature if a thermometer is taken from a room where the temperature is 68° to the outdoors, where after $\frac{1}{2}$ minute and 1 minute the thermometer reads 53° and 42°, respectively.

32. When an object is removed from a furnace and placed in an environment with a constant temperature of 90°, its core temperature is 1500°. One hour after it is removed, the core temperature is 1120°. Find the core temperature 5 hours after the unit is removed from the furnace.

33. Atmospheric pressure P (measured in millimeters of mercury) decreases exponentially with increasing altitude x (measured in meters). The pressure is 760 millimeters of mercury at sea level ($x = 0$) and 672.71 millimeters of mercury at an altitude of 1000 meters. Find the pressure at an altitude of 3000 meters.

34. The standard normal probability density function is

$$f(x) = \frac{1}{\sqrt{2\pi}} e^{-x^2/2}.$$

If x is chosen at random from a population with this density, then the probability that x lies in the interval $[a, b]$ is given by

$$P(a \le x \le b) = \int_a^b f(x)\, dx.$$

Estimate the following probabilities using Simpson's Rule with $n = 6$.
   (a) $P(0 \le x \le 1)$ (b) $P(0 \le x \le 2)$

35. Because of a slump in the economy, a company finds that its annual revenues have dropped from $742,000 in 1988 to $632,000 in 1990. If the revenue is following an exponential pattern of decline, what is the expected revenue for 1991? (Let $t = 0$ represent 1988.)

6.7 Inverse Trigonometric Functions and Differentiation

Inverse trigonometric functions ▪ Derivatives of inverse trigonometric functions ▪ Review of basic differentiation formulas

$y = \sin x$
Domain: $[-\pi/2,\ \pi/2]$
Range: $[-1,\ 1]$

FIGURE 6.26

This section begins with a rather startling statement: *None of the six basic trigonometric functions has an inverse.* This statement is true because all six trigonometric functions are periodic, and hence not one-to-one. In this section we will examine these six functions to see whether we can redefine their domains in such a way that they will have inverses on the *restricted domains*.

For example, in Section 6.3, we demonstrated that the sine function is increasing (and therefore is one-to-one) on the interval $[-\pi/2,\ \pi/2]$, as shown in Figure 6.26. On this interval we define the inverse of the *restricted* sine function to be

$$y = \arcsin x \quad \text{if and only if} \quad \sin y = x$$

where $-1 \le x \le 1$ and $-\pi/2 \le \arcsin x \le \pi/2$.

Under suitable restrictions, each of the six trigonometric functions is one-to-one and so possesses an inverse, as indicated in the following definition. (The term iff is used to represent the phrase "if and only if.")

DEFINITION OF INVERSE TRIGONOMETRIC FUNCTIONS

Function	Domain	Range
$y = \mathbf{arcsin}\ x$ iff $\sin y = x$	$-1 \le x \le 1$	$-\dfrac{\pi}{2} \le y \le \dfrac{\pi}{2}$
$y = \mathbf{arccos}\ x$ iff $\cos y = x$	$-1 \le x \le 1$	$0 \le y \le \pi$
$y = \mathbf{arctan}\ x$ iff $\tan y = x$	$-\infty < x < \infty$	$-\dfrac{\pi}{2} < y < \dfrac{\pi}{2}$
$y = \mathbf{arccot}\ x$ iff $\cot y = x$	$-\infty < x < \infty$	$0 < y < \pi$
$y = \mathbf{arcsec}\ x$ iff $\sec y = x$	$\lvert x \rvert \ge 1$	$0 \le y \le \pi,\ y \ne \dfrac{\pi}{2}$
$y = \mathbf{arccsc}\ x$ iff $\csc y = x$	$\lvert x \rvert \ge 1$	$-\dfrac{\pi}{2} \le y \le \dfrac{\pi}{2},\ y \ne 0$

REMARK   The term arcsin x is read as the "inverse sine of x" or sometimes the "angle whose sine is x." An alternate notation for the inverse sine function is $\sin^{-1} x$.

The graphs of these six inverse functions are shown in Figure 6.27 on page 372. (Compare these to the graphs of the six trigonometric functions given in Section 1.6.)

Domain: $[-1, 1]$
Range: $[-\pi/2, \pi/2]$

$y = \arcsin x$

Domain: $(-\infty, -1]$ and $[1, \infty)$
Range: $[-\pi/2, 0)$ and $(0, \pi/2]$

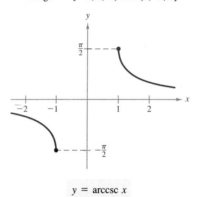

$y = \text{arccsc } x$

Domain: $(-\infty, \infty)$
Range: $(-\pi/2, \pi/2)$

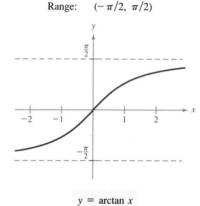

$y = \arctan x$

Domain: $[-1, 1]$
Range: $[0, \pi]$

$y = \arccos x$

Domain: $(-\infty, -1]$ and $[1, \infty)$
Range: $[0, \pi/2)$ and $(\pi/2, \pi]$

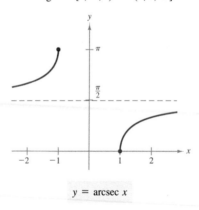

$y = \text{arcsec } x$

Domain: $(-\infty, \infty)$
Range: $(0, \pi)$

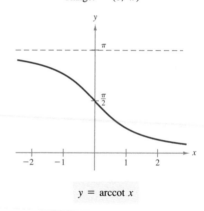

$y = \text{arccot } x$

FIGURE 6.27

When evaluating inverse trigonometric functions, remember that they denote *angles in radian measure*.

EXAMPLE 1 Evaluating inverse trigonometric functions

Evaluate the following.

(a) $\arcsin \left(-\dfrac{1}{2}\right)$ (b) $\arccos 0$

(c) $\arctan \sqrt{3}$ (d) $\arcsin (0.3)$

SOLUTION

(a) By definition, $y = \arcsin\left(-\frac{1}{2}\right)$ implies that $\sin y = -\frac{1}{2}$. In the interval $[-\pi/2,\ \pi/2]$, we choose $y = -\pi/6$. Therefore,

$$\arcsin\left(-\frac{1}{2}\right) = -\frac{\pi}{6}.$$

(b) By definition, $y = \arccos 0$ implies that $\cos y = 0$. In the interval $[0,\ \pi]$, we choose $y = \pi/2$. Therefore,

$$\arccos 0 = \frac{\pi}{2}.$$

(c) By definition, $y = \arctan \sqrt{3}$ implies that $\tan y = \sqrt{3}$. In the interval $(-\pi/2,\ \pi/2)$, we choose $y = \pi/3$. Therefore,

$$\arctan \sqrt{3} = \frac{\pi}{3}.$$

(d) By using a calculator set in *radian mode*, we obtain

$$\arcsin (0.3) \approx 0.3047.$$

Inverse functions possess the properties

$$f(f^{-1}(x)) = x \qquad \text{and} \qquad f^{-1}(f(x)) = x.$$

When applying these properties to inverse trigonometric functions, remember that the trigonometric functions possess inverses only in restricted domains. For *x*-values outside these domains, these two properties do not hold. For example,

$$\arcsin (\sin \pi) = \arcsin 0 = 0 \neq \pi.$$

INVERSE PROPERTIES

If $-1 \le x \le 1$ and $-\pi/2 \le y \le \pi/2$, then

$$\sin (\arcsin x) = x \qquad \text{and} \qquad \arcsin (\sin y) = y.$$

If $-\pi/2 < y < \pi/2$, then

$$\tan (\arctan x) = x \qquad \text{and} \qquad \arctan (\tan y) = y.$$

If $|x| \ge 1$ and $0 \le y < \pi/2$ or $\pi/2 < y \le \pi$, then

$$\sec (\operatorname{arcsec} x) = x \qquad \text{and} \qquad \operatorname{arcsec} (\sec y) = y.$$

REMARK   Similar properties hold for the other three inverse trigonometric functions.

Notice how we use one of these inverse properties to solve the equation in the next example.

EXAMPLE 2   Solving an equation

Solve for x in the equation arctan $(2x - 3) = \pi/4$.

SOLUTION

$$\arctan (2x - 3) = \frac{\pi}{4}$$

$$\tan [\arctan (2x - 3)] = \tan \frac{\pi}{4}$$

$$2x - 3 = 1$$

$$x = 2$$

There are some important types of problems in calculus in which we evaluate expressions like sec (arctan x). To solve this type of problem, it helps to use right triangles, as demonstrated in the next example.

EXAMPLE 3   Using right triangles

(a) Given $y = \arcsin x$, where $0 < y < \pi/2$, find cos y.
(b) Given $y = \text{arcsec} (\sqrt{5}/2)$, find tan y.

SOLUTION

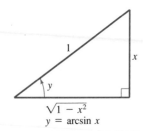

$$\sqrt{1 - x^2}$$
$$y = \arcsin x$$

FIGURE 6.28

(a) Since y is the angle whose sine is x, we form a right triangle having an acute angle $y = \arcsin x$, as shown in Figure 6.28. Therefore,

$$\cos y = \cos (\arcsin x) \qquad \cos y = \frac{\text{adj.}}{\text{hyp.}}$$

$$= \sqrt{1 - x^2}.$$

It can also be shown that for $-\pi/2 < y \leq 0$, $y = \arcsin x$ implies $\cos y = \sqrt{1 - x^2}$.

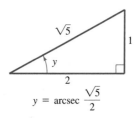

$$y = \text{arcsec} \frac{\sqrt{5}}{2}$$

FIGURE 6.29

(b) Since y is the angle whose secant is $\sqrt{5}/2$, we can sketch this angle as part of a triangle, as shown in Figure 6.29. Therefore,

$$\tan y = \tan \left[\text{arcsec} \left(\frac{\sqrt{5}}{2} \right) \right] \qquad \tan y = \frac{\text{opp.}}{\text{adj.}}$$

$$= \frac{1}{2}.$$

Derivatives of inverse trigonometric functions

In Section 6.1 we saw that the derivative of the *transcendental* function $f(x) = \ln x$ is the *algebraic* function $f'(x) = 1/x$. We will now see that the derivatives of the inverse trigonometric functions also are algebraic, even though the inverse trigonometric functions are themselves transcendental.

The following theorem lists the derivative of each of the six inverse trigonometric functions. Note that the derivatives of arccos u, arccot u, and arccsc u are the *negatives* of the derivatives of arcsin u, arctan u, and arcsec u, respectively.

**THEOREM 6.18
DERIVATIVES OF THE INVERSE
TRIGONOMETRIC FUNCTIONS**

Let u be a differentiable function of x.

$$\frac{d}{dx}[\arcsin u] = \frac{u'}{\sqrt{1 - u^2}} \qquad \frac{d}{dx}[\arccos u] = \frac{-u'}{\sqrt{1 - u^2}}$$

$$\frac{d}{dx}[\arctan u] = \frac{u'}{1 + u^2} \qquad \frac{d}{dx}[\text{arccot } u] = \frac{-u'}{1 + u^2}$$

$$\frac{d}{dx}[\text{arcsec } u] = \frac{u'}{|u|\sqrt{u^2 - 1}} \qquad \frac{d}{dx}[\text{arccsc } u] = \frac{-u'}{|u|\sqrt{u^2 - 1}}$$

PROOF   We prove the first of these formulas and leave the proofs of the others as an exercise (see Exercise 67). Let $f(x) = \sin x$ and $f^{-1}(x) = g(x) = \arcsin x$. Then, by using Theorem 6.10, we have

$$\frac{d}{dx}[g(x)] = \frac{1}{f'[g(x)]} = \frac{1}{\cos(\arcsin x)}.$$

Now, by using the result of Example 3(a), we have

$$\frac{d}{dx}[\arcsin x] = \frac{1}{\cos(\arcsin x)} = \frac{1}{\sqrt{1 - x^2}}.$$

If u is a differentiable function of x, then the Chain Rule gives us

$$\frac{d}{dx}[\arcsin u] = \frac{u'}{\sqrt{1 - u^2}}, \quad \text{where } u' = \frac{du}{dx}.$$

There is no common agreement regarding the definition of arcsec x (or arccsc x) for negative values of x. When we defined the range of the arcsecant, we chose to preserve the reciprocal identity

$$\text{arcsec } x = \arccos \frac{1}{x}.$$

For example, to evaluate arcsec x when $x = -2$, we write arcsec $(-2) =$ arccos $(-0.5) \approx 2.09$.

One of the consequences of the definition of the inverse secant function given in this text is that its graph has a positive slope at every x-value in its domain. (See Figure 6.27.) This accounts for the absolute value sign in the formula for the derivative of arcsec x.

EXAMPLE 4 Differentiating inverse trigonometric functions

(a) $\dfrac{d}{dx}[\arctan(3x)] = \dfrac{3}{1+(3x)^2}$ $u = 3x$

$= \dfrac{3}{1+9x^2}$

(b) $\dfrac{d}{dx}[\arcsin\sqrt{x}] = \dfrac{(1/2)x^{-1/2}}{\sqrt{1-x}}$ $u = \sqrt{x}$

$= \dfrac{1}{2\sqrt{x}\sqrt{1-x}}$

$= \dfrac{1}{2\sqrt{x-x^2}}$

(c) $\dfrac{d}{dx}[\text{arcsec }e^{2x}] = \dfrac{2e^{2x}}{e^{2x}\sqrt{(e^{2x})^2-1}}$ $u = e^{2x}$

$= \dfrac{2e^{2x}}{e^{2x}\sqrt{e^{4x}-1}}$

$= \dfrac{2}{\sqrt{e^{4x}-1}}$

Note that we omit the absolute value sign because $e^{2x} > 0$.

EXAMPLE 5 A derivative that can be simplified

Differentiate $y = \arcsin x + x\sqrt{1-x^2}$.

SOLUTION

$y' = \dfrac{1}{\sqrt{1-x^2}} + x\left(\dfrac{1}{2}\right)(-2x)(1-x^2)^{-1/2} + \sqrt{1-x^2}$

$= \dfrac{1}{\sqrt{1-x^2}} - \dfrac{x^2}{\sqrt{1-x^2}} + \sqrt{1-x^2}$

$= \sqrt{1-x^2} + \sqrt{1-x^2}$

$= 2\sqrt{1-x^2}$

EXAMPLE 6 Graphing an inverse trigonometric function

Sketch the graph of $y = (\arctan x)^2$.

SOLUTION

From the derivative

$$y' = 2(\arctan x)\left(\frac{1}{1 + x^2}\right) = \frac{2 \arctan x}{1 + x^2}$$

we can see that the only critical number is $x = 0$. By the First Derivative Test, this value corresponds to a relative minimum. From the second derivative

$$y'' = \frac{(1 + x^2)\left(\dfrac{2}{1 + x^2}\right) - (2 \arctan x)(2x)}{(1 + x^2)^2} = \frac{2(1 - 2x \arctan x)}{(1 + x^2)^2}$$

we see that points of inflection occur when $2x \arctan x = 1$. Using Newton's Method, we find that these points occur when $x \approx \pm 0.765$. Finally, the graph has a horizontal asymptote at $y = \pi^2/4$ since

$$\lim_{x \to \pm\infty} (\arctan x)^2 = \frac{\pi^2}{4}.$$

The graph is shown in Figure 6.30.

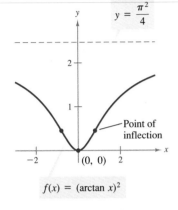

FIGURE 6.30

$f(x) = (\arctan x)^2$

We conclude this section with an example describing an application involving the derivative of an inverse trigonometric function.

EXAMPLE 7   An application

A photographer is taking a picture of a four-foot painting hung in an art gallery. The camera lens is one foot below the lower edge of the painting, as shown in Figure 6.31. How far should the camera be from the painting to maximize the angle subtended by the camera lens?

SOLUTION

In Figure 6.31, we let β be the angle we wish to maximize. Then we have

$$\beta = \theta - \alpha = \text{arccot} \, \frac{x}{5} - \text{arccot} \, x.$$

Differentiating, we have

$$\frac{d\beta}{dx} = \frac{-1/5}{1 + (x^2/25)} - \frac{-1}{1 + x^2}$$

$$= \frac{-5}{25 + x^2} + \frac{1}{1 + x^2}$$

$$= \frac{4(5 - x^2)}{(25 + x^2)(1 + x^2)}.$$

Since $d\beta/dx = 0$ when $x = \sqrt{5}$, we conclude from the First Derivative Test that this distance yields a maximum value for β.

FIGURE 6.31

Galileo Galilei

Review of basic differentiation formulas

In the 1600s, Europe was ushered into the scientific age by such great thinkers as Descartes, Galileo, Huygens, Newton, and Kepler. These men believed that nature is governed by basic laws—laws that can, for the most part, be written in terms of mathematical equations. One of the most influential publications during this period was written by Galileo Galilei (1564–1642). It is called *Dialogue on the Great World Systems,* and it has become a classic description of modern scientific thought.

As mathematics developed during the past few hundred years, a list of elementary functions has proven sufficient for modeling most phenomena in physics, chemistry, biology, engineering, economics, and a variety of other fields. An **elementary function** is one that can be formed as the sum, product, quotient, or composition of functions from the following list.

Algebraic functions	*Transcendental functions*
Polynomial functions	Logarithmic functions
Rational functions	Exponential functions
Functions involving radicals	Trigonometric functions
	Inverse trigonometric functions

There are isolated cases of the use of other types of functions, but this list is fairly comprehensive. Moreover, with the differentiation formulas introduced in this section, we can differentiate *any* elementary function. For convenience, we summarize these differentiation rules here.

BASIC DIFFERENTIATION RULES FOR ELEMENTARY FUNCTIONS

1. $\frac{d}{dx}[cu] = cu'$

2. $\frac{d}{dx}[u \pm v] = u' \pm v'$

3. $\frac{d}{dx}[uv] = uv' + vu'$

4. $\frac{d}{dx}\left[\frac{u}{v}\right] = \frac{vu' - uv'}{v^2}$

5. $\frac{d}{dx}[c] = 0$

6. $\frac{d}{dx}[u^n] = nu^{n-1}u'$

7. $\frac{d}{dx}[x] = 1$

8. $\frac{d}{dx}[|u|] = \frac{u}{|u|}(u'), \ u \neq 0$

9. $\frac{d}{dx}[\ln u] = \frac{u'}{u}$

10. $\frac{d}{dx}[e^u] = e^u u'$

11. $\frac{d}{dx}[\sin u] = (\cos u)u'$

12. $\frac{d}{dx}[\cos u] = -(\sin u)u'$

13. $\frac{d}{dx}[\tan u] = (\sec^2 u)u'$

14. $\frac{d}{dx}[\cot u] = -(\csc^2 u)u'$

15. $\frac{d}{dx}[\sec u] = (\sec u \tan u)u'$

16. $\frac{d}{dx}[\csc u] = -(\csc u \cot u)u'$

17. $\frac{d}{dx}[\arcsin u] = \frac{u'}{\sqrt{1 - u^2}}$

18. $\frac{d}{dx}[\arccos u] = \frac{-u'}{\sqrt{1 - u^2}}$

19. $\frac{d}{dx}[\arctan u] = \frac{u'}{1 + u^2}$

20. $\frac{d}{dx}[\text{arccot } u] = \frac{-u'}{1 + u^2}$

21. $\frac{d}{dx}[\text{arcsec } u] = \frac{u'}{|u|\sqrt{u^2 - 1}}$

22. $\frac{d}{dx}[\text{arccsc } u] = \frac{-u'}{|u|\sqrt{u^2 - 1}}$

EXERCISES for Section 6.7

In Exercises 1–10, evaluate the given expression.

1. $\arcsin \dfrac{1}{2}$ **2.** $\arcsin 0$

3. $\arccos \dfrac{1}{2}$ **4.** $\arccos 0$

5. $\arctan \dfrac{\sqrt{3}}{3}$ **6.** $\text{arccot}\,(-1)$

7. $\text{arccsc}\,\sqrt{2}$ **8.** $\arcsin\,(-0.39)$
9. $\text{arcsec}\,1.269$ **10.** $\arctan\,(-3)$

In Exercises 11–16, evaluate the given expression without a calculator. [Hint: Make a sketch of a right triangle, as illustrated in Example 3.]

11. (a) $\sin\left(\arcsin \dfrac{1}{2}\right)$ (b) $\cos\left(2\arcsin \dfrac{1}{2}\right)$

12. (a) $\tan\left(\arccos \dfrac{\sqrt{2}}{2}\right)$ (b) $\cos\left(\arcsin \dfrac{5}{13}\right)$

13. (a) $\sin\left(\arctan \dfrac{3}{4}\right)$ (b) $\sec\left(\arcsin \dfrac{4}{5}\right)$

14. (a) $\tan\,(\text{arccot}\,2)$ (b) $\cos\,(\text{arcsec}\,\sqrt{5})$

15. (a) $\cot\left[\arcsin\left(-\dfrac{1}{2}\right)\right]$ (b) $\csc\left[\arctan\left(-\dfrac{5}{12}\right)\right]$

16. (a) $\sec\left[\arctan\left(-\dfrac{3}{5}\right)\right]$ (b) $\tan\left[\arcsin\left(-\dfrac{5}{6}\right)\right]$

In Exercises 17–26, write the given expression in algebraic form.

17. $\tan\,(\arctan x)$ **18.** $\sin\,(\arccos x)$
19. $\cos\,(\arcsin 2x)$ **20.** $\sec\,(\arctan 3x)$
21. $\sin\,(\text{arcsec}\,x)$ **22.** $\cos\,(\text{arccot}\,x)$

23. $\tan\left(\text{arcsec}\,\dfrac{x}{3}\right)$ **24.** $\sec\,[\arcsin\,(x-1)]$

25. $\csc\left(\arctan \dfrac{x}{\sqrt{2}}\right)$ **26.** $\cos\left(\arcsin \dfrac{x-h}{r}\right)$

In Exercises 27 and 28, fill in the blank.

27. $\arctan \dfrac{9}{x} = \arcsin\,(\blacksquare)$

28. $\arcsin \dfrac{\sqrt{36-x^2}}{6} = \arccos\,(\blacksquare)$

In Exercises 29 and 30, verify each identity.

29. (a) $\text{arccsc}\,x = \arcsin \dfrac{1}{x},\ |x| \ge 1$

(b) $\text{arccot}\,x = \arctan \dfrac{1}{x},\ x > 0$

30. (a) $\arcsin\,(-x) = -\arcsin x,\ |x| \le 1$
(b) $\arccos\,(-x) = \pi - \arccos x,\ |x| \le 1$

In Exercises 31–34, sketch the graph of the function.

31. $f(x) = \arcsin\,(x-1)$ **32.** $f(x) = \arctan x + \dfrac{\pi}{2}$

33. $f(x) = \text{arcsec}\,2x$ **34.** $f(x) = \arccos \dfrac{x}{4}$

In Exercises 35–38, solve the given equation for x.

35. $\arcsin\,(3x - \pi) = \dfrac{1}{2}$ **36.** $\arctan 2x = -1$

37. $\arcsin \sqrt{2x} = \arccos \sqrt{x}$
38. $\arccos x = \text{arcsec}\,x$

In Exercises 39–58, find the derivative of the given function.

39. $f(x) = \arcsin 2x$ **40.** $f(x) = \arcsin x^2$
41. $f(x) = 2\arcsin\,(x-1)$ **42.** $f(x) = \arccos \sqrt{x}$

43. $f(x) = 3\arccos \dfrac{x}{2}$ **44.** $f(x) = \arctan \sqrt{x}$

45. $f(x) = \arctan 5x$ **46.** $f(x) = x\arctan x$

47. $f(x) = \arccos \dfrac{1}{x}$ **48.** $f(x) = \text{arcsec}\,2x$

49. $f(x) = \arcsin x + \arccos x$
50. $f(x) = \text{arcsec}\,x + \text{arccsc}\,x$
51. $h(t) = \sin\,(\arccos t)$ **52.** $g(t) = \tan\,(\arcsin t)$

53. $f(t) = \dfrac{1}{\sqrt{6}}\arctan \dfrac{\sqrt{6}t}{2}$

54. $f(x) = \dfrac{1}{2}\left(\dfrac{1}{2}\ln \dfrac{x+1}{x-1} - \arctan x\right)$

55. $f(x) = \dfrac{1}{2}\left(\dfrac{1}{2}\ln \dfrac{x+1}{x-1} + \arctan x\right)$

56. $f(x) = \dfrac{1}{2}(x\sqrt{1-x^2} + \arcsin x)$
57. $f(x) = x\arcsin x + \sqrt{1-x^2}$
58. $f(x) = x\arctan 2x - \dfrac{1}{4}\ln\,(1+4x^2)$

In Exercises 59 and 60, find the point of inflection of the graph of the given function.

59. $f(x) = \arcsin x$ **60.** $f(x) = \text{arccot}\,2x$

In Exercises 61 and 62, find any relative extrema of the given function.

61. $f(x) = \text{arcsec}\,x - x$ **62.** $f(x) = \arcsin x - 2x$

In Exercises 63 and 64, find the point of intersection of the graphs of the given functions.

63. $y = \arccos x$, $y = \arctan x$
64. $y = \arcsin x$, $y = \arccos x$

65. A small boat is being pulled toward a dock that is 10 feet above the water. The rope is being pulled in at a rate of 1.5 feet per second. Find the rate at which the angle the rope makes with the horizontal is changing when 20 feet of rope is out.

66. An observer is standing 300 feet from the point at which a balloon is released. The balloon rises at a rate of 5 feet per second. How fast is the angle of elevation of the observer's line of sight increasing when the balloon is 100 feet high?

67. Verify the following differentiation formulas.

(a) $\dfrac{d}{dx}[\arctan u] = \dfrac{u'}{1 + u^2}$

(b) $\dfrac{d}{dx}[\text{arcsec } u] = \dfrac{u'}{|u|\sqrt{u^2 - 1}}$

(c) $\dfrac{d}{dx}[\arccos u] = \dfrac{-u'}{\sqrt{1 - u^2}}$

(d) $\dfrac{d}{dx}[\text{arccot } u] = \dfrac{-u'}{1 + u^2}$

(e) $\dfrac{d}{dx}[\text{arccsc } u] = \dfrac{-u'}{|u|\sqrt{u^2 - 1}}$

68. Show that the function

$$f(x) = \arcsin\left(\frac{x - 2}{2}\right) - 2 \arcsin \frac{\sqrt{x}}{2}$$

is constant for $0 \le x \le 4$.

6.8 Inverse Trigonometric Functions: Integration and Completing the Square

Integrals involving inverse trigonometric functions ▪ Completing the square ▪ Review of basic integration formulas

The derivatives of the six inverse trigonometric functions occur in three pairs. In each pair the derivative of one function is the negative of the other. For example,

$$\frac{d}{dx}[\arcsin x] = \frac{1}{\sqrt{1 - x^2}} \quad \text{and} \quad \frac{d}{dx}[\arccos x] = -\frac{1}{\sqrt{1 - x^2}}.$$

When listing the *antiderivative* that corresponds to each of the inverse trigonometric functions, we need use only one member from each pair. For example, we choose to use arcsin x as the antiderivative of $1/\sqrt{1 - x^2}$, rather than $-\arccos x$. The next theorem gives one antiderivative formula for each of the three pairs.

THEOREM 6.19
INTEGRALS INVOLVING INVERSE TRIGONOMETRIC FUNCTIONS

Let u be a differentiable function of x, and let $a > 0$.

$$\int \frac{du}{\sqrt{a^2 - u^2}} = \arcsin \frac{u}{a} + C$$

$$\int \frac{du}{a^2 + u^2} = \frac{1}{a} \arctan \frac{u}{a} + C$$

$$\int \frac{du}{u\sqrt{u^2 - a^2}} = \frac{1}{a} \text{arcsec} \frac{|u|}{a} + C$$

PROOF   We prove the first formula and leave the remaining proofs as an exercise (see Exercise 52). Let $y = \arcsin (u/a)$. Then

$$y' = \frac{1}{\sqrt{1 - (u/a)^2}}\left(\frac{u'}{a}\right) = \frac{u'}{a\sqrt{(a^2 - u^2)/a^2}} = \frac{u'}{\sqrt{a^2 - u^2}}.$$

Consequently, by the definition of an antiderivative, we have

$$\int \frac{du}{\sqrt{a^2 - u^2}} = \int \frac{u'}{\sqrt{a^2 - u^2}}\, dx = \arcsin \frac{u}{a} + C.$$

EXAMPLE 1   Integration with inverse trigonometric functions

(a) $\displaystyle\int \frac{dx}{\sqrt{4 - x^2}} = \arcsin \frac{x}{2} + C$

(b) $\displaystyle\int \frac{dx}{2 + 9x^2} = \frac{1}{3} \int \frac{3\, dx}{(\sqrt{2})^2 + (3x)^2}$ $\quad u = 3x,\ a = \sqrt{2}$

$\displaystyle\qquad\qquad = \frac{1}{3\sqrt{2}} \arctan \frac{3x}{\sqrt{2}} + C$ ▭

The integrals in Example 1 are fairly straightforward applications of integration formulas. Unfortunately, this is not typical. The inverse trigonometric integration formulas can be disguised in many ways. In this section we give some typical ways in which these integrals can be disguised, and we illustrate techniques for removing the disguise.

EXAMPLE 2   Integration by substitution

Evaluate

$$\int \frac{dx}{\sqrt{e^{2x} - 1}}.$$

SOLUTION

Because this integral doesn't fit any of the three inverse trigonometric formulas as it stands, we try the substitution $u = e^x$.

$$u = e^x \quad \Longrightarrow \quad du = e^x\, dx \quad \Longrightarrow \quad dx = \frac{du}{e^x} = \frac{du}{u}$$

Now, we have

$$\int \frac{dx}{\sqrt{(e^x)^2 - 1}} = \int \frac{du/u}{\sqrt{u^2 - 1}} = \int \frac{du}{u\sqrt{u^2 - 1}}.$$

This integral now fits the arcsecant formula, and we have

$$\int \frac{dx}{\sqrt{e^{2x} - 1}} = \operatorname{arcsec} \frac{|u|}{1} + C = \operatorname{arcsec} e^x + C.$$ ▭

A useful integration technique is to rewrite the integrand as the sum of two quotients.

EXAMPLE 3   Rewriting the integrand as the sum of two quotients

Evaluate

$$\int \frac{x + 2}{\sqrt{4 - x^2}} \, dx.$$

SOLUTION

This integral does not appear to fit any of our integration formulas, but by splitting the integrand into two parts we obtain

$$\int \frac{x + 2}{\sqrt{4 - x^2}} \, dx = \int \frac{x}{\sqrt{4 - x^2}} \, dx + \int \frac{2}{\sqrt{4 - x^2}} \, dx.$$

Now, the first integral on the right can be evaluated with the Power Rule, and the second integral will yield an inverse sine function. Therefore, we write

$$\int \frac{x + 2}{\sqrt{4 - x^2}} \, dx = -\frac{1}{2} \int (4 - x^2)^{-1/2}(-2x) \, dx + 2 \int \frac{1}{\sqrt{4 - x^2}} \, dx$$

$$= -\frac{1}{2} \left[\frac{(4 - x^2)^{1/2}}{1/2} \right] + 2 \arcsin \frac{x}{2} + C$$

$$= -\sqrt{4 - x^2} + 2 \arcsin \frac{x}{2} + C. \qquad \blacksquare$$

Another disguised use of integration formulas occurs when the integrand is an improper rational function. (That is, the integrand is a rational function in which the degree of the numerator is greater than or equal to the degree of the denominator.) For such functions, we use long division to write the integrand as the sum of a polynomial and a proper rational function.

EXAMPLE 4   Integrating an improper rational function

Evaluate

$$\int \frac{3x^3 - 2}{x^2 + 4} \, dx.$$

SOLUTION

Since the degree of the numerator is greater than the degree of the denominator, we divide to obtain

$$\int \frac{3x^3 - 2}{x^2 + 4}\, dx = \int \left(3x - \frac{12x + 2}{x^2 + 4} \right) dx$$

$$= \int 3x\, dx - 6 \int \frac{2x}{x^2 + 4}\, dx - 2 \int \frac{dx}{x^2 + 4}$$

$$= \frac{3x^2}{2} - 6 \ln (x^2 + 4) - \arctan \frac{x}{2} + C. \qquad \blacksquare$$

Completing the square

Completing the square helps when quadratic functions are involved in the integrand. For example, the quadratic $x^2 + bx + c$ can be written as

$$x^2 + bx + c = x^2 + bx + \left(\frac{b}{2} \right)^2 - \left(\frac{b}{2} \right)^2 + c = \left(x + \frac{b}{2} \right)^2 + \left(\frac{4c - b^2}{4} \right).$$

Thus, we have written $x^2 + bx + c$ as the sum or difference of two squares. Notice that the key step in completing the square is to *add and subtract* $(b/2)^2$.

EXAMPLE 5   Completing the square

Evaluate

$$\int \frac{dx}{x^2 - 4x + 7}.$$

SOLUTION

By completing the square, we obtain

$$x^2 - 4x + 7 = (x^2 - 4x + 4) - 4 + 7 = (x - 2)^2 + 3 = u^2 + a^2.$$

Therefore,

$$\int \frac{dx}{x^2 - 4x + 7} = \int \frac{dx}{(x - 2)^2 + 3} \qquad u = x - 2,\ a = \sqrt{3}$$

$$= \frac{1}{\sqrt{3}} \arctan \frac{x - 2}{\sqrt{3}} + C. \qquad \blacksquare$$

In Example 5, we completed the square for a polynomial whose leading coefficient is 1. If the leading coefficient is not 1, we suggest factoring before completing the square, as demonstrated in the next example.

EXAMPLE 6 Completing the square when the leading coefficient is not 1

Evaluate

$$\int \frac{dx}{2x^2 - 8x + 10}.$$

SOLUTION

To complete the square, we write

$$2x^2 - 8x + 10 = 2(x^2 - 4x + 5)$$
$$= 2(x^2 - 4x + 4 - 4 + 5)$$
$$= 2[(x - 2)^2 + 1]$$
$$= 2(u^2 + a^2).$$

Therefore, we have

$$\int \frac{dx}{2x^2 - 8x + 10} = \frac{1}{2} \int \frac{dx}{(x - 2)^2 + 1}$$
$$= \frac{1}{2} \int \frac{du}{u^2 + 1} \qquad u = x - 2, a = 1$$
$$= \frac{1}{2} \arctan u + C$$
$$= \frac{1}{2} \arctan (x - 2) + C.$$

In the next example, notice how we complete the square when the coefficient of x^2 is negative.

EXAMPLE 7 Completing the square for a negative leading coefficient

Find the area of the region bounded by the graph of $f(x) = 1/\sqrt{3x - x^2}$, the x-axis, and the lines $x = \frac{3}{2}$ and $x = \frac{9}{4}$.

SOLUTION

From Figure 6.32, we see that the area is given by

$$\text{area} = \int_{3/2}^{9/4} \frac{1}{\sqrt{3x - x^2}}\, dx.$$

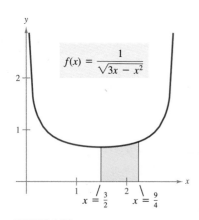

$$f(x) = \frac{1}{\sqrt{3x - x^2}}$$

$$x = \frac{3}{2} \qquad x = \frac{9}{4}$$

FIGURE 6.32

To complete the square, we write

$$3x - x^2 = -(x^2 - 3x) = -\left[x^2 - 3x + \left(\frac{3}{2}\right)^2 - \left(\frac{3}{2}\right)^2 \right]$$

$$= \left(\frac{3}{2}\right)^2 - \left(x - \frac{3}{2}\right)^2 = a^2 - u^2.$$

Therefore,

$$\int_{3/2}^{9/4} \frac{dx}{\sqrt{3x - x^2}} = \int_{3/2}^{9/4} \frac{dx}{\sqrt{(3/2)^2 - [x - (3/2)]^2}}$$

$$= \left[\arcsin \frac{x - (3/2)}{3/2} \right]_{3/2}^{9/4}$$

$$= \arcsin \frac{1}{2} - \arcsin 0 = \frac{\pi}{6} - 0 = \frac{\pi}{6}.$$

Review of basic integration formulas

We now have completed our list of the **basic integration formulas.** It is important to memorize each formula in the list.

BASIC INTEGRATION FORMULAS

1. $\int kf(u) \, du = k \int f(u) \, du$

2. $\int [f(u) \pm g(u)] \, du = \int f(u) \, du \pm \int g(u) \, du$

3. $\int du = u + C$

4. $\int u^n \, du = \dfrac{u^{n+1}}{n + 1} + C, \, n \neq -1$

5. $\int \dfrac{du}{u} = \ln |u| + C$

6. $\int e^u \, du = e^u + C$

7. $\int \sin u \, du = -\cos u + C$

8. $\int \cos u \, du = \sin u + C$

9. $\int \tan u \, du = -\ln |\cos u| + C$

10. $\int \cot u \, du = \ln |\sin u| + C$

11. $\int \sec u \, du = \ln |\sec u + \tan u| + C$

12. $\int \csc u \, du = -\ln |\csc u + \cot u| + C$

13. $\int \sec^2 u \, du = \tan u + C$

14. $\int \csc^2 u \, du = -\cot u + C$

15. $\int \sec u \tan u \, du = \sec u + C$

16. $\int \csc u \cot u \, du = -\csc u + C$

17. $\int \dfrac{du}{\sqrt{a^2 - u^2}} = \arcsin \dfrac{u}{a} + C$

18. $\int \dfrac{du}{a^2 + u^2} = \dfrac{1}{a} \arctan \dfrac{u}{a} + C$

19. $\int \dfrac{du}{u\sqrt{u^2 - a^2}} = \dfrac{1}{a} \operatorname{arcsec} \dfrac{|u|}{a} + C$

You can learn quite a bit about the nature of integration by comparing this list to the summary of differentiation formulas given in the previous section. For differentiation, we now have formulas that allow us to differentiate *any* elementary function. For integration, this is certainly not true. The integration formulas listed above are primarily those that we happened upon when we were developing differentiation formulas. We do not have a formula for the antiderivative of a general product or quotient, the natural logarithmic function, or the inverse trigonometric functions. More importantly, we cannot apply any of the formulas in this list unless we can create the proper *du* corresponding to the *u* in the formula. The point is that we need to work more on integration techniques, which we will do in Chapter 8. To give you a better feeling for the integration problems that we *can* and *cannot* do with the techniques and formulas we now have, we conclude this section with two examples.

EXAMPLE 8   Comparing integration problems

Evaluate as many of the following integrals as you can using the formulas and techniques we have studied so far in the text.

(a) $\displaystyle\int \frac{dx}{x\sqrt{x^2 - 1}}$ (b) $\displaystyle\int \frac{x\,dx}{\sqrt{x^2 - 1}}$ (c) $\displaystyle\int \frac{dx}{\sqrt{x^2 - 1}}$

SOLUTION

(a) We *can* evaluate this integral (it fits the arcsecant formula).

$$\int \frac{dx}{x\sqrt{x^2 - 1}} = \operatorname{arcsec} |x| + C$$

(b) We *can* also evaluate this integral (it fits the Power Rule).

$$\int \frac{x\,dx}{\sqrt{x^2 - 1}} = \frac{1}{2}\int (x^2 - 1)^{-1/2}(2x)\,dx$$

$$= \frac{1}{2}\left(\frac{(x^2 - 1)^{1/2}}{1/2}\right) + C$$

$$= \sqrt{x^2 - 1} + C$$

(c) We *cannot* evaluate this integral using our present techniques. (You should scan the list of Basic Integration Formulas to verify this conclusion.)

EXAMPLE 9   Comparing integration problems

Evaluate as many of the following integrals as you can using the formulas and techniques we have studied so far in the text.

(a) $\displaystyle\int \frac{dx}{x \ln x}$ (b) $\displaystyle\int \frac{\ln x\,dx}{x}$ (c) $\displaystyle\int \ln x\,dx$

SOLUTION

(a) We *can* evaluate this integral (it fits the log formula).

$$\int \frac{dx}{x \ln x} = \int \frac{1/x}{\ln x}\, dx = \ln |\ln x| + C$$

(b) We *can* also evaluate this integral (it fits the Power Rule).

$$\int \frac{\ln x}{x}\, dx = \int \left(\frac{1}{x}\right)(\ln x)^1\, dx = \frac{(\ln x)^2}{2} + C$$

(c) We *cannot* evaluate this integral using our present techniques.   ▭

REMARK   Note in Examples 8 and 9 that it is the *simplest* function that we cannot yet integrate.

EXERCISES for Section 6.8

In Exercises 1–26, evaluate the given integral.

1. $\displaystyle\int_0^{1/6} \frac{1}{\sqrt{1 - 9x^2}}\, dx$

2. $\displaystyle\int_0^1 \frac{dx}{\sqrt{4 - x^2}}$

3. $\displaystyle\int_0^{\sqrt{3}/2} \frac{1}{1 + 4x^2}\, dx$

4. $\displaystyle\int_{\sqrt{3}}^3 \frac{1}{9 + x^2}\, dx$

5. $\displaystyle\int \frac{1}{x\sqrt{4x^2 - 1}}\, dx$

6. $\displaystyle\int \frac{1}{4 + (x - 1)^2}\, dx$

7. $\displaystyle\int \frac{x^3}{x^2 + 1}\, dx$

8. $\displaystyle\int \frac{x^4 - 1}{x^2 + 1}\, dx$

9. $\displaystyle\int \frac{1}{\sqrt{1 - (x + 1)^2}}\, dx$

10. $\displaystyle\int \frac{t}{t^4 + 16}\, dt$

11. $\displaystyle\int \frac{t}{\sqrt{1 - t^4}}\, dt$

12. $\displaystyle\int \frac{1}{x\sqrt{x^4 - 4}}\, dx$

13. $\displaystyle\int \frac{\arctan x}{1 + x^2}\, dx$

14. $\displaystyle\int \frac{1}{(x - 1)\sqrt{(x - 1)^2 - 4}}\, dx$

15. $\displaystyle\int_0^{1/\sqrt{2}} \frac{\arcsin x}{\sqrt{1 - x^2}}\, dx$

16. $\displaystyle\int_0^{1/\sqrt{2}} \frac{\arccos x}{\sqrt{1 - x^2}}\, dx$

17. $\displaystyle\int_{-1/2}^0 \frac{x}{\sqrt{1 - x^2}}\, dx$

18. $\displaystyle\int_{-\sqrt{3}}^0 \frac{x}{1 + x^2}\, dx$

19. $\displaystyle\int \frac{e^x}{\sqrt{1 - e^{2x}}}\, dx$

20. $\displaystyle\int \frac{\cos x}{\sqrt{4 - \sin^2 x}}\, dx$

21. $\displaystyle\int \frac{1}{9 + (x - 3)^2}\, dx$

22. $\displaystyle\int \frac{x + 1}{x^2 + 1}\, dx$

23. $\displaystyle\int \frac{1}{\sqrt{x}(1 + x)}\, dx$

24. $\displaystyle\int_1^2 \frac{1}{3 + (x - 2)^2}\, dx$

25. $\displaystyle\int_{\pi/2}^\pi \frac{\sin x}{1 + \cos^2 x}\, dx$

26. $\displaystyle\int \frac{e^{2x}}{4 + e^{4x}}\, dx$

In Exercises 27–40, evaluate the given integral. (Complete the square, if necessary.)

27. $\displaystyle\int_0^2 \frac{dx}{x^2 - 2x + 2}$

28. $\displaystyle\int_{-3}^{-1} \frac{dx}{x^2 + 6x + 13}$

29. $\displaystyle\int \frac{2x}{x^2 + 6x + 13}\, dx$

30. $\displaystyle\int \frac{2x - 5}{x^2 + 2x + 2}\, dx$

31. $\displaystyle\int \frac{1}{\sqrt{-x^2 - 4x}}\, dx$

32. $\displaystyle\int \frac{1}{\sqrt{-x^2 + 2x}}\, dx$

33. $\displaystyle\int \frac{x + 2}{\sqrt{-x^2 - 4x}}\, dx$

34. $\displaystyle\int \frac{x - 1}{\sqrt{x^2 - 2x}}\, dx$

35. $\displaystyle\int_2^3 \frac{2x - 3}{\sqrt{4x - x^2}}\, dx$

36. $\displaystyle\int \frac{1}{(x - 1)\sqrt{x^2 - 2x}}\, dx$

37. $\displaystyle\int \frac{x}{x^4 + 2x^2 + 2}\, dx$ **38.** $\displaystyle\int \frac{x}{\sqrt{9 + 8x^2 - x^4}}\, dx$

39. $\displaystyle\int \frac{1}{\sqrt{-16x^2 + 16x - 3}}\, dx$

40. $\displaystyle\int \frac{1}{(x - 1)\sqrt{9x^2 - 18x + 5}}\, dx$

In Exercises 41–44, use substitution to evaluate the given integral.

41. $\displaystyle\int \frac{\sqrt{x - 1}}{x}\, dx$ **42.** $\displaystyle\int \frac{\sqrt{x - 2}}{x + 1}\, dx$

43. $\displaystyle\int \sqrt{e^t - 3}\, dt$ **44.** $\displaystyle\int \frac{1}{t\sqrt{t} + \sqrt{t}}\, dt$

In Exercises 45–48, find the area of the region bounded by the graphs of the given equations.

45. $y = \dfrac{1}{1 + x^2},\ y = 0,\ x = 0,\ x = 1$

46. $y = \dfrac{1}{\sqrt{4 - x^2}},\ y = 0,\ x = 0,\ x = 1$

47. $y = \dfrac{1}{x^2 - 2x + 5},\ y = 0,\ x = 1,\ x = 3$

48. $y = \dfrac{1}{\sqrt{3 + 2x - x^2}},\ y = 0,\ x = 0,\ x = 2$

49. An object is projected upward with an initial velocity of 500 feet per second. If the air resistance is proportional to the square of the velocity, we obtain the equation

$$\frac{dv}{dt} = -(32 + kv^2)$$

where 32 feet per second is the acceleration due to gravity and k is a constant. Find the velocity (on the upward flight) as a function of time by solving the equation

$$\int \frac{dv}{32 + kv^2} = -\int dt.$$

50. A weight of mass m is attached to a spring and oscillates with simple harmonic motion (see figure). By Hooke's Law, we can determine that

$$\int \frac{dy}{\sqrt{A^2 - y^2}} = \int \sqrt{\frac{k}{m}}\, dt$$

where A is the maximum displacement, t is the time, and k is a constant. Find y as a function of t, given that $y = 0$ when $t = 0$.

FIGURE FOR 50

51. (a) Show that

$$\int_0^1 \frac{4}{1 + x^2}\, dx = \pi.$$

(b) Approximate the number π using Simpson's Rule (with $n = 6$) and the integral of part (a).

52. Verify the following rules by differentiating.

(a) $\displaystyle\int \frac{du}{a^2 + u^2} = \frac{1}{a}\arctan\frac{u}{a} + C$

(b) $\displaystyle\int \frac{du}{u\sqrt{u^2 - a^2}} = \frac{1}{a}\operatorname{arcsec}\frac{|u|}{a} + C$

In Exercises 53–56, determine which of the given integrals can be evaluated using the Basic Integration Formulas we have studied so far.

53. (a) $\displaystyle\int \frac{1}{\sqrt{1 - x^2}}\, dx$ **(b)** $\displaystyle\int \frac{x}{\sqrt{1 - x^2}}\, dx$

(c) $\displaystyle\int \frac{1}{x\sqrt{1 - x^2}}\, dx$

54. (a) $\displaystyle\int e^{x^2}\, dx$ **(b)** $\displaystyle\int x\, e^{x^2}\, dx$

(c) $\displaystyle\int \frac{1}{x^2}e^{1/x}\, dx$

55. (a) $\displaystyle\int \sqrt{x - 1}\, dx$ **(b)** $\displaystyle\int x\sqrt{x - 1}\, dx$

(c) $\displaystyle\int \frac{x}{\sqrt{x - 1}}\, dx$

56. (a) $\displaystyle\int \frac{1}{1 + x^4}\, dx$ **(b)** $\displaystyle\int \frac{x}{1 + x^4}\, dx$

(c) $\displaystyle\int \frac{x^3}{1 + x^4}\, dx$

6.9 Hyperbolic Functions

Hyperbolic functions ▪ Differentiation and integration of hyperbolic functions ▪ Inverse hyperbolic functions ▪ Differentiation and integration involving inverse hyperbolic functions

Johann Heinrich Lambert

In this section we look briefly at a special class of exponential functions called the **hyperbolic functions.** The first person to publish a comprehensive study on hyperbolic functions was Johann Heinrich Lambert (1728–1777), a Swiss-German mathematician and colleague of Euler.

The name, *hyperbolic function,* arose from comparing the area of a semicircular region, as shown in Figure 6.33, to the area of a region under a hyperbola, as shown in Figure 6.34. The integral for the semicircular region involves an *inverse trigonometric (circular) function:*

$$\int_{-1}^{1} \sqrt{1 - x^2}\,dx = \frac{1}{2}\left[x\sqrt{1 - x^2} + \arcsin x \right]_{-1}^{1} = \frac{\pi}{2} \approx 1.571.$$

The integral for the hyperbolic region involves an inverse hyperbolic function:

$$\int_{-1}^{1} \sqrt{1 + x^2}\,dx = \frac{1}{2}\left[x\sqrt{1 + x^2} + \sinh^{-1} x \right]_{-1}^{1} \approx 2.296.$$

This is only one of many ways in which the hyperbolic functions are similar to the trigonometric functions.

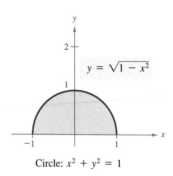

Circle: $x^2 + y^2 = 1$

FIGURE 6.33

Hyperbola: $-x^2 + y^2 = 1$

FIGURE 6.34

DEFINITION OF THE HYPERBOLIC FUNCTIONS			

$$\sinh x = \frac{e^x - e^{-x}}{2} \qquad \cosh x = \frac{e^x + e^{-x}}{2} \qquad \tanh x = \frac{\sinh x}{\cosh x}$$

$$\operatorname{csch} x = \frac{1}{\sinh x}, x \neq 0 \qquad \operatorname{sech} x = \frac{1}{\cosh x} \qquad \coth x = \frac{1}{\tanh x}, x \neq 0$$

REMARK We read $\sinh x$ as "the hyperbolic sine of x," $\cosh x$ as "the hyperbolic cosine of x," and so on.

The graphs of the six hyperbolic functions and their domains and ranges are shown in Figure 6.35. Note that the graphs of sinh x and cosh x can be obtained by *addition of ordinates* using the exponential functions $f(x) = \frac{1}{2}e^x$ and $g(x) = \frac{1}{2}e^{-x}$.

Domain: $(-\infty, \infty)$
Range: $(-\infty, \infty)$

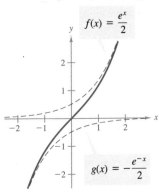

$y = \sinh x$

Domain: $(-\infty, \infty)$
Range: $[1, \infty)$

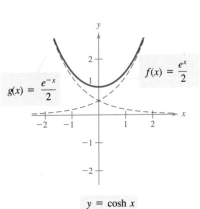

$y = \cosh x$

Domain: $(-\infty, \infty)$
Range: $(-1, 1)$

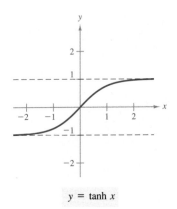

$y = \tanh x$

Domain: $(-\infty, 0)$ and $(0, \infty)$
Range: $(-\infty, 0)$ and $(0, \infty)$

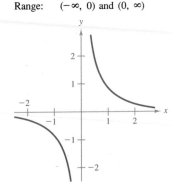

$y = \operatorname{csch} x = \dfrac{1}{\sinh x}$

Domain: $(-\infty, \infty)$
Range: $(0, 1]$

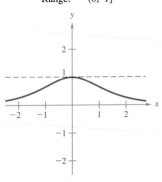

$y = \operatorname{sech} x = \dfrac{1}{\cosh x}$

Domain: $(-\infty, 0)$ and $(0, \infty)$
Range: $(-\infty, -1)$ and $(1, \infty)$

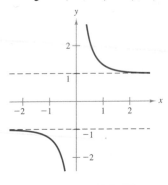

$y = \coth x = \dfrac{1}{\tanh x}$

FIGURE 6.35

Many of the trigonometric identities have corresponding *hyperbolic identities*. For instance,

$$\cosh^2 x - \sinh^2 x = \left(\frac{e^x + e^{-x}}{2}\right)^2 - \left(\frac{e^x - e^{-x}}{2}\right)^2$$

$$= \frac{e^{2x} + 2 + e^{-2x}}{4} - \frac{e^{2x} - 2 + e^{-2x}}{4} = \frac{4}{4} = 1$$

and

$$2\sinh x \cosh x = 2\left(\frac{e^x - e^{-x}}{2}\right)\left(\frac{e^x + e^{-x}}{2}\right) = \frac{e^{2x} - e^{-2x}}{2} = \sinh 2x.$$

In the following list, notice the similarity of hyperbolic identities to trigonometric identities.

HYPERBOLIC IDENTITIES

$$\cosh^2 x - \sinh^2 x = 1 \qquad \sinh(x + y) = \sinh x \cosh y + \cosh x \sinh y$$

$$\tanh^2 x + \operatorname{sech}^2 x = 1 \qquad \sinh(x - y) = \sinh x \cosh y - \cosh x \sinh y$$

$$\coth^2 x - \operatorname{csch}^2 x = 1 \qquad \cosh(x + y) = \cosh x \cosh y + \sinh x \sinh y$$

$$\cosh(x - y) = \cosh x \cosh y - \sinh x \sinh y$$

$$\sinh^2 x = \frac{-1 + \cosh 2x}{2} \qquad \sinh 2x = 2 \sinh x \cosh x$$

$$\cosh^2 x = \frac{1 + \cosh 2x}{2} \qquad \cosh 2x = \cosh^2 x + \sinh^2 x$$

Since the hyperbolic functions are written in terms of e^x and e^{-x}, we can easily derive formulas for their derivatives. In the following theorem, we list these derivatives with the corresponding integration formulas.

THEOREM 6.20
DERIVATIVES AND INTEGRALS OF HYPERBOLIC FUNCTIONS

Let u be a differentiable function of x.

$$\frac{d}{dx}[\sinh u] = (\cosh u)u' \qquad \int \cosh u \, du = \sinh u + C$$

$$\frac{d}{dx}[\cosh u] = (\sinh u)u' \qquad \int \sinh u \, du = \cosh u + C$$

$$\frac{d}{dx}[\tanh u] = (\operatorname{sech}^2 u)u' \qquad \int \operatorname{sech}^2 u \, du = \tanh u + C$$

$$\frac{d}{dx}[\coth u] = -(\operatorname{csch}^2 u)u' \qquad \int \operatorname{csch}^2 u \, du = -\coth u + C$$

$$\frac{d}{dx}[\operatorname{sech} u] = -(\operatorname{sech} u \tanh u)u' \qquad \int \operatorname{sech} u \tanh u \, du = -\operatorname{sech} u + C$$

$$\frac{d}{dx}[\operatorname{csch} u] = -(\operatorname{csch} u \coth u)u' \qquad \int \operatorname{csch} u \coth u \, du = -\operatorname{csch} u + C$$

PROOF We give proofs for the derivatives of the hyperbolic sine and tangent, and leave the proofs of two of the remaining formulas as exercises (see Exercises 75 and 76).

$$\frac{d}{dx}[\sinh x] = \frac{d}{dx}\left[\frac{e^x - e^{-x}}{2}\right] = \frac{e^x + e^{-x}}{2} = \cosh x$$

$$\frac{d}{dx}[\tanh x] = \frac{d}{dx}\left[\frac{\sinh x}{\cosh x}\right] = \frac{\cosh x(\cosh x) - \sinh x(\sinh x)}{\cosh^2 x}$$

$$= \frac{1}{\cosh^2 x} = \operatorname{sech}^2 x$$

EXAMPLE 1 Differentiation of hyperbolic functions

(a) $\dfrac{d}{dx}[\sinh (x^2 - 3)] = 2x \cosh (x^2 - 3)$

(b) $\dfrac{d}{dx}[\ln (\cosh x)] = \dfrac{\sinh x}{\cosh x} = \tanh x$

(c) $\dfrac{d}{dx}[x \sinh x - \cosh x] = x \cosh x + \sinh x - \sinh x = x \cosh x$ ▭

EXAMPLE 2 Finding relative extrema

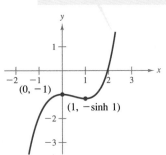

$y = (x - 1) \cosh x - \sinh x$

FIGURE 6.36

Find the relative extrema of $f(x) = (x - 1) \cosh x - \sinh x$.

SOLUTION

Letting $f'(x) = 0$, we have

$$f'(x) = (x - 1) \sinh x + \cosh x - \cosh x$$
$$= (x - 1) \sinh x = 0.$$

Thus, the critical numbers are $x = 1$ and $x = 0$. By the Second Derivative Test, we can verify that the point $(0, -1)$ yields a relative maximum and the point $(1, -\sinh 1)$ yields a relative minimum, as shown in Figure 6.36. ▭

When a uniform flexible cable, such as a telephone wire, is suspended from two points, it takes the shape of a **catenary**, as shown in Figure 6.37. The equation for such a curve is shown in Example 3.

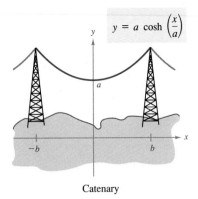

$y = a \cosh\left(\dfrac{x}{a}\right)$

FIGURE 6.37

Catenary

EXAMPLE 3 An application of hyperbolic functions

Power cables are suspended between two towers, forming the catenary in Figure 6.37. The equation for this catenary is

$$y = a \cosh \frac{x}{a}.$$

The distance between the two towers is $2b$. Find the slope of the catenary at the point where the cable meets the right-hand tower.

SOLUTION

Differentiating, we have

$$y' = a\left(\frac{1}{a}\right) \sinh \frac{x}{a} = \sinh \frac{x}{a}.$$

At b, the slope (from the left) is given by $m = \sinh (b/a)$.

EXAMPLE 4 Integrating a hyperbolic function

Evaluate $\int \cosh 2x \sinh^2 2x\, dx$.

SOLUTION

If we let $u = \sinh 2x$, then by the Power Rule it follows that

$$\int \cosh 2x \sinh^2 2x\, dx = \frac{1}{2} \int (\sinh 2x)^2 (2 \cosh 2x)\, dx$$

$$= \frac{1}{2} \left[\frac{(\sinh 2x)^3}{3} \right] + C$$

$$= \frac{\sinh^3 2x}{6} + C.$$

Unlike trigonometric functions, hyperbolic functions are *not* periodic. In fact, by looking back at Figure 6.35, we can see that four of the six hyperbolic functions are actually one-to-one (the hyperbolic sine, tangent, cosecant, and cotangent). Thus, we can apply Theorem 6.8 to conclude that these four functions have inverse functions. The other two (the hyperbolic cosine and secant) are one-to-one if their domains are restricted to the positive real numbers, and for this restricted domain they also have inverse functions. Since the hyperbolic functions are defined in terms of exponential functions, it is not surprising to find that the inverse hyperbolic functions can be written in terms of logarithmic functions, as shown in Theorem 6.21.

THEOREM 6.21
INVERSE HYPERBOLIC
FUNCTIONS

Domain

$\sinh^{-1} x = \ln (x + \sqrt{x^2 + 1})$ $(-\infty, \infty)$

$\cosh^{-1} x = \ln (x + \sqrt{x^2 - 1})$ $[1, \infty)$

$\tanh^{-1} x = \dfrac{1}{2} \ln \dfrac{1 + x}{1 - x}$ $(-1, 1)$

$\coth^{-1} x = \dfrac{1}{2} \ln \dfrac{x + 1}{x - 1}$ $(-\infty, -1) \cup (1, \infty)$

$\operatorname{sech}^{-1} x = \ln \dfrac{1 + \sqrt{1 - x^2}}{x}$ $(0, 1]$

$\operatorname{csch}^{-1} x = \ln \dfrac{1 + \sqrt{1 + x^2}}{|x|}$ $(-\infty, 0) \cup (0, \infty)$

PROOF   The proof of this theorem is a straightforward application of the properties of the exponential and logarithmic functions. For example, if

$$f(x) = \sinh x = \frac{e^x - e^{-x}}{2} \quad \text{and} \quad g(x) = \ln(x + \sqrt{x^2 + 1}),$$

then

$$f(g(x)) = \frac{(x + \sqrt{x^2 + 1}) - [1/(x + \sqrt{x^2 + 1})]}{2}$$

$$= \frac{2x(x + \sqrt{x^2 + 1})}{2(x + \sqrt{x^2 + 1})} = x.$$

A similar argument can show that $g(f(x)) = x$, and we can conclude that g is the inverse function of f.

FIGURE 6.38

The graphs of the inverse hyperbolic functions are shown in Figure 6.38.

Domain: $[1, \infty)$
Range: $[0, \infty)$

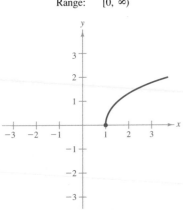

$y = \cosh^{-1} x$

Domain: $(-1, 1)$
Range: $(-\infty, \infty)$

$y = \tanh^{-1} x$

Domain: $(-\infty, \infty)$
Range: $(-\infty, \infty)$

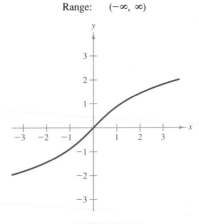

$y = \sinh^{-1} x$

Domain: $(0, 1]$
Range: $[0, \infty)$

$y = \operatorname{sech}^{-1} x$

Domain: $(-\infty, -1)$ and $(1, \infty)$
Range: $(-\infty, 0)$ and $(0, \infty)$

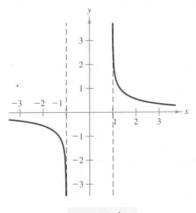

$y = \coth^{-1} x$

Domain: $(-\infty, 0)$ and $(0, \infty)$
Range: $(-\infty, 0)$ and $(0, \infty)$

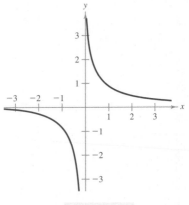

$y = \operatorname{csch}^{-1} x$

The inverse hyperbolic secant can be used to define a curve called a *tractrix* or *pursuit curve*. This is described in the following example.

EXAMPLE 5 An application: the tractrix

A person is holding a rope that is tied to a boat, as shown in Figure 6.39. As the person walks along the dock, the boat travels along a **tractrix**, given by the equation

$$y = a \operatorname{sech}^{-1} \frac{x}{a} - \sqrt{a^2 - x^2}$$

where a is the length of the rope. If $a = 20$ feet, find the distance the person must walk to bring the boat 5 feet from the dock.

SOLUTION

In Figure 6.39, we see that the distance the person has walked is given by

$$y_1 = y + \sqrt{20^2 - x^2} = \left(20 \operatorname{sech}^{-1} \frac{x}{20} - \sqrt{20^2 - x^2} \right) + \sqrt{20^2 - x^2}$$

$$= 20 \operatorname{sech}^{-1} \frac{x}{20}.$$

When $x = 5$, this distance is

$$y_1 = 20 \operatorname{sech}^{-1} \frac{5}{20} = 20 \ln \frac{1 + \sqrt{1 - (1/4)^2}}{1/4}$$

$$= 20 \ln (4 + \sqrt{15}) \approx 41.27 \text{ ft.}$$

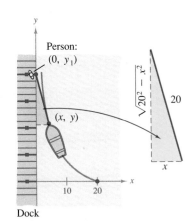

Person:
$(0, y_1)$

(x, y)

$\sqrt{20^2 - x^2}$

20

10   20

Dock

$$y = 20 \operatorname{sech}^{-1} \left(\frac{x}{20} \right) - \sqrt{20^2 - x^2}$$

FIGURE 6.39

The derivatives of the inverse hyperbolic functions resemble the derivatives of the inverse trigonometric functions. We list these in Theorem 6.22 with the corresponding integration formulas (listed in logarithmic form).

THEOREM 6.22
DIFFERENTIATION AND INTEGRATION INVOLVING INVERSE HYPERBOLIC FUNCTIONS

Let u be a differentiable function of x.

$$\frac{d}{dx}[\sinh^{-1} u] = \frac{u'}{\sqrt{u^2 + 1}}$$

$$\frac{d}{dx}[\cosh^{-1} u] = \frac{u'}{\sqrt{u^2 - 1}}$$

$$\frac{d}{dx}[\tanh^{-1} u] = \frac{u'}{1 - u^2}$$

$$\frac{d}{dx}[\coth^{-1} u] = \frac{u'}{1 - u^2}$$

$$\frac{d}{dx}[\operatorname{sech}^{-1} u] = \frac{-u'}{u\sqrt{1 - u^2}}$$

$$\frac{d}{dx}[\operatorname{csch}^{-1} u] = \frac{-u'}{|u|\sqrt{1 + u^2}}$$

$$\int \frac{du}{\sqrt{u^2 \pm a^2}} = \ln (u + \sqrt{u^2 \pm a^2}) + C$$

$$\int \frac{du}{a^2 - u^2} = \frac{1}{2a} \ln \left| \frac{a + u}{a - u} \right| + C$$

$$\int \frac{du}{u\sqrt{a^2 \pm u^2}} = -\frac{1}{a} \ln \frac{a + \sqrt{a^2 \pm u^2}}{|u|} + C$$

PROOF   As with the previous theorem, we can prove these differentiation and integration formulas with a straightforward application of the properties of exponential and logarithmic functions. For example,

$$\frac{d}{dx}[\tanh^{-1} x] = \frac{d}{dx}\left[\frac{1}{2} \ln \frac{1 + x}{1 - x}\right]$$

$$= \frac{1}{2}\left[\frac{d}{dx}[\ln (1 + x) - \ln (1 - x)]\right]$$

$$= \frac{1}{2}\left[\frac{1}{1 + x} - \frac{-1}{1 - x}\right]$$

$$= \frac{1}{2}\left[\frac{2}{1 - x^2}\right] = \frac{1}{1 - x^2}.$$

In Exercises 77–79, you are asked to verify some of the other formulas given in this theorem.

EXAMPLE 6   Differentiation of inverse hyperbolic functions

For the tractrix given in Example 5, show that the boat is always pointing toward the person.

SOLUTION

For a point (x, y) on the tractrix, the slope of the graph gives us the direction of the boat, as shown in Figure 6.39. This slope is given by

$$y' = \frac{d}{dx}\left[20 \, \text{sech}^{-1} \frac{x}{20} - \sqrt{20^2 - x^2}\right]$$

$$= -20\left(\frac{1}{20}\right)\left(\frac{1}{(x/20)\sqrt{1 - (x/20)^2}}\right) - \left(\frac{1}{2}\right)\left(\frac{-2x}{\sqrt{20^2 - x^2}}\right)$$

$$= \frac{-20^2}{x\sqrt{20^2 - x^2}} + \frac{x}{\sqrt{20^2 - x^2}}$$

$$= -\frac{\sqrt{20^2 - x^2}}{x}.$$

However, from Figure 6.39, we can see that the slope of the line segment connecting the point $(0, y_1)$ with the point (x, y) is also

$$m = -\frac{\sqrt{20^2 - x^2}}{x}.$$

Thus, we can conclude that the boat is always pointing toward the person. (It is because of this property that a tractrix is called a *pursuit curve*.) ▭

EXAMPLE 7   Integration using inverse hyperbolic functions

$$\int \frac{dx}{x\sqrt{4-9x^2}} = \int \frac{3\ dx}{(3x)\sqrt{4-9x^2}} \qquad a = 2,\ u = 3x$$

$$= \int \frac{du}{u\sqrt{a^2-u^2}}$$

$$= -\frac{1}{a} \ln \frac{a+\sqrt{a^2-u^2}}{|u|} + C$$

$$= -\frac{1}{2} \ln \frac{2+\sqrt{4-9x^2}}{|3x|} + C$$ ▭

EXAMPLE 8   Integration using inverse hyperbolic functions

Evaluate

$$\int_0^1 \frac{dx}{5-4x^2}.$$

SOLUTION

Since the denominator is of the form $a^2 - u^2$, we let $u = 2x$ and $a = \sqrt{5}$.
Then $du = 2\ dx$, and we have

$$\int_0^1 \frac{dx}{5-4x^2} = \frac{1}{2} \int_0^1 \frac{2\ dx}{(\sqrt{5})^2 - (2x)^2}$$

$$= \frac{1}{2} \frac{1}{2\sqrt{5}} \left[\ln \frac{\sqrt{5}+2x}{\sqrt{5}-2x} \right]_0^1$$

$$= \frac{1}{4\sqrt{5}} \ln \frac{\sqrt{5}+2}{\sqrt{5}-2} \approx 0.3228.$$ ▭

EXERCISES for Section 6.9

In Exercises 1–6, evaluate the given function. If the function value is not a rational number, give the answer to three-decimal-place accuracy.

1. (a) sinh 3   (b) tanh (-2)
2. (a) cosh 0   (b) sech 1
3. (a) csch (ln 2)   (b) coth (ln 5)
4. (a) $\sinh^{-1} 0$ (b) $\tanh^{-1} 0$
5. (a) $\cosh^{-1} 2$ (b) $\mathrm{sech}^{-1}\ \dfrac{2}{3}$
6. (a) $\mathrm{csch}^{-1}\ 2$ (b) $\coth^{-1} 3$

In Exercises 7–12, verify the given identity.

7. $\tanh^2 x + \mathrm{sech}^2 x = 1$
8. $\cosh^2 x = \dfrac{1 + \cosh 2x}{2}$
9. $\sinh (x + y) = \sinh x \cosh y + \cosh x \sinh y$
10. $\sinh 2x = 2 \sinh x \cosh x$
11. $\sinh 3x = 3 \sinh x + 4 \sinh^3 x$
12. $\cosh x + \cosh y = 2 \cosh \dfrac{x+y}{2} \cosh \dfrac{x-y}{2}$

In Exercises 13–34, find y' and simplify.

13. $y = \sinh (1 - x^2)$

14. $y = \coth 3x$

15. $y = \ln (\sinh x)$

16. $y = \ln (\cosh x)$

17. $y = \ln \left(\tanh \dfrac{x}{2} \right)$

18. $y = x \sinh x - \cosh x$

19. $y = \dfrac{1}{4} \sinh 2x - \dfrac{x}{2}$

20. $y = x - \coth x$

21. $y = \arctan (\sinh x)$

22. $y = e^{\sinh x}$

23. $y = x^{\cosh x}$

24. $y = \text{sech}^2 \, 3x$

25. $y = (\cosh x - \sinh x)^2$

26. $y = \text{sech} \, (x + 1)$

27. $y = \cosh^{-1} (3x)$

28. $y = \tanh^{-1} \dfrac{x}{2}$

29. $y = \sinh^{-1} (\tan x)$

30. $y = \text{sech}^{-1} (\cos 2x),$

$$0 < x < \frac{x}{4}$$

31. $y = \coth^{-1} (\sin 2x)$

32. $y = (\text{csch}^{-1} x)^2$

33. $y = 2x \sinh^{-1} (2x) - \sqrt{1 + 4x^2}$

34. $y = x \tanh^{-1} x + \ln \sqrt{1 - x^2}$

In Exercises 35–62, evaluate the integral.

35. $\displaystyle\int \sinh (1 - 2x) \, dx$

36. $\displaystyle\int \dfrac{\cosh \sqrt{x}}{\sqrt{x}} \, dx$

37. $\displaystyle\int \cosh^2 (x - 1) \sinh (x - 1) \, dx$

38. $\displaystyle\int \dfrac{\sinh x}{1 + \sinh^2 x} \, dx$

39. $\displaystyle\int \dfrac{\cosh x}{\sinh x} \, dx$

40. $\displaystyle\int \text{sech}^2 (2x - 1) \, dx$

41. $\displaystyle\int x \, \text{csch}^2 \dfrac{x^2}{2} \, dx$

42. $\displaystyle\int \text{sech}^3 x \tanh x \, dx$

43. $\displaystyle\int \dfrac{\text{csch} \, (1/x) \coth (1/x)}{x^2} \, dx$

44. $\displaystyle\int \sinh^2 x \, dx$

45. $\displaystyle\int_0^4 \dfrac{1}{25 - x^2} \, dx$

46. $\displaystyle\int_0^4 \dfrac{1}{\sqrt{25 - x^2}} \, dx$

47. $\displaystyle\int_0^{\sqrt{2}/4} \dfrac{2}{\sqrt{1 - 4x^2}} \, dx$

48. $\displaystyle\int \dfrac{2}{x\sqrt{1 + 4x^2}} \, dx$

49. $\displaystyle\int \dfrac{x}{x^4 + 1} \, dx$

50. $\displaystyle\int \dfrac{\cosh x}{\sqrt{9 - \sinh^2 x}} \, dx$

51. $\displaystyle\int \dfrac{1}{\sqrt{1 + e^{2x}}} \, dx$

52. $\displaystyle\int \dfrac{e^x}{1 - e^{2x}} \, dx$

53. $\displaystyle\int \dfrac{1}{\sqrt{x}\sqrt{1 + x}} \, dx$

54. $\displaystyle\int \dfrac{\sqrt{x}}{\sqrt{1 + x^3}} \, dx$

55. $\displaystyle\int \dfrac{1}{(x - 1)\sqrt{x^2 - 2x + 2}} \, dx$

56. $\displaystyle\int \dfrac{-1}{4x - x^2} \, dx$

57. $\displaystyle\int \dfrac{1}{1 - 4x - 2x^2} \, dx$

58. $\displaystyle\int \dfrac{1}{(x + 1)\sqrt{2x^2 + 4x + 8}} \, dx$

59. $\displaystyle\int \dfrac{1}{\sqrt{80 + 8x - 16x^2}} \, dx$

60. $\displaystyle\int \dfrac{1}{(x - 1)\sqrt{-4x^2 + 8x - 1}} \, dx$

61. $\displaystyle\int \dfrac{x^3 - 21x}{5 + 4x - x^2} \, dx$

62. $\displaystyle\int \dfrac{1 - 2x}{4x - x^2} \, dx$

In Exercises 63 and 64, find any relative extrema and points of inflection of the function. Sketch the graph.

63. $f(x) = x \cosh x - \sinh x$

64. $f(x) = x - \tanh x$

In Exercises 65 and 66, show that the given equation satisfies the differential equation.

65. $y = a \sinh x, \quad y''' - y' = 0$

66. $y = a \cosh x, \quad y'' - y = 0$

In Exercises 67 and 68, use the equation of the tractrix

$$y = a \, \text{sech}^{-1} \dfrac{x}{a} - \sqrt{a^2 - x^2}, \quad a > 0.$$

67. Find dy/dx.

68. Let L be the tangent line at the point P to the tractrix. If L intersects the y-axis at the point Q, show that the distance between P and Q is a.

69. Suppose that two chemicals A and B combine in a 3-to-1 ratio to form a compound. The amount of compound x being produced at any time t is proportional to the unchanged amounts of A and B remaining in the solution. Thus, if 3 kilograms of A is mixed with 2 kilograms of B, we have

$$\dfrac{dx}{dt} = k\left(3 - \dfrac{3x}{4}\right)\left(2 - \dfrac{x}{4}\right) = \dfrac{3k}{16}(x^2 - 12x + 32).$$

If 1 kilogram of the compound is formed after 10 minutes, find the amount formed after 20 minutes by solving the integral

$$\int \dfrac{3k}{16} \, dt = \int \dfrac{dx}{x^2 - 12x + 32}.$$

70. Electric wires suspended between two towers form a catenary (see figure) modeled by the equation

$$y = 60 \cosh \dfrac{x}{60}, \quad -60 \le x \le 60$$

where x and y are measured in feet. Find the length of the suspended cable if the towers are 120 feet apart.

FIGURE FOR 70

71. Repeat Exercise 70, if the equation for the catenary is given by

$$y = a \cosh \frac{x}{a}, \quad -b \le x \le b$$

where a is measured in feet and the distance between the towers is $2b$ feet.

72. A barn is 100 feet long and 40 feet wide (see figure). A cross section of the roof is the inverted catenary

$$y = 31 - 20 \cosh \frac{x}{20}.$$

FIGURE FOR 72

Find the number of cubic feet of storage space in the barn.

73. In the Chapter 6 Application, we introduced the following equation for the catenary for the Gateway Arch,

$$y = 757.71 - 127.71 \cosh \frac{x}{127.71}.$$

Show that the height of the Gateway Arch is the same as the distance between its two legs.

74. Use the formula given in Exercise 73 to find the arc length of the Gateway Arch.

In Exercises 75–79, verify the given derivative formula.

75. $\dfrac{d}{dx}[\tanh x] = \text{sech}^2 x$

76. $\dfrac{d}{dx}[\text{sech } x] = -\text{sech } x \tanh x$

77. $\dfrac{d}{dx}[\cosh^{-1} x] = \dfrac{1}{\sqrt{x^2 - 1}}$

78. $\dfrac{d}{dx}[\sinh^{-1} x] = \dfrac{1}{\sqrt{x^2 + 1}}$

79. $\dfrac{d}{dx}[\text{sech}^{-1} x] = \dfrac{-1}{x\sqrt{1 - x^2}}$

REVIEW EXERCISES for Chapter 6

In Exercises 1–6, (a) find the inverse of the given function, (b) sketch the graphs of $f(x)$ and $f^{-1}(x)$ on the same axes, and (c) verify that $f^{-1}[f(x)] = f[f^{-1}(x)] = x$.

1. $f(x) = \dfrac{1}{2}x - 3$

2. $f(x) = 5x - 7$

3. $f(x) = \sqrt{x + 1}$

4. $f(x) = x^3 + 2$

5. $f(x) = x^2 - 5, x \ge 0$

6. $f(x) = \sqrt[3]{x + 1}$

In Exercises 7 and 8, the function does not have an inverse. Give a restriction on the domain so that the restricted function has an inverse, and then find the inverse.

7. $f(x) = 2(x - 4)^2$

8. $f(x) = |x - 2|$

In Exercises 9–12, solve the given equation for x.

9. $e^{\ln x} = 3$

10. $\ln x + \ln (x - 3) = 0$

11. $\log_3 x + \log_3 (x - 1) - \log_3 (x - 2) = 2$

12. $\log_x 125 = 3$

In Exercises 13–40, find dy/dx.

13. $y = \ln \sqrt{x}$

14. $y = \ln \dfrac{x(x - 1)}{x - 2}$

15. $y = x\sqrt{\ln x}$

16. $y = \ln [x(x^2 - 2)^{2/3}]$

17. $y \ln x + y^2 = 0$

18. $\ln (x + y) = x$

19. $y = \dfrac{1}{b^2}\left[\ln (a + bx) + \dfrac{a}{a + bx}\right]$

20. $y = \dfrac{1}{b^2}[a + bx - a \ln (a + bx)]$

21. $y = -\dfrac{1}{a} \ln \dfrac{a + bx}{x}$

22. $y = -\dfrac{1}{ax} + \dfrac{b}{a^2} \ln \dfrac{a + bx}{x}$

23. $y = \ln (e^{-x^2})$

24. $y = \ln \dfrac{e^x}{1 + e^x}$

25. $y = x^2 e^x$

26. $y = e^{-x^2/2}$

27. $y = \sqrt{e^{2x} + e^{-2x}}$

28. $y = x^{2x+1}$

29. $y = 3^{x-1}$

30. $y = (4e)^x$

31. $ye^x + xe^y = xy$

32. $y = \dfrac{x^2}{e^x}$

33. $\cos x^2 = xe^y$

34. $y = \dfrac{1}{2}e^{\sin 2x}$

35. $y = \tan(\arcsin x)$

36. $y = \arctan(x^2 - 1)$

37. $y = x \operatorname{arcsec} x$

38. $y = \dfrac{1}{2} \arctan e^{2x}$

39. $y = x(\arcsin x)^2 - 2x + 2\sqrt{1 - x^2} \arcsin x$

40. $y = \sqrt{x^2 - 4} - 2 \operatorname{arcsec} \dfrac{x}{2}, \quad 0 < x < \dfrac{\pi}{4}$

41. Find the derivative of each of the following, given that a is constant.
  (a) $y = x^a$ (b) $y = a^x$
  (c) $y = x^x$ (d) $y = a^a$

42. Show that

$$y = e^x(a \cos 3x + b \sin 3x)$$

satisfies the differential equation $y'' - 2y' + 10y = 0$.

In Exercises 43–70, evaluate the integral.

43. $\displaystyle\int \dfrac{1}{7x - 2}\, dx$

44. $\displaystyle\int \dfrac{x}{x^2 - 1}\, dx$

45. $\displaystyle\int \dfrac{\sin x}{1 + \cos x}\, dx$

46. $\displaystyle\int \dfrac{\ln \sqrt{x}}{x}\, dx$

47. $\displaystyle\int \dfrac{x^2 + 3}{x}\, dx$

48. $\displaystyle\int \dfrac{x + 2}{2x + 3}\, dx$

49. $\displaystyle\int_1^4 \dfrac{x + 1}{x}\, dx$

50. $\displaystyle\int_1^e \dfrac{\ln x}{x}\, dx$

51. $\displaystyle\int_0^{\pi/3} \sec\theta\, d\theta$

52. $\displaystyle\int_0^{\pi/4} \tan\left(\dfrac{\pi}{4} - x\right) dx$

53. $\displaystyle\int xe^{-3x^2}\, dx$

54. $\displaystyle\int \dfrac{e^{1/x}}{x^2}\, dx$

55. $\displaystyle\int \dfrac{e^{4x} - e^{2x} + 1}{e^x}\, dx$

56. $\displaystyle\int \dfrac{e^{2x} - e^{-2x}}{e^{2x} + e^{-2x}}\, dx$

57. $\displaystyle\int \dfrac{e^x}{e^x - 1}\, dx$

58. $\displaystyle\int x^2 e^{x^3 + 1}\, dx$

59. $\displaystyle\int xe^{-x^2/2}\, dx$

60. $\displaystyle\int \dfrac{x - 1}{3x^2 - 6x - 1}\, dx$

61. $\displaystyle\int \dfrac{e^{-2x}}{1 + e^{-2x}}\, dx$

62. $\displaystyle\int \dfrac{\tan(1/x)}{x^2}\, dx$

63. $\displaystyle\int \dfrac{1}{e^{2x} + e^{-2x}}\, dx$

64. $\displaystyle\int \dfrac{1}{3 + 25x^2}\, dx$

65. $\displaystyle\int \dfrac{x}{\sqrt{1 - x^4}}\, dx$

66. $\displaystyle\int \dfrac{1}{16 + x^2}\, dx$

67. $\displaystyle\int \dfrac{x}{16 + x^2}\, dx$

68. $\displaystyle\int \dfrac{\arcsin x}{\sqrt{1 - x^2}}\, dx$

69. $\displaystyle\int \dfrac{\arctan(x/2)}{4 + x^2}\, dx$

70. $\displaystyle\int \dfrac{4 - x}{\sqrt{4 - x^2}}\, dx$

In Exercises 71 and 72, sketch the graph of the region whose area is given by the integral and find the area.

71. $\displaystyle\int_0^{\pi/3} \tan x\, dx$

72. $\displaystyle\int_0^1 \dfrac{1}{x + 1}\, dx$

In Exercises 73 and 74, find the area of the region bounded by the graphs of the equations.

73. $y = xe^{-x^2},\ y = 0,\ x = 0,\ x = 4$

74. $y = 3e^{-x/2},\ y = 0,\ x = 0,\ x = 4$

In Exercises 75 and 76, use Simpson's Rule to approximate the definite integral.

75. $\displaystyle\int_0^1 e^{x^3}\, dx,\quad n = 4$

76. $\displaystyle\int_0^1 e^{-x^2}\, dx,\quad n = 4$

77. A deposit of $500 earns interest at the rate of 5 percent compounded continuously. Find its value after each of the following time periods.
  (a) 1 year
  (b) 10 years
  (c) 100 years

78. A deposit earns interest at the rate of r percent compounded continuously and doubles in value in 10 years. Find r.

79. How large a deposit, at 7 percent interest compounded continuously, must be made to obtain a balance of $10,000 in 15 years?

80. A deposit of $2500 is made in a savings account at an annual interest rate of 12 percent compounded continuously. Find the average balance in this account during the first five years.

81. A population is growing continuously at the rate of $2\frac{1}{2}$ percent per year. Find the time necessary for the population to (a) double and (b) triple in size.

82. Under ideal conditions, air pressure decreases continuously with height above sea level at a rate proportional to the pressure at that height. If the barometer reads 30 inches at sea level and 15 inches at 18,000 feet, find the barometric pressure at 35,000 feet.

In Exercises 83 and 84, use the following model for human memory.

$$p(t) = 80e^{-0.5t} + 20$$

where $p(t)$ is the percentage retained after t weeks (see figure).

83. At what rate is information being retained after (a) 1 week and (b) 2 weeks?

84. Find the average percentage retained during (a) the first 2 weeks and (b) the second 2 weeks.

$p(t) = 80e^{-0.5t} + 20$

FIGURE FOR 83 and 84

In Exercises 85 and 86, use the **exponential density function** given by $P = (1/\mu)e^{-t/\mu}$, where t is the time in minutes and μ is the average time between successive events. The definite integral

$$\int_{t_1}^{t_2} \frac{1}{\mu} e^{-t/\mu}\, dt$$

gives the probability that the elapsed time before the next occurrence lies between t_1 and t_2 units of time.

85. Trucks arrive at a terminal at an average rate of 3 per hour ($\mu = 20$ minutes is the average time between arrivals). If a truck has just arrived, find the probability that the next arrival will occur
   (a) within 10 minutes
   (b) within 30 minutes
   (c) between 15 and 30 minutes
   (d) within 60 minutes

86. The average time between incoming calls at a switchboard is 3 minutes. Complete the accompanying table to show the probabilities of the various ranges of time elapsed between incoming calls.

Time elapsed between calls in minutes	0–2	2–4	4–6	6–8	8–10
Probability					

87. Two numbers between 0 and 10 are chosen at random. The probability that their product is less than n ($0 < n < 100$) is given by

$$P = \frac{1}{100}\left(n + \int_{n/10}^{10} \frac{n}{x}\, dx\right).$$

(a) What is the probability that the product is less than 25?
(b) What is the probability that the product is less than 50?

88. A solution of a certain drug contains 500 units per milliliter when it is prepared. After 40 days it contains 300 units per milliliter. Assuming that the rate of decomposition is proportional to the amount present, find an equation giving the amount A after t days.

89. An object is projected horizontally with an initial velocity of v_0 feet per second. Assuming that the air resistance is proportional to its velocity, show that the distance the object travels from its starting point in t seconds is given by

$$s = \frac{v_0}{k}(1 - e^{-kt})$$

where k is a constant.

90. Assume that an object falling from rest encounters air resistance that is proportional to its velocity. If the acceleration due to gravity is considered to be -32 feet per second per second, the net change in velocity is given by

$$\frac{dv}{dt} = kv - 32.$$

Find the velocity of the object as a function of time.

91. Solve Exercise 90 for an object that has an initial velocity of v_0 feet per second.

92. For the falling object in Exercise 90, find the limit of the velocity as t approaches infinity.

93. Integrate the velocity function found in Exercise 90 to find the position function s.

94. A certain automobile gets 28 miles per gallon of gasoline for speeds up to 50 miles per hour. Over 50 miles per hour the number of miles per gallon drops at the rate of 12 percent for each 10 miles per hour.
   (a) If s is the speed and y is the miles per gallon, find y as a function of s by solving the differential equation

$$\frac{dy}{ds} = -0.012y, \quad s > 50.$$

   (b) Use the function in part (a) to complete the following table.

Speed	50	55	60	65	70
Miles per gallon					

Chapter 7 Application

American football, in its modern form, is a twentieth-century invention. In the 1800s a rough, soccer-like game was played with a "round football." In 1905, at the request of President Theodore Roosevelt, the Intercollegiate Athletic Association (which became the NCAA in 1910) was formed. With the introduction of the forward pass in 1906, the shape of the ball was altered to make it easier to grip.

Volume of a Football

A regulation-size football has the following dimensions.

Length: 11.00–11.25 in.
Girth: 21.25–21.50 in.

A mathematical model that falls within these dimensions is given by revolving the graph of

$$f(x) = -0.0944x^2 + 3.4, \quad -5.5 \leq x \leq 5.5$$

about the *x*-axis. The resulting solid is called a *solid of revolution* and its volume is given by

$$\text{volume} = \pi \int_{-5.5}^{5.5} [f(x)]^2 \, dx$$

$$= \pi \int_{-5.5}^{5.5} [-0.0944x^2 + 3.4]^2 \, dx$$

$$\approx 232 \text{ cubic inches.}$$

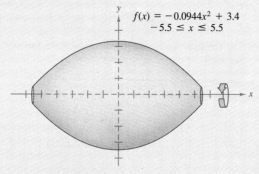

$f(x) = -0.0944x^2 + 3.4$
$-5.5 \leq x \leq 5.5$

Football-shaped solid formed by revolving a parabolic segment about the *x*-axis.

Chapter Overview

Integration has a wide variety of applications. For each of the applications presented in this chapter, we will begin with a known formula, such as the area of a rectangular region, the volume of a circular disc, or the work done by a constant force. Then we will show how the limit of a sum gives rise to new formulas that involve integration.

The chapter begins by showing how to use a definite integral to find the area of a region bounded by two curves. Section 7.2 presents a technique for finding the volume of a solid of revolution by the *disc method*. The section closes by showing that the disc method is a special case of using integration to find the volume of a solid with known cross sections. Section 7.3 presents an alternate way to find the volume of a solid of revolution—using the *shell method*.

Section 7.4 discusses the length of a curve and the surface area of a surface of revolution. Section 7.5 shows how integration can be used to find the work done by a variable force, and Section 7.6 shows how integration can be used to find fluid pressure. For instance, integration can be used to find the water pressure against the side of a dam.

The last section in the chapter discusses moments, centers of mass, and centroids. All three of these concepts have many uses in engineering and physics, and several are discussed in the section.

See Exercise 43, Section 7.2.

Applications
of Integration

7

7.1 Area of a Region Between Two Curves

Area of a region between two curves ▪ Points of intersection of two curves ▪ Representative elements of integration

With a few modifications we can extend the application of definite integrals from the area of a region *under* a curve to the area of a region *between* two curves. If, as in Figure 7.1, the graphs of both f and g lie above the x-axis, we can geometrically interpret the area of the region between the graphs as the area of the region under the graph of g subtracted from the area of the region under the graph of f as shown in Figure 7.2.

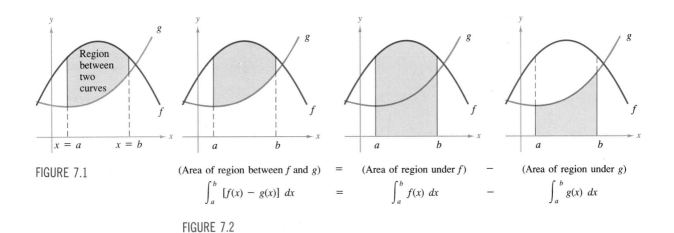

FIGURE 7.1

Region between two curves

$x = a$ $x = b$

(Area of region between f and g) $= \int_a^b [f(x) - g(x)]\, dx$

(Area of region under f) $= \int_a^b f(x)\, dx$

$-$

(Area of region under g) $- \int_a^b g(x)\, dx$

FIGURE 7.2

Although the graphs of f and g are shown above the x-axis in Figure 7.1, this is not necessary. The same integrand $[f(x) - g(x)]$ can be used as long as f and g are continuous and $g(x) \leq f(x)$ on the interval $[a, b]$. This result is summarized in the following theorem.

403

THEOREM 7.1
AREA OF A REGION BETWEEN
TWO CURVES

If f and g are continuous on $[a, b]$ and $g(x) \leq f(x)$ for all x in $[a, b]$, then the area of the region bounded by the graphs of f and g and the vertical lines $x = a$ and $x = b$ is

$$A = \int_a^b [f(x) - g(x)] \, dx.$$

PROOF

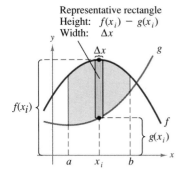

FIGURE 7.3

We partition the interval $[a, b]$ into n subintervals, each of width Δx, and sketch a **representative rectangle** of width Δx and height $f(x_i) - g(x_i)$, where x_i is in the ith interval, as shown in Figure 7.3. The area of this representative rectangle is

$$\Delta A_i = (\text{height})(\text{width}) = [f(x_i) - g(x_i)]\Delta x.$$

By adding the areas of the n rectangles and taking the limit as $\|\Delta\| \to 0$ ($n \to \infty$), we have

$$\lim_{n \to \infty} \sum_{i=1}^{n} [f(x_i) - g(x_i)]\Delta x.$$

Since f and g are continuous on $[a, b]$, $f - g$ is also continuous on this interval and the limit exists. Therefore, the area A of the given region is

$$A = \lim_{n \to \infty} \sum_{i=1}^{n} [f(x_i) - g(x_i)]\Delta x = \int_a^b [f(x) - g(x)] \, dx.$$

It is important to realize that the area formula in Theorem 7.1 depends *only* on the continuity of f and g and the assumption that $g(x) \leq f(x)$. The graphs of f and g can be placed anywhere with respect to the x-axis, as illustrated in Figure 7.4.

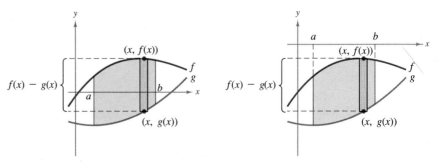

FIGURE 7.4   Height of representative rectangle is $f(x) - g(x)$ regardless of the relative position of the x-axis.

Representative rectangles are used throughout this chapter in various applications of integration. A vertical rectangle (of width Δx) implies integration with respect to x, while a horizontal rectangle (of width Δy) implies integration with respect to y.

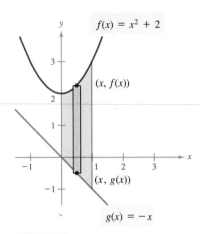

$f(x) = x^2 + 2$

$(x, f(x))$

$(x, g(x))$

$g(x) = -x$

FIGURE 7.5

EXAMPLE 1 Finding the area of a region between two curves

Find the area of the region bounded by the graphs of $y = x^2 + 2$, $y = -x$, $x = 0$, and $x = 1$.

SOLUTION

If we let $g(x) = -x$ and $f(x) = x^2 + 2$, then $g(x) \le f(x)$ for all x in $[0, 1]$, as shown in Figure 7.5. Thus, the area of the representative rectangle is

$$\Delta A = [f(x) - g(x)]\Delta x = [(x^2 + 2) - (-x)]\Delta x.$$

Therefore, by Theorem 7.1, we find the area of the region to be

$$A = \int_a^b [f(x) - g(x)]\, dx = \int_0^1 [(x^2 + 2) - (-x)]\, dx$$

$$= \left[\frac{x^3}{3} + \frac{x^2}{2} + 2x\right]_0^1$$

$$= \frac{1}{3} + \frac{1}{2} + 2 = \frac{17}{6}.$$

In Example 1, the graphs of $f(x) = x^2 + 2$ and $g(x) = -x$ do not intersect, and the values of a and b are given explicitly. A more common problem involves the area of a region bounded by two *intersecting* graphs, where the values of a and b must be calculated.

EXAMPLE 2 A region lying between two intersecting graphs

Find the area of the region bounded by the graphs of $f(x) = 2 - x^2$ and $g(x) = x$.

SOLUTION

From Figure 7.6, we see that the graphs of f and g have two points of intersection. To find the x-coordinates of these points, we set $f(x)$ and $g(x)$ equal to each other and solve for x.

$$2 - x^2 = x \qquad \text{Set } f(x) \text{ equal to } g(x)$$

$$-x^2 - x + 2 = 0$$

$$-(x + 2)(x - 1) = 0 \qquad x = -2 \text{ and } x = 1$$

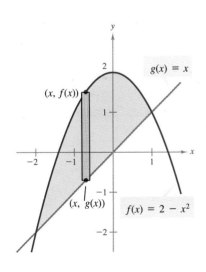

$g(x) = x$

$(x, f(x))$

$(x, g(x))$

$f(x) = 2 - x^2$

FIGURE 7.6

Thus, we have $a = -2$ and $b = 1$. Since $g(x) \le f(x)$ on the interval $[-2, 1]$, the representative rectangle has an area of

$$\Delta A = [f(x) - g(x)]\Delta x = [(2 - x^2) - x]\Delta x$$

and the area of the region is

$$A = \int_{-2}^{1} [(2 - x^2) - x]\, dx = \left[-\frac{x^3}{3} - \frac{x^2}{2} + 2x \right]_{-2}^{1}$$

$$= \left(-\frac{1}{3} - \frac{1}{2} + 2 \right) - \left(\frac{8}{3} - 2 - 4 \right)$$

$$= \frac{9}{2}.$$

EXAMPLE 3 A region lying between two intersecting graphs

The sine and cosine curves intersect infinitely many times bounding regions of equal areas, as shown in Figure 7.7. Find the area of one of these regions.

SOLUTION

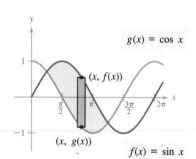

$g(x) = \cos x$

$(x, f(x))$

$(x, g(x))$

$f(x) = \sin x$

FIGURE 7.7

To find the points of intersection of the graphs of $f(x) = \sin x$ and $g(x) = \cos x$, we set the two functions equal to each other and solve for x.

$$\sin x = \cos x \qquad \text{Set } f(x) \text{ equal to } g(x)$$

$$\frac{\sin x}{\cos x} = 1$$

$$\tan x = 1 \qquad x = \frac{\pi}{4}, \frac{5\pi}{4}; \quad 0 \le x \le 2\pi$$

Thus, $a = \pi/4$ and $b = 5\pi/4$. Since $\sin x \ge \cos x$ on the interval $[\pi/4, 5\pi/4]$, the area of the region is

$$A = \int_{\pi/4}^{5\pi/4} [\sin x - \cos x]\, dx = \left[-\cos x - \sin x \right]_{\pi/4}^{5\pi/4}$$

$$= \left(\frac{\sqrt{2}}{2} + \frac{\sqrt{2}}{2} \right) - \left(-\frac{\sqrt{2}}{2} - \frac{\sqrt{2}}{2} \right)$$

$$= 2\sqrt{2}.$$

If two curves intersect at *more* than two points, then to find the area of the region between the curves, we must find all points of intersection and check to see which curve is above the other in each interval determined by these points.

EXAMPLE 4 Curves that intersect at more than two points

Find the area of the region between the graphs of $f(x) = 3x^3 - x^2 - 10x$ and $g(x) = -x^2 + 2x$.

SOLUTION

Solving for x in the equation $f(x) = g(x)$ produces

$$f(x) - g(x) = (3x^3 - x^2 - 10x) - (-x^2 + 2x) = 0$$
$$3x^3 - 12x = 0$$
$$3x(x^2 - 4) = 0.$$

Thus, the two graphs intersect when $x = -2$, 0, and 2. In Figure 7.8, we see that $g(x) \leq f(x)$ in the interval $[-2, 0]$. However, the two graphs switch at the origin, and $f(x) \leq g(x)$ in the interval $[0, 2]$. Hence, we need two integrals—one for the interval $[-2, 0]$ and one for $[0, 2]$.

$$A = \int_{-2}^{0} [f(x) - g(x)]\, dx + \int_{0}^{2} [g(x) - f(x)]\, dx$$

$$= \int_{-2}^{0} (3x^3 - 12x)\, dx + \int_{0}^{2} (-3x^3 + 12x)\, dx$$

$$= \left[\frac{3x^4}{4} - 6x^2\right]_{-2}^{0} + \left[\frac{-3x^4}{4} + 6x^2\right]_{0}^{2}$$

$$= -(12 - 24) + (-12 + 24)$$

$$= 24$$

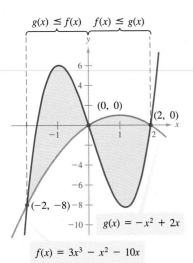

$g(x) \leq f(x)$ $f(x) \leq g(x)$

$(0, 0)$

$(2, 0)$

$(-2, -8)$

$g(x) = -x^2 + 2x$

$f(x) = 3x^3 - x^2 - 10x$

FIGURE 7.8

If the graph of a function of y is a boundary of a region, it is often convenient to use representative rectangles that are *horizontal* and find the area by integrating with respect to y. In general, to determine the area between two curves, we use

$$A = \int_{x_1}^{x_2} \underbrace{[(\text{top curve}) - (\text{bottom curve})]}_{\text{in variable } x}\, dx \qquad \text{Vertical rectangles}$$

$$A = \int_{y_1}^{y_2} \underbrace{[(\text{right curve}) - (\text{left curve})]}_{\text{in variable } y}\, dy \qquad \text{Horizontal rectangles}$$

where (x_1, y_1) and (x_2, y_2) are either adjacent points of intersection of the two curves involved or points on the specified boundary lines.

EXAMPLE 5   Horizontal representative rectangles

Find the area of the region bounded by the graphs of $x = 3 - y^2$ and $y = x - 1$.

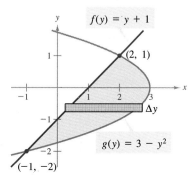

Horizontal rectangles
(Integration with respect to y)

FIGURE 7.9

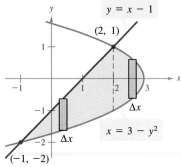

Vertical rectangles
(Integration with respect to x)

FIGURE 7.10

SOLUTION

We consider $g(y) = 3 - y^2$ and $f(y) = y + 1$. These two curves intersect when $y = -2$ and $y = 1$, as shown in Figure 7.9. Since $f(y) \le g(y)$ on this interval, we have

$$\Delta A = [g(y) - f(y)]\Delta y = [(3 - y^2) - (y + 1)]\Delta y.$$

Hence, the area is

$$A = \int_{-2}^{1} [(3 - y^2) - (y + 1)]\, dy$$

$$= \int_{-2}^{1} (-y^2 - y + 2)\, dy = \left[\frac{-y^3}{3} - \frac{y^2}{2} + 2y \right]_{-2}^{1}$$

$$= \left(-\frac{1}{3} - \frac{1}{2} + 2 \right) - \left(\frac{8}{3} - 2 - 4 \right) = \frac{9}{2}.$$

REMARK Note in Example 5 that by integrating with respect to y we need only one integral. If we had integrated with respect to x, we would have needed two integrals, as shown in Figure 7.10.

In this section, we developed the integration formula for the area between two curves by using a rectangle as the *representative element*. For each new application in the remaining sections of this chapter, we will construct an appropriate representative element using precalculus formulas you already know. Each integration formula then will be obtained by summing these representative elements.

| Known precalculus formula | $\longrightarrow$ | Representative element | $\longrightarrow$ | New integration formula |

For example, in this section we developed the area formula as follows.

| $A = $ (height)(width) | $\longrightarrow$ | $\Delta A = [f(x) - g(x)]\,\Delta x$ | $\longrightarrow$ | $A = \displaystyle\int_{a}^{b} [f(x) - g(x)]\, dx$ |

EXAMPLE 6 An application

The total consumption of petroleum fuel for transportation in the United States from 1960 to 1979 followed a growth pattern described by the equation

$$f(t) = 0.000433t^2 + 0.0962t + 2.76, \quad -10 \le t \le 9$$

where $f(t)$ is measured in billions of barrels and t in years, with $t = 0$ corresponding to January 1, 1970. When crude oil prices increased dramatically in the late 1970s, the growth pattern for consumption changed and began following the pattern described by the model

$$g(t) = -0.00831t^2 + 0.152t + 2.81, \quad 9 \le t \le 16$$

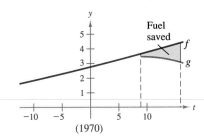

f: Pre-1979 consumption rate
g: Post-1979 consumption rate

FIGURE 7.11

as shown in Figure 7.11. Find the total amount of fuel saved from 1979 through 1985 as a result of fuel being consumed at the post-1979 rate rather than at the pre-1979 rate.

SOLUTION

Since the graph of the pre-1979 model lies above the post-1979 graph on the interval [9, 16], the amount of gasoline saved is given by the following integral.

$$\int_9^{16} \overbrace{[(0.000433t^2 + 0.0962t + 2.76)}^{f(t)} - \overbrace{(-0.00831t^2 + 0.152t + 2.81)]}^{g(t)} \, dt$$

$$= \int_9^{16} (0.008743t^2 - 0.0558t - 0.05) \, dt$$

$$= \left[\frac{0.008743t^3}{3} - \frac{0.0558t^2}{2} - 0.05t \right]_9^{16}$$

$$\approx 4.58 \text{ billion barrels}$$

Therefore, approximately 4.58 billion barrels of fuel were saved. (At 42 gallons per barrel, about 200 billion gallons were saved!)

EXERCISES for Section 7.1

In Exercises 1–6, find the area of the given region.

1. $f(x) = x^2 - 6x$
$g(x) = 0$

2. $f(x) = x^2 + 2x + 1$
$g(x) = 2x + 5$

3. $f(x) = x^2 - 4x + 3$
$g(x) = -x^2 + 2x + 3$

4. $f(x) = x^2$
$g(x) = x^3$

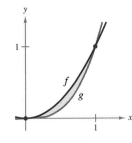

5. $f(x) = 3(x^3 - x)$
$g(x) = 0$

6. $f(x) = (x - 1)^3$
$g(x) = x - 1$

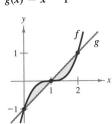

In Exercises 7–36, sketch the region bounded by the graphs of the given functions and find the area of the region.

7. $f(x) = x^2 - 4x$, $g(x) = 0$
8. $f(x) = 3 - 2x - x^2$, $g(x) = 0$
9. $f(x) = x^2 + 2x + 1$, $g(x) = 3x + 3$
10. $f(x) = -x^2 + 4x + 2$, $g(x) = x + 2$
11. $y = x$, $y = 2 - x$, $y = 0$
12. $y = \frac{1}{x^2}$, $y = 0$, $x = 1$, $x = 5$
13. $f(x) = 3x^2 + 2x$, $g(x) = 8$
14. $f(x) = x(x^2 - 3x + 3)$, $g(x) = x^2$
15. $f(x) = x^3 - 2x + 1$, $g(x) = -2x$, $x = 1$

16. $f(x) = \sqrt[3]{x}$, $g(x) = x$

17. $f(x) = \sqrt{3x + 1}$, $g(x) = x + 1$

18. $f(x) = x^2 + 5x - 6$, $g(x) = 6x - 6$

19. $y = x^2 - 4x + 3$, $y = 3 + 4x - x^2$

20. $y = x^4 - 2x^2$, $y = 2x^2$

21. $f(y) = y^2$, $g(y) = y + 2$

22. $f(y) = y(2 - y)$, $g(y) = -y$

23. $f(y) = y^2 + 1$, $g(y) = 0$, $y = -1$, $y = 2$

24. $f(y) = \dfrac{y}{\sqrt{16 - y^2}}$, $g(y) = 0$, $y = 3$

25. $f(x) = \dfrac{1}{x}$, $g(x) = -x^2 + 4x - 2$, $x > 0$

26. $f(x) = 3^x$, $g(x) = 2x + 1$

27. $f(x) = \dfrac{1}{1 + x^2}$, $g(x) = \dfrac{x^2}{2}$

28. $f(x) = 2$, $g(x) = \sec x$, $-\dfrac{\pi}{3} \le x \le \dfrac{\pi}{3}$

29. $f(x) = 2 \sin x$, $g(x) = \tan x$, $-\dfrac{\pi}{3} \le x \le \dfrac{\pi}{3}$

30. $f(x) = \sin 2x$, $g(x) = \cos x$, $\dfrac{\pi}{6} \le x \le \dfrac{5\pi}{6}$

31. $f(x) = 2 \sin x + \sin 2x$, $y = 0$, $0 \le x \le \pi$

32. $f(x) = 2 \sin x + \cos 2x$, $y = 0$, $0 \le x \le \pi$

33. $f(x) = xe^{-x^2}$, $y = 0$, $0 \le x \le 1$

34. $f(x) = \dfrac{1}{x^2}e^{1/x}$, $y = 0$, $1 \le x \le 3$

35. $f(x) = \dfrac{6x}{x^2 + 1}$, $y = 0$, $0 \le x \le 3$

36. $f(x) = \dfrac{4}{x}$, $x = 0$, $y = 1$, $y = 4$

In Exercises 37–40, use integration to find the area of the triangle having the given vertices.

37. $(0, 0)$, $(4, 0)$, $(4, 4)$

38. $(0, 0)$, $(4, 0)$, $(6, 4)$

39. $(0, 0)$, $(a, 0)$, (b, c)

40. $(2, -3)$, $(4, 6)$, $(6, 1)$

In Exercises 41 and 42, find b so that the line $y = b$ divides the region bounded by the graphs of the two equations into two regions of equal area.

41. $y = 9 - x^2$, $y = 0$

42. $y = 9 - |x|$, $y = 0$

43. The graphs of $y = x^4 - 2x^2 + 1$ and $y = 1 - x^2$ intersect at three points. However, the area between the curves *can* be found by a single integral. Explain why this is so, and write an integral for this area.

44. The area of the region bounded by the graphs of $y = x^3$ and $y = x$ *cannot* be found by the single integral

$$\int_{-1}^{1} (x^3 - x)\, dx.$$

Explain why this is so. Use symmetry to write a single integral that does represent the area.

In Exercises 45 and 46, find the area of the region bounded by the graph of the function and the tangent line to the graph at the specified point.

Function	Point
45. $f(x) = x^3$	$(1, 1)$
46. $f(x) = \dfrac{1}{x^2 + 1}$	$\left(1, \dfrac{1}{2}\right)$

In Exercises 47 and 48, use a computer or calculator and Simpson's Rule (with $n = 4$) to approximate the area of the region bounded by the graphs of the given equations.

47. $y = \sqrt{1 + x^3}$, $y = \dfrac{1}{2}x + 2$, $x = 0$

48. $y = \sqrt{x}e^x$, $y = 0$, $x = 0$, $x = 1$

In Exercises 49 and 50, evaluate the given limit and sketch the graph of the region whose area is given by the limit.

49. $\displaystyle\lim_{\|\Delta\| \to 0} \sum_{i=1}^{n} (x_i - x_i^2)\, \Delta x$
    where $x_i = i/n$ and $\Delta x = 1/n$

50. $\displaystyle\lim_{\|\Delta\| \to 0} \sum_{i=1}^{n} (4 - x_i^2)\, \Delta x$
    where $x_i = -2 + (4i/n)$ and $\Delta x = 4/n$

In Exercises 51 and 52, two models R_1 and R_2 are given for revenue (in billions of dollars) for a large corporation. The model R_1 gives projected annual revenues from 1990 to 1995, with $t = 0$ corresponding to 1990, and R_2 gives projected revenues if there is a decrease in growth of corporate sales over the period. Approximate the total reduction in revenue if corporate sales are actually closer to the model R_2.

51. $R_1 = 7.21 + 0.58t$
    $R_2 = 7.21 + 0.45t$

52. $R_1 = 7.21 + 0.26t + 0.02t^2$
    $R_2 = 7.21 + 0.1t + 0.01t^2$

In Exercises 53–56, find the consumer surplus and producer surplus for the given supply and demand curves. The consumer surplus and producer surplus are represented by the areas shown in the accompanying figure.

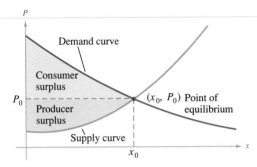

Demand function	Supply function
53. $p_1(x) = 50 - 0.5x$	$p_2(x) = 0.125x$
54. $p_1(x) = 1000 - 0.4x^2$	$p_2(x) = 42x$
55. $p_1(x) = \dfrac{10,000}{\sqrt{x + 100}}$	$p_2(x) = 100\sqrt{0.05x + 10}$
56. $p_1(x) = \sqrt{25 - 0.1x}$	$p_2(x) = \sqrt{9 + 0.1x} - 2$

FIGURE FOR 53–56

In Exercises 57 and 58, use Simpson's Rule on a computer or calculator (with $n = 10$) to approximate the area of the region bounded by the graphs of the given equations.

57. $y = \sqrt{\dfrac{x^3}{4 - x}}, \; y = 0, \; x = 3$

58. $y = x\sqrt{\dfrac{4 - x}{4 + x}}, \; y = 0, \; x = 4$

7.2 Volume: The Disc Method

Solid of revolution ▪ The disc method ▪ The washer method ▪ Solids with known cross sections

In Chapter 5, we mentioned that area is only *one* of the many applications of the definite integral. Another important application is its use in finding the volume of a three-dimensional solid. In this section we consider a particular type of three-dimensional solid—one whose cross sections are similar. We begin with solid figures having *circular* (or annular) cross sections. Such solids are called **solids of revolution** and are used commonly in engineering and manufacturing. Some examples are axles, funnels, pills, bottles, and pistons.

If a region in the plane is revolved about a line, the resulting solid is called a **solid of revolution,** and the line is called the **axis of revolution.** The simplest such solid is a right circular cylinder or **disc,** which is formed by revolving a rectangle about an axis adjacent to one of the sides of the rectangle as shown in Figure 7.12. The volume of such a disc is given by

volume of disc $= \pi R^2 w$

where R is the radius of the disc and w is the width.

FIGURE 7.12

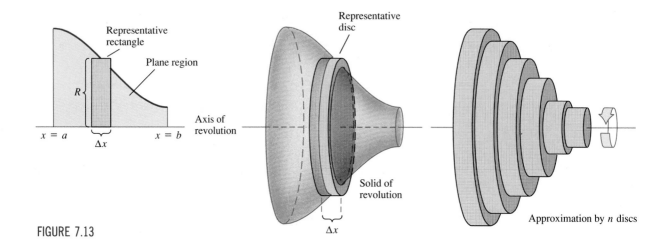

FIGURE 7.13

To see how to use the volume of a disc to find the volume of a general solid of revolution, consider a solid of revolution formed by revolving the plane region in Figure 7.13 about the indicated axis. To determine the volume of this solid, we consider a representative rectangle in the plane region. When this rectangle is revolved about the axis of revolution, it generates a representative disc whose volume is

$$\Delta V = \pi R^2 \Delta x.$$

If we approximate the volume of the solid by n such discs of width Δx and radius $R(x_i)$, we have

$$\text{volume of solid} \approx \sum_{i=1}^{n} \pi [R(x_i)]^2 \Delta x = \pi \sum_{i=1}^{n} [R(x_i)]^2 \Delta x.$$

By taking the limit as $\|\Delta\| \to 0$ $(n \to \infty)$, we have

$$\text{volume of solid} = \lim_{n \to \infty} \pi \sum_{i=1}^{n} [R(x_i)]^2 \Delta x = \pi \int_{a}^{b} [R(x)]^2 \, dx.$$

Schematically, the disc method looks like this.

Known precalculus formula	*Representative element*	*New integration formula*
volume of disc $V = \pi R^2 w$	$\Delta V = \pi [R(x_i)]^2 \Delta x$	solid of revolution $V = \pi \int_{a}^{b} [R(x)]^2 \, dx$

A similar formula can be derived if the axis of revolution is vertical.

THE DISC METHOD

To find the volume of a solid of revolution with the **disc method,** use one of the following as indicated in Figure 7.14.

Horizontal axis of revolution	*Vertical axis of revolution*
volume $= V = \pi \displaystyle\int_a^b [R(x)]^2 \, dx$	volume $= V = \pi \displaystyle\int_c^d [R(y)]^2 \, dy$

Horizontal axis of revolution

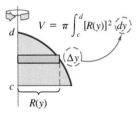

Vertical axis of revolution

FIGURE 7.14

REMARK   In Figure 7.14 note that we can determine the variable of integration by placing a representative rectangle in the *plane* region "perpendicular" to the axis of revolution. If the width of the rectangle is Δx, we integrate with respect to x, and if the width of the rectangle is Δy, we integrate with respect to y.

The simplest application of the disc method involves a plane region bounded by the graph of f and the x-axis. If the axis of revolution is the x-axis, then the radius $R(x)$ is simply $f(x)$, as shown in Example 1.

EXAMPLE 1   Finding the volume of a solid of revolution: the disc method

Find the volume of the solid formed by revolving the region bounded by the graph of $f(x) = \sqrt{\sin x}$ and the x-axis ($0 \le x \le \pi$) about the x-axis.

SOLUTION

From the representative rectangle in Figure 7.15, we see that the radius of this solid is given by

$$R(x) = f(x) = \sqrt{\sin x}$$

and it follows that its volume is

$$V = \pi \int_a^b [R(x)]^2 \, dx = \pi \int_0^\pi (\sqrt{\sin x})^2 \, dx = \pi \int_0^\pi \sin x \, dx$$

$$= -\pi \cos x \Big]_0^\pi = \pi(1 + 1) = 2\pi.$$

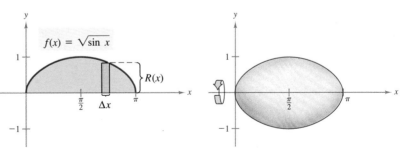

FIGURE 7.15   Plane region   Solid of revolution

REMARK   Note in Example 1 that the problem was solved *without* referring to the three-dimensional portion of the sketch in Figure 7.15. In general, to set up an integral to find the volume of a solid of revolution, a sketch of the plane region is more useful than a sketch of the solid, since the radius is more easily visualized in the plane region.

EXAMPLE 2   Finding the volume of a solid of revolution: the disc method

Find the volume of the solid formed by revolving the region bounded by $f(x) = 2 - x^2$ and $g(x) = 1$ about the line $y = 1$, as shown in Figure 7.16.

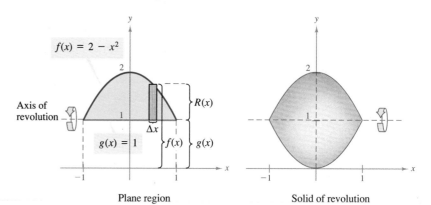

FIGURE 7.16

Plane region   Solid of revolution

SOLUTION

We begin by finding the points of intersection of f and g.

$$2 - x^2 = 1 \qquad \text{Set } f(x) \text{ equal to } g(x)$$
$$x^2 = 1$$
$$x = \pm 1$$

Now, to find the radius, we subtract $g(x)$ from $f(x)$ to obtain

$$R(x) = f(x) - g(x) = (2 - x^2) - 1 = 1 - x^2.$$

Finally, we integrate between -1 and 1 to find the volume.

$$V = \pi \int_a^b [R(x)]^2 \, dx = \pi \int_{-1}^1 (1 - x^2)^2 \, dx$$
$$= \pi \int_{-1}^1 (1 - 2x^2 + x^4) \, dx$$
$$= \left[\pi \left(x - \frac{2x^3}{3} + \frac{x^5}{5} \right) \right]_{-1}^1 = \frac{16\pi}{15} \approx 3.35 \quad \blacksquare$$

FIGURE 7.17

The washer method

The disc method can be extended to cover solids of revolution with a hole by replacing the representative disc with a representative **washer.** The washer is formed by revolving a rectangle about an axis, shown in Figure 7.17. If r

and R are the inner and outer radii of the washer and w is the width of the washer, then the volume is given by

$$\text{volume of washer} = \pi(R^2 - r^2)w.$$

Now, suppose that a region is bounded by an **outer radius** $R(x)$ and an **inner radius** $r(x)$, as shown in Figure 7.18. If this region is revolved about its axis of revolution, then the volume of the resulting solid is given by

$$V = \pi \int_a^b [R(x)]^2 \, dx - \pi \int_a^b [r(x)]^2 \, dx = \pi \int_a^b ([R(x)]^2 - [r(x)]^2) \, dx.$$

Note that the integral involving the inner radius represents the volume of the hole and is *subtracted* from the integral involving the outer radius.

FIGURE 7.18

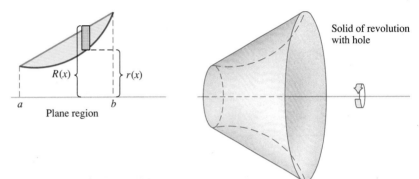

Plane region

Solid of revolution
with hole

EXAMPLE 3   The washer method

Find the volume of the solid formed by revolving the region bounded by the graphs of $y = \sqrt{x}$ and $y = x^2$ about the x-axis, as shown in Figure 7.19.

SOLUTION

From Figure 7.19, we have

$$R(x) = \sqrt{x} \qquad \text{Outer radius}$$
$$r(x) = x^2. \qquad \text{Inner radius}$$

Now, integrating between 0 and 1, we have

$$V = \pi \int_a^b ([R(x)]^2 - [r(x)]^2) \, dx = \pi \int_0^1 (x - x^4) \, dx$$
$$= \pi \left[\frac{x^2}{2} - \frac{x^5}{5} \right]_0^1$$
$$= \frac{3\pi}{10}.$$

In each example so far, the axis of revolution has been *horizontal* and we integrated with respect to x. In the next example, the axis of revolution is *vertical* and (using the disc method) we must integrate with respect to y. In this example, we need two separate integrals to compute the volume.

Plane region

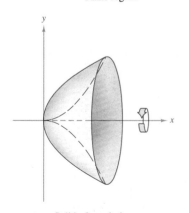

Solid of revolution

FIGURE 7.19

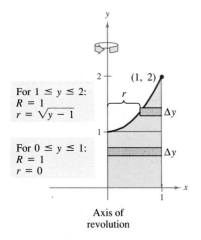

For $1 \le y \le 2$:
$R = 1$
$r = \sqrt{y - 1}$

For $0 \le y \le 1$:
$R = 1$
$r = 0$

Axis of revolution

FIGURE 7.20

EXAMPLE 4   Two-integral case, integrating with respect to *y*

Find the volume of the solid formed by revolving the region bounded by the graphs of $y = x^2 + 1$, $y = 0$, $x = 0$, and $x = 1$ about the y-axis, as shown in Figure 7.20.

SOLUTION

We use two integrals because we have no single formula to represent the inner radius r. When $0 \le y \le 1$, then $r = 0$, but when $1 \le y \le 2$, r is determined by the equation $y = x^2 + 1$, which implies that $r = \sqrt{y - 1}$. Since $R = 1$, the volume is

$$V = \pi \int_0^1 (1^2 - 0^2) \, dy + \pi \int_1^2 [1^2 - (\sqrt{y - 1})^2] \, dy$$

$$= \pi \int_0^1 1 \, dy + \pi \int_1^2 (2 - y) \, dy$$

$$= \pi y \Big]_0^1 + \pi \left[2y - \frac{y^2}{2} \right]_1^2$$

$$= \pi \left(1 + 4 - 2 - 2 + \frac{1}{2} \right) = \frac{3\pi}{2}.$$

Note that the first integral $\pi \int_0^1 1 \, dy$ represents the volume of a right circular cylinder of radius 1 and height 1. This portion of the volume could have been determined without using calculus.   ▭

EXAMPLE 5   An application

A manufacturer drills a hole through the center of a metal sphere of radius 5 inches, as shown in Figure 7.21(a). The hole has a radius of 3 inches. What is the volume of the resulting metal ring?

SOLUTION

(a) Solid of revolution

We imagine the ring to be generated by a segment of the circle whose equation is $x^2 + y^2 = 25$, as shown in Figure 7.21(b). Since the radius of the hole is 3 inches, we let $y = 3$ and solve the equation $x^2 + y^2 = 25$ to determine that the limits of integration are $x = \pm 4$. Thus, we have

$$r(x) = 3 \quad \text{and} \quad R(x) = \sqrt{25 - x^2}$$

and the volume is given by

$$V = \pi \int_a^b ([R(x)]^2 - [r(x)]^2) \, dx$$

$$= \pi \int_{-4}^4 [(\sqrt{25 - x^2})^2 - (3)^2] \, dx = \pi \int_{-4}^4 (16 - x^2) \, dx$$

$$= \pi \left[16x - \frac{x^3}{3} \right]_{-4}^4 = \frac{256\pi}{3} \text{ in}^3.$$

(b) Plane region

FIGURE 7.21

Solids with known cross sections

With the disc method, we can find the volume of a solid having a circular cross section whose area is $\Delta A = \pi R^2$. We can generalize this method to solids of any shape, as long as we know a formula for the area of an arbitrary cross section. Some common cross sections are squares, rectangles, triangles, semicircles, and trapezoids.

VOLUMES OF SOLIDS WITH KNOWN CROSS SECTIONS	1. For cross sections of area $A(x)$, taken perpendicular to the x-axis, $$\text{volume} = \int_a^b A(x)\, dx. \qquad \text{(See Figure 7.22)}$$ 2. For cross sections of area $A(y)$, taken perpendicular to the y-axis, $$\text{volume} = \int_c^d A(y)\, dy. \qquad \text{(See Figure 7.22.)}$$

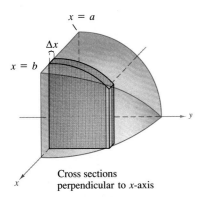

Cross sections
perpendicular to x-axis

Cross sections
perpendicular to y-axis

FIGURE 7.22

EXAMPLE 6   Triangular cross sections

Find the volume of the solid whose base is the area bounded by the lines

$$f(x) = 1 - \frac{x}{2}, \qquad g(x) = -1 + \frac{x}{2}, \qquad \text{and} \qquad x = 0$$

and whose cross sections perpendicular to the x-axis are equilateral triangles, as shown in Figure 7.23.

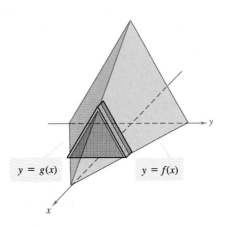

$y = g(x)$ $y = f(x)$

Cross sections are equilateral triangles.

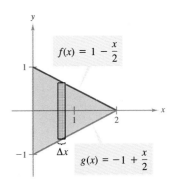

$f(x) = 1 - \dfrac{x}{2}$

$g(x) = -1 + \dfrac{x}{2}$

Triangular base in xy-plane

FIGURE 7.23

SOLUTION

For each triangular cross section, we have

$$\text{base} = \left(1 - \frac{x}{2}\right) - \left(-1 + \frac{x}{2}\right) = 2 - x \qquad \text{Length of base}$$

$$\text{area} = \frac{\sqrt{3}}{4}(\text{base})^2 \qquad\qquad \text{Area of equilateral triangle}$$

$$A(x) = \frac{\sqrt{3}}{4}(2 - x)^2. \qquad\qquad \text{Area of cross section}$$

Since x ranges from 0 to 2, the volume of the solid is

$$V = \int_a^b A(x)\ dx = \int_0^2 \frac{\sqrt{3}}{4}(2 - x)^2\ dx$$

$$= -\frac{\sqrt{3}}{4}\left[\frac{(2 - x)^3}{3}\right]_0^2$$

$$= \frac{2\sqrt{3}}{3}. \qquad\qquad\blacksquare$$

EXAMPLE 7 An application to geometry

Prove that the volume of a pyramid with a square base is $V = \frac{1}{3}hB$, where h is the height of the pyramid and B is the area of the base.

SOLUTION

In Figure 7.24, we intersect the pyramid with a plane parallel to the base at height y to form a square cross section whose sides are of length b'. Using similar triangles, we can show that

$$\frac{b'}{b} = \frac{h - y}{h} \qquad \text{or} \qquad b' = \frac{b}{h}(h - y)$$

where b is the length of the sides of the base of the pyramid. Thus,

$$A(y) = (b')^2 = \frac{b^2}{h^2}(h - y)^2.$$

Integrating between 0 and h, we have

$$\text{volume} = \int_0^h \frac{b^2}{h^2}(h - y)^2\ dy = \frac{b^2}{h^2}\int_0^h (h - y)^2\ dy$$

$$= -\left(\frac{b^2}{h^2}\right)\frac{(h - y)^3}{3}\Bigg]_0^h = \frac{b^2}{h^2}\left(\frac{h^3}{3}\right) = \frac{b^2 h}{3} = \frac{1}{3}hB. \qquad\blacksquare$$

An interesting generalization of the formula in Example 7 is that the volume of *any cone* can be determined by the formula $V = \frac{1}{3}hB$, where B is the area of the base and h is the height of the cone, as shown in Figure 7.25.

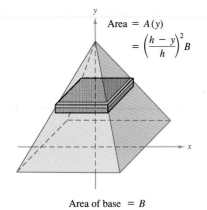

Area $= A(y)$

$= \left(\dfrac{h - y}{h}\right)^2 B$

Area of base $= B$

$h - y$

h

$\frac{1}{2}b'$

y

$\frac{1}{2}b$

FIGURE 7.24

FIGURE 7.25

Square base
of area B

Circular base
of area B

Irregular base
of area B

EXERCISES for Section 7.2

In Exercises 1–8, find the volume of the solid formed by revolving the given region about the x-axis.

1. $y = -x + 1$

2. $y = 4 - x^2$

3. $y = \sqrt{4 - x^2}$

4. $y = x^2$

5. $y = \sqrt{x}$

6. $y = \sqrt{4 - x^2}$

7. $y = x^2,\ y = x^3$

8. $y = 2,\ y = 4 - \dfrac{x^2}{4}$

In Exercises 9–12, find the volume of the solid formed by revolving the given region about the y-axis.

9. $y = x^2$

10. $y = \sqrt{16 - x^2}$

11. $y = x^{2/3}$

12. $x = -y^2 + 4y$

In Exercises 13–16, find the volume of the solid generated by revolving the region bounded by the graphs of the given equations about the indicated lines.

13. $y = \sqrt{x},\ y = 0,\ x = 4$
   (a) the x-axis   (b) the y-axis
   (c) the line $x = 4$ (d) the line $x = 6$

14. $y = 2x^2,\ y = 0,\ x = 2$
   (a) the y-axis   (b) the x-axis
   (c) the line $y = 8$ (d) the line $x = 2$

15. $y = x^2,\ y = 4x - x^2$
   (a) the x-axis   (b) the line $y = 6$

16. $y = 6 - 2x - x^2,\ y = x + 6$
   (a) the x-axis   (b) the line $y = 3$

In Exercises 17–20, find the volume of the solid generated by revolving the region bounded by the graphs of the given equations about the line $y = 4$.

17. $y = x$, $y = 3$, $x = 0$
18. $y = x^2$, $y = 4$
19. $y = \dfrac{1}{x}$, $y = 0$, $x = 1$, $x = 4$
20. $y = \sec x$, $y = 0$, $0 \le x \le \dfrac{\pi}{3}$

In Exercises 21–24, find the volume of the solid generated by revolving the region bounded by the graphs of the given equations about the line $x = 6$.

21. $y = x$, $y = 0$, $y = 4$, $x = 6$
22. $y = 6 - x$, $y = 0$, $y = 4$, $x = 0$
23. $x = y^2$, $x = 4$
24. $xy = 6$, $y = 2$, $y = 6$, $x = 6$

In Exercises 25–32, find the volume of the solid generated by revolving the region bounded by the graphs of the given equations about the x-axis.

25. $y = \dfrac{1}{\sqrt{x + 1}}$, $y = 0$, $x = 0$, $x = 3$
26. $y = x\sqrt{4 - x^2}$, $y = 0$
27. $y = \dfrac{1}{x}$, $y = 0$, $x = 1$, $x = 4$
28. $y = \dfrac{3}{x + 1}$, $y = 0$, $x = 0$, $x = 8$
29. $y = e^{-x}$, $y = 0$, $x = 0$, $x = 1$
30. $y = e^{x/2}$, $y = 0$, $x = 0$, $x = 4$
31. $y = \sqrt{\sin x}$, $y = 0$, $x = 0$, $x = \dfrac{\pi}{2}$
32. $y = \sqrt{\cos x}$, $y = 0$, $x = 0$, $x = \dfrac{\pi}{2}$

33. The region bounded by the parabola $y = 4x - x^2$ and the x-axis is revolved about the x-axis. Find the volume of the resulting solid.
34. If the equation of the parabola in Exercise 33 were changed to $y = 4 - x^2$, would the volume of the solid generated be different? Why or why not?
35. The upper half of the ellipse $9x^2 + 25y^2 = 225$ is revolved about the x-axis to form a prolate spheroid (shaped like a football). Find the volume of the spheroid.
36. The right half of the ellipse $9x^2 + 25y^2 = 225$ is revolved about the y-axis to form an oblate spheroid (shaped like an M&M candy). Find the volume of the spheroid.
37. If the portion of the line $y = \frac{1}{2}x$ lying in the first quadrant is revolved about the x-axis, a cone is generated. Find the volume of the cone extending from $x = 0$ to $x = 6$.

38. Use the disc method to verify that the volume of a right circular cone is $\frac{1}{3}\pi r^2 h$, where r is the radius of the base and h is the height.
39. Use the disc method to verify that the volume of a sphere of radius r is $\frac{4}{3}\pi r^3$.
40. A sphere of radius r is cut by a plane h ($h < r$) units above the equator. Find the volume of the solid (spherical segment) above the plane.
41. A cone with a base of radius r and height H is cut by a plane parallel to and h units above the base. Find the volume of the solid (frustum of a cone) below the plane.
42. The region bounded by $y = \sqrt{x}$, $y = 0$, $x = 0$, and $x = 4$ is revolved about the x-axis. (a) Find the value of x in the interval $[0, 4]$ that divides the solid into two parts of equal volume. (b) Find the values of x in the interval $[0, 4]$ that divide the solid into three parts of equal volume.
43. In the Chapter 7 Application, we showed that a solid of revolution having the form of a regulation-size football can be generated by revolving the graph of

$$f(x) = -0.0944x^2 + 3.4, \quad -5.5 \le x \le 5.5$$

about the x-axis. (The x and y values are measured in inches.) Show that the volume of this solid is approximately 232 cubic inches.
44. A tank on a water tower is a sphere of radius 50 feet. Determine the depth of the water when the tank is filled to 21.6 percent of its total capacity.
45. A tank on the wing of a jet is formed by revolving the region bounded by the graph of $y = \frac{1}{8}x^2\sqrt{2 - x}$ and the x-axis about the x-axis (see figure), where x and y are measured in meters. Find the volume of the tank.

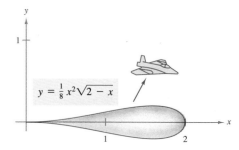

$$y = \tfrac{1}{8} x^2 \sqrt{2 - x}$$

FIGURE FOR 45

In Exercises 46 and 47, use Simpson's Rule (with $n = 4$) to approximate the volume of the solid generated by revolving the region bounded by the graphs of the given equations about the x-axis.

46. $y = \ln x$, $y = 0$, $x = 1$, $x = 3$
47. $y = e^{-x^2}$, $y = 0$, $x = 0$, $x = 2$

48. Match each of the following integrals with the solid whose volume it represents, and give the dimensions of each solid.
(a) right circular cylinder   (b) ellipsoid
(c) sphere   (d) right circular cone
(e) torus

___ $\pi \int_0^h \left(\frac{rx}{h}\right)^2 dx$

___ $\pi \int_0^h r^2\, dx$

___ $\pi \int_{-r}^r (\sqrt{r^2 - x^2})^2\, dx$

___ $\pi \int_{-b}^b \left(a\sqrt{1 - \frac{x^2}{b^2}}\right)^2 dx$

___ $\pi \int_{-r}^r [(R + \sqrt{r^2 - x^2})^2 - (R - \sqrt{r^2 - x^2})^2]\, dx$

49. Find the volume of the solid whose base is bounded by the circle $x^2 + y^2 = 4$, with the indicated cross sections taken perpendicular to the x-axis.
(a) squares   (b) equilateral triangles

(c) semicircles   (d) isosceles right triangles

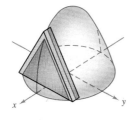

50. Find the volume of the solid whose base is bounded by the graphs of $y = x + 1$ and $y = x^2 - 1$, with the indicated cross sections taken perpendicular to the x-axis.
(a) squares   (b) rectangles of height 1

(c) semiellipses of   (d) equilateral triangles
    height 2 ($A = \pi ab$)

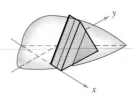

51. The base of a solid is bounded by $y = x^3$, $y = 0$, and $x = 1$. Find the volume of the solid for the following cross sections (taken perpendicular to the y-axis).
(a) squares
(b) semicircles
(c) equilateral triangles
(d) trapezoids for which $h = b_1 = \frac{1}{2}b_2$, where b_1 and b_2 are the lengths of the upper and lower bases
(e) semiellipses whose heights are twice the lengths of their bases

52. Find the volume of the solid of intersection (the solid common to both) of the two right circular cylinders of radius r whose axes meet at right angles (see figure).

Two intersecting cylinders

Solid of intersection   FIGURE FOR 52

53. A wedge is cut from a right circular cylinder of radius r inches by a plane through the diameter of the base, making an angle of $45°$ with the plane of the base (see figure). Find the volume of the wedge.

FIGURE FOR 53

54. *Cavalieri's Theorem* Prove that if two solids have equal altitudes and all plane sections parallel to their bases and at equal distances from their bases have equal areas, then the solids have the same volume.

Area of R_1 = area of R_2

FIGURE FOR 54

In Exercises 55 and 56, use Simpson's Rule on a computer or calculator (with $n = 10$) to approximate the volume of the solid generated when the region bounded by the graphs of the given equations is revolved about the x-axis.

55. $y = e^{x/2} + e^{-x/2}$, $y = 0$, $x = -1$, $x = 2$

56. $y = \dfrac{8x}{9 + x^2}$, $y = 0$, $x = 0$, $x = 5$

7.3 Volume: The Shell Method

The shell method ▪ Comparison of disc and shell methods

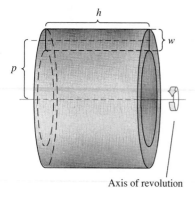

FIGURE 7.26

In this section, we look at an alternate method for finding the volume of a solid of revolution, a method that uses cylindrical shells. We will compare the advantages of the two methods later in this section.

To introduce the **shell method,** we consider a representative rectangle as shown in Figure 7.26, where

  w = width of the rectangle

  h = height of the rectangle

  p = distance between axis of revolution and *center* of the rectangle.

When this rectangle is revolved about its axis of revolution, it forms a cylindrical shell (or tube) of thickness w. To find the volume of this shell we consider two cylinders. The radius of the larger cylinder corresponds to the outer radius of the shell, and the radius of the smaller cylinder corresponds to the inner radius of the shell. Since p is the average radius of the shell, we know the outer radius is $p + (w/2)$ and the inner radius is $p - (w/2)$. Thus, the volume of the shell is given by the difference.

volume of shell = (volume of cylinder) − (volume of hole)

$$= \pi\left(p + \frac{w}{2}\right)^2 h - \pi\left(p - \frac{w}{2}\right)^2 h$$

$$= \pi\left(p^2 + pw + \frac{w^2}{4} - p^2 + pw - \frac{w^2}{4}\right)h$$

$$= \pi 2phw = 2\pi(\text{average radius})(\text{height})(\text{thickness})$$

We use this formula to find the volume of a solid of revolution as follows. Assume that the plane region in Figure 7.27 is revolved about a line to form the indicated solid. If we consider a horizontal rectangle of width Δy, then as the plane region is revolved about a line parallel to the x-axis, the rectangle generates a representative shell whose volume is

$$\Delta V = 2\pi[p(y)h(y)]\,\Delta y.$$

Now, if we approximate the volume of the solid by n such shells of thickness Δy, height $h(y_i)$, and average radius $p(y_i)$, we have

$$\text{volume of solid} \approx \sum_{i=1}^{n} 2\pi[p(y_i)h(y_i)]\,\Delta y = 2\pi \sum_{i=1}^{n} [p(y_i)h(y_i)]\,\Delta y.$$

By taking the limit as $\|\Delta\| \to 0$ $(n \to \infty)$, we have

$$\text{volume of solid} = \lim_{n \to \infty} 2\pi \sum_{i=1}^{n} [p(y_i)h(y_i)]\,\Delta y = 2\pi \int_{c}^{d} [p(y)h(y)]\,dy.$$

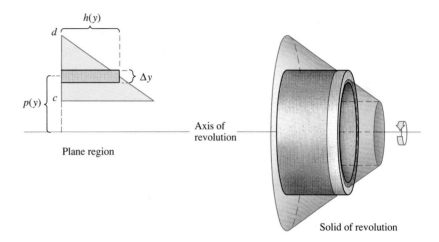

FIGURE 7.27

THE SHELL METHOD

To find the volume of a solid of revolution with the **shell method,** use one of the following, as shown in Figure 7.28.

Horizontal axis of revolution	Vertical axis of revolution
volume $= V = 2\pi \int_{c}^{d} p(y)h(y)\,dy$	volume $= V = 2\pi \int_{a}^{b} p(x)h(x)\,dx$

FIGURE 7.28

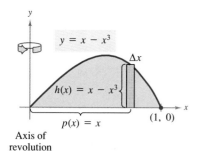

$y = x - x^3$

Δx

$h(x) = x - x^3$

$p(x) = x$ (1, 0)

Axis of
revolution

FIGURE 7.29

EXAMPLE 1   Using the shell method to find volume

Find the volume of the solid of revolution formed by revolving the region bounded by $y = x - x^3$ and the x-axis $(0 \le x \le 1)$ about the y-axis.

SOLUTION

Since the axis of revolution is vertical, we use a vertical representative rectangle, as shown in Figure 7.29. The width Δx indicates that x is the variable of integration. The distance from the center of the rectangle to the axis of revolution is $p(x) = x$, and the height of the rectangle is $h(x) = x - x^3$. Since x ranges from 0 to 1, the volume of the solid is

$$V = 2\pi \int_a^b p(x)h(x)\,dx = 2\pi \int_0^1 x(x - x^3)\,dx = 2\pi \int_0^1 (-x^4 + x^2)\,dx$$

$$= 2\pi \left[-\frac{x^5}{5} + \frac{x^3}{3} \right]_0^1 = \frac{4\pi}{15}.$$

EXAMPLE 2   Using the shell method to find volume

Find the volume of the solid of revolution formed by revolving the region bounded by the graph of $x = e^{-y^2}$ and the y-axis $(0 \le y \le 1)$ about the x-axis.

SOLUTION

Since the axis of revolution is horizontal, we use a horizontal representative rectangle, as shown in Figure 7.30. The width Δy indicates that y is the variable of integration. The distance from the center of the rectangle to the axis of revolution is $p(y) = y$, and the height of the rectangle is $h(y) = e^{-y^2}$. Since y ranges from 0 to 1, the volume of the solid is

$$V = 2\pi \int_c^d p(y)h(y)\,dy = 2\pi \int_0^1 ye^{-y^2}\,dy$$

$$= \left[-\pi e^{-y^2} \right]_0^1$$

$$= \pi\left(1 - \frac{1}{e} \right)$$

$$\approx 1.986.$$

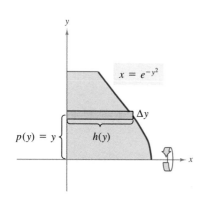

y

$x = e^{-y^2}$

Δy

$p(y) = y$ $h(y)$

x

FIGURE 7.30

Comparison of disc and shell methods

The disc and shell methods can be distinguished as follows. For the disc method, the representative rectangle is always *perpendicular* to the axis of revolution, whereas for the shell method, the representative rectangle is always *parallel* to the axis of revolution, as shown in Figure 7.31.

$$V = \pi \int_c^d (R^2 - r^2)\,(dy)$$

(a) Vertical axis
of revolution

$$V = \pi \int_a^b (R^2 - r^2)\,(dx)$$

Horizontal axis
of revolution

Disc Method: Representative rectangle is
perpendicular to the axis of revolution.

$$V = 2\pi \int_a^b ph\,(dx)$$

(b) Vertical axis
of revolution

$$V = 2\pi \int_c^d ph\,(dy)$$

Horizontal axis
of revolution

Shell Method: Representative rectangle is
parallel to the axis of revolution.

FIGURE 7.31

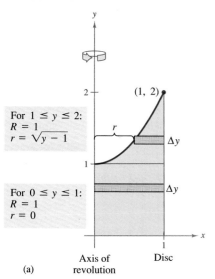

For $1 \le y \le 2$:
$R = 1$
$r = \sqrt{y - 1}$

For $0 \le y \le 1$:
$R = 1$
$r = 0$

(a)

Axis of   Disc
revolution

(b)

Axis of   Shell
revolution

$p(x) = x$

$h(x) = x^2 + 1$

FIGURE 7.32

Often, one method is more convenient to use than the other. The following example illustrates a case in which the shell method is preferable.

EXAMPLE 3   Shell method preferable

Find the volume of the solid formed by revolving the region bounded by the graphs of $y = x^2 + 1$, $y = 0$, $x = 0$, and $x = 1$ about the y-axis.

SOLUTION

In Example 4 in the previous section, we found that the disc method requires two integrals to determine the volume of this solid. See Figure 7.32(a).

$$V = \pi \int_0^1 (1^2 - 0^2)\,dy + \pi \int_1^2 [1^2 - (\sqrt{y - 1})^2]\,dy \qquad \text{Disc method}$$

$$= \pi \int_0^1 1\,dy + \pi \int_1^2 (2 - y)\,dy$$

$$= \pi y \Big]_0^1 + \pi \left[2y - \frac{y^2}{2} \right]_1^2$$

$$= \pi \left(1 + 4 - 2 - 2 + \frac{1}{2} \right) = \frac{3\pi}{2}$$

From Figure 7.32(b), we can see that the shell method requires only one integral to find the volume.

$$V = 2\pi \int_a^b p(x)h(x)\,dx \qquad \text{Shell method}$$

$$= 2\pi \int_0^1 x(x^2 + 1)\,dx$$

$$= 2\pi \left[\frac{x^4}{4} + \frac{x^2}{2} \right]_0^1$$

$$= 2\pi \left(\frac{3}{4} \right) = \frac{3\pi}{2}$$

FIGURE 7.33

FIGURE 7.34

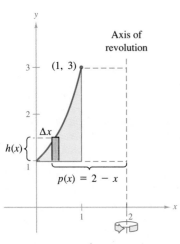

$$h(x) = (x^3 + x + 1) - 1$$
$$p(x) = 2 - x$$

FIGURE 7.35

EXAMPLE 4 Disc method preferable

A pontoon is to be made in the shape shown in Figure 7.33. The pontoon is designed by rotating the graph of

$$y = 1 - \frac{x^2}{16}, \quad -4 \leq x \leq 4$$

about the *x*-axis, where *x* and *y* are measured in feet. Find the volume of the pontoon.

SOLUTION

In Figure 7.34 we compare the shell and disc methods. Using the shell method, we see that Δy is the width of the rectangle, $p(y) = y$, and $h(y) = 2x$. Solving for *x* in the equation $y = 1 - (x^2/16)$, we obtain

$$x = 4\sqrt{1 - y} \quad \text{or} \quad h(y) = 2x = 8\sqrt{1 - y}.$$

Hence, the shell method results in the integral

$$V = 2\pi \int_0^1 8y\sqrt{1 - y} \, dy. \qquad \text{Shell method}$$

Although this is not a particularly difficult integral, its evaluation does require a *u*-substitution. By contrast, the disc method yields the relatively simple integral

$$V = \pi \int_{-4}^4 \left(1 - \frac{x^2}{16}\right)^2 dx = \pi \int_{-4}^4 \left(1 - \frac{x^2}{8} + \frac{x^4}{256}\right) dx \qquad \text{Disc method}$$

$$= \pi \left[x - \frac{x^3}{24} + \frac{x^5}{1280}\right]_{-4}^4$$

$$= \frac{64\pi}{15} \approx 13.4 \text{ ft}^3.$$

Note that for the shell method in Example 4 we had to solve for *x* in terms of *y* in the equation $y = 1 - (x^2/16)$. Sometimes solving for *x* is very difficult (or even impossible). In such cases we must use a vertical rectangle (of width Δx), thus making *x* the variable of integration. The position (horizontal or vertical) of the axis of revolution then determines the method to be used. This is illustrated in Example 5.

EXAMPLE 5 Shell method necessary

Find the volume of the solid formed by revolving the region bounded by the graphs of $y = x^3 + x + 1$, $y = 1$, and $x = 1$ about the line $x = 2$, as shown in Figure 7.35.

SOLUTION

In the equation $y = x^3 + x + 1$, we cannot easily solve for x in terms of y. (See the discussion at the end of Section 4.8.) Therefore, the variable of integration must be x, and we choose a vertical representative rectangle. Since the rectangle is parallel to the axis of revolution, we use the shell method and obtain

$$V = 2\pi \int_a^b p(x)h(x)\, dx = 2\pi \int_0^1 (2 - x)(x^3 + x + 1 - 1)\, dx$$

$$= 2\pi \int_0^1 (-x^4 + 2x^3 - x^2 + 2x)\, dx$$

$$= 2\pi \left[-\frac{x^5}{5} + \frac{x^4}{2} - \frac{x^3}{3} + x^2 \right]_0^1$$

$$= 2\pi \left(-\frac{1}{5} + \frac{1}{2} - \frac{1}{3} + 1 \right) = \frac{29\pi}{15}.$$

EXERCISES for Section 7.3

In Exercises 1–20, use the shell method to find the volume of the solid generated by revolving the given plane region about the indicated line.

1. $y = x$

2. $y = 1 - x$

3. $y = x$

4. $y = 2 - x$

5. $y = \sqrt{x}$

6. $y = x^2 + 4$

7. $y = x^2$, $y = 0$, $x = 2$, about the y-axis
8. $y = x^2$, $y = 0$, $x = 4$, about the y-axis
9. $y = x^2$, $y = 4x - x^2$, about the y-axis
10. $y = x^2$, $y = 4x - x^2$, about the line $x = 2$
11. $y = x^2$, $y = 4x - x^2$, about the line $x = 4$

12. $y = \dfrac{1}{x}$, $x = 1$, $x = 2$, $y = 0$, about the x-axis

13. $y = 4x - x^2$, $y = 0$, about the line $x = 5$
14. $x + y^2 = 9$, $x = 0$, about the x-axis
15. $y = 4x - x^2$, $x = 0$, $y = 4$, about the y-axis
16. $y = 4 - x^2$, $y = 0$, about the y-axis
17. $y = \sqrt{x}$, $y = 0$, $x = 4$, about the line $x = 6$
18. $y = 2x$, $y = 4$, $x = 0$, about the y-axis

19. $y = \dfrac{1}{\sqrt{2\pi}} e^{-x^2/2}$, $y = 0$, $x = 0$, $x = 1$, about the y-axis

20. $y = \begin{cases} \dfrac{\sin x}{x}, & x > 0 \\ 1, & x = 0 \end{cases}$, $y = 0$, $x = 0$, $x = \pi$

about the y-axis

In Exercises 21–24, use the disc or shell method to find the volume of the solid generated by revolving the region bounded by the graphs of the given equations about the specified line.

21. $y = x^3$, $y = 0$, $x = 2$
  (a) the x-axis   (b) the y-axis
  (c) the line $x = 4$ (d) the line $y = 8$

22. $y = \dfrac{1}{x^2}$, $y = 0$, $x = 1$, $x = 4$
  (a) the x-axis   (b) the y-axis
  (c) the line $x = 4$

23. $x^{1/2} + y^{1/2} = a^{1/2}$, $x = 0$, $y = 0$
  (a) the x-axis   (b) the y-axis
  (c) the line $x = a$
24. $x^{2/3} + y^{2/3} = a^{2/3}$ (hypocycloid)
  (a) the x-axis   (b) the y-axis

25. A solid is generated by revolving the region bounded by $y = \frac{1}{2}x^2$ and $y = 2$ about the y-axis. A hole, centered along the axis of revolution, is drilled through this solid so that one-quarter of the volume is removed. Find the diameter of the hole.

26. A solid is generated by revolving the region bounded by $y = \sqrt{9 - x^2}$ and $y = 0$ about the y-axis. A hole, centered along the axis of revolution, is drilled through this solid so that one-third of the volume is removed. Find the diameter of the hole.

27. Let a sphere of radius r be cut by a plane, thus forming a segment of height h. Show that the volume of this segment is $\frac{1}{3}\pi h^2(3r - h)$.

In Exercises 28 and 29, use Simpson's Rule (with $n = 4$) to approximate the volume of the solid. The solid is generated by revolving the region bounded by the graphs of the given equations about the y-axis.

28. $y = \sqrt{1 - x^3}$, $y = 0$, $x = 0$
29. $x^{4/3} + y^{4/3} = 1$

30. A pond is approximately circular with a diameter of 400 feet (see figure). Starting at the center, the depth of the water is measured every 25 feet and recorded in the accompanying table. Use Simpson's Rule to approximate the volume of water in the pond.

x	0	25	50	75	100
Depth	20	19	19	17	15

x	125	150	175	200
Depth	14	10	6	0

31. A storage shed has a circular base of diameter 80 feet (see figure). Starting at the center, the interior height is measured every 10 feet and recorded in the accompanying table. Use Simpson's Rule to approximate the volume of the building.

x	0	10	20	30	40
Height	50	46	40	20	0

FIGURE FOR 30

FIGURE FOR 31

32. Match each of the following integrals with the solid whose volume it represents and give the dimensions of each solid.

$\underline{} 2\pi \int_0^r hx \, dx$

$\underline{} 2\pi \int_0^r hx\left(1 - \frac{x}{r}\right) dx$

$\underline{} 2\pi \int_0^r 2x\sqrt{r^2 - x^2} \, dx$

$\underline{} 2\pi \int_0^b 2ax\sqrt{1 - \frac{x^2}{b^2}} \, dx$

$\underline{} 2\pi \int_{-r}^r (R - x)(2\sqrt{r^2 - x^2}) \, dx$

(a) right circular cone
(b) torus
(c) sphere
(d) right circular cylinder
(e) ellipsoid

33. A **torus** is formed by revolving the circle $x^2 + y^2 = 1$ about the vertical line $x = 2$, as shown in the figure. Find the volume of this "doughnut-shaped" solid. [Hint: The integral $\int_{-1}^1 \sqrt{1 - x^2} \, dx$ represents the area of a semicircle.]

34. A hole is cut through the center of a sphere of radius r. The height of the remaining spherical ring is h, as shown in the figure. Show that the volume of the ring is $V = \pi h^3/6$. [Note that the volume is independent of r.]

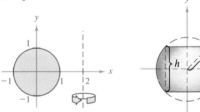

FIGURE FOR 33

FIGURE FOR 34

In Exercises 35 and 36, use Simpson's Rule on a computer or calculator (with $n = 10$) to approximate the volume of the solid generated when the region bounded by the graphs of the given equations is revolved about the y-axis.

35. $y = \sqrt[3]{(x - 2)^2(x - 6)^2}$, $y = 0$, $x = 2$, $x = 6$
36. $y = \dfrac{2}{1 + e^{1/x}}$, $y = 0$, $x = 1$, $x = 3$

7.4 Arc Length and Surfaces of Revolution

Arc length ▪ Surface of revolution ▪ Area of a surface of revolution

Christian Huygens

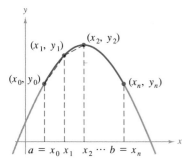

FIGURE 7.36

In this section the summation character of the definite integral is used to find the arc length of a plane curve and the area of a surface of revolution. In both cases, we approximate an arc (a segment of a curve) by straight line segments whose lengths are given by the familiar distance formula

$$d = \sqrt{(x_2 - x_1)^2 + (y_2 - y_1)^2}.$$

If a segment of a curve has a finite arc length, we say that it is **rectifiable.** The problem of calculating the length of a curve has motivated a wide range of mathematical work. Some early contributions to the problem were made by the Dutch mathematician Christian Huygens (1629–1695), who invented the pendulum clock. Another pioneer in the work of rectifiable curves was James Gregory (1638–1675), a Scottish mathematician. Both men played important roles in the early development of calculus.

We will see in developing the formula for arc length that a sufficient condition for the graph of a function f to be rectifiable between $(a, f(a))$ and $(b, f(b))$ is that f' is continuous on $[a, b]$. Such a function is said to be **continuously differentiable** on $[a, b]$ and its graph on the interval $[a, b]$ is called a **smooth curve.**

Assume that a function given by $y = f(x)$ is continuously differentiable on the interval $[a, b]$, and let s denote the length of its graph on this interval. We approximate the graph of f by n line segments whose endpoints are determined by the partition

$$a = x_0 < x_1 < x_2 < \cdots < x_n = b$$

as shown in Figure 7.36.

By letting $\Delta x_i = x_i - x_{i-1}$ and $\Delta y_i = y_i - y_{i-1}$, we approximate the total arc length by

$$s \approx \sum_{i=1}^{n} \sqrt{(\Delta x_i)^2 + (\Delta y_i)^2}.$$

Taking the limit as $\|\Delta\| \to 0$, $(n \to \infty)$, we obtain

$$s = \lim_{n \to \infty} \sum_{i=1}^{n} \sqrt{(\Delta x_i)^2 + (\Delta y_i)^2} = \lim_{n \to \infty} \sum_{i=1}^{n} \sqrt{1 + \left(\frac{\Delta y_i}{\Delta x_i}\right)^2} (\Delta x_i).$$

Since $f'(x)$ exists for each x in (x_{i-1}, x_i), the Mean Value Theorem guarantees the existence of c_i in (x_{i-1}, x_i) such that

$$f(x_i) - f(x_{i-1}) = f'(c_i)(x_i - x_{i-1}) \quad \Longrightarrow \quad \frac{\Delta y_i}{\Delta x_i} = f'(c_i).$$

Moreover, since f' is continuous on $[a, b]$, we know that $\sqrt{1 + [f'(x)]^2}$ is also continuous (and hence integrable) on $[a, b]$ and we have

$$s = \lim_{n \to \infty} \sum_{i=1}^{n} \sqrt{1 + [f'(c_i)]^2} (\Delta x_i) = \int_a^b \sqrt{1 + [f'(x)]^2} \, dx.$$

We call s the **arc length** of f between a and b.

DEFINITION OF ARC LENGTH

If the function given by $y = f(x)$ represents a smooth curve on the interval $[a, b]$, then the **arc length** of f between a and b is given by

$$s = \int_a^b \sqrt{1 + [f'(x)]^2}\, dx.$$

Similarly, for a smooth curve given by $x = g(y)$, the **arc length** of g between c and d is given by

$$s = \int_c^d \sqrt{1 + [g'(y)]^2}\, dy.$$

Definite integrals representing arc length often are very difficult to evaluate. In this section we present a few examples. In the next chapter, with more advanced integration techniques, we will be able to tackle more difficult arc length problems.

Since the definition for arc length can be applied to a linear function, we should check to see that this new definition agrees with the standard distance formula for the length of a line segment. We do this in Example 1.

EXAMPLE 1 The length of a line segment

Find the arc length from (x_1, y_1) to (x_2, y_2) on the graph of $f(x) = mx + b$, as shown in Figure 7.37.

SOLUTION

Since

$$m = f'(x) = \frac{y_2 - y_1}{x_2 - x_1}$$

it follows that

$$s = \int_{x_1}^{x_2} \sqrt{1 + [f'(x)]^2}\, dx$$

$$= \int_{x_1}^{x_2} \sqrt{1 + \left(\frac{y_2 - y_1}{x_2 - x_1}\right)^2}\, dx$$

$$= \left[\sqrt{\frac{(x_2 - x_1)^2 + (y_2 - y_1)^2}{(x_2 - x_1)^2}}\,(x) \right]_{x_1}^{x_2}$$

$$= \sqrt{\frac{(x_2 - x_1)^2 + (y_2 - y_1)^2}{(x_2 - x_1)^2}}\,(x_2 - x_1)$$

$$= \sqrt{(x_2 - x_1)^2 + (y_2 - y_1)^2}$$

which is the formula for the distance between two points in the plane.

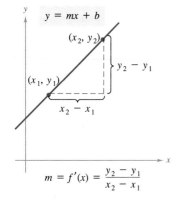

$$m = f'(x) = \frac{y_2 - y_1}{x_2 - x_1}$$

FIGURE 7.37

EXAMPLE 2 Finding arc length

Find the arc length of the graph of

$$f(x) = \frac{x^3}{6} + \frac{1}{2x}$$

on the interval $\left[\frac{1}{2}, 2\right]$, as shown in Figure 7.38.

SOLUTION

Since

$$f'(x) = \frac{3x^2}{6} - \frac{1}{2x^2} = \frac{1}{2}\left(x^2 - \frac{1}{x^2}\right)$$

the arc length is

$$s = \int_a^b \sqrt{1 + \left(\frac{dy}{dx}\right)^2}\, dx$$

$$= \int_{1/2}^2 \sqrt{1 + \left[\frac{1}{2}\left(x^2 - \frac{1}{x^2}\right)\right]^2}\, dx$$

$$= \int_{1/2}^2 \sqrt{\frac{1}{4}\left(x^4 + 2 + \frac{1}{x^4}\right)}\, dx$$

$$= \int_{1/2}^2 \frac{1}{2}\left(x^2 + \frac{1}{x^2}\right) dx$$

$$= \left[\frac{1}{2}\left(\frac{x^3}{3} - \frac{1}{x}\right)\right]_{1/2}^2$$

$$= \frac{1}{2}\left(\frac{13}{6} + \frac{47}{24}\right) = \frac{99}{48} = \frac{33}{16}.$$

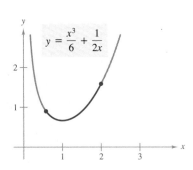

FIGURE 7.38

EXAMPLE 3 Finding arc length

Find the arc length of the graph of $(y - 1)^3 = x^2$ on the interval $[0, 8]$, as shown in Figure 7.39.

SOLUTION

We can solve for either x or y in the equation $(y - 1)^3 = x^2$. For this example we choose to solve for x, and we obtain

$$x = \pm(y - 1)^{3/2}.$$

By choosing the positive value for x, we have

$$\frac{dx}{dy} = \frac{3}{2}(y - 1)^{1/2}.$$

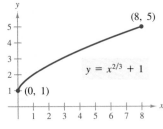

FIGURE 7.39

Therefore, the arc length is

$$s = \int_c^d \sqrt{1 + \left(\frac{dx}{dy}\right)^2}\, dy = \int_1^5 \sqrt{1 + \left[\frac{3}{2}(y-1)^{1/2}\right]^2}\, dy$$

$$= \int_1^5 \sqrt{\frac{9}{4}y - \frac{5}{4}}\, dy$$

$$= \frac{1}{2}\int_1^5 \sqrt{9y - 5}\, dy$$

$$= \frac{1}{18}\left[\frac{(9y-5)^{3/2}}{3/2}\right]_1^5 = \frac{1}{27}(40^{3/2} - 4^{3/2})$$

$$= \frac{8}{27}(10^{3/2} - 1) \approx 9.0734.$$ ▭

EXAMPLE 4   Finding arc length

Find the arc length of the graph of $y = \ln(\cos x)$ from $x = 0$ to $x = \pi/4$, as shown in Figure 7.40.

SOLUTION

Since

$$\frac{dy}{dx} = -\frac{\sin x}{\cos x} = -\tan x$$

the arc length is given by

$$s = \int_a^b \sqrt{1 + \left(\frac{dy}{dx}\right)^2}\, dx = \int_0^{\pi/4} \sqrt{1 + \tan^2 x}\, dx$$

$$= \int_0^{\pi/4} \sqrt{\sec^2 x}\, dx$$

$$= \int_0^{\pi/4} \sec x\, dx$$

$$= \left[\ln|\sec x + \tan x|\right]_0^{\pi/4}$$

$$= \ln(\sqrt{2} + 1) - \ln 1 \approx 0.8814.$$ ▭

$y = \ln(\cos x)$

FIGURE 7.40

EXAMPLE 5   An application of arc length

An electric cable is hung between two towers that are 200 feet apart, as shown in Figure 7.41. The cable takes the shape of a **catenary** whose equation is

$$y = 75(e^{x/150} + e^{-x/150}) = 150 \cosh \frac{x}{150}.$$

Find the arc length of the cable between the two towers.

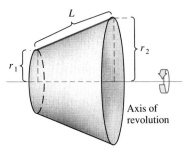

Catenary:
$$y = 150 \cosh \frac{x}{150}$$

150

−100   100

FIGURE 7.41

SOLUTION

Since

$$y' = \frac{1}{2}(e^{x/150} - e^{-x/150})$$

we have

$$(y')^2 = \frac{1}{4}(e^{x/75} - 2 + e^{-x/75})$$

and

$$1 + (y')^2 = \frac{1}{4}(e^{x/75} + 2 + e^{-x/75}) = \left[\frac{1}{2}(e^{x/150} + e^{-x/150})\right]^2.$$

Therefore, the arc length of the cable is given by

$$s = \int_a^b \sqrt{1 + (y')^2}\, dx = \frac{1}{2}\int_{-100}^{100}(e^{x/150} + e^{-x/150})\, dx$$

$$= \left[75(e^{x/150} - e^{-x/150})\right]_{-100}^{100}$$

$$= 150(e^{2/3} - e^{-2/3}) \approx 215 \text{ ft.}$$

Surfaces of revolution

In Sections 7.2 and 7.3 integration was used to calculate the volume of a solid of revolution. We now look at a procedure for finding the area of a surface of revolution.

DEFINITION OF SURFACE OF REVOLUTION

If the graph of a continuous function is revolved about a line, the resulting surface is called a **surface of revolution.**

L

r_2

r_1

Axis of revolution

FIGURE 7.42

To find the area of a surface of revolution, we use the formula for the lateral surface area of the frustum of a right circular cone. Consider the line segment in Figure 7.42, where

L = length of line segment

r_1 = radius at left end of line segment

r_2 = radius at right end of line segment.

When the line segment is revolved about its axis of revolution, it forms a frustum of a right circular cone, with

$$S = 2\pi r L \qquad \text{Lateral surface area of frustum}$$

$$r = \frac{1}{2}(r_1 + r_2). \qquad \text{Average radius of frustum}$$

(In Exercise 38 you are asked to verify this formula for S.)

Now, suppose the graph of a function f, having a continuous derivative on the interval $[a, b]$, is revolved about the x-axis to form a surface of revolution, as shown in Figure 7.43. Let Δ be a partition of $[a, b]$, with subintervals of width Δx_i. Then the line segment of length

$$\Delta L_i = \sqrt{\Delta x_i^2 + \Delta y_i^2}$$

generates a frustum of a cone.

By the Intermediate Value Theorem, a point d_i exists such that $r_i = f(d_i)$ is the average radius of this frustum. Finally, the lateral surface area, ΔS_i, of the frustum is given by

$$\Delta S_i = 2\pi r_i \Delta L_i = 2\pi f(d_i)\sqrt{\Delta x_i^2 + \Delta y_i^2} = 2\pi f(d_i)\sqrt{1 + \left(\frac{\Delta y_i}{\Delta x_i}\right)^2}\,\Delta x_i.$$

FIGURE 7.43

By the Mean Value Theorem, a point c_i exists in (x_{i-1}, x_i) such that

$$f'(c_i) = \frac{f(x_i) - f(x_{i-1})}{x_i - x_{i-1}} = \frac{\Delta y_i}{\Delta x_i}.$$

Therefore, $\Delta S_i = 2\pi f(d_i)\sqrt{1 + [f'(c_i)]^2}\,\Delta x_i$, and the total surface area can be approximated by

$$S \approx 2\pi \sum_{i=1}^{n} f(d_i)\sqrt{1 + [f'(c_i)]^2}\,\Delta x_i.$$

It can be shown that the limit of the right side as $\|\Delta\| \to 0$ (or $n \to \infty$) is

$$S = 2\pi \int_a^b f(x)\sqrt{1 + [f'(x)]^2}\,dx.$$

In a similar manner, it follows that if the graph of f is revolved about the y-axis, then S is given by

$$S = 2\pi \int_a^b x\sqrt{1 + [f'(x)]^2}\,dx.$$

In both formulas for S, we can regard the products $2\pi f(x)$ and $2\pi x$ as the circumference of the circle traced by a point (x, y) on the graph of f as it is revolved about the x- or y-axis (Figure 7.44). In one case the radius is $r = f(x)$, and in the other case the radius is $r = x$. Moreover, by appropriately adjusting r, we can generalize this formula for surface area to cover *any* horizontal or vertical axis of revolution, as indicated in the following definition.

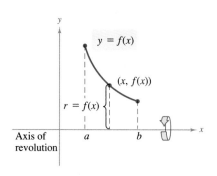

FIGURE 7.44

DEFINITION OF THE AREA OF A SURFACE OF REVOLUTION	If $y = f(x)$ has a continuous derivative on the interval $[a, b]$, then the area S of the surface of revolution formed by revolving the graph of f about a horizontal or vertical axis is $$S = 2\pi \int_a^b r(x)\sqrt{1 + [f'(x)]^2}\, dx$$ where $r(x)$ is the distance between the graph of f and the axis of revolution.

REMARK   If $x = g(y)$ on the interval $[c, d]$, then the surface area is

$$S = 2\pi \int_c^d r(y)\sqrt{1 + [g'(y)]^2}\, dy$$

where $r(y)$ is the distance between the graph of g and the axis of revolution.

EXAMPLE 6   The area of a surface of revolution

Find the area of the surface formed by revolving the graph of $f(x) = x^3$ on the interval $[0, 1]$ about the x-axis, as shown in Figure 7.45.

SOLUTION

The distance between the x-axis and the graph of f is $r(x) = f(x)$, and since $f'(x) = 3x^2$, the surface area is given by

$$\begin{aligned}
S = 2\pi \int_a^b r(x)\sqrt{1 + [f'(x)]^2}\, dx &= 2\pi \int_0^1 x^3\sqrt{1 + (3x^2)^2}\, dx \\
&= \frac{2\pi}{36} \int_0^1 (36x^3)(1 + 9x^4)^{1/2}\, dx \\
&= \frac{\pi}{18}\left[\frac{(1 + 9x^4)^{3/2}}{3/2}\right]_0^1 \\
&= \frac{\pi}{27}(10^{3/2} - 1) \approx 3.563.
\end{aligned}$$

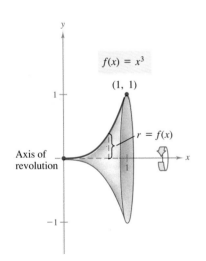

y

$f(x) = x^3$

(1, 1)

$r = f(x)$

Axis of revolution

x

FIGURE 7.45

EXAMPLE 7   The area of a surface of revolution

Find the area of the surface formed by revolving the graph of $f(x) = x^2$ on the interval $[0, \sqrt{2}]$ about the y-axis, as shown in Figure 7.46.

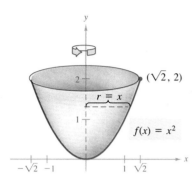

FIGURE 7.46

SOLUTION

In this case, the distance between the graph of f and the y-axis is $r(x) = x$, and since $f'(x) = 2x$, the surface area is

$$S = 2\pi \int_a^b r(x)\sqrt{1 + [f'(x)]^2}\, dx = 2\pi \int_0^{\sqrt{2}} x\sqrt{1 + (2x)^2}\, dx$$

$$= \frac{2\pi}{8} \int_0^{\sqrt{2}} (1 + 4x^2)^{1/2}(8x)\, dx$$

$$= \frac{2\pi}{8} \left[\frac{(1 + 4x^2)^{3/2}}{3/2} \right]_0^{\sqrt{2}}$$

$$= \frac{\pi}{6}[(1 + 8)^{3/2} - (1)^{3/2}] = \frac{13\pi}{3}.$$

EXERCISES for Section 7.4

In Exercises 1 and 2, find the distance between the given points by (a) using the distance formula and (b) determining the equation of the line through the points and using the formula for arc length.

1. (0, 0), (5, 12)   **2.** (1, 2), (7, 10)

In Exercises 3–8, find the arc length of the graph of the given function over the indicated interval.

Function	Interval
3. $y = \frac{2}{3}x^{3/2} + 1$	[0, 1]
4. $y = x^{3/2} - 1$	[0, 4]
5. $y = \frac{x^4}{8} + \frac{1}{4x^2}$	[1, 2]
6. $y = \frac{3}{2}x^{2/3}$	[1, 8]
7. $y = \frac{x^5}{10} + \frac{1}{6x^3}$	[1, 2]
8. $y = \frac{1}{2}(e^x + e^{-x})$	[0, 2]

In Exercises 9–14, find a definite integral that represents the arc length of the curve over the indicated interval. (Do not evaluate the integral.)

Function	Interval
9. $y = x^2 + x - 2$	[-2, 1]
10. $y = \frac{1}{x + 1}$	[0, 1]
11. $y = 4 - x^2$	[0, 2]
12. $y = \cos x$	$\left[-\frac{\pi}{2}, \frac{\pi}{2}\right]$

Function	Interval
13. $x = e^{-y}$	[0, 2]
14. $x = \sqrt{a^2 - y^2}$	$\left[0, \frac{a}{2}\right]$

In Exercises 15–18, use Simpson's Rule (with $n = 4$) to approximate the arc length of the function over the indicated interval.

Function	Interval
15. $y = \frac{1}{x}$	[1, 3]
16. $y = x^2$	[0, 1]
17. $y = \sin x$	[0, π]
18. $y = \ln x$	[1, 5]

19. A fleeing object leaves the origin and moves up the y-axis (see figure). At the same time, a pursuer leaves the point (1, 0) and moves always toward the fleeing object. If the pursuer's speed is twice that of the fleeing object, the equation of the path is

$$y = \frac{1}{3}(x^{3/2} - 3x^{1/2} + 2).$$

How far has the fleeing object traveled when it is caught? Show that the pursuer has traveled twice as far.

20. A barn is 100 feet long and 40 feet wide (see figure). A cross section of the roof is the inverted catenary

$$y = 31 - 10(e^{x/20} + e^{-x/20}).$$

Find the number of square feet of roofing on the barn. [Hint: Find the arc length of the catenary and multiply by the length of the barn.]

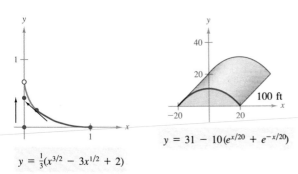

$y = \frac{1}{3}(x^{3/2} - 3x^{1/2} + 2)$

FIGURE FOR 19

$y = 31 - 10(e^{x/20} + e^{-x/20})$

FIGURE FOR 20

21. Find the arc length from $(0, 3)$ clockwise to $(2, \sqrt{5})$ along the circle $x^2 + y^2 = 9$.

22. Find the arc length from $(-3, 4)$ clockwise to $(4, 3)$ along the circle $x^2 + y^2 = 25$. Show that the result is one-fourth of the circumference of the circle.

In Exercises 23–26, find the area of the surface of revolution generated by revolving the given plane curve about the x-axis.

Function	Interval
23. $y = \dfrac{x^3}{3}$	$[0, 3]$
24. $y = \sqrt{x}$	$[1, 4]$
25. $y = \dfrac{x^3}{6} + \dfrac{1}{2x}$	$[1, 2]$
26. $y = \dfrac{x}{2}$	$[0, 6]$

In Exercises 27 and 28, find the area of the surface of revolution generated by revolving the given plane curve over the indicated interval about the y-axis.

Function	Interval
27. $y = \sqrt[3]{x} + 2$	$[1, 8]$
28. $y = 4 - x^2$	$[0, 2]$

In Exercises 29 and 30, find the definite integral that represents the surface area of the solid of revolution. (Do not evaluate the integral.)

Function	Interval
29. $y = \sin x$ revolved about the x-axis	$[0, \pi]$
30. $y = \ln x$ revolved about the y-axis	$[1, e]$

31. A right circular cone is generated by revolving the region bounded by $y = hx/r$, $y = h$, and $x = 0$ about the y-axis. Verify that the lateral surface area of the cone is

$$S = \pi r \sqrt{r^2 + h^2}.$$

32. A sphere of radius r is generated by revolving the graph of $y = \sqrt{r^2 - x^2}$ about the x-axis. Verify that the surface area of the sphere is $4\pi r^2$.

33. Find the area of the zone of a sphere formed by revolving the graph of $y = \sqrt{9 - x^2}$, $0 \le x \le 2$, about the y-axis.

34. Find the area of the zone of a sphere formed by revolving the graph of $y = \sqrt{r^2 - x^2}$, $0 \le x \le a$, about the y-axis. Assume that $a < r$.

35. An ornamental light bulb is designed by revolving the graph of

$$y = \frac{1}{3}x^{1/2} - x^{3/2}, \quad 0 \le x \le \frac{1}{3}$$

about the x-axis, where x and y are measured in feet (see figure). Find the surface area of the bulb and use the result to approximate the amount of glass needed to make the bulb. (Assume the glass is 0.015 inch thick.)

$y = \frac{1}{3} x^{1/2} - x^{3/2}$

FIGURE FOR 35

36. Given a circular sector with radius L and central angle θ (see figure) show that its area is given by

$$S = \frac{1}{2}L^2\theta.$$

37. By joining the straight line edges of the sector of Exercise 36, a right circular cone is formed (see figure) and the lateral surface area of the cone is the same as the area of the sector. Show that the area is given by

$$S = \pi r L$$

where r is the radius of the base of the cone. [Hint: The arc length of the sector equals the circumference of the base of the cone.]

FIGURE FOR 36

FIGURE FOR 37

38. Use the result of Exercise 37 to verify that the formula for the lateral surface area of the frustum of a cone with slant height L and radii r_1 and r_2 (see figure) is given by

$$S = \pi(r_1 + r_2)L.$$

[Note: This formula was used to develop the integral for finding the surface area of a surface of revolution.]

Axis of
revolution

FIGURE FOR 38

In Exercises 39 and 40, use Simpson's Rule on a computer or calculator (with $n = 10$) to approximate the surface area of a vase generated by revolving the region bounded by the graphs of the given equations about the y-axis.

39. $x = \dfrac{y}{27}(2y^2 - 19y + 51),\ x = 0,\ y = 0.5,\ y = 6$

40. $x = \dfrac{18}{y^2 + 9},\ x = 0,\ y = 0,\ y = 6$

7.5 Work

Work done by a constant force ▪ Work done by a variable force

The concept of work is important to scientists and engineers for determining the energy needed to perform various jobs. For instance, it is useful to know the amount of work done when a crane lifts a steel girder, when a spring is compressed, when a rocket is propelled into the air, or when a truck pulls a load along a highway.

In general, we say that **work** is done by a force when it moves an object. If the force applied to the object is *constant*, we have the following definition of work.

DEFINITION OF WORK DONE BY A CONSTANT FORCE	If an object is moved a distance D in the direction of an applied constant force F, then the **work** W done by the force is defined as $W = FD$.

There are many types of forces—centrifugal, electromotive, and gravitational, to name a few. A **force** can be thought of as a *push* or a *pull*; a force changes the state of rest or state of motion of a body. For gravitational forces on the earth, we commonly use units of measure corresponding to the weight of an object, as illustrated in the following example.

EXAMPLE 1   Work done by a constant force

Determine the work done in lifting a 150-pound object 4 feet.

SOLUTION

The magnitude of the required force F is the weight of the object, as shown in Figure 7.47. Thus, the work done in lifting the object 4 feet is

$$\text{work} = W = FD = 150(4) = 600 \text{ ft} \cdot \text{lb}.$$

REMARK In the U.S. system of measurement, work is typically expressed in foot-pounds (ft · lb), inch-pounds, or foot-tons. In the centimeter-gram-second (C-G-S) system, the basic unit of force is the **dyne**—the force required to produce an acceleration of one centimeter per second per second in a gram mass. In this system work is typically expressed in dyne-centimeters (erg) or newton-meters (joule), where one joule $= 10^7$ ergs.

FIGURE 7.47

Work done by a variable force

In Example 1, the force involved is *constant*. If a *variable* force is applied to an object, then calculus is needed to determine the work done, because the amount of force changes as the object changes positions. For instance, the force required to compress a spring increases as the spring is compressed.

Suppose that an object is moved along a straight line from $x = a$ to $x = b$ by a continuously varying force $F(x)$. Let Δ be a partition that divides the interval $[a, b]$ into n subintervals determined by

$$a = x_0 < x_1 < x_2 < \cdots < x_n = b$$

and let $\Delta x_i = x_i - x_{i-1}$. For each i, choose c_i such that $x_{i-1} \le c_i \le x_i$. Then at c_i the force is given by $F(c_i)$. Since F is continuous, we conclude (assuming that Δx_i is small) that the work done in moving the object through the ith subinterval is approximated by the increment $\Delta W_i = F(c_i) \Delta x_i$, as shown in Figure 7.48. By adding the work done in each subinterval, we can approximate the total work done as the object moves from a to b by

$$W \approx \sum_{i=1}^{n} \Delta W_i = \sum_{i=1}^{n} F(c_i) \Delta x_i.$$

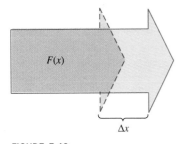

FIGURE 7.48

Taking the limit of this sum as $\|\Delta\| \to 0$, $(n \to \infty)$, we have

$$W = \lim_{n \to \infty} \sum_{i=1}^{n} F(c_i) \Delta x_i = \int_a^b F(x) \, dx.$$

We therefore define work done by a variable force as follows.

DEFINITION OF WORK DONE BY A VARIABLE FORCE

If an object is moved along a straight line by a continuously varying force $F(x)$, then the **work** W done by the force as the object is moved from $x = a$ to $x = b$ is given by

$$W = \lim_{n \to \infty} \sum_{i=1}^{n} \Delta W_i = \int_a^b F(x) \, dx.$$

Emilie de Breteuil

The remaining examples in this section use some well-known physical laws. The discovery of many of these laws occurred during the same period in which calculus was being developed. In fact, during the seventeenth and eighteenth centuries, there was little difference between physicists and mathematicians. One such physicist-mathematician was Emilie de Breteuil (1706–1749). Breteuil was instrumental in synthesizing the work of many other scientists including Newton, Leibniz, Huygens, Kepler, and Descartes. Her physics text *Institutions* was widely used for many years.

The following three laws of physics were developed by Robert Hooke (1635–1703), Isaac Newton (1642–1727), and Charles Coulomb (1736–1806). Although Hooke and Coulombe are known primarily for their work in physics, they were also gifted mathematicians.

1. **Hooke's Law:** The force F required to compress or stretch a spring (within its elastic limits) is proportional to the distance d that the spring is compressed or stretched from its original length. That is,

$$F = kd$$

where the constant of proportionality k depends on the specific nature of the spring.

2. **Law of Universal Gravitation:** The force F of attraction between two particles of mass m_1 and m_2 is proportional to the product of the masses and inversely proportional to the square of the distance d between the two particles. That is,

$$F = k\frac{m_1 m_2}{d^2}.$$

(If m_1 and m_2 are given in grams and d in centimeters, then F will be in dynes for a value of $k = 6.670 \times 10^{-8}$.)

3. **Coulomb's Law:** The force between two charges q_1 and q_2 in a vacuum is proportional to the product of the charges and inversely proportional to the square of the distance d between the two charges. That is,

$$F = k\frac{q_1 q_2}{d^2}.$$

(If q_1 and q_2 are given in electrostatic units and d in centimeters, then F will be in dynes for a value of $k = 1$.)

EXAMPLE 2 An application of Hooke's Law

A force of 750 pounds compresses a spring 3 inches from its natural length of 15 inches. Find the work done in compressing the spring an additional 3 inches.

SOLUTION

Using Hooke's Law, the force $F(x)$ required to compress the spring x units (from its natural length) is given by $F(x) = kx$. Using the given data, we have

$$F(3) = 750 = (k)(3)$$

Natural length $(F = 0)$

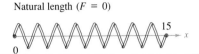

and thus $k = 250$ and $F(x) = 250x$, as shown in Figure 7.49. To find the increment of work, we assume that the force required to compress the spring over a small increment Δx is nearly constant. Thus, the increment of work is

Compressed 3 inches $(F = 750)$

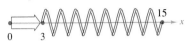

$$\Delta W = (\text{force})(\text{distance increment}) = (250x)\Delta x.$$

Now, since the spring is compressed from $x = 3$ to $x = 6$ inches less than its natural length, the work required is

Compressed x inches $(F = 250x)$

FIGURE 7.49

$$W = \int_3^6 250x \, dx = 125x^2 \Big]_3^6 = 4500 - 1125 = 3375 \text{ in} \cdot \text{lb}.$$

Note that we do not integrate from $x = 0$ to $x = 6$ because we were asked to determine the work done in compressing the spring an *additional* 3 inches (not including the first 3 inches). ◻

EXAMPLE 3 An application involving gravitational force

If a space module weighs 15 tons on the surface of the earth, how much work is done in propelling the module to a height of 800 miles above the earth, as shown in Figure 7.50? (Do not consider the effect of air resistance or the weight of the propellant.)

SOLUTION

Since the weight of a body varies inversely as the square of its distance from the center of the earth, the force $F(x)$ exerted by gravity is

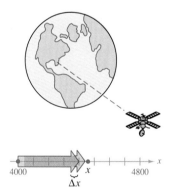

Space module is moved 800 miles above the earth.

FIGURE 7.50

$$F(x) = \frac{C}{x^2}.$$

Since the module weighs 15 tons on the surface of the earth and the radius of the earth is approximately 4000 miles, we obtain

$$15 = \frac{C}{(4000)^2} \quad \Longrightarrow \quad C = 240{,}000{,}000.$$

Thus, the increment of work is

$$\Delta W = (\text{force})(\text{distance increment})$$

$$= \frac{240{,}000{,}000}{x^2} \Delta x.$$

Finally, since the module is propelled from $x = 4000$ to $x = 4800$ miles, the total work done is

$$W = \int_{4000}^{4800} \frac{240{,}000{,}000}{x^2}\,dx = \frac{-240{,}000{,}000}{x}\Bigg]_{4000}^{4800}$$

$$= -50{,}000 + 60{,}000 = 10{,}000 \text{ mile-tons}$$

$$\approx 1.056 \times 10^{11} \text{ ft} \cdot \text{lb}.$$

The solutions to Examples 2 and 3 conform to our development of work as the summation of increments in the form

$$\Delta W = (\text{force})(\text{distance increment}) = F(\Delta x).$$

An equally useful way to formulate the increment of work is

$$\Delta W = (\text{force increment})(\text{distance}) = (\Delta F)x.$$

This second interpretation of ΔW is useful in problems involving the movement of nonrigid substances such as fluids or chains, as seen in the next two examples.

EXAMPLE 4   Work required to move a liquid

A spherical tank of radius 8 feet is half full of oil that weighs 50 pounds per cubic foot. Find the work required to pump the oil out through a hole in the top of the tank.

SOLUTION

We consider the oil to be subdivided into discs of thickness Δy and radius x, as shown in Figure 7.51. Since the increment of force for each disc is given by its weight, we have

$$\Delta F = \text{weight} = \left(\frac{50\ \text{lb}}{\text{ft}^3}\right)(\text{volume}) = 50(\pi x^2 \Delta y).$$

Now, for a circle of radius 8 and center at $(0, 8)$, we have

$$x^2 + (y - 8)^2 = 8^2$$

$$x^2 = 16y - y^2$$

and we can write the force increment as

$$\Delta F = 50(\pi x^2 \Delta y) = 50\pi(16y - y^2)\Delta y.$$

In Figure 7.51, we see that a disc y feet from the bottom of the tank must be moved a distance of $(16 - y)$ feet. Therefore, the increment of work is

$$\Delta W = \Delta F(16 - y) = 50\pi(16y - y^2)\Delta y(16 - y)$$

$$= 50\pi(256y - 32y^2 + y^3)\Delta y.$$

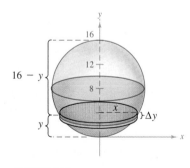

FIGURE 7.51

Finally, since the tank is half full, y ranges from 0 to 8, and the work required to empty the tank is

$$W = \int_0^8 50\pi(256y - 32y^2 + y^3) \, dy = 50\pi\left[128y^2 - \frac{32}{3}y^3 + \frac{y^4}{4}\right]_0^8$$

$$= 50\pi\left(\frac{11264}{3}\right) \approx 589,782 \text{ ft} \cdot \text{lb.}$$

EXAMPLE 5   Work done in lifting a chain

FIGURE 7.52

A 20-foot chain, weighing 5 pounds per foot, is lying coiled on the ground. How much work is required to raise one end of the chain to a height of 20 feet so that it is fully extended, as shown in Figure 7.52?

SOLUTION

Imagine that the chain is divided into small sections, each of length Δy. Then the weight of each section is the increment of force

$$\Delta F = (\text{weight}) = \left(\frac{5 \text{ lb}}{\text{ft}}\right)(\text{length}) = 5\Delta y.$$

Since a typical section (initially on the ground) is raised to a height of y, we conclude that the increment of work is

$$\Delta W = (\text{force increment})(\text{distance}) = (5\Delta y)y = 5y\Delta y.$$

Finally, since y ranges from 0 to 20, the total work is

$$W = \int_0^{20} 5y \, dy = \frac{5y^2}{2}\Bigg]_0^{20} = \frac{5(400)}{2} = 1000 \text{ ft} \cdot \text{lb.}$$

x Work done by expanding gas

FIGURE 7.53

In the next example we consider a piston of radius r in a cylindrical casing, as shown in Figure 7.53. As the gas in the cylinder expands, the piston moves and work is done. If p represents the pressure of the gas (in pounds per square foot) against the piston head and V represents the volume of the gas (in cubic feet), then the work increment involved in moving the piston Δx feet is

$$\Delta W = (\text{force})(\text{distance increment}) = F(\Delta x) = p(\pi r^2)\Delta x = p\Delta V.$$

Thus, as the volume of the gas expands from V_0 to V_1, the work done in moving the piston is

$$W = \int_{V_0}^{V_1} p \, dV.$$

Assuming the pressure of the gas to be inversely proportional to its volume, we have $p = k/V$ and the integral for work becomes

$$W = \int_{V_0}^{V_1} \frac{k}{V} \, dV.$$

EXAMPLE 6 Work done by a gas

A quantity of gas with an initial volume of 1 cubic foot and pressure of 500 pounds per square foot expands to a volume of 2 cubic feet. Find the work done by the gas. (Assume the pressure is inversely proportional to the volume.)

SOLUTION

Since $p = k/V$ and $p = 500$ when $V = 1$, we have $k = 500$. Thus, the work is

$$W = \int_{V_0}^{V_1} \frac{k}{V}\, dV = \int_1^2 \frac{500}{V}\, dV = 500 \ln |V| \Big]_1^2$$
$$= 500 \ln 2 \approx 346.6 \text{ ft} \cdot \text{lb}.$$

EXERCISES for Section 7.5

1. Determine the work done in lifting a 100-pound bag of sugar 10 feet.

2. Determine the work done by a hoist in lifting a 2400-pound car 6 feet.

3. A force of 25 pounds is required to slide a cement block on a plank in a construction project. The plank is 12 feet long. Determine the work done in sliding the block along the length of the plank.

4. The locomotive of a freight train pulls its cars with a constant force of 9 tons while traveling at a constant rate of 55 miles per hour on a level track. How many foot-pounds of work does the locomotive do in a distance of one-half mile?

5. A force of 5 pounds compresses a 15-inch spring a total of 4 inches. How much work is done in compressing the spring 7 inches?

6. How much work is done in compressing the spring in Exercise 5 from a length of 10 inches to a length of 6 inches?

7. A force of 60 pounds stretches a spring 1 foot. How much work is done in stretching the spring from 9 inches to 15 inches?

8. A force of 200 pounds stretches a spring 2 feet on a mechanical device for driving fence posts. Find the work done in stretching the spring the required 2 feet.

9. A force of 15 pounds stretches a spring 6 inches in an exercise machine. Find the work done in stretching the spring 1 foot from its natural position.

10. An overhead garage door has two springs, one on each side of the door. A force of 15 pounds is required to stretch each spring 1 foot. Because of the pulley system, the springs only stretch one-half the distance the door travels. Find the work done by the pair of springs if the door moves a total of 8 feet and the springs are at their natural length when the door is open.

11. A rectangular tank with a base 4 feet by 5 feet and a height of 4 feet is filled with water (see figure). (The water weighs 62.4 pounds per cubic foot.) How much work is done in pumping water out over the top edge in order to empty
    (a) half of the tank?
    (b) all of the tank?

5 ft   4 ft   FIGURE FOR 11

12. Repeat Exercise 11 for a tank filled with gasoline that weighs 42 pounds per cubic foot.

13. A cylindrical water tank 12 feet high with a radius of 8 feet is buried so that the top of the tank is 3 feet below ground level (see figure). How much work is done in pumping a full tank of water up to ground level?

14. Suppose the tank in Exercise 13 is located on a tower so that the bottom of the tank is 20 feet above the level of a stream (see figure). How much work is done in filling the tank half full of water through a hole in the bottom, using water from the stream?

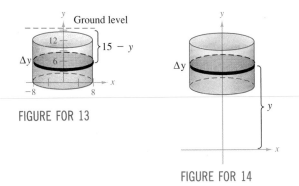

FIGURE FOR 13

FIGURE FOR 14

15. A hemispherical tank of radius 6 feet is positioned so that its base is circular. How much work is required to fill the tank with water through a hole in the base if the water source is at the base?

16. Suppose the tank in Exercise 15 is inverted and the top 2 feet of water is pumped out through a hole in the top. How much work is done?

17. An open tank has the shape of a right circular cone (see figure). The tank is 8 feet across the top and is 6 feet high. How much work is done in emptying the tank by pumping the water over the top edge?

FIGURE FOR 17

18. If water is pumped in through the bottom of the tank in Exercise 17, how much work is done to fill the tank
(a) to a depth of 2 feet?
(b) from a depth of 4 feet to a depth of 6 feet?

19. A cylindrical gasoline tank 3 feet in diameter and 4 feet long is carried on the back of a truck and is used to fuel tractors in the field. The axis of the tank is horizontal. Find the work done to pump the entire contents of the full tank into a tractor if the opening on the tractor tank is 5 feet above the top of the tank in the truck. Assume gasoline weighs 42 pounds per cubic foot. (Hint: Evaluate one integral by a geometric formula and the other by observing that the integrand is an odd function.)

20. The top of a cylindrical storage tank for gasoline at a service station is 4 feet below ground level. The axis of the tank is horizontal and its diameter and length are 5 feet and 12 feet, respectively. Find the work done in pumping the entire contents of the full tank to a height of 3 feet above ground level. (Hint: Evaluate one integral by a geometric formula and by observing that the integrand is an odd function.)

21. Neglecting air resistance and the weight of the propellant, determine the work done in propelling a 4-ton satellite to a height of (a) 200 miles and (b) 400 miles above the earth.

22. Use the information from Exercise 21 to write the work W of the propulsion system as a function of the height h of the satellite above the earth. Find the limit (if it exists) of W as h approaches infinity.

23. Neglecting air resistance, determine the work done in propelling a 10-ton satellite to a height of (a) 11,000 miles and (b) 22,000 miles above the earth.

24. If a lunar module weighs 12 tons on the surface of the earth, how much work is done in propelling the module from the surface of the moon to a height of 50 miles? Consider the radius of the moon to be 1100 miles and its force of gravity to be one-sixth that of the earth's.

25. Two electrons repel each other with a force that varies inversely as the square of the distance between them. If one electron is fixed at the point $(2, 4)$, find the work done in moving a second electron from $(-2, 4)$ to $(1, 4)$.

26. The force generated by a press in a manufacturing process is given by

$$F(x) = \frac{e^{x^2} - 1}{100}, \quad 0 \le x \le 4$$

where x is the distance in feet the press moves. Use Simpson's Rule (with $n = 8$) to approximate the work done through one cycle of the press.

In Exercises 27–30, consider a 15-foot chain hanging from a winch 15 feet above ground level. Find the work done by the winch in winding up the required amount of chain, if the chain weighs 3 pounds per foot.

27. Wind up the entire chain.

28. Wind up one-third of the chain.

29. Run the winch until the bottom of the chain is at the 10-foot level.

30. Wind up the entire chain with a 100-pound load attached.

In Exercises 31 and 32, consider a 15-foot hanging chain that weighs 3 pounds per foot. Find the work done in lifting the chain vertically to the required position.

31. Take the bottom of the chain and raise it to the 15-foot level, leaving the chain doubled and still hanging vertically (see figure).

FIGURE FOR 31

32. Repeat Exercise 31 raising the bottom of the chain to the 12-foot level.

In Exercises 33 and 34, consider a demolition crane with a 500-pound ball suspended from a 40-foot cable that weighs 1 pound per foot.

33. Find the work required to wind up 15 feet of the apparatus.
34. Find the work required to wind up all 40 feet of the apparatus.

In Exercises 35 and 36, find the work done by the gas for the given volumes and pressures. Assume the pressure is inversely proportional to the volume. (Use Example 6 as a model.)

35. A quantity of gas with an initial volume of 2 cubic feet and pressure of 1000 pounds per square foot expands to a volume of 3 cubic feet.
36. A quantity of gas with an initial volume of 1 cubic foot and pressure of 2000 pounds per square foot expands to a volume of 4 cubic feet.

37. In a manufacturing process an object is moved linearly 5 feet with a variable force given by

$$F(x) = 1000[1.8 - \ln (x + 1)], \quad 0 \le x \le 5$$

where F is given in pounds and x gives the position of the unit in feet. Use Simpson's Rule on a computer or calculator (with $n = 10$) to approximate the work done to move one unit through the given process.
38. Repeat Exercise 37 for $F(x) = 100x\sqrt{125 - x^3}$.

7.6 Fluid Pressure and Fluid Force

Fluid pressure ▪ Force exerted by a fluid

Swimmers know that the deeper an object is submerged in a fluid, the greater the pressure on the object. We define **pressure** to be the force per unit of area over the surface of a body. For example, since a ten-foot column of water (one-inch square) weighs 4.3 pounds, we say that the *fluid* pressure at a depth of 10 feet of water is 4.3 pounds per square inch.* At 20 feet, this would increase to 8.6 pounds per square inch, and in general the pressure is proportional to the depth of the object in the fluid, as indicated in the following definition.

DEFINITION OF FLUID PRESSURE

The **pressure** on an object at depth h in a liquid is given by

$$\text{pressure} = P = wh$$

where w is the weight of the liquid per unit of volume.

Listed below are some commonly used weights of fluids in pounds per cubic foot.

Water	62.4
Ethyl alcohol	49.4
Gasoline	41.0–43.0
Kerosene	51.2
Mercury	849.0
Glycerin	78.6
Sea water	64.0

*The total pressure on an object in ten feet of water would also include the pressure due to the earth's atmosphere. At sea level, atmospheric pressure is approximately 14.7 pounds per square inch.

Blaise Pascal

When calculating fluid pressure, we use an important (and rather surprising) physical law called **Pascal's Principle,** named after the French mathematician Blaise Pascal (1623–1662). Pascal is well known for his work in many areas of mathematics and physics and also for his influence on Leibniz. Although much of Pascal's work in calculus was intuitive and lacked the rigor of modern mathematics, he nevertheless anticipated many important results. Pascal's Principle states that the pressure exerted by a fluid at a depth h is transmitted equally *in all directions*. For example, in Figure 7.54, the pressure at the indicated depth is the same for all three objects.

Since fluid pressure is given in terms of force per unit area $(P = F/A)$, the fluid force on a *submerged horizontal* surface of area A is given by

fluid force $= F = PA =$ (pressure)(area).

EXAMPLE 1   Fluid force on a horizontal surface

Find the fluid force on a rectangular metal sheet 3 feet by 4 feet that is submerged in 6 feet of water, as shown in Figure 7.55.

SOLUTION

Since the weight of water is 62.4 pounds per cubic foot and the sheet is submerged in 6 feet of water, the fluid pressure is

$$P = (62.4)(6) = 374.4 \text{ lb/ft}^2.$$

Now, since the total area of the sheet is $A = (3)(4) = 12$ square feet, the fluid force is given by

$$F = PA = (374.4)(12) = 4492.8 \text{ lb}.$$

REMARK   The answer to Example 1 is independent of the size of the body of water. The fluid force would be the same in a swimming pool or a lake.

In Example 1, the fact that the sheet is rectangular and horizontal means that we do not need the methods of calculus to solve the problem. We now look at a surface that is submerged vertically in a fluid. This problem is more difficult because the pressure is not constant over the surface.

Suppose a vertical plate is submerged in a fluid of weight w (per unit of volume), as shown in Figure 7.56. We wish to determine the total force against *one side* of this region from depth c to depth d. First, we subdivide the interval $[c, d]$ into n subintervals, each of width Δy. Now consider the representative rectangle of width Δy and length $L(y_i)$, where y_i is in the ith subinterval. The force against this representative rectangle is

$$\Delta F_i = w(\text{depth})(\text{area}) = wh(y_i)L(y_i)\Delta y.$$

The force against n such rectangles is

$$\sum_{i=1}^{n} \Delta F_i = w \sum_{i=1}^{n} h(y_i)L(y_i)\Delta y.$$

FIGURE 7.54

FIGURE 7.55

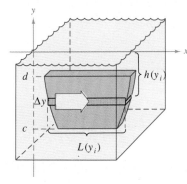

FIGURE 7.56

Note that w is considered to be constant and is factored out of the summation. Therefore, taking the limit as $\|\Delta\| \to 0$ ($n \to \infty$), we find the total force against the region to be

$$F = w \lim_{n \to \infty} \sum_{i=1}^{n} h(y_i)L(y_i)\Delta y = w \int_c^d h(y)L(y)\,dy.$$

DEFINITION OF FORCE EXERTED BY A FLUID

The **force F exerted by a fluid** of constant weight w (per unit of volume) against a submerged vertical plane region from $y = c$ to $y = d$ is given by

$$F = w \int_c^d h(y)L(y)\,dy$$

where $h(y)$ is the depth of the fluid at y and $L(y)$ is the horizontal length of the region at y.

(a)

Water gate in a dam

(b)

FIGURE 7.57

EXAMPLE 2 Fluid force on a vertical surface

A vertical gate in a dam has the shape of an isosceles trapezoid 8 feet across the top and 6 feet across the bottom, with a height of 5 feet, as shown in Figure 7.57(a). What is the fluid force against the gate if the top of the gate is 4 feet below the surface of the water?

SOLUTION

In setting up a mathematical model for this problem, we are at liberty to locate the x- and y-axes in a number of different ways. A convenient approach is to let the y-axis bisect the gate and place the x-axis at the surface of the water, as shown in Figure 7.57(b). Thus, the depth of the water at y is

$$\text{depth} = h(y) = -y.$$

To find the length $L(y)$ of the region at y, we find the equation of the line forming the right side of the gate. Since this line passes through the two points $(3, -9)$ and $(4, -4)$, its equation is

$$y - (-9) = \frac{-4 - (-9)}{4 - 3}(x - 3) = 5(x - 3)$$

$$y = 5x - 24 \quad \Longrightarrow \quad x = \frac{y + 24}{5}.$$

From Figure 7.57(b) we can see that the length of the region at y is given by

$$\text{length} = 2x = \frac{2}{5}(y + 24) = L(y).$$

Finally, by integrating from $y = -9$ to $y = -4$, we find the fluid force to be

$$F = w \int_c^d h(y)L(y) \, dy = 62.4 \int_{-9}^{-4} (-y)\left(\frac{2}{5}\right)(y + 24) \, dy$$

$$= -62.4\left(\frac{2}{5}\right) \int_{-9}^{-4} (y^2 + 24y) \, dy$$

$$= -62.4\left(\frac{2}{5}\right)\left[\frac{y^3}{3} + 12y^2\right]_{-9}^{-4}$$

$$= -62.4\left(\frac{2}{5}\right)\left(\frac{-1675}{3}\right)$$

$$= 13{,}936 \text{ lb.}$$

REMARK In Example 2, we let the x-axis coincide with the surface of the water. This was convenient, but arbitrary. In choosing a coordinate system to represent a physical situation, you should consider various possibilities. Often, you can simplify the calculations in a problem by locating the coordinate system so as to take advantage of special characteristics of the problem, such as symmetry.

EXAMPLE 3 Fluid force on a vertical surface

A circular observation window on a marine science ship has a radius of 1 foot, and the center of the window is 8 feet below water level, as shown in Figure 7.58(a). What is the fluid force on the window?

SOLUTION

To take advantage of symmetry, we locate a coordinate system so that the origin coincides with the center of the window, as shown in Figure 7.58(b). The depth at y is then given by

$$\text{depth} = h(y) = 8 - y.$$

The horizontal length of the window is $2x$, and we can use the equation for the circle, $x^2 + y^2 = 1$, to solve for x as follows.

$$\text{length} = 2x = 2\sqrt{1 - y^2} = L(y)$$

Finally, since y ranges from -1 to 1, we have, using the weight of sea water as 64 pounds per cubic foot,

$$F = w \int_c^d h(y)L(y) \, dy = 64 \int_{-1}^{1} (8 - y)(2)\sqrt{1 - y^2} \, dy.$$

Initially it looks as if this integral would be difficult to solve. However, if we break the integral into two parts and apply symmetry, the solution is simple:

$$F = 64(16) \int_{-1}^{1} \sqrt{1 - y^2} \, dy - 64(2) \int_{-1}^{1} y\sqrt{1 - y^2} \, dy.$$

(a)

(b)

FIGURE 7.58

The second integral is zero (since the integrand is odd and the limits of integration are symmetric to the origin). Moreover, by recognizing that the first integral represents the area of a semicircle of radius 1, we have

$$F = 64(16)\left(\frac{\pi}{2}\right) - 64(2)(0)$$

$$= 512\pi \approx 1608.5 \text{ lb.}$$

EXAMPLE 4 Fluid force on a vertical surface

(a)

(b)

FIGURE 7.59

A swimming pool is 2 feet deep at one end and 10 feet deep at the other, as shown in Figure 7.59(a). The pool is 40 feet long and 30 feet wide with vertical sides. Find the fluid force against one of the 40-foot sides.

SOLUTION

By placing the x- and y-axes as shown in Figure 7.59(b), we see that the depth at y is

$$\text{depth} = h(y) = 10 - y.$$

The equation representing the base of the side is given by

$$y = mx + b = \frac{1}{5}x \implies x = 5y.$$

However, we must note that the equation $x = 5y$ is only valid for $0 \le y \le 8$. When y is between 8 and 10, x is a constant 40. Thus, the length of the side of the pool is

$$\text{length} = L(y) = \begin{cases} 5y, & 0 \le y \le 8 \\ 40, & 8 \le y \le 10. \end{cases}$$

Therefore, the fluid force on the side of the pool is given by the two integrals

$$F = \int_0^8 62.4(10 - y)(5y)\, dy + \int_8^{10} 62.4(10 - y)(40)\, dy$$

$$= 312 \int_0^8 (10y - y^2)\, dy + 2496 \int_8^{10} (10 - y)\, dy$$

$$= 312\left[5y^2 - \frac{y^3}{3}\right]_0^8 + 2496\left[10y - \frac{y^2}{2}\right]_8^{10}$$

$$= 312\left(\frac{448}{3}\right) + 2496(2) = 51{,}584 \text{ lb.}$$

REMARK   In Example 4, note that the 30-foot width of the swimming pool is a "red herring." That is, the width of the pool is unnecessary information, and the fluid force against the 40-foot sides can be determined without knowing the width of the pool.

EXERCISES for Section 7.6

In Exercises 1 and 2, find the fluid force on the top side of the metal sheet of given area submerged horizontally in 5 feet of water.

1. 3 square feet

2. 18 square feet

In Exercises 3 and 4, find the **buoyant force** of a rectangular solid of given dimensions submerged in water so that the top side is parallel to the surface of the water. The buoyant force is the difference between the fluid forces on the top and bottom sides of the solid.

3.

4.

In Exercises 5–10, find the fluid force on the indicated vertical side of a tank. Assume that the tank is full of water.

5. Rectangle

6. Triangle

7. Trapezoid

8. Semicircle

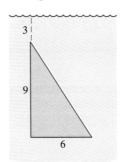

9. Parabola

$$y = x^2$$

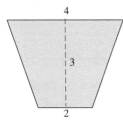

10. Semiellipse

$$y = -\frac{1}{2}\sqrt{36 - 9x^2}$$

In Exercises 11–14, find the fluid force on the given vertical plates submerged in water.

11. Square

12. Square

13. Triangle

14. Rectangle

In Exercises 15–18, the given vertical plate is the side of a form for poured concrete that weighs 140.7 pounds per cubic foot. Determine the fluid force on the plate.

15. Rectangle

16. Rectangle

17. Semiellipse

$$y = -\frac{3}{4}\sqrt{16 - x^2}$$

18. Triangle

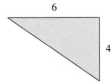

19. A cylindrical gasoline tank is placed so that the axis of the cylinder is horizontal. Find the fluid force on a circular end of the tank if the tank is half full, assuming that the diameter is 3 feet and the gasoline weighs 42 pounds per cubic foot.

20. Repeat Exercise 19 for a tank that is full. (Evaluate one integral by a geometric formula and the other by observing that the integrand is an odd function.)

21. A circular plate of radius r feet is submerged vertically in a tank of fluid that weighs w pounds per cubic foot. The center of the circle is k feet below the surface of the fluid. Show that the fluid force on the surface of the circle is given by

$$F = wk(\pi r^2).$$

(Evaluate one integral by a geometric formula and the other by observing that the integrand is an odd function.)

22. A rectangular plate of height h feet and base b feet is submerged vertically in a tank of fluid that weighs w pounds per cubic foot. The center is k feet below the surface of the fluid, where $h \le k/2$. Show that the fluid force on the surface of the rectangle is given by

$$F = wkhb.$$

23. A porthole on a vertical side of a submarine (submerged in sea water) is 1 foot square. Find the fluid force on the porthole, assuming that the center of the square is 15 feet below the surface.

24. Repeat Exercise 23 for a circular porthole that has a diameter of 1 foot. The center is 15 feet below the surface.

25. A swimming pool is 20 feet wide, 40 feet long, 4 feet deep at one end, and 8 feet deep at the other. The bottom is an inclined plane. Find the fluid force on each of the vertical walls.

26. The vertical cross section of an irrigation canal is modeled by

$$f(x) = \frac{5x^2}{x^2 + 4}$$

where x is measured in feet and $x = 0$ corresponds to the center of the canal. Use Simpson's Rule (with $n = 6$) to approximate the fluid force against a vertical gate used to stop the flow of water if the water is 3 feet deep.

In Exercises 27 and 28, use Simpson's Rule on a computer or calculator (with $n = 10$) to approximate the fluid force on the vertical plate bounded by the x-axis and the top half of the graph of the given equation. Assume that the base of the plate is 12 feet beneath the surface of the water.

27. $x^{2/3} + y^{2/3} = 4^{2/3}$

28. $\dfrac{x^2}{28} + \dfrac{y^2}{16} = 1$

7.7   Moments, Centers of Mass, and Centroids

Mass ▪ Moments ▪ Center of mass ▪ Centroid ▪ Theorem of Pappus

In this section we look at several important applications of integration that are related to **mass.** Mass is a measure of a body's resistance to changes in motion, and is independent of the particular gravitational system in which the body is located. However, because so many applications involving mass occur on the earth's surface, we tend to equate an object's mass with its *weight.* This is not technically correct. Weight is a type of force and as such is dependent on gravity. Force and mass are related by the equation

$$\text{force} = (\text{mass})(\text{acceleration}).$$

In Table 7.1 we list some commonly used measures of mass and force, together with their conversion factors.

TABLE 7.1

System of measurement	*Measure of mass*	*Measure of force*
U.S.	slug	pound = (slug)(ft/sec^2)
International	kilogram	newton = (kilogram)(m/sec^2)
C-G-S	gram	dyne = (gram)(cm/sec^2)

Conversions:	
1 pound = 4.448 newtons	1 slug = 14.59 kilograms
1 newton = 0.2248 pound	1 kilogram = 0.06854 slug
1 dyne = 0.02248 pound	1 gram = 0.00006854 slug

EXAMPLE 1 Mass on the surface of the earth

Find the mass (in slugs) of an object whose weight at sea level is one pound.

SOLUTION

Using 32 feet per second per second as the acceleration due to gravity, we have

$$\text{mass} = \frac{\text{force}}{\text{acceleration}} = \frac{1 \text{ lb}}{32 \text{ ft/sec}^2} = 0.03125 \frac{\text{lb}}{\text{ft/sec}^2} = 0.03125 \text{ slug}.$$

Because many applications occur on the earth's surface, this amount of mass is called a **pound mass.** ▭

The moment of a mass

We shall consider two types of moments of a mass—the **moment about a point** and the **moment about a line.** To define these two moments, we consider an idealized situation in which a mass m is concentrated at a point. If x is the distance between this point-mass and another point P, then the **moment of m about the point P** is given by

$$\text{moment} = mx$$

and x is called the **length of the moment arm.**

The concept of moment can be demonstrated simply by a seesaw, as illustrated by Figure 7.60. Suppose a child of mass 20 kilograms sits 2 meters to the left of fulcrum P, and an older child of mass 30 kilograms sits 2 meters to the right of P. From experience, we know that the seesaw would begin to rotate clockwise, moving the larger child down. This rotation occurs because the moment produced by the child on the left is less than the moment produced by the child on the right:

left moment = (20)(2) = 40 kg · m

right moment = (30)(2) = 60 kg · m.

20 kg   30 kg

P

2 m   2 m

FIGURE 7.60

To balance the seesaw, the two moments must be equal. For example, if the larger child moved to a position $\frac{4}{3}$ meters from the fulcrum, then the seesaw would balance, since both children would produce a moment of 40 kilogram-meters.

To generalize this situation we introduce a coordinate line on which the origin corresponds to the fulcrum, as shown in Figure 7.61. Suppose several point masses are located on the x-axis. The measure of the tendency of this system to rotate about the origin is called the **moment about the origin,** and it is defined to be the sum of the n products $m_i x_i$.

$$M_0 = m_1 x_1 + m_2 x_2 + \cdots + m_n x_n$$

If M_0 is zero, then the system is said to be in **equilibrium.**

For a system that is not in equilibrium, we define the **center of mass** to be the point $\bar{x}$ at which the fulcrum could be relocated to attain equilibrium. If the system were translated $\bar{x}$ units, each coordinate x_i would become $(x_i - \bar{x})$, and since the moment of the translated system is zero, we have

$$\sum_{i=1}^{n} m_i(x_i - \bar{x}) = \sum_{i=1}^{n} m_i x_i - \sum_{i=1}^{n} m_i \bar{x} = 0.$$

Solving for $\bar{x}$, we have

$$\bar{x} = \frac{\displaystyle\sum_{i=1}^{n} m_i x_i}{\displaystyle\sum_{i=1}^{n} m_i} = \frac{\text{moment of system about origin}}{\text{total mass of system}}.$$

FIGURE 7.61

If $x_1 m_1 + x_2 m_2 + \cdots + x_n m_n = 0$, then the system is in equilibrium.

DEFINITION OF THE MOMENT AND CENTER OF MASS OF A LINEAR SYSTEM

Let the point masses $m_1, m_2, \ldots, m_n$ be located at $x_1, x_2, \ldots, x_n$, respectively.

1. The **moment about the origin** is $M_0 = m_1 x_1 + m_2 x_2 + \cdots + m_n x_n$.
2. The **center of mass** is $\bar{x} = \dfrac{M_0}{m}$

where $m = m_1 + m_2 + \cdots + m_n$ is the **total mass** of the system.

EXAMPLE 2 The center of mass of a linear system

Find the center of mass of the linear system shown in Figure 7.62.

$$m_1 = 10, \quad m_2 = 15, \quad m_3 = 5, \quad m_4 = 10$$
$$x_1 = -5, \quad x_2 = 0, \quad x_3 = 4, \quad x_4 = 7$$

FIGURE 7.62

SOLUTION

The moment about the origin is given by

$$M_0 = m_1x_1 + m_2x_2 + m_3x_3 + m_4x_4$$
$$= 10(-5) + 15(0) + 5(4) + 10(7) = -50 + 0 + 20 + 70 = 40.$$

Since the total mass of the system is $m = 10 + 15 + 5 + 10 = 40$, the center of mass is

$$\bar{x} = \frac{M_0}{m} = \frac{40}{40} = 1.$$

Rather than define the moment of a *mass*, we could define the moment of a *force*. In this context, the center of mass is called the **center of gravity.** Suppose that a system of point masses m_1, m_2, . . . , m_n is located at $x_1, x_2, . . . , x_n$. Then, since force = (mass)(acceleration), the total force of the system is

$$F = m_1a + m_2a + \cdots + m_na = ma.$$

The **torque** (moment) about the origin is given by

$$T_0 = (m_1a)x_1 + (m_2a)x_2 + \cdots + (m_na)x_n = M_0a$$

and the **center of gravity** is

$$\frac{T_0}{F} = \frac{M_0a}{ma} = \frac{M_0}{m} = \bar{x}.$$

Therefore, the center of gravity and the center of mass have the same location.

Two-dimensional systems

We can extend the concept of moment to two dimensions by considering a system of masses located in the *xy*-plane at the points (x_1, y_1), (x_2, y_2), . . . , (x_n, y_n) as shown in Figure 7.63. Rather than defining a single moment (with respect to the origin), we define two moments—one with respect to the *x*-axis and one with respect to the *y*-axis.

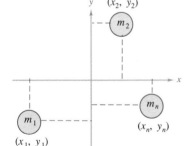

FIGURE 7.63

DEFINITION OF THE MOMENTS AND CENTER OF MASS OF A TWO-DIMENSIONAL SYSTEM

Let the point masses m_1, m_2, . . . , m_n be located at (x_1, y_1), (x_2, y_2), . . . , and (x_n, y_n), respectively.

1. The **moment about the y-axis** is $M_y = m_1x_1 + m_2x_2 + \cdots + m_nx_n$.
2. The **moment about the x-axis** is $M_x = m_1y_1 + m_2y_2 + \cdots + m_ny_n$.
3. The **center of mass** $(\bar{x}, \bar{y})$ (or **center of gravity**) is given by

$$\bar{x} = \frac{M_y}{m} \quad \text{and} \quad \bar{y} = \frac{M_x}{m}$$

where $m = m_1 + m_2 + \cdots + m_n$ is the **total mass** of the system.

The moment of a system of masses in the plane can be taken about any horizontal or vertical line. In general, the moment about a line is the sum of the product of the masses and the *directed distances* from the points to the line.

$$\text{Moment} = m_1(y_1 - b) + m_2(y_2 - b) + \cdots + m_n(y_n - b)$$
Horizontal line $y = b$

$$\text{Moment} = m_1(x_1 - a) + m_2(x_2 - a) + \cdots + m_n(x_n - a)$$
Vertical line $x = a$

EXAMPLE 3   The center of mass of a two-dimensional system

Find the center of mass of a system of point masses $m_1 = 6$, $m_2 = 3$, $m_3 = 2$, and $m_4 = 9$, located at $(3, -2)$, $(0, 0)$, $(-5, 3)$, and $(4, 2)$, as shown in Figure 7.64.

SOLUTION

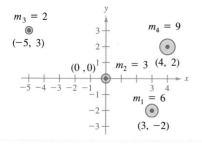

FIGURE 7.64

$$m = 6 \quad + 3 \quad + 2 \quad + 9 \quad = 20 \qquad \text{Mass}$$

$$M_y = 6(3) \quad + 3(0) + 2(-5) + 9(4) = 44 \qquad \text{Moment about } y\text{-axis}$$

$$M_x = 6(-2) + 3(0) + 2(3) \quad + 9(2) = 12 \qquad \text{Moment about } x\text{-axis}$$

Therefore,

$$\bar{x} = \frac{M_y}{m} = \frac{44}{20} = \frac{11}{5} \qquad \text{and} \qquad \bar{y} = \frac{M_x}{m} = \frac{12}{20} = \frac{3}{5}$$

and we conclude that the center of mass is $\left(\frac{11}{5}, \frac{3}{5}\right)$.

The center of mass of a planar lamina

In the preceding discussion we assumed the total mass of a system to be distributed at discrete points in the plane (or on a line). We now consider a thin flat plate of material of uniform density called a **planar lamina. Density** is a measure of mass per unit of volume, such as grams per cubic centimeter. We denote density by ρ, the lowercase Greek letter rho. Intuitively, we think of the center of mass $(\bar{x}, \bar{y})$ of a lamina as its balancing point. For example, the center of mass of a circular lamina is located at the center of the circle, and the center of mass of a rectangular lamina is located at the center of the rectangle, as shown in Figure 7.65.

FIGURE 7.65

Consider an irregularly shaped planar lamina of uniform density ρ, bounded by the graphs of $y = f(x)$, $y = g(x)$, and $a \le x \le b$, as shown in Figure 7.66. The mass of this region is given by

$$m = (\text{density})(\text{area}) = \rho \int_a^b [f(x) - g(x)]\, dx = \rho A$$

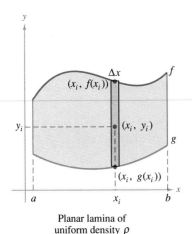

Planar lamina of
uniform density ρ

FIGURE 7.66

where A is the area of the region. To find the center of mass of this lamina, we partition the interval $[a, b]$ into n equal subintervals of width Δx. If x_i is the center of the ith subinterval, then we can approximate the portion of the lamina lying in the ith subinterval by a rectangle whose height is $h = f(x_i) - g(x_i)$. Since the density of the rectangle is ρ, we know that its mass is

$$m_i = (\text{density})(\text{area}) = \rho\underbrace{[f(x_i) - g(x_i)]}_{\text{height}}\underbrace{\Delta x}_{\text{width}}.$$
$$\underbrace{}_{\text{density}}$$

Now, considering this mass to be located at the center (x_i, y_i) of the rectangle, we know that the directed distance from the x-axis to (x_i, y_i) is $y_i = [f(x_i) + g(x_i)]/2$. Thus, the moment of m_i about the x-axis is

$$\text{moment} = (\text{mass})(\text{distance})$$
$$= m_i y_i = \rho[f(x_i) - g(x_i)]\Delta x \left[\frac{f(x_i) + g(x_i)}{2}\right].$$

Summing these moments and taking the limit as $n \to \infty$, we have the moment about the x-axis defined as

$$M_x = \rho \int_a^b \left[\frac{f(x) + g(x)}{2}\right][f(x) - g(x)]\, dx.$$

For the moment about the y-axis, the directed distance from the y-axis to (x_i, y_i) is x_i and we have

$$M_y = \rho \int_a^b x[f(x) - g(x)]\, dx.$$

**DEFINITION OF MOMENTS
AND CENTER OF MASS
OF A PLANAR LAMINA**

Let f and g be continuous functions such that $f(x) \geq g(x)$ on $[a, b]$ and consider the planar lamina of uniform density ρ bounded by the graphs of $y = f(x)$, $y = g(x)$, $a \leq x \leq b$.

1. The **moments about the x- and y-axes** are

$$M_x = \rho \int_a^b \left[\frac{f(x) + g(x)}{2}\right][f(x) - g(x)]\, dx$$

$$M_y = \rho \int_a^b x[f(x) - g(x)]\, dx.$$

2. The **center of mass** $(\bar{x}, \bar{y})$ is given by $\bar{x} = \dfrac{M_y}{m}$ and $\bar{y} = \dfrac{M_x}{m}$

where $m = \rho \int_a^b [f(x) - g(x)]\, dx$ is the mass of the lamina.

REMARK   Note that the integrals for both moments can be formed by inserting into the integral for mass the directed distance from the axis (or line) about which the moment is taken to the center of the representative rectangle.

EXAMPLE 4 The center of mass of a planar lamina

Find the center of mass of the lamina of uniform density ρ bounded by the graph of $f(x) = 4 - x^2$ and the x-axis.

SOLUTION

First, since the center of mass lies on the axis of symmetry, we know that $\bar{x} = 0$. Moreover, the mass of the lamina is given by

$$m = \rho \int_{-2}^{2} (4 - x^2) \, dx = \rho \left[4x - \frac{x^3}{3} \right]_{-2}^{2} = \frac{32\rho}{3}.$$

FIGURE 7.67

To find the moment about the x-axis, we place a representative rectangle in the region, as shown in Figure 7.67. The distance from the x-axis to the center of this rectangle is

$$y_i = \frac{f(x)}{2} = \frac{4 - x^2}{2}.$$

Since the mass of the representative rectangle is

$$\rho f(x) \Delta x = \rho (4 - x^2) \Delta x$$

we have

$$M_x = \rho \int_{-2}^{2} \frac{4 - x^2}{2} (4 - x^2) \, dx = \frac{\rho}{2} \int_{-2}^{2} (16 - 8x^2 + x^4) \, dx$$

$$= \frac{\rho}{2} \left[16x - \frac{8x^3}{3} + \frac{x^5}{5} \right]_{-2}^{2} = \frac{256\rho}{15}$$

and $\bar{y}$ is given by

$$\bar{y} = \frac{M_x}{m} = \frac{256\rho/15}{32\rho/3} = \frac{8}{5}.$$

Center of mass:
$\left(0, \frac{8}{5} \right)$

$y = 4 - x^2$

Center of mass is
balancing point.

FIGURE 7.68

Thus, the center of mass (the balancing point) of the lamina is $\left(0, \frac{8}{5} \right)$, as shown in Figure 7.68.

The density ρ in Example 4 is a common factor of both the moments and the mass, and as such cancels out of the quotients representing the coordinates of the center of mass. Thus, the center of mass of a lamina of *uniform* density depends only on the shape of the lamina and not on its density. For this reason, the point $(\bar{x}, \bar{y})$ is sometimes called the center of mass of a *region* in the plane, or the **centroid** of the region. In other words, to find the centroid of a region in the plane, we simply assume that the region has a constant density of $\rho = 1$, and compute the corresponding center of mass.

EXAMPLE 5 The centroid of a plane region

Find the centroid of the region bounded by the graphs of $f(x) = 4 - x^2$ and $g(x) = x + 2$.

$f(x) = 4 - x^2$

$g(x) = x + 2$

(1, 3)

$\dfrac{f(x) + g(x)}{2}$

$f(x) - g(x)$

(−2, 0)

FIGURE 7.69

SOLUTION

The two graphs intersect at the points $(-2, 0)$ and $(1, 3)$, as shown in Figure 7.69. Thus, the area of the region is given by

$$A = \int_{-2}^{1} [f(x) - g(x)]\, dx = \int_{-2}^{1} (2 - x - x^2)\, dx = \frac{9}{2}.$$

The centroid $(\bar{x}, \bar{y})$ of the region has coordinates

$$\bar{x} = \frac{1}{A} \int_{-2}^{1} x[(4 - x^2) - (x + 2)]\, dx = \frac{2}{9} \int_{-2}^{1} (-x^3 - x^2 + 2x)\, dx$$

$$= \frac{2}{9}\left[-\frac{x^4}{4} - \frac{x^3}{3} + x^2 \right]_{-2}^{1} = -\frac{1}{2}$$

$$\bar{y} = \frac{1}{A} \int_{-2}^{1} \left[\frac{(4 - x^2) + (x + 2)}{2} \right][(4 - x^2) - (x + 2)]\, dx$$

$$= \frac{2}{9}\left(\frac{1}{2} \right) \int_{-2}^{1} (-x^2 + x + 6)(-x^2 - x + 2)\, dx$$

$$= \frac{1}{9} \int_{-2}^{1} (x^4 - 9x^2 - 4x + 12)\, dx$$

$$= \frac{1}{9}\left[\frac{x^5}{5} - 3x^3 - 2x^2 + 12x \right]_{-2}^{1} = \frac{12}{5}.$$

Thus, the centroid of the region is $(\bar{x}, \bar{y}) = \left(-\frac{1}{2}, \frac{12}{5} \right)$.

For simple plane regions you may be able to find the centroid without resorting to integration. Example 6 presents such a case.

EXAMPLE 6   The centroid of a simple plane region

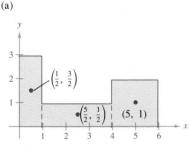

(a)

$\left(\frac{1}{2}, \frac{3}{2}\right)$

$\left(\frac{5}{2}, \frac{1}{2}\right)$ (5, 1)

FIGURE 7.70

Find the centroid of the region shown in Figure 7.70(a).

SOLUTION

By superimposing a coordinate system on the region, as indicated in Figure 7.70(b), we locate the centroids of the three rectangles as

$$\left(\frac{1}{2}, \frac{3}{2} \right), \quad \left(\frac{5}{2}, \frac{1}{2} \right), \quad \text{and} \quad (5, 1).$$

Now, we can calculate the centroid as follows.

$$A = \text{area of region} = 3 + 3 + 4 = 10$$

$$\bar{x} = \frac{(1/2)(3) + (5/2)(3) + (5)(4)}{10} = \frac{29}{10} = 2.9$$

$$\bar{y} = \frac{(3/2)(3) + (1/2)(3) + (1)(4)}{10} = \frac{10}{10} = 1$$

Thus, the centroid of the region is $(2.9, 1)$.

The final topic in this section is a useful theorem credited to Pappus of Alexandria (ca. 300 A.D.), a Greek mathematician whose eight-volume *Mathematical Collection* is a record of much of classical Greek mathematics. We delay the proof of this theorem until Section 15.4 (Exercise 43).

THEOREM 7.2 **THE THEOREM OF PAPPUS**	Let R be a region in a plane and let L be a line in the same plane such that L does not intersect the interior of R, as shown in Figure 7.71. If r is the distance between the centroid of R and the line, then the volume V of the solid of revolution formed by revolving R about the line is given by $$V = 2\pi r A$$ where A is the area of R. (Note that $2\pi r$ is the distance traveled by the centroid as the region is revolved about the line.)

Volume $= 2\pi r$(area of region R)

FIGURE 7.71

The Theorem of Pappus can be used to find the volume of a torus, as shown in the following example. Recall that a **torus** is a doughnut-shaped solid formed by revolving a circular region about a line that lies in the same plane as the circle (but does not intersect the circle).

EXAMPLE 7 Finding volume by the Theorem of Pappus

Find the volume of the torus formed by revolving the circular region bounded by $(x - 2)^2 + y^2 = 1$ about the y-axis, as shown in Figure 7.72(a).

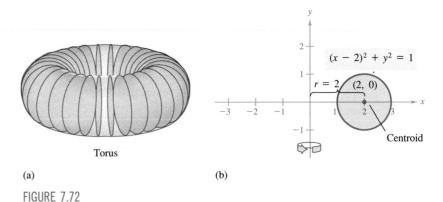

Torus

(a)   (b)

FIGURE 7.72

SOLUTION

From Figure 7.72(b) we see that the centroid of the circular region is $(2, 0)$. Thus, the distance between the centroid and the axis of revolution is $r = 2$. Since the area of the circular region is $A = \pi$, the volume of the torus is given by

$$V = 2\pi r A = 2\pi(2)(\pi) = 4\pi^2 \approx 39.5.$$

EXERCISES for Section 7.7

In Exercises 1–4, find the center of the given point masses lying on the x-axis.

1. $m_1 = 6$, $m_2 = 3$, $m_3 = 5$
$x_1 = -5$, $x_2 = 1$, $x_3 = 3$

2. $m_1 = 7$, $m_2 = 4$, $m_3 = 3$, $m_4 = 8$
$x_1 = -3$, $x_2 = -2$, $x_3 = 5$, $x_4 = 6$

3. $m_1 = 1$, $m_2 = 1$, $m_3 = 1$, $m_4 = 1$, $m_5 = 1$
$x_1 = 7$, $x_2 = 8$, $x_3 = 12$, $x_4 = 15$, $x_5 = 18$

4. $m_1 = 12$, $m_2 = 1$, $m_3 = 6$, $m_4 = 3$, $m_5 = 11$
$x_1 = -3$, $x_2 = -2$, $x_3 = -1$, $x_4 = 0$, $x_5 = 4$

5. Notice in Exercise 3 that $\bar{x}$ is the arithmetic mean of the x-coordinates. Translate each point mass to the right 5 units and compare the resulting center of mass with that obtained in Exercise 3.

6. Translate each point mass in Exercise 4 to the left 3 units and compare the resulting center of mass with that obtained in Exercise 4.

In Exercises 7–10, find the center of mass of the given system of point masses.

7.

m_i	5	1	3
(x_i, y_i)	(2, 2)	(−3, 1)	(1, −4)

8.

m_i	10	2	5
(x_i, y_i)	(1, −1)	(5, 5)	(−4, 0)

9.

m_i	3	4
(x_i, y_i)	(−2, −3)	(−1, 0)

m_i	2	1	6
(x_i, y_i)	(7, 1)	(0, 0)	(−3, 0)

10.

m_i	4	2	5/2	5
(x_i, y_i)	(2, 3)	(−1, 5)	(6, 8)	(2, −2)

In Exercises 11–14, introduce an appropriate coordinate system and find the coordinates of the center of mass of the given planar lamina.

11.

12.

13.

14.

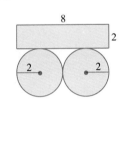

15. Suppose that the circular lamina in Exercise 11 has twice the density of the square lamina, and find the resulting center of mass.

16. Suppose that the square lamina in Exercise 11 has twice the density of the circular lamina, and find the resulting center of mass.

In Exercises 17–28, find M_x, M_y, and $(\bar{x}, \bar{y})$ for the laminas of uniform density ρ bounded by the graphs of the given equations.

17. $y = \sqrt{x}$, $y = 0$, $x = 4$
18. $y = x^2$, $y = 0$, $x = 4$
19. $y = \sqrt{x}$, $y = x$, $x \geq 0$
20. $y = x^2$, $y = x^3$
21. $y = -x^2 + 4x + 2$, $y = x + 2$
22. $y = \sqrt{3x + 1}$, $y = x + 1$
23. $x = 4 - y^2$, $x = 0$
24. $x = 2y - y^2$, $x = 0$
25. $x = -y$, $x = 2y - y^2$
26. $x = y + 2$, $x = y^2$
27. $y = x^{2/3}$, $y = 0$, $x = 8$
28. $y = x^{2/3}$, $y = 4$

In Exercises 29–34, find the centroid of the region bounded by the graphs of the given equations.

29. $y = \sqrt{1 - x^2}$, $y = 0$
30. $y = x^3$, $y = x$, $0 \le x \le 1$
31. $y = x^2$, $y = x$
32. $y = \dfrac{1}{x}$, $y = 0$, $1 \le x \le 4$
33. $y = 2x + 4$, $y = 0$, $0 \le x \le 3$
34. $y = x^2 - 4$, $y = 0$

35. Find the centroid of the triangular region with vertices $(-a, 0)$, $(a, 0)$, and (b, c), as shown in the figure. Show that it is the point of intersection of the medians of the triangle.

36. Find the centroid of the **parabolic spandrel** shown in the figure.

Parabolic spandrel

FIGURE FOR 35   FIGURE FOR 36

In Exercises 37 and 38, use Simpson's Rule on a computer or calculator (with $n = 10$) to approximate the centroid of the region bounded by the graphs of the given equations.

37. $y = 10x\sqrt{125 - x^3}$, $y = 0$, $x = 0$
38. $y = xe^{-x/2}$, $y = 0$, $x = 0$, $x = 4$

39. The prefabricated end section of a building is modeled by the region bounded by the graphs of $y =$ $5\sqrt[3]{400 - x^2}$ and $y = 0$. Use Simpson's Rule on a computer or calculator (with $n = 10$) to approximate the y-coordinate of the centroid of the region.

40. Use Simpson's Rule on a computer or calculator (with $n = 8$) to approximate the y-coordinate of the centroid of the region bounded by the graphs $y = 8/(x^2 + 4)$, $y = 0$, $x = -2$, and $x = 2$. The graph of the curve is called the **Witch of Agnesi,** and the exact y-coordinate of the centroid is $(\pi + 2)/2\pi$.

In Exercises 41–44, use the Theorem of Pappus to find the volume of the solid of revolution.

41. Torus formed by revolving the circle $(x - 5)^2 + y^2 = 16$ about the y-axis.
42. Torus formed by revolving the circle $x^2 + (y - 3)^2 = 4$ about the x-axis.
43. Solid formed by revolving the region bounded by the graphs of $y = x$, $y = 4$, and $x = 0$ about the x-axis.
44. Solid formed by revolving the region bounded by the graphs of $y = \sqrt{x - 1}$, $y = 0$, and $x = 5$ about the y-axis.

In Exercises 45 and 46, use the **Second Theorem of Pappus,** which is stated as follows. If a segment of a plane curve C is revolved about an axis that does not intersect the curve (except possibly at its endpoints), then the area S of the resulting surface of revolution is given by the product of the length of C times the distance d traveled by the centroid of C.

45. A sphere is formed by revolving the graph of $y = \sqrt{r^2 - x^2}$ about the x-axis. Use the formula for surface area, $S = 4\pi r^2$, to find the centroid of the semicircle $y = \sqrt{r^2 - x^2}$.
46. A torus is formed by revolving the graph of $(x - 1)^2 + y^2 = 1$ about the y-axis. Find the surface area of the torus.

REVIEW EXERCISES for Chapter 7

In Exercises 1–18, sketch the region bounded by the graphs of the given equations and determine the area of the region.

1. $y = \dfrac{1}{x^2}$, $y = 0$, $x = 1$, $x = 5$
2. $y = \dfrac{1}{x^2}$, $y = 4$, $x = 5$
3. $y = \dfrac{1}{x^2 + 1}$, $y = 0$, $x = -1$, $x = 1$
4. $y = 1 - \dfrac{x}{2}$, $y = x - 2$, $y = 1$
5. $x = y^2 - 2y$, $x = 0$
6. $x = y^2 - 2y$, $x = -1$, $y = 0$
7. $y = x$, $y = x^3$
8. $x = y^2 + 1$, $x = y + 3$
9. $y = x^2 - 8x + 3$, $y = 3 + 8x - x^2$
10. $y = x^2 - 4x + 3$, $y = x^3$, $x = 0$
11. $y = \sqrt{x - 1}$, $y = 2$, $y = 0$, $x = 0$
12. $y = \sqrt{x - 1}$, $y = \dfrac{x - 1}{2}$
13. $\sqrt{x} + \sqrt{y} = 1$, $y = 0$, $x = 0$
14. $y = x^4 - 2x^2$, $y = 2x^2$
15. $y = e^x$, $y = e^2$, $x = 0$.
16. $y = \csc x$, $y = 2$ (one region)
17. $y = \sin x$, $y = \cos x$, $\dfrac{\pi}{4} \le x \le \dfrac{5\pi}{4}$
18. $x = \cos y$, $x = \dfrac{1}{2}$, $\dfrac{\pi}{3} \le y \le \dfrac{7\pi}{3}$

In Exercises 19–26, find the volume of the solid generated by revolving the plane region bounded by the given equations about the indicated line.

19. $y = x$, $y = 0$, $x = 4$
    (a) the x-axis   (b) the y-axis
    (c) the line $x = 4$ (d) the line $x = 6$

20. $y = \sqrt{x}$, $y = 2$, $x = 0$
    (a) the x-axis   (b) the line $y = 2$
    (c) the y-axis   (d) the line $x = -1$

21. $\dfrac{x^2}{16} + \dfrac{y^2}{9} = 1$
    (a) the y-axis (oblate spheroid)
    (b) the x-axis (prolate spheroid)

22. $\dfrac{x^2}{a^2} + \dfrac{y^2}{b^2} = 1$
    (a) the y-axis (oblate spheroid)
    (b) the x-axis (prolate spheroid)

23. $y = \dfrac{1}{x^4 + 1}$, $y = 0$, $x = 0$, $x = 1$,
    revolved about the y-axis

24. $y = \dfrac{1}{\sqrt{1 + x^2}}$, $y = 0$, $x = -1$, $x = 1$,
    revolved about the x-axis

25. $y = -x^2 + 6x - 5$, $y = 0$
    (a) the x-axis   (b) the y-axis

26. $y = e^{-x}$, $y = 0$, $x = 0$, $x = 1$,
    revolved about the x-axis

27. Find the work done in stretching a spring from its natural length of 10 inches to a length of 15 inches, if a force of 4 pounds is needed to stretch it 1 inch from its natural position.

28. Find the work done in stretching a spring from its natural length of 9 inches to double that length, if a force of 50 pounds is required to hold the spring at double its natural length.

29. A water well has an 8-inch casing (diameter) and is 175 feet deep. If the water is 25 feet from the top of the well, determine the amount of work done in pumping it dry, assuming that no water enters the well while it is being pumped.

30. Repeat Exercise 29, assuming that water enters the well at the rate of 4 gallons per minute and the pump works at the rate of 12 gallons per minute. How many gallons are pumped in this case?

31. A chain 10 feet long weighs 5 pounds per foot and is suspended from a platform 20 feet above the ground. How much work is required to raise the entire chain to the 20-foot level?

32. A windlass, 200 feet above ground level on the top of a building, uses a cable weighing 4 pounds per foot. Find the work done in winding up the cable if
    (a) one end is at ground level.
    (b) there is a 300-pound load attached to the end of the cable.

33. A swimming pool is 5 feet deep at one end and 10 feet deep at the other, and the bottom is an inclined plane. The length and width of the pool are 40 feet and 20 feet, respectively. If the pool is full of water, what is the fluid force on each of the vertical walls?

34. Show that the fluid force against any vertical region in a liquid is the product of the weight per cubic volume of the liquid, the area of the region, and the depth of the centroid of the region.

35. Using the result of Exercise 34, find the fluid force on one side of a vertical circular plate of radius 4 feet that is submerged in water so that its center is 5 feet below the surface.

36. How much must the water level be raised to double the fluid force on one side of the plate in Exercise 35?

In Exercises 37–40, find the centroid of the region bounded by the graphs of the given equations.

37. $\sqrt{x} + \sqrt{y} = \sqrt{a}$, $x = 0$, $y = 0$
38. $y = x^2$, $y = 2x + 3$
39. $y = a^2 - x^2$, $y = 0$
40. $y = x^{2/3}$, $y = \dfrac{1}{2}x$

41. Use integration to show that the arc length of the circle $x^2 + y^2 = 4$ from $(-\sqrt{3}, 1)$ clockwise to $(\sqrt{3}, 1)$ is one-third the circumference of the circle.

42. Find the length of the graph of
$$y = \frac{1}{6}x^3 + \frac{1}{2x}$$
from $x = 1$ to $x = 3$.

43. Use integration to find the lateral surface area of a right circular cone of height 4 and radius 3.

44. A gasoline tank is an oblate spheroid generated by revolving the region bounded by the graph of
$$\frac{x^2}{16} + \frac{y^2}{9} = 1$$
about the y-axis, where x and y are measured in feet. Find the depth of the gasoline in the tank when it is filled to one-fourth its capacity.

45. Find the area of the region bounded by $y = x\sqrt{x + 1}$ and $y = 0$.

46. The region defined in Exercise 45 is revolved around the x-axis. Find the volume of the solid generated.

47. Find the volume of the solid generated by revolving the region defined in Exercise 45 about the y-axis.

48. The region bounded by $y = 2\sqrt{x}$, $y = 0$, and $x = 3$ is revolved around the x-axis. Find the surface area of the solid generated.

49. Find the arc length of the graph of $f(x) = \frac{4}{5}x^{5/4}$ from $x = 0$ to $x = 4$.

50. Find the volume of the solid generated by revolving the region bounded by $y = 1/(1 + \sqrt{x - 2})$, $y = 0$, $x = 2$, and $x = 6$ about the y-axis.

Chapter 8 Application

The space shuttle Columbia, after four test flights, made its first operational flight on November 16, 1982. It stayed in orbit five days. The second of the space shuttles, the *Challenger*, made its maiden flight in April of 1983. On September 29, 1988, the *Discovery* was launched. It returned four days later to Edwards Air Force Base in California.

Velocity of a Rocket

If a rocket whose initial mass is m (including fuel) is fired vertically at time $t = 0$, then its velocity at time t is given by

$$v = gt + u \ln \frac{m}{m - rt}$$

where u is the expulsion speed of the fuel, r is the rate at which the fuel is consumed, and $g = -32$ is the acceleration due to gravity.

The position equation for the rocket can be found by integrating the velocity with respect to t to obtain the equation

$$s = \frac{gt^2}{2} + u\left[t + \left(t - \frac{m}{r}\right) \ln \frac{m}{m - rt}\right].$$

For example, consider a rocket for which $m = 40,000$ lb, $u = 10,000$ ft/sec, and $r = 300$ lb/sec. If the rocket contains 36,000 pounds of fuel, then it will accelerate for 120 seconds and have a velocity of

$$v = -32t + 10,000 \ln \frac{40,000}{40,000 - 300t}, \quad 0 \le t \le 120$$

where v is measured in feet per second. Thus, after 120 seconds, this rocket will be traveling at 19,186 ft/sec and will have attained a height of 662,590 feet.

See Exercise 73, Section 8.2.

Chapter Overview

This chapter begins by reviewing the nineteen basic integration formulas developed in earlier chapters. Then, in Sections 8.2–8.6, we look at a variety of integration techniques: (1) integration by parts, (2) integration of powers of trigonometric functions, (3) trigonometric substitution, (4) partial fractions, (5) integration by tables, and (6) integration of rational functions of sine and cosine. Success in these new techniques depends on mastery of the basic integration formulas given in Section 8.1.

In Section 8.7, we revisit a problem that was introduced in Chapter 2—determination of a limit involving an indeterminate form. To do this, we present a new technique called **L'Hôpital's Rule.**

In the last section, we introduce the concept of an **improper integral.** Improper integrals come in two forms: (1) one or both of the limits of integration can be infinite such as in the improper integral

$$\int_1^\infty \frac{1}{x} \, dx$$

and (2) the integrand can have a finite number of infinite discontinuities such as in the improper integral

$$\int_0^1 \frac{1}{x^2} \, dx.$$

8

Integration Techniques, L'Hôpital's Rule, and Improper Integrals

8.1 Basic Integration Formulas

Basic integration formulas ▪ Fitting integrands to basic formulas

In this chapter we study several integration techniques that greatly expand the set of integrals to which the basic integration formulas can be applied.

BASIC INTEGRATION FORMULAS

1. $\displaystyle\int kf(u) \, du = k \int f(u) \, du$

2. $\displaystyle\int [f(u) \pm g(u)] \, du = \int f(u) \, du \pm \int g(u) \, du$

3. $\displaystyle\int du = u + C$

4. $\displaystyle\int u^n \, du = \frac{u^{n+1}}{n+1} + C, \, n \neq -1$

5. $\displaystyle\int \frac{du}{u} = \ln|u| + C$

6. $\displaystyle\int e^u \, du = e^u + C$

7. $\displaystyle\int \sin u \, du = -\cos u + C$

8. $\displaystyle\int \cos u \, du = \sin u + C$

9. $\displaystyle\int \tan u \, du = -\ln|\cos u| + C$

10. $\displaystyle\int \cot u \, du = \ln|\sin u| + C$

11. $\displaystyle\int \sec u \, du = \ln|\sec u + \tan u| + C$

12. $\displaystyle\int \csc u \, du = -\ln|\csc u + \cot u| + C$

13. $\displaystyle\int \sec^2 u \, du = \tan u + C$

14. $\displaystyle\int \csc^2 u \, du = -\cot u + C$

15. $\displaystyle\int \sec u \tan u \, du = \sec u + C$

16. $\displaystyle\int \csc u \cot u \, du = -\csc u + C$

17. $\displaystyle\int \frac{du}{\sqrt{a^2 - u^2}} = \arcsin \frac{u}{a} + C$

18. $\displaystyle\int \frac{du}{a^2 + u^2} = \frac{1}{a} \arctan \frac{u}{a} + C$

19. $\displaystyle\int \frac{du}{u\sqrt{u^2 - a^2}} = \frac{1}{a} \text{arcsec} \frac{|u|}{a} + C$

A major step in solving any integration problem is recognizing the proper basic integration formula to be used. To do this, you must memorize the basic formulas and practice using them. In this first section we review the use of basic integration formulas.

One of the challenging things about integration is the fact that slight differences in the integrand can lead to very different solution techniques.

EXAMPLE 1 A comparison of three similar integrals

Evaluate the following integrals.

(a) $\int \dfrac{4}{x^2 + 9}\, dx$ (b) $\int \dfrac{4x}{x^2 + 9}\, dx$ (c) $\int \dfrac{4x^2}{x^2 + 9}\, dx$

SOLUTION

(a) Considering the Arctangent Rule, we let $u = x$ and $a = 3$. Then

$$\int \frac{4}{x^2 + 9}\, dx = 4 \int \frac{1}{x^2 + 3^2}\, dx = 4\left[\frac{1}{3} \arctan \frac{x}{3}\right] + C$$

$$= \frac{4}{3} \arctan \frac{x}{3} + C.$$

(b) Here the Arctangent Rule does not apply because the numerator contains a factor of x. Considering the Log Rule, we let $u = x^2 + 9$. Then, $du = 2x\, dx$, and we have

$$\int \frac{4x}{x^2 + 9}\, dx = 2 \int \frac{2x\, dx}{x^2 + 9} = 2 \int \frac{du}{u}$$

$$= 2 \ln |u| + C$$

$$= 2 \ln (x^2 + 9) + C.$$

(c) Since the degree of the numerator is equal to the degree of the denominator, we first divide to obtain

$$\frac{4x^2}{x^2 + 9} = 4 - \frac{36}{x^2 + 9}.$$

Thus,

$$\int \frac{4x^2}{x^2 + 9}\, dx = \int \left(4 - \frac{36}{x^2 + 9}\right) dx = \int 4\, dx - 36 \int \frac{1}{x^2 + 9}\, dx$$

$$= 4x - 36\left[\frac{1}{3} \arctan \frac{x}{3}\right] + C$$

$$= 4x - 12 \arctan \frac{x}{3} + C. \quad \blacksquare$$

Notice in part (c) of Example 1 that some preliminary algebra was required before applying the rules for integration, and that subsequently more than one formula was needed to evaluate the resulting integral. This is also the case in the next example.

EXAMPLE 2 Using two basic formulas to solve a single integral

Evaluate

$$\int \frac{x+3}{\sqrt{4-x^2}}\,dx.$$

SOLUTION

We rewrite the integral as the sum of two integrals and then apply the Power Rule and the Arcsine Rule as follows.

$$\int \frac{x+3}{\sqrt{4-x^2}}\,dx = \int \frac{x}{\sqrt{4-x^2}}\,dx + \int \frac{3}{\sqrt{4-x^2}}\,dx$$

$$= -\frac{1}{2}\int \underbrace{(4-x^2)^{-1/2}}_{u^n}\underbrace{(-2x)\,dx}_{du} + 3\int \frac{1}{\underbrace{\sqrt{2^2-x^2}}_{a=2,\,u=x}}\,dx$$

$$= -\frac{1}{2}\int u^{-1/2}\,du + 3\int \frac{1}{\sqrt{a^2-u^2}}\,du$$

$$= -\frac{1}{2}\left(\frac{u^{1/2}}{1/2}\right) + 3\arcsin\frac{u}{a} + C$$

$$= -(4-x^2)^{1/2} + 3\arcsin\frac{x}{2} + C$$

When the integrand is a quotient whose numerator is a sum (or difference), you should consider breaking the quotient into two or more parts, as in Example 2. The next example further demonstrates this useful procedure.

EXAMPLE 3 Breaking a quotient into two parts

Evaluate

$$\int \frac{1+\cos(e^{-2x})}{e^{2x}}\,dx.$$

SOLUTION

As it stands, the integrand does not appear to fit any basic formula. However, since the numerator consists of a sum, we break the integrand into two fractions and proceed as follows.

$$\int \frac{1 + \cos e^{-2x}}{e^{2x}} \, dx = \int \frac{dx}{e^{2x}} + \int \frac{\cos e^{-2x}}{e^{2x}} \, dx$$

$$= \int \underbrace{e^{-2x}}_{u \, = \, -2x} \, dx + \int (\cos e^{-2x}) \underbrace{(e^{-2x})}_{u \, = \, e^{-2x}} \, dx$$

$$= -\frac{1}{2} \int e^{-2x}(-2) \, dx - \frac{1}{2} \int (\cos e^{-2x})(-2e^{-2x}) \, dx$$

$$= -\frac{1}{2} \int e^u \, du - \frac{1}{2} \int \cos u \, du$$

$$= -\frac{1}{2} e^u - \frac{1}{2} \sin u + C$$

$$= -\frac{1}{2} e^{-2x} - \frac{1}{2} \sin e^{-2x} + C$$

Formulas 17, 18, and 19 all have expressions involving the sum or difference of two squares:

$$a^2 - [f(x)]^2 \qquad a^2 + [f(x)]^2 \qquad [f(x)]^2 - a^2.$$

When you encounter such an expression, consider the substitution $u = f(x)$, as shown in the next example.

EXAMPLE 4   A substitution involving $a^2 - u^2$

Evaluate

$$\int \frac{x^2}{\sqrt{16 - x^6}} \, dx.$$

SOLUTION

The radical in the denominator can be written in the form $\sqrt{a^2 - u^2} = \sqrt{4^2 - (x^3)^2}$, and we try the substitution $u = x^3$. Then $du = 3x^2 \, dx$, and we have

$$\int \frac{x^2}{\sqrt{16 - x^6}} \, dx = \frac{1}{3} \int \frac{3x^2 \, dx}{\sqrt{16 - (x^3)^2}}$$

$$= \frac{1}{3} \int \frac{du}{\sqrt{4^2 - u^2}} = \frac{1}{3} \arcsin \frac{u}{4} + C$$

$$= \frac{1}{3} \arcsin \frac{x^3}{4} + C.$$

Also, when applying Formulas 17, 18, and 19, recall from Section 6.8 that completing the square can be helpful. For instance,

$$\int \frac{dx}{x^2 + 2x + 5} = \int \frac{dx}{(x + 1)^2 + 2^2}$$

$$= \frac{1}{2} \arctan \frac{x + 1}{2} + C.$$

Surprisingly, two of the most commonly overlooked integration formulas are the Log Rule and the Power Rule. Notice in the next two examples how these two integration formulas can be disguised.

EXAMPLE 5 A disguised form of the Log Rule

Evaluate

$$\int \frac{1}{1 + e^x} \, dx.$$

SOLUTION

The integral does not appear to fit any of the basic formulas. However, the quotient form suggests the Log Rule. If we let $u = 1 + e^x$, then $du = e^x \, dx$. We can obtain the required du by adding and subtracting e^x in the numerator as follows.

$$\int \frac{1}{1 + e^x} \, dx = \int \frac{1 + e^x - e^x}{1 + e^x} \, dx \qquad \text{Add and subtract } e^x \text{ in numerator}$$

$$= \int \left(\frac{1 + e^x}{1 + e^x} - \frac{e^x}{1 + e^x} \right) dx$$

$$= \int dx - \int \frac{e^x \, dx}{1 + e^x}$$

$$= \int dx - \int \frac{du}{u}$$

$$= x - \ln |u| + C$$

$$= x - \ln (1 + e^x) + C$$

In this chapter, you will see that there is often more than one way to solve an integration problem. For instance, in Example 5, we could multiply the numerator and denominator by e^{-x} to obtain an integral of the form $-\int (u'/u) \, dx$. See if you can get the same answer by this procedure. (Be careful; the answer will appear in a different form.)

EXAMPLE 6   A disguised form of the Power Rule

Evaluate $\int (\cot x)[\ln (\sin x)] \, dx$.

SOLUTION

Again, this integral does not appear to fit any of the basic formulas. However, considering the two primary choices for u ($u = \cot x$ or $u = \ln \sin x$), we see that the second choice is the appropriate one, since

$$u = \ln \sin x \quad \Longrightarrow \quad du = \frac{\cos x}{\sin x} \, dx = \cot x \, dx.$$

Thus, we have

$$\int (\cot x)[\ln (\sin x)] \, dx = \int u \, du = \frac{u^2}{2} + C$$

$$= \frac{1}{2}[\ln (\sin x)]^2 + C.$$

REMARK   In Example 6, try *checking* that the derivative of $\frac{1}{2}[\ln (\sin x)]^2 + C$ is the integrand of the original integral.

Trigonometric identities can often be used to fit integrals to one of the basic integration formulas. The last example in this review section demonstrates this procedure.

EXAMPLE 7   Using trigonometric identities

Evaluate $\int \tan^2 2x \, dx$.

SOLUTION

By looking over the list of basic integration formulas, we see that $\tan^2 u$ is not on the list, but $\sec^2 u$ is on the list. This observation suggests the trigonometric identity $\tan^2 u = \sec^2 u - 1$. If we let $u = 2x$, then $du = 2 \, dx$ and

$$\int \tan^2 2x \, dx = \frac{1}{2} \int \tan^2 u \, du = \frac{1}{2} \int (\sec^2 u - 1) \, du$$

$$= \frac{1}{2} \int \sec^2 u \, du - \frac{1}{2} \int du = \frac{1}{2} \tan u - \frac{u}{2} + C$$

$$= \frac{1}{2} \tan 2x - x + C.$$

We conclude with a summary of some common procedures for fitting integrands to the basic integration formulas.

PROCEDURES FOR FITTING INTEGRANDS TO BASIC FORMULAS	Technique	Example
	Expand (numerator)	$(1 + e^x)^2 = 1 + 2e^x + e^{2x}$
	Separate numerator	$\dfrac{1 + x}{x^2 + 1} = \dfrac{1}{x^2 + 1} + \dfrac{x}{x^2 + 1}$
	Complete the square	$\dfrac{1}{\sqrt{2x - x^2}} = \dfrac{1}{\sqrt{1 - (x - 1)^2}}$
	Divide if rational function is improper	$\dfrac{x^2}{x^2 + 1} = 1 - \dfrac{1}{x^2 + 1}$
	Add and subtract terms in numerator	$\dfrac{2x}{x^2 + 2x + 1} = \dfrac{2x + 2 - 2}{x^2 + 2x + 1}$ $= \dfrac{2x + 2}{x^2 + 2x + 1} - \dfrac{2}{(x + 1)^2}$
	Use trigonometric identities	$\cot^2 x = \csc^2 x - 1$
	Multiply and divide by Pythagorean conjugate	$\dfrac{1}{1 + \sin x} = \left(\dfrac{1}{1 + \sin x}\right)\left(\dfrac{1 - \sin x}{1 - \sin x}\right)$ $= \dfrac{1 - \sin x}{1 - \sin^2 x} = \dfrac{1 - \sin x}{\cos^2 x}$ $= \sec^2 x - \dfrac{\sin x}{\cos^2 x}$

REMARK   Watch out for this common error when fitting integrands to basic fomulas.

$$\frac{1}{x^2 + 1} \neq \frac{1}{x^2} + \frac{1}{1} \qquad \text{Do } not \text{ separate denominators}$$

EXERCISES for Section 8.1

In Exercises 1–50, evaluate the indefinite integral.

1. $\int (3x - 2)^4 \, dx$

2. $\int \dfrac{2}{(t - 9)^2} \, dt$

3. $\int (-2x + 5)^{3/2} \, dx$

4. $\int x\sqrt{4 - 2x^2} \, dx$

5. $\int \left[v + \dfrac{1}{(3v - 1)^3}\right] dv$

6. $\int \dfrac{2t - 1}{t^2 - t + 2} \, dt$

7. $\int \dfrac{t^2 - 3}{-t^3 + 9t + 1} \, dt$

8. $\int \dfrac{2x}{x - 4} \, dx$

9. $\int \dfrac{x^2}{x - 1} \, dx$

10. $\int \dfrac{x + 1}{\sqrt{x^2 + 2x - 4}} \, dx$

11. $\int \left(\dfrac{1}{3x - 1} - \dfrac{1}{3x + 1}\right) dx$

12. $\int \dfrac{e^x}{1 + e^x} \, dx$

13. $\int t \sin t^2 \, dt$

14. $\int \sec 4u \, du$

15. $\int \cos x \, e^{\sin x} \, dx$

16. $\int e^{5x} \, dx$

17. $\int \dfrac{(1 + e^t)^2}{e^t} \, dt$

18. $\int \dfrac{1 + \sin x}{\cos x} \, dx$

19. $\int \sec 3x \tan 3x \, dx$

20. $\int \dfrac{1}{\sqrt{x}(1 - 2\sqrt{x})} \, dx$

21. $\int \dfrac{2}{e^{-x} + 1} \, dx$

22. $\int \dfrac{1}{2e^x - 3} \, dx$

23. $\int \dfrac{1}{1 - \cos x} \, dx$

24. $\int \dfrac{1}{\sec x - 1} \, dx$

25. $\int \dfrac{2t - 1}{t^2 + 4} \, dt$

26. $\int \dfrac{2}{(2t - 1)^2 + 4} \, dt$

27. $\int \dfrac{3}{t^2 + 1} \, dt$

28. $\int \dfrac{3}{\sqrt{1 - t^2}} \, dt$

29. $\int \dfrac{1}{x\sqrt{x^2 - 4}} \, dx$

30. $\int \dfrac{-2x}{\sqrt{x^2 - 4}} \, dx$

31. $\int \dfrac{-1}{\sqrt{1 - (2t - 1)^2}} \, dt$

32. $\int \dfrac{1}{4 + 3x^2} \, dx$

33. $\int \csc \pi x \cot \pi x \, dx$

34. $\int \frac{1}{x\sqrt{4x^2 - 1}} \, dx$

35. $\int \frac{t}{\sqrt{1 - t^4}} \, dt$

36. $\int \frac{\sin x}{\sqrt{\cos x}} \, dx$

37. $\int \frac{\sec^2 x}{4 + \tan^2 x} \, dx$

38. $\int \tan^2 2x \, dx$

39. $\int \frac{\tan (2/t)}{t^2} \, dt$

40. $\int \frac{e^{1/t}}{t^2} \, dt$

41. $\int (1 + 2x^2)^2 \, dx$

42. $\int (2 - x^2)^2 \, dx$

43. $\int x\left(1 + \frac{1}{x}\right)^3 \, dx$

44. $\int (x + x^2)^3 \, dx$

45. $\int (1 + e^x)^2 \, dx$

46. $\int \left(\frac{e^x + e^{-x}}{2}\right)^2 \, dx$

47. $\int \frac{3}{\sqrt{6x - x^2}} \, dx$

48. $\int \frac{1}{(x - 1)\sqrt{4x^2 - 8x + 3}} \, dx$

49. $\int \frac{4}{4x^2 + 4x + 65} \, dx$

50. $\int \frac{1}{\sqrt{2 - 2x - x^2}} \, dx$

In Exercises 51–60, evaluate the definite integral.

51. $\int_0^1 xe^{-x^2} \, dx$

52. $\int_0^\pi \sin^2 t \cos t \, dt$

53. $\int_1^e \frac{1 - \ln x}{x} \, dx$

54. $\int_1^2 \frac{x - 2}{x} \, dx$

55. $\int_0^4 \frac{2x}{\sqrt{x^2 + 9}} \, dx$

56. $\int_0^{\pi/4} \cos 2x \, dx$

57. $\int_{\pi/4}^{\pi/2} \cot x \, dx$

58. $\int_0^4 \frac{1}{\sqrt{25 - x^2}} \, dx$

59. $\int_1^2 \frac{1}{2x\sqrt{4x^2 - 1}} \, dx$

60. $\int_0^{2/\sqrt{3}} \frac{1}{4 + 9x^2} \, dx$

In Exercises 61 and 62, find the area of the region bounded by the graph(s) of the given equation(s).

61. $y^2 = x^2(1 - x^2)$

62. $y = \sin 2x, \ y = 0, \ x = 0, \ x = \dfrac{\pi}{2}$

63. The graphs of $f(x) = x$ and $g(x) = ax^2$ intersect at the points $(0, 0)$ and $(1/a, 1/a)$. Find a so that the area of the region bounded by the graphs of these two functions is $\frac{2}{3}$.

64. Find the volume of the solid generated by revolving the region bounded by $y = e^{-x^2}$, $y = 0$, $x = 0$, and $x = 1$ about the y-axis.

65. The region bounded by $y = e^{-x^2}$, $y = 0$, $x = 0$, and $x = b$ is revolved around the y-axis. Find b so that the volume of the generated solid is $\frac{4}{3}$ cubic units.

66. Compute the average value of each of the functions over the indicated interval.

(a) $f(x) = \sin nx, \ 0 \leq x \leq \pi/n$, n is a positive integer

(b) $f(x) = \dfrac{1}{1 + x^2}, \ -3 \leq x \leq 3$

67. Find the x-coordinate of the centroid of the region bounded by the graphs of

$$y = \frac{5}{\sqrt{25 - x^2}}, \ y = 0, \ x = 0, \ x = 4.$$

68. Find the area of the surface formed by revolving the graph of $y = 2\sqrt{x}$ on the interval $[0, 9]$ about the x-axis.

In Exercises 69 and 70, use Simpson's Rule on a computer or calculator (with $n = 12$) to approximate the arc length of the curve over the indicated interval.

	Equation	Interval
69.	$y = \tan \pi x$	$[0, \frac{1}{4}]$
70.	$y = x^{2/3}$	$[1, 8]$

8.2 Integration by Parts

Integration by parts ▪ Tabular method

The first new integration technique we present in this chapter is called **integration by parts.** This technique applies to a wide variety of functions and is particularly useful for integrands involving a *product* of algebraic and transcendental functions. For instance, integration by parts works well with integrals like $\int x \ln x \, dx$, $\int x^2 e^x \, dx$, and $\int e^x \sin x \, dx$.

Integration by parts is based on the formula for the derivative of a product

$$\frac{d}{dx}[uv] = u\frac{dv}{dx} + v\frac{du}{dx} = uv' + vu'$$

where both u and v are differentiable functions of x. If u' and v' are continuous, then we can integrate both sides of this equation to obtain

$$uv = \int uv' \, dx + \int vu' \, dx = \int u \, dv + \int v \, du.$$

By rewriting this equation, we obtain the following theorem.

**THEOREM 8.1
INTEGRATION BY PARTS**

If u and v are functions of x and have continuous derivatives, then

$$\int u \, dv = uv - \int v \, du.$$

This formula expresses the original integrand in terms of another integral. Depending on the choices for u and dv, it may be easier to evaluate the second integral than the original one. Since the choices of u and dv are critical in the integration by parts process, we provide the following guidelines.

**GUIDELINES FOR INTEGRATION
BY PARTS**

1. Try letting dv be the most complicated portion of the integrand that fits a basic integration formula. Then u will be the remaining factor(s) of the integrand.
2. Try letting u be the portion of the integrand whose derivative is a simpler function than u. Then dv will be the remaining factor(s) of the integrand.

EXAMPLE 1 Integration by parts

Evaluate $\int xe^x \, dx$.

SOLUTION

To apply integration by parts, we want to write the integral in the form $\int u \, dv$. There are several ways to do this.

$$\int \underbrace{(x)}_{u}\underbrace{(e^x \, dx)}_{dv}, \qquad \int \underbrace{(e^x)}_{u}\underbrace{(x \, dx)}_{dv}, \qquad \int \underbrace{(1)}_{u}\underbrace{(xe^x \, dx)}_{dv}, \qquad \int \underbrace{(xe^x)}_{u}\underbrace{(dx)}_{dv},$$

Following our guidelines, we choose the first option because the derivative of $u = x$ is simpler than x *and* $dv = e^x \, dx$ is the most complicated portion of the integrand that fits a basic integration formula. Thus, we have

$$dv = e^x \, dx \quad \Longrightarrow \quad v = \int dv = \int e^x \, dx = e^x$$
$$u = x \quad \Longrightarrow \quad du = dx.$$

Now, by the integration by parts formula, we have

$$\int u \, dv = uv - \int v \, du$$
$$\int x \, e^x \, dx = xe^x - \int e^x \, dx = xe^x - e^x + C.$$

Note in Example 1 that it is not necessary to include a constant integration when solving $v = \int e^x \, dx = e^x + C_1$. To illustrate this, we replace v by $v + C_1$ in the general formula to obtain

$$\int u \, dv = u(v + C_1) - \int (v + C_1) \, du = uv + C_1 u - \int C_1 \, du - \int v \, du$$

$$= uv + C_1 u - C_1 u - \int v \, du = uv - \int v \, du.$$

EXAMPLE 2 Integration by parts

Evaluate $\int x^2 \ln x \, dx$.

SOLUTION

In this case x^2 is more easily integrated than $\ln x$. Furthermore, the derivative of $\ln x$ is simpler than $\ln x$. Therefore, we let $dv = x^2 \, dx$.

$$dv = x^2 \, dx \implies v = \int x^2 \, dx = \frac{x^3}{3}$$

$$u = \ln x \implies du = \frac{1}{x} \, dx$$

Therefore, we have

$$\int x^2 \ln x \, dx = \frac{x^3}{3} \ln x - \int \left(\frac{x^3}{3}\right)\left(\frac{1}{x}\right) dx$$

$$= \frac{x^3}{3} \ln x - \frac{1}{3} \int x^2 \, dx$$

$$= \frac{x^3}{3} \ln x - \frac{x^3}{9} + C.$$

One unusual application of integration by parts involves integrands consisting of a single factor, such as $\int \ln x \, dx$ or $\int \arcsin x \, dx$. In such cases, we let $dv = dx$, as illustrated in the next example.

EXAMPLE 3 An integrand with a single term

Evaluate $\int_0^1 \arcsin x \, dx$.

SOLUTION

Letting $dv = dx$, we have

$$dv = dx \implies v = \int dx = x$$

$$u = \arcsin x \implies du = \frac{1}{\sqrt{1 - x^2}} \, dx.$$

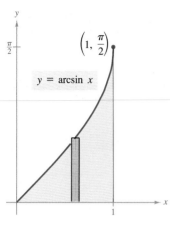

FIGURE 8.1

Therefore, we have

$$\int \arcsin x \, dx = x \arcsin x - \int \frac{x}{\sqrt{1 - x^2}} \, dx$$

$$= x \arcsin x + \frac{1}{2} \int (1 - x^2)^{-1/2}(-2x) \, dx$$

$$= x \arcsin x + \sqrt{1 - x^2} + C.$$

Now, using this antiderivative, we evaluate the definite integral as follows.

$$\int_0^1 \arcsin x \, dx = \left[x \arcsin x + \sqrt{1 - x^2} \right]_0^1 = \frac{\pi}{2} - 1$$

The area represented by this definite integral is shown in Figure 8.1. ▭

It may happen that an integral requires repeated application of the integration by parts formula. This is demonstrated in the next example.

EXAMPLE 4 Repeated application of integration by parts

Evaluate $\int x^2 \sin x \, dx$.

SOLUTION

We may consider x^2 and $\sin x$ to be equally easy to integrate. However, the derivative of x^2 becomes simpler, whereas the derivative of $\sin x$ does not. Therefore, we let $u = x^2$ and write

$$dv = \sin x \, dx \quad \Longrightarrow \quad v = \int \sin x \, dx = -\cos x$$

$$u = x^2 \quad \Longrightarrow \quad du = 2x \, dx$$

and it follows that

$$\int x^2 \sin x \, dx = -x^2 \cos x + \int 2x \cos x \, dx.$$

Now, we apply integration by parts to the new integral. We let $u = 2x$ and write

$$dv = \cos x \, dx \quad \Longrightarrow \quad v = \int \cos x \, dx = \sin x$$

$$u = 2x \quad \Longrightarrow \quad du = 2 \, dx$$

and it follows that

$$\int 2x \cos x \, dx = 2x \sin x - \int 2 \sin x \, dx = 2x \sin x + 2 \cos x + C.$$

Combining these two results, we have

$$\int x^2 \sin x \, dx = -x^2 \cos x + 2x \sin x + 2 \cos x + C. \quad ▭$$

When making repeated applications of integration by parts, you need to be careful not to interchange the substitutions in successive applications. For instance, in Example 4 our first substitution was $u = x^2$ and $dv = \sin x \, dx$. If, in the second application, we had switched the substitution to

$$dv = 2x \, dx \quad \Longrightarrow \quad v = \int 2x \, dx = x^2$$

$$u = \cos x \quad \Longrightarrow \quad du = -\sin x \, dx$$

we would have obtained

$$\int x^2 \sin x \, dx = -x^2 \cos x + \int 2x \cos x \, dx$$

$$= -x^2 \cos x + x^2 \cos x + \int x^2 \sin x \, dx$$

$$= \int x^2 \sin x \, dx$$

thus undoing the previous integration and returning to the *original* integral.

When making repeated applications of integration by parts, you should also watch for the appearance of a *constant multiple* of the original integral. This is illustrated in the next example.

EXAMPLE 5   Repeated application of integration by parts

Evaluate $\int e^x \cos 2x \, dx$.

SOLUTION

Our guidelines fail to help with a choice of u and dv, so we arbitrarily choose $dv = e^x \, dx$ and $u = \cos 2x$. (You might try verifying that the choice of $dv = \cos 2x \, dx$ and $u = e^x$ works equally well.)

$$dv = e^x \, dx \quad \Longrightarrow \quad v = \int e^x \, dx = e^x$$

$$u = \cos 2x \quad \Longrightarrow \quad du = -2 \sin 2x \, dx$$

Thus, it follows that

$$\int e^x \cos 2x \, dx = e^x \cos 2x + 2 \int e^x \sin 2x \, dx.$$

Making the same type of substitutions for the next application of integration by parts, we have

$$dv = e^x \, dx \quad \Longrightarrow \quad v = \int e^x \, dx = e^x$$

$$u = \sin 2x \quad \Longrightarrow \quad du = 2 \cos 2x \, dx$$

and it follows that

$$\int e^x \sin 2x \, dx = e^x \sin 2x - 2 \int e^x \cos 2x \, dx.$$

Therefore, we have

$$\int e^x \cos 2x \, dx = e^x \cos 2x + 2e^x \sin 2x - 4 \int e^x \cos 2x \, dx.$$

Now, since the right-hand integral is a constant multiple of the original integral, we add it to the left side of the equation to obtain

$$5 \int e^x \cos 2x \, dx = e^x \cos 2x + 2e^x \sin 2x$$

$$\int e^x \cos 2x \, dx = \frac{1}{5}e^x \cos 2x + \frac{2}{5}e^x \sin 2x + C. \qquad \text{Divide by 5} \quad \blacksquare$$

The integral in the next example is an important one. In Section 8.4 (Example 7) we will see that it is used to find the arc length of a parabolic segment.

EXAMPLE 6   Integration by parts

Evaluate $\int \sec^3 x \, dx$.

SOLUTION

The most complicated portion of the integrand that can be easily integrated is $\sec^2 x$. Letting $dv = \sec^2 x \, dx$ and $u = \sec x$, we have

$$dv = \sec^2 x \, dx \quad \Longrightarrow \quad v = \int \sec^2 x \, dx = \tan x$$

$$u = \sec x \quad \Longrightarrow \quad du = \sec x \tan x \, dx.$$

Therefore, we have

$$\int \sec^3 x \, dx = \sec x \tan x - \int \sec x \tan^2 x \, dx$$

$$= \sec x \tan x - \int \sec x (\sec^2 x - 1) \, dx$$

$$= \sec x \tan x - \int \sec^3 x \, dx + \int \sec x \, dx$$

$$2 \int \sec^3 x \, dx = \sec x \tan x + \int \sec x \, dx \qquad \text{Collect like integrals}$$

$$\int \sec^3 x \, dx = \frac{1}{2}\sec x \tan x + \frac{1}{2} \ln |\sec x + \tan x| + C. \qquad \blacksquare$$

Since we developed the integration by parts formula from the Product Rule for derivatives, we would expect many of our examples of this technique to involve a product. However, integration by parts is also useful in cases where the integrand is a quotient, as demonstrated in the next example.

EXAMPLE 7 An integrand involving a quotient

Evaluate

$$\int \frac{xe^x}{(x+1)^2}\,dx.$$

SOLUTION

Since $1/(x+1)^2$ is easily integrated, we make the following choices.

$$dv = \frac{dx}{(x+1)^2} \quad\Longrightarrow\quad v = \int \frac{dx}{(x+1)^2} = -\frac{1}{x+1}$$

$$u = xe^x \quad\Longrightarrow\quad du = (xe^x + e^x)\,dx = e^x(x+1)\,dx$$

Thus, we have

$$\int \frac{xe^x}{(x+1)^2}\,dx = xe^x\!\left(\frac{-1}{x+1}\right) - \int (x+1)e^x\!\left(\frac{-1}{x+1}\right)dx$$

$$= -\frac{xe^x}{x+1} + \int e^x\,dx$$

$$= -\frac{xe^x}{x+1} + e^x + C = \frac{e^x}{x+1} + C.$$

EXAMPLE 8 An application of integration by parts

Find the centroid of the region bounded by the graph of $y = \sin x$ and the x-axis, $0 \le x \le \pi/2$.

SOLUTION

Using the formulas presented in Section 7.7, together with Figure 8.2, we have

$$\text{area} = A = \int_0^{\pi/2} \sin x\,dx = -\cos x\,\Big]_0^{\pi/2} = 1$$

$$\bar{y} = \frac{1}{A}\int_0^{\pi/2} \frac{\sin x}{2}(\sin x)\,dx$$

$$= \frac{1}{4}\int_0^{\pi/2}(1 - \cos 2x)\,dx \qquad \text{Half-angle formula}$$

$$= \frac{1}{4}\left[x - \frac{\sin 2x}{2}\right]_0^{\pi/2} = \frac{\pi}{8}$$

$$\bar{x} = \frac{1}{A}\int_0^{\pi/2} x \sin x\,dx.$$

Now, using *integration by parts* on this integral, we let $dv = \sin x\,dx$, $u = x$, and obtain $v = -\cos x$ and $du = dx$. Thus, we have

$$\int x \sin x\,dx = -x \cos x + \int \cos x\,dx = -x \cos x + \sin x + C.$$

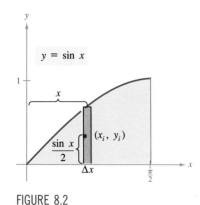

FIGURE 8.2

Now, we determine $\bar{x}$ to be

$$\bar{x} = \left[-x \cos x + \sin x \right]_0^{\pi/2} = 1.$$

Therefore, we conclude that the centroid of the region is $(1, \pi/8)$. ▭

As you gain experience in using integration by parts, your skill in determining u and dv will increase. In the following summary, we list several common integrals with suggestions for the choice of u and dv.

SUMMARY OF COMMON INTEGRALS USING INTEGRATION BY PARTS

1. $\int x^n e^{ax}\, dx,\qquad \int x^n \sin ax\, dx,\qquad \int x^n \cos ax\, dx$

    Let $u = x^n$ and $dv = e^{ax}\, dx$, $\sin ax\, dx$, or $\cos ax\, dx$. (Examples 1, 4)

2. $\int x^n \ln x\, dx,\qquad \int x^n \arcsin ax\, dx,\qquad \int x^n \arctan ax\, dx$

    Let $u = \ln x$, $\arcsin ax$, or $\arctan ax$ and $dv = x^n\, dx$. (Examples 2, 3)

3. $\int e^{ax} \sin bx\, dx,\qquad \int e^{ax} \cos bx\, dx$

    Let $u = \sin bx$ or $\cos bx$ and $dv = e^{ax}\, dx$. (Example 5)

Tabular method for repeated use of integration by parts

In problems involving repeated applications of integration by parts, a tabular method, illustrated in Example 9, can help to organize work. The method works well for integrals of the form

$$\int x^n \sin ax\, dx,\qquad \int x^n \cos ax\, dx,\qquad \text{and}\qquad \int x^n e^{ax}\, dx.$$

EXAMPLE 9   Using the Tabular Method

Evaluate $\int x^2 \sin 4x\, dx$.

SOLUTION

We begin as usual by letting $u = x^2$ and $dv = v'\, dx = \sin 4x\, dx$. Next, we create a table consisting of three columns as follows.

Alternate signs	u and its derivative	v' and its antiderivatives
+	x^2	$\sin 4x$
−	$2x$	$-\dfrac{1}{4}\cos 4x$
+	2	$-\dfrac{1}{16}\sin 4x$
−	0	$\dfrac{1}{64}\cos 4x$

Differentiate until you obtain 0 as a derivative.

Finally, the solution is given by multiplying the *signed* products of the diagonal entries to obtain

$$\int x^2 \sin 4x \, dx = -\frac{1}{4}x^2 \cos 4x + \frac{1}{8}x \sin 4x + \frac{1}{32} \cos 4x + C. \quad \blacksquare$$

EXERCISES for Section 8.2

In Exercises 1–34, evalute the given integral. (Note: Solve by the simplest method—not all require integration by parts.)

1. $\int xe^{2x} \, dx$ **2.** $\int x^2 e^{2x} \, dx$

3. $\int xe^{x^2} \, dx$ **4.** $\int x^2 e^{x^3} \, dx$

5. $\int xe^{-2x} \, dx$ **6.** $\int \frac{x}{e^x} \, dx$

7. $\int x^3 e^x \, dx$ **8.** $\int \frac{e^{1/t}}{t^2} \, dt$

9. $\int x^3 \ln x \, dx$ **10.** $\int \ln x \, dx$

11. $\int t \ln (t + 1) \, dt$ **12.** $\int \frac{1}{x(\ln x)^3} \, dx$

13. $\int (\ln x)^2 \, dx$ **14.** $\int \ln 3x \, dx$

15. $\int \frac{(\ln x)^2}{x} \, dx$ **16.** $\int \frac{\ln x}{x^2} \, dx$

17. $\int \frac{xe^{2x}}{(2x + 1)^2} \, dx$ **18.** $\int \frac{x^3 e^{x^2}}{(x^2 + 1)^2} \, dx$

19. $\int (x^2 - 1)e^x \, dx$ **20.** $\int \frac{\ln 2x}{x^2} \, dx$

21. $\int x\sqrt{x - 1} \, dx$ **22.** $\int x^2\sqrt{x - 1} \, dx$

23. $\int \frac{x^2}{\sqrt{2 + 3x}} \, dx$ **24.** $\int \frac{x}{\sqrt{2 + 3x}} \, dx$

25. $\int x \cos x \, dx$ **26.** $\int x^2 \cos x \, dx$

27. $\int x \sec^2 x \, dx$ **28.** $\int \theta \sec \theta \tan \theta \, d\theta$

29. $\int \arcsin 2x \, dx$ **30.** $\int \arccos x \, dx$

31. $\int \arctan x \, dx$ **32.** $\int \arctan \frac{x}{2} \, dx$

33. $\int e^{2x} \sin x \, dx$ **34.** $\int e^x \cos 2x \, dx$

In Exercises 35–40, evaluate the definite integral.

35. $\int_0^\pi x \sin 2x \, dx$ **36.** $\int_0^1 x \arcsin x^2 \, dx$

37. $\int_0^1 e^x \sin x \, dx$ **38.** $\int_0^1 x^2 e^x \, dx$

39. $\int_0^{\pi/2} x \cos x \, dx$ **40.** $\int_0^1 \ln (1 + x^2) \, dx$

In Exercises 41–46, use the tabular method for repeated applications of integration by parts to evaluate the given integral.

41. $\int x^2 e^{2x} \, dx$ **42.** $\int x^4 e^{-x} \, dx$

43. $\int x^3 \sin x \, dx$ **44.** $\int x^3 \cos 2x \, dx$

45. $\int x \sec^2 x \, dx$ **46.** $\int x^2(x - 2)^{3/2} \, dx$

47. Integrate $\int 2x\sqrt{2x - 3} \, dx$
  (a) by parts, letting $dv = \sqrt{2x - 3} \, dx$
  (b) by substitution, letting $u = \sqrt{2x - 3}$.

48. Integrate $\int x\sqrt{4 + x} \, dx$
  (a) by parts, letting $dv = \sqrt{4 + x} \, dx$
  (b) by substitution, letting $u = \sqrt{4 + x}$.

49. Integrate

$$\int \frac{x^3}{\sqrt{4 + x^2}} \, dx$$

  (a) by parts, letting $dv = \frac{x}{\sqrt{4 + x^2}} \, dx$
  (b) by substitution, letting $u = \sqrt{4 + x^2}$.

50. Integrate $\int x\sqrt{4 - x} \, dx$
  (a) by parts, letting $dv = \sqrt{4 - x} \, dx$
  (b) by substitution, letting $u = \sqrt{4 - x}$.

In Exercises 51–56, use integration by parts to verify the given formula. (Assume n is a positive integer.)

51. $\int x^n \sin x \, dx = -x^n \cos x + n \int x^{n-1} \cos x \, dx$

52. $\int x^n \cos x \, dx = x^n \sin x - n \int x^{n-1} \sin x \, dx$

53. $\int x^n \ln x \, dx = \frac{x^{n+1}}{(n + 1)^2}[-1 + (n + 1) \ln x] + C$

54. $\int x^n e^{ax} \, dx = \frac{x^n e^{ax}}{a} - \frac{n}{a} \int x^{n-1} e^{ax} \, dx$

55. $\int e^{ax} \sin bx \, dx = \frac{e^{ax}(a \sin bx - b \cos bx)}{a^2 + b^2} + C$

56. $\int e^{ax} \cos bx \, dx = \frac{e^{ax}(a \cos bx + b \sin bx)}{a^2 + b^2} + C$

In Exercises 57–60, evaluate the given integral by using the appropriate formula from Exercises 51–56.

57. $\int x^3 \ln x\, dx$

58. $\int x^2 \cos x\, dx$

59. $\int e^{2x} \cos 3x\, dx$

60. $\int x^3 e^{2x}\, dx$

In Exercises 61–64, find the area of the region bounded by the graphs of the given equations.

61. $y = xe^{-x}$, $y = 0$, $x = 4$

62. $y = \frac{1}{9}xe^{-x/3}$, $y = 0$, $x = 0$, $x = 3$

63. $y = e^{-x} \sin \pi x$, $y = 0$, $x = 0$, $x = 1$

64. $y = x \sin x$, $y = 0$, $x = 0$, $x = \pi$

65. Given the region bounded by the graphs of $y = \ln x$, $y = 0$, and $x = e$, find the following.
(a) the area of the region
(b) the volume of the solid generated by revolving the region about the x-axis
(c) the volume of the solid generated by revolving the region about the y-axis
(d) the centroid of the region

66. Find the volume of the solid generated by revolving the region bounded by $y = e^x$, $y = 0$, $x = 0$, and $x = 1$ about the y-axis.

67. A model for the ability M of a child to memorize, measured on a scale from 0 to 10, is given by

$$M = 1 + 1.6t \ln t, \quad 0 < t \le 4$$

where t is the child's age in years. Find the average value of this function
(a) between the child's first and second birthdays.
(b) between the child's third and fourth birthdays.

68. A company sells a seasonal product, and the model for the daily revenue from the product is

$$R = 410.5t^2 e^{-t/30} + 25,000, \quad 0 \le t \le 365$$

where t is the time in days.
(a) Find the average daily receipts during the first quarter, $0 \le t \le 91$.
(b) Find the average daily receipts during the fourth quarter, $274 \le t \le 365$.

In Exercises 69 and 70, find the present value P of a continuous income flow of $c(t)$ dollars per year if

$$P = \int_0^{t_1} c(t)e^{-rt}\, dt$$

where t_1 is time in years and r is the annual interest rate compounded continuously.

69. $c(t) = 100,000 + 4000t$, $r = 9\%$, $t_1 = 10$
70. $c(t) = 30,000 + 500t$, $r = 7\%$, $t_1 = 5$

71. A string stretched between two points $(0, 0)$ and $(0, 2)$ is plucked by displacing the string h units at its midpoint. The motion of the string is modeled by a **Fourier Sine Series** whose coefficients are given by

$$b_n = h \int_0^1 x \sin \frac{n\pi x}{2}\, dx + h \int_1^2 (-x + 2) \sin \frac{n\pi x}{2}\, dx.$$

Evaluate b_n.

72. A damping force affects the vibration of a spring so that the displacement of the spring is given by

$$y = e^{-4t} (\cos 2t + 5 \sin 2t).$$

Find the average value of y on the interval from $t = 0$ to $t = \pi$.

73. In the Chapter 8 Application, we developed the following equation for the velocity of a rocket,

$$v(t) = -32t + 10,000 \ln \frac{40,000}{40,000 - 300t}$$

where v is measured in ft/sec. Use integration by parts to find the position function for the rocket. (The initial height is $s(0) = 0$.) What is the height of the rocket when $t = 120$ seconds?

74. The velocity (in ft/sec) of a rocket whose initial mass is m (including fuel) is given by

$$v = gt + u \ln \frac{m}{m - rt}$$

where u is the expulsion speed of the fuel, r is the rate at which the fuel is consumed, and $g = -32$ is the acceleration due to gravity. Find the position equation for a rocket for which $m = 50,000$ lb, $u = 12,000$ ft/sec, and $r = 400$ lb/sec. What is the height of the rocket when $t = 100$ seconds? (Assume it was fired from ground level and is moving straight up.)

75. Find the fallacy in the following argument that $0 = 1$.

$$dv = dx \implies v = x$$

$$u = \frac{1}{x} \implies du = -\frac{1}{x^2}\, dx$$

$$0 + \int \frac{dx}{x} = \left(\frac{1}{x}\right)(x) - \int \left(-\frac{1}{x^2}\right)(x)\, dx$$

$$= 1 + \int \frac{dx}{x}$$

Hence $0 = 1$.

76. Is there a fallacy in the following evaluation of $\int \ln (x + 5)\, dx$?

$$dv = dx \quad \Longrightarrow \quad v = x + 5$$

$$u = \ln (x + 5) \quad \Longrightarrow \quad du = \frac{1}{x + 5}\, dx$$

$$\int \ln (x + 5)\, dx = (x + 5) \ln (x + 5) - \int dx$$

$$= (x + 5) \ln (x + 5) - x + C.$$

In Exercises 77 and 78, use Simpson's Rule on a computer or calculator (with $n = 12$) to approximate the area of the region bounded by the graphs of the given equations.

77. $y = \sqrt[3]{x}\sqrt{4 - x},\ y = 0,\ x = 0$

78. $y = \sin (\pi x)^2,\ y = 0,\ x = 0,\ x = 1$

8.3 Trigonometric Integrals

Integrals involving powers of sine and cosine ▪ Integrals involving powers of secant and tangent ▪ Integrals involving sine-cosine products with different angles

In this section we introduce techniques for evaluating integrals of the form

$$\int \sin^m x \cos^n x\, dx \quad \text{and} \quad \int \sec^m x \tan^n x\, dx$$

where either m or n is a positive integer. To find antiderivatives for these forms, we try to break them into combinations of trigonometric integrals to which we can apply the Power Rule. For instance, we can evaluate $\int \sin^5 x \cos x\, dx$ with the Power Rule by letting $u = \sin x$. Then, $du = \cos x\, dx$ and we have

$$\int \sin^5 x \cos x\, dx = \int u^5\, du = \frac{u^6}{6} + C = \frac{\sin^6 x}{6} + C.$$

Similarly, to evaluate $\int \sec^4 x \tan x\, dx$ by the Power Rule, we let $u = \sec x$. Then, $du = \sec x \tan x\, dx$ and we have

$$\int \sec^4 x \tan x\, dx = \int \sec^3 x\, (\sec x \tan x)\, dx = \frac{\sec^4 x}{4} + C.$$

To break up $\int \sin^m x \cos^n x\, dx$ into forms to which we can apply the Power Rule, we use the identities

$$\sin^2 x + \cos^2 x = 1 \qquad \text{Pythagorean identity}$$

$$\sin^2 x = \frac{1 - \cos 2x}{2} \qquad \text{Half-angle identity for } \sin^2 x$$

$$\cos^2 x = \frac{1 + \cos 2x}{2} \qquad \text{Half-angle identity for } \cos^2 x$$

as indicated in the following guidelines.

INTEGRALS INVOLVING SINE AND COSINE

1. If the power of the sine is odd and positive, save one sine factor and convert the remaining factors to cosine. Then, expand and integrate.

$$\int \sin^{\overset{\frown}{2k+1}} x \cos^n x \, dx = \int \overset{\text{convert to cosine}}{(\sin^2 x)^k} \cos^n x \overset{\text{save for } du}{\sin x} \, dx$$

$$= \int (1 - \cos^2 x)^k \cos^n x \sin x \, dx$$

2. If the power of the cosine is odd and positive, save one cosine factor and convert the remaining factors to sine. Then, expand and integrate.

$$\int \sin^m x \cos^{\overset{\frown}{2k+1}} x \, dx = \int \sin^m x \overset{\text{convert to sine}}{(\cos^2 x)^k} \overset{\text{save for } du}{\cos x} \, dx$$

$$= \int \sin^m x (1 - \sin^2 x)^k \cos x \, dx$$

3. If the powers of *both* the sine and cosine are even and nonnegative, make repeated use of the identities

$$\sin^2 x = \frac{1 - \cos 2x}{2} \quad \text{and} \quad \cos^2 x = \frac{1 + \cos 2x}{2}$$

to convert the integrand to odd powers of the cosine. Then, proceed as in case 2.

EXAMPLE 1   Power of sine is odd and positive

Evaluate $\int \sin^3 x \cos^4 x \, dx$.

SOLUTION

Expecting to use the Power Rule with $u = \cos x$, we *save one sine factor* to form du and convert the remaining sine factors to cosines, as follows.

$$\int \sin^3 x \cos^4 x \, dx = \int \sin^2 x \cos^4 x \, (\sin x) \, dx \qquad \text{Save } \sin x$$

$$= \int (1 - \cos^2 x) \cos^4 x \sin x \, dx \qquad \text{Identity for } \sin^2 x$$

$$= \int (\cos^4 x - \cos^6 x) \sin x \, dx$$

$$= \int \cos^4 x \sin x \, dx - \int \cos^6 x \sin x \, dx$$

$$= -\int \cos^4 x \, (-\sin x) \, dx + \int \cos^6 x \, (-\sin x) \, dx$$

$$= -\frac{\cos^5 x}{5} + \frac{\cos^7 x}{7} + C \qquad \text{Power Rule}$$

EXAMPLE 2 Power of cosine is odd and positive

Evaluate $\int \sin^2 x \cos^5 x \, dx$.

SOLUTION

Since the power of the cosine is odd and positive, we have

$$\int \sin^2 x \cos^5 x \, dx = \int \sin^2 x \cos^4 x \cos x \, dx \qquad \text{Save } \cos x$$

$$= \int \sin^2 x \, (\cos^2 x)^2 \cos x \, dx$$

$$= \int \sin^2 x \, (1 - \sin^2 x)^2 \cos x \, dx \qquad \text{Identity for } \cos^2 x$$

$$= \int \sin^2 x \, (1 - 2\sin^2 x + \sin^4 x) \cos x \, dx$$

$$= \int [\sin^2 x - 2\sin^4 x + \sin^6 x] \cos x \, dx$$

$$= \frac{\sin^3 x}{3} - \frac{2\sin^5 x}{5} + \frac{\sin^7 x}{7} + C. \qquad \text{Power Rule}$$

In Examples 1 and 2, *both* of the powers m and n happened to be positive integers. However, the method will work as long as either m or n is odd and positive. For instance, in the next example the power of the sine is 3, but the power of the cosine is $-\frac{1}{2}$.

EXAMPLE 3 Power of sine is odd and positive

Evaluate

$$\int_0^{\pi/3} \frac{\sin^3 x}{\sqrt{\cos x}} \, dx.$$

SOLUTION

Since the power of the sine is odd and positive, we have

$$\int_0^{\pi/3} \frac{\sin^3 x}{\sqrt{\cos x}} \, dx = \int_0^{\pi/3} \frac{(1 - \cos^2 x)(\sin x)}{\sqrt{\cos x}} \, dx$$

$$= \int_0^{\pi/3} [(\cos x)^{-1/2} \sin x - (\cos x)^{3/2} \sin x] \, dx$$

$$= \int_0^{\pi/3} [-(\cos x)^{-1/2} (-\sin x) + (\cos x)^{3/2} (-\sin x)] \, dx$$

$$= \left[-\frac{(\cos x)^{1/2}}{1/2} + \frac{(\cos x)^{5/2}}{5/2} \right]_0^{\pi/3}$$

$$= \left(-\frac{2}{\sqrt{2}} + \frac{2}{5(2^{5/2})} + 2 - \frac{2}{5} \right) = \frac{32 - 19\sqrt{2}}{20} \approx 0.256.$$

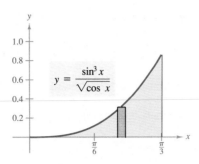

$$y = \frac{\sin^3 x}{\sqrt{\cos\ x}}$$

FIGURE 8.3

The region whose area is represented by this integral is shown in Figure 8.3.

EXAMPLE 4   Powers of sine and cosine are even and nonnegative

---

Evaluate $\int \cos^4 x\, dx$.

SOLUTION

Since m and n are both even and nonnegative ($m = 0$), we replace $\cos^4 x$ by $[(1 + \cos 2x)/2]^2$. Then, we have

$$\int \cos^4 x\, dx = \int \left(\frac{1 + \cos 2x}{2}\right)^2 dx \qquad \text{Identity for } \cos^2 x$$

$$= \int \left[\frac{1}{4} + \frac{\cos 2x}{2} + \frac{\cos^2 2x}{4}\right] dx$$

$$= \int \left[\frac{1}{4} + \frac{\cos 2x}{2} + \frac{1}{4}\left(\frac{1 + \cos 4x}{2}\right)\right] dx \qquad \text{Identity for } \cos^2 2x$$

$$= \frac{3}{8}\int dx + \frac{1}{4}\int 2\cos 2x\, dx + \frac{1}{32}\int 4\cos 4x\, dx$$

$$= \frac{3x}{8} + \frac{\sin 2x}{4} + \frac{\sin 4x}{32} + C.$$

In Example 4, suppose that we were to evaluate the *definite* integral from 0 to $\pi/2$. We would obtain

$$\int_0^{\pi/2} \cos^4 x\, dx = \left[\frac{3x}{8} + \frac{\sin 2x}{4} + \frac{\sin 4x}{32}\right]_0^{\pi/2} = \frac{3\pi}{16}.$$

Note that the only term that contributes to the solution is $3x/8$. This observation is generalized in the following formulas developed by John Wallis (1616–1703). Wallis did much of his work prior to Newton and Leibniz, and he influenced the thinking of both of these men.

John Wallis

WALLIS'S FORMULAS

If n is odd $(n \geq 3)$, then

$$\int_0^{\pi/2} \cos^n x \, dx = \left(\frac{2}{3}\right)\left(\frac{4}{5}\right)\left(\frac{6}{7}\right) \cdots \left(\frac{n-1}{n}\right).$$

If n is even $(n \geq 2)$, then

$$\int_0^{\pi/2} \cos^n x \, dx = \left(\frac{1}{2}\right)\left(\frac{3}{4}\right)\left(\frac{5}{6}\right) \cdots \left(\frac{n-1}{n}\right)\left(\frac{\pi}{2}\right).$$

These formulas are also valid if $\cos^n x$ is replaced by $\sin^n x$. (You are asked to prove both formulas in Exercises 85 and 86.)

Integrals involving powers of secant and tangent

To help you evaluate integrals of the form $\int \sec^m x \tan^n x \, dx$ we provide the following guidelines.

INTEGRALS INVOLVING SECANTS AND TANGENTS

1. If the power of the secant is even and positive, save a secant squared factor and convert the remaining factors to tangents. Then, expand and integrate.

$$\int \sec^{2k} x \tan^n x \, dx = \int (\sec^2 x)^{k-1} \tan^n x \sec^2 x \, dx$$

$$= \int (1 + \tan^2 x)^{k-1} \tan^n x \sec^2 x \, dx$$

2. If the power of the tangent is odd and positive, save a secant-tangent factor and convert the remaining factors to secants. Then, expand and integrate.

$$\int \sec^m x \tan^{2k+1} x \, dx = \int \sec^{m-1} x (\tan^2 x)^k \sec x \tan x \, dx$$

$$= \int \sec^{m-1} x (\sec^2 x - 1)^k \sec x \tan x \, dx$$

3. If there are no secant factors and the power of the tangent is even and positive, convert a tangent squared factor to secants; then expand and repeat if necessary.

$$\int \tan^n x \, dx = \int \tan^{n-2} x (\tan^2 x) \, dx = \int \tan^{n-2} x (\sec^2 x - 1) \, dx$$

$$= \int \tan^{n-2} x (\sec^2 x) \, dx - \int \tan^{n-2} x \, dx$$

4. If the integral is of the from $\int \sec^m x \, dx$, where m is odd and positive, use integration by parts as illustrated in Example 6 in the previous section.
5. If none of the first four cases apply, try converting to sines and cosines.

EXAMPLE 5 Power of tangent is odd and positive

Evaluate

$$\int \frac{\tan^3 x}{\sqrt{\sec x}}\, dx.$$

SOLUTION

Since the power of the tangent is odd and positive, we write

$$\int \frac{\tan^3 x}{\sqrt{\sec x}}\, dx = \int (\sec x)^{-1/2} \tan^3 x\, dx$$

$$= \int (\sec x)^{-3/2} (\tan^2 x)(\sec x \tan x)\, dx \qquad \text{Save } \sec x \tan x$$

$$= \int (\sec x)^{-3/2} (\sec^2 x - 1)(\sec x \tan x)\, dx$$

$$= \int [(\sec x)^{1/2} - (\sec x)^{-3/2}](\sec x \tan x)\, dx$$

$$= \frac{2}{3}(\sec x)^{3/2} + 2(\sec x)^{-1/2} + C. \qquad \text{Power Rule}$$

EXAMPLE 6 Power of secant is even and positive

Evaluate $\int \sec^4 3x \tan^3 3x\, dx$.

SOLUTION

Since the power of the secant is even, we save a secant squared factor to create du. If $u = \tan 3x$, then

$$du = 3 \sec^2 3x\, dx$$

and we write

$$\int \sec^4 3x \tan^3 3x\, dx = \int \sec^2 3x \tan^3 3x\, (\sec^2 3x)\, dx \qquad \text{Save } \sec^2 3x$$

$$= \int (1 + \tan^2 3x) \tan^3 3x\, (\sec^2 3x)\, dx$$

$$= \frac{1}{3} \int [\tan^3 3x + \tan^5 3x](3 \sec^2 3x)\, dx$$

$$= \frac{1}{3}\left[\frac{\tan^4 3x}{4} + \frac{\tan^6 3x}{6} \right] + C. \qquad \text{Power Rule}$$

REMARK   In Example 6, the power of the tangent is odd and positive. Thus, we could have evaluated the integral with the procedure described in case 2. In Exercise 61, you are asked to show that the results obtained by these two procedures differ only by a constant.

EXAMPLE 7   Power of tangent is even

Evaluate

$$\int_0^{\pi/4} \tan^4 x \, dx.$$

SOLUTION

Since there are no secant factors, we convert a tangent squared factor to secants.

$$\int \tan^4 x \, dx = \int \tan^2 x \, (\tan^2 x) \, dx = \int \tan^2 x \, (\sec^2 x - 1) \, dx$$

$$= \int \tan^2 x \, \sec^2 x \, dx - \int \tan^2 x \, dx$$

$$= \int \tan^2 x \, \sec^2 x \, dx - \int (\sec^2 x - 1) \, dx$$

$$= \frac{\tan^3 x}{3} - \tan x + x + C$$

Thus, the definite integral (representing the area shown in Figure 8.4) has the following value.

$$\int_0^{\pi/4} \tan^4 x \, dx = \left[\frac{\tan^3 x}{3} - \tan x + x \right]_0^{\pi/4} = \frac{\pi}{4} - \frac{2}{3} \approx 0.119 \qquad \blacksquare$$

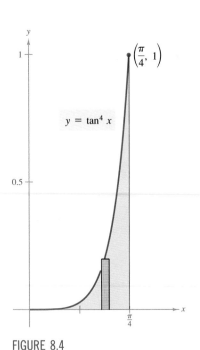

$y = \tan^4 x$

$\left(\frac{\pi}{4}, 1 \right)$

FIGURE 8.4

For integrals involving powers of cotangents and cosecants, we follow a strategy similar to that used for powers of tangents and secants, as illustrated in the next example.

EXAMPLE 8   Powers of cotangent and cosecant

Evaluate $\int \csc^4 x \, \cot^4 x \, dx$.

SOLUTION

Since the power of the cosecant is even, we write

$$\int \csc^4 x \cot^4 x \, dx = \int \csc^2 x \cot^4 x \, (\csc^2 x) \, dx$$

$$= \int (1 + \cot^2 x) \cot^4 x \, (\csc^2 x) \, dx$$

$$= -\int (\cot^4 x + \cot^6 x)(-\csc^2 x) \, dx$$

$$= -\left(\frac{\cot^5 x}{5} + \frac{\cot^7 x}{7}\right) + C$$

$$= -\frac{\cot^5 x}{5} - \frac{\cot^7 x}{7} + C.$$

EXAMPLE 9   Converting to sines and cosines

Evaluate

$$\int \frac{\sec x}{\tan^2 x} \, dx.$$

SOLUTION

Since the first two cases do not apply, we try converting the integrand to sines and cosines. In this case, we are able to integrate the resulting powers of sine and cosine as follows.

$$\int \frac{\sec x}{\tan^2 x} \, dx = \int \left(\frac{1}{\cos x}\right)\left(\frac{\cos x}{\sin x}\right)^2 \, dx = \int (\sin x)^{-2} (\cos x) \, dx$$

$$= -(\sin x)^{-1} + C = -\csc x + C$$

Integrals involving sine-cosine products with different angles

Occasionally, we encounter integrals involving the product of sines and cosines of two *different* angles. In such instances we use the following product-to-sum identities.

$$\sin mx \sin nx = \frac{1}{2}(\cos [(m - n)x] - \cos [(m + n)x])$$

$$\sin mx \cos nx = \frac{1}{2}(\sin [(m - n)x] + \sin [(m + n)x])$$

$$\cos mx \cos nx = \frac{1}{2}(\cos [(m - n)x] + \cos [(m + n)x])$$

EXAMPLE 10 Using product-to-sum identities

Evaluate $\int \sin 5x \cos 4x \, dx$.

SOLUTION

Considering the second product-to-sum identity, we write

$$\int \sin 5x \cos 4x \, dx = \frac{1}{2} \int (\sin 9x + \sin x) \, dx$$

$$= \frac{1}{2} \left[-\frac{\cos 9x}{9} - \cos x \right] + C$$

$$= -\frac{\cos 9x}{18} - \frac{\cos x}{2} + C.$$

As you review this section, concentrate on the *general pattern* followed in evaluating trigonometric integrals.

EXERCISES for Section 8.3

In Exercises 1–50, evaluate the given integral.

1. $\int \cos^3 x \sin x \, dx$

2. $\int \cos^3 x \sin^2 x \, dx$

3. $\int \sin^5 2x \cos 2x \, dx$

4. $\int \sin^3 x \, dx$

5. $\int \sin^5 x \cos^2 x \, dx$

6. $\int \cos^3 \frac{x}{3} \, dx$

7. $\int \cos^2 3x \, dx$

8. $\int \sin^2 2x \, dx$

9. $\int \sin^4 \pi x \, dx$

10. $\int \cos^4 \frac{x}{2} \, dx$

11. $\int \sin^2 x \cos^2 x \, dx$

12. $\int \sin^2 \frac{x}{2} \cos^2 \frac{x}{2} \, dx$

13. $\int x \sin^2 x \, dx$ (Integration by parts)

14. $\int x^2 \sin^2 x \, dx$ (Integration by parts)

15. $\int \sec 3x \, dx$

16. $\int \sec^2 (2x - 1) \, dx$

17. $\int \sec^4 5x \, dx$

18. $\int \sec^6 \frac{x}{2} \, dx$

19. $\int \sec^3 \pi x \, dx$ (Integration by parts)

20. $\int \sec^5 \pi x \, dx$ (Integration by parts)

21. $\int \tan^3 (1 - x) \, dx$

22. $\int \tan^2 x \, dx$

23. $\int \tan^5 \frac{x}{4} \, dx$

24. $\int \tan^3 \frac{\pi x}{2} \sec^2 \frac{\pi x}{2} \, dx$

25. $\int \sec^2 x \tan x \, dx$

26. $\int \csc^2 3x \cot 3x \, dx$

27. $\int \tan^2 x \sec^2 x \, dx$

28. $\int \tan^5 2x \sec^2 2x \, dx$

29. $\int \sec^5 \pi x \tan \pi x \, dx$

30. $\int \sec^4 (1 - x) \tan (1 - x) \, dx$

31. $\int \sec^6 4x \tan 4x \, dx$

32. $\int \sec^2 \frac{x}{2} \tan \frac{x}{2} \, dx$

33. $\int \sec^3 x \tan x \, dx$

34. $\int \tan^3 3x \, dx$

35. $\int \tan^3 3x \sec 3x \, dx$

36. $\int \sqrt{\tan x} \sec^4 x \, dx$

37. $\int \cot^3 2x \, dx$

38. $\int \tan^4 \frac{x}{2} \sec^4 \frac{x}{2} \, dx$

39. $\int \csc^4 \theta \, d\theta$

40. $\int \tan^3 t \sec^3 t \, dt$

41. $\int \frac{\cot^2 t}{\csc t} \, dt$

42. $\int \frac{\cot^3 t}{\csc t} \, dt$

43. $\int \sin 3x \cos 2x \, dx$

44. $\int \cos 3\theta \cos (-2\theta) \, d\theta$

45. $\int \sin \theta \sin 3\theta \, d\theta$

46. $\int x \tan^2 x^2 \, dx$

47. $\int \frac{1}{\sec x \tan x} \, dx$

48. $\int \frac{\sin^2 x - \cos^2 x}{\cos x} \, dx$

49. $\int (\tan^4 t - \sec^4 t) \, dt$

50. $\int \frac{1 - \sec t}{\cos t - 1} \, dt$

In Exercises 51–60, evaluate the given definite integral.

51. $\int_{-\pi}^{\pi} \sin^2 x \, dx$

52. $\int_{0}^{\pi/4} \tan^2 x \, dx$

53. $\int_{0}^{\pi/4} \tan^3 x \, dx$

54. $\int_{0}^{\pi/4} \sec^2 t \sqrt{\tan t} \, dt$

55. $\int_{0}^{\pi/2} \frac{\cos t}{1 + \sin t} \, dt$

56. $\int_{-\pi}^{\pi} \sin 3\theta \cos \theta \, d\theta$

57. $\int_{0}^{\pi/4} \sin 2\theta \sin 3\theta \, d\theta$

58. $\int_{0}^{\pi/2} (1 - \cos \theta)^2 \, d\theta$

59. $\int_{-\pi/2}^{\pi/2} \cos^3 x \, dx$

60. $\int_{-\pi/2}^{\pi/2} (\sin^2 x + 1) \, dx$

In Exercises 61 and 62, find the indefinite integral in two different ways and show that the results differ only by a constant.

61. $\int \sec^4 3x \tan^3 3x \, dx$

62. $\int \sec^2 x \tan x \, dx$

In Exercises 63 and 64, find the area of the region bounded by the graphs of the given equations.

63. $y = \sin^2 \pi x, \; y = 0, \; x = 0, \; x = 1$

64. $y = \tan^2 x, \; y = \dfrac{4x}{\pi}, \; x = -\dfrac{\pi}{4}, \; x = \dfrac{\pi}{4}$

In Exercises 65 and 66, find the volume of the solid generated by revolving the region bounded by the graphs of the given equations about the x-axis.

65. $y = \tan x, \; y = 0, \; x = -\dfrac{\pi}{4}, \; x = \dfrac{\pi}{4}$

66. $y = \sin x \cos^2 x, \; y = 0, \; x = 0, \; x = \dfrac{\pi}{2}$

In Exercises 67 and 68, for the region bounded by the graphs of the given equations, find (a) the volume of the solid formed by revolving the region about the x-axis and (b) the centroid of the region.

67. $y = \sin x, \; y = 0, \; x = 0, \; x = \pi$

68. $y = \cos x, \; y = 0, \; x = 0, \; x = \dfrac{\pi}{2}$

In Exercises 69–74, verify Wallis's Formulas for the given integral.

69. $\int_{0}^{\pi/2} \cos^3 x \, dx = \dfrac{2}{3}$

70. $\int_{0}^{\pi/2} \cos^5 x \, dx = \dfrac{8}{15}$

71. $\int_{0}^{\pi/2} \cos^7 x \, dx = \dfrac{16}{35}$

72. $\int_{0}^{\pi/2} \sin^2 x \, dx = \dfrac{\pi}{4}$

73. $\int_{0}^{\pi/2} \sin^4 x \, dx = \dfrac{3\pi}{16}$

74. $\int_{0}^{\pi/2} \sin^6 x \, dx = \dfrac{5\pi}{32}$

In Exercises 75–78, use integration by parts to verify the given reduction formula.

75. $\int \sin^n x \, dx = -\dfrac{\sin^{n-1} x \cos x}{n} + \dfrac{n-1}{n} \int \sin^{n-2} x \, dx$

76. $\int \cos^n x \, dx = \dfrac{\cos^{n-1} x \sin x}{n} + \dfrac{n-1}{n} \int \cos^{n-2} x \, dx$

77. $\int \cos^m x \sin^n x \, dx = -\dfrac{\cos^{m+1} x \sin^{n-1} x}{m+n}$
$\qquad + \dfrac{n-1}{m+n} \int \cos^m x \sin^{n-2} x \, dx$

78. $\int \sec^n x \, dx = \dfrac{1}{n-1} \sec^{n-2} x \tan x$
$\qquad + \dfrac{n-2}{n-1} \int \sec^{n-2} x \, dx$

In Exercises 79–84, find the indefinite integral by using the appropriate formula from Exercises 75–78.

79. $\int \sin^5 x \, dx$

80. $\int \cos^4 x \, dx$

81. $\int \cos^6 x \, dx$

82. $\int \sin^4 x \cos^2 x \, dx$

83. $\int \sin^2 \pi x \cos^2 \pi x \, dx$

84. $\int \sin^4 \dfrac{2\pi x}{5} \, dx$

85. (Wallis's Formula) Use the result of Exercise 76 to prove that if n is odd $(n \geq 3)$, then
$$\int_{0}^{\pi/2} \cos^n x \, dx = \left(\frac{2}{3}\right)\left(\frac{4}{5}\right)\left(\frac{6}{7}\right) \cdots \left(\frac{n-1}{n}\right).$$

86. (Wallis's Formula) Use the result of Exercise 76 to prove that if n is even $(n \geq 2)$, then
$$\int_{0}^{\pi/2} \cos^n x \, dx = \left(\frac{1}{2}\right)\left(\frac{3}{4}\right)\left(\frac{5}{6}\right) \cdots \left(\frac{n-1}{n}\right)\left(\frac{\pi}{2}\right).$$

In Exercises 87 and 88, use Simpson's Rule on a computer or calculator (with $n = 12$) to approximate the volume of the solid generated by revolving the region bounded by the graphs of the given equations about the y-axis.

87. $y = \sqrt{x} \sin \dfrac{\pi x}{2}, \; y = 0, \; x = 0, \; x = 2$

88. $y = \dfrac{10}{\sqrt{x} e^x}, \; y = 0, \; x = 1, \; x = 4$

89. The **inner product** of two functions f and g on $[a, b]$ is given by $\langle f, g \rangle = \int_{a}^{b} f(x)g(x) \, dx$. Two distinct functions f and g are said to be **orthogonal** if $\langle f, g \rangle = 0$.
  (a) Show that the functions given by $f_n(x) = \cos nx$, $n = 0, 1, 2, \ldots$ form an orthogonal family on $[0, \pi]$.
  (b) Show that the functions given by
$$f_n(x) = \sin nx, \quad n = 1, 2, 3, \ldots$$
$$g_n(x) = \cos nx, \quad n = 0, 1, 2, \ldots$$
form an orthogonal family on $[-\pi, \pi]$.

8.4 Trigonometric Substitution

Trigonometric substitution

Now that we can evaluate integrals involving powers of trigonometric functions, we can use the method of **trigonometric substitution** to evaluate integrals involving the radicals

$$\sqrt{a^2 - u^2}, \qquad \sqrt{a^2 + u^2}, \qquad \text{and} \qquad \sqrt{u^2 - a^2}.$$

Our objective with trigonometric substitution is to eliminate the radical in the integrand. We do this with the Pythagorean identities

$$\cos^2 \theta = 1 - \sin^2 \theta$$
$$\sec^2 \theta = 1 + \tan^2 \theta$$
$$\tan^2 \theta = \sec^2 \theta - 1.$$

For example, if $a > 0$, we let $u = a \sin \theta$, where $-\pi/2 \leq \theta \leq \pi/2$. Then

$$\sqrt{a^2 - u^2} = \sqrt{a^2 - a^2 \sin^2 \theta} = \sqrt{a^2(1 - \sin^2 \theta)}$$
$$= \sqrt{a^2 \cos^2 \theta} = a \cos \theta.$$

Note that $\cos \theta \geq 0$, since $-\pi/2 \leq \theta \leq \pi/2$.

TRIGONOMETRIC SUBSTITUTION ($a > 0$)

1. For integrals involving $\sqrt{a^2 - u^2}$, let $u = a \sin \theta$. Then $\sqrt{a^2 - u^2} = a \cos \theta$ where $-\pi/2 \leq \theta \leq \pi/2$.

2. For integrals involving $\sqrt{a^2 + u^2}$, let $u = a \tan \theta$. Then $\sqrt{a^2 + u^2} = a \sec \theta$ where $-\pi/2 < \theta < \pi/2$.

3. For integrals involving $\sqrt{u^2 - a^2}$, let $u = a \sec \theta$. Then $\sqrt{u^2 - a^2} = \pm a \tan \theta$ where $0 \leq \theta < \pi/2$ or $\pi/2 < \theta \leq \pi$. Use the positive value if $u > a$ and the negative value if $u < -a$.

EXAMPLE 1 Trigonometric substitution: $u = a \sin \theta$

Evaluate

$$\int \frac{dx}{x^2\sqrt{9-x^2}}.$$

SOLUTION

First, we observe that none of the basic integration formulas in Section 8.1 apply. To use trigonometric substitution, we observe that $\sqrt{9-x^2}$ is of the form $\sqrt{a^2-u^2}$, with the corresponding triangle shown in Figure 8.5. Hence, we use the substitution

$$x = a \sin \theta = 3 \sin \theta$$

which implies that

$$dx = 3 \cos \theta \, d\theta, \qquad \sqrt{9-x^2} = 3 \cos \theta, \qquad \text{and} \qquad x^2 = 9 \sin^2 \theta.$$

Therefore,

$$
\begin{aligned}
\int \frac{dx}{x^2\sqrt{9-x^2}} &= \int \frac{3 \cos \theta \, d\theta}{(9 \sin^2 \theta)(3 \cos \theta)} \\
&= \frac{1}{9} \int \frac{d\theta}{\sin^2 \theta} \\
&= \frac{1}{9} \int \csc^2 \theta \, d\theta \\
&= -\frac{1}{9} \cot \theta + C \\
&= -\frac{1}{9}\left(\frac{\sqrt{9-x^2}}{x}\right) + C \qquad \text{Substitute for } \cot \theta \\
&= -\frac{\sqrt{9-x^2}}{9x} + C.
\end{aligned}
$$

Note that we used the triangle in Figure 8.5 to convert from θ's back to x's as follows.

$$\cot \theta = \frac{\text{adj}}{\text{opp}} = \frac{\sqrt{9-x^2}}{x}$$

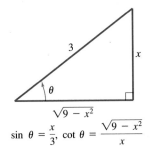

$\sin \theta = \dfrac{x}{3}, \ \cot \theta = \dfrac{\sqrt{9-x^2}}{x}$

FIGURE 8.5

In Section 6.9, we saw how the inverse hyperbolic functions can be used to evaluate the integrals

$$\int \frac{du}{\sqrt{u^2 \pm a^2}}, \qquad \int \frac{du}{a^2 - u^2}, \qquad \text{and} \qquad \int \frac{du}{u\sqrt{a^2 \pm u^2}}.$$

You can also evaluate these integrals using trigonometric substitution. This is illustrated in the next example.

EXAMPLE 2 Trigonometric substitution: $u = a \tan \theta$

Evaluate

$$\int \frac{dx}{\sqrt{4x^2 + 1}}.$$

SOLUTION

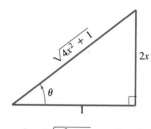

$\sec \theta = \sqrt{4x^2 + 1}$, $\tan \theta = 2x$

FIGURE 8.6

We let $u = 2x$, $a = 1$, and $2x = \tan \theta$, as shown in Figure 8.6. Then,

$$dx = \frac{1}{2} \sec^2 \theta \, d\theta \qquad \text{and} \qquad \sqrt{4x^2 + 1} = \sec \theta .$$

Therefore,

$$\int \frac{1}{\sqrt{4x^2 + 1}} \, dx = \frac{1}{2} \int \frac{\sec^2 \theta \, d\theta}{\sec \theta} = \frac{1}{2} \int \sec \theta \, d\theta$$

$$= \frac{1}{2} \ln |\sec \theta + \tan \theta| + C$$

$$= \frac{1}{2} \ln |\sqrt{4x^2 + 1} + 2x| + C.$$

We can extend the use of trigonometric substitution to cover integrals involving expressions such as $(a^2 - u^2)^{n/2}$ by writing the expression as

$$(a^2 - u^2)^{n/2} = (\sqrt{a^2 - u^2})^n.$$

This procedure is demonstrated in the next example.

EXAMPLE 3 Trigonometric substitution: rational powers

Evaluate

$$\int \frac{dx}{(x^2 + 1)^{3/2}}.$$

SOLUTION

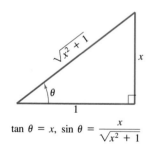

$\tan \theta = x$, $\sin \theta = \dfrac{x}{\sqrt{x^2 + 1}}$

FIGURE 8.7

We write $(x^2 + 1)^{3/2}$ in the form $(\sqrt{x^2 + 1})^3$, and let $a = 1$ and $u = x = \tan \theta$, as shown in Figure 8.7. Then,

$$dx = \sec^2 \theta \, d\theta \qquad \text{and} \qquad \sqrt{x^2 + 1} = \sec \theta.$$

Therefore,

$$\int \frac{dx}{(x^2 + 1)^{3/2}} = \int \frac{dx}{(\sqrt{x^2 + 1})^3} = \int \frac{\sec^2 \theta \, d\theta}{\sec^3 \theta}$$

$$= \int \frac{d\theta}{\sec \theta} = \int \cos \theta \, d\theta = \sin \theta + C$$

$$= \frac{x}{\sqrt{x^2 + 1}} + C. \qquad \text{Substitute for } \sin \theta$$

For definite integrals, it is often convenient to determine the integration limits for θ and thus avoid converting back to x. You might want to review this procedure in Section 5.5, Examples 9 and 10.

EXAMPLE 4 Converting the limits of integration

Evaluate

$$\int_{\sqrt{3}}^{2} \frac{\sqrt{x^2 - 3}}{x} \, dx.$$

SOLUTION

Since $\sqrt{x^2 - 3}$ has the form $\sqrt{u^2 - a^2}$, we consider $u = x$, $a = \sqrt{3}$, and let $x = \sqrt{3} \sec \theta$, as shown in Figure 8.8. Then,

$$dx = \sqrt{3} \sec \theta \tan \theta \, d\theta \quad \text{and} \quad \sqrt{x^2 - 3} = \sqrt{3} \tan \theta.$$

To determine the upper and lower limits of integration, we use the substitution $x = \sqrt{3} \sec \theta$ as follows.

Lower limit	*Upper limit*
When $x = \sqrt{3}$, $\sec \theta = 1$	When $x = 2$, $\sec \theta = \dfrac{2}{\sqrt{3}}$
and $\theta = 0$.	and $\theta = \dfrac{\pi}{6}$.

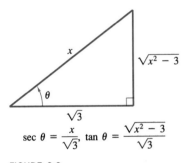

$\sec \theta = \dfrac{x}{\sqrt{3}}$, $\tan \theta = \dfrac{\sqrt{x^2 - 3}}{\sqrt{3}}$

FIGURE 8.8

Therefore, we have

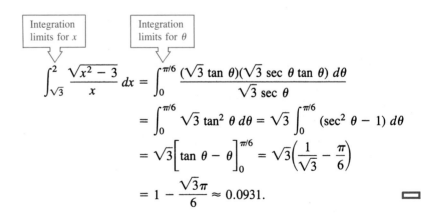

Integration limits for x

Integration limits for θ

$$\int_{\sqrt{3}}^{2} \frac{\sqrt{x^2 - 3}}{x} \, dx = \int_{0}^{\pi/6} \frac{(\sqrt{3} \tan \theta)(\sqrt{3} \sec \theta \tan \theta) \, d\theta}{\sqrt{3} \sec \theta}$$

$$= \int_{0}^{\pi/6} \sqrt{3} \tan^2 \theta \, d\theta = \sqrt{3} \int_{0}^{\pi/6} (\sec^2 \theta - 1) \, d\theta$$

$$= \sqrt{3} \left[\tan \theta - \theta \right]_{0}^{\pi/6} = \sqrt{3} \left(\frac{1}{\sqrt{3}} - \frac{\pi}{6} \right)$$

$$= 1 - \frac{\sqrt{3}\pi}{6} \approx 0.0931.$$

REMARK In Example 4, we leave as an exercise the alternate procedure of converting back to variable x and evaluating the antiderivative at the original limits of integration. (See Exercise 47.)

When using trigonometric substitution to evaluate definite integrals, you must be careful to check that the values of θ lie in the intervals discussed at the beginning of this section. For instance, if in Example 4 we had been asked to evaluate the definite integral

$$\int_{-2}^{-\sqrt{3}} \frac{\sqrt{x^2 - 3}}{x}\, dx$$

then using $u = x$ and $a = \sqrt{3}$ in the interval $[-2, -\sqrt{3}]$ would imply that $u < -a$. Thus, when determining the upper and lower limits of integration, we would have to choose θ such that $\pi/2 < \theta \le \pi$. In this case the integral would be evaluated as follows.

$$\int_{-2}^{-\sqrt{3}} \frac{\sqrt{x^2 - 3}}{x}\, dx = \int_{5\pi/6}^{\pi} \frac{(-\sqrt{3}\tan\theta)(\sqrt{3}\sec\theta\tan\theta)\,d\theta}{\sqrt{3}\sec\theta}$$

$$= \int_{5\pi/6}^{\pi} -\sqrt{3}\tan^2\theta\,d\theta$$

$$= -\sqrt{3}\int_{5\pi/6}^{\pi} (\sec^2\theta - 1)\,d\theta$$

$$= -\sqrt{3}\Big[\tan\theta - \theta\Big]_{5\pi/6}^{\pi}$$

$$= -\sqrt{3}\Big[(0 - \pi) - \left(-\frac{1}{\sqrt{3}} - \frac{5\pi}{6}\right)\Big]$$

$$= -1 + \frac{\sqrt{3}\pi}{6} \approx -0.0931$$

We can further expand the range of problems to which trigonometric substitution applies by employing the technique of completing the square. (To review this procedure, see Examples 5, 6, and 7 in Section 6.8.) We further demonstrate the technique of completing the square in the next example.

EXAMPLE 5 Completing the square

Evaluate

$$\int \frac{dx}{(x^2 - 4x)^{3/2}}, \quad x > 4.$$

SOLUTION

Completing the square, we have

$$x^2 - 4x = x^2 - 4x + 4 - 4 = (x - 2)^2 - 2^2 = u^2 - a^2.$$

We let $a = 2$ and $u = x - 2 = 2\sec\theta$ as shown in Figure 8.9. (Note that $x > 4$ implies that $u > a$.) Then

$$dx = 2\sec\theta\tan\theta\,d\theta \quad \text{and} \quad \sqrt{(x - 2)^2 - 2^2} = 2\tan\theta.$$

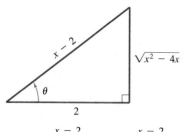

$$\sec\theta = \frac{x - 2}{2}, \quad \csc\theta = \frac{x - 2}{\sqrt{x^2 - 4x}}$$

FIGURE 8.9

Therefore,

$$\int \frac{dx}{(x^2 - 4x)^{3/2}} = \int \frac{2 \sec \theta \tan \theta \, d\theta}{(2 \tan \theta)^3} = \frac{2}{8} \int \frac{\sec \theta}{\tan^2 \theta} \, d\theta$$

$$= \frac{1}{4} \int (\sin \theta)^{-2}(\cos \theta) \, d\theta = -\frac{1}{4} \csc \theta + C$$

$$= -\frac{x - 2}{4\sqrt{x^2 - 4x}} + C. \qquad \text{Substitute for csc } \theta$$

In Section 8.5, we will encounter integrals involving rational functions. Some of these important integral forms can be solved using trigonometric substitution as illustrated in the next example.

EXAMPLE 6 Trigonometric substitution for rational functions

Evaluate

$$\int \frac{dx}{(x^2 + 1)^2}.$$

SOLUTION

Letting $a = 1$ and $u = x = \tan \theta$ as shown in Figure 8.10, we have

$$dx = \sec^2 \theta \, d\theta \qquad \text{and} \qquad x^2 + 1 = \sec^2 \theta.$$

Therefore,

$$\int \frac{dx}{(x^2 + 1)^2} = \int \frac{\sec^2 \theta \, d\theta}{\sec^4 \theta}$$

$$= \int \cos^2 \theta \, d\theta$$

$$= \frac{1}{2} \int (1 + \cos 2\theta) \, d\theta$$

$$= \frac{\theta}{2} + \frac{\sin 2\theta}{4} + C$$

$$= \frac{\theta}{2} + \frac{1}{2} \sin \theta \cos \theta + C$$

$$= \frac{\arctan x}{2} + \frac{1}{2}\left(\frac{x}{\sqrt{x^2 + 1}}\right)\left(\frac{1}{\sqrt{x^2 + 1}}\right) + C$$

$$= \frac{1}{2}\left(\arctan x + \frac{x}{x^2 + 1}\right) + C.$$

$\tan \theta = x$, $\sin \theta = \dfrac{x}{\sqrt{x^2 + 1}}$,

$\cos \theta = \dfrac{1}{\sqrt{x^2 + 1}}$

FIGURE 8.10

Trigonometric substitution can be used to evaluate the three integrals listed in the following theorem. We will have occasion to use these integrals several times in the remainder of the text, and when we do, we will simply refer to this theorem. If you prefer not to memorize these formulas, you should remember that they can be generated by trigonometric substitution.

**THEOREM 8.2
SPECIAL INTEGRATION
FORMULAS ($a > 0$)**

1. $\int \sqrt{a^2 - u^2}\, du = \frac{1}{2}\left(a^2 \arcsin \frac{u}{a} + u\sqrt{a^2 - u^2}\right) + C$

2. $\int \sqrt{u^2 - a^2}\, du = \frac{1}{2}\left(u\sqrt{u^2 - a^2} - a^2 \ln |u + \sqrt{u^2 - a^2}|\right) + C, \quad u > a$

3. $\int \sqrt{u^2 + a^2}\, du = \frac{1}{2}\left(u\sqrt{u^2 + a^2} + a^2 \ln |u + \sqrt{u^2 + a^2}|\right) + C$

PROOF We outline the proof of part 1 and leave the second and third parts as an exercise (see Exercise 65). For the first integral, we let $u = a \sin \theta$. Then $du = a \cos \theta\, d\theta$ and $\sqrt{a^2 - u^2} = a \cos \theta$ and we have

$$\int \sqrt{a^2 - u^2}\, du = \int a^2 \cos^2 \theta\, d\theta = a^2 \int \frac{1 + \cos 2\theta}{2}\, d\theta$$

$$= \frac{a^2}{2}\left(\theta + \frac{1}{2} \sin 2\theta\right) + C = \frac{a^2}{2}(\theta + \sin \theta \cos \theta) + C.$$

From Figure 8.11, it follows that

$$\int \sqrt{a^2 - u^2}\, du = \frac{a^2}{2}\left[\arcsin \frac{u}{a} + \left(\frac{u}{a}\right)\left(\frac{\sqrt{a^2 - u^2}}{a}\right)\right] + C$$

$$= \frac{a^2}{2}\left[\frac{a^2 \arcsin (u/a) + u\sqrt{a^2 - u^2}}{a^2}\right] + C$$

$$= \frac{1}{2}\left[a^2 \arcsin \frac{u}{a} + u\sqrt{a^2 - u^2}\right] + C.$$

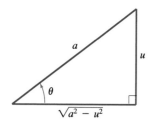

$\sin \theta = \dfrac{u}{a}, \; \cos \theta = \dfrac{\sqrt{a^2 - u^2}}{a}$

FIGURE 8.11

There are many practical applications involving trigonometric substitution, as illustrated in the last two examples.

EXAMPLE 7 An application involving arc length

Find the arc length of the graph of $f(x) = \frac{1}{2}x^2$ from $x = 0$ to $x = 1$, as shown in Figure 8.12.

SOLUTION

From the formula for arc length s, we have

$$s = \int_0^1 \sqrt{1 + [f'(x)]^2}\, dx = \int_0^1 \sqrt{1 + x^2}\, dx.$$

Letting $a = 1$ and $x = \tan \theta$, we have

$$dx = \sec^2 \theta\, d\theta \quad \text{and} \quad \sqrt{1 + x^2} = \sec \theta.$$

Moreover, the upper and lower limits are as follows.

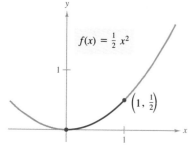

$f(x) = \frac{1}{2}x^2$

$\left(1, \frac{1}{2}\right)$

FIGURE 8.12

Lower limit	Upper limit
When $x = \tan\theta = 0$,	When $x = \tan\theta = 1$,
$\theta = 0$.	$\theta = \dfrac{\pi}{4}$.

Therefore, the arc length is given by

$$s = \int_0^1 \sqrt{1 + x^2}\, dx = \int_0^{\pi/4} \sec^3\theta\, d\theta \qquad \text{Example 6, Section 8.2}$$

$$= \frac{1}{2}\left[\sec\theta\tan\theta + \ln\left|\sec\theta + \tan\theta\right|\right]_0^{\pi/4}$$

$$= \frac{1}{2}[\sqrt{2} + \ln(\sqrt{2} + 1)] \approx 1.148.$$

REMARK Try using Theorem 8.2 to evaluate the integral in Example 7.

EXAMPLE 8 A comparison of two fluid forces

A sealed barrel of oil (weighing 48 pounds per cubic foot) is floating in sea water (weighing 64 pounds per cubic foot), as shown in Figures 8.13 and 8.14. (Note that the barrel is not completely full of oil—on its side, the top 0.2 feet of the barrel is empty.) Compare the fluid force against one end of the barrel from the inside and from the outside.

SOLUTION

In Figure 8.14 we locate the coordinate system with the origin at the center of the circle given by $x^2 + y^2 = 1$. Then, to find the fluid force against an end of the barrel *from the inside*, we integrate between -1 and 0.8 (using a weight of $w = 48$) to obtain

$$F_{\text{inside}} = 48 \int_{-1}^{0.8} (0.8 - y)(2)\sqrt{1 - y^2}\, dy$$

$$= 76.8 \int_{-1}^{0.8} \sqrt{1 - y^2}\, dy - 96 \int_{-1}^{0.8} y\sqrt{1 - y^2}\, dy.$$

To find the fluid force *from the outside*, we integrate between -1 and 0.4 (using a weight of $w = 64$) to obtain

$$F_{\text{outside}} = 64 \int_{-1}^{0.4} (0.4 - y)(2)\sqrt{1 - y^2}\, dy$$

$$= 51.2 \int_{-1}^{0.4} \sqrt{1 - y^2}\, dy - 128 \int_{-1}^{0.4} y\sqrt{1 - y^2}\, dy.$$

We leave the details of integration for you to complete in Exercises 61 and 62. Intuitively, would you say that the force from the oil (the inside) or the force from the sea water (the outside) is greater? By evaluating these two integrals, you can determine that

$$F_{\text{inside}} \approx 121.3 \text{ lb} \qquad \text{and} \qquad F_{\text{outside}} \approx 93.0 \text{ lb.}$$

FIGURE 8.13

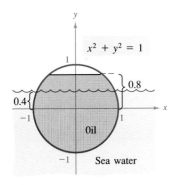

FIGURE 8.14

EXERCISES for Section 8.4

In Exercises 1–4, evaluate the indefinite integral using the substitution $x = 5 \sin \theta$.

1. $\displaystyle \int \frac{1}{(25 - x^2)^{3/2}}\, dx$

2. $\displaystyle \int \frac{1}{x^2\sqrt{25 - x^2}}\, dx$

3. $\displaystyle \int \frac{\sqrt{25 - x^2}}{x}\, dx$

4. $\displaystyle \int \frac{x^2}{\sqrt{25 - x^2}}\, dx$

In Exercises 5–8, evaluate the indefinite integral using the substitution $x = 2 \sec \theta$.

5. $\displaystyle \int \frac{1}{\sqrt{x^2 - 4}}\, dx$

6. $\displaystyle \int \frac{\sqrt{x^2 - 4}}{x}\, dx$

7. $\displaystyle \int x^3\sqrt{x^2 - 4}\, dx$

8. $\displaystyle \int \frac{x^3}{\sqrt{x^2 - 4}}\, dx$

In Exercises 9–12, evaluate the indefinite integral using the substitution $x = \tan \theta$.

9. $\displaystyle \int x\sqrt{1 + x^2}\, dx$

10. $\displaystyle \int \frac{x^3}{\sqrt{1 + x^2}}\, dx$

11. $\displaystyle \int \frac{1}{(1 + x^2)^2}\, dx$

12. $\displaystyle \int \frac{x^2}{(1 + x^2)^2}\, dx$

In Exercises 13–44, evaluate the indefinite integral.

13. $\displaystyle \int \frac{x}{\sqrt{x^2 + 9}}\, dx$

14. $\displaystyle \int \frac{1}{\sqrt{25 - x^2}}\, dx$

15. $\displaystyle \int_0^2 \sqrt{16 - 4x^2}\, dx$

16. $\displaystyle \int_0^2 x\sqrt{16 - 4x^2}\, dx$

17. $\displaystyle \int \frac{1}{\sqrt{x^2 - 9}}\, dx$

18. $\displaystyle \int \frac{t}{(1 - t^2)^{3/2}}\, dt$

19. $\displaystyle \int_0^{\sqrt{3}/2} \frac{t^2}{(1 - t^2)^{3/2}}\, dt$

20. $\displaystyle \int_0^{\sqrt{3}/2} \frac{1}{(1 - t^2)^{5/2}}\, dt$

21. $\displaystyle \int \frac{\sqrt{1 - x^2}}{x^4}\, dx$

22. $\displaystyle \int \frac{\sqrt{4x^2 + 9}}{x^4}\, dx$

23. $\displaystyle \int \frac{1}{x\sqrt{4x^2 + 9}}\, dx$

24. $\displaystyle \int \frac{1}{(x^2 + 3)^{3/2}}\, dx$

25. $\displaystyle \int \frac{x}{(x^2 + 3)^{3/2}}\, dx$

26. $\displaystyle \int \frac{1}{x\sqrt{4x^2 + 16}}\, dx$

27. $\displaystyle \int e^{2x}\sqrt{1 + e^{2x}}\, dx$

28. $\displaystyle \int \frac{1}{\sqrt{4x - x^2}}\, dx$

29. $\displaystyle \int (x + 1)\sqrt{x^2 + 2x + 2}\, dx$

30. $\displaystyle \int \frac{x^2}{\sqrt{2x - x^2}}\, dx$

31. $\displaystyle \int \frac{x}{\sqrt{x^2 + 4x + 8}}\, dx$

32. $\displaystyle \int \frac{x}{\sqrt{x^2 - 6x + 5}}\, dx$

33. $\displaystyle \int \frac{x^2}{\sqrt{x^2 + 10x + 9}}\, dx$

34. $\displaystyle \int (x^2 + 2x + 11)^{3/2}\, dx$

35. $\displaystyle \int e^x\sqrt{1 - e^{2x}}\, dx$

36. $\displaystyle \int \frac{\sqrt{1 - x}}{\sqrt{x}}\, dx$

37. $\displaystyle \int \frac{1}{4 + 4x^2 + x^4}\, dx$

38. $\displaystyle \int \frac{x^3 + x + 1}{x^4 + 2x^2 + 1}\, dx$

39. $\displaystyle \int \sqrt{4 + 9x^2}\, dx$

40. $\displaystyle \int \sqrt{1 + x^2}\, dx$

41. $\displaystyle \int \frac{x^2}{\sqrt{x^2 - 1}}\, dx$

42. $\displaystyle \int x^2\sqrt{x^2 - 4}\, dx$

43. $\displaystyle \int \text{arcsec } 2x\, dx$

44. $\displaystyle \int x \arcsin x\, dx$

In Exercises 45 and 46, evaluate the given integral using (a) the given integration limits and (b) the limits obtained by trigonometric substitution.

45. $\displaystyle \int_0^3 \frac{x^3}{\sqrt{x^2 + 9}}\, dx$

46. $\displaystyle \int_0^{5/3} \sqrt{25 - 9x^2}\, dx$

47. Find the value of the definite integral of Example 4 by converting back to the variable x and evaluating the antiderivative at the original limits.

48. The field strength H of a magnet of length $2L$ on a particle r units from the center of the magnet is given by

$$H = \frac{2mL}{(r^2 + L^2)^{3/2}}$$

where $\pm m$ are the poles of the magnet (see figure). Find the average field strength as the particle moves from 0 to R units from the center by evaluating the integral

$$\frac{1}{R}\int_0^R \frac{2mL}{(r^2 + L^2)^{3/2}}\, dr.$$

FIGURE FOR 48

In Exercises 49 and 50, find the fluid force on the circular observation window of radius 1 foot in a vertical wall of a large water-filled tank at a fish hatchery.

49. The center of the window is 3 feet below the surface of the water (see figure).

$x^2 + y^2 = 1$ FIGURE FOR 49

50. The center of the window is d feet below the surface of the water $(d > 1)$.

In Exercises 51 and 52, use trigonometric substitution to find the area enclosed by the graph of the given conic.

51. $x^2 + y^2 = r^2$ **52.** $\dfrac{x^2}{2^2} + \dfrac{y^2}{3^2} = 1$

In Exercises 53 and 54, find the volume of the torus generated by revolving the region bounded by the graph of the given circle about the y-axis.

53. $(x - 3)^2 + y^2 = 1$ (see figure)
54. $(x - h)^2 + y^2 = r^2$, $h > r$

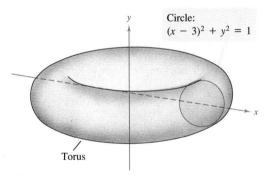

Circle:
$(x - 3)^2 + y^2 = 1$

Torus

FIGURE FOR 53

In Exercises 55 and 56, find the arc length of the given plane curve over the indicated interval.

Function	Interval
55. $y = \ln x$	$[1, 5]$
56. $y = x^2$	$[0, 3]$

In Exercises 57 and 58, find the distance that a projectile travels when it follows the path given by the graph of the equation.

Function	Interval
57. $y = x - 0.005x^2$	$[0, 200]$
58. $y = x - \dfrac{x^2}{72}$	$[0, 72]$

59. Find the surface area of the solid generated by revolving the region bounded by the graphs of $y = x^2$, $y = 0$, $x = 0$, and $x = \sqrt{2}$ about the x-axis.

60. Find the centroid of the region bounded by the graphs of $y = 3/\sqrt{x^2 + 9}$, $y = 0$, $x = -4$, and $x = 4$.

61. Evaluate the first integral for the fluid force given in Example 8.

$$F = 48 \int_{-1}^{0.8} (0.8 - y)(2)\sqrt{1 - y^2}\, dy$$

62. Evaluate the second integral for the fluid force given in Example 8.

$$F = 64 \int_{-1}^{0.4} (0.4 - y)(2)\sqrt{1 - y^2}\, dy$$

In Exercises 63 and 64, use Simpson's Rule on a computer or calculator (with $n = 12$) to approximate the centroid of the region bounded by the graphs of the given equations.

63. $y = \dfrac{4}{\ln x}$, $y = 0$, $x = 2$, $x = 10$

64. $y = 2 \cos x^2$, $y = 0$, $x = 0$, $x = 1$

65. Use trigonometric substitution to verify the second and third integration formulas given in Theorem 8.2.

66. Prove that the area of the region enclosed by the ellipse

$$\frac{x^2}{a^2} + \frac{y^2}{b^2} = 1, \ a > 0, b > 0$$

is πab.

8.5 Partial Fractions

Partial fractions (linear factors) ▪ Partial fractions (quadratic factors)

John Bernoulli

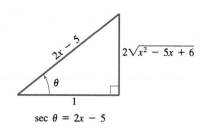

sec $\theta = 2x - 5$

FIGURE 8.15

This section examines a procedure for decomposing a rational function into simpler rational functions to which we can apply the basic integration formulas. We call this procedure the **method of partial fractions.** This technique was introduced in 1702 by John Bernoulli (1667–1748), a Swiss mathematician who was instrumental in the early development of calculus. John Bernoulli was a professor at the University of Basel and taught many outstanding students, the most famous of whom was Leonhard Euler.

To introduce the method of partial fractions, let's consider the integral of $1/(x^2 - 5x + 6)$. To evaluate this integral *without* partial fractions, we complete the square and use trigonometric substitution (see Figure 8.15) to obtain the following.

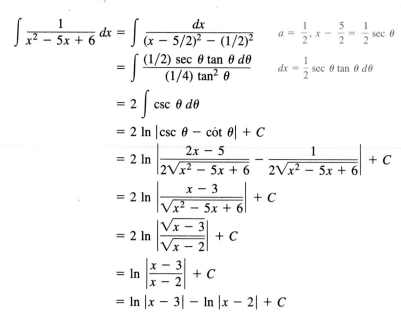

$$\int \frac{1}{x^2 - 5x + 6}\, dx = \int \frac{dx}{(x - 5/2)^2 - (1/2)^2} \qquad a = \frac{1}{2},\ x - \frac{5}{2} = \frac{1}{2}\sec\theta$$

$$= \int \frac{(1/2)\sec\theta\tan\theta\, d\theta}{(1/4)\tan^2\theta} \qquad dx = \frac{1}{2}\sec\theta\tan\theta\, d\theta$$

$$= 2\int \csc\theta\, d\theta$$

$$= 2\ln|\csc\theta - \cot\theta| + C$$

$$= 2\ln\left|\frac{2x - 5}{2\sqrt{x^2 - 5x + 6}} - \frac{1}{2\sqrt{x^2 - 5x + 6}}\right| + C$$

$$= 2\ln\left|\frac{x - 3}{\sqrt{x^2 - 5x + 6}}\right| + C$$

$$= 2\ln\left|\frac{\sqrt{x - 3}}{\sqrt{x - 2}}\right| + C$$

$$= \ln\left|\frac{x - 3}{x - 2}\right| + C$$

$$= \ln|x - 3| - \ln|x - 2| + C$$

Now, to see the benefit of the partial fraction method, suppose we had observed that

$$\frac{1}{x^2 - 5x + 6} = \frac{1}{x - 3} - \frac{1}{x - 2}. \qquad \text{Partial fraction decomposition}$$

Then we could evaluate the integral easily as follows.

$$\int \frac{1}{x^2 - 5x + 6}\, dx = \int \left(\frac{1}{x - 3} - \frac{1}{x - 2}\right) dx$$

$$= \ln|x - 3| - \ln|x - 2| + C$$

This method is clearly preferable to trigonometric substitution. However, its use depends on the ability to factor the denominator, $x^2 - 5x + 6$, and to find the **partial fractions** $1/(x - 3)$ and $-1/(x - 2)$.

Recall from algebra that every polynomial with real coefficients can be factored into linear and irreducible quadratic factors.* For instance, the polynomial $x^5 + x^4 - x - 1$ can be written as

$$x^5 + x^4 - x - 1 = (x - 1)(x + 1)^2(x^2 + 1)$$

where $(x - 1)$ is a linear factor, $(x + 1)^2$ is a repeated linear factor, and $(x^2 + 1)$ is an irreducible quadratic factor. Using this factorization, we can write the partial fraction decomposition of the rational expression

$$\frac{N(x)}{x^5 + x^4 - x - 1}$$

where $N(x)$ is a polynomial of degree less than 5, as follows.

$$\frac{N(x)}{(x - 1)(x + 1)^2(x^2 + 1)} = \frac{A}{x - 1} + \frac{B}{x + 1} + \frac{C}{(x + 1)^2} + \frac{Dx + E}{x^2 + 1}$$

We summarize the steps for partial fraction decomposition as follows.

DECOMPOSITION OF $N(x)/D(x)$ INTO PARTIAL FRACTIONS

1. **Divide if improper:** If $N(x)/D(x)$ is an improper fraction (that is, if the degree of the numerator is greater than or equal to the degree of the denominator), then divide the denominator into the numerator to obtain

   $$\frac{N(x)}{D(x)} = (\text{a polynomial}) + \frac{N_1(x)}{D(x)}$$

   where the degree of $N_1(x)$ is less than the degree of $D(x)$. Then apply steps 2, 3, and 4 to the proper rational expression $N_1(x)/D(x)$.
2. **Factor denominator:** Completely factor the denominator into factors of the form

   $$(px + q)^m$$

   and

   $$(ax^2 + bx + c)^n$$

   where $ax^2 + bx + c$ is irreducible.
3. **Linear factors:** For *each* factor of the form $(px + q)^m$, the partial fraction decomposition must include the following sum of m fractions.

   $$\frac{A_1}{(px + q)} + \frac{A_2}{(px + q)^2} + \cdots + \frac{A_m}{(px + q)^m}$$

4. **Quadratic factors:** For *each* factor of the form $(ax^2 + bx + c)^n$, the partial fraction decomposition must include the following sum of n fractions.

   $$\frac{B_1x + C_1}{ax^2 + bx + c} + \frac{B_2x + C_2}{(ax^2 + bx + c)^2} + \cdots + \frac{B_nx + C_n}{(ax^2 + bx + c)^n}$$

---

*For a review of factorization techniques, see *Precalculus*, 2nd edition, by Larson and Hostetler (Lexington, Mass.: D. C. Heath and Company, 1989).

Linear factors

Algebraic techniques for determining the constants in the numerators are demonstrated in the following examples.

EXAMPLE 1   Distinct linear factors

Write the partial fraction decomposition for

$$\frac{1}{x^2 - 5x + 6}.$$

SOLUTION

Since

$$x^2 - 5x + 6 = (x - 3)(x - 2)$$

we include one partial fraction for each factor and write

$$\frac{1}{x^2 - 5x + 6} = \frac{A}{x - 3} + \frac{B}{x - 2}$$

where A and B are to be determined. Multiplying this equation by the lowest common denominator, $(x - 3)(x - 2)$, leads to the **basic equation**

$$1 = A(x - 2) + B(x - 3). \qquad \text{Basic equation}$$

Since this equation is to be true for all x, we can substitute *convenient* values for x to obtain equations in A and B. These values are the ones that make particular factors zero.

To solve for A, we let $x = 3$ and obtain

$$1 = A(3 - 2) + B(3 - 3) \qquad \text{Let } x = 3 \text{ in basic equation.}$$
$$1 = A(1) + B(0)$$
$$A = 1.$$

To solve for B, we let $x = 2$ and obtain

$$1 = A(2 - 2) + B(2 - 3) \qquad \text{Let } x = 2 \text{ in basic equation.}$$
$$1 = A(0) + B(-1)$$
$$B = -1.$$

Therefore, the decomposition is

$$\frac{1}{x^2 - 5x + 6} = \frac{1}{x - 3} - \frac{1}{x - 2}$$

as indicated at the beginning of this section.   ▭

REMARK   The substitutions for x in Example 1 were chosen for their convenience in determining values for A and B. We chose $x = 2$ so as to eliminate the term $A(x - 2)$, and $x = 3$ was chosen to eliminate the term $B(x - 3)$. The goal is to make *convenient* substitutions whenever possible.

EXAMPLE 2   Repeated linear factors

Evaluate

$$\int \frac{5x^2 + 20x + 6}{x^3 + 2x^2 + x} \, dx.$$

SOLUTION

Since

$$x^3 + 2x^2 + x = x(x^2 + 2x + 1) = x(x + 1)^2$$

we include one fraction for *each power* of x and $(x + 1)$ and write

$$\frac{5x^2 + 20x + 6}{x(x + 1)^2} = \frac{A}{x} + \frac{B}{x + 1} + \frac{C}{(x + 1)^2}.$$

Multiplying by the least common denominator, $x(x + 1)^2$, leads to the *basic equation*

$$5x^2 + 20x + 6 = A(x + 1)^2 + Bx(x + 1) + Cx. \qquad \text{Basic equation}$$

To solve for A, let $x = 0$. This eliminates the B and C terms and yields

$$6 = A(1) + 0 + 0$$
$$A = 6.$$

To solve for C, let $x = -1$. This eliminates the A and B terms and yields

$$5 - 20 + 6 = 0 + 0 - C$$
$$C = 9.$$

We have exhausted the most convenient choices for x, so to find the value of B we use *any other value* for x along with the calculated values of A and C. Thus, using $x = 1$, $A = 6$, and $C = 9$, we have

$$5 + 20 + 6 = A(4) + B(2) + C$$
$$= 6(4) + 2B + 9$$
$$B = -1.$$

Therefore, it follows that

$$\int \frac{5x^2 + 20x + 6}{x(x + 1)^2} \, dx = \int \left(\frac{6}{x} - \frac{1}{x + 1} + \frac{9}{(x + 1)^2} \right) dx$$

$$= 6 \ln |x| - \ln |x + 1| + 9 \frac{(x + 1)^{-1}}{-1} + C$$

$$= \ln \left| \frac{x^6}{x + 1} \right| - \frac{9}{x + 1} + C.$$

REMARK   Note that it is necessary to make as many substitutions for x as there are unknowns (A, B, C, . . .) to be determined. For instance, in Example 2 we made three substitutions ($x = -1$, $x = 0$, and $x = 1$) to solve for C, A, and B, respectively.

Quadratic factors

When using the method of partial fractions with *linear* factors, a convenient choice of x immediately yields a value for one of the coefficients. With *quadratic* factors, a system of linear equations will typically have to be solved, regardless of the choice of x.

EXAMPLE 3 Distinct linear and quadratic factors

Evaluate

$$\int \frac{2x^3 - 4x - 8}{(x^2 - x)(x^2 + 4)}\, dx.$$

SOLUTION

Since

$$(x^2 - x)(x^2 + 4) = x(x - 1)(x^2 + 4)$$

we include one partial fraction for each factor and write

$$\frac{2x^3 - 4x - 8}{x(x - 1)(x^2 + 4)} = \frac{A}{x} + \frac{B}{x - 1} + \frac{Cx + D}{x^2 + 4}.$$

Multiplying by the least common denominator, $x(x - 1)(x^2 + 4)$, yields the *basic equation*

$$2x^3 - 4x - 8 = A(x - 1)(x^2 + 4) + Bx(x^2 + 4) + (Cx + D)(x)(x - 1).$$

To solve for A, let $x = 0$ and obtain

$$-8 = A(-1)(4) + 0 + 0 \quad \Longrightarrow \quad A = 2.$$

To solve for B, let $x = 1$ and obtain

$$-10 = 0 + B(5) + 0 \quad \Longrightarrow \quad B = -2.$$

At this point C and D are yet to be determined. We can find these remaining constants by choosing two other values for x and solving the resulting system of linear equations.

If $x = -1$, then, since $A = 2$ and $B = -2$, we have

$$-6 = (2)(-2)(5) + (-2)(-1)(5) + (-C + D)(-1)(-2)$$
$$2 = -C + D.$$

If $x = 2$, then we have

$$0 = (2)(1)(8) + (-2)(2)(8) + (2C + D)(2)(1)$$
$$8 = 2C + D.$$

Solving this system of two equations with two unknowns, we have

$$\begin{array}{rl} -C + D = 2 & \\ \underline{2C + D = 8} & \\ -3C \quad\quad = -6 & \text{Subtract second equation from first} \end{array}$$

which yields $C = 2$. Consequently, $D = 4$, and it follows that

$$\int \frac{2x^3 - 4x - 8}{x(x - 1)(x^2 + 4)} \, dx$$

$$= \int \left(\frac{2}{x} - \frac{2}{x - 1} + \frac{2x}{x^2 + 4} + \frac{4}{x^2 + 4} \right) dx$$

$$= 2 \ln |x| - 2 \ln |x - 1| + \ln (x^2 + 4) + 2 \arctan \frac{x}{2} + C.$$

When integrating rational expressions, keep in mind that for *improper* rational expressions like

$$\frac{N(x)}{D(x)} = \frac{2x^3 + x^2 - 7x + 7}{x^2 + x - 2}$$

you first must divide to obtain

$$\frac{N(x)}{D(x)} = 2x - 1 + \frac{-2x + 5}{x^2 + x - 2}.$$

The proper rational expression is then decomposed into its partial fractions by the usual methods.

In Examples 1, 2, and 3, we began the solution of the basic equation by substituting values of x that made the linear factors zero. This method works well when the partial fraction decomposition involves *only* linear factors. However, if the decomposition involves a quadratic factor, then an alternate procedure is often more convenient. Both methods are outlined in the following summary.

GUIDELINES FOR SOLVING THE BASIC EQUATION

Linear factors

1. Substitute the *roots* of the distinct linear factors into the basic equation.
2. For repeated linear factors use the coefficients determined in part 1 to rewrite the basic equation. Then substitute *other* convenient values of x and solve for the remaining coefficients.

Quadratic factors

1. Expand the basic equation.
2. Collect terms according to powers of x.
3. Equate the coefficients of like powers to obtain a system of linear equations involving A, B, C, and so on.
4. Solve the system of linear equations.

The second procedure for solving the basic equation is demonstrated in the next two examples.

EXAMPLE 4   Repeated quadratic factors

Evaluate

$$\int \frac{8x^3 + 13x}{(x^2 + 2)^2}\, dx.$$

SOLUTION

We include one partial fraction for each power of $(x^2 + 2)$ and write

$$\frac{8x^3 + 13x}{(x^2 + 2)^2} = \frac{Ax + B}{x^2 + 2} + \frac{Cx + D}{(x^2 + 2)^2}.$$

Multiplying by the least common denominator, $(x^2 + 2)^2$, yields the *basic equation*

$$8x^3 + 13x = (Ax + B)(x^2 + 2) + Cx + D.$$

Expanding the basic equation and collecting like terms, we have

$$8x^3 + 13x = Ax^3 + 2Ax + Bx^2 + 2B + Cx + D$$
$$8x^3 + 13x = Ax^3 + Bx^2 + (2A + C)x + (2B + D).$$

Now, we can equate the coefficients of like terms on opposite sides of the equation.

$$8 = A$$
$$13 = 2A + C$$
$$8x^3 + 0x^2 + 13x + 0 = Ax^3 + Bx^2 + (2A + C)x + (2B + D)$$
$$0 = 2B + D$$
$$0 = B$$

Using the known values $A = 8$ and $B = 0$, we have

$$13 = 2A + C = 2(8) + C \implies C = -3$$
$$0 = 2B + D = 2(0) + D \implies D = 0.$$

Finally, we conclude that

$$\int \frac{8x^3 + 13x}{(x^2 + 2)^2}\, dx = \int \left(\frac{8x}{x^2 + 2} + \frac{-3x}{(x^2 + 2)^2} \right) dx$$

$$= 4 \ln (x^2 + 2) + \frac{3}{2(x^2 + 2)} + C.$$

EXAMPLE 5   Repeated quadratic factors

Evaluate

$$\int \frac{x^2}{(x^2 + 1)^2}\, dx.$$

SOLUTION

We write

$$\frac{x^2}{(x^2 + 1)^2} = \frac{Ax + B}{x^2 + 1} + \frac{Cx + D}{(x^2 + 1)^2}$$

and multiply by $(x^2 + 1)^2$ to obtain

$$x^2 = (Ax + B)(x^2 + 1) + Cx + D$$
$$= Ax^3 + Bx^2 + (A + C)x + (B + D).$$

Therefore, by equating coefficients we obtain

$$A = 0, \quad B = 1, \quad C = 0, \quad D = -1$$

and we have

$$\int \frac{x^2}{(x^2 + 1)^2}\, dx = \int \left[\frac{1}{x^2 + 1} - \frac{1}{(x^2 + 1)^2} \right] dx.$$

The first integral can be evaluated by the Arctangent Rule, and the second integral was solved in Section 8.4 (see Example 6). Therefore, we have

$$\int \frac{x^2\, dx}{(x^2 + 1)^2} = \int \frac{dx}{x^2 + 1} - \int \frac{dx}{(x^2 + 1)^2}$$

$$= \arctan x - \frac{1}{2}\left(\arctan x + \frac{x}{x^2 + 1} \right) + C$$

$$= \frac{1}{2}\left(\arctan x - \frac{x}{x^2 + 1} \right) + C.$$

The final example shows how to combine the two methods effectively to solve a basic equation.

EXAMPLE 6   Linear and quadratic factors

Evaluate

$$\int \frac{3x + 4}{x^3 - 2x - 4}\, dx.$$

SOLUTION

Since

$$x^3 - 2x - 4 = (x - 2)(x^2 + 2x + 2)$$

we write

$$\frac{3x + 4}{x^3 - 2x - 4} = \frac{A}{x - 2} + \frac{Bx + C}{x^2 + 2x + 2}$$

$$3x + 4 = A(x^2 + 2x + 2) + (Bx + C)(x - 2) \qquad \text{Basic equation}$$

$$3x + 4 = (A + B)x^2 + (2A - 2B + C)x + (2A - 2C).$$

If $x = 2$, then in the basic equation we have $10 = 10A$ and $A = 1$. Now, from the expanded equation, we have the system

$$A + B \qquad = 0$$
$$2A - 2B + C = 3$$
$$2A \qquad - 2C = 4.$$

Using $A = 1$, we have $B = -1$ and $C = -1$. Therefore,

$$\int \frac{3x + 4}{x^3 - 2x - 4}\, dx = \int \left(\frac{A}{x - 2} + \frac{Bx + C}{x^2 + 2x + 2} \right) dx$$

$$= \int \left(\frac{1}{x - 2} + \frac{-x - 1}{x^2 + 2x + 2} \right) dx$$

$$= \ln |x - 2| - \frac{1}{2} \ln (x^2 + 2x + 2) + C. \qquad \square$$

Before we conclude this section, here are a few things you should remember. First, it is not necessary to use the partial fractions technique on all rational functions. For instance, the following integral is evaluated more easily by the Log Rule.

$$\int \frac{x^2 + 1}{x^3 + 3x - 4}\, dx = \frac{1}{3} \int \frac{3x^2 + 3}{x^3 + 3x - 4}\, dx = \frac{1}{3} \ln |x^3 + 3x - 4| + C$$

Second, if the integrand is not in reduced form, reducing it may eliminate the need for partial fractions, as shown in the following integral.

$$\int \frac{x^2 - x - 2}{x^3 - 2x - 4}\, dx = \int \frac{(x + 1)(x - 2)}{(x - 2)(x^2 + 2x + 2)}\, dx = \int \frac{x + 1}{x^2 + 2x + 2}\, dx$$

$$= \frac{1}{2} \ln |x^2 + 2x + 2| + C$$

Finally, partial fractions can be used with some quotients involving transcendental functions. For instance, the substitution $u = \sin x$ allows us to write

$$\int \frac{\cos x}{\sin x(\sin x - 1)}\, dx = \int \frac{du}{u(u - 1)} = \int \left(-\frac{1}{u} + \frac{1}{u - 1} \right) du$$

$$= -\ln |u| + \ln |u - 1| + C = \ln \left| \frac{\sin x - 1}{\sin x} \right| + C.$$

EXERCISES for Section 8.5

In Exercises 1–30, evaluate the indefinite integral.

1. $\displaystyle\int \frac{1}{x^2 - 1}\, dx$

2. $\displaystyle\int \frac{1}{4x^2 - 9}\, dx$

3. $\displaystyle\int \frac{3}{x^2 + x - 2}\, dx$

4. $\displaystyle\int \frac{x + 1}{x^2 + 4x + 3}\, dx$

5. $\displaystyle\int \frac{5 - x}{2x^2 + x - 1}\, dx$

6. $\displaystyle\int \frac{3x^2 - 7x - 2}{x^3 - x}\, dx$

7. $\displaystyle\int \frac{x^2 + 12x + 12}{x^3 - 4x}\, dx$

8. $\displaystyle\int \frac{x^3 - x + 3}{x^2 + x - 2}\, dx$

9. $\displaystyle\int \frac{2x^3 - 4x^2 - 15x + 5}{x^2 - 2x - 8}\, dx$

10. $\displaystyle\int \frac{x + 2}{x^2 - 4x}\, dx$

11. $\displaystyle\int \frac{4x^2 + 2x - 1}{x^3 + x^2}\, dx$

12. $\displaystyle\int \frac{2x - 3}{(x - 1)^2}\, dx$

13. $\int \dfrac{x^4}{(x-1)^3}\,dx$

14. $\int \dfrac{4x^2-1}{(2x)(x^2+2x+1)}\,dx$

15. $\int \dfrac{3x}{x^2-6x+9}\,dx$

16. $\int \dfrac{6x^2+1}{x^2(x-1)^3}\,dx$

17. $\int \dfrac{x^2-1}{x^3+x}\,dx$

18. $\int \dfrac{x}{x^3-1}\,dx$

19. $\int \dfrac{x^2}{x^4-2x^2-8}\,dx$

20. $\int \dfrac{2x^2+x+8}{(x^2+4)^2}\,dx$

21. $\int \dfrac{x}{16x^4-1}\,dx$

22. $\int \dfrac{x^2-4x+7}{x^3-x^2+x+3}\,dx$

23. $\int \dfrac{x^2+x+2}{(x^2+2)^2}\,dx$

24. $\int \dfrac{x^3}{(x^2-4)^2}\,dx$

25. $\int \dfrac{x^2+5}{x^3-x^2+x+3}\,dx$

26. $\int \dfrac{x^2+x+3}{x^4+6x^2+9}\,dx$

27. $\int \dfrac{6x^2-3x+14}{x^3-2x^2+4x-8}\,dx$

28. $\int \dfrac{x(2x-9)}{x^3-6x^2+12x-8}\,dx$

29. $\int \dfrac{2x^2-2x+3}{x^3-x^2-x-2}\,dx$

30. $\int \dfrac{x^3-6x^2+x-4}{1-x^4}\,dx$

In Exercises 31–36, evaluate the definite integral.

31. $\int_0^1 \dfrac{3}{2x^2+5x+2}\,dx$

32. $\int_3^4 \dfrac{1}{x^2-4}\,dx$

33. $\int_1^2 \dfrac{x+1}{x(x^2+1)}\,dx$

34. $\int_1^5 \dfrac{x-1}{x^2(x+1)}\,dx$

35. $\int_2^3 \dfrac{x^2-x+2}{x^3-x^2+x-1}\,dx$

36. $\int_0^1 \dfrac{x^2-x}{x^2+x+1}\,dx$

In Exercises 37–42, find the indefinite integral by using the indicated substitution.

37. $\int \dfrac{\sin x}{\cos x(\cos x-1)}\,dx$; let $u=\cos x$

38. $\int \dfrac{\sin x}{\cos x+\cos^2 x}\,dx$; let $u=\cos x$

39. $\int \dfrac{3\cos x}{\sin^2 x+\sin x-2}\,dx$; let $u=\sin x$

40. $\int \dfrac{\sec^2 x}{\tan x(\tan x+1)}\,dx$; let $u=\tan x$

41. $\int \dfrac{e^x}{(e^x-1)(e^x+4)}\,dx$; let $u=e^x$

42. $\int \dfrac{e^x}{(e^{2x}+1)(e^x-1)}\,dx$; let $u=e^x$

In Exercises 43–46, use the method of partial fractions to verify the given indefinite integral.

43. $\int \dfrac{1}{x(a+bx)}\,dx=\dfrac{1}{a}\ln\left|\dfrac{x}{a+bx}\right|+C$

44. $\int \dfrac{1}{a^2-x^2}\,dx=\dfrac{1}{2a}\ln\left|\dfrac{a+x}{a-x}\right|+C$

45. $\int \dfrac{x}{(a+bx)^2}\,dx=\dfrac{1}{b^2}\left(\dfrac{a}{a+bx}+\ln|a+bx|\right)+C$

46. $\int \dfrac{1}{x^2(a+bx)}\,dx=-\dfrac{1}{ax}-\dfrac{b}{a^2}\ln\left|\dfrac{x}{a+bx}\right|+C$

47. Find the area of the region bounded by the graphs of $y=7/(16-x^2)$ and $y=1$.

48. Find the centroid of the region bounded by the graphs of $y=2x/(x^2+1)$, $y=0$, $x=0$, and $x=3$.

49. Find the volume of the solid generated by revolving the region in Exercise 48 about the x-axis.

50. In Section 6.6 the exponential growth equation was derived from the assumption that the rate of growth is proportional to the existing quantity. In practice, there often exists some upper limit L past which growth cannot occur. In such cases, we assume the rate of growth to be proportional not only to the existing quantity, but also to the difference between the existing quantity y and the upper limit L. That is,

$$\dfrac{dy}{dt}=ky(L-y).$$

In integral form we can express this relationship as

$$\int \dfrac{dy}{y(L-y)}=\int k\,dt.$$

(a) Use partial fractions to evaluate the integral on the left, and solve for y as a function of t, where y_0 is the initial quantity.

(b) The graph of the function y is called a **logistics curve** (see figure). Show that the rate of growth is a maximum at the point of inflection, and that this occurs when $y=L/2$.

Logistics curve

FIGURE FOR 50

51. A single infected individual enters a community of n susceptible individuals. Let x be the number of newly infected individuals at time t. The common **epidemic model** assumes that the disease spreads at a rate proportional to the product of the total number infected and the number not yet infected. Thus

$$\frac{dx}{dt} = k(x + 1)(n - x)$$

and we obtain

$$\int \frac{1}{(x + 1)(n - x)} \, dx = \int k \, dt.$$

Solve for x as a function of t.

52. In a chemical reaction, one unit of compound Y and one unit of compound Z are converted into a single unit of compound X. If x is the amount of compound X formed, and the rate of formation of X is proportional

to the product of the amounts of unconverted compounds Y and Z, then

$$\frac{dx}{dt} = k(y_0 - x)(z_0 - x)$$

where y_0 and z_0 are the initial amounts of substances Y and Z, respectively. From the above equation we obtain

$$\int \frac{1}{(y_0 - x)(z_0 - x)} \, dx = \int k \, dt.$$

Do the two integrations and solve for x in terms of t.

53. Use the result of Exercise 52 to find x as $t \to \infty$ if (a) $y_0 < z_0$ and (b) $y_0 > z_0$. If $y_0 = z_0$, what is the limit of x as $t \to \infty$.

In Exercises 54 and 55, use Simpson's Rule on a computer or calculator (with $n = 12$) to approximate the definite integral.

54. $\displaystyle\int_1^4 \frac{4}{\sqrt{x} + \sqrt[3]{x}} \, dx$

55. $\displaystyle\int_{-\pi}^{\pi} \sqrt{4 + 3 \sin^2 x} \, dx$

8.6 Integration by Tables and Other Integration Techniques

Integration by tables ▪ Reduction formulas ▪ Substitution for rational functions of sine and cosine

So far in this chapter we have discussed a number of integration techniques to use with the basic integration formulas. Certainly we have not considered every method for finding antiderivatives, but we have considered the most important ones.

But merely knowing *how* to use the various techniques is not enough. You also need to know *when* to use them. Integration is first and foremost a problem of recognition. That is, you must recognize which formula or technique to apply to obtain an antiderivative. Frequently, a slight alteration of an integrand will require a different integration technique, as shown below.

$$\int x \ln x \, dx = \frac{x^2}{2} \ln x - \frac{x^2}{4} + C \qquad \text{Integration by parts}$$

$$\int \frac{\ln x}{x} \, dx = \frac{(\ln x)^2}{2} + C \qquad \text{Power Rule}$$

$$\int \frac{1}{x \ln x} \, dx = \ln |\ln x| + C \qquad \text{Log Rule}$$

Integration by tables requires considerable thought and insight and often involves substitution. Many people find a table of integrals to be a valuable supplement to the integration techniques discussed in this chapter. A table of common integrals can be found in Appendix C of this text.

Each integration formula in the table in this text can be developed using one or more of the techniques we have studied. We encourage you to verify several of the formulas. For instance, Formula 4

$$\int \frac{u}{(a + bu)^2} \, du = \frac{1}{b^2} \left(\frac{a}{a + bu} + \ln |a + bu| \right) + C \qquad \text{Formula 4}$$

can be verified using the method of partial fractions, and Formula 19

$$\int \frac{\sqrt{a + bu}}{u}\, du = 2\sqrt{a + bu} + a \int \frac{du}{u\sqrt{a + bu}} \qquad \text{Formula 19}$$

can be verified using integration by parts.

The next four examples demonstrate the use of integration tables. Note that the integrals in the tables in this text are classified according to forms involving the following: u^n, $(a + bu)$, $(a + bu + cu^2)$, $\sqrt{a + bu}$, $(a^2 \pm u^2)$, $\sqrt{u^2 \pm a^2}$, $\sqrt{a^2 - u^2}$, trigonometric functions, inverse trigonometric functions, exponential functions, and logarithmic functions.

EXAMPLE 1 Integration by tables

Evaluate

$$\int \frac{dx}{x\sqrt{x - 1}}.$$

SOLUTION

Since the expression inside the radical is linear, we consider forms involving $\sqrt{a + bu}$.

$$\int \frac{du}{u\sqrt{a + bu}} = \frac{2}{\sqrt{-a}} \arctan \sqrt{\frac{a + bu}{-a}} + C \qquad \text{Formula 17 } (a < 0)$$

We let $a = -1$, $b = 1$, and $u = x$. Then $du = dx$, and we write

$$\int \frac{dx}{x\sqrt{x - 1}} = 2 \arctan \sqrt{x - 1} + C.$$

EXAMPLE 2 Integration by tables

Evaluate $\int x\sqrt{x^4 - 9}\, dx$.

SOLUTION

Since the radical has the form $\sqrt{u^2 - a^2}$, we consider the following formula.

$$\int \sqrt{u^2 - a^2}\, du = \frac{1}{2}(u\sqrt{u^2 - a^2} - a^2 \ln |u + \sqrt{u^2 - a^2}|) + C \quad \text{Formula 26}$$

We let $u = x^2$ and $a = 3$. Then $du = 2x\, dx$, and we write

$$\int x\sqrt{x^4 - 9}\, dx = \frac{1}{2} \int \sqrt{(x^2)^2 - 3^2}\, (2x)\, dx$$

$$= \frac{1}{4}(x^2\sqrt{x^4 - 9} - 9 \ln |x^2 + \sqrt{x^4 - 9}|) + C.$$

EXAMPLE 3   Integration by tables

Evaluate

$$\int \frac{x}{1 + e^{-x^2}} \, dx.$$

SOLUTION

Of the forms involving e^u, we consider the following.

$$\int \frac{du}{1 + e^u} = u - \ln(1 + e^u) + C \qquad \text{Formula 84}$$

We let $u = -x^2$. Then, $du = -2x \, dx$, and we write

$$\int \frac{x}{1 + e^{-x^2}} \, dx = -\frac{1}{2} \int \frac{-2x \, dx}{1 + e^{-x^2}}$$

$$= -\frac{1}{2}[-x^2 - \ln(1 + e^{-x^2})] + C$$

$$= \frac{1}{2}[x^2 + \ln(1 + e^{-x^2})] + C.$$

EXAMPLE 4   Integration by tables

Evaluate

$$\int \frac{\sin 2x}{2 + \cos x} \, dx.$$

SOLUTION

Substituting $2 \sin x \cos x$ for $\sin 2x$, we have

$$\int \frac{\sin 2x}{2 + \cos x} \, dx = 2 \int \frac{\sin x \cos x}{2 + \cos x} \, dx.$$

A check of the forms involving $\sin u$ or $\cos u$ shows that none of those listed applies. Therefore, we consider forms involving $a + bu$. For example,

$$\int \frac{u \, du}{a + bu} = \frac{1}{b^2}(bu - a \ln|a + bu|) + C. \qquad \text{Formula 3}$$

We let $a = 2$, $b = 1$, and $u = \cos x$. Then, $du = -\sin x \, dx$ and we write

$$2 \int \frac{\sin x \cos x}{2 + \cos x} \, dx = -2 \int \frac{\cos x \, (-\sin x \, dx)}{2 + \cos x}$$

$$= -2(\cos x - 2 \ln|2 + \cos x|) + C$$

$$= -2 \cos x + 4 \ln|2 + \cos x| + C.$$

Reduction formulas

You will notice that a number of integrals in the integration tables have the form

$$\int f(x)\ dx = g(x) + \int h(x)\ dx.$$

Such integration formulas are called **reduction formulas** because they reduce a given integral to the sum of a function and a simpler integral. We demonstrate the use of a reduction formula in the next two examples.

EXAMPLE 5   Use of a reduction formula

Evaluate

$$\int x^3 \sin x\ dx.$$

SOLUTION

From the integration table, we have the following three formulas.

$$\int u \sin u\ du = \sin u - u \cos u + C \qquad\qquad \text{Formula 52}$$

$$\int u^n \sin u\ du = -u^n \cos u + n \int u^{n-1} \cos u\ du \qquad \text{Formula 54}$$

$$\int u^n \cos u\ du = u^n \sin u - n \int u^{n-1} \sin u\ du \qquad \text{Formula 55}$$

Using Formula 54 followed by Formula 55, we have

$$\int x^3 \sin x\ dx = -x^3 \cos x + 3 \int x^2 \cos x\ dx$$

$$= -x^3 \cos x + 3\left(x^2 \sin x - 2 \int x \sin x\ dx \right).$$

Now, by Formula 52, we have

$$\int x^3 \sin x\ dx = -x^3 \cos x + 3x^2 \sin x + 6x \cos x - 6 \sin x + C.$$

EXAMPLE 6   Use of a reduction formula

Evaluate

$$\int \frac{\sqrt{3 - 5x}}{2x}\ dx.$$

SOLUTION

From the integration table, we have the following two formulas.

$$\int \frac{du}{u\sqrt{a + bu}} = \frac{1}{\sqrt{a}} \ln \left| \frac{\sqrt{a + bu} - \sqrt{a}}{\sqrt{a + bu} + \sqrt{a}} \right| + C \qquad \text{Formula 17 } (0 < a)$$

$$\int \frac{\sqrt{a + bu}}{u} \, du = 2\sqrt{a + bu} + a \int \frac{du}{u\sqrt{a + bu}} \qquad \text{Formula 19}$$

Using Formula 19 with $a = 3$, $b = -5$, and $u = x$, it follows that

$$\frac{1}{2} \int \frac{\sqrt{3 - 5x}}{x} \, dx = \frac{1}{2} \left(2\sqrt{3 - 5x} + 3 \int \frac{dx}{x\sqrt{3 - 5x}} \right)$$

$$= \sqrt{3 - 5x} + \frac{3}{2} \int \frac{dx}{x\sqrt{3 - 5x}}.$$

Now, by Formula 17, with $a = 3$, $b = -5$, and $u = x$, we conclude that

$$\int \frac{\sqrt{3 - 5x}}{2x} \, dx = \sqrt{3 - 5x} + \frac{3}{2} \left(\frac{1}{\sqrt{3}} \ln \left| \frac{\sqrt{3 - 5x} - \sqrt{3}}{\sqrt{3 - 5x} + \sqrt{3}} \right| \right) + C$$

$$= \sqrt{3 - 5x} + \frac{\sqrt{3}}{2} \ln \left| \frac{\sqrt{3 - 5x} - \sqrt{3}}{\sqrt{3 - 5x} + \sqrt{3}} \right| + C. \quad \square$$

Rational functions of sine and cosine

Example 4 involves a rational expression of $\sin x$ and $\cos x$. If you are unable to find an integral of this form in the integration tables, use the following special substitution to convert the trigonometric expression to a standard rational expression.

SUBSTITUTION FOR RATIONAL FUNCTIONS OF SINE AND COSINE

For integrals involving rational functions of sine and cosine, the substitution

$$u = \frac{\sin x}{1 + \cos x} = \tan \frac{x}{2}$$

yields

$$\cos x = \frac{1 - u^2}{1 + u^2}, \qquad \sin x = \frac{2u}{1 + u^2}, \qquad \text{and} \qquad dx = \frac{2 \, du}{1 + u^2}.$$

PROOF   From the substitution for u, it follows that

$$u^2 = \frac{\sin^2 x}{(1 + \cos x)^2} = \frac{1 - \cos^2 x}{(1 + \cos x)^2} = \frac{1 - \cos x}{1 + \cos x}.$$

Solving for $\cos x$ in this equation, we have

$$\cos x = \frac{1 - u^2}{1 + u^2}.$$

To find sin x, we write $u = \sin x/(1 + \cos x)$ as

$$\sin x = u(1 + \cos x) = u\left(1 + \frac{1 - u^2}{1 + u^2}\right) = \frac{2u}{1 + u^2}.$$

Finally, to find dx, we consider $u = \tan (x/2)$. Then $\arctan u = x/2$ and

$$dx = \frac{2\ du}{1 + u^2}.$$

EXAMPLE 7   Substitution for rational functions of sine and cosine

Evaluate

$$\int \frac{dx}{1 + \sin x - \cos x}.$$

SOLUTION

Let $u = \sin x/(1 + \cos x)$. Then

$$\int \frac{dx}{1 + \sin x - \cos x} = \int \frac{2\ du/(1 + u^2)}{1 + [2u/(1 + u^2)] - [(1 - u^2)/(1 + u^2)]}$$

$$= \int \frac{2\ du}{(1 + u^2) + 2u - (1 - u^2)} = \int \frac{2\ du}{2u + 2u^2}$$

$$= \int \frac{du}{u(1 + u)} = \int \frac{1}{u}\ du - \int \frac{1}{1 + u}\ du \qquad \text{Partial fractions}$$

$$= \ln |u| - \ln |1 + u| + C = \ln \left|\frac{u}{1 + u}\right| + C$$

$$= \ln \left|\frac{\sin x/(1 + \cos x)}{1 + [\sin x/(1 + \cos x)]}\right| + C$$

$$= \ln \left|\frac{\sin x}{1 + \sin x + \cos x}\right| + C.$$ ▭

EXERCISES for Section 8.6

In Exercises 1–52, use the integration tables at the end of the text to evaluate the given integral.

1. $\displaystyle \int \frac{x^2}{1 + x}\ dx$

2. $\displaystyle \int \frac{x}{\sqrt{1 + x}}\ dx$

3. $\displaystyle \int \frac{1}{x^2\sqrt{1 - x^2}}\ dx$

4. $\displaystyle \int x \sin x\ dx$

5. $\displaystyle \int x^2 \ln x\ dx$

6. $\displaystyle \int \operatorname{arcsec} 2x\ dx$

7. $\displaystyle \int \frac{1}{x^2\sqrt{x^2 - 4}}\ dx$

8. $\displaystyle \int \frac{\sqrt{x^2 - 4}}{x}\ dx$

9. $\displaystyle \int xe^{x^2}\ dx$

10. $\displaystyle \int \frac{x}{\sqrt{9 - x^4}}\ dx$

11. $\displaystyle \int \frac{2x}{(1 - 3x)^2}\ dx$

12. $\displaystyle \int \frac{1}{x^2 + 2x + 2}\ dx$

13. $\displaystyle \int e^x \arccos e^x\ dx$

14. $\displaystyle \int \frac{\theta^2}{1 - \sin \theta^3}\ d\theta$

15. $\displaystyle \int x^3 \ln x\ dx$

16. $\displaystyle \int \cot^3 \theta\ d\theta$

17. $\displaystyle \int \frac{x^2}{(3x - 5)^2}\ dx$

18. $\displaystyle \int \frac{1}{2x^2(2x - 1)^2}\ dx$

19. $\displaystyle \int \frac{x}{1 - \sec x^2}\ dx$

20. $\displaystyle \int \frac{e^x}{1 - \tan e^x}\ dx$

21. $\displaystyle\int \frac{\cos x}{1 + \sin^2 x}\, dx$

22. $\displaystyle\int \frac{1}{t[1 + (\ln t)^2]}\, dt$

23. $\displaystyle\int \frac{1}{1 + e^{2x}}\, dx$

24. $\displaystyle\int \frac{1}{\sqrt{x}(1 + 2\sqrt{x})}\, dx$

25. $\displaystyle\int \frac{\cos \theta}{3 + 2 \sin \theta + \sin^2 \theta}\, d\theta$

26. $\displaystyle\int x^2\sqrt{2 + 9x^2}\, dx$

27. $\displaystyle\int \frac{1}{x^2\sqrt{2 + 9x^2}}\, dx$

28. $\displaystyle\int \sqrt{3 + x^2}\, dx$

29. $\displaystyle\int e^x\sqrt{1 + e^{2x}}\, dx$

30. $\displaystyle\int \frac{1}{\sqrt{x}(x - 4)^{3/2}}\, dx$

31. $\displaystyle\int \sin^4 2x\, dx$

32. $\displaystyle\int \frac{\cos^3 \sqrt{x}}{\sqrt{x}}\, dx$

33. $\displaystyle\int \frac{1}{\sqrt{x}(1 - \cos \sqrt{x})}\, dx$

34. $\displaystyle\int \frac{1}{1 - \tan 5x}\, dx$

35. $\displaystyle\int t^4 \cos t\, dt$

36. $\displaystyle\int \sqrt{x}\, \arctan x^{3/2}\, dx$

37. $\displaystyle\int x\, \text{arcsec}\, (x^2 + 1)\, dx$

38. $\displaystyle\int (\ln x)^3\, dx$

39. $\displaystyle\int \frac{\ln x}{x(3 + 2 \ln x)}\, dx$

40. $\displaystyle\int \frac{e^x}{(1 - e^{2x})^{3/2}}\, dx$

41. $\displaystyle\int \frac{\sqrt{2 - 2x - x^2}}{x + 1}\, dx$

42. $\displaystyle\int \frac{1}{(x^2 - 6x + 10)^2}\, dx$

43. $\displaystyle\int \frac{x}{x^4 - 6x^2 + 10}\, dx$

44. $\displaystyle\int (2x - 3)^2\sqrt{(2x - 3)^2 + 4}\, dx$

45. $\displaystyle\int \frac{x}{\sqrt{x^4 - 6x^2 + 5}}\, dx$

46. $\displaystyle\int \frac{\cos x}{\sqrt{\sin^2 x + 1}}\, dx$

47. $\displaystyle\int \frac{x^3}{\sqrt{4 - x^2}}\, dx$

48. $\displaystyle\int \sqrt{\frac{3 - x}{3 + x}}\, dx$

49. $\displaystyle\int \frac{1}{x^{3/2}\sqrt{1 - x}}\, dx$

50. $\displaystyle\int x\sqrt{x^2 + 2x}\, dx$

51. $\displaystyle\int \frac{e^{3x}}{(1 + e^x)^3}\, dx$

52. $\displaystyle\int \sec^5 \theta\, d\theta$

In Exercises 53–58, verify the given integration formula.

53. $\displaystyle\int \frac{u^2}{(a + bu)^2}\, du =$
$$\frac{1}{b^3}\left(bu - \frac{a^2}{a + bu} - 2a \ln |a + bu|\right) + C$$

54. $\displaystyle\int \frac{u^n}{\sqrt{a + bu}}\, du =$
$$\frac{2}{(2n + 1)b}\left(u^n\sqrt{a + bu} - na \int \frac{u^{n-1}}{\sqrt{a + bu}}\, du\right)$$

55. $\displaystyle\int \frac{1}{(u^2 \pm a^2)^{3/2}}\, du = \frac{\pm u}{a^2\sqrt{u^2 \pm a^2}} + C$

56. $\displaystyle\int u^n \cos u\, du = u^n \sin u - n \int u^{n-1} \sin u\, du$

57. $\displaystyle\int \arctan u\, du = u \arctan u - \ln \sqrt{1 + u^2} + C$

58. $\displaystyle\int (\ln u)^n\, du = u(\ln u)^n - n \int (\ln u)^{n-1}\, du$

In Exercises 59–68, evaluate the given integral.

59. $\displaystyle\int \frac{1}{2 - 3 \sin \theta}\, d\theta$

60. $\displaystyle\int \frac{\sin \theta}{1 + \cos^2 \theta}\, d\theta$

61. $\displaystyle\int_0^{\pi/2} \frac{1}{1 + \sin \theta + \cos \theta}\, d\theta$

62. $\displaystyle\int_0^{\pi/2} \frac{1}{3 - 2 \cos \theta}\, d\theta$

63. $\displaystyle\int \frac{\sin \theta}{3 - 2 \cos \theta}\, d\theta$

64. $\displaystyle\int \frac{\sin \theta}{1 + \sin \theta}\, d\theta$

65. $\displaystyle\int \frac{\cos \sqrt{\theta}}{\sqrt{\theta}}\, d\theta$

66. $\displaystyle\int \frac{\sin \theta}{(\cos \theta)(1 + \sin \theta)}\, d\theta$

67. $\displaystyle\int \frac{1}{\sin \theta \tan \theta}\, d\theta$

68. $\displaystyle\int \frac{1}{\sec \theta - \tan \theta}\, d\theta$

69. A hydraulic cylinder on an industrial machine pushes a steel block a distance of x feet ($0 \le x \le 5$), where the variable force required is

$$F(x) = 2000xe^{-x} \text{ lb.}$$

Find the work done in pushing the block the full 5 feet through the machine.

70. Repeat Exercise 69, using a force of

$$F(x) = \frac{500x}{\sqrt{26 - x^2}} \text{ lb.}$$

8.7 Indeterminate Forms and L'Hôpital's Rule

Indeterminate forms ▪ L'Hôpital's Rule

In Chapter 2, we described the forms $0/0$ and ∞/∞ as *indeterminate*, because they do not guarantee that a limit exists, nor do they indicate what the limit is, if one does exist. When we encountered one of these indeterminate forms, we attempted to rewrite the expression by using various algebraic techniques, as illustrated by the examples in Table 8.1.

TABLE 8.1   Indeterminate Forms

Indeterminate form	Limit	Algebraic technique
$0/0$	$\displaystyle\lim_{x \to -1} \frac{2x^2 - 2}{x + 1} = \lim_{x \to -1} 2(x - 1) = -4$	Divide numerator and denominator by $(x + 1)$.
∞/∞	$\displaystyle\lim_{x \to \infty} \frac{3x^2 - 1}{2x^2 + 1} = \lim_{x \to \infty} \frac{3 - (1/x^2)}{2 + (1/x^2)} = \frac{3}{2}$	Divide numerator and denominator by x^2.

Occasionally, we can extend these algebraic techniques to find limits of transcendental functions. For instance, the limit

$$\lim_{x \to 0} \frac{e^{2x} - 1}{e^x - 1}$$

produces the indeterminate form $0/0$. Factoring and then dividing, we have

$$\lim_{x \to 0} \frac{e^{2x} - 1}{e^x - 1} = \lim_{x \to 0} \frac{(e^x + 1)(e^x - 1)}{e^x - 1}$$

$$= \lim_{x \to 0} (e^x + 1) = 2.$$

However, not all indeterminate forms can be evaluated by algebraic manipulation. This is particularly true when *both* algebraic and transcendental functions are involved. For instance, the limit

$$\lim_{x \to 0} \frac{e^{2x} - 1}{x}$$

produces the indeterminate form $0/0$. Dividing the numerator and denominator by x to obtain

$$\lim_{x \to 0} \left[\frac{e^{2x}}{x} - \frac{1}{x} \right]$$

merely produces another indeterminate form, $\infty - \infty$. Of course, we could use a calculator to estimate this limit, as shown in Table 8.2. From the table, the limit appears to be 2. (This limit will be verified in Example 1.)

TABLE 8.2

x	-1.0	-0.1	-0.01	-0.001	0	0.001	0.01	0.1	1.0
$\dfrac{e^{2x} - 1}{x}$	0.865	1.813	1.980	1.998	?	2.002	2.020	2.214	6.389

Guillaume L'Hôpital

To find the limit illustrated in Table 8.2, we use a theorem called **L'Hôpital's Rule.** This theorem states that under certain conditions the limit of the quotient $f(x)/g(x)$ is determined by the limit of $f'(x)/g'(x)$. This theorem is named after the French mathematician Guillaume Francois Antoine De L'Hôpital (1661–1704), who published the first calculus text in 1696. To prove this theorem, we use a more general result called the **Extended Mean-Value Theorem,** which states the following. If f and g are differentiable on an open interval (a, b) and continuous on $[a, b]$ such that $g(x) \neq 0$ for any x in (a, b), then there exists a point c in (a, b) such that

$$\frac{f'(c)}{g'(c)} = \frac{f(b) - f(a)}{g(b) - g(a)}.$$

We prove the Extended Mean-Value Theorem and L'Hôpital's Rule in Appendix A.

THEOREM 8.3
L'HÔPITAL'S RULE

Let f and g be functions that are differentiable on an open interval (a, b) containing c, except possibly at c itself. If the limit of $f(x)/g(x)$ as x approaches c produces the indeterminate form $0/0$ or ∞/∞, then

$$\lim_{x \to c} \frac{f(x)}{g(x)} = \lim_{x \to c} \frac{f'(x)}{g'(x)}$$

provided the limit on the right exists (or is infinite).

REMARK   The indeterminate form ∞/∞ comes in four forms: ∞/∞, $(-\infty)/\infty$, $\infty/(-\infty)$, and $(-\infty)/(-\infty)$. L'Hôpital's Rule can be applied to each of these forms.

Students occasionally use L'Hôpital's Rule incorrectly by applying the Quotient Rule to $f(x)/g(x)$. Be sure you see that the rule involves $f'(x)/g'(x)$, not the derivative of $f(x)/g(x)$.

EXAMPLE 1   Indeterminate form $0/0$

Evaluate

$$\lim_{x \to 0} \frac{e^{2x} - 1}{x}.$$

SOLUTION

Since direct substitution results in the indeterminate form 0/0, we apply L'Hôpital's Rule to obtain

$$\lim_{x \to 0} \frac{e^{2x} - 1}{x} = \lim_{x \to 0} \frac{\frac{d}{dx}[e^{2x} - 1]}{\frac{d}{dx}[x]} = \lim_{x \to 0} \frac{2e^{2x}}{1} = 2.$$

REMARK   In writing the string of equations in Example 1, we actually do not know that the first limit is equal to the second until we have shown that the second limit exists. In other words, if the second limit had not existed, it would not have been permissible to apply L'Hôpital's Rule.

Another form of L'Hôpital's Rule states that if the limit of $f(x)/g(x)$ as x approaches ∞ (or $-\infty$) produces the indeterminate form 0/0 or ∞/∞, then

$$\lim_{x \to \infty} \frac{f(x)}{g(x)} = \lim_{x \to \infty} \frac{f'(x)}{g'(x)}$$

provided the limit on the right exists. We illustrate this form of L'Hôpital's Rule in Example 2.

EXAMPLE 2   Indeterminate form ∞/∞

Evaluate

$$\lim_{x \to \infty} \frac{\ln x}{x}.$$

SOLUTION

Since direct substitution results in the indeterminate form ∞/∞, we apply L'Hôpital's Rule to obtain

$$\lim_{x \to \infty} \frac{\ln x}{x} = \lim_{x \to \infty} \frac{\frac{d}{dx}[\ln x]}{\frac{d}{dx}[x]} = \lim_{x \to \infty} \frac{1}{x} = 0.$$

Occasionally it is necessary to apply L'Hôpital's Rule more than once to remove an indeterminate form. This is illustrated in the next example.

EXAMPLE 3   Applying L'Hôpital's Rule more than once

Evaluate

$$\lim_{x \to -\infty} \frac{x^2}{e^{-x}}.$$

SOLUTION

Since direct substitution results in the indeterminate form ∞/∞, we apply L'Hôpital's Rule to obtain

$$\lim_{x \to -\infty} \frac{x^2}{e^{-x}} = \lim_{x \to -\infty} \frac{\dfrac{d}{dx}[x^2]}{\dfrac{d}{dx}[e^{-x}]} = \lim_{x \to -\infty} \frac{2x}{-e^{-x}}.$$

This limit yields the indeterminate form $(-\infty)/(-\infty)$, so we apply L'Hôpital's Rule again to obtain

$$\lim_{x \to -\infty} \frac{2x}{-e^{-x}} = \lim_{x \to -\infty} \frac{\dfrac{d}{dx}[2x]}{\dfrac{d}{dx}[-e^{-x}]} = \lim_{x \to -\infty} \frac{2}{e^{-x}} = 0. \qquad \blacksquare$$

In addition to the forms $0/0$ and ∞/∞, there are other indeterminate forms such as $0 \cdot \infty$, 1^∞, ∞^0, 0^0, and $\infty - \infty$. For example, consider the following four limits that lead to the indeterminate form $0 \cdot \infty$.

$$\underbrace{\lim_{x \to 0} (x)\left(\frac{1}{x}\right)}_{\text{limit is } 1}, \qquad \underbrace{\lim_{x \to 0} (x)\left(\frac{2}{x}\right)}_{\text{limit is } 2}, \qquad \underbrace{\lim_{x \to \infty} (x)\left(\frac{1}{e^x}\right)}_{\text{limit is } 0}, \qquad \underbrace{\lim_{x \to \infty} (e^x)\left(\frac{1}{x}\right)}_{\text{limit is } \infty}$$

Since each limit is different, it is clear that the form $0 \cdot \infty$ is indeterminate in the sense that it does not determine the value (or even the existence) of the limit. The following examples indicate methods for evaluating these forms. Basically, we attempt to convert each of these forms to those for which L'Hôpital's Rule is applicable.

EXAMPLE 4   Indeterminate form $0 \cdot \infty$

Evaluate

$$\lim_{x \to \infty} e^{-x}\sqrt{x}.$$

SOLUTION

Since direct substitution produces the indeterminate form $0 \cdot \infty$, we rewrite the limit to fit the form $0/0$ or ∞/∞. In this case, we choose the second form and write

$$\lim_{x \to \infty} e^{-x}\sqrt{x} = \lim_{x \to \infty} \frac{\sqrt{x}}{e^x}.$$

Now, by L'Hôpital's Rule, we have

$$\lim_{x \to \infty} \frac{\sqrt{x}}{e^x} = \lim_{x \to \infty} \frac{1/(2\sqrt{x})}{e^x} = \lim_{x \to \infty} \frac{1}{2\sqrt{x}\, e^x} = 0.$$

If rewriting a limit in one of the forms $0/0$ or ∞/∞ does not seem to work, try the other form. For instance, in Example 4 we could have written the limit as

$$\lim_{x \to \infty} e^{-x}\sqrt{x} = \lim_{x \to \infty} \frac{e^{-x}}{x^{-1/2}}$$

which yields the indeterminate form $0/0$. However, in applying L'Hôpital's Rule to this limit, we obtain

$$\lim_{x \to \infty} \frac{e^{-x}}{x^{-1/2}} = \lim_{x \to \infty} \frac{-e^{-x}}{-1/(2x^{3/2})}$$

which also yields the indeterminate form $0/0$. Moreover, since the quotient seems to be getting more complicated, we abandon this approach and try the ∞/∞ form, as shown in Example 4.

The indeterminate forms 1^∞, ∞^0, and 0^0 arise from limits of functions that have a variable base and a variable exponent. When we encountered this type of function in Section 6.5 we used logarithmic differentiation to find the derivative. We use a similar procedure when taking limits, as indicated in the next example. (Note that this example serves as a proof of Theorem 6.16.)

EXAMPLE 5   Indeterminate form 1^∞

Evaluate

$$\lim_{x \to \infty} \left(1 + \frac{1}{x}\right)^x.$$

SOLUTION

Since direct substitution yields the indeterminate form 1^∞, we proceed as follows. We assume the limit exists, and we represent it by

$$y = \lim_{x \to \infty} \left(1 + \frac{1}{x}\right)^x.$$

Now, taking the natural logarithm of both sides, we have

$$\ln y = \ln \left[\lim_{x \to \infty} \left(1 + \frac{1}{x} \right)^x \right].$$

Using the fact that the natural logarithmic function is continuous, we write

$$\ln y = \lim_{x \to \infty} \left[x \ln \left(1 + \frac{1}{x} \right) \right] \qquad \text{Indeterminate form: } \infty \cdot 0$$

$$= \lim_{x \to \infty} \left[\frac{\ln [1 + (1/x)]}{1/x} \right] \qquad \text{Indeterminate form: } 0/0$$

$$= \lim_{x \to \infty} \left[\frac{(-1/x^2)(1/[1 + (1/x)])}{-1/x^2} \right] \qquad \text{L'Hôpital's Rule}$$

$$= \lim_{x \to \infty} \frac{1}{1 + (1/x)} = 1.$$

Finally, since $\ln y = 1$, we know that $y = e$ and we conclude that

$$\lim_{x \to \infty} \left(1 + \frac{1}{x} \right)^x = e. \qquad \blacksquare$$

L'Hôpital's Rule can also be applied to one-sided limits, as demonstrated in Examples 6 and 7.

EXAMPLE 6 Indeterminate form 0^0

Evaluate

$$\lim_{x \to 0^+} (\sin x)^x.$$

SOLUTION

Since direct substitution produces the indeterminate form 0^0, we proceed as follows.

$$y = \lim_{x \to 0^+} (\sin x)^x \qquad \text{Indeterminate form } 0^0$$

$$\ln y = \ln \left[\lim_{x \to 0^+} (\sin x)^x \right] \qquad \text{Take log of both sides}$$

$$= \lim_{x \to 0^+} [\ln (\sin x)^x] \qquad \text{Continuity}$$

$$= \lim_{x \to 0^+} [x \ln (\sin x)] \qquad \text{Indeterminate form } 0 \cdot (-\infty)$$

$$= \lim_{x \to 0^+} \frac{\ln (\sin x)}{1/x} \qquad \text{Indeterminate form } -\infty/\infty$$

$$= \lim_{x \to 0^+} \frac{\cot x}{-1/x^2} \qquad \text{L'Hôpital's Rule}$$

$$= \lim_{x \to 0^+} \frac{-x^2}{\tan x} \qquad \text{Indeterminate form } 0/0$$

$$= \lim_{x \to 0^+} \frac{-2x}{\sec^2 x} = 0 \qquad \text{L'Hôpital's Rule}$$

Now, since $\ln y = 0$, we conclude that $y = e^0 = 1$, and it follows that

$$\lim_{x \to 0^+} (\sin x)^x = 1.$$

REMARK   When evaluating complicated limits like the one in Example 6, it is helpful to check the reasonableness of the solution with a calculator. For instance, the calculations shown in Table 8.3 are consistent with our conclusion that $(\sin x)^x$ approaches 1 as x approaches 0 from the right.

TABLE 8.3

x	1.0	0.1	0.01	0.001	0.0001	0.00001
$(\sin x)^x$	0.8415	0.7942	0.9550	0.9931	0.9991	0.9999

EXAMPLE 7   Indeterminate form $\infty - \infty$

Evaluate

$$\lim_{x \to 1^+} \left(\frac{1}{\ln x} - \frac{1}{x - 1} \right).$$

SOLUTION

Since direct substitution yields the indeterminate form $\infty - \infty$, we try to rewrite the expression to produce a form to which we can apply L'Hôpital's Rule. In this case, we combine the two fractions to obtain

$$\lim_{x \to 1^+} \left(\frac{1}{\ln x} - \frac{1}{x - 1} \right) = \lim_{x \to 1^+} \left[\frac{x - 1 - \ln x}{(x - 1) \ln x} \right].$$

Now, since direct substitution produces the indeterminate form $0/0$, we can apply L'Hôpital's Rule to obtain

$$\lim_{x \to 1^+} \left(\frac{1}{\ln x} - \frac{1}{x - 1} \right) = \lim_{x \to 1^+} \left[\frac{1 - (1/x)}{(x - 1)(1/x) + \ln x} \right]$$

$$= \lim_{x \to 1^+} \left[\frac{x - 1}{x - 1 + x \ln x} \right].$$

This limit also yields the indeterminate form $0/0$, so we apply L'Hôpital's Rule again to obtain

$$\lim_{x \to 1^+} \left(\frac{1}{\ln x} - \frac{1}{x - 1} \right) = \lim_{x \to 1^+} \left[\frac{1}{1 + x(1/x) + \ln x} \right] = \frac{1}{2}.$$

We have identified the forms $0/0$, ∞/∞, $\infty - \infty$, $0 \cdot \infty$, 0^0, 1^∞, and ∞^0 as *indeterminate*. There are similar forms that you should recognize as *determinate*, such as

$$\infty + \infty \to \infty \qquad -\infty - \infty \to -\infty \qquad 0^\infty \to 0 \qquad 0^{-\infty} \to \infty .$$

(You are asked to verify two of these in Exercises 43 and 44.)

In each of the examples so far in this section, we have used L'Hôpital's Rule to find a limit that exists. L'Hôpital's Rule can also be used to conclude that a limit is infinite, and this is demonstrated in the last example.

EXAMPLE 8   An infinite limit

Evaluate

$$\lim_{x \to \infty} \frac{e^x}{x}.$$

SOLUTION

Direct substitution produces the indeterminate form ∞/∞, therefore we apply L'Hôpital's Rule to obtain

$$\lim_{x \to \infty} \frac{e^x}{x} = \lim_{x \to \infty} \frac{e^x}{1} = \lim_{x \to \infty} e^x = \infty.$$

Now, since $e^x \to \infty$ as $x \to \infty$, we conclude that the limit of e^x/x as $x \to \infty$ is also infinite.

As a final comment, we remind you that L'Hôpital's Rule can be applied only to quotients leading to the indeterminate forms $0/0$ or ∞/∞. For instance, the following application of L'Hôpital's Rule is *incorrect*.

$$\lim_{x \to 0} \frac{e^x}{x} \overset{?}{=} \lim_{x \to 0} \frac{e^x}{1} = 1 \quad \text{— Incorrect use of L'Hôpital's Rule}$$

The reason this application is incorrect is that, even though the limit of the denominator is 0, the limit of the numerator is 1, which means that the hypotheses of L'Hôpital's Rule have not been satisfied.

EXERCISES for Section 8.7

In Exercises 1–34, evaluate each limit, using L'Hôpital's Rule if necessary.

1. $\displaystyle \lim_{x \to 2} \frac{x^2 - x - 2}{x - 2}$

2. $\displaystyle \lim_{x \to -1} \frac{x^2 - x - 2}{x + 1}$

3. $\displaystyle \lim_{x \to 0} \frac{\sqrt{4 - x^2} - 2}{x}$

4. $\displaystyle \lim_{x \to 2^-} \frac{\sqrt{4 - x^2}}{x - 2}$

5. $\displaystyle \lim_{x \to 0} \frac{e^x - (1 - x)}{x}$

6. $\displaystyle \lim_{x \to 0^+} \frac{e^x - (1 + x)}{x^3}$

7. $\displaystyle \lim_{x \to 0^+} \frac{e^x - (1 + x)}{x^n}$, $\quad n = 1, 2, 3, \ldots$

8. $\displaystyle \lim_{x \to 1} \frac{\ln x}{x^2 - 1}$

9. $\displaystyle \lim_{x \to \infty} \frac{\ln x}{x}$

10. $\displaystyle \lim_{x \to \infty} \frac{e^x}{x}$

11. $\displaystyle \lim_{x \to \infty} \frac{3x^2 - 2x + 1}{2x^2 + 3}$

12. $\displaystyle \lim_{x \to \infty} \frac{x - 1}{x^2 + 2x + 3}$

13. $\displaystyle \lim_{x \to \infty} \frac{x^2 + 2x + 3}{x - 1}$

14. $\displaystyle \lim_{x \to \infty} \frac{x^2}{e^x}$

15. $\displaystyle \lim_{x \to 0^+} x^2 \ln x$

16. $\displaystyle \lim_{x \to 0} \left(\frac{1}{x} - \frac{1}{x^2} \right)$

17. $\displaystyle \lim_{x \to 2} \left(\frac{8}{x^2 - 4} - \frac{x}{x - 2} \right)$

18. $\displaystyle \lim_{x \to 2} \left(\frac{1}{x^2 - 4} - \frac{\sqrt{x - 1}}{x^2 - 4} \right)$

19. $\lim\limits_{x \to \infty} \dfrac{x}{\sqrt{x^2 + 1}}$

20. $\lim\limits_{x \to 1^+} \left(\dfrac{3}{\ln x} - \dfrac{2}{x - 1} \right)$

21. $\lim\limits_{x \to 0^+} x^{1/x}$

22. $\lim\limits_{x \to 0^+} (e^x + x)^{1/x}$

23. $\lim\limits_{x \to \infty} x^{1/x}$

24. $\lim\limits_{x \to \infty} \left(1 + \dfrac{1}{x} \right)^x$

25. $\lim\limits_{x \to \infty} (1 + x)^{1/x}$

26. $\lim\limits_{x \to \pi} \dfrac{\sin x}{x - \pi}$

27. $\lim\limits_{x \to 0} \dfrac{\sin 2x}{\sin 3x}$

28. $\lim\limits_{x \to 0} \dfrac{\sin ax}{\sin bx}$

29. $\lim\limits_{x \to 0} x \csc x$

30. $\lim\limits_{x \to 0} x^2 \cot x$

31. $\lim\limits_{x \to \infty} \left(x \sin \dfrac{1}{x} \right)$

32. $\lim\limits_{x \to \infty} x \tan \dfrac{1}{x}$

33. $\lim\limits_{x \to 0} \dfrac{\arcsin x}{x}$

34. $\lim\limits_{x \to 1} \dfrac{\arctan x - (\pi/4)}{x - 1}$

In Exercises 35–40, use L'Hôpital's Rule to determine the comparative rates of increase of the functions.

$$f(x) = x^m$$
$$g(x) = e^{nx}$$
$$h(x) = (\ln x)^n$$

where $n > 0$, $m > 0$, and $x \to \infty$. The limits obtained in these exercises suggest that $(\ln x)^n$ tends toward infinity more slowly than x^m, which in turn tends toward infinity more slowly than e^{nx}.

35. $\lim\limits_{x \to \infty} \dfrac{x^2}{e^{5x}}$

36. $\lim\limits_{x \to \infty} \dfrac{x^3}{e^{2x}}$

37. $\lim\limits_{x \to \infty} \dfrac{(\ln x)^3}{x}$

38. $\lim\limits_{x \to \infty} \dfrac{(\ln x)^2}{x^3}$

39. $\lim\limits_{x \to \infty} \dfrac{(\ln x)^n}{x^m}$, where $0 < n, m$

40. $\lim\limits_{x \to \infty} \dfrac{x^m}{e^{nx}}$, where $0 < n, m$

41. Complete the following table to show that x eventually "overpowers" $(\ln x)^4$.

x	10	10^2	10^4	10^6	10^8	10^{10}
$\dfrac{(\ln x)^4}{x}$						

42. Complete the following table to show that e^x eventually "overpowers" x^5.

x	1	5	10	20	30	40	50	100
$\dfrac{e^x}{x^5}$								

43. Prove that if $f(x) \geq 0$, $\lim\limits_{x \to a} f(x) = 0$, and $\lim\limits_{x \to a} g(x) = \infty$, then

$$\lim\limits_{x \to a} f(x)^{g(x)} = 0.$$

44. Prove that if $f(x) \geq 0$, $\lim\limits_{x \to a} f(x) = 0$, and $\lim\limits_{x \to a} g(x) = -\infty$, then

$$\lim\limits_{x \to a} f(x)^{g(x)} = \infty.$$

In Exercises 45–48, find any asymptotes and relative extrema that may exist and sketch the graph of the function. [Hint: Some of the limits required in finding asymptotes have been found in preceding exercises.]

45. $y = x^{1/x}$, $x > 0$

46. $y = x^x$, $x > 0$

47. $y = 2xe^{-x}$

48. $y = \dfrac{\ln x}{x}$

In Exercises 49–52, L'Hôpital's Rule is used *incorrectly*. Describe the error.

49. $\lim\limits_{x \to 0} \dfrac{e^{2x} - 1}{e^x} = \lim\limits_{x \to 0} \dfrac{2e^{2x}}{e^x} = \lim\limits_{x \to 0} 2e^x = 2$

50. $\lim\limits_{x \to 0} \dfrac{\sin \pi x - 1}{x} = \lim\limits_{x \to 0} \dfrac{\pi \cos \pi x}{1} = \pi$

51. $\lim\limits_{x \to \infty} x \cos \dfrac{1}{x} = \lim\limits_{x \to \infty} \dfrac{\cos (1/x)}{1/x}$

$$= \lim\limits_{x \to \infty} \dfrac{[-\sin (1/x)](1/x^2)}{-1/x^2} = 0$$

52. $\lim\limits_{x \to \infty} \dfrac{e^{-x}}{1 + e^{-x}} = \lim\limits_{x \to \infty} \dfrac{-e^{-x}}{-e^{-x}} = \lim\limits_{x \to \infty} 1 = 1$

53. Show that L'Hôpital's Rule fails for the limit

$$\lim\limits_{x \to \infty} \dfrac{x}{\sqrt{x^2 + 1}}.$$

What is this limit?

54. The formula for the amount A in a savings account compounded n times a year for t years at an interest rate of r and an initial deposit of P is

$$A = P\left(1 + \dfrac{r}{n} \right)^{nt}.$$

Use L'Hôpital's Rule to show that the limiting formula as the number of compoundings per year becomes infinite is

$$A = Pe^{rt}.$$

(Note that this limiting formula is used for continuous compounding of interest.)

55. The velocity of an object falling through a resisting medium such as air or water is given by

$$v = \frac{32}{k}\left(1 - e^{-kt} + \frac{v_0 k e^{-kt}}{32}\right)$$

where v_0 is the initial velocity, t is the time, and k is the resistance constant of the medium. Use L'Hôpital's Rule to find the formula for the velocity of a falling body in a vacuum by fixing v_0 and t and letting k approach zero. (Assume the downward direction is positive.)

56. The Gamma Function $\Gamma(n)$ is defined in terms of the integral of the function

$$f(x) = x^{n-1}e^{-x}, \quad n > 0.$$

Show that for any fixed value of n, the limit of $f(x)$ as x approaches infinity is zero.

In Exercises 57–60, use a computer or graphics calculator, to (a) graph the given function, and (b) find the required limit (if it exists).

57. $\lim_{x \to 3} \dfrac{x - 3}{\ln (2x - 5)}$

58. $\lim_{x \to 0^+} (\sin x)^x$

59. $\lim_{x \to \infty} (\sqrt{x^2 + 5x + 2} - x)$

60. $\lim_{x \to \infty} \dfrac{x^3}{e^{2x}}$

61. Use a computer or graphics calculator to sketch the graph of

$$f(x) = \frac{x^k - 1}{k}$$

for $k = 1, 0.1, 0.01$. Then evaluate the limit

$$\lim_{k \to 0^+} \frac{x^k - 1}{k}.$$

8.8 Improper Integrals

Improper integrals (infinite limits of integration) ▪ Improper integrals (infinite discontinuities)

The definition of the definite integral $\int_a^b f(x)\, dx$ requires that the interval $[a, b]$ be finite. Furthermore, the Fundamental Theorem of Calculus, by which we have been evaluating definite integrals, requires that f be continuous on $[a, b]$. In this section we discuss a limit procedure for evaluating integrals that do not satisfy these requirements because either (1) one or both of the limits of integration are infinite, or (2) f has a finite number of infinite discontinuities in the interval $[a, b]$. Integrals that possess either property are called **improper integrals.** Note that we say that a function f has an **infinite discontinuity** at c if, *from the right or left,*

$$\lim_{x \to c} f(x) = \infty \quad \text{or} \quad \lim_{x \to c} f(x) = -\infty.$$

EXAMPLE 1 Examples of improper integrals

(a) The integrals

$$\int_1^\infty \frac{dx}{x} \quad \text{and} \quad \int_{-\infty}^\infty \frac{dx}{x^2 + 1}$$

are improper because one or both of the limits of integration are infinite.

(b) The integrals

$$\int_1^5 \frac{dx}{\sqrt{x - 1}} \quad \text{and} \quad \int_{-2}^2 \frac{dx}{(x + 1)^2}$$

are improper because the integrands have infinite discontinuities somewhere in the interval of integration. $1/\sqrt{x - 1}$ has an infinite discontinuity at $x = 1$, and $1/(x + 1)^2$ has an infinite discontinuity at $x = -1$.

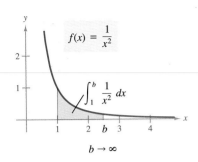

$f(x) = \dfrac{1}{x^2}$

$\displaystyle\int_1^b \dfrac{1}{x^2}\,dx$

$b \to \infty$

FIGURE 8.16

Improper integrals with infinite limits of integration

To get an idea of how we can evaluate an improper integral, consider the integral

$$\int_1^b \frac{dx}{x^2} = \left[-\frac{1}{x} \right]_1^b = -\frac{1}{b} + 1 = 1 - \frac{1}{b}$$

which we can interpret as the area of the shaded region shown in Figure 8.16. Taking the limit as $b \to \infty$, we have

$$\int_1^\infty \frac{dx}{x^2} = \lim_{b \to \infty} \left[\int_1^b \frac{dx}{x^2} \right] = \lim_{b \to \infty} \left(1 - \frac{1}{b} \right) = 1$$

and we can interpret this improper integral as the area of the *unbounded* region between the graph of $f(x) = 1/x^2$ and the x-axis (to the right of $x = 1$). We define improper integrals with infinite limits of integration as follows.

DEFINITION OF IMPROPER INTEGRALS WITH INFINITE LIMITS OF INTEGRATION

1. If f is continuous on the interval $[a, \infty)$, then

$$\int_a^\infty f(x)\,dx = \lim_{b \to \infty} \int_a^b f(x)\,dx.$$

2. If f is continuous on the interval $(-\infty, b]$, then

$$\int_{-\infty}^b f(x)\,dx = \lim_{a \to -\infty} \int_a^b f(x)\,dx.$$

3. If f is continuous on the interval $(-\infty, \infty)$, then

$$\int_{-\infty}^\infty f(x)\,dx = \int_{-\infty}^c f(x)\,dx + \int_c^\infty f(x)\,dx$$

where c is any real number.

In each case, if the limit exists, then the improper integral is said to **converge;** otherwise, the improper integral **diverges.** In the third case, the improper integral on the left diverges if either of the improper integrals on the right diverges.

EXAMPLE 2   An improper integral that diverges

Evaluate

$$\int_1^\infty \frac{dx}{x}.$$

SOLUTION

$$\int_1^\infty \frac{dx}{x} = \lim_{b \to \infty} \int_1^b \frac{dx}{x} = \lim_{b \to \infty} \left[\ln x \right]_1^b = \lim_{b \to \infty} (\ln b - 0) = \infty$$

FIGURE 8.17

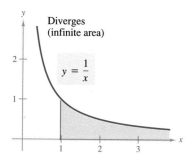

FIGURE 8.18

A comparison of the improper integral $\int_1^\infty (1/x^2)\, dx$, which converges to 1, and $\int_1^\infty (1/x)\, dx$, which diverges, suggests the rather unpredictable nature of improper integrals. The functions $f(x) = 1/x^2$ and $g(x) = 1/x$ have similar graphs, as shown in Figures 8.17 and 8.18. Yet, the shaded region shown in Figure 8.17 has a finite area, whereas the shaded region shown in Figure 8.18 has an infinite area.

EXAMPLE 3 Evaluating improper integrals

(a) $\displaystyle\int_0^\infty e^{-x}\, dx = \lim_{b\to\infty} \int_0^b e^{-x}\, dx = \lim_{b\to\infty}\left[-e^{-x}\right]_0^b = \lim_{b\to\infty}(-e^{-b} + 1) = 1$

(b) $\displaystyle\int_0^\infty \frac{1}{x^2 + 1}\, dx = \lim_{b\to\infty} \int_0^b \frac{1}{x^2 + 1}\, dx = \lim_{b\to\infty}\left[\arctan x\right]_0^b$

$$= \lim_{b\to\infty} \arctan b = \frac{\pi}{2}$$

EXAMPLE 4 An improper integral that diverges

Evaluate

$$\int_0^\infty \sin x\, dx.$$

SOLUTION

$$\int_0^\infty \sin x\, dx = \lim_{b\to\infty} \int_0^b \sin x\, dx = \lim_{b\to\infty}\left[-\cos x\right]_0^b = \lim_{b\to\infty}(1 - \cos b)$$

Since $\cos b$ does not approach a limit as b approaches infinity, we conclude that the given improper integral diverges.

REMARK   Note that the improper integrals in Examples 2 and 4 diverge for different reasons. In Example 2, the improper integral $\int_1^\infty (1/x)\, dx$ diverges because the region (to the right of 1) between $y = 1/x$ and the x-axis has an infinite area. In Example 4, the improper integral $\int_0^\infty \sin x\, dx$ diverges because of the oscillation of the sine function. This type of divergence is sometimes called *divergence by oscillation*.

In the following example, we illustrate a technique for using L'Hôpital's Rule to evaluate an improper integral.

EXAMPLE 5 Using L'Hôpital's Rule with an improper integral

Evaluate

$$\int_1^\infty (1 - x)e^{-x}\, dx.$$

SOLUTION

Using integration by parts, with $dv = e^{-x} \, dx$ and $u = (1 - x)$, we have

$$\int u \, dv \quad = \quad uv \quad - \quad \int v \, du$$

$$\int (1 - x)e^{-x} \, dx = -e^{-x}(1 - x) - \int e^{-x} \, dx$$
$$= -e^{-x} + xe^{-x} + e^{-x} + C = xe^{-x} + C.$$

Therefore, we have

$$\int_1^\infty (1 - x)e^{-x} \, dx = \lim_{b \to \infty} \left[xe^{-x} \right]_1^b = \left(\lim_{b \to \infty} \frac{b}{e^b} \right) - \frac{1}{e}.$$

Now, using L'Hôpital's Rule on the right-hand limit, we have

$$\lim_{b \to \infty} \frac{b}{e^b} = \lim_{b \to \infty} \frac{1}{e^b} = 0.$$

Finally, we conclude that

$$\int_1^\infty (1 - x)e^{-x} \, dx = -\frac{1}{e}.$$

(See Figure 8.19.)

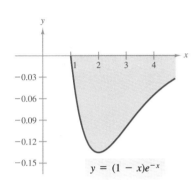

FIGURE 8.19

$y = (1 - x)e^{-x}$

EXAMPLE 6   An integral with infinite upper and lower limits of integration

Evaluate

$$\int_{-\infty}^\infty \frac{e^x}{1 + e^{2x}} \, dx.$$

SOLUTION

Note that the integrand is continuous on $(-\infty, \infty)$. To evaluate this integral, we break it into two parts, choosing $c = 0$ as a convenient value.

$$\int_{-\infty}^\infty \frac{e^x}{1 + e^{2x}} \, dx = \int_{-\infty}^0 \frac{e^x}{1 + e^{2x}} \, dx + \int_0^\infty \frac{e^x}{1 + e^{2x}} \, dx$$
$$= \lim_{b \to -\infty} \left[\arctan e^x \right]_b^0 + \lim_{b \to \infty} \left[\arctan e^x \right]_0^b$$
$$= \lim_{b \to -\infty} \left[\frac{\pi}{4} - \arctan e^b \right] + \lim_{b \to \infty} \left[\arctan e^b - \frac{\pi}{4} \right]$$
$$= \frac{\pi}{4} - 0 + \frac{\pi}{2} - \frac{\pi}{4}$$
$$= \frac{\pi}{2}$$

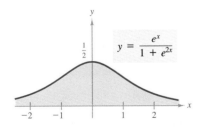

FIGURE 8.20

$y = \dfrac{e^x}{1 + e^{2x}}$

(See Figure 8.20.)

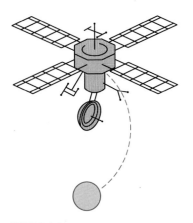

FIGURE 8.21

EXAMPLE 7 An application involving an improper integral

In Example 3 of Section 7.5, we determined that it would require 10,000 mile-tons of work to propel a 15-ton space module to a height of 800 miles above the earth. How much work is required to propel this same module an unlimited distance away from the earth's surface? (See Figure 8.21.)

SOLUTION

At first you might think that an infinite amount of work would be required. But if this were the case, then it would be impossible to send rockets into outer space. Since this has been done, the work required must be finite, and we can determine the work in the following manner. Using the integral of Example 3, Section 7.5, we replace the upper bound of 4800 miles by ∞ and write

$$W = \int_{4000}^{\infty} \frac{240{,}000{,}000}{x^2}\, dx$$

$$= \lim_{b \to \infty} \left[-\frac{240{,}000{,}000}{x} \right]_{4000}^{b}$$

$$= \lim_{b \to \infty} \left[-\frac{240{,}000{,}000}{b} + \frac{240{,}000{,}000}{4000} \right]$$

$$= 60{,}000 \text{ mile-tons}$$

$$\approx 6.336 \times 10^{11} \text{ ft} \cdot \text{lb.}$$

Improper integrals with infinite discontinuities

The second basic type of improper integral is one that has an infinite discontinuity *at or between* the limits of integration.

DEFINITION OF IMPROPER INTEGRALS WITH AN INFINITE DISCONTINUITY

1. If f is continuous on the interval $[a, b)$ and has an infinite discontinuity at b, then

$$\int_a^b f(x)\, dx = \lim_{c \to b^-} \int_a^c f(x)\, dx.$$

2. If f is continuous on the interval $(a, b]$ and has an infinite discontinuity at a, then

$$\int_a^b f(x)\, dx = \lim_{c \to a^+} \int_c^b f(x)\, dx.$$

3. If f is continuous on the interval $[a, b]$, except for some c in (a, b) at which f has an infinite discontinuity, then

$$\int_a^b f(x)\, dx = \int_a^c f(x)\, dx + \int_c^b f(x)\, dx.$$

In each case, if the limit exists, then the improper integral is said to **converge;** otherwise, the improper integral **diverges.** In the third case, the improper integral on the left diverges if either of the improper integrals on the right diverges.

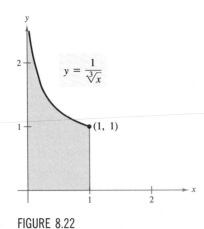

FIGURE 8.22

EXAMPLE 8 An improper integral with an infinite discontinuity

Evaluate

$$\int_0^1 \frac{dx}{\sqrt[3]{x}}.$$

SOLUTION

The integrand has an infinite discontinuity at $x = 0$, as shown in Figure 8.22. To evaluate this integral, we write

$$\int_0^1 x^{-1/3} \, dx = \lim_{b \to 0^+} \left[\frac{x^{2/3}}{2/3} \right]_b^1$$

$$= \lim_{b \to 0^+} \frac{3}{2}(1 - b^{2/3}) = \frac{3}{2}.$$

EXAMPLE 9 An improper integral that diverges

Evaluate

$$\int_0^2 \frac{dx}{x^3}.$$

SOLUTION

Since the integrand has an infinite discontinuity at $x = 0$, we write

$$\int_0^2 \frac{dx}{x^3} = \lim_{b \to 0^+} \left[-\frac{1}{2x^2} \right]_b^2 = \lim_{b \to 0^+} \left[-\frac{1}{8} + \frac{1}{2b^2} \right] = \infty.$$

Thus, we conclude that this improper integral diverges.

EXAMPLE 10 An improper integral with an interior discontinuity

Evaluate

$$\int_{-1}^2 \frac{dx}{x^3}.$$

SOLUTION

This integral is improper because the integrand has an infinite discontinuity at the interior point $x = 0$, as shown in Figure 8.23. Thus, we write

$$\int_{-1}^2 \frac{dx}{x^3} = \int_{-1}^0 \frac{dx}{x^3} + \int_0^2 \frac{dx}{x^3}.$$

From Example 9 we know that the second integral diverges. Therefore, the original improper integral also diverges.

FIGURE 8.23

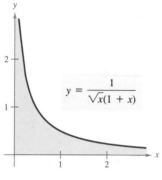

$$y = \frac{1}{\sqrt{x}(1 + x)}$$

FIGURE 8.24

REMARK   Remember to check for infinite discontinuities at interior points as well as endpoints when determining whether an integral is improper. For instance, if we had not recognized that the integral in Example 10 was improper, we would have obtained the *incorrect* result

$$\int_{-1}^{2} \frac{dx}{x^3} \stackrel{?}{=} \left[\frac{-1}{2x^2}\right]_{-1}^{2} = -\frac{1}{8} + \frac{1}{2} = \frac{3}{8}. \qquad \text{Incorrect evaluation}$$

The integral in the next example is improper for *two* reasons. One limit of integration is infinite, and the integrand has an infinite discontinuity at the other limit of integration, as shown in Figure 8.24.

EXAMPLE 11   A doubly improper integral

Evaluate

$$\int_{0}^{\infty} \frac{dx}{\sqrt{x}(x + 1)}.$$

SOLUTION

To evaluate this integral, we split it at a convenient point (say, $x = 1$) and write

$$\int_{0}^{\infty} \frac{dx}{\sqrt{x}(x + 1)} = \int_{0}^{1} \frac{dx}{\sqrt{x}(x + 1)} + \int_{1}^{\infty} \frac{dx}{\sqrt{x}(x + 1)}$$

$$= \lim_{b \to 0^+} \left[2 \arctan \sqrt{x}\right]_{b}^{1} + \lim_{c \to \infty} \left[2 \arctan \sqrt{x}\right]_{1}^{c}$$

$$= 2\left(\frac{\pi}{4}\right) - 0 + 2\left(\frac{\pi}{2}\right) - 2\left(\frac{\pi}{4}\right) = \pi. \qquad \square$$

EXAMPLE 12   An improper integral involving L'Hôpital's Rule

Evaluate

$$\int_{0}^{1} \ln x \, dx.$$

SOLUTION

Using integration by parts, with $dv = dx$ and $u = \ln x$, we have

$$\int_{0}^{1} \ln x \, dx = \lim_{b \to 0^+} \left[x \ln x - x\right]_{b}^{1} = \lim_{b \to 0^+} [0 - 1 - b \ln b + b].$$

By L'Hôpital's Rule, we have

$$\lim_{b \to 0^+} b \ln b = \lim_{b \to 0^+} \frac{\ln b}{1/b} = \lim_{b \to 0^+} \frac{1/b}{-1/b^2} = \lim_{b \to 0^+} - b = 0.$$

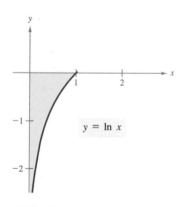

$$y = \ln x$$

FIGURE 8.25

Finally, we conclude that

$$\int_0^1 \ln x \, dx = \lim_{b \to 0^+} [0 - 1 - b \ln b + b] = 0 - 1 - 0 + 0 = -1.$$

(See Figure 8.25.)

EXAMPLE 13   An application involving arc length

Use the formula for arc length to show that the circumference of the circle $x^2 + y^2 = 1$ is 2π.

SOLUTION

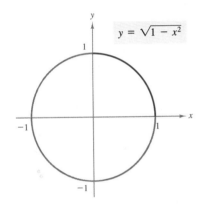

$y = \sqrt{1 - x^2}$

FIGURE 8.26

To simplify the work, we consider the quarter circle given by $y = \sqrt{1 - x^2}$, where $0 \le x \le 1$. The arc length of this quarter circle is given by

$$s = \int_0^1 \sqrt{1 + (y')^2} \, dx$$

$$= \int_0^1 \sqrt{1 + \left(\frac{-x}{\sqrt{1 - x^2}} \right)^2} \, dx$$

$$= \int_0^1 \frac{dx}{\sqrt{1 - x^2}}.$$

This integral is improper, since it has an infinite discontinuity at $x = 1$. Thus, we write

$$s = \int_0^1 \frac{dx}{\sqrt{1 - x^2}} = \lim_{b \to 1^-} \left[\arcsin x \right]_0^b = \frac{\pi}{2} - 0 = \frac{\pi}{2}.$$

Finally, multiplying by 4, we conclude that the circumference of the circle is $4s = 2\pi$, as shown in Figure 8.26.

We conclude this section with a useful theorem describing the convergence or divergence of a common type of improper integral. The proof of this theorem is left as an exercise (see Exercise 33).

THEOREM 8.4 **A SPECIAL TYPE OF IMPROPER** **INTEGRAL**	$\displaystyle \int_1^\infty \frac{dx}{x^p} = \begin{cases} \dfrac{1}{p - 1}, & \text{if } p > 1 \\ \text{diverges}, & \text{if } p \le 1 \end{cases}$

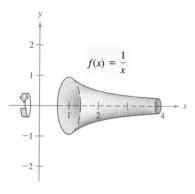

FIGURE 8.27

EXAMPLE 14   An application involving a solid of revolution

The solid formed by revolving the *unbounded* region lying between the graph of $f(x) = 1/x$ and the x-axis ($x \geq 1$) is called **Gabriel's Horn.** (See Figure 8.27.) Show that this solid has a finite volume and an infinite surface area.

SOLUTION

Using the disc method, with Theorem 8.4, we determine the volume to be

$$V = \pi \int_1^\infty \left(\frac{1}{x}\right)^2 dx$$

$$= \pi\left(\frac{1}{2-1}\right) = \pi.$$

The surface area is given by

$$S = 2\pi \int_1^\infty f(x)\sqrt{1 + [f'(x)]^2}\, dx$$

$$= 2\pi \int_1^\infty \frac{1}{x}\sqrt{1 + \frac{1}{x^4}}\, dx$$

$$= 2\pi \int_1^\infty \frac{\sqrt{1 + x^4}}{x^3}\, dx.$$

Now, using the substitution $u = x^2$ together with Formula 30 in the integration table, we have

$$S = \lim_{b \to \infty} \pi \left[\frac{-\sqrt{1 + x^4}}{x^2} + \ln\left|x^2 + \sqrt{1 + x^4}\right|\right]_1^b.$$

We leave it up to you to show that this limit is infinite.

EXERCISES for Section 8.8

In Exercises 1–32, determine the divergence or convergence of the given improper integral. Evaluate the integral if it converges.

1. $\displaystyle\int_0^4 \frac{1}{\sqrt{x}}\, dx$

2. $\displaystyle\int_3^4 \frac{1}{\sqrt{x-3}}\, dx$

3. $\displaystyle\int_0^2 \frac{1}{(x-1)^{2/3}}\, dx$

4. $\displaystyle\int_0^2 \frac{1}{(x-1)^2}\, dx$

5. $\int_0^\infty e^{-x}\,dx$

6. $\int_{-\infty}^0 e^{2x}\,dx$

7. $\int_{-\infty}^0 xe^{-2x}\,dx$

8. $\int_0^\infty xe^{-x}\,dx$

9. $\int_0^\infty x^2 e^{-x}\,dx$

10. $\int_0^\infty (x-1)e^{-x}\,dx$

11. $\int_1^\infty \dfrac{1}{x^2}\,dx$

12. $\int_1^\infty \dfrac{1}{\sqrt{x}}\,dx$

13. $\int_0^\infty e^{-x}\cos x\,dx$

14. $\int_0^\infty e^{-ax}\sin bx\,dx,\ a>0$

15. $\int_{-\infty}^\infty \dfrac{1}{1+x^2}\,dx$

16. $\int_0^\infty \dfrac{x^3}{(x^2+1)^2}\,dx$

17. $\int_0^\infty \dfrac{1}{e^x+e^{-x}}\,dx$

18. $\int_0^\infty \dfrac{e^x}{1+e^x}\,dx$

19. $\int_0^\infty \cos \pi x\,dx$

20. $\int_0^\infty \sin \dfrac{x}{2}\,dx$

21. $\int_0^1 \dfrac{1}{x^2}\,dx$

22. $\int_0^1 \dfrac{1}{x}\,dx$

23. $\int_0^8 \dfrac{1}{\sqrt[3]{8-x}}\,dx$

24. $\int_0^e \ln x\,dx$

25. $\int_0^1 x\ln x\,dx$

26. $\int_0^{\pi/2} \sec \theta\,d\theta$

27. $\int_0^{\pi/2} \tan \theta\,d\theta$

28. $\int_0^2 \dfrac{1}{\sqrt{4-x^2}}\,dx$

29. $\int_2^4 \dfrac{1}{\sqrt{x^2-4}}\,dx$

30. $\int_0^2 \dfrac{1}{4-x^2}\,dx$

31. $\int_0^2 \dfrac{1}{\sqrt[3]{x-1}}\,dx$

32. $\int_0^2 \dfrac{1}{(x-1)^{4/3}}\,dx$

33. Determine all values of p for which the following integral converges.

$$\int_1^\infty \dfrac{1}{x^p}\,dx$$

34. Determine all values of p for which the following integral converges.

$$\int_0^1 \dfrac{1}{x^p}\,dx$$

35. Use mathematical induction to verify that the following integral converges for any positive integer n.

$$\int_0^\infty x^n e^{-x}\,dx$$

36. Given continuous functions f and g such that $0 \le f(x) \le g(x)$ on the interval (a, ∞), prove the following.
(a) If $\int_a^\infty g(x)\,dx$ converges, then $\int_a^\infty f(x)\,dx$ converges.
(b) If $\int_a^\infty f(x)\,dx$ diverges, then $\int_a^\infty g(x)\,dx$ diverges.

In Exercises 37–46, use the results of Exercises 33–36 to determine whether the improper integral converges.

37. $\int_0^1 \dfrac{1}{x^3}\,dx$

38. $\int_0^1 \dfrac{1}{\sqrt[3]{x}}\,dx$

39. $\int_1^\infty \dfrac{1}{x^3}\,dx$

40. $\int_0^\infty x^4 e^{-x}\,dx$

41. $\int_1^\infty \dfrac{1}{x^2+5}\,dx$

42. $\int_2^\infty \dfrac{1}{\sqrt{x-1}}\,dx$

43. $\int_2^\infty \dfrac{1}{\sqrt[3]{x(x-1)}}\,dx$

44. $\int_1^\infty \dfrac{1}{\sqrt{x}(x+1)}\,dx$

45. $\int_0^\infty e^{-x^2}\,dx$

46. $\int_2^\infty \dfrac{1}{\sqrt{x}\ln x}\,dx$

47. Given the region bounded by the graphs of $y = 1/x^2$ and $y = 0$ ($x \ge 1$), find the following.
(a) the area of the region
(b) the volume of the solid generated by revolving the region about the x-axis
(c) the volume of the solid generated by revolving the region about the y-axis

48. Given the region bounded by the graphs of $y = e^{-x}$ and $y = 0$ ($x \ge 0$), find the following.
(a) the area of the region
(b) the volume of the solid generated by revolving the region about the y-axis

49. Sketch the graph of the hypocycloid of four cusps

$$x^{2/3} + y^{2/3} = 1$$

and find its perimeter.

50. The region bounded by

$$(x-2)^2 + y^2 = 1$$

is revolved about the y-axis to form a torus. Find the surface area of this torus.

51. The **Gamma Function** $\Gamma(n)$ is defined by

$$\Gamma(n) = \int_0^\infty x^{n-1}e^{-x}\,dx, \quad n > 0.$$

Evaluate this function at $n = 1, 2$, and 3. Then use integration by parts to show that $\Gamma(n+1) = n\Gamma(n)$.

52. A 5-ton rocket is fired from the surface of the earth into outer space. How much work is required to overcome the earth's gravitational force? How far has the rocket traveled when half the total work has occurred?

In Exercises 53 and 54, use the following definitions. A nonnegative function f is called a **probability density function** if

$$\int_{-\infty}^{\infty} f(t)\, dt = 1.$$

The **probability** that t lies between a and b is given by

$$P(a \le t \le b) = \int_{a}^{b} f(t)\, dt.$$

The **expected value** of t is given by

$$E(t) = \int_{-\infty}^{\infty} t f(t)\, dt.$$

53. For

$$f(t) = \begin{cases} \dfrac{1}{7} e^{-t/7}, & t \ge 0 \\ 0, & t < 0 \end{cases}$$

(a) show that f is a probability density function.
(b) find $P(0 \le t \le 5)$.
(c) find $E(t)$.

54. For

$$f(t) = \begin{cases} \dfrac{2}{5} e^{-2t/5}, & t \ge 0 \\ 0, & t < 0 \end{cases}$$

(a) show that f is a probability density function.
(b) find $P(0 \le t \le 3)$.
(c) find $E(t)$.

In Exercises 55 and 56, find the capitalized cost C of an asset (a) for $n = 5$ years, (b) for $n = 10$ years, and (c) forever. The **capitalized cost** is given by

$$C = C_0 + \int_{0}^{n} c(t) e^{-rt}\, dt$$

where C_0 is the original investment, t is the time in years, r is the annual rate compounded continuously, and $c(t)$ is annual cost of maintenance.

55. $C_0 = \$650{,}000$, $c(t) = \$25{,}000$, $r = 0.12$
56. $C_0 = \$650{,}000$, $c(t) = \$25{,}000(1 + 0.08t)$, $r = 0.12$

57. Find the value of the following integral used in electromagnetic theory.

$$P = k \int_{1}^{\infty} \frac{1}{(a^2 + x^2)^{3/2}}\, dx$$

REVIEW EXERCISES for Chapter 8

In Exercises 1–40, evaluate the given integral.

1. $\displaystyle\int \frac{x^2}{x^2 + 2x - 15}\, dx$

2. $\displaystyle\int \frac{\sqrt{x^2 - 9}}{x}\, dx$

3. $\displaystyle\int \frac{1}{1 - \sin\theta}\, d\theta$

4. $\displaystyle\int x^2 \sin 2x\, dx$

5. $\displaystyle\int e^{2x} \sin 3x\, dx$

6. $\displaystyle\int (x^2 - 1) e^x\, dx$

7. $\displaystyle\int \frac{\ln(2x)}{x^2}\, dx$

8. $\displaystyle\int 2x\sqrt{2x - 3}\, dx$

9. $\displaystyle\int \sqrt{4 - x^2}\, dx$

10. $\displaystyle\int \frac{\sqrt{4 - x^2}}{2x}\, dx$

11. $\displaystyle\int \frac{-12}{x^2\sqrt{4 - x^2}}\, dx$

12. $\displaystyle\int \tan\theta \sec^4\theta\, d\theta$

13. $\displaystyle\int \sec^4 \frac{x}{2}\, dx$

14. $\displaystyle\int \sec\theta \cos 2\theta\, d\theta$

15. $\displaystyle\int \frac{9}{x^2 - 9}\, dx$

16. $\displaystyle\int \frac{\sec^2\theta}{\tan\theta(\tan\theta - 1)}\, d\theta$

17. $\displaystyle\int \frac{x^2 + 2x}{x^3 - x^2 + x - 1}\, dx$

18. $\displaystyle\int \frac{4x - 2}{3(x - 1)^2}\, dx$

19. $\displaystyle\int \frac{3x^3 + 4x}{(x^2 + 1)^2}\, dx$

20. $\displaystyle\int \sqrt{\frac{x - 2}{x + 2}}\, dx$

21. $\displaystyle\int \frac{16}{\sqrt{16 - x^2}}\, dx$

22. $\displaystyle\int \frac{\sin\theta}{1 + 2\cos^2\theta}\, d\theta$

23. $\displaystyle\int \frac{e^x}{4 + e^{2x}}\, dx$

24. $\displaystyle\int \frac{x}{x^2 - 4x + 8}\, dx$

25. $\displaystyle\int \frac{x}{x^2 + 4x + 8}\, dx$

26. $\displaystyle\int \frac{3}{2x\sqrt{9x^2 - 1}}\, dx$

27. $\displaystyle\int \theta \sin\theta \cos\theta\, d\theta$

28. $\displaystyle\int \frac{\csc\sqrt{2x}}{\sqrt{x}}\, dx$

29. $\displaystyle\int (\sin\theta + \cos\theta)^2\, d\theta$

30. $\displaystyle\int \cos 2\theta\,(\sin\theta + \cos\theta)^2\, d\theta$

31. $\displaystyle\int \frac{x^{1/4}}{1 + x^{1/2}}\, dx$

32. $\displaystyle\int \sqrt{1 - \cos x}\, dx$

33. $\displaystyle\int \sqrt{1 + \cos x}\, dx$

34. $\displaystyle\int \ln\sqrt{x^2 - 1}\, dx$

35. $\int \ln (x^2 + x) \, dx$　　**36.** $\int x \arcsin 2x \, dx$

37. $\int \cos x \ln (\sin x) \, dx$　　**38.** $\int e^x \arctan e^x \, dx$

39. $\int \dfrac{x^4 + 2x^2 + x + 1}{(x^2 + 1)^2} \, dx$

40. $\int \sqrt{1 + \sqrt{x}} \, dx$

In Exercises 41–44, evaluate the given integral using the indicated methods.

41. $\int \dfrac{1}{x^2 \sqrt{4 + x^2}} \, dx$

  (a) trigonometric substitution
  (b) substitution: $x = 2/u$

42. $\int \dfrac{1}{x \sqrt{4 + x^2}} \, dx$

  (a) trigonometric substitution
  (b) substitution: $u^2 = 4 + x^2$

43. $\int \dfrac{x^3}{\sqrt{4 + x^2}} \, dx$

  (a) trigonometric substitution
  (b) substitution: $u^2 = 4 + x^2$
  (c) by parts: $dv = (x/\sqrt{4 + x^2}) \, dx$

44. $\int x \sqrt{4 + x} \, dx$

  (a) trigonometric substitution
  (b) substitution: $u^2 = 4 + x$
  (c) substitution: $u = 4 + x$
  (d) by parts: $dv = \sqrt{4 + x} \, dx$

In Exercises 45 and 46, evaluate each definite integral to two-decimal-place accuracy.

45. (a) $\int_0^1 e^x \, dx$　　(b) $\int_0^1 x e^x \, dx$

  (c) $\int_0^1 x e^{x^2} \, dx$　　(d) $\int_0^1 e^{x^2} \, dx$

46. (a) $\int_0^{\pi/2} \cos x \, dx$　　(b) $\int_0^{\pi/2} \cos^2 x \, dx$

  (c) $\int_0^{\pi/2} \cos x^2 \, dx$　　(d) $\int_0^{\pi/2} \cos \sqrt{x} \, dx$

In Exercises 47–56, use L'Hôpital's Rule to evaluate the given limit.

47. $\lim\limits_{x \to 1} \dfrac{(\ln x)^2}{x - 1}$　　**48.** $\lim\limits_{x \to k} \dfrac{\sqrt[3]{x} - \sqrt[3]{k}}{x - k}$

49. $\lim\limits_{x \to \infty} \dfrac{e^{2x}}{x^2}$　　**50.** $\lim\limits_{x \to 1^+} \left(\dfrac{2}{\ln x} - \dfrac{2}{x - 1} \right)$

51. $\lim\limits_{x \to \infty} (\ln x)^{2/x}$　　**52.** $\lim\limits_{x \to 1} (x - 1)^{\ln x}$

53. $\lim\limits_{n \to \infty} 1000 \left(1 + \dfrac{0.09}{n} \right)^n$　　**54.** $\lim\limits_{x \to \infty} x e^{-x^2}$

55. $\lim\limits_{x \to 0} \csc 3x \tan \pi x$　　**56.** $\lim\limits_{x \to 0} \dfrac{\sin \pi x}{\sin 2 \pi x}$

In Exercises 57 and 58, approximate to two decimal places the arc length of the curve over the given interval.

Function	Interval
57. $y = \sin x$	$[0, \pi]$
58. $y = \sin^2 x$	$[0, \pi]$

In Exercises 59 and 60, find the centroid of the region bounded by the graphs of the given equations.

59. $y = \sqrt{1 - x^2}, \; y = 0$

60. $(x - 1)^2 + y^2 = 1, \; (x - 4)^2 + y^2 = 4$

61. Approximate

$$\int_2^\infty \dfrac{1}{x^5 - 1} \, dx$$

using the inequality

$$\dfrac{1}{x^5} + \dfrac{1}{x^{10}} + \dfrac{1}{x^{15}} < \dfrac{1}{x^5 - 1} < \dfrac{1}{x^5} + \dfrac{1}{x^{10}} + \dfrac{2}{x^{15}}$$

for $x \geq 2$.

62. Let

$$I_n = \int \dfrac{x^{2n-1}}{(x^2 + 1)^{n+3}} \, dx, \quad n \geq 1.$$

Prove that

$$I_n = \left(\dfrac{n - 1}{n + 2} \right) I_{n-1}$$

and then evaluate the following.

  (a) $\int_0^\infty \dfrac{x}{(x^2 + 1)^4} \, dx$

  (b) $\int_0^\infty \dfrac{x^3}{(x^2 + 1)^5} \, dx$

  (c) $\int_0^\infty \dfrac{x^5}{(x^2 + 1)^6} \, dx$

63. Verify the reduction formula

$$\int \tan^n x \, dx = \dfrac{1}{n - 1} \tan^{n-1} x - \int \tan^{n-2} x \, dx.$$

64. Show that

$$\int_x^1 \dfrac{1}{1 + t^2} \, dt = \int_1^{1/x} \dfrac{1}{1 + t^2} \, dt$$

by evaluating each integral and then using the identity

$$\arctan x + \arctan \dfrac{1}{x} = \dfrac{\pi}{2}, \quad x > 0.$$

Chapter 9 Application

A golf ball, after it is hit, continuously encounters air resistance, and its speed decreases. This means that, even on a calm day, the path of a golf ball is not parabolic. Typically, a golf ball that is hit with an iron will travel from 70–90 yards (9-iron) to 165–190 yards (2-iron).

Trajectories and Air Resistance

Neglecting air resistance, a projectile fired from ground level follows a parabolic trajectory given by

$$y = (\tan \theta)x + \frac{gx^2}{2v_0{}^2 \cos^2 \theta}$$

where v_0 is the initial speed, θ is the angle of projection, and $g = -32$ ft/sec is the acceleration due to gravity. If *air resistance is considered*, the trajectory is given by

$$y = \left(\tan \theta - \frac{g}{kv_0 \cos \theta}\right)x - \frac{g}{k^2} \ln \left(1 - \frac{kx}{v_0 \cos \theta}\right)$$

where k is the drag factor caused by the air resistance. Using the power series representation

$$\ln (1 + x) = x - \frac{x^2}{2} + \frac{x^3}{3} - \frac{x^4}{4} + \cdots, \qquad -1 < x < 1$$

we can rewrite the equation for the trajectory (considering air resistance) as

$$y = (\tan \theta)x + \frac{gx^2}{2v_0{}^2 \cos^2 \theta} + \frac{kgx^3}{3v_0{}^3 \cos^3 \theta} + \frac{k^2gx^4}{4v_0{}^4 \cos^4 \theta} + \cdots.$$

Note that the first two terms of this power series give the equation for the parabolic trajectory. Thus, when k is small (little air resistance), the trajectory is nearly parabolic.

$$v_0 = 48, \ \theta = 45, \ k = 0.25$$

See Exercise 45, Section 9.10.

Chapter Overview

This chapter is divided into two basic parts. The first six sections discuss **infinite sequences** and **infinite series.** In particular, we are interested in the convergence or divergence of infinite sequences and series. For example,

$$\frac{1}{2}, \frac{1}{4}, \frac{1}{8}, \frac{1}{16}, \cdots, \frac{1}{2^n}, \cdots$$

is an infinite sequence that converges to 0. By adding the terms of this sequence, we obtain an infinite series that converges to 1. That is,

$$\sum_{n=1}^{\infty} \left(\frac{1}{2}\right)^n = \frac{1}{2} + \frac{1}{4} + \frac{1}{8} + \frac{1}{16} + \cdots = 1.$$

The last four sections in the chapter discuss **Taylor** and **Maclaurin polynomials** and power series. The nth Maclaurin polynomial for a function f is given by

$$P_n(x) = f(0) + f'(0)x + \frac{f''(0)}{2!}x^2 + \cdots + \frac{f^{(n)}(0)}{n!}x^n.$$

This polynomial can be used as an approximation for the function f—good approximations occur when x is close to zero or when n is large.

Taylor or Maclaurin "polynomials" with an infinite number of terms are called Taylor or Maclaurin **series,** and often these power series are precisely equal to the function f. In such cases, we say that the series **converges** to $f(x)$.

Infinite Series

9.1 Sequences

Sequences ▪ Limit of a sequence ▪ Pattern recognition for sequences ▪ Monotonic sequences ▪ Bounded sequences

In mathematics the word "sequence" is used in much the same way as in ordinary English. When we say that a collection of objects or events is *in sequence* we usually mean that the collection is ordered so that it has an identified first member, second member, third member, and so on.

We define a sequence mathematically as a *function whose domain is the set of positive integers*. Although a sequence is a function, we usually represent sequences by subscript notation, rather than the standard function notation. For instance, in the sequence

$$1, \quad 2, \quad 3, \quad 4, \quad \ldots, \quad n, \quad \ldots$$
$$\downarrow \quad \downarrow \quad \downarrow \quad \downarrow \qquad \quad \downarrow$$
$$a_1, \quad a_2, \quad a_3, \quad a_4, \quad \ldots, \quad a_n, \quad \ldots$$

1 is mapped onto a_1, 2 is mapped onto a_2, and so on. We call a_n the **nth term** of the sequence and we denote the sequence by $\{a_n\}$.

DEFINITION OF A SEQUENCE
A **sequence** $\{a_n\}$ is a function whose domain is the set of positive integers. The functional values $a_1, a_2, a_3, \ldots, a_n, \ldots$ are called the **terms** of the sequence.

REMARK   Occasionally, it is convenient to begin a sequence with a_0, so that the terms of the sequence become

$$a_0, a_1, a_2, a_3, a_4, \ldots, a_n, \ldots.$$

EXAMPLE 1 Listing the terms of a sequence

(a) For the sequence $\{a_n\} = \{3 + (-1)^n\}$, the first four terms are

$$3 + (-1)^1, \; 3 + (-1)^2, \; 3 + (-1)^3, \; 3 + (-1)^4, \; \ldots$$
$$2, \, 4, \, 2, \, 4, \; \ldots.$$

(b) For the sequence $\{b_n\} = \{2n/(1 + n)\}$, the first four terms are

$$\frac{2 \cdot 1}{1 + 1}, \; \frac{2 \cdot 2}{1 + 2}, \; \frac{2 \cdot 3}{1 + 3}, \; \frac{2 \cdot 4}{1 + 4}, \; \ldots$$
$$\frac{2}{2}, \; \frac{4}{3}, \; \frac{6}{4}, \; \frac{8}{5}, \; \ldots.$$

(c) For the sequence $\{c_n\} = \{n^2/(2^n - 1)\}$, the first four terms are

$$\frac{1^2}{2^1 - 1}, \; \frac{2^2}{2^2 - 1}, \; \frac{3^2}{2^3 - 1}, \; \frac{4^2}{2^4 - 1}, \; \ldots$$
$$\frac{1}{1}, \; \frac{4}{3}, \; \frac{9}{7}, \; \frac{16}{15}, \; \ldots. \qquad \blacksquare$$

In this chapter our primary interest is in sequences whose terms approach a limiting value. Such sequences are said to **converge.** For instance, the sequence $\{1/2^n\}$

$$\frac{1}{2}, \; \frac{1}{4}, \; \frac{1}{8}, \; \frac{1}{16}, \; \frac{1}{32}, \; \ldots$$

converges to 0 as indicated in the following definition.

DEFINITION OF THE LIMIT OF A SEQUENCE

The **limit** of a sequence $\{a_n\}$ is L, written as

$$\lim_{n \to \infty} a_n = L$$

if for each $\varepsilon > 0$, there exists $M > 0$ such that $|a_n - L| < \varepsilon$ whenever $n > M$. Sequences that have a (finite) limit are said to **converge,** and sequences that do not have a limit are said to **diverge.**

Graphically, this definition says that eventually (for $n > M$) the terms of a sequence that converges to L will lie within the band between the lines

$$y = L + \varepsilon \qquad \text{and} \qquad y = L - \varepsilon$$

as illustrated in Figure 9.1.

If a sequence $\{a_n\}$ agrees with a function f at every positive integer, and if $f(x)$ approaches a limit as $x \to \infty$, then the sequence must converge to the same limit. This result is stated formally in the following theorem.

For $n > M$ the terms of the sequence all lie within ε units of L.

FIGURE 9.1

THEOREM 9.1 **LIMIT OF A SEQUENCE**	Let f be a function of a real variable such that $$\lim_{x \to \infty} f(x) = L.$$ If $\{a_n\}$ is a sequence such that $f(n) = a_n$ for every positive integer n, then $$\lim_{n \to \infty} a_n = L.$$

EXAMPLE 2   Finding the limit of a sequence

Find the limit of the sequence whose nth term is

$$a_n = \left(1 + \frac{1}{n}\right)^n$$

SOLUTION

We know from Theorem 6.16 that

$$\lim_{x \to \infty} \left(1 + \frac{1}{x}\right)^x = e.$$

Therefore, we can apply Theorem 9.1 to conclude that

$$\lim_{n \to \infty} a_n = \lim_{n \to \infty} \left(1 + \frac{1}{n}\right)^n = e.$$

The following properties of limits of sequences parallel those given for limits of functions of a real variable in Section 4.5.

THEOREM 9.2 **PROPERTIES OF LIMITS** **OF SEQUENCES**	If $\lim\limits_{n \to \infty} a_n = L$ and $\lim\limits_{n \to \infty} b_n = K$, then the following properties are true. 1. $\lim\limits_{n \to \infty} (a_n \pm b_n) = L \pm K$ 2. $\lim\limits_{n \to \infty} ca_n = cL,$ c is any real number 3. $\lim\limits_{n \to \infty} (a_n b_n) = LK$ 4. $\lim\limits_{n \to \infty} \dfrac{a_n}{b_n} = \dfrac{L}{K},$ $b_n \neq 0$ and $K \neq 0$

EXAMPLE 3   Determining the convergence or divergence of a sequence

Determine the convergence or divergence of the following sequences.

(a) $\{a_n\} = \{3 + (-1)^n\}$ (b) $\{b_n\} = \left\{\dfrac{n}{1 - 2n}\right\}$

SOLUTION

(a) Since the sequence $\{a_n\} = \{3 + (-1)^n\}$ has terms

$$2, 4, 2, 4, \ldots$$

that oscillate between 2 and 4, the limit does not exist, and we conclude that the sequence diverges.

(b) For $\{b_n\}$, we can divide the numerator and denominator by n to obtain

$$\lim_{n \to \infty} \frac{n}{1 - 2n} = \lim_{n \to \infty} \left[\frac{1}{(1/n) - 2} \right] = -\frac{1}{2}$$

and we conclude that the sequence converges to $-\frac{1}{2}$. ▭

Theorem 9.1 opens up the possibility of using L'Hôpital's Rule to determine the limit of a sequence, as demonstrated in the next example.

EXAMPLE 4   Using L'Hôpital's Rule to determine convergence

Show that the sequence whose nth term is $a_n = n^2/(2^n - 1)$ converges.

SOLUTION

We consider the function of a real variable

$$f(x) = \frac{x^2}{2^x - 1}.$$

Then, applying L'Hôpital's Rule twice, we have

$$\lim_{x \to \infty} \frac{x^2}{2^x - 1} = \lim_{x \to \infty} \frac{2x}{(\ln 2)2^x} = \lim_{x \to \infty} \frac{2}{(\ln 2)^2 2^x} = 0.$$

Since $f(n) = a_n$ for every positive integer, we can apply Theorem 9.1 to conclude that

$$\lim_{n \to \infty} \frac{n^2}{2^n - 1} = 0.$$ ▭

To simplify some of the formulas developed in this chapter, we use the symbol $n!$ (read "n factorial"), as given in the following definition.

DEFINITION OF n FACTORIAL

Let n be a positive integer; then **n factorial** is given by

$$n! = 1 \cdot 2 \cdot 3 \cdot 4 \cdots (n - 1) \cdot n.$$

Zero factorial is given by $0! = 1$.

REMARK   From this definition, we see that $0! = 1$, $1! = 1$, $2! = 1 \cdot 2 = 2$, $3! = 1 \cdot 2 \cdot 3 = 6$, and so on. Factorials follow the same conventions for order of operation as exponents. That is, just as $2x^3$ and $(2x)^3$ imply different orders of operations, $2n!$ and $(2n)!$ imply the following orders:

$$2n! = 2(n!) = 2(1 \cdot 2 \cdot 3 \cdot 4 \cdots n)$$

and

$$(2n)! = 1 \cdot 2 \cdot 3 \cdot 4 \cdots n \cdot (n + 1) \cdots 2n.$$

Another useful limit theorem that can be rewritten for sequences is the Squeeze Theorem of Section 2.3.

THEOREM 9.3 **SQUEEZE THEOREM** **FOR SEQUENCES**	If $$\lim_{n \to \infty} a_n = L = \lim_{n \to \infty} b_n$$ and there exists an integer N such that $a_n \leq c_n \leq b_n$ for all $n > N$, then $$\lim_{n \to \infty} c_n = L.$$

The usefulness of the Squeeze Theorem is seen in Example 5.

EXAMPLE 5   Using the Squeeze Theorem

Show that the following sequence converges and find its limit.

$$\{c_n\} = \left\{ (-1)^n \frac{1}{n!} \right\}$$

SOLUTION

To apply the Squeeze Theorem, we must find two convergent sequences that can be related to the given factorial sequence. Two possibilities are $a_n = -1/2^n$ and $b_n = 1/2^n$, both of which converge to zero. By comparing the term $n!$ with 2^n, we see that

$$n! = 1 \cdot 2 \cdot 3 \cdot 4 \cdot 5 \cdot 6 \cdots n = 24 \cdot 5 \cdot 6 \cdots n$$

and

$$2^n = 2 \cdot 2 \cdot 2 \cdot 2 \cdot 2 \cdot 2 \cdots 2 = 16 \cdot \underbrace{2 \cdot 2 \cdots 2}_{n - 4 \text{ factors}}.$$

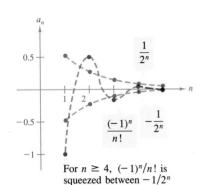

For $n \geq 4$, $(-1)^n/n!$ is squeezed between $-1/2^n$ and $1/2^n$.

FIGURE 9.2

This implies that for $n \geq 4$, $2^n < n!$, and we have

$$\frac{-1}{2^n} \leq (-1)^n \frac{1}{n!} \leq \frac{1}{2^n}$$

as illustrated in Figure 9.2. Therefore, by the Squeeze Theorem it follows that

$$\lim_{n \to \infty} (-1)^n \frac{1}{n!} = 0.$$

Table 9.1 lists the first six terms of each of the three sequences $\{a_n\}$, $\{b_n\}$, and $\{c_n\}$.

TABLE 9.1

n	1	2	3	4	5	6
$a_n = -\dfrac{1}{2^n}$	$-\dfrac{1}{2}$	$-\dfrac{1}{4}$	$-\dfrac{1}{8}$	$-\dfrac{1}{16}$	$-\dfrac{1}{32}$	$-\dfrac{1}{64}$
$b_n = (-1)^n \dfrac{1}{n!}$	$-\dfrac{1}{1}$	$\dfrac{1}{2}$	$-\dfrac{1}{6}$	$\dfrac{1}{24}$	$-\dfrac{1}{120}$	$\dfrac{1}{720}$
$c_n = \dfrac{1}{2^n}$	$\dfrac{1}{2}$	$\dfrac{1}{4}$	$\dfrac{1}{8}$	$\dfrac{1}{16}$	$\dfrac{1}{32}$	$\dfrac{1}{64}$

In Example 5 the sequence $\{c_n\}$ has both positive and negative terms. For this sequence, it happens that the sequence of absolute values, $\{|c_n|\}$, also converges to zero. You can show this by the Squeeze Theorem using the inequality

$$0 \leq \frac{1}{n!} \leq \frac{1}{2^n}, \quad n \geq 4.$$

In such cases it is often convenient to consider the sequence of positive terms, then apply Theorem 9.4, which tells us that if the positive term sequence converges to zero, the original signed sequence also must converge to zero.

THEOREM 9.4
ABSOLUTE VALUE THEOREM

For the sequence $\{a_n\}$, if

$$\lim_{n \to \infty} |a_n| = 0 \quad \text{then} \quad \lim_{n \to \infty} a_n = 0.$$

PROOF   Consider the two sequences $\{|a_n|\}$ and $\{-|a_n|\}$. Since both of these sequences converge to 0 and since

$$-|a_n| \leq a_n \leq |a_n|$$

we can apply the Squeeze Theorem and conclude that $\{a_n\}$ converges to 0.

The result in Example 5 suggests something about the rate at which $n!$ increases as $n \to \infty$. From Figure 9.2 we can see that both $1/2^n$ and $1/n!$ approach zero as $n \to \infty$. Yet $1/n!$ approaches 0 so much faster than $1/2^n$ does that

$$\lim_{n \to \infty} \frac{1/n!}{1/2^n} = \lim_{n \to \infty} \frac{2^n}{n!} = 0.$$

In fact, it can be shown that for any fixed number k,

$$\lim_{n \to \infty} \frac{k^n}{n!} = 0.$$

This means that *the factorial function grows faster than any exponential function.* We will find this fact to be very useful as we work with limits of sequences.

Pattern recognition for sequences

Sometimes the first several terms of a sequence are listed without the nth term, or the terms may be generated by some rule that does not explicitly identify the nth term of the sequence. In such cases, we are required to discover a *pattern* in the sequence and to describe the nth term. Once the nth term is specified, we can discuss the convergence or divergence of the sequence. This is demonstrated in the next two examples.

EXAMPLE 6   Finding the nth term of a sequence

Find a sequence $\{a_n\}$ whose first five terms are

$$\frac{2}{1}, \frac{4}{3}, \frac{8}{5}, \frac{16}{7}, \frac{32}{9}, \dots$$

and then determine whether the particular sequence you have chosen converges or diverges.

SOLUTION

First, we note that the numerators are successive powers of 2, and the denominators form the sequence of positive odd integers. Then, by comparing a_n with n, we have the following pattern.

$$1, \quad 2, \quad 3, \quad 4, \quad 5, \quad \dots$$
$$\downarrow \quad \downarrow \quad \downarrow \quad \downarrow \quad \downarrow$$
$$\frac{2}{1}, \frac{4}{3}, \frac{8}{5}, \frac{16}{7}, \frac{32}{9}, \dots$$
$$\frac{2^1}{1}, \frac{2^2}{3}, \frac{2^3}{5}, \frac{2^4}{7}, \frac{2^5}{9}, \dots$$

Using the fact that the odd integer denominators are generated by $2n - 1$, we can conclude that $a_n = 2^n/(2n - 1)$. Now, using L'Hôpital's Rule to evaluate the limit of $f(x) = 2^x/(2x - 1)$, we obtain

$$\lim_{x \to \infty} \frac{2^x}{2x - 1} = \lim_{x \to \infty} \frac{2^x(\ln 2)}{2} = \infty \quad \Longrightarrow \quad \lim_{n \to \infty} \frac{2^n}{2n - 1} = \infty.$$

Hence, the sequence *diverges*.

Without a specific rule for generating the terms of a sequence or some knowledge of the context in which the terms of the sequence are obtained, it is not possible to determine the convergence or divergence of the sequence merely from its first several terms. For instance, although the first three terms of the four sequences given next are identical, the first two sequences converge to 0, the third sequence converges to $\frac{1}{9}$, and the fourth one diverges.

$$\{a_n\}: \frac{1}{2}, \frac{1}{4}, \frac{1}{8}, \frac{1}{16}, \cdots, \frac{1}{2^n}, \cdots$$

$$\{b_n\}: \frac{1}{2}, \frac{1}{4}, \frac{1}{8}, \frac{1}{15}, \cdots, \frac{6}{(n + 1)(n^2 - n + 6)}, \cdots$$

$$\{c_n\}: \frac{1}{2}, \frac{1}{4}, \frac{1}{8}, \frac{7}{62}, \cdots, \frac{n^2 - 3n + 3}{9n^2 - 25n + 18}, \cdots$$

$$\{d_n\}: \frac{1}{2}, \frac{1}{4}, \frac{1}{8}, 0, \cdots, \frac{-n(n + 1)(n - 4)}{6(n^2 + 3n - 2)}, \cdots$$

The process of determining an nth term from the pattern observed in the first several terms of a sequence is an example of *inductive reasoning*.

EXAMPLE 7   Finding the nth term of a sequence

Determine an nth term for a sequence whose first five terms are

$$-\frac{2}{1}, \frac{8}{2}, -\frac{26}{6}, \frac{80}{24}, -\frac{242}{120}, \cdots$$

and then decide if your sequence converges or diverges.

SOLUTION

We observe that the numerators are one less than 3^n. Hence, we reason that the numerators are given by the rule $3^n - 1$. If we factor the denominators, we have

$$1 = 1$$
$$2 = 1 \cdot 2$$
$$6 = 1 \cdot 2 \cdot 3$$
$$24 = 1 \cdot 2 \cdot 3 \cdot 4$$
$$120 = 1 \cdot 2 \cdot 3 \cdot 4 \cdot 5 \cdots.$$

This suggests that the denominators are represented by $n!$. Finally, since the signs alternate, we can write the nth term as

$$a_n = (-1)^n \left(\frac{3^n - 1}{n!} \right).$$

From Theorem 9.4 and the discussion about the growth of $n!$, it follows that

$$\lim_{n \to \infty} |a_n| = \lim_{n \to \infty} \frac{3^n - 1}{n!} = 0 = \lim_{n \to \infty} a_n$$

and we conclude that $\{a_n\}$ converges to 0.

Monotonic sequences

So far we have determined the convergence of a sequence by finding its limit. Even if we cannot determine the limit of a particular sequence, it still may be useful to know whether the sequence converges. Theorem 9.5 identifies a test for convergence of sequences without involving the determination of the limit. First, we look at some preliminary definitions.

DEFINITION OF A MONOTONIC SEQUENCE

A sequence $\{a_n\}$ is **monotonic** if its terms are nondecreasing

$$a_1 \leq a_2 \leq a_3 \leq \cdots \leq a_n \leq \cdots$$

or if its terms are nonincreasing

$$a_1 \geq a_2 \geq a_3 \geq \cdots \geq a_n \geq \cdots.$$

EXAMPLE 8 Determining if a sequence is monotonic

Determine whether the sequence having the given nth term is monotonic.

(a) $a_n = 3 + (-1)^n$ (b) $b_n = \dfrac{2n}{1 + n}$ (c) $c_n = \dfrac{n^2}{2^n - 1}$

SOLUTION

(a) This sequence alternates between 2 and 4. Therefore, it is not monotonic.
(b) This sequence is monotonic because each successive term is larger than its predecessor. To see this, we compare the terms b_n and b_{n+1} as follows. (Note that because n is positive we can multiply both sides of the inequality by $(1 + n)$ and $(2 + n)$ without reversing the inequality sign.)

$$b_n = \frac{2n}{1 + n} \overset{?}{<} \frac{2(n + 1)}{1 + (n + 1)} = b_{n+1}$$
$$2n(2 + n) \overset{?}{<} (1 + n)(2n + 2)$$
$$4n + 2n^2 \overset{?}{<} 2 + 4n + 2n^2$$
$$0 < 2$$

Since the final inequality is valid, we can reverse the steps to conclude that the original inequality is also valid.

(c) This sequence is not monotonic since the second term is larger than the first term, and smaller than the third. (Note that if we drop the first term, the remaining sequence c_2, c_3, c_4, . . . is monotonic.) Figure 9.3 graphically illustrates these three sequences. ▱

REMARK   In Example 8(b), another way to see that the sequence is monotonic is to consider the derivative of the corresponding differentiable function $f(x) = 2x/(1 + x)$. Since

$$f'(x) = \frac{2}{(1 + x)^2}$$

is positive for all x, we can conclude that f is increasing. This implies that $\{a_n\}$ is increasing.

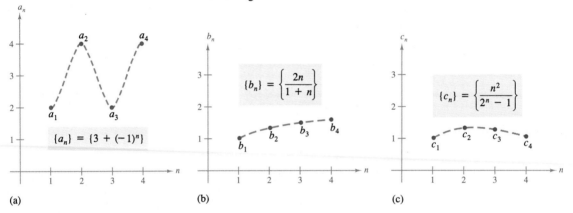

FIGURE 9.3

DEFINITION OF A BOUNDED SEQUENCE

A sequence $\{a_n\}$ is **bounded** if there is a positive real number M such that $|a_n| \leq M$ for all n. We call M an **upper bound** for the sequence.

In Figure 9.3, all three sequences are bounded. That is,

$$|3 + (-1)^n| \leq 4, \qquad \left|\frac{2n}{1 + n}\right| \leq 2, \qquad \text{and} \qquad \left|\frac{n^2}{2^n - 1}\right| \leq \frac{4}{3}.$$

One important property of the real numbers is that they are **complete.** Informally, this means that there are no holes or gaps on the real number line. (The set of rational numbers does not have the completeness property.) The completeness axiom for real numbers can be used to conclude that if a sequence has an upper bound, then it must have a **least upper bound** (an upper bound that is smaller than all other upper bounds for the sequence). For example, the least upper bound of the sequence $\{a_n\} = \{n/(n + 1)\}$,

$$\frac{1}{2}, \frac{2}{3}, \frac{3}{4}, \frac{4}{5}, \cdot \cdot \cdot, \frac{n}{n + 1}, \cdot \cdot \cdot$$

is 1. We use the completeness axiom in the proof of Theorem 9.5.

**THEOREM 9.5
BOUNDED MONOTONIC
SEQUENCES**

If a sequence $\{a_n\}$ is bounded and monotonic, then it converges.

PROOF

We give the proof for a nondecreasing sequence, as shown in Figure 9.4, and leave the nonincreasing case as an exercise (see Exercise 72). For the sake of simplicity, we assume that each term in the sequence is positive. Then, since the sequence is bounded, there must exist an upper bound M such that

$$a_1 \le a_2 \le a_3 \le \cdots \le a_n \le \cdots \le M.$$

From the completeness axiom, it follows that there is a least upper bound L such that

$$a_1 \le a_2 \le a_3 \le \cdots \le a_n \le \cdots \le L.$$

Now, we will show that $\{a_n\}$ converges to L. For $\varepsilon > 0$, it follows that $L - \varepsilon < L$, and therefore $L - \varepsilon$ cannot be an upper bound for the sequence. Consequently, at least one term of $\{a_n\}$ is greater than $L - \varepsilon$. That is, $L - \varepsilon < a_N$ for some positive integer N. Since the terms of $\{a_n\}$ are nondecreasing we conclude that $a_N \le a_n$ for $n > N$. We now know that

$$L - \varepsilon < a_N < a_n \le L < L + \varepsilon, \quad \text{for every } n > N.$$

It follows that $|a_n - L| < \varepsilon$ for $n > N$, which by definition means that $\{a_n\}$ converges to L.

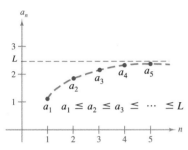

Every bounded nondecreasing sequence converges.

FIGURE 9.4

EXAMPLE 9 Bounded and monotonic sequences

(a) The sequence

$$\{a_n\} = \left\{\frac{1}{n}\right\}$$

is both bounded and monotonic and thus, by Theorem 9.5, must converge.

(b) The divergent sequence

$$\{b_n\} = \left\{\frac{n^2}{n + 1}\right\}$$

is monotonic, but not bounded.

(c) The divergent sequence

$$\{c_n\} = \{(-1)^n\}$$

is bounded, but not monotonic.

EXERCISES for Section 9.1

In Exercises 1–8, write out the first five terms of the sequence with the given nth term.

1. $a_n = 2^n$

2. $a_n = \dfrac{n}{n + 1}$

3. $a_n = \left(-\dfrac{1}{2}\right)^n$

4. $a_n = \sin\dfrac{n\pi}{2}$

5. $a_n = \dfrac{3^n}{n!}$

6. $a_n = 5 - \dfrac{1}{n} + \dfrac{1}{n^2}$

7. $a_n = \dfrac{(-1)^{n(n+1)/2}}{n^2}$

8. $a_n = \dfrac{3n!}{(n - 1)!}$

In Exercises 9–24, write an expression for the nth term of the sequence.

9. $1, 4, 7, 10, \ldots$

10. $3, 7, 11, 15, \ldots$

11. $-1, 2, 7, 14, 23, \ldots$

12. $1, \dfrac{1}{4}, \dfrac{1}{9}, \dfrac{1}{16}, \ldots$

13. $\dfrac{2}{3}, \dfrac{3}{4}, \dfrac{4}{5}, \dfrac{5}{6}, \ldots$

14. $2, \dfrac{3}{3}, \dfrac{4}{5}, \dfrac{5}{7}, \dfrac{6}{9}, \ldots$

15. $2, -1, \dfrac{1}{2}, -\dfrac{1}{4}, \dfrac{1}{8}, \ldots$

16. $\dfrac{1}{2}, \dfrac{1}{3}, \dfrac{2}{9}, \dfrac{4}{27}, \dfrac{8}{81}, \ldots$

17. $2, 1 + \dfrac{1}{2}, 1 + \dfrac{1}{3}, 1 + \dfrac{1}{4}, 1 + \dfrac{1}{5}, \ldots$

18. $1 + \dfrac{1}{2}, 1 + \dfrac{3}{4}, 1 + \dfrac{7}{8}, 1 + \dfrac{15}{16}, 1 + \dfrac{31}{32}, \ldots$

19. $\dfrac{1}{2 \cdot 3}, \dfrac{2}{3 \cdot 4}, \dfrac{3}{4 \cdot 5}, \dfrac{4}{5 \cdot 6}, \ldots$

20. $1, \dfrac{1}{2}, \dfrac{1}{6}, \dfrac{1}{24}, \dfrac{1}{120}, \ldots$

21. $1, \dfrac{1}{1 \cdot 3}, \dfrac{1}{1 \cdot 3 \cdot 5}, \dfrac{1}{1 \cdot 3 \cdot 5 \cdot 7}, \ldots$

22. $2, -4, 6, -8, 10, \ldots$

23. $1, -1, -1, 1, 1, -1, -1, \ldots$

24. $1, x, \dfrac{x^2}{2}, \dfrac{x^3}{6}, \dfrac{x^4}{24}, \dfrac{x^5}{120}, \ldots$

In Exercises 25–46, determine the convergence or divergence of the sequence with the given nth term. If the sequence converges, find its limit.

25. $a_n = \dfrac{n + 1}{n}$

26. $a_n = \dfrac{1}{n^{3/2}}$

27. $a_n = (-1)^n \left(\dfrac{n}{n + 1}\right)$

28. $a_n = \dfrac{n - 1}{n} - \dfrac{n}{n - 1}$
$n \geq 2$

29. $a_n = \dfrac{3n^2 - n + 4}{2n^2 + 1}$

30. $a_n = \dfrac{\sqrt{n}}{\sqrt{n} + 1}$

31. $a_n = \dfrac{n^2 - 1}{n + 1}$

32. $a_n = 1 + (-1)^n$

33. $a_n = \dfrac{1 + (-1)^n}{n}$

34. $a_n = \dfrac{\ln(n^2)}{n}$

35. $a_n = \cos\dfrac{n\pi}{2}$

36. $a_n = \dfrac{n}{\sqrt{n^2 + 1}}$

37. $a_n = \dfrac{3^n}{4^n}$

38. $a_n = \dfrac{(n - 2)!}{n!}$

39. $a_n = f^{(n-1)}(2), f(x) = \ln x$

40. $a_n = \dfrac{n^2}{2n + 1} - \dfrac{n^2}{2n - 1}$

41. $a_n = 3 - \dfrac{1}{2^n}$

42. $a_n = \dfrac{n!}{n^n}$

43. $a_n = \left(1 + \dfrac{k}{n}\right)^n$

44. $a_n = 2^{1/n}$

45. $a_n = \dfrac{n^p}{e^n} (p > 0)$

46. $a_n = n \sin\dfrac{1}{n}$

In Exercises 47–56, determine if the sequence with the given nth term is monotonic.

47. $a_n = 4 - \dfrac{1}{n}$

48. $a_n = \dfrac{4n}{n + 1}$

49. $a_n = \dfrac{\cos n}{n}$

50. $a_n = ne^{-n/2}$

51. $a_n = (-1)^n \left(\dfrac{1}{n}\right)$

52. $a_n = \left(-\dfrac{2}{3}\right)^n$

53. $a_n = \left(\dfrac{2}{3}\right)^n$

54. $a_n = \left(\dfrac{3}{2}\right)^n$

55. $a_n = \sin\dfrac{n\pi}{6}$

56. $a_n = \dfrac{n}{2^{n+2}}$

In Exercises 57–60, use Theorem 9.5 to show that the sequence with the given nth term converges, and find the limit.

57. $a_n = 5 + \dfrac{1}{n}$

58. $a_n = 3 - \dfrac{4}{n}$

59. $a_n = \dfrac{1}{3}\left(1 - \dfrac{1}{3^n}\right)$

60. $a_n = 4 + \dfrac{1}{2^n}$

61. Consider the sequence $\{A_n\}$, whose nth term is given by

$$A_n = P\left(1 + \dfrac{r}{12}\right)^n$$

where P is the principal, A_n is the amount at compound interest after n months, and r is the annual percentage rate.

(a) Is $\{A_n\}$ a convergent sequence?

(b) Find the first ten terms of the sequence if $P = \$9,000$ and $r = 0.115$.

62. A deposit of $100 is made each month in an account that earns 12-percent interest compounded monthly. The balance in the account after n months is given by

$$A_n = 100(101)[(1.01)^n - 1].$$

(a) Compute the first 6 terms of the sequence $\{A_n\}$.
(b) Find the balance after 5 years by computing the 60th term of the sequence.
(c) Find the balance after 20 years by computing the 240th term of the sequence.

63. A government program that currently costs taxpayers $2.5 billion per year is to be cut back by 20 percent per year.
(a) Write an expression for the amount budgeted for this program after n years.
(b) Compute the budgets for the first 4 years.
(c) Determine the convergence or divergence of the sequence of reduced budgets. If the sequence converges, find its limit.

64. If the average price of a new car increases $5\frac{1}{2}$ percent per year and the average price is currently $11,000, then the average price after n years is

$$P_n = \$11,000(1.055)^n.$$

Compute the average price for the first 5 years of increases.

65. Consider an idealized population with the characteristic that each member of the population produces 1 offspring at the end of every time period. If each member has a life span of 3 time periods and the population begins with 10 newborn members, then the following table gives the population during the first 5 time periods.

Age bracket	Time period				
	1	2	3	4	5
0–1	10	10	20	40	70
1–2		10	10	20	40
2–3			10	10	20
Total	10	20	40	70	130

The sequence for the total population has the property that

$$S_n = S_{n-1} + S_{n-2} + S_{n-3}, \quad n > 3.$$

Find the total population during the next 5 time periods.

66. Consider the sequence $\{x_n\}$ defined by

$$x_n = x_{n-1} - \frac{f(x_{n-1})}{f'(x_{n-1})}.$$

If $f(x) = x^2 + x - 1$ and $x_1 = 0.5$, find the next three terms of the sequence.

67. Compute the first 6 terms of the sequence $\{a_n\} = \{\sqrt[n]{n}\}$. If the sequence converges, find its limit.

68. Show that $\{r^n/(1 - r)\}$ diverges if $|r| \geq 1$, and converges to 0 if $|r| < 1$.

69. Prove that if $\{s_n\}$ converges to L and $L > 0$, then there exists a number N such that $s_n > 0$ for $n > N$.

70. If $\{s_n\}$ is a convergent sequence, show that

$$\lim_{n \to \infty} s_{n-1} = \lim_{n \to \infty} s_n.$$

71. In the study of the progeny of rabbits, **Fibonacci** (ca. 1175–ca. 1250) encountered the now famous sequence bearing his name. It is defined recursively by

$$a_{n+2} = a_n + a_{n+1}, \quad \text{where } a_1 = 1 \text{ and } a_2 = 1.$$

(a) Write out the first 12 terms of the sequence.
(b) Write out the first 10 terms of the sequence defined by

$$b_n = \frac{a_{n+1}}{a_n}, \quad \text{for } n \geq 1.$$

(c) Using the definition of part (b) show that

$$b_n = 1 + \frac{1}{b_{n-1}}.$$

(d) If

$$\lim_{n \to \infty} b_n = \rho$$

use the results of part (b) and Exercise 70 to show that

$$\rho = 1 + \frac{1}{\rho}.$$

Solve this equation for ρ. (ρ is called the *golden ratio*.)

72. Complete the proof of Theorem 9.5.

9.2 Series and Convergence

Infinite series ▪ Convergent and divergent series ▪ Telescoping series ▪ Geometric series ▪ *n*th-Term Test for Divergence

One important application of infinite sequences is in representing infinite summations. Informally, if $\{a_n\}$ is an infinite sequence, then

$$\sum_{n=1}^{\infty} a_n = a_1 + a_2 + a_3 + \cdots + a_n + \cdots$$

is called an **infinite series** (or simply a **series**). The numbers $a_1, a_2, a_3, \ldots$ are called the **terms** of the series. For some series it is convenient to begin the index at $n = 0$. As a typesetting convention, it is common to represent an infinite series as simply $\Sigma \, a_n$. In such cases the starting point for the index ($n = 0$ or $n = 1$) must be taken from the context of the statement.

To find the sum of an infinite series, we consider the following **sequence of partial sums.**

$$S_1 = a_1$$
$$S_2 = a_1 + a_2$$
$$S_3 = a_1 + a_2 + a_3$$
$$\vdots$$
$$S_n = a_1 + a_2 + a_3 + \cdots + a_n$$

If this sequence converges, then we say that the series also converges and has the sum indicated in the following definition.

DEFINITION OF CONVERGENT AND DIVERGENT SERIES

For the infinite series $\Sigma \, a_n$, the **nth partial sum** is given by

$$S_n = a_1 + a_2 + \cdots + a_n.$$

If the sequence of partial sums $\{S_n\}$ converges to S, then we say that the series $\Sigma \, a_n$ **converges.** We call S the **sum of the series** and write

$$S = a_1 + a_2 + \cdots + a_n + \cdots.$$

If $\{S_n\}$ diverges, then we say that the series also **diverges.**

As you study this chapter, you will see that there are two basic questions involving infinite series. Does a series converge or does it diverge? If a series converges, what is its sum? These questions are not always easy to answer, especially the second one. We begin with some simple examples.

EXAMPLE 1 Convergent and divergent series

(a) The series

$$\sum_{n=1}^{\infty} \frac{1}{2^n} = \frac{1}{2} + \frac{1}{4} + \frac{1}{8} + \frac{1}{16} + \cdots$$

has the following partial sums.

$$S_1 = \frac{1}{2}$$

$$S_2 = \frac{1}{2} + \frac{1}{4} = \frac{3}{4}$$

$$S_3 = \frac{1}{2} + \frac{1}{4} + \frac{1}{8} = \frac{7}{8}$$

$$\vdots$$

$$S_n = \frac{1}{2} + \frac{1}{4} + \frac{1}{8} + \cdots + \frac{1}{2^n} = \frac{2^n - 1}{2^n}$$

Since

$$\lim_{n \to \infty} \frac{2^n - 1}{2^n} = 1$$

we conclude that the series converges and its sum is 1.

(b) The nth partial sum of the series

$$\sum_{n=1}^{\infty} \left(\frac{1}{n} - \frac{1}{n + 1} \right) = \left(1 - \frac{1}{2} \right) + \left(\frac{1}{2} - \frac{1}{3} \right) + \left(\frac{1}{3} - \frac{1}{4} \right) + \cdots$$

is given by

$$S_n = 1 - \frac{1}{n + 1}$$

and since the limit of S_n is 1, the series converges and its sum is 1.

(c) The series

$$\sum_{n=1}^{\infty} 1 = 1 + 1 + 1 + 1 + \cdots$$

diverges because $S_n = n$ and the sequence of partial sums diverges.

The series in part (b) of Example 1 is called a **telescoping series.** That is, it is of the form

$$(b_1 - b_2) + (b_2 - b_3) + (b_3 - b_4) + (b_4 - b_5) + \cdots .$$

Note that b_2 is cancelled by the second term, b_3 is cancelled by the third term, and so on. Since the nth partial sum of this series is $S_n = b_1 - b_n$, it follows that a telescoping series will converge if and only if b_n approaches a finite number as $n \to \infty$. Moreover, if the series converges, then its sum is

$$S = b_1 - \lim_{n \to \infty} b_n.$$

In Example 2, we find the limit of a series by writing the series in *telescoping form*.

EXAMPLE 2 Writing a series in telescoping form

Find the sum of the following series.

$$\sum_{n=1}^{\infty} \frac{2}{4n^2 - 1}$$

SOLUTION

Using partial fractions, we find that

$$a_n = \frac{2}{4n^2 - 1} = \frac{2}{(2n - 1)(2n + 1)} = \frac{1}{2n - 1} - \frac{1}{2n + 1}.$$

From this telescoping form, we see that the nth partial sum is

$$S_n = \left(\frac{1}{1} - \frac{1}{3}\right) + \left(\frac{1}{3} - \frac{1}{5}\right) + \cdots + \left(\frac{1}{2n - 1} - \frac{1}{2n + 1}\right)$$

$$= 1 - \frac{1}{2n + 1}.$$

Thus, the series converges and its sum is 1. That is,

$$\sum_{n=1}^{\infty} \frac{2}{4n^2 - 1} = \lim_{n \to \infty} S_n = \lim_{n \to \infty} \left(1 - \frac{1}{2n + 1}\right) = 1.$$

Geometric series

The series given in part (a) of Example 1 is called a **geometric series.** This important type of series is defined as follows.

DEFINITION OF GEOMETRIC SERIES	The series given by $$\sum_{n=0}^{\infty} ar^n = a + ar + ar^2 + \cdots + ar^n + \cdots, \quad a \neq 0$$ is called a **geometric series** with ratio r.

The conditions for the convergence or divergence of a geometric series are given in the following theorem.

THEOREM 9.6 CONVERGENCE OF A GEOMETRIC SERIES	A geometric series with ratio r diverges if $\|r\| \geq 1$. If $0 < \|r\| < 1$, then the series converges to the sum $$\sum_{n=0}^{\infty} ar^n = \frac{a}{1 - r}, \quad 0 < \|r\| < 1.$$

PROOF   It is easy to see that the series diverges if $r = \pm 1$. If $r \neq \pm 1$, then

$$S_n = a + ar + ar^2 + \cdots + ar^{n-1}.$$

Multiplication by r yields

$$rS_n = ar + ar^2 + ar^3 + \cdots + ar^n.$$

By subtracting the second equation from the first, we obtain $S_n - rS_n = a - ar^n$. Therefore, $S_n(1 - r) = a(1 - r^n)$, and the nth partial sum is

$$S_n = \frac{a}{1 - r}(1 - r^n).$$

Now, since $0 < |r| < 1$, it follows that $r^n \to 0$ as $n \to \infty$, and we obtain

$$\lim_{n \to \infty} S_n = \lim_{n \to \infty} \left[\frac{a}{1 - r}(1 - r^n) \right] = \frac{a}{1 - r}[\lim_{n \to \infty} (1 - r^n)] = \frac{a}{1 - r}$$

which means the series *converges* and its sum is $a/(1 - r)$. We leave it to you to show that the series diverges if $|r| > 1$.

EXAMPLE 3   A convergent geometric series

The geometric series

$$\sum_{n=0}^{\infty} \frac{3}{2^n} = \sum_{n=0}^{\infty} 3\left(\frac{1}{2}\right)^n = 3(1) + 3\left(\frac{1}{2}\right) + 3\left(\frac{1}{2}\right)^2 + \cdots$$

has a ratio of $r = \frac{1}{2}$ with $a = 3$. Since $0 < |r| < 1$, the series converges and its sum is

$$S = \frac{a}{1 - r} = \frac{3}{1 - (1/2)} = 6. \qquad \square$$

EXAMPLE 4   A divergent geometric series

The geometric series

$$\sum_{n=0}^{\infty} \left(\frac{3}{2}\right)^n = 1 + \frac{3}{2} + \frac{9}{4} + \frac{27}{8} + \cdots$$

has a ratio of $r = \frac{3}{2}$. Since $|r| \geq 1$, the series diverges. $\qquad \square$

The formula for the sum of a geometric series can be used to write a repeating decimal as the ratio of two integers, as demonstrated in the next example.

EXAMPLE 5   A geometric series for a repeating decimal

Use a geometric series to express $0.080\overline{808}$ as the ratio of two integers.

SOLUTION

For

$$0.080808 \ldots = \frac{8}{10^2} + \frac{8}{10^4} + \frac{8}{10^6} + \frac{8}{10^8} + \cdots = \sum_{n=0}^{\infty} \left(\frac{8}{10^2}\right)\left(\frac{1}{10^2}\right)^n$$

we have $a = 8/10^2$ and $r = 1/10^2$. Thus,

$$0.080808 \ldots = \frac{a}{1-r} = \frac{8/10^2}{1-(1/10^2)} = \frac{8}{99}.$$

Try dividing 8 by 99 on a calculator to see that it produces 0.080808 . . . .

▭

The convergence of a series is not affected by removing a finite number of terms from the beginning of the series. For instance, the geometric series

$$\sum_{n=4}^{\infty} \left(\frac{1}{2}\right)^n \quad \text{and} \quad \sum_{n=0}^{\infty} \left(\frac{1}{2}\right)^n$$

both converge. Furthermore, because the sum of the second series is $a/(1-r) = 2$, we can conclude that the sum of the first series is

$$S = 2 - \left[\left(\frac{1}{2}\right)^0 + \left(\frac{1}{2}\right)^1 + \left(\frac{1}{2}\right)^2 + \left(\frac{1}{2}\right)^3\right]$$

$$= 2 - \frac{15}{8} = \frac{1}{8}.$$

Properties of series

The following properties are direct consequences of the corresponding properties of limits of sequences.

THEOREM 9.7
PROPERTIES OF INFINITE SERIES

If $\Sigma a_n = A$, $\Sigma b_n = B$, and c is a real number, then the following series converge to the indicated sums.

1. $\sum_{n=1}^{\infty} ca_n = cA$ 2. $\sum_{n=1}^{\infty} (a_n + b_n) = A + B$

3. $\sum_{n=1}^{\infty} (a_n - b_n) = A - B$

The following theorem tells us that if a series converges, then the limit of its nth term must be zero.

THEOREM 9.8
LIMIT OF nTH TERM OF A CONVERGENT SERIES

If the series Σa_n converges, then the sequence $\{a_n\}$ converges to 0.

PROOF Assume that

$$\sum_{n=1}^{\infty} a_n = \lim_{n \to \infty} S_n = L.$$

Then, because $S_n = S_{n-1} + a_n$ and

$$\lim_{n \to \infty} S_n = \lim_{n \to \infty} S_{n-1} = L$$

it follows that

$$L = \lim_{n \to \infty} S_n = \lim_{n \to \infty} (S_{n-1} + a_n) = \lim_{n \to \infty} S_{n-1} + \lim_{n \to \infty} a_n = L + \lim_{n \to \infty} a_n$$

which implies that $\{a_n\}$ converges to 0.

The contrapositive of Theorem 9.8 provides a useful test for *divergence*. This **nth-Term Test for Divergence** tells us that if the limit of the nth term of a series does *not* converge to 0, then the series must diverge.

THEOREM 9.9 ***n*TH-TERM TEST FOR DIVERGENCE**	If the sequence $\{a_n\}$ does not converge to 0, then the series $\Sigma\, a_n$ diverges.

REMARK   Be sure you see that this theorem does *not* state that the series $\Sigma\, a_n$ converges if $\{a_n\}$ converges to 0.

EXAMPLE 6   Using the *n*th-Term Test for Divergence

(a) For the series

$$\sum_{n=0}^{\infty} 2^n$$

we have

$$\lim_{n \to \infty} 2^n = \infty.$$

Thus, the limit of the nth term is not zero, and we conclude that the series *diverges*.

(b) For the series

$$\sum_{n=1}^{\infty} \frac{n!}{2n! + 1}$$

we have

$$\lim_{n \to \infty} \frac{n!}{2n! + 1} = \frac{1}{2}.$$

Thus, the limit of the nth term is not zero, and we conclude that the series *diverges*.

(c) For the series

$$\sum_{n=1}^{\infty} \frac{1}{n}$$

we have

$$\lim_{n \to \infty} \frac{1}{n} = 0.$$

Because the limit of the nth term is zero, the nth-Term Test for Divergence does *not* apply and we draw no conclusions about convergence or divergence. (In the next section, we will see that this particular series diverges.) ▭

Applications

EXAMPLE 7 An application of geometric series

A ball is dropped from a height of 6 feet and begins bouncing as shown in Figure 9.5. The height of each bounce is three-fourths the height of the previous bounce. Find the total vertical distance traveled by the ball.

SOLUTION

When the ball hits the ground for the first time, it has traveled a distance of $D_1 = 6$. For subsequent bounces, we let D_n be the distance traveled up *and* down. For example, D_2 and D_3 are as follows.

$$D_2 = \underbrace{6\left(\frac{3}{4}\right)}_{\text{up}} + \underbrace{6\left(\frac{3}{4}\right)}_{\text{down}} = 12\left(\frac{3}{4}\right)$$

$$D_3 = \underbrace{6\left(\frac{3}{4}\right)\left(\frac{3}{4}\right)}_{\text{up}} + \underbrace{6\left(\frac{3}{4}\right)\left(\frac{3}{4}\right)}_{\text{down}} = 12\left(\frac{3}{4}\right)^2$$

By continuing this process, we have a total vertical distance of

$$D = 6 + 12\left(\frac{3}{4}\right) + 12\left(\frac{3}{4}\right)^2 + 12\left(\frac{3}{4}\right)^3 + \cdots.$$

To find the sum of this series, we rewrite the first term in geometric form as $6 = -6 + 12 = -6 + 12\left(\frac{3}{4}\right)^0$. Then we obtain

$$D = -6 + 12\left(\frac{3}{4}\right)^0 + 12\left(\frac{3}{4}\right)^1 + 12\left(\frac{3}{4}\right)^2 + 12\left(\frac{3}{4}\right)^3 + \cdots$$

$$= -6 + \sum_{n=0}^{\infty} 12\left(\frac{3}{4}\right)^n = -6 + \frac{12}{1 - (3/4)} = -6 + 48 = 42 \text{ ft.} \ \ ▭$$

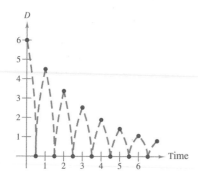

FIGURE 9.5

EXERCISES for Section 9.2

In Exercises 1–6, find the first five terms of the sequence of partial sums.

1. $1 + \dfrac{1}{4} + \dfrac{1}{9} + \dfrac{1}{16} + \dfrac{1}{25} + \cdots$

2. $\dfrac{1}{2 \cdot 3} + \dfrac{2}{3 \cdot 4} + \dfrac{3}{4 \cdot 5} + \dfrac{4}{5 \cdot 6} + \dfrac{5}{6 \cdot 7} + \cdots$

3. $3 - \dfrac{9}{2} + \dfrac{27}{4} - \dfrac{81}{8} + \dfrac{243}{16} - \cdots$

4. $\dfrac{1}{1} + \dfrac{1}{3} + \dfrac{1}{5} + \dfrac{1}{7} + \dfrac{1}{9} + \dfrac{1}{11} + \cdots$

5. $\displaystyle\sum_{n=1}^{\infty} \dfrac{3}{2^{n-1}}$

6. $\displaystyle\sum_{n=1}^{\infty} \dfrac{(-1)^{n+1}}{n!}$

In Exercises 7–18, verify that the infinite series diverges.

7. $\dfrac{1}{2} + \dfrac{2}{3} + \dfrac{3}{4} + \dfrac{4}{5} + \cdots$

8. $3 - \dfrac{9}{2} + \dfrac{27}{4} - \dfrac{81}{8} + \dfrac{243}{16} - \cdots$

9. $\displaystyle\sum_{n=1}^{\infty} \dfrac{3n}{n+1} = \dfrac{3}{2} + \dfrac{6}{3} + \dfrac{9}{4} + \dfrac{12}{5} + \cdots$

10. $\displaystyle\sum_{n=1}^{\infty} \dfrac{n}{2n+3} = \dfrac{1}{5} + \dfrac{2}{7} + \dfrac{3}{9} + \dfrac{4}{11} + \cdots$

11. $\displaystyle\sum_{n=1}^{\infty} \dfrac{n^2}{n^2+1}$

12. $\displaystyle\sum_{n=1}^{\infty} \dfrac{n}{\sqrt{n^2+1}}$

13. $\displaystyle\sum_{n=0}^{\infty} 3\left(\dfrac{3}{2}\right)^n$

14. $\displaystyle\sum_{n=0}^{\infty} \left(\dfrac{4}{3}\right)^n$

15. $\displaystyle\sum_{n=0}^{\infty} 1000(1.055)^n$

16. $\displaystyle\sum_{n=0}^{\infty} 2(-1.03)^n$

17. $\displaystyle\sum_{n=1}^{\infty} \dfrac{2^n+1}{2^{n+1}}$

18. $\displaystyle\sum_{n=1}^{\infty} \dfrac{n!}{2^n}$

In Exercises 19–24, verify that the infinite series converges.

19. $2 + \dfrac{3}{2} + \dfrac{9}{8} + \dfrac{27}{32} + \dfrac{81}{128} + \cdots$

20. $2 - 1 + \dfrac{1}{2} - \dfrac{1}{4} + \dfrac{1}{8} - \cdots$

21. $\displaystyle\sum_{n=0}^{\infty} (0.9)^n = 1 + 0.9 + 0.81 + 0.729 + \cdots$

22. $\displaystyle\sum_{n=0}^{\infty} (-0.6)^n = 1 - 0.6 + 0.36 - 0.216 + \cdots$

23. $\displaystyle\sum_{n=1}^{\infty} \dfrac{1}{n(n+1)}$ (Use partial fractions.)

24. $\displaystyle\sum_{n=1}^{\infty} \dfrac{1}{n(n+2)}$ (Use partial fractions.)

In Exercises 25–40, find the sum of the convergent series.

25. $\displaystyle\sum_{n=0}^{\infty} \left(\dfrac{1}{2}\right)^n$

26. $\displaystyle\sum_{n=0}^{\infty} 2\left(\dfrac{2}{3}\right)^n$

27. $\displaystyle\sum_{n=0}^{\infty} \left(\dfrac{-1}{2}\right)^n$

28. $\displaystyle\sum_{n=0}^{\infty} 2\left(\dfrac{-2}{3}\right)^n$

29. $1 + 0.1 + 0.01 + 0.001 + \cdots$

30. $8 + 6 + \dfrac{9}{2} + \dfrac{27}{8} + \cdots$

31. $3 - 1 + \dfrac{1}{3} - \dfrac{1}{9} + \cdots$

32. $4 - 2 + 1 - \dfrac{1}{2} + \cdots$

33. $\displaystyle\sum_{n=4}^{\infty} 3\left(\dfrac{5}{8}\right)^n$

34. $\displaystyle\sum_{n=5}^{\infty} 2\left(\dfrac{-3}{4}\right)^n$

35. $\displaystyle\sum_{n=2}^{\infty} \dfrac{1}{n^2-1}$

36. $\displaystyle\sum_{n=1}^{\infty} \dfrac{1}{n(n+1)}$

37. $\displaystyle\sum_{n=1}^{\infty} \dfrac{4}{n(n+2)}$

38. $\displaystyle\sum_{n=1}^{\infty} \dfrac{1}{(2n+1)(2n+3)}$

39. $\displaystyle\sum_{n=0}^{\infty} \left(\dfrac{1}{2^n} - \dfrac{1}{3^n}\right)$

40. $\displaystyle\sum_{n=1}^{\infty} [(0.7)^n + (0.9)^n]$

In Exercises 41–44, express the repeated decimal as a geometric series, and write its sum as the ratio of two integers.

41. $0.6\overline{66}$

42. $0.23\overline{23}$

43. $0.075\overline{75}$

44. $0.215\overline{15}$

In Exercises 45–56, determine the convergence or divergence of the series.

45. $\displaystyle\sum_{n=1}^{\infty} \dfrac{n+10}{10n+1}$

46. $\displaystyle\sum_{n=0}^{\infty} \dfrac{4}{2^n}$

47. $\displaystyle\sum_{n=1}^{\infty} \left(\dfrac{1}{n} - \dfrac{1}{n+2}\right)$

48. $\displaystyle\sum_{n=1}^{\infty} \dfrac{n+1}{2n-1}$

49. $\displaystyle\sum_{n=1}^{\infty} \dfrac{3n-1}{2n+1}$

50. $\displaystyle\sum_{n=0}^{\infty} \dfrac{1}{4^n}$

51. $\displaystyle\sum_{n=0}^{\infty} (1.075)^n$

52. $\displaystyle\sum_{n=1}^{\infty} \dfrac{2^n}{100}$

53. $\displaystyle\sum_{n=2}^{\infty} \dfrac{n}{\ln n}$

54. $\displaystyle\sum_{n=1}^{\infty} \dfrac{2^n}{n^2}$

55. $\displaystyle\sum_{n=1}^{\infty} \left(1 + \dfrac{k}{n}\right)^n$

56. $\displaystyle\sum_{n=1}^{\infty} \dfrac{1}{n(n+3)}$

57. A company producing a new product estimates the annual sales to be 8000 units. Suppose that in any given year 10 percent of the units (regardless of age) will become inoperative. How many units will be in use after n years?

58. Repeat Exercise 57 with the assumption that 25 percent of the units will become inoperative each year.

59. A ball is dropped from a height of 16 feet. Each time it drops h feet, it rebounds vertically $0.81h$ feet. Find the total distance traveled by the ball.

60. Find the total time it takes for the ball in Exercise 59 to come to rest.

61. Find the fraction of the total area of the square that is eventually shaded if the pattern of shading shown in the figure is continued. (Note that each side of the shaded corner squares is one-fourth that of the square in which it is placed.)

FIGURE FOR 61

In Exercises 62–66, use the formula for the nth partial sum of a geometric series:

$$\sum_{i=0}^{n-1} ar^i = \frac{a(1 - r^n)}{1 - r}.$$

62. A deposit of $100 is made each month for 5 years in an account that pays 10-percent interest compounded monthly. What is the balance A in the account at the end of the 5 years?

$$A = 100\left(1 + \frac{0.10}{12}\right) + \cdots + 100\left(1 + \frac{0.10}{12}\right)^{60}$$

63. A deposit of $50 is made each month in an account that pays 12-percent interest compounded monthly. What is the balance in the account at the end of 10 years?

64. A deposit of P dollars is made each month for t years in an account that pays interest at an annual rate of r, compounded monthly. Let $N = 12t$ be the total number of deposits, and show that the balance in the account after t years is

$$A = P\left[\left(1 + \frac{r}{12}\right)^N - 1\right]\left(1 + \frac{12}{r}\right).$$

65. Use the formula in Exercise 64 to find the amount in an account earning 9-percent interest compounded monthly after deposits of $50 have been made monthly for 40 years.

66. The number of direct ancestors a person has had is given by

$$2 + 2^2 + 2^3 + 2^4 + \cdots + 2^n + \cdots.$$

This formula is valid *provided* the person has had no common ancestors. [A common ancestor is one to whom you are related in more than one way. For example, one of your great-grandmothers on your father's side might also be one of your great-grandmothers on your mother's side (see figure). How many of your direct ancestors have lived since the year 1 A.D.? Assume that the average time between generations was 30 years (resulting in 66 generations) so that the total is given by

$$2 + 2^2 + 2^3 + 2^4 + \cdots + 2^{66}.$$

Considering your total, is it reasonable to assume that you have had no common ancestors in the past 2000 years?

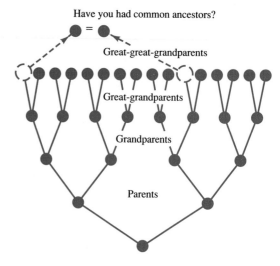

FIGURE FOR 66

67. Prove that $0.75 = 0.749999 \ldots .$

68. Prove that every decimal with a repeating pattern of digits is a rational number.

69. Show that the series

$$\sum_{n=1}^{\infty} a_n$$

can be written in the telescoping form

$$\sum_{n=1}^{\infty} [(c - S_{n-1}) - (c - S_n)]$$

where $S_0 = 0$ and S_n is the nth partial sum.

70. Let $\Sigma\, a_n$ be a convergent series, and let

$$R_N = a_{N+1} + a_{N+2} + \cdots$$

be the remainder of the series after the first N terms. Prove that

$$\lim_{N \to \infty} R_N = 0.$$

71. Find two divergent series $\Sigma\, a_n$ and $\Sigma\, b_n$ such that $\Sigma\,(a_n + b_n)$ converges.

72. Given two infinite series $\Sigma\, a_n$ and $\Sigma\, b_n$ such that $\Sigma\, a_n$ converges and $\Sigma\, b_n$ diverges, prove that $\Sigma\,(a_n + b_n)$ diverges.

9.3 The Integral Test and *p*-Series

The Integral Test ▪ *p*-Series ▪ Harmonic series

In this and the following section, we look at several convergence tests that apply to series with *positive* terms.

Since a definite integral is defined as the limit of a sum, it seems logical to expect that we might be able to use integrals to test for convergence or divergence of an infinite series. This is indeed the case, and we show how this is done in the proof of the following theorem.

**THEOREM 9.10
INTEGRAL TEST**

If f is positive, continuous, and decreasing for $x \geq 1$ and $a_n = f(n)$, then

$$\sum_{n=1}^{\infty} a_n \quad \text{and} \quad \int_1^{\infty} f(x)\, dx$$

either both converge or both diverge.

PROOF   We begin by partitioning the interval $[1, n]$ into $n - 1$ unit intervals, as shown in Figure 9.6. The total area of the inscribed rectangles is given by

$$\sum_{i=2}^{n} f(i) = f(2) + f(3) + \cdots + f(n). \qquad \text{Inscribed area}$$

Similarly, the total area of the circumscribed rectangles is

$$\sum_{i=1}^{n-1} f(i) = f(1) + f(2) + \cdots + f(n-1). \qquad \text{Circumscribed area}$$

FIGURE 9.6

The exact area under the graph of f from $x = 1$ to $x = n$, $\int_1^n f(x)\, dx$, lies between the inscribed and circumscribed areas, and we have

$$\sum_{i=2}^{n} f(i) \le \int_1^n f(x)\, dx \le \sum_{i=1}^{n-1} f(i).$$

Using the nth partial sum, $S_n = f(1) + f(2) + \cdots + f(n)$, we can write this inequality as

$$S_n - f(1) \le \int_1^n f(x)\, dx \le S_{n-1}.$$

Now, assuming that $\int_1^\infty f(x)\, dx$ converges to L, it follows that for $n \ge 1$

$$S_n - f(1) \le L \implies S_n \le L + f(1).$$

Consequently, $\{S_n\}$ is bounded and monotonic, and by Theorem 9.5 it converges. Thus, $\Sigma\, a_n$ converges.

For the other direction of the proof, we assume the improper integral diverges. Then $\int_1^n f(x)\, dx$ approaches infinity as $n \to \infty$, and the inequality

$$S_{n-1} \ge \int_1^n f(x)\, dx$$

implies that $\{S_n\}$ diverges. Thus, $\Sigma\, a_n$ diverges.

We noted earlier that the convergence or divergence of $\Sigma\, a_n$ is not affected by deleting the first N terms. Similarly, if the conditions for the Integral Test are satisfied for all $x \ge N > 1$, we simply use the integral $\int_N^\infty f(x)\, dx$ to test for convergence or divergence. (This is illustrated in Example 4.)

EXAMPLE 1   Using the Integral Test

Apply the Integral Test to the series

$$\sum_{n=1}^{\infty} \frac{n}{n^2 + 1}.$$

SOLUTION

Since $f(x) = x/(x^2 + 1)$ satisfies the conditions for the Integral Test (check this), we integrate to obtain

$$\int_1^\infty \frac{x}{x^2 + 1}\, dx = \frac{1}{2} \int_1^\infty \frac{2x}{x^2 + 1}\, dx$$

$$= \frac{1}{2} \lim_{b \to \infty} \left[\ln (x^2 + 1) \right]_1^b$$

$$= \frac{1}{2} \lim_{b \to \infty} [\ln (b^2 + 1) - \ln 2] = \infty.$$

Thus, the series *diverges*.

EXAMPLE 2   Using the Integral Test

Apply the Integral Test to the series

$$\sum_{n=1}^{\infty} \frac{1}{n^2 + 1}.$$

SOLUTION

Since $f(x) = 1/(x^2 + 1)$ satisfies the conditions for the Integral Test, we integrate to obtain

$$\int_1^{\infty} \frac{1}{x^2 + 1}\, dx = \lim_{b \to \infty} \left[\arctan x \right]_1^b$$

$$= \lim_{b \to \infty} [\arctan b - \arctan 1] = \frac{\pi}{4}.$$

Thus, the series *converges*.

p-Series

In the remainder of this section, we investigate a second type of series that has a simple arithmetic test for convergence or divergence.

DEFINITION OF *p*-SERIES

A series of the form

$$\sum_{n=1}^{\infty} \frac{1}{n^p} = \frac{1}{1^p} + \frac{1}{2^p} + \frac{1}{3^p} + \cdots$$

is called a **p-series,** where p is a positive constant. For $p = 1$, the series

$$\sum_{n=1}^{\infty} \frac{1}{n} = 1 + \frac{1}{2} + \frac{1}{3} + \cdots$$

is called the **harmonic series.**

REMARK   A **general harmonic series** is of the form $\Sigma\, [1/(an + b)]$. In music, strings of the same material, diameter, and tension, whose lengths form a harmonic series, produce harmonic tones.

The Integral Test is convenient for establishing the convergence or divergence of p-series. This is shown in the proof of Theorem 9.11.

**THEOREM 9.11
CONVERGENCE OF *p*-SERIES**

The p-series

$$\sum_{n=1}^{\infty} \frac{1}{n^p} = \frac{1}{1^p} + \frac{1}{2^p} + \frac{1}{3^p} + \frac{1}{4^p} + \cdots$$

1. converges if $p > 1$, and
2. diverges if $0 < p \le 1$.

PROOF The proof follows from the Integral Test and from Theorem 8.4, which states that

$$\int_1^\infty \frac{1}{x^p}\, dx$$

converges if $p > 1$, and diverges if $0 < p \leq 1$.

EXAMPLE 3 Convergent and divergent p-series

(a) From Theorem 9.11, it follows that the *harmonic series*

$$\sum_{n=1}^\infty \frac{1}{n} = \frac{1}{1} + \frac{1}{2} + \frac{1}{3} + \cdots \qquad p = 1$$

diverges.

(b) From Theorem 9.11, it follows that the p-series

$$\sum_{n=1}^\infty \frac{1}{n^2} = \frac{1}{1^2} + \frac{1}{2^2} + \frac{1}{3^2} + \cdots \qquad p = 2$$

converges. ▭

The sum of the series in part (b) of Example 3 can be shown to be $\pi^2/6$. (This was shown by Leonhard Euler, but the proof is too difficult to present here.) Be sure you see that the Integral Test does not tell us that the sum of the series is given by the value of the integral. For instance, the sum of the series $\Sigma \,(1/n^2)$ is given by

$$\sum_{n=1}^\infty \frac{1}{n^2} = \frac{\pi^2}{6} \approx 1.645$$

but the value of the corresponding improper integral is

$$\int_1^\infty \frac{1}{x^2}\, dx = 1.$$

EXAMPLE 4 A comparison to the harmonic series

Determine whether the following series converges or diverges.

$$\sum_{n=2}^\infty \frac{1}{n \ln n}$$

SOLUTION

This series is similar to the divergent harmonic series. If its terms were larger than those of the harmonic series, we would expect it to diverge. However, since its terms are smaller, we are not sure what to expect. Using the Integral Test, with $f(x) = 1/(x \ln x)$, we see that the series diverges.

$$\int_2^\infty \frac{1}{x \ln x}\, dx = \int_2^\infty \frac{1/x}{\ln x}\, dx$$

$$= \lim_{b\to\infty}\left[\ln(\ln x)\right]_2^b$$

$$= \lim_{b\to\infty}[\ln(\ln b) - \ln(\ln 2)] = \infty$$

EXERCISES for Section 9.3

In Exercises 1–10, determine the convergence or divergence of the given series using the Integral Test.

1. $\displaystyle\sum_{n=1}^\infty \frac{1}{n+1}$ **2.** $\displaystyle\sum_{n=1}^\infty e^{-n}$

3. $\displaystyle\sum_{n=1}^\infty ne^{-n}$ **4.** $\displaystyle\sum_{n=1}^\infty e^{-n}\cos n$

5. $\dfrac{1}{2}+\dfrac{1}{5}+\dfrac{1}{10}+\dfrac{1}{17}+\dfrac{1}{26}+\cdots$

6. $\dfrac{1}{3}+\dfrac{1}{5}+\dfrac{1}{7}+\dfrac{1}{9}+\dfrac{1}{11}+\cdots$

7. $\dfrac{\ln 2}{2}+\dfrac{\ln 3}{3}+\dfrac{\ln 4}{4}+\dfrac{\ln 5}{5}+\dfrac{\ln 6}{6}+\cdots$

8. $\dfrac{1}{4}+\dfrac{2}{7}+\dfrac{3}{12}+\cdots+\dfrac{n}{n^2+3}+\cdots$

9. $\displaystyle\sum_{n=1}^\infty \frac{n^{k-1}}{n^k+c}$, k is a positive integer

10. $\displaystyle\sum_{n=1}^\infty n^k e^{-n}$, k is a positive integer

In Exercises 11–20, determine the convergence or divergence of the given p-series.

11. $\displaystyle\sum_{n=1}^\infty \frac{1}{n^3}$ **12.** $\displaystyle\sum_{n=1}^\infty \frac{1}{n^{1/3}}$

13. $\displaystyle\sum_{n=1}^\infty \frac{1}{\sqrt[5]{n}}$ **14.** $\displaystyle\sum_{n=1}^\infty \frac{1}{n^{4/3}}$

15. $1+\dfrac{1}{\sqrt2}+\dfrac{1}{\sqrt3}+\dfrac{1}{\sqrt4}+\cdots$

16. $1+\dfrac{1}{4}+\dfrac{1}{9}+\dfrac{1}{16}+\dfrac{1}{25}+\cdots$

17. $1+\dfrac{1}{2\sqrt2}+\dfrac{1}{3\sqrt3}+\dfrac{1}{4\sqrt4}+\dfrac{1}{5\sqrt5}+\cdots$

18. $1+\dfrac{1}{\sqrt[3]4}+\dfrac{1}{\sqrt[3]9}+\dfrac{1}{\sqrt[3]{16}}+\dfrac{1}{\sqrt[3]{25}}+\cdots$

19. $\displaystyle\sum_{n=1}^\infty \frac{1}{n^{1.04}}$ **20.** $\displaystyle\sum_{n=1}^\infty \frac{1}{n^\pi}$

In Exercises 21–32, determine the convergence or divergence of the given series.

21. $\displaystyle\sum_{n=1}^\infty \frac{1}{2n-1}$ **22.** $\displaystyle\sum_{n=2}^\infty \frac{1}{n\sqrt{n^2-1}}$

23. $\displaystyle\sum_{n=1}^\infty \frac{1}{n\sqrt[4]{n}}$ **24.** $3\displaystyle\sum_{n=1}^\infty \frac{1}{n^{0.95}}$

25. $\displaystyle\sum_{n=0}^\infty \left(\frac{2}{3}\right)^n$ **26.** $\displaystyle\sum_{n=1}^\infty \frac{n}{\sqrt{n^2+1}}$

27. $\displaystyle\sum_{n=0}^\infty (1.075)^n$ **28.** $\displaystyle\sum_{n=1}^\infty \left(\frac{1}{n^2}-\frac{1}{n^3}\right)$

29. $\displaystyle\sum_{n=1}^\infty \left(1+\frac{1}{n}\right)^n$ **30.** $\displaystyle\sum_{n=2}^\infty \ln n$

31. $\displaystyle\sum_{n=2}^\infty \frac{1}{n(\ln n)^3}$ **32.** $\displaystyle\sum_{n=2}^\infty \frac{\ln n}{n^3}$

In Exercises 33 and 34, find the positive values of p for which the series converges.

33. $\displaystyle\sum_{n=2}^\infty \frac{1}{n(\ln n)^p}$ **34.** $\displaystyle\sum_{n=2}^\infty \frac{\ln n}{n^p}$

35. Use a computer to find the smallest value of k such that

$$\sum_{n=1}^k \frac{1}{n} > 4.$$

36. The **Riemann zeta function** for real numbers is

$$\zeta(x) = \sum_{n=1}^\infty n^{-x}.$$

Find the domain of this function.

In the remaining exercises of this set use the following result. Let f be a positive, continuous, and decreasing function for $x \ge 1$, such that $a_n = f(n)$. If the series

$$\sum_{n=1}^\infty a_n$$

converges to S, then the remainder $R_N = S - S_N$ is bounded by

$$0 \le R_N \le \int_N^\infty f(x)\, dx.$$

In Exercises 37–42, approximate the sum of the convergent series using the indicated number of terms. Include an estimate of the maximum error for your approximation.

37. $\sum_{n=1}^{\infty} \dfrac{1}{n^4}$, six terms

38. $\sum_{n=1}^{\infty} \dfrac{1}{n^5}$, four terms

39. $\sum_{n=1}^{\infty} \dfrac{1}{n^2 + 1}$, ten terms

40. $\sum_{n=1}^{\infty} \dfrac{1}{(n + 1)[\ln (n + 1)]^3}$, ten terms

41. $\sum_{n=1}^{\infty} ne^{-n^2}$, four terms

42. $\sum_{n=1}^{\infty} e^{-n}$, four terms

In Exercises 43–46, find N so that $R_N \le 0.001$ for the given convergent series.

43. $\sum_{n=1}^{\infty} \dfrac{1}{n^4}$

44. $\sum_{n=1}^{\infty} \dfrac{1}{n^{3/2}}$

45. $\sum_{n=1}^{\infty} e^{-5n}$

46. $\sum_{n=1}^{\infty} \dfrac{1}{n^2 + 1}$

9.4 Comparisons of Series

Direct Comparison Test ▪ Limit Comparison Test

In the previous two sections we developed tests for convergence or divergence of special types of series. In each case the terms of the series were quite simple, and the series *had to possess* special characteristics in order for the convergence tests to be applied. The slightest deviation from these special characteristics could make a test nonapplicable. For example, notice in the following pairs that the second series cannot be tested by the same convergence test as the first series even though it is similar to the first.

1. $\sum_{n=0}^{\infty} \dfrac{1}{2^n}$ is geometric, but $\sum_{n=0}^{\infty} \dfrac{n}{2^n}$ is not.

2. $\sum_{n=1}^{\infty} \dfrac{1}{n^3}$ is a *p*-series, but $\sum_{n=1}^{\infty} \dfrac{1}{n^3 + 1}$ is not.

3. $a_n = \dfrac{n}{(n^2 + 3)^2}$ is easily integrated, but $b_n = \dfrac{n^2}{(n^2 + 3)^2}$ is not.

In this section we discuss two additional tests for positive term series. These two tests greatly expand the variety of series we are able to test for convergence or divergence by allowing us to *compare* one series having similar but more complicated terms to a simpler series whose convergence or divergence is known.

**THEOREM 9.12
DIRECT COMPARISON TEST**

Let $0 \le a_n \le b_n$ for all n.

1. If $\sum_{n=1}^{\infty} b_n$ converges, then $\sum_{n=1}^{\infty} a_n$ converges.

2. If $\sum_{n=1}^{\infty} a_n$ diverges, then $\sum_{n=1}^{\infty} b_n$ diverges.

PROOF   To prove the first property, let

$$L = \sum_{n=1}^{\infty} b_n$$

and

$$S_n = a_1 + a_2 + \cdots + a_n.$$

Since $0 \le a_n \le b_n$, we know that the sequence

$$S_1, S_2, S_3, \ldots$$

is nondecreasing and bounded above by L; hence it must converge. Moreover, since

$$\lim_{n \to \infty} S_n = \sum_{n=1}^{\infty} a_n$$

it follows that $\Sigma \, a_n$ converges. The second property is logically equivalent to the first.

REMARK   As stated, the Direct Comparison Test requires that $0 \le a_n \le b_n$ for all n. Since the convergence of a series is not dependent on its first several terms, we could modify the test to require only that $0 \le a_n \le b_n$ for all n greater than some integer N.

EXAMPLE 1   Using the Direct Comparison Test

Determine the convergence or divergence of

$$\sum_{n=1}^{\infty} \frac{1}{2 + 3^n}.$$

SOLUTION

This series resembles

$$\sum_{n=1}^{\infty} \frac{1}{3^n}. \qquad \text{Convergent geometric series}$$

Term-by-term comparison yields

$$a_n = \frac{1}{2 + 3^n} < \frac{1}{3^n} = b_n, \quad n \ge 1.$$

Thus, it follows by the Direct Comparison Test that the series converges.

EXAMPLE 2 Using the Direct Comparison Test

Determine the convergence or divergence of

$$\sum_{n=1}^{\infty} \frac{1}{2 + \sqrt{n}}.$$

SOLUTION

This series resembles

$$\sum_{n=1}^{\infty} \frac{1}{n^{1/2}}.$$ Divergent p-series

Term-by-term comparison yields

$$\frac{1}{2 + \sqrt{n}} \leq \frac{1}{\sqrt{n}}, \quad n \geq 1$$

which *does not* meet the requirements for divergence. (Remember that if term-by-term comparison reveals a series that is *smaller* than a divergent series, then the Direct Comparison Test tells us nothing.) Still expecting the series to diverge, we compare the given series to

$$\sum_{n=1}^{\infty} \frac{1}{n}.$$ Divergent harmonic series

In this case, term-by-term comparison yields

$$a_n = \frac{1}{n} \leq \frac{1}{2 + \sqrt{n}} = b_n, \quad n \geq 4$$

and, by the Direct Comparison Test, the given series diverges. ▭

REMARK To verify the last inequality in Example 2, try showing that $2 + \sqrt{n} \leq n$ whenever $n \geq 4$.

When using the Direct Comparison Test, remember that $0 \leq a_n \leq b_n$ for both parts of the theorem. Informally, the test says the following about two series with nonnegative terms.

1. If the "larger" series $\Sigma\, b_n$ converges, then the "smaller" series $\Sigma\, a_n$ must also converge.
2. If the "smaller" series $\Sigma\, a_n$ diverges, then the "larger" series $\Sigma\, b_n$ must also diverge.

Limit Comparison Test

Often a given series closely resembles a p-series or a geometric series, yet we cannot establish the term-by-term comparison necessary to apply the Direct Comparison Test. Under these circumstances we may be able to apply a second comparison test, called the **Limit Comparison Test.**

THEOREM 9.13 LIMIT COMPARISON TEST	Suppose $a_n > 0$, $b_n > 0$, and

$$\lim_{n \to \infty} \left(\frac{a_n}{b_n} \right) = L$$

where L is *finite and positive*. Then the two series $\Sigma\, a_n$ and $\Sigma\, b_n$ either both converge or both diverge.

PROOF   Since $a_n > 0$, $b_n > 0$, and $(a_n/b_n) \to L$ as $n \to \infty$, there exists $N > 0$ such that for $n \geq N$, $0 < (a_n/b_n) < (L + 1)$. This implies that

$$0 < a_n < (L + 1)b_n.$$

Hence, by the Direct Comparison Test, the convergence of $\Sigma\, b_n$ implies the convergence of $\Sigma\, a_n$. Similarly, the fact that $(b_n/a_n) \to (1/L)$ as $n \to \infty$ can be used to show that the convergence of $\Sigma\, a_n$ implies the convergence of $\Sigma\, b_n$.

REMARK   As with the Direct Comparison Test, the Limit Comparison Test could be modified to require only that a_n and b_n be positive for all n greater than some integer N.

The versatility of the Limit Comparison Test is seen in the next example, where we show that a general harmonic series diverges.

EXAMPLE 3   Using the Limit Comparison Test

Show that the following general harmonic series diverges.

$$\sum_{n=1}^{\infty} \frac{1}{an + b}, \quad a > 0$$

SOLUTION

By comparison to

$$\sum_{n=1}^{\infty} \frac{1}{n} \qquad \text{Divergent harmonic series}$$

we have

$$\lim_{n \to \infty} \frac{1/(an + b)}{1/n} = \lim_{n \to \infty} \frac{n}{an + b} = \frac{1}{a}.$$

Since this limit is greater than 0, we conclude from the Limit Comparison Test that the given series diverges.

The Limit Comparison Test works well for comparing "messy" algebraic series to a *p*-series. In choosing an appropriate *p*-series, we must choose one with an *n*th term of the same magnitude as the *n*th term of the given series.

Given series	*Comparison series*	*Conclusion*
$\displaystyle\sum_{n=1}^{\infty} \frac{1}{3n^2 - 4n + 5}$	$\displaystyle\sum_{n=1}^{\infty} \frac{1}{n^2}$	Both series converge.
$\displaystyle\sum_{n=1}^{\infty} \frac{1}{\sqrt{3n - 2}}$	$\displaystyle\sum_{n=1}^{\infty} \frac{1}{\sqrt{n}}$	Both series diverge.
$\displaystyle\sum_{n=1}^{\infty} \frac{n^2 - 10}{4n^5 + n^3}$	$\displaystyle\sum_{n=1}^{\infty} \frac{n^2}{n^5} = \sum_{n=1}^{\infty} \frac{1}{n^3}$	Both series converge.
$\displaystyle\sum_{n=1}^{\infty} \frac{\sqrt{n}}{\sqrt{n^3 + 1}}$	$\displaystyle\sum_{n=1}^{\infty} \frac{n^{1/2}}{n^{3/2}} = \sum_{n=1}^{\infty} \frac{1}{n}$	Both series diverge.

In other words, when choosing a series for comparison, we disregard all but the *highest powers of n* in both the numerator and the denominator.

EXAMPLE 4 Using the Limit Comparison Test

Determine the convergence or divergence of

$$\sum_{n=1}^{\infty} \frac{4\sqrt{n} - 1}{n^2 + 2\sqrt{n}}.$$

SOLUTION

Disregarding all but the highest powers of n in the numerator and the denominator, we compare the series to

$$\sum_{n=1}^{\infty} \frac{\sqrt{n}}{n^2} = \sum_{n=1}^{\infty} \frac{1}{n^{3/2}}. \qquad \text{Convergent } p\text{-series}$$

Since

$$\lim_{n \to \infty} \frac{a_n}{b_n} = \lim_{n \to \infty} \left(\frac{4\sqrt{n} - 1}{n^2 + 2\sqrt{n}} \right)\left(\frac{n^{3/2}}{1} \right)$$

$$= \lim_{n \to \infty} \frac{4n^2 - n^{3/2}}{n^2 + 2\sqrt{n}} = 4$$

we conclude by the Limit Comparison Test that the given series converges.

EXAMPLE 5 Using the Limit Comparison Test

Determine the convergence or divergence of

$$\sum_{n=1}^{\infty} \frac{n2^n + 5}{4n^3 + 3n}.$$

SOLUTION

A reasonable comparison would be to the series

$$\sum_{n=1}^{\infty} \frac{2^n}{n^2}.$$

Note that this series diverges by the nth-Term Test, since

$$\lim_{n\to\infty} \frac{2^n}{n^2} \neq 0.$$

From the limit

$$\lim_{n\to\infty} \frac{a_n}{b_n} = \lim_{n\to\infty} \left(\frac{n2^n + 5}{4n^3 + 3n}\right)\left(\frac{n^2}{2^n}\right)$$

$$= \lim_{n\to\infty} \frac{1 + [5/(2^n n)]}{4 + (3/n^2)} = \frac{1}{4}$$

we conclude that the given series diverges. ▭

EXERCISES for Section 9.4

In Exercises 1–12, use the Direct Comparison Test to determine the convergence or divergence of the series.

1. $\sum_{n=1}^{\infty} \dfrac{1}{n^2 + 1}$

2. $\sum_{n=1}^{\infty} \dfrac{1}{3n^2 + 2}$

3. $\sum_{n=2}^{\infty} \dfrac{1}{n - 1}$

4. $\sum_{n=2}^{\infty} \dfrac{1}{\sqrt{n} - 1}$

5. $\sum_{n=0}^{\infty} \dfrac{1}{3^n + 1}$

6. $\sum_{n=0}^{\infty} \dfrac{2^n}{3^n + 5}$

7. $\sum_{n=2}^{\infty} \dfrac{\ln n}{n + 1}$

8. $\sum_{n=1}^{\infty} \dfrac{1}{\sqrt{n^3 + 1}}$

9. $\sum_{n=0}^{\infty} \dfrac{1}{n!}$

10. $\sum_{n=1}^{\infty} \dfrac{1}{3\sqrt[4]{n} - 1}$

11. $\sum_{n=0}^{\infty} e^{-n^2}$

12. $\sum_{n=0}^{\infty} \dfrac{4^n}{3^n - 1}$

In Exercises 13–26, use the Limit Comparison Test to determine the convergence or divergence of the series.

13. $\sum_{n=1}^{\infty} \dfrac{n}{n^2 + 1}$

14. $\sum_{n=2}^{\infty} \dfrac{1}{\sqrt{n^2 - 1}}$

15. $\sum_{n=0}^{\infty} \dfrac{1}{\sqrt{n^2 + 1}}$

16. $\sum_{n=1}^{\infty} \dfrac{1}{2^n - 5}$

17. $\sum_{n=1}^{\infty} \dfrac{2n^2 - 1}{3n^5 + 2n + 1}$

18. $\sum_{n=1}^{\infty} \dfrac{5n - 3}{n^2 - 2n + 5}$

19. $\sum_{n=1}^{\infty} \dfrac{n + 3}{n(n + 2)}$

20. $\sum_{n=1}^{\infty} \dfrac{1}{n(n^2 + 1)}$

21. $\sum_{n=1}^{\infty} \dfrac{1}{n\sqrt{n^2 + 1}}$

22. $\sum_{n=1}^{\infty} \dfrac{n}{(n + 1)2^{n-1}}$

23. $\sum_{n=1}^{\infty} \dfrac{n^{k-1}}{n^k + 1}, \; k > 2$

24. $\sum_{n=1}^{\infty} \dfrac{1}{n + \sqrt{n^2 + 1}}$

25. $\sum_{n=1}^{\infty} \sin \dfrac{1}{n}$

26. $\sum_{n=1}^{\infty} \tan \dfrac{1}{n}$

In Exercises 27–34, test for convergence or divergence, using each of the seven tests at least once.
(a) nth-Term Test
(b) Geometric Series Test
(c) p-Series Test
(d) telescoping series
(e) Integral Test
(f) Direct Comparison Test
(g) Limit Comparison Test

27. $\sum_{n=1}^{\infty} \dfrac{\sqrt{n}}{n}$

28. $\sum_{n=0}^{\infty} 5\left(\dfrac{-1}{5}\right)^n$

29. $\sum_{n=1}^{\infty} \dfrac{1}{3^n + 2}$

30. $\sum_{n=4}^{\infty} \dfrac{1}{3n^2 - 2n - 15}$

31. $\sum_{n=1}^{\infty} \dfrac{n}{2n + 3}$

32. $\sum_{n=1}^{\infty} \left(\dfrac{1}{n + 1} - \dfrac{1}{n + 2}\right)$

33. $\sum_{n=1}^{\infty} \dfrac{n}{(n^2 + 1)^2}$

34. $\sum_{n=1}^{\infty} \dfrac{3}{n(n + 3)}$

35. Use the Limit Comparison Test with the harmonic series to show that the series $\Sigma \, a_n$ (where $0 < a_n < a_{n-1}$) diverges if

$$\lim_{n\to\infty} na_n \neq 0.$$

36. Prove that if $P(n)$ and $Q(n)$ are polynomials of degree j and k, respectively, then the series

$$\sum_{n=1}^{\infty} \frac{P(n)}{Q(n)}$$

converges if $j < k - 1$, and diverges if $j \geq k - 1$.

In Exercises 37–40, use the polynomial test as given in Exercise 36 to determine whether the series converges or diverges.

37. $\dfrac{1}{2} + \dfrac{2}{5} + \dfrac{3}{10} + \dfrac{4}{17} + \dfrac{5}{26} + \cdots$

38. $\dfrac{1}{3} + \dfrac{1}{8} + \dfrac{1}{15} + \dfrac{1}{24} + \dfrac{1}{35} + \cdots$

39. $\displaystyle\sum_{n=1}^{\infty} \frac{1}{n^3 + 1}$ **40.** $\displaystyle\sum_{n=1}^{\infty} \frac{n^2}{n^3 + 1}$

In Exercises 41 and 42, use the divergence test used in Exercise 35 to show that each series diverges.

41. $\displaystyle\sum_{n=1}^{\infty} \frac{n^3}{5n^4 + 3}$ **42.** $\displaystyle\sum_{n=2}^{\infty} \frac{1}{\ln n}$

43. Prove that if the nonnegative series

$$\sum_{n=1}^{\infty} a_n \quad \text{and} \quad \sum_{n=1}^{\infty} b_n$$

converge, then so does the series

$$\sum_{n=1}^{\infty} a_n b_n.$$

44. Use the result of Exercise 43 to prove that if the nonnegative series

$$\sum_{n=1}^{\infty} a_n$$

converges, then so does the series

$$\sum_{n=1}^{\infty} a_n^2.$$

9.5 Alternating Series

Alternating series ▪ Alternating series remainder ▪ Absolute and conditional convergence

Most of the results we have studied so far have been restricted to series with positive terms. In this and the following section we consider series that contain both positive and negative terms. The simplest such series is an **alternating series,** whose terms alternate in sign. For example, the geometric series

$$\sum_{n=0}^{\infty} \left(-\frac{1}{2}\right)^n = \sum_{n=0}^{\infty} (-1)^n \frac{1}{2^n} = 1 - \frac{1}{2} + \frac{1}{4} - \frac{1}{8} + \frac{1}{16} - \cdots$$

is an *alternating geometric series* with $r = -\frac{1}{2}$.

Alternating series occur in two ways:

$$\sum_{n=1}^{\infty} (-1)^n a_n = -a_1 + a_2 - a_3 + a_4 - \cdots, \quad a_n > 0$$

and

$$\sum_{n=1}^{\infty} (-1)^{n-1} a_n = a_1 - a_2 + a_3 - a_4 + \cdots, \quad a_n > 0.$$

In one case the *odd* terms are negative, and in the other case the *even* terms are negative. The conditions for the convergence of an alternating series are given in the following theorem.

THEOREM 9.14
ALTERNATING SERIES TEST

If $a_n > 0$, then the alternating series

$$\sum_{n=1}^{\infty} (-1)^n a_n$$

and

$$\sum_{n=1}^{\infty} (-1)^{n-1} a_n$$

converge, provided that the following two conditions are met.

1. $a_{n+1} \le a_n$, for all n
2. $\lim_{n \to \infty} a_n = 0$

PROOF   The proof follows the same pattern for either form of the alternating series. In this proof, we use the form

$$\sum_{n=1}^{\infty} (-1)^{n-1} a_n = a_1 - a_2 + a_3 - a_4 + \cdots.$$

For this series, the partial sum (where $2n$ is even)

$$S_{2n} = (a_1 - a_2) + (a_3 - a_4) + (a_5 - a_6) + \cdots + (a_{2n-1} - a_{2n})$$

has all nonnegative terms, and therefore $\{S_{2n}\}$ is a nondecreasing sequence. But we can also write

$$S_{2n} = a_1 - (a_2 - a_3) - (a_4 - a_5) - \cdots - (a_{2n-2} - a_{2n-1}) - a_{2n}$$

which implies that $S_{2n} \le a_1$ for every integer n. Thus $\{S_{2n}\}$ is a bounded, nondecreasing sequence that converges to some value L. Now, since $S_{2n-1} = S_{2n} - a_{2n}$ and $a_{2n} \to 0$, we have

$$\lim_{n \to \infty} S_{2n-1} = \lim_{n \to \infty} S_{2n} - \lim_{n \to \infty} a_{2n} = L - \lim_{n \to \infty} a_{2n} = L.$$

Since both S_{2n} and S_{2n-1} converge to the same limit L, it follows that $\{S_n\}$ also converges to L. Consequently, the given alternating series converges.

REMARK   The first condition in the Alternating Series Test can be modified to require only that $0 < a_{n+1} \le a_n$ for all n greater than some integer N.

EXAMPLE 1   Using the Alternating Series Test

Determine the convergence or divergence of

$$\sum_{n=1}^{\infty} \frac{n}{(-2)^{n-1}} = \frac{1}{1} - \frac{2}{2} + \frac{3}{4} - \frac{4}{8} + \cdots.$$

SOLUTION

To apply the Alternating Series Test, we note that for $n \geq 1$,

$$\frac{1}{2} \leq \frac{n}{n+1}$$

which implies that

$$\frac{2^{n-1}}{2^n} \leq \frac{n}{n+1}$$

$$(n+1)\, 2^{n-1} \leq n 2^n$$

$$\frac{n+1}{2^n} \leq \frac{n}{2^{n-1}}.$$

Hence, $a_{n+1} = (n+1)/2^n \leq n/2^{n-1} = a_n$ for all n. Furthermore, by L'Hôpital's Rule,

$$\lim_{x \to \infty} \frac{x}{2^{x-1}} = \lim_{x \to \infty} \frac{1}{2^{x-1}(\ln 2)} = 0 \quad \Longrightarrow \quad \lim_{n \to \infty} \frac{n}{2^{n-1}} = 0.$$

Therefore, by the Alternating Series Test, the given series converges. ◻

EXAMPLE 2   Using the Alternating Series Test

Determine whether the following series converge or diverge.

(a) $\displaystyle \sum_{n=1}^{\infty} \frac{(-1)^n n}{\ln 2n}$ (b) $\displaystyle \sum_{n=1}^{\infty} (-1)^{n+1} \left(\frac{3n+2}{4n^2-3} \right)$

SOLUTION

(a) By L'Hôpital's Rule, we have

$$\lim_{x \to \infty} \frac{x}{\ln 2x} = \lim_{x \to \infty} \frac{1}{1/x} = \lim_{x \to \infty} x = \infty.$$

Thus, $\{a_n\}$ does not converge to zero, and the Alternating Series Test does not apply. However, by the nth-Term Test for Divergence, we can conclude that the series diverges.

(b) Sometimes it is convenient to use differentiation to establish that $a_{n+1} \leq a_n$. In this case, we let

$$f(x) = \frac{3x+2}{4x^2-3}.$$

Then the derivative

$$f'(x) = \frac{-12x^2 - 16x - 9}{(4x^2-3)^2}$$

is always negative. Hence, f is a decreasing function, and it follows that $a_{n+1} \leq a_n$ for $n \geq 1$. Furthermore, since

$$\lim_{n \to \infty} \frac{3n+2}{4n^2-3} = 0$$

the series converges by the Alternating Series Test. ◻

For a convergent alternating series, the partial sum S_N can be a useful approximation for the sum S of the series. Just how close S_N is to S is stated in the following theorem.

THEOREM 9.15
ALTERNATING SERIES REMAINDER

If a convergent alternating series satisfies the condition $a_{n+1} \le a_n$, then the absolute value of the remainder R_N involved in approximating the sum S by S_N is less than (or equal to) the first neglected term. That is,

$$|S - S_N| = |R_N| \le a_{N+1}.$$

PROOF   The series obtained by deleting the first N terms of the given series satisfies the conditions of the Alternating Series Test and has a sum of R_N.

$$R_N = S - S_N = \sum_{n=1}^{\infty} (-1)^{n-1} a_n - \sum_{n=1}^{N} (-1)^{n-1} a_n$$
$$= (-1)^N a_{N+1} + (-1)^{N+1} a_{N+2} + (-1)^{N+2} a_{N+3} + \cdots$$
$$= (-1)^N (a_{N+1} - a_{N+2} + a_{N+3} - \cdots)$$
$$|R_N| = a_{N+1} - a_{N+2} + a_{N+3} - a_{N+4} + a_{N+5} - \cdots$$
$$= a_{N+1} - (a_{N+2} - a_{N+3}) - (a_{N+4} - a_{N+5}) - \cdots$$
$$\le a_{N+1}$$

Consequently, $|S - S_N| = |R_N| \le a_{N+1}$, which establishes the theorem.

EXAMPLE 3   Approximating the sum of an alternating series

Approximate the sum of the following series by its first six terms.

$$\sum_{n=1}^{\infty} (-1)^{n-1} \left(\frac{1}{n!} \right) = \frac{1}{1!} - \frac{1}{2!} + \frac{1}{3!} - \frac{1}{4!} + \frac{1}{5!} - \frac{1}{6!} + \cdots$$

SOLUTION

The Alternating Series Test establishes that the series converges because

$$\frac{1}{(n+1)!} \le \frac{1}{n!} \quad \text{and} \quad \lim_{n \to \infty} \frac{1}{n!} = 0.$$

Now, the sum of the first six terms is

$$S_6 = 1 - \frac{1}{2} + \frac{1}{6} - \frac{1}{24} + \frac{1}{120} - \frac{1}{720} \approx 0.63194$$

and, by the Alternating Series Remainder, we have

$$|S - S_6| = |R_6| \le a_7 = \frac{1}{5040} \approx 0.0002.$$

Therefore, the sum S lies between $0.63194 - 0.0002$ and $0.63194 + 0.0002$, and we have

$$0.63174 \leq S \leq 0.63214.$$

(Later, we will see that the actual sum is $(e - 1)/e \approx 0.63212$.)

Absolute and conditional convergence

Occasionally, a series may have both positive and negative terms and not be an alternating series. For instance, the series

$$\sum_{n=1}^{\infty} \frac{\sin n}{n^2} = \frac{\sin 1}{1} + \frac{\sin 2}{4} + \frac{\sin 3}{9} + \cdots$$

has both positive and negative terms, yet it is not an alternating series. One way to obtain some information about the convergence of this series is to investigate the convergence of the series

$$\sum_{n=1}^{\infty} \left| \frac{\sin n}{n^2} \right|.$$

By direct comparison, we have $|\sin n| \leq 1$ for all n, so

$$\left| \frac{\sin n}{n^2} \right| \leq \frac{1}{n^2}, \quad n \geq 1.$$

Thus, by the Direct Comparison Test, the series $\Sigma |(\sin n)/n^2|$ converges. But the question still is "Does the original series converge?" The next theorem tells us that the answer is yes.

THEOREM 9.16 ABSOLUTE CONVERGENCE	If the series $\Sigma |a_n|$ converges, then the series Σa_n also converges.

PROOF Since $0 \leq a_n + |a_n| \leq 2|a_n|$ for all n, the series

$$\sum_{n=1}^{\infty} (a_n + |a_n|)$$

converges by comparison to the convergent series

$$\sum_{n=1}^{\infty} 2|a_n|.$$

Furthermore, since $a_n = (a_n + |a_n|) - |a_n|$, we can write

$$\sum_{n=1}^{\infty} a_n = \sum_{n=1}^{\infty} (a_n + |a_n|) - \sum_{n=1}^{\infty} |a_n|$$

where both series on the right converge. Hence it follows that Σa_n converges.

The converse of Theorem 9.16 is not true. For instance, the **alternating harmonic series**

$$\sum_{n=1}^{\infty} \frac{(-1)^{n+1}}{n} = \frac{1}{1} - \frac{1}{2} + \frac{1}{3} - \frac{1}{4} + \cdots$$

converges by the Alternating Series Test. Yet the harmonic series diverges. We call this type of convergence **conditional.**

DEFINITION OF ABSOLUTE AND CONDITIONAL CONVERGENCE	1. $\sum a_n$ is **absolutely convergent** if $\sum \lvert a_n \rvert$ converges. 2. $\sum a_n$ is **conditionally convergent** if $\sum a_n$ converges but $\sum \lvert a_n \rvert$ diverges.

EXAMPLE 4   Absolute and conditional convergence

Determine whether the following series are convergent or divergent. If convergent, classify the series as absolutely or conditionally convergent.

(a) $\displaystyle\sum_{n=1}^{\infty} \frac{(-1)^{n(n+1)/2}}{3^n} = -\frac{1}{3} - \frac{1}{9} + \frac{1}{27} + \frac{1}{81} - \cdots$

(b) $\displaystyle\sum_{n=1}^{\infty} \frac{(-1)^n}{\ln(n+1)} = -\frac{1}{\ln 2} + \frac{1}{\ln 3} - \frac{1}{\ln 4} + \frac{1}{\ln 5} - \cdots$

(c) $\displaystyle\sum_{n=0}^{\infty} \frac{(-1)^n n!}{2^n} = \frac{0!}{2^0} - \frac{1!}{2^1} + \frac{2!}{2^2} - \frac{3!}{2^3} + \cdots$

(d) $\displaystyle\sum_{n=1}^{\infty} \frac{(-1)^n}{\sqrt{n}} = -\frac{1}{\sqrt{1}} + \frac{1}{\sqrt{2}} - \frac{1}{\sqrt{3}} + \frac{1}{\sqrt{4}} - \cdots$

SOLUTION

(a) This is *not* an alternating series, since the signs change in pairs. However, we note that

$$\sum_{n=1}^{\infty} \left\lvert \frac{(-1)^{n(n+1)/2}}{3^n} \right\rvert = \sum_{n=1}^{\infty} \frac{1}{3^n}$$

is a convergent geometric series with $r = \frac{1}{3} < 1$. Consequently, by Theorem 9.16, we can conclude that the given series is *absolutely* convergent, hence convergent.

(b) In this case, the Alternating Series Test indicates that the given series converges. However, the series

$$\sum_{n=1}^{\infty} \left\lvert \frac{(-1)^n}{\ln(n+1)} \right\rvert = \frac{1}{\ln 2} + \frac{1}{\ln 3} + \frac{1}{\ln 4} + \cdots$$

diverges by direct comparison with the terms of the harmonic series. Therefore, we conclude that the given series is *conditionally* convergent.

(c) This is an alternating series, but the Alternating Series Test does not apply because the limit of the nth term is not zero. However, by the nth-Term Test for Divergence, the series diverges.

(d) In this case, the Alternating Series Test tells us that the series converges. Moreover, since the series

$$\sum_{n=1}^{\infty} \left| \frac{(-1)^n}{\sqrt{n}} \right| = \frac{1}{\sqrt{1}} + \frac{1}{\sqrt{2}} + \frac{1}{\sqrt{3}} + \frac{1}{\sqrt{4}} + \cdots$$

diverges, we conclude that the given series is *conditionally* convergent.

EXERCISES for Section 9.5

In Exercises 1–22, use the Alternating Series Test to determine the convergence or divergence of the series.

1. $\displaystyle\sum_{n=1}^{\infty} \frac{(-1)^{n+1}}{n}$

2. $\displaystyle\sum_{n=1}^{\infty} \frac{(-1)^{n+1}n}{2n-1}$

3. $\displaystyle\sum_{n=1}^{\infty} \frac{(-1)^{n+1}}{2n-1}$

4. $\displaystyle\sum_{n=2}^{\infty} \frac{(-1)^n}{\ln n}$

5. $\displaystyle\sum_{n=1}^{\infty} \frac{(-1)^n n^2}{n^2+1}$

6. $\displaystyle\sum_{n=1}^{\infty} \frac{(-1)^{n+1}n}{n^2+1}$

7. $\displaystyle\sum_{n=1}^{\infty} \frac{(-1)^n}{\sqrt{n}}$

8. $\displaystyle\sum_{n=1}^{\infty} \frac{(-1)^{n+1}n^3}{n^3+6}$

9. $\displaystyle\sum_{n=1}^{\infty} \frac{(-1)^{n+1}(n+1)}{\ln(n+1)}$

10. $\displaystyle\sum_{n=1}^{\infty} \frac{(-1)^{n+1}\ln(n+1)}{n+1}$

11. $\displaystyle\sum_{n=0}^{\infty} (-1)^n e^{-n}$

12. $\displaystyle\sum_{n=1}^{\infty} \frac{(-1)^{n+1}\sqrt{n+2}}{\sqrt{n(n+2)}}$

13. $\displaystyle\sum_{n=1}^{\infty} \sin\frac{(2n-1)\pi}{2}$

14. $\displaystyle\sum_{n=1}^{\infty} \cos n\pi$

15. $\displaystyle\sum_{n=1}^{\infty} \frac{1}{n}\sin\frac{(2n-1)\pi}{2}$

16. $\displaystyle\sum_{n=1}^{\infty} \frac{1}{n}\cos n\pi$

17. $\displaystyle\sum_{n=0}^{\infty} \frac{(-1)^n}{n!}$

18. $\displaystyle\sum_{n=0}^{\infty} \frac{(-1)^n}{(2n)!}$

19. $\displaystyle\sum_{n=1}^{\infty} \frac{(-1)^{n+1}\sqrt{n}}{n+2}$

20. $\displaystyle\sum_{n=1}^{\infty} \frac{(-1)^{n+1}\sqrt{n}}{\sqrt[3]{n}}$

21. $\displaystyle\sum_{n=1}^{\infty} \frac{2(-1)^{n+1}}{e^n - e^{-n}} = \sum_{n=1}^{\infty} (-1)^{n+1}\operatorname{csch} n$

22. $\displaystyle\sum_{n=1}^{\infty} \frac{2(-1)^{n+1}}{e^n + e^{-n}} = \sum_{n=1}^{\infty} (-1)^{n+1}\operatorname{sech} n$

In Exercises 23–38, examine the series for conditional convergence or absolute convergence.

23. $\displaystyle\sum_{n=1}^{\infty} \frac{(-1)^{n+1}}{(n+1)^2}$

24. $\displaystyle\sum_{n=1}^{\infty} \frac{(-1)^{n+1}}{n+1}$

25. $\displaystyle\sum_{n=1}^{\infty} \frac{(-1)^{n+1}}{\sqrt{n}}$

26. $\displaystyle\sum_{n=1}^{\infty} \frac{(-1)^{n+1}}{n\sqrt{n}}$

27. $\displaystyle\sum_{n=2}^{\infty} \frac{(-1)^n}{\ln n}$

28. $\displaystyle\sum_{n=0}^{\infty} (-1)^n e^{-n^2}$

29. $\displaystyle\sum_{n=2}^{\infty} \frac{(-1)^n n}{n^3 - 1}$

30. $\displaystyle\sum_{n=1}^{\infty} \frac{(-1)^{n+1}}{n^{1.5}}$

31. $\displaystyle\sum_{n=0}^{\infty} \frac{(-1)^n}{(2n+1)!}$

32. $\displaystyle\sum_{n=0}^{\infty} \frac{(-1)^n}{\sqrt[3]{n+1}}$

33. $\displaystyle\sum_{n=0}^{\infty} \frac{\cos n\pi}{n+1}$

34. $\displaystyle\sum_{n=1}^{\infty} (-1)^{n+1}\arctan n$

35. $\displaystyle\sum_{n=1}^{\infty} \frac{\cos n}{n^2}$

36. $\displaystyle\sum_{n=1}^{\infty} \frac{\cos n}{n\sqrt{n}}$

37. $\displaystyle\sum_{n=1}^{\infty} \frac{\sin[(2n-1)\pi/2]}{\sqrt{n}}$

38. $\displaystyle\sum_{n=1}^{\infty} \frac{\sin[(2n-1)\pi/2]}{n}$

In Exercises 39–46, approximate the sum of the series with an error of less than 0.001. (Use Theorem 9.15.)

39. $\displaystyle\sum_{n=1}^{\infty} \frac{(-1)^{n+1}}{2n^3 - 1}$

40. $\displaystyle\sum_{n=1}^{\infty} \frac{(-1)^{n+1}}{n^4}$

41. $\displaystyle\sum_{n=0}^{\infty} \frac{(-1)^n}{n!}$ $\left(\text{the sum is }\dfrac{1}{e}\right)$

42. $\displaystyle\sum_{n=0}^{\infty} \frac{(-1)^n}{2^n n!}$ $\left(\text{the sum is }\dfrac{1}{\sqrt{e}}\right)$

43. $\displaystyle\sum_{n=0}^{\infty} \frac{(-1)^n}{(2n+1)!}$ (the sum is sin 1)

44. $\displaystyle\sum_{n=0}^{\infty} \frac{(-1)^n}{(2n)!}$ (the sum is cos 1)

45. $\displaystyle\sum_{n=1}^{\infty} \frac{(-1)^{n+1}}{n2^n}$ $\left(\text{the sum is }\ln\dfrac{3}{2}\right)$

46. $\displaystyle\sum_{n=1}^{\infty} \frac{(-1)^{n+1}}{n4^n}$ $\left(\text{the sum is }\ln\dfrac{5}{4}\right)$

In Exercises 47 and 48, find the number of terms necessary to approximate the sum of the series with an error less than 0.001. (Use Theorem 9.15.)

47. $\displaystyle\sum_{n=0}^{\infty} \frac{(-1)^n}{2n+1}$ $\left(\text{the sum is }\dfrac{\pi}{4}\right)$

48. $\displaystyle\sum_{n=1}^{\infty} \frac{(-1)^{n+1}}{n^2}$ $\left(\text{the sum is }\dfrac{\pi^2}{12}\right)$

49. Prove that the alternating p-series

$$\sum_{n=1}^{\infty} (-1)^n \left(\frac{1}{n^p}\right)$$

converges if $p > 0$.

50. Prove that if $\Sigma\, |a_n|$ converges, then $\Sigma\, a_n^2$ converges.

51. Find all values of x for which the series

$$\sum_{n=1}^{\infty} \frac{x^n}{n}$$

(a) converges absolutely and
(b) converges conditionally.

52. Determine the error in the following argument that $0 = 1$.

$$0 = 0 + 0 + 0 + \cdots$$
$$= (1 - 1) + (1 - 1) + (1 - 1) + \cdots$$
$$= 1 + (-1 + 1) + (-1 + 1) + \cdots$$
$$= 1 + 0 + 0 + \cdots$$
$$= 1$$

9.6 The Ratio and Root Tests

Ratio Test ▪ Root Test ▪ Summary of tests for convergence and divergence

We begin this section with a test for absolute convergence—the **Ratio Test.**

**THEOREM 9.17
RATIO TEST**

Let $\Sigma\, a_n$ be a series with nonzero terms.

1. $\Sigma\, a_n$ converges if

$$\lim_{n\to\infty} \left| \frac{a_{n+1}}{a_n} \right| < 1. \qquad \text{Absolute convergence}$$

2. $\Sigma\, a_n$ diverges if

$$\lim_{n\to\infty} \left| \frac{a_{n+1}}{a_n} \right| > 1.$$

3. The Ratio Test is inconclusive if

$$\lim_{n\to\infty} \left| \frac{a_{n+1}}{a_n} \right| = 1.$$

PROOF   For the proof of the first part, we let

$$\lim_{n\to\infty} \left| \frac{a_{n+1}}{a_n} \right| = r < 1$$

and choose R such that $0 \le r < R < 1$. By the definition of the limit of a sequence, there exists some $N > 0$ such that $|a_{n+1}/a_n| < R$ for all $n > N$. Therefore, we can write the following inequalities.

$$|a_{N+1}| < |a_N|R$$
$$|a_{N+2}| < |a_{N+1}|R < |a_N|R^2$$
$$|a_{N+3}| < |a_{N+2}|R < |a_{N+1}|R^2 < |a_N|R^3$$
$$\vdots$$

The geometric series

$$\sum_{n=1}^{\infty} a_N R^n = a_N R + a_N R^2 + \cdots + a_N R^n + \cdots$$

converges, so, by the Direct Comparison Test, the series

$$\sum_{n=1}^{\infty} |a_{N+n}| = |a_{N+1}| + |a_{N+2}| + \cdots + |a_{N+n}| + \cdots$$

also converges. This in turn implies that the series $\Sigma |a_n|$ converges, since discarding a finite number of terms ($n = N - 1$) does not affect convergence. Consequently, by Theorem 9.16, the series Σa_n converges.

The proof of the second part is similar, except that we choose R such that

$$\lim_{n \to \infty} \left| \frac{a_{n+1}}{a_n} \right| = r > R > 1$$

and show that there exists some $M > 0$ such that $|a_{M+n}| > |a_M| R^n$.

The fact that the Ratio Test fails to give us any useful information when $|a_{n+1}/a_n| \to 1$ can be seen by comparing the two series

$$\sum_{n=1}^{\infty} \frac{1}{n} \quad \text{and} \quad \sum_{n=1}^{\infty} \frac{1}{n^2}.$$

The first series diverges and the second one converges, but in both cases

$$\lim_{n \to \infty} \left| \frac{a_{n+1}}{a_n} \right| = 1.$$

Although the Ratio Test is not a cure for all ills related to tests for convergence, it is particularly useful for series that *converge rapidly*. Series involving factorials or exponentials are frequently of this type.

EXAMPLE 1 Using the Ratio Test

Determine the convergence or divergence of

$$\sum_{n=0}^{\infty} \frac{2^n}{n!}.$$

SOLUTION

Since $a_n = 2^n/n!$, we have

$$\lim_{n \to \infty} \left| \frac{a_{n+1}}{a_n} \right| = \lim_{n \to \infty} \left[\frac{2^{n+1}}{(n+1)!} \div \frac{2^n}{n!} \right] = \lim_{n \to \infty} \left[\frac{2^{n+1}}{(n+1)!} \cdot \frac{n!}{2^n} \right]$$

$$= \lim_{n \to \infty} \frac{2}{n+1} = 0.$$

Therefore, the series converges.

EXAMPLE 2 Using the Ratio Test

Determine whether the following series converge or diverge.

(a) $\displaystyle\sum_{n=0}^{\infty} \frac{n^2 2^{n+1}}{3^n}$ (b) $\displaystyle\sum_{n=0}^{\infty} \frac{n^n}{n!}$

SOLUTION

(a) Since

$$\lim_{n \to \infty} \left| \frac{a_{n+1}}{a_n} \right| = \lim_{n \to \infty} \left[(n+1)^2 \left(\frac{2^{n+2}}{3^{n+1}} \right) \left(\frac{3^n}{n^2 2^{n+1}} \right) \right]$$

$$= \lim_{n \to \infty} \frac{2(n+1)^2}{3n^2} = \frac{2}{3} < 1$$

we conclude by the Ratio Test that the series converges.

(b) Since

$$\lim_{n \to \infty} \left| \frac{a_{n+1}}{a_n} \right| = \lim_{n \to \infty} \left[\frac{(n+1)^{n+1}}{(n+1)!} \left(\frac{n!}{n^n} \right) \right]$$

$$= \lim_{n \to \infty} \left[\frac{(n+1)^{n+1}}{(n+1)} \left(\frac{1}{n^n} \right) \right]$$

$$= \lim_{n \to \infty} \frac{(n+1)^n}{n^n} = \lim_{n \to \infty} \left(1 + \frac{1}{n} \right)^n = e > 1$$

we conclude by the Ratio Test that the series diverges.

EXAMPLE 3   A failure of the Ratio Test

Determine the convergence or divergence of

$$\sum_{n=1}^{\infty} (-1)^n \frac{\sqrt{n}}{n+1}.$$

SOLUTION

Using the Ratio Test, we have

$$\lim_{n \to \infty} \left| \frac{a_{n+1}}{a_n} \right| = \lim_{n \to \infty} \left[\left(\frac{\sqrt{n+1}}{n+2} \right) \left(\frac{n+1}{\sqrt{n}} \right) \right]$$

$$= \lim_{n \to \infty} \left[\sqrt{\frac{n+1}{n}} \left(\frac{n+1}{n+2} \right) \right]$$

$$= \sqrt{1}(1) = 1.$$

Thus, as the Ratio Test gives us no useful information, we use the Alternating Series Test. To show that $a_{n+1} \le a_n$, we let $f(x) = \sqrt{x}/(x+1)$. Then the derivative is

$$f'(x) = \frac{-x+1}{2\sqrt{x}(x+1)^2}$$

and since this is negative for $x > 1$, we know that f is a decreasing function. Also, by L'Hôpital's Rule,

$$\lim_{x \to \infty} \frac{\sqrt{x}}{x+1} = \lim_{x \to \infty} \frac{1/(2\sqrt{x})}{1} = \lim_{x \to \infty} \frac{1}{2\sqrt{x}} = 0.$$

Therefore, the series converges.

REMARK   Note that the series in Example 3 is *conditionally convergent*, since $\Sigma\, |a_n|$ diverges (by the Limit Comparison Test with $\Sigma\, 1/\sqrt{n}$), but $\Sigma\, a_n$ converges.

The Root Test

The next test for convergence or divergence of series works especially well for series involving an *n*th power. The proof of this theorem is similar to that given for the Ratio Test, and we leave it as an exercise (see Exercise 51).

THEOREM 9.18 **ROOT TEST**	1. $\Sigma\, a_n$ converges if $$\lim_{n\to\infty} \sqrt[n]{|a_n|} < 1.$$ 2. $\Sigma\, a_n$ diverges if $$\lim_{n\to\infty} \sqrt[n]{|a_n|} > 1.$$ 3. The Root Test is inconclusive if $$\lim_{n\to\infty} \sqrt[n]{|a_n|} = 1.$$

EXAMPLE 4   Using the Root Test

Determine the convergence or divergence of

$$\sum_{n=1}^{\infty} \frac{e^{2n}}{n^n}.$$

SOLUTION

Since

$$\lim_{n\to\infty} \sqrt[n]{|a_n|} = \lim_{n\to\infty} \sqrt[n]{\frac{e^{2n}}{n^n}} = \lim_{n\to\infty} \frac{e^{2n/n}}{n^{n/n}}$$

$$= \lim_{n\to\infty} \frac{e^2}{n} = 0 < 1$$

we conclude by the Root Test that the series converges.

EXAMPLE 5   Using the Root Test

Determine the convergence or divergence of

$$\sum_{n=1}^{\infty} \frac{n^3}{3^n}.$$

SOLUTION

First, we have

$$\lim_{n \to \infty} \sqrt[n]{|a_n|} = \lim_{n \to \infty} \sqrt[n]{\frac{n^3}{3^n}} = \lim_{n \to \infty} \frac{n^{3/n}}{3}.$$

Since the limit in the numerator yields the indeterminate form ∞^0, we apply L'Hôpital's Rule as follows.

$$y = \lim_{x \to \infty} x^{3/x}$$

$$\ln y = \ln \left(\lim_{x \to \infty} x^{3/x} \right)$$

$$= \lim_{x \to \infty} (\ln x^{3/x})$$

$$= \lim_{x \to \infty} \left(\frac{3}{x} \ln x \right)$$

$$= \lim_{x \to \infty} \frac{3 \ln x}{x}$$

$$= \lim_{x \to \infty} \frac{3/x}{1} = 0 \qquad \text{L'Hôpital's Rule}$$

Now, since $\ln y = 0$, we conclude that

$$y = \lim_{x \to \infty} x^{3/x} = 1$$

from which it follows that

$$\lim_{n \to \infty} \frac{n^{3/n}}{3} = \frac{1}{3} < 1.$$

Thus, by the Root Test, the given series converges.

In the last five sections we have discussed ten different tests for determining the convergence or divergence of an infinite series. Skill in choosing and applying the various tests will come only with practice. We summarize in Table 9.2 the various tests we have studied. Below is a useful checklist for choosing an appropriate test.

1. Does the nth term approach zero? If not, the series diverges.
2. Is the series one of the special types—geometric, p-series, telescoping, alternating?
3. Can the Integral, Root, or Ratio Test be applied?
4. Can the series be compared favorably to one of the special types?

In some instances more than one test is applicable. However, your objective should be to learn to choose the most efficient way to test a series.

TABLE 9.2 Summary of Tests for Series

Test	Series	Converges	Diverges	Comment
nth-Term	$\displaystyle\sum_{n=1}^{\infty} a_n$		$\displaystyle\lim_{n\to\infty} a_n \neq 0$	This test cannot be used to show convergence.
Geometric Series	$\displaystyle\sum_{n=0}^{\infty} ar^n$	$\|r\| < 1$	$\|r\| \geq 1$	Sum: $S = \dfrac{a}{1-r}$
Telescoping Series	$\displaystyle\sum_{n=1}^{\infty} (b_n - b_{n+1})$	$\displaystyle\lim_{n\to\infty} b_n = L$		Sum: $S = b_1 - L$
p-Series	$\displaystyle\sum_{n=1}^{\infty} \dfrac{1}{n^p}$	$p > 1$	$p \leq 1$	
Alternating Series	$\displaystyle\sum_{n=1}^{\infty} (-1)^{n-1}a_n$	$0 < a_{n+1} \leq a_n$ and $\displaystyle\lim_{n\to\infty} a_n = 0$		Remainder: $\|R_N\| \leq a_{N+1}$
Integral (f is continuous, positive, and decreasing)	$\displaystyle\sum_{n=1}^{\infty} a_n,\ a_n = f(n) \geq 0$	$\displaystyle\int_1^{\infty} f(x)\,dx$ converges	$\displaystyle\int_1^{\infty} f(x)\,dx$ diverges	Remainder: $0 < R_N < \displaystyle\int_N^{\infty} f(x)\,dx$
Root	$\displaystyle\sum_{n=1}^{\infty} a_n$	$\displaystyle\lim_{n\to\infty} \sqrt[n]{\|a_n\|} < 1$	$\displaystyle\lim_{n\to\infty} \sqrt[n]{\|a_n\|} > 1$	Test is inconclusive if $\displaystyle\lim_{n\to\infty} \sqrt[n]{\|a_n\|} = 1$.
Ratio	$\displaystyle\sum_{n=1}^{\infty} a_n$	$\displaystyle\lim_{n\to\infty} \left\|\dfrac{a_{n+1}}{a_n}\right\| < 1$	$\displaystyle\lim_{n\to\infty} \left\|\dfrac{a_{n+1}}{a_n}\right\| > 1$	Test is inconclusive if $\displaystyle\lim_{n\to\infty} \left\|\dfrac{a_{n+1}}{a_n}\right\| = 1$.
Direct Comparison ($a_n, b_n > 0$)	$\displaystyle\sum_{n=1}^{\infty} a_n$	$0 \leq a_n \leq b_n$ and $\displaystyle\sum_{n=1}^{\infty} b_n$ converges	$0 \leq b_n \leq a_n$ and $\displaystyle\sum_{n=1}^{\infty} b_n$ diverges	
Limit Comparison ($a_n, b_n > 0$)	$\displaystyle\sum_{n=1}^{\infty} a_n$	$\displaystyle\lim_{n\to\infty} \dfrac{a_n}{b_n} = L > 0$ and $\displaystyle\sum_{n=1}^{\infty} b_n$ converges	$\displaystyle\lim_{n\to\infty} \dfrac{a_n}{b_n} = L > 0$ and $\displaystyle\sum_{n=1}^{\infty} b_n$ diverges	

EXERCISES for Section 9.6

In Exercises 1–20, use the Ratio Test to test for convergence or divergence of the series.

1. $\displaystyle\sum_{n=0}^{\infty} \dfrac{n!}{3^n}$

2. $\displaystyle\sum_{n=1}^{\infty} n\left(\dfrac{2}{3}\right)^n$

3. $\displaystyle\sum_{n=0}^{\infty} \dfrac{3^n}{n!}$

4. $\displaystyle\sum_{n=1}^{\infty} n\left(\dfrac{3}{2}\right)^n$

5. $\displaystyle\sum_{n=1}^{\infty} \dfrac{n}{2^n}$

6. $\displaystyle\sum_{n=1}^{\infty} \dfrac{n^2}{2^n}$

7. $\displaystyle\sum_{n=1}^{\infty} \dfrac{2^n}{n^2}$

8. $\displaystyle\sum_{n=1}^{\infty} \dfrac{(-1)^{n+1}(n+2)}{n(n+1)}$

9. $\displaystyle\sum_{n=0}^{\infty} \dfrac{(-1)^n 2^n}{n!}$

10. $\displaystyle\sum_{n=1}^{\infty} \dfrac{(-1)^{n-1}(3/2)^n}{n^2}$

11. $\displaystyle\sum_{n=0}^{\infty} \frac{n!}{n3^n}$

12. $\displaystyle\sum_{n=1}^{\infty} \frac{(2n)!}{n^5}$

13. $\displaystyle\sum_{n=0}^{\infty} \frac{4^n}{n!}$

14. $\displaystyle\sum_{n=1}^{\infty} \frac{n^n}{n!}$

15. $\displaystyle\sum_{n=0}^{\infty} \frac{3^n}{(n+1)^n}$

16. $\displaystyle\sum_{n=0}^{\infty} \frac{(n!)^2}{(3n)!}$

17. $\displaystyle\sum_{n=0}^{\infty} \frac{4^n}{3^n + 1}$

18. $\displaystyle\sum_{n=0}^{\infty} \frac{(-1)^n 2^{4n}}{(2n+1)!}$

19. $\displaystyle\sum_{n=0}^{\infty} \frac{(-1)^{n+1} n!}{1 \cdot 3 \cdot 5 \cdots (2n+1)}$

20. $\displaystyle\sum_{n=1}^{\infty} \frac{(-1)^n 2 \cdot 4 \cdot 6 \cdots (2n)}{2 \cdot 5 \cdot 8 \cdots (3n-1)}$

In Exercises 21–30, use the Root Test to test for convergence or divergence of the series.

21. $\displaystyle\sum_{n=1}^{\infty} \left(\frac{n}{2n+1}\right)^n$

22. $\displaystyle\sum_{n=1}^{\infty} \left(\frac{2n}{n+1}\right)^n$

23. $\displaystyle\sum_{n=2}^{\infty} \frac{(-1)^n}{(\ln n)^n}$

24. $\displaystyle\sum_{n=1}^{\infty} \left(\frac{-2n}{3n+1}\right)^{3n}$

25. $\displaystyle\sum_{n=1}^{\infty} (2\sqrt[n]{n} + 1)^n$

26. $\displaystyle\sum_{n=1}^{\infty} (\sqrt[n]{n} - 1)^n$

27. $\displaystyle\sum_{n=2}^{\infty} \left(\frac{\ln n}{n}\right)^n$

28. $\displaystyle\sum_{n=0}^{\infty} e^{-n}$

29. $\displaystyle \frac{1}{(\ln 3)^3} + \frac{1}{(\ln 4)^4} + \frac{1}{(\ln 5)^5} + \frac{1}{(\ln 6)^6} + \cdots$

30. $\displaystyle 1 + \frac{2}{3} + \frac{3}{3^2} + \frac{4}{3^3} + \frac{5}{3^4} + \frac{6}{3^5} + \cdots$

In Exercises 31–50, test for convergence or divergence using any appropriate test from this chapter. Identify the test used.

31. $\displaystyle\sum_{n=1}^{\infty} \frac{(-1)^{n+1} 5}{n}$

32. $\displaystyle\sum_{n=1}^{\infty} \frac{5}{n}$

33. $\displaystyle\sum_{n=1}^{\infty} \frac{3}{n\sqrt{n}}$

34. $\displaystyle\sum_{n=1}^{\infty} \left(\frac{\pi}{4}\right)^n$

35. $\displaystyle\sum_{n=1}^{\infty} \frac{2n}{n+1}$

36. $\displaystyle\sum_{n=1}^{\infty} \frac{n}{2n^2 + 1}$

37. $\displaystyle\sum_{n=1}^{\infty} \frac{(-1)^n 3^{n-2}}{2^n}$

38. $\displaystyle\sum_{n=1}^{\infty} \frac{10}{3\sqrt{n^3}}$

39. $\displaystyle\sum_{n=1}^{\infty} \frac{10n + 3}{n2^n}$

40. $\displaystyle\sum_{n=1}^{\infty} \frac{2^n}{4n^2 - 1}$

41. $\displaystyle\sum_{n=1}^{\infty} (-1)^n \ln\left(\frac{n+2}{n}\right)$

42. $\displaystyle\sum_{n=1}^{\infty} \frac{1}{\sqrt{n} + 2}$

43. $\displaystyle\sum_{n=1}^{\infty} \frac{\cos n}{2^n}$

44. $\displaystyle\sum_{n=2}^{\infty} \frac{(-1)^n}{n \ln n}$

45. $\displaystyle\sum_{n=1}^{\infty} \frac{n7^n}{n!}$

46. $\displaystyle\sum_{n=1}^{\infty} \frac{\ln n}{n^2}$

47. $\displaystyle\sum_{n=1}^{\infty} \frac{(-1)^n 3^{n-1}}{n!}$

48. $\displaystyle\sum_{n=1}^{\infty} \frac{(-1)^n 3^n}{n2^n}$

49. $\displaystyle\sum_{n=1}^{\infty} \frac{(-3)^n}{3 \cdot 5 \cdot 7 \cdots (2n+1)}$

50. $\displaystyle\sum_{n=1}^{\infty} \frac{3 \cdot 5 \cdot 7 \cdots (2n+1)}{18^n(2n-1)n!}$

51. Prove Theorem 9.18. [Hint for part 1: If the limit equals $r < 1$, choose a real number R such that $r < R < 1$. By the definition of the limit there exists some $N > 0$ such that $\sqrt[n]{|a_n|} < R$ for $n > N$.]

52. Show that the Ratio Test is inconclusive for a p-series.

9.7 Taylor Polynomials and Approximations

Polynomial approximations of elementary functions ▪ Taylor and Maclaurin polynomials ▪ Remainder of a Taylor polynomial

The remaining four sections of this chapter discuss Taylor polynomials and Taylor series.

Polynomial approximations of elementary functions

The goal of this section is to show how special types of polynomials can be used as approximations for other elementary functions. To find a polynomial function P that approximates another function f, we begin by choosing a number c in the domain of f at which we require that f and P have the same value. That is,

$$P(c) = f(c). \qquad \text{Graphs of } f \text{ and } P \text{ pass through } (c, f(c))$$

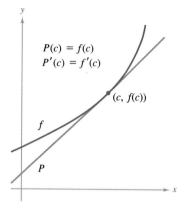

FIGURE 9.7

We say that the approximating polynomial is **expanded about c** or **centered at c.** Geometrically, the requirement that $P(c) = f(c)$ means that the graph of P passes through the point $(c, f(c))$. Of course, there are many polynomials whose graph passes through the point $(c, f(c))$. Our task is to find a polynomial whose graph resembles the graph of f near this point. One way to do this is to impose the additional requirement that the slope of the polynomial function be the same as the slope of the graph of f at the point $(c, f(c))$. That is, we require that

$$P'(c) = f'(c). \qquad \text{Graphs of } f \text{ and } P \text{ have same slope at } (c, f(c))$$

With these two requirements, we can obtain a simple linear approximation of f, as shown in Figure 9.7. This procedure is demonstrated in Example 1.

EXAMPLE 1   First-degree polynomial approximation of $f(x) = e^x$

For the function $f(x) = e^x$, find a first-degree polynomial function

$$P_1(x) = a_1 x + a_0$$

whose value and slope agree with the value and slope of f at $x = 0$.

SOLUTION

Since $f(x) = e^x$ and $f'(x) = e^x$, the value and slope of f, at $x = 0$, are given by

$$f(0) = e^0 = 1 \qquad \text{and} \qquad f'(0) = e^0 = 1.$$

Now, since $P_1(x) = a_1 x + a_0$, we can impose the condition that $P_1(0) = f(0)$ to conclude that $a_0 = 1$. Moreover, since $P_1'(x) = a_1$, we can use the condition that $P_1'(0) = f'(0)$ to conclude that $a_1 = 1$. Therefore, we have

$$P_1(x) = x + 1.$$

Figure 9.8 shows the graphs of $P_1(x) = x + 1$ and $f(x) = e^x$. ▭

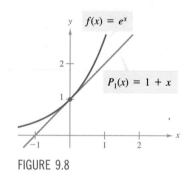

FIGURE 9.8

From Figure 9.8 we can see that at points near $(0, 1)$, the graph of $P_1(x) = x + 1$ is reasonably close to the graph of $f(x) = e^x$. However, as we move away from $(0, 1)$, the graphs move farther from each other and the approximation is not good. To improve the approximation, we can impose yet another requirement—that the values of the second derivatives of P and f agree when $x = 0$. The polynomial of least degree that satisfies all three requirements $P_2(0) = f(0)$, $P_2'(0) = f'(0)$, and $P_2''(0) = f''(0)$ can be shown to be

$$P_2(x) = 1 + x + \frac{1}{2}x^2.$$

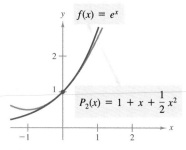

FIGURE 9.9

Moreover, from Figure 9.9, we can see that P_2 is a better approximation to f than P. If we continue this pattern, requiring that the values of $P_n(x)$ and its first n derivatives match those of $f(x) = e^x$ at $x = 0$, we obtain the following.

$$P_n(x) = 1 + x + \frac{1}{2}x^2 + \frac{1}{3!}x^3 + \cdots + \frac{1}{n!}x^n \approx e^x$$

EXAMPLE 2 Third-degree polynomial approximation of $f(x) = e^x$

Construct a table comparing the values of the polynomial

$$P_3(x) = 1 + x + \frac{1}{2}x^2 + \frac{1}{3!}x^3$$

to $f(x) = e^x$ for several values of x near 0.

SOLUTION

Using a calculator, we obtain the results shown in Table 9.3. Note that for $x = 0$, the two functions have the same value, but that as x moves farther away from 0, the accuracy of the approximating polynomial $P_3(x)$ decreases.

TABLE 9.3

x	-1.0	-0.2	-0.1	0.0	0.1	0.2	1.0
e^x	0.367879	0.818731	0.904837	1.000000	1.105171	1.221403	2.718282
$P_3(x)$	0.333333	0.818667	0.904833	1.000000	1.105167	1.221333	2.666667

Taylor and Maclaurin polynomials

The polynomial approximation for $f(x) = e^x$ given in Example 2 is expanded about $c = 0$. For expansions about an arbitrary value of c, it is convenient to write the polynomial in the form

$$P_n(x) = a_0 + a_1(x - c) + a_2(x - c)^2 + a_3(x - c)^3 + \cdots + a_n(x - c)^n.$$

In this form, repeated differentiation produces

$$P_n'(x) = a_1 + 2a_2(x - c) + 3a_3(x - c)^2 + \cdots + na_n(x - c)^{n-1}$$
$$P_n''(x) = 2a_2 + 2 \cdot 3a_3(x - c) + \cdots + n(n - 1)a_n(x - c)^{n-2}$$
$$P_n'''(x) = 2 \cdot 3a_3 + \cdots + n(n - 1)(n - 2)a_n(x - e)^{n-3}$$
$$\vdots$$
$$P_n^{(n)}(x) = n(n - 1)(n - 2) \cdots (2)(1)a_n.$$

Brook Taylor

Letting $x = c$, we then obtain

$$P_n(c) = a_0, \quad P_n'(c) = a_1, \quad P_n''(c) = 2a_2, \quad \ldots, \quad P_n^{(n)}(c) = n!a_n$$

and because the value of f and its first n derivatives must agree with the value of P_n and its first n derivatives at $x = c$, it follows that

$$f(c) = a_0, \quad f'(c) = a_1, \quad \frac{f''(c)}{2!} = a_2, \quad \ldots, \quad \frac{f^{(n)}(c)}{n!} = a_n.$$

With these coefficients we obtain the following definition of **Taylor polynomials,** named after the English mathematician Brook Taylor (1685–1731). Although Taylor was not the first to seek polynomial approximations of transcendental functions, his published account in 1715 was the first comprehensive work on the subject.

DEFINITION OF nTH TAYLOR POLYNOMIAL AND MACLAURIN POLYNOMIAL

If f has n derivatives at c, then the polynomial

$$P_n(x) = f(c) + f'(c)(x - c) + \frac{f''(c)}{2!}(x - c)^2 + \cdots + \frac{f^{(n)}(c)}{n!}(x - c)^n$$

is called the **nth Taylor polynomial for f at c.** If $c = 0$, then

$$P_n(x) = f(0) + f'(0)x + \frac{f''(0)}{2!}x^2 + \frac{f'''(0)}{3!}x^3 + \cdots + \frac{f^{(n)}(0)}{n!}x^n$$

is called the **nth Maclaurin polynomial for f.**

REMARK   The nth Maclaurin polynomial for f is named after the English mathematician, Colin Maclaurin (1698–1746).

EXAMPLE 3   A Maclaurin polynomial for $f(x) = e^x$

From our discussion earlier in this section, the nth Maclaurin polynomial for $f(x) = e^x$ is given by

$$P_n(x) = 1 + x + \frac{1}{2}x^2 + \frac{1}{3!}x^3 + \cdots + \frac{1}{n!}x^n.$$ ▢

EXAMPLE 4   Finding Taylor polynomials for $\ln x$

Find the Taylor polynomials P_0, P_1, P_2, P_3, and P_4 for $f(x) = \ln x$ centered at $c = 1$.

SOLUTION

Expanding about $c = 1$ yields the following.

$$f(x) = \ln x \qquad f(1) = 0$$

$$f'(x) = \frac{1}{x} \qquad f'(1) = 1$$

$$f''(x) = -\frac{1}{x^2} \qquad f''(1) = -1$$

$$f'''(x) = \frac{2!}{x^3} \qquad f'''(1) = 2$$

$$f^{(4)}(x) = -\frac{3!}{x^4} \qquad f^{(4)}(1) = -6$$

Therefore, the Taylor polynomials are as follows.

$$P_0(x) = f(1) = 0$$
$$P_1(x) = P_0(x) + f'(1)(x - 1) = (x - 1)$$
$$P_2(x) = P_1(x) + \frac{f''(1)}{2!}(x - 1)^2 = (x - 1) - \frac{1}{2}(x - 1)^2$$
$$P_3(x) = P_2(x) + \frac{f'''(1)}{3!}(x - 1)^3 = (x - 1) - \frac{1}{2}(x - 1)^2 + \frac{1}{3}(x - 1)^3$$
$$P_4(x) = P_3(x) + \frac{f^{(4)}(1)}{4!}(x - 1)^4$$

$$= (x - 1) - \frac{1}{2}(x - 1)^2 + \frac{1}{3}(x - 1)^3 - \frac{1}{4}(x - 1)^4$$

Figure 9.10 compares the graphs of P_1, P_2, P_3, and P_4 to the graph of f. Note that near $x = 1$ the graphs are nearly indistinguishable. ▭

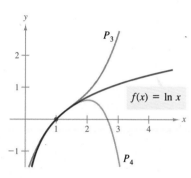

FIGURE 9.10

EXAMPLE 5 Finding Maclaurin polynomials for cos x

Find the Maclaurin polynomials P_0, P_2, P_4, and P_6 for $f(x) = \cos x$. Use $P_6(x)$ to approximate the value of $\cos (0.1)$.

SOLUTION

Expanding about $c = 0$ yields the following.

$$f(x) = \cos x \qquad f(0) = 1$$
$$f'(x) = -\sin x \qquad f'(0) = 0$$
$$f''(x) = -\cos x \qquad f''(0) = -1$$
$$f'''(x) = \sin x \qquad f'''(0) = 0$$

Through repeated differentiation, we can see that the pattern 1, 0, -1, 0 continues, and we obtain the following Maclaurin polynomials.

$$P_0(x) = f(0) = 1$$

$$P_2(x) = P_0(x) + f'(0)x + \frac{f''(0)}{2!}x^2 = 1 - \frac{1}{2!}x^2$$

$$P_4(x) = P_2(x) + \frac{f'''(0)}{3!}x^3 + \frac{f^{(4)}(0)}{4!}x^4 = 1 - \frac{1}{2!}x^2 + \frac{1}{4!}x^4$$

$$P_6(x) = P_4(x) + \frac{f^{(5)}(0)}{5!}x^5 + \frac{f^{(6)}(0)}{6!}x^6 = 1 - \frac{1}{2!}x^2 + \frac{1}{4!}x^4 - \frac{1}{6!}x^6$$

Using $P_6(x)$, we obtain the approximation $\cos(0.1) \approx 0.995004165$, which coincides with the calculator value to nine decimal places. ▭

Note in Example 5 that the Maclaurin polynomials for $\cos x$ have only even powers of x. Similarly, the Maclaurin polynomials for $g(x) = \sin x$ have only odd powers of x (see Exercise 5). This is not generally true for the Taylor polynomials for $\sin x$ and $\cos x$ expanded about $c \neq 0$, as we will see in the next example.

EXAMPLE 6   Finding a Taylor polynomial for sin x

Find the third Taylor polynomial for $f(x) = \sin x$, expanded about $c = \pi/6$.

SOLUTION

Expanding about $c = \pi/6$ yields the following.

$$f(x) = \sin x \qquad f\left(\frac{\pi}{6}\right) = \frac{1}{2}$$

$$f'(x) = \cos x \qquad f'\left(\frac{\pi}{6}\right) = \frac{\sqrt{3}}{2}$$

$$f''(x) = -\sin x \qquad f''\left(\frac{\pi}{6}\right) = -\frac{1}{2}$$

$$f'''(x) = -\cos x \qquad f'''\left(\frac{\pi}{6}\right) = -\frac{\sqrt{3}}{2}$$

Thus, the third Taylor polynomial for $f(x) = \sin x$, expanded about $c = \pi/6$, is as follows.

$$P_3(x) = f\left(\frac{\pi}{6}\right) + f'\left(\frac{\pi}{6}\right)\left(x - \frac{\pi}{6}\right) + \frac{f''\left(\frac{\pi}{6}\right)}{2!}\left(x - \frac{\pi}{6}\right)^2 + \frac{f'''\left(\frac{\pi}{6}\right)}{3!}\left(x - \frac{\pi}{6}\right)^3$$

$$= \frac{1}{2} + \frac{\sqrt{3}}{2}\left(x - \frac{\pi}{6}\right) - \frac{1}{2 \cdot 2!}\left(x - \frac{\pi}{6}\right)^2 - \frac{\sqrt{3}}{2 \cdot 3!}\left(x - \frac{\pi}{6}\right)^3 \qquad ▭$$

To approximate the value of a function at a specific point, we can use Taylor or Maclaurin polynomials. For instance, to approximate the value of $\ln(1.1)$, we can use Taylor polynomials for $f(x) = \ln x$ expanded about $c = 1$, as in Example 4. Or we can use Maclaurin polynomials, as shown in the next example.

EXAMPLE 7   Approximation using Maclaurin polynomials

Use a fourth Maclaurin polynomial to approximate the value of ln (1.1).

SOLUTION

Since 1.1 is closer to 1 than to 0, we consider Maclaurin polynomials for the function $g(x) = \ln (1 + x)$.

$$
\begin{aligned}
g(x) &= \ln (1 + x) & g(0) &= 0 \\
g'(x) &= (1 + x)^{-1} & g'(0) &= 1 \\
g''(x) &= -(1 + x)^{-2} & g''(0) &= -1 \\
g'''(x) &= 2(1 + x)^{-3} & g'''(0) &= 2 \\
g^{(4)}(x) &= -6(1 + x)^{-4} & g^{(4)}(0) &= -6
\end{aligned}
$$

Note that we get the same coefficients as we did in Example 4. Therefore, the fourth Maclaurin polynomial for $g(x) = \ln (1 + x)$ is

$$P_4(x) = g(0) + g'(0)x + \frac{g''(0)}{2!}x^2 + \frac{g'''(0)}{3!}x^3 + \frac{g^{(4)}(0)}{4!}x^4$$

$$= x - \frac{1}{2}x^2 + \frac{1}{3}x^3 - \frac{1}{4}x^4.$$

Consequently,

$$\ln (1.1) = \ln (1 + 0.1) \approx P_4(0.1)$$

$$= (0.1) - \frac{1}{2}(0.1)^2 + \frac{1}{3}(0.1)^3 - \frac{1}{4}(0.1)^4$$

$$\approx 0.0953083.$$

Check to see that the fourth Taylor polynomial (from Example 4), evaluated at $x = 1.1$, yields the same result.

Table 9.4 illustrates the accuracy of the Taylor polynomial approximation to the calculator value of ln (1.1). We can see that as n becomes larger, $P_n(1.1)$ is closer to the calculator value of 0.0953102.

TABLE 9.4   Approximations of ln (1.1), Using Taylor Polynomials

n	1	2	3	4
$P_n(1.1)$	0.1000000	0.0950000	0.0953333	0.0953083

On the other hand, Table 9.5 illustrates that as we move away from the expansion point $c = 1$, the accuracy of the approximation decreases.

TABLE 9.5 Fourth Taylor Polynomial Approximations of ln x

x	1.0	1.1	1.5	1.75	2.0
ln x	0.0000000	0.0953102	0.4054651	0.5596158	0.6931472
$P_4(x)$	0.0000000	0.0953083	0.4010417	0.5302734	0.5833333

Tables 9.4 and 9.5 illustrate two very important points about the accuracy of Taylor (or Maclaurin) polynomials for use in approximations.

1. The approximation is usually better at x-values close to c than at x-values far from c.
2. The approximation is usually better for higher degree Taylor (or Maclaurin) polynomials than for those of lower degree.

The remainder of a Taylor polynomial

An approximation technique is of little value without some idea of its accuracy. To measure the accuracy of approximating a functional value $f(x)$ by the Taylor polynomial $P_n(x)$, we use the concept of a **remainder, $R_n(x)$**, defined as follows.

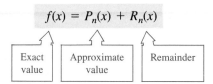

$$f(x) = P_n(x) + R_n(x)$$

Exact value   Approximate value   Remainder

Thus, $R_n(x) = f(x) - P_n(x)$, and we call the absolute value of $R_n(x)$ the **error** associated with the approximation. That is,

$$\text{error} = |R_n(x)| = |f(x) - P_n(x)|.$$

The next theorem gives a general procedure for estimating the remainder associated with a Taylor polynomial. This important theorem is called **Taylor's Theorem** and the remainder given in the theorem is called the **Lagrange form of the remainder.**

THEOREM 9.19
TAYLOR'S THEOREM

If a function f is differentiable through order $n + 1$ in an interval I containing c, then for each x in I, there exists z between x and c such that

$$f(x) = f(c) + f'(c)(x - c) + \frac{f''(c)}{2!}(x - c)^2 + \cdots + \frac{f^{(n)}(c)}{n!}(x - c)^n + R_n(x)$$

where

$$R_n(x) = \frac{f^{(n+1)}(z)}{(n + 1)!}(x - c)^{n+1}.$$

PROOF To find $R_n(x)$ we fix x in I ($x \neq c$) and write

$$R_n(x) = f(x) - P_n(x)$$

where $P_n(x)$ is the nth Taylor polynomial for $f(x)$. Then we let g be a function of t defined by

$$g(t) = f(x) - f(t) - f'(t)(x - t) - \cdots - \frac{f^{(n)}(t)}{n!}(x - t)^n - R_n(x)\frac{(x - t)^{n+1}}{(x - c)^{n+1}}.$$

The reason for defining g this way is that differentiation with respect to t has a telescoping effect. For example, we have

$$\frac{d}{dt}[-f(t) - f'(t)(x - t)] = -f'(t) + f'(t) - f''(t)(x - t)$$

$$= -f''(t)(x - t).$$

The result is that the derivative $g'(t)$ simplifies to

$$g'(t) = -\frac{f^{(n+1)}(t)}{n!}(x - t)^n + (n + 1)R_n(x)\frac{(x - t)^n}{(x - c)^{n+1}}$$

for all t between c and x. Moreover, for a fixed x

$$g(c) = f(x) - [P_n(x) + R_n(x)] = f(x) - f(x) = 0$$

and

$$g(x) = f(x) - f(x) - 0 - \cdots - 0 = f(x) - f(x) = 0.$$

Therefore, g satisfies the conditions of Rolle's Theorem, and it follows that there is a number z between c and x such that $g'(z) = 0$. Substituting z for t in the equation for $g'(t)$ and then solving for $R_n(x)$, we obtain

$$g'(z) = -\frac{f^{(n+1)}(z)}{n!}(x - z)^n + (n + 1)R_n(x)\frac{(x - z)^n}{(x - c)^{n+1}} = 0$$

$$R_n(x) = \frac{f^{(n+1)}(z)}{(n + 1)!}(x - c)^{n+1}.$$

Finally, since $g(c) = 0$, we have

$$0 = f(x) - f(c) - f'(c)(x - c) - \cdots - \frac{f^{(n)}(c)}{n!}(x - c)^n - R_n(x)$$

$$f(x) = f(c) + f'(c)(x - c) + \cdots + \frac{f^{(n)}(c)}{n!}(x - c)^n + R_n(x).$$

When applying Taylor's Theorem, we do not expect to be able to find the exact value of z. (If we could do that, an approximation usually would not have been necessary.) Rather, we try to find bounds for $f^{(n+1)}(z)$ from which we are able to tell how large the remainder $R_n(x)$ is. This is demonstrated in Example 8.

EXAMPLE 8   Determining the accuracy of an approximation

The third Maclaurin polynomial for $\sin x$ is given by

$$P_3(x) = x - \frac{x^3}{3!}.$$

Use Taylor's Theorem to approximate $\sin (0.1)$ by $P_3(0.1)$ and determine the accuracy of the approximation.

SOLUTION

Using Taylor's Theorem, we have

$$\sin x = x - \frac{x^3}{3!} + R_3(x) = x - \frac{x^3}{3!} + \frac{f^{(4)}(z)}{4!}x^4$$

where $0 < z < 0.1$. Therefore,

$$\sin (0.1) \approx 0.1 - \frac{(0.1)^3}{3!} \approx 0.1 - 0.000167 = 0.099833.$$

Since $f^{(4)}(z) = \sin z$ and the sine function is increasing on the interval $[0, 0.1]$, it follows that $0 < \sin z < 1$ and we have

$$0 \le R_3(0.1) = \frac{\sin z}{4!}(0.1)^4 < \frac{0.0001}{4!} \approx 0.000004$$

and we conclude that

$$0.099833 \le \sin (0.1) \le 0.099833 + R_3(x)$$
$$0.099833 \le \sin (0.1) \le 0.099837.$$

EXAMPLE 9   Approximating a functional value to a desired accuracy

Determine the degree of the Taylor polynomial $P_n(x)$ expanded about $c = 1$ that should be used to approximate $\ln (1.2)$ so that the error is less than 0.001.

SOLUTION

Following the pattern of Example 4, we see that the $(n + 1)$st derivative of $f(x) = \ln x$ is given by

$$f^{(n+1)}(x) = (-1)^{n+1}\frac{n!}{x^{n+1}}.$$

Using Taylor's Theorem, we know that the error $|R_n(1.2)|$ is given by

$$|R_n(1.2)| = \left| \frac{f^{(n+1)}(z)}{(n + 1)!}(1.2 - 1)^{n+1} \right| = \frac{n!}{z^{n+1}}\left(\frac{1}{(n + 1)!} \right)(0.2)^{n+1}$$

$$= \frac{(0.2)^{n+1}}{z^{n+1}(n + 1)}$$

where $1 < z < 1.2$. In this interval, $|R_n(1.2)|$ is largest when $z = 1$; thus we are seeking a value of n such that

$$\frac{(0.2)^{n+1}}{(1)^{n+1}(n+1)} < 0.001 \implies 1000 < (n+1)5^{n+1}.$$

By trial and error, we can determine that the smallest value of n satisfying this inequality is $n = 3$. Thus, we would need the third Taylor polynomial to achieve the desired accuracy in approximating $\ln (1.2)$. ▭

EXERCISES for Section 9.7

In Exercises 1–14, find the Maclaurin polynomial of degree n for the given function.

1. $f(x) = e^{-x}$, $n = 3$
2. $f(x) = e^{-x}$, $n = 5$
3. $f(x) = e^{2x}$, $n = 4$
4. $f(x) = e^{3x}$, $n = 4$
5. $f(x) = \sin x$, $n = 5$
6. $f(x) = \sin \pi x$, $n = 3$
7. $f(x) = xe^x$, $n = 4$
8. $f(x) = x^2 e^{-x}$, $n = 4$
9. $f(x) = \dfrac{1}{x+1}$, $n = 4$
10. $f(x) = \dfrac{1}{x^2+1}$, $n = 4$
11. $f(x) = \tan x$, $n = 3$
12. $f(x) = \sec x$, $n = 2$
13. $f(x) = 2 - 3x^3 + x^4$, $n = 4$
14. $f(x) = 3 - 2x^2 + x^3$, $n = 3$

In Exercises 15–18, find the Taylor polynomial of degree n centered at c.

15. $f(x) = \dfrac{1}{x}$, $n = 4$, $c = 1$
16. $f(x) = \sqrt{x}$, $n = 4$, $c = 4$
17. $f(x) = \ln x$, $n = 4$, $c = 1$
18. $f(x) = x^2 \cos x$, $n = 2$, $c = \pi$

19. Use the Maclaurin polynomials $P_1(x)$, $P_3(x)$, and $P_5(x)$ for $f(x) = \sin x$ to complete the following table. (See Exercise 5.)

x	0	0.25	0.50	0.75	1.00
$\sin x$	0	0.2474	0.4794	0.6816	0.8415
$P_1(x)$					
$P_3(x)$					
$P_5(x)$					

20. Use the Taylor polynomials $P_1(x)$ and $P_4(x)$ for $f(x) = \ln x$ centered at 1 to complete the following table. (See Exercise 17.)

x	1.00	1.25	1.50	1.75	2.00
$\ln x$	0	0.2231	0.4055	0.5596	0.6931
$P_1(x)$					
$P_4(x)$					

In Exercises 21–24, approximate the function at the given value of x, using the polynomial found in the indicated exercise.

21. $f(x) = e^{-x}$, $f\left(\dfrac{1}{2}\right)$, Exercise 1

22. $f(x) = x^2 e^{-x}$, $f\left(\dfrac{1}{4}\right)$, Exercise 8

23. $f(x) = x^2 \cos x$, $f\left(\dfrac{7\pi}{8}\right)$, Exercise 18

24. $f(x) = \sqrt{x}$, $f(5)$, Exercise 16

25. Compare the Maclaurin polynomials of degree 4 for the functions

$$f(x) = e^x \quad \text{and} \quad g(x) = xe^x$$

What is the relationship between them?

26. Differentiate the Maclaurin polynomial of degree 5 for $f(x) = \sin x$ and compare the result with the Maclaurin polynomial of degree 4 for $g(x) = \cos x$.

In Exercises 27 and 28, determine the values of x for which the given function can be replaced by the Taylor Polynomial if the error cannot exceed 0.001.

27. $f(x) = e^x \approx 1 + x + \dfrac{x^2}{2!} + \dfrac{x^3}{3!}$, $x < 0$

28. $f(x) = \sin x \approx x - \dfrac{x^3}{3!}$

29. Estimate the error in approximating e by the fifth-degree polynomial

$$1 + x + \frac{x^2}{2!} + \frac{x^3}{3!} + \frac{x^4}{4!} + \frac{x^5}{5!}.$$

30. What degree Maclaurin polynomial for $\ln(x + 1)$ should be used to guarantee finding $\ln 1.5$ if the error cannot exceed 0.0001?

In Exercises 31 and 32,
(a) Find the Taylor polynomial $P_3(x)$ of degree three for for $f(x)$.
(b) Complete the accompanying table for $f(x)$ and $P_3(x)$.
(c) Sketch the graphs of $f(x)$ and $P_3(x)$ on the same axes.

31. $f(x) = \arcsin x$ **32.** $f(x) = \arctan x$

x	-1	-0.75	-0.50	-0.25
$f(x)$				
$P_3(x)$				

x	0	0.25	0.50	0.75	1
$f(x)$					
$P_3(x)$					

In Exercises 33 and 34, use a computer or graphics calculator to sketch the graph of the given function over the specified interval and sketch the graph of each of the Maclaurin polynomials. Place all of the graphs on the same set of coordinate axes.

33. $f(x) = \sin x, \quad [-\pi, \pi]$
(a) $P_1(x)$ (b) $P_3(x)$
(c) $P_5(x)$ (d) $P_7(x)$
34. $f(x) = e^x, \quad [-1, 1]$
(a) $P_1(x)$ (b) $P_3(x)$
(c) $P_5(x)$ (d) $P_7(x)$

35. Prove that if f is an odd function then the nth Maclaurin polynomial for the function will contain only terms with odd powers of x.
36. Prove that if f is an even function then the nth Maclaurin polynomial for the function will contain only terms with even powers of x.
37. Let $P_n(x)$ be the nth Taylor polynomial for f at c. Prove that $P_n(c) = f(c)$ and $P^{(k)}(c) = f^{(k)}(c)$ for $1 \le k \le n$.

9.8 Power Series

Power series ▪ Radius of convergence ▪ Interval of convergence ▪ Endpoint convergence ▪ Differentiation and integration of power series

In Section 9.7, we looked at the problem of approximating functions by Taylor polynomials. For instance, the function $f(x) = e^x$ can be *approximated* by its third-degree Taylor polynomial as follows.

$$e^x \approx 1 + x + \frac{x^2}{2} + \frac{x^3}{3!}$$

In the last three sections of this chapter, we will see that several important types of functions, including $f(x) = e^x$, can be represented *exactly* by a **power series**. For example, the power series representation for e^x is

$$e^x = 1 + x + \frac{x^2}{2} + \frac{x^3}{3!} + \cdots + \frac{x^n}{n!} + \cdots.$$

For each real number x, we will show that the infinite series on the right converges to the number e^x. Before doing that, we look at some preliminary results dealing with power series—beginning with the following definition.

DEFINITION OF POWER SERIES

If x is a variable, then an infinite series of the form

$$\sum_{n=0}^{\infty} a_n x^n = a_0 + a_1 x + a_2 x^2 + a_3 x^3 + \cdots + a_n x^n + \cdots$$

is called a **power series.** More generally, we call a series of the form

$$\sum_{n=0}^{\infty} a_n (x - c)^n = a_0 + a_1 (x - c) + a_2 (x - c)^2 + \cdots + a_n (x - c)^n + \cdots$$

a **power series centered at c,** where c is a constant.

REMARK   To simplify the notation for power series, we agree that $(x - c)^0 = 1$, even if $x = c$.

EXAMPLE 1   Power series

(a) The following power series is centered at 0.

$$\sum_{n=0}^{\infty} \frac{x^n}{n!} = 1 + x + \frac{x^2}{2} + \frac{x^3}{3!} + \cdots$$

(b) The following power series is centered at 1.

$$\sum_{n=1}^{\infty} \frac{1}{n}(x - 1)^n = (x - 1) + \frac{1}{2}(x - 1)^2 + \frac{1}{3}(x - 1)^3 + \cdots \quad \blacksquare$$

A power series in x can be viewed as a function of x

$$f(x) = \sum_{n=0}^{\infty} a_n (x - c)^n$$

where the *domain of f* is the set of all x for which the power series converges. The determination of the domain of a power series is our primary concern in this section. Of course, every power series converges at its center c, since

$$f(c) = \sum_{n=0}^{\infty} a_n (c - c)^n$$

$$= a_0(1) + 0 + 0 + \cdots + 0 + \cdots$$

$$= a_0.$$

Thus, c always lies in the domain of f. The following important theorem (which we state without proof) tells us that the domain of a power series can take three basic forms: a single point, an interval centered at c, or the entire real line.

THEOREM 9.20
CONVERGENCE OF
A POWER SERIES

For a power series centered at c, precisely one of the following is true.

1. The series converges only at c.
2. There exists a real number $R > 0$ such that the series converges (absolutely) for $|x - c| < R$, and diverges for $|x - c| > R$.
3. The series converges for all x.

The number R is called the **radius of convergence** of the power series. If the series converges only at c, then we say that the radius of convergence is $R = 0$, and if the series converges for all x, then we say that the radius of convergence is $R = \infty$. The set of all values of x for which the power series converges is called the **interval of convergence** of the power series.

To determine the radius of convergence of a power series, we use the Ratio Test, as shown in Examples 2, 3, and 4.

EXAMPLE 2 Finding the radius of convergence

Find the radius of convergence of the power series

$$\sum_{n=0}^{\infty} n! x^n.$$

SOLUTION

For $x = 0$, we obtain

$$f(0) = \sum_{n=0}^{\infty} n! 0^n = 1 + 0 + 0 + \cdots = 1.$$

For any fixed value of x such that $|x| > 0$, let $u_n = n! x^n$. Then

$$\lim_{n \to \infty} \left| \frac{u_{n+1}}{u_n} \right| = \lim_{n \to \infty} \left| \frac{(n + 1)! x^{n+1}}{n! x^n} \right| = |x| \lim_{n \to \infty} (n + 1) = \infty.$$

Therefore, by the Ratio Test, we conclude that the series diverges for $|x| > 0$, and converges only at its center, 0. Hence, the radius of convergence is $R = 0$. ▭

EXAMPLE 3 Finding the radius of convergence

Find the radius of convergence of the power series

$$\sum_{n=0}^{\infty} 3(x - 2)^n.$$

SOLUTION

For $x \neq 2$, we let $u_n = 3(x - 2)^n$. Then,

$$\lim_{n \to \infty} \left| \frac{u_{n+1}}{u_n} \right| = \lim_{n \to \infty} \left| \frac{3(x - 2)^{n+1}}{3(x - 2)^n} \right| = \lim_{n \to \infty} |x - 2| = |x - 2|.$$

By the Ratio Test, it follows that the series converges if $|x - 2| < 1$ and diverges if $|x - 2| > 1$. Therefore, the radius of convergence of the series is $R = 1$.   ▭

EXAMPLE 4 Finding the radius of convergence

Find the radius of convergence of the power series

$$\sum_{n=0}^{\infty} \frac{(-1)^n x^{2n+1}}{(2n + 1)!} .$$

SOLUTION

If we let $u_n = (-1)^n x^{2n+1}/(2n + 1)!$, then

$$\lim_{n \to \infty} \left| \frac{u_{n+1}}{u_n} \right| = \lim_{n \to \infty} \left| \frac{[(-1)^{n+1} x^{2n+3}]/(2n + 3)!}{[(-1)^n x^{2n+1}]/(2n + 1)!} \right|$$

$$= \lim_{n \to \infty} \frac{x^2}{(2n + 3)(2n + 2)}.$$

For any *fixed* value for x, this limit is 0, and by the Ratio Test, the series converges for all x. Therefore, the radius of convergence is $R = \infty$.   ▭

Endpoint convergence

Note that for a power series whose radius of convergence is a finite number R, Theorem 9.20 says nothing about the convergence at the *endpoints* of the interval of convergence. Each endpoint must be tested separately for convergence or divergence. As a result, the interval of convergence of a power series can take any one of the six forms shown in Figure 9.11.

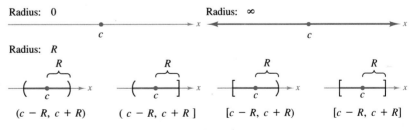

Intervals of Convergence

FIGURE 9.11

EXAMPLE 5 Finding the interval of convergence

Find the interval of convergence of the power series

$$\sum_{n=1}^{\infty} \frac{x^n}{n}.$$

SOLUTION

Letting $u_n = x^n/n$, we have

$$\lim_{n \to \infty} \left| \frac{u_{n+1}}{u_n} \right| = \lim_{n \to \infty} \left| \frac{x^{n+1}/(n+1)}{x^n/n} \right| = \lim_{n \to \infty} \left| \frac{nx}{n+1} \right| = |x|.$$

Therefore, by the Ratio Test, the radius of convergence is $R = 1$. Moreover, because the series is centered at 0, it will converge in the interval $(-1, 1)$, and we proceed to test for convergence at the endpoints. When $x = 1$, we obtain the *divergent* harmonic series

$$\sum_{n=1}^{\infty} \frac{1}{n} = \frac{1}{1} + \frac{1}{2} + \frac{1}{3} + \cdots .$$ Diverges when $x = 1$

When $x = -1$, we obtain the *convergent* alternating harmonic series

$$\sum_{n=1}^{\infty} \frac{(-1)^n}{n} = -1 + \frac{1}{2} - \frac{1}{3} + \frac{1}{4} - \cdots .$$ Converges when $x = -1$

Therefore, the interval of convergence for the given series is $[-1, 1)$, as shown in Figure 9.12.   ▭

Interval: $[-1, 1)$
Radius: $R = 1$

FIGURE 9.12

EXAMPLE 6 Finding the interval of convergence

Find the interval of convergence of the power series

$$\sum_{n=0}^{\infty} \frac{(-1)^n(x+1)^n}{2^n}.$$

SOLUTION

Since

$$\lim_{n \to \infty} \left| \frac{u_{n+1}}{u_n} \right| = \lim_{n \to \infty} \left| \frac{(-1)^{n+1}(x+1)^{n+1}/2^{n+1}}{(-1)^n(x+1)^n/2^n} \right|$$

$$= \lim_{n \to \infty} \left| \frac{2^n(x+1)}{2^{n+1}} \right| = \left| \frac{x+1}{2} \right|$$

the radius of convergence is $R = 2$. Since the series is centered at $x = -1$, it will converge in the interval $(-3, 1)$. Furthermore, at the endpoints we have

$$\sum_{n=0}^{\infty} \frac{(-1)^n(-2)^n}{2^n} = \sum_{n=0}^{\infty} \frac{2^n}{2^n} = \sum_{n=0}^{\infty} 1$$ Diverges when $x = -3$

Interval: $(-3, 1)$
Radius: $R = 2$

FIGURE 9.13

and

$$\sum_{n=0}^{\infty} \frac{(-1)^n (2)^n}{2^n} = \sum_{n=0}^{\infty} (-1)^n \qquad \text{Diverges when } x = 1$$

both of which diverge. Thus, the interval of convergence is $(-3, 1)$, as shown in Figure 9.13.

EXAMPLE 7 Finding the interval of convergence

Find the interval of convergence of the power series

$$\sum_{n=1}^{\infty} \frac{x^n}{n^2}.$$

SOLUTION

Since

$$\lim_{n \to \infty} \left| \frac{u_{n+1}}{u_n} \right| = \lim_{n \to \infty} \left| \frac{x^{n+1}/(n+1)^2}{x^n/n^2} \right| = \lim_{n \to \infty} \left| \frac{n^2 x}{(n+1)^2} \right| = |x|$$

the radius of convergence is $R = 1$. Moreover, since the series is centered at $x = 0$, it will converge in the interval $(-1, 1)$ and we proceed to test for convergence at the endpoints. When $x = 1$, we obtain the *convergent p*-series

$$\sum_{n=1}^{\infty} \frac{1}{n^2} = \frac{1}{1^2} + \frac{1}{2^2} + \frac{1}{3^2} + \frac{1}{4^2} + \cdots. \qquad \text{Converges when } x = 1$$

When $x = -1$, we obtain the *convergent* alternating series

$$\sum_{n=1}^{\infty} \frac{(-1)^n}{n^2} = -\frac{1}{1^2} + \frac{1}{2^2} - \frac{1}{3^2} + \frac{1}{4^2} - \cdots. \qquad \text{Converges when } x = -1$$

Therefore, the interval of convergence for the given series is $[-1, 1]$.

Differentiation and integration of power series

Power series representation of functions has played an important role in the development of calculus. In fact, much of Newton's work with differentiation and integration was done in the context of power series—especially his work with complicated algebraic functions and transcendental functions. Euler, Lagrange, Leibniz, and the Bernoullis all used power series extensively in calculus. One of the earliest mathematicians to work with power series was a Scotsman, James Gregory (1638–1675). He developed a power series method for interpolating table values—a method that was later used by Brook Taylor in the development of Taylor polynomials and Taylor series.

Once we have defined a function with a power series, it is natural to wonder how we can determine the characteristics of the function. Is it continuous? Differentiable? Integrable? Theorem 9.21, which we state without proof, answers these questions.

James Gregory

THEOREM 9.21 PROPERTIES OF A FUNCTION DEFINED BY A POWER SERIES	If the function given by

$$f(x) = \sum_{n=0}^{\infty} a_n(x - c)^n = a_0 + a_1(x - c) + a_2(x - c)^2 + a_3(x - c)^3 + \cdots$$

has a radius of convergence of $R > 0$, then on the interval $(c - R, c + R)$ f is continuous, differentiable, and integrable. Moreover, the derivative and antiderivative of f are as follows.

1. $f'(x) = a_1 + 2a_2(x - c) + 3a_3(x - c)^2 + \cdots$

2. $\displaystyle\int f(x) \, dx = C + a_0(x - c) + a_1\frac{(x - c)^2}{2} + a_2\frac{(x - c)^3}{3} + \cdots$

REMARK   Theorem 9.21 tells us that, in many ways, a function defined by a power series behaves like a polynomial. It is continuous in its interval of convergence, and both its derivative and antiderivative can be determined by differentiating and integrating each term of the given power series.

The *radius of convergence* for the series obtained by differentiating or integrating a power series is the same as for the original power series. For example, suppose the power series

$$f(x) = a_0 + a_1(x - c) + a_2(x - c)^2 + a_3(x - c)^3 + \cdots$$

has a finite radius of convergence of $R \neq 0$. Then, by the Ratio Test, we know that

$$\lim_{n \to \infty} \left| \frac{a_{n+1}(x - c)^{n+1}}{a_n(x - c)^n} \right| = \left| \frac{x - c}{R} \right|.$$

For the series

$$f'(x) = a_1 + 2a_2(x - c) + 3a_3(x - c)^2 + \cdots$$

it follows that the radius of convergence is also R since

$$\lim_{n \to \infty} \left| \frac{(n + 1)a_{n+1}(x - c)^n}{na_n(x - c)^{n-1}} \right| = \lim_{n \to \infty} \left(\frac{n + 1}{n} \right) \left| \frac{a_{n+1}(x - c)}{a_n} \right|$$

$$= \left| \frac{x - c}{R} \right|.$$

However, the *interval of convergence* may differ due to the behavior at the endpoints. This is demonstrated in Example 8.

EXAMPLE 8   Intervals of convergence for $f(x)$, $f'(x)$, and $\int f(x) \, dx$

Find the intervals of convergence of (a) $\int f(x) \, dx$, (b) $f(x)$, and (c) $f'(x)$, where

$$f(x) = \sum_{n=1}^{\infty} \frac{x^n}{n} = x + \frac{x^2}{2} + \frac{x^3}{3} + \cdots.$$

SOLUTION

By Theorem 9.21 we have

$$f'(x) = \sum_{n=1}^{\infty} x^{n-1} = 1 + x + x^2 + x^3 + \cdots$$

and

$$\int f(x)\, dx = C + \sum_{n=1}^{\infty} \frac{x^{n+1}}{n(n+1)} = C + \frac{x^2}{1 \cdot 2} + \frac{x^3}{2 \cdot 3} + \frac{x^4}{3 \cdot 4} + \cdots.$$

By the Ratio Test we know that each series has a radius of convergence of $R = 1$. Now, considering the interval $(-1, 1)$, we have the following.

(a) For $\int f(x)\, dx$, the series

$$\sum_{n=1}^{\infty} \frac{x^{n+1}}{n(n+1)} \qquad \text{Interval of convergence: } [-1, 1]$$

converges for $x = \pm 1$, and its interval of convergence is $[-1, 1]$.

(b) For $f(x)$, the series

$$\sum_{n=1}^{\infty} \frac{x^n}{n} \qquad \text{Interval of convergence: } [-1, 1)$$

converges for $x = -1$ and diverges for $x = 1$, and its interval of convergence is $[-1, 1)$.

(c) For $f'(x)$, the series

$$\sum_{n=1}^{\infty} x^{n-1} \qquad \text{Interval of convergence: } (-1, 1)$$

diverges for $x = \pm 1$ and its interval of convergence is $(-1, 1)$.   ▭

REMARK   From Example 8, it appears that of the three series, the one for the derivative, $f'(x)$, is the least likely to converge at the endpoints. In fact, it can be shown that if the series for $f'(x)$ converges at the endpoints $x = c \pm R$, then the series for $f(x)$ will also converge there (see Exercise 38).

EXERCISES for Section 9.8

In Exercises 1–6, find the radius of convergence of the power series.

1. $\sum_{n=0}^{\infty} (-1)^n \dfrac{x^n}{n+1}$

2. $\sum_{n=0}^{\infty} (4x)^n$

3. $\sum_{n=1}^{\infty} \dfrac{(2x)^n}{n^2}$

4. $\sum_{n=0}^{\infty} \dfrac{(-1)^n x^n}{2^n}$

5. $\sum_{n=0}^{\infty} \dfrac{(2x)^n}{n!}$

6. $\sum_{n=0}^{\infty} \dfrac{(2n)!x^n}{n!}$

In Exercises 7–30, find the interval of convergence of the power series. (Be sure to include a check for convergence at the endpoints of the interval.)

7. $\sum_{n=0}^{\infty} \left(\dfrac{x}{2}\right)^n$

8. $\sum_{n=0}^{\infty} \left(\dfrac{x}{k}\right)^n$

9. $\sum_{n=1}^{\infty} \dfrac{(-1)^n x^n}{n}$

10. $\sum_{n=0}^{\infty} (-1)^{n+1} n x^n$

11. $\sum_{n=0}^{\infty} \dfrac{x^n}{n!}$

12. $\sum_{n=0}^{\infty} \dfrac{(3x)^n}{(2n)!}$

13. $\sum_{n=0}^{\infty} (2n)! \left(\dfrac{x}{2}\right)^n$

14. $\sum_{n=0}^{\infty} \dfrac{(-1)^n x^n}{(n+1)(n+2)}$

15. $\sum_{n=1}^{\infty} \dfrac{(-1)^{n+1} x^n}{4^n}$

16. $\sum_{n=0}^{\infty} \dfrac{(-1)^n n! (x-4)^n}{3^n}$

17. $\sum_{n=1}^{\infty} \dfrac{(-1)^{n+1}(x-5)^n}{n5^n}$

18. $\sum_{n=0}^{\infty} \dfrac{(x-2)^{n+1}}{(n+1)3^{n+1}}$

19. $\displaystyle\sum_{n=0}^{\infty} \frac{(-1)^{n+1}(x-1)^{n+1}}{n+1}$ **20.** $\displaystyle\sum_{n=1}^{\infty} \frac{(-1)^{n+1}(x-c)^n}{nc^n}$

21. $\displaystyle\sum_{n=1}^{\infty} \frac{(x-c)^{n-1}}{c^{n-1}}, \; 0 < c$ **22.** $\displaystyle\sum_{n=1}^{\infty} \frac{(-1)^{n+1}x^{2n-1}}{2n-1}$

23. $\displaystyle\sum_{n=1}^{\infty} \frac{n}{n+1}(-2x)^{n-1}$ **24.** $\displaystyle\sum_{n=0}^{\infty} \frac{(-1)^n x^{2n}}{n!}$

25. $\displaystyle\sum_{n=0}^{\infty} \frac{x^{2n+1}}{(2n+1)!}$ **26.** $\displaystyle\sum_{n=1}^{\infty} \frac{n!x^n}{(2n)!}$

27. $\displaystyle\sum_{n=1}^{\infty} \frac{k(k+1)(k+2)\cdots(k+n-1)x^n}{n!}, \quad k \ge 1$

28. $\displaystyle\sum_{n=1}^{\infty} \left(\frac{2\cdot 4\cdot 6\cdots 2n}{3\cdot 5\cdot 7\cdots(2n+1)}\right)x^{2n+1}$

29. $\displaystyle\sum_{n=1}^{\infty} \frac{(-1)^{n+1}3\cdot 7\cdot 11\cdots(4n-1)(x-3)^n}{4^n}$

30. $\displaystyle\sum_{n=1}^{\infty} \frac{n!(x-c)^n}{1\cdot 3\cdot 5\cdots(2n-1)}$

In Exercises 31–34, find the interval of convergence of (a) $f(x)$, (b) $f'(x)$, (c) $f''(x)$, and (d) $\int f(x)\,dx$. Include a check for convergence at the endpoints.

31. $f(x) = \displaystyle\sum_{n=0}^{\infty} \left(\frac{x}{2}\right)^n$

32. $f(x) = \displaystyle\sum_{n=1}^{\infty} \frac{(-1)^{n+1}(x-5)^n}{n5^n}$

33. $f(x) = \displaystyle\sum_{n=0}^{\infty} \frac{(-1)^{n+1}(x-1)^{n+1}}{n+1}$

34. $f(x) = \displaystyle\sum_{n=1}^{\infty} \frac{(-1)^{n+1}(x-1)^n}{n}$

35. Let

$$f(x) = \sum_{n=0}^{\infty} \frac{(-1)^n x^{2n+1}}{(2n+1)!}$$

and

$$g(x) = \sum_{n=0}^{\infty} \frac{(-1)^n x^{2n}}{(2n)!}.$$

(a) Find the interval of convergence of f and g.
(b) Show that $f'(x) = g(x)$.
(c) Show that $g'(x) = -f(x)$.
(d) Identify the functions f and g.

36. Let

$$f(x) = \sum_{n=0}^{\infty} \frac{x^{2n+1}}{(2n+1)!}$$

and

$$g(x) = \sum_{n=0}^{\infty} \frac{x^{2n}}{(2n)!}.$$

(a) Find the interval of convergence of f and g.
(b) Show that $f'(x) = g(x)$.
(c) Show that $g'(x) = f(x)$.
(d) Identify the functions f and g. (See Section 6.9.)

37. Let

$$f(x) = \sum_{n=0}^{\infty} \frac{x^n}{n!}.$$

(a) Find the interval of convergence of f.
(b) Show that $f'(x) = f(x)$.
(c) Show that $f(0) = 1$.
(d) Identify the function f.

38. Prove that if the series for f' converges at the endpoints of its interval of convergence, then the series for f does also.

9.9 Representation of Functions by Power Series

Geometric power series ▪ Operations with power series

Joseph Fourier

In this section and the next we look at several techniques for finding a power series to represent a given function. Some of the early work in representing functions by power series was done by the French mathematician Joseph Fourier (1768–1830). Fourier's work is important in the history of calculus, partly because it forced eighteenth century mathematicians to question the then prevailing narrow concept of a function. Both Cauchy and Dirichlet were motivated by Fourier's work with series, and in 1837 Dirichlet published the general definition of function that is used today.

Geometric power series

Consider the function given by $f(x) = 1/(1-x)$. The form of f closely resembles the sum of a geometric series

$$\sum_{n=0}^{\infty} ar^n = \frac{a}{1-r}, \quad |r| < 1.$$

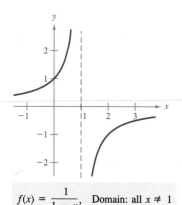

$f(x) = \dfrac{1}{1-x}$, Domain: all $x \neq 1$

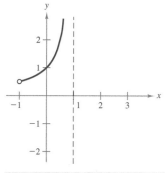

$f(x) = \displaystyle\sum_{n=0}^{\infty} x^n$, Domain: $-1 < x < 1$

FIGURE 9.14

In other words, if we let $a = 1$ and $r = x$, then a power series representation for $1/(1 - x)$, centered at 0, is

$$\frac{1}{1-x} = \sum_{n=0}^{\infty} x^n = 1 + x + x^2 + x^3 + \cdots, \quad |x| < 1.$$

Of course, this series represents $f(x) = 1/(1 - x)$ only on the interval $(-1, 1)$, whereas f is defined for all $x \neq 1$, as shown in Figure 9.14. To represent f on another interval, we must develop a different series. For instance, to obtain the power series centered at -1, we write

$$\frac{1}{1-x} = \frac{1}{2 - (x + 1)}$$

$$= \frac{1/2}{1 - [(x + 1)/2]} = \frac{a}{1 - r}$$

which implies that $a = \frac{1}{2}$ and $r = (x + 1)/2$. Thus, for $|x + 1| < 2$, we have

$$\frac{1}{1-x} = \sum_{n=0}^{\infty} \left(\frac{1}{2}\right)\left(\frac{x + 1}{2}\right)^n$$

$$= \frac{1}{2}\left[1 + \frac{(x + 1)}{2} + \frac{(x + 1)^2}{4} + \frac{(x + 1)^3}{8} + \cdots\right], \quad |x + 1| < 2$$

which converges on the interval $(-3, 1)$.

EXAMPLE 1 Finding a geometric power series centered at 0

Find a power series for $f(x) = 4/(x + 2)$, centered at 0.

SOLUTION

Writing $f(x)$ in the form $a/(1 - r)$, we have

$$\frac{4}{2 + x} = \frac{2}{1 - (-x/2)} = \frac{a}{1 - r}$$

which implies that $a = 2$ and $r = -x/2$. Therefore, the power series for $f(x)$ is given by

$$\frac{4}{x + 2} = \sum_{n=0}^{\infty} ar^n$$

$$= \sum_{n=0}^{\infty} 2\left(-\frac{x}{2}\right)^n$$

$$= 2\left(1 - \frac{x}{2} + \frac{x^2}{4} - \frac{x^3}{8} + \cdots\right).$$

This power series converges when $|-x/2| < 1$, which implies that the interval of convergence is $(-2, 2)$. ▭

Another way to determine a power series for a rational function such as the one in Example 1 is to use long division. For instance, by dividing $2 + x$ into 4, we obtain the following.

$$
\begin{array}{r}
2 - x + \tfrac{1}{2}x^2 - \tfrac{1}{4}x^3 + \cdots \\
2 + x \overline{\smash{)}4} \\
\underline{4 + 2x} \\
- 2x \\
\underline{- 2x - x^2} \\
x^2 \\
\underline{x^2 + \tfrac{1}{2}x^3} \\
- \tfrac{1}{2}x^3 \\
\underline{- \tfrac{1}{2}x^3 - \tfrac{1}{4}x^4}
\end{array}
$$

EXAMPLE 2   Finding a geometric power series centered at 1

Find a power series for $f(x) = 1/x$, centered at 1.

SOLUTION

Writing $f(x)$ in the form $a/(1 - r)$, we have

$$\frac{1}{x} = \frac{1}{1 - (-x + 1)} = \frac{a}{1 - r}$$

which implies that $a = 1$ and $r = 1 - x = -(x - 1)$. Therefore, the power series for $f(x)$ is given by

$$\frac{1}{x} = \sum_{n=0}^{\infty} ar^n$$

$$= \sum_{n=0}^{\infty} [-(x - 1)]^n$$

$$= 1 - (x - 1) + (x - 1)^2 - (x - 1)^3 + \cdots.$$

This power series converges when $|x - 1| < 1$, which implies that the interval of convergence is $(0, 2)$.

Operations with power series

The versatility of geometric power series will be shown later in this section, following a discussion of power series operations. These operations, used with differentiation and integration, give us a means for developing power series for a variety of elementary functions. For instance, we will find power series for functions as diverse as

$$f(x) = \frac{3x - 1}{x^2 - 1}, \qquad f(x) = \ln x, \qquad \text{and} \qquad f(x) = \arctan x.$$

We summarize the basic operations with power series in the following theorem. For simplicity, we state the results for a series centered at 0.

THEOREM 9.22 OPERATIONS WITH POWER SERIES	Given $f(x) = \Sigma\, a_n x^n$ and $g(x) = \Sigma\, b_n x^n$, the following properties are true.

$$1.\ f(kx) = \sum_{n=0}^{\infty} a_n k^n x^n \qquad\qquad 2.\ f(x^N) = \sum_{n=0}^{\infty} a_n x^{nN}$$

$$3.\ f(x) \pm g(x) = \sum_{n=0}^{\infty} (a_n \pm b_n)x^n \qquad 4.\ f(x)g(x) = \left(\sum_{n=0}^{\infty} a_n x^n\right)\left(\sum_{n=0}^{\infty} b_n x^n\right)$$

The operations described in Theorem 9.22 can change the interval of convergence for the resulting series. For example, in the following addition, the interval of convergence for the sum is the *intersection* of the intervals of convergence of the two original series.

$$\underbrace{\sum_{n=0}^{\infty} x^n}_{(-1,\,1)} + \underbrace{\sum_{n=0}^{\infty} \left(\frac{x}{2}\right)^n}_{(-2,\,2)} = \underbrace{\sum_{n=0}^{\infty} \left(1 + \frac{1}{2^n}\right)x^n}_{(-1,\,1)}$$

$$\cap$$

EXAMPLE 3 Adding two power series

Find a power series, centered at 0, for

$$f(x) = \frac{3x - 1}{x^2 - 1}.$$

SOLUTION

Using partial fractions, we can write $f(x)$ as

$$\frac{3x - 1}{x^2 - 1} = \frac{2}{x + 1} + \frac{1}{x - 1}.$$

Now, using two geometric power series, we write

$$\frac{2}{x + 1} = \frac{2}{1 - (-x)} = \sum_{n=0}^{\infty} 2(-1)^n x^n, \quad |x| < 1$$

and

$$\frac{1}{x - 1} = \frac{-1}{1 - x} = -\sum_{n=0}^{\infty} x^n, \quad |x| < 1.$$

Finally, adding these two power series produces

$$\frac{3x - 1}{x^2 - 1} = \sum_{n=0}^{\infty} [2(-1)^n - 1]x^n = 1 - 3x + x^2 - 3x^3 + x^4 - \cdots .$$

The interval of convergence for this power series is $(-1, 1)$.

EXAMPLE 4   Finding a power series by integration

Find a power series for $f(x) = \ln x$, centered at 1.

SOLUTION

From Example 2 we know that

$$\frac{1}{x} = \sum_{n=0}^{\infty} (-1)^n (x-1)^n. \qquad (0, 2)$$

By integration, we have

$$\ln x = \int \frac{1}{x}\, dx + C = C + \sum_{n=0}^{\infty} (-1)^n \frac{(x-1)^{n+1}}{n+1}.$$

To determine C, we let $x = 1$ and conclude that $C = 0$. Therefore,

$$\ln x = \sum_{n=0}^{\infty} (-1)^n \frac{(x-1)^{n+1}}{n+1}$$

$$= \frac{(x-1)}{1} - \frac{(x-1)^2}{2} + \frac{(x-1)^3}{3} - \frac{(x-1)^4}{4} + \cdots. \qquad (0, 2]$$

Note that the series converges at $x = 2$. This is consistent with our observation in the previous section that integration of a power series may alter the convergence at the endpoints of the interval of convergence. ▭

In Section 9.7, we approximated $\ln (1.1)$ using the fourth Taylor polynomial for the natural logarithmic function.

$$\ln x \approx (x-1) - \frac{(x-1)^2}{2} + \frac{(x-1)^3}{3} - \frac{(x-1)^4}{4}$$

$$\ln (1.1) \approx (0.1) - \frac{1}{2}(0.1)^2 + \frac{1}{3}(0.1)^3 - \frac{1}{4}(0.1)^4 \approx 0.0953083$$

We now know from Example 4 that this polynomial represents the first four terms of the power series for $\ln x$. Moreover, using the Alternating Series Remainder, we can determine that the error in this approximation is less than

$$|R_3| \leq |a_4| = \frac{1}{5}(0.1)^5 \approx 0.000002.$$

During the seventeenth and eighteenth centuries, mathematical tables for logarithms and values of other transcendental functions were computed in this manner. The use for such numerical techniques is far from outdated, since it is precisely by such means that modern calculating devices are programmed to evaluate transcendental functions.

EXAMPLE 5   Finding a power series by integration

Find a power series for $f(x) = \arctan x$, centered at 0.

SOLUTION

Since $D_x[\arctan x] = 1/(1 + x^2)$, we use the series

$$f(x) = \frac{1}{1 + x} = \sum_{n=0}^{\infty} (-1)^n x^n. \qquad (-1, 1)$$

Now, substituting x^2 for x, we obtain

$$f(x^2) = \frac{1}{1 + x^2} = \sum_{n=0}^{\infty} (-1)^n x^{2n}.$$

Finally, by integrating, we obtain

$$\arctan x = \int \frac{1}{1 + x^2} \, dx + C$$

$$= C + \sum_{n=0}^{\infty} (-1)^n \frac{x^{2n+1}}{2n + 1}$$

$$= \sum_{n=0}^{\infty} (-1)^n \frac{x^{2n+1}}{2n + 1} \qquad \text{Let } x = 0, \text{ then } C = 0.$$

$$= x - \frac{x^3}{3} + \frac{x^5}{5} - \frac{x^7}{7} + \cdots. \qquad (-1, 1)$$

It can be shown that the power series developed for $\arctan x$ in Example 5 also converges (to $\arctan x$) for $x = \pm 1$. For instance, when $x = 1$, we can write

$$\arctan 1 = 1 - \frac{1}{3} + \frac{1}{5} - \frac{1}{7} + \cdots = \frac{\pi}{4}.$$

However, this series (developed by James Gregory in 1671) does not give us a practical way of approximating π because it converges so slowly that hundreds of terms would have to be used to obtain reasonable accuracy. Example 6 shows how to use *two* different arctangent series to obtain a very good approximation of π using only a few terms. This approximation was developed by John Machin in 1706.

Series that can be used to approximate π have interested mathematicians for the past 300 years. An amazing series for approximating $1/\pi$ was developed by the Indian mathematician Srinivasa Ramanujan (1887–1920) in 1914. Each successive term of Ramanujan's series adds roughly eight more correct digits to the value of $1/\pi$.

Srinivasa Ramanujan

EXAMPLE 6 Approximating π with a series

Use the trigonometric identity

$$4 \arctan \frac{1}{5} - \arctan \frac{1}{239} = \frac{\pi}{4}$$

to approximate the number π.

SOLUTION

By using only five terms from each of the series for arctan $(1/5)$ and arctan $(1/239)$, we obtain

$$4\left(4 \arctan \frac{1}{5} - \arctan \frac{1}{239}\right) \approx 3.141593$$

which agrees with the decimal representation of π with an error less than 0.000001.

EXERCISES for Section 9.9

In Exercises 1–14, find a power series for the given function, centered at c, and determine the interval of convergence.

1. $f(x) = \dfrac{1}{2 - x}$, $c = 0$

2. $f(x) = \dfrac{3}{4 - x}$, $c = 0$

3. $f(x) = \dfrac{1}{2 - x}$, $c = 5$

4. $f(x) = \dfrac{3}{4 - x}$, $c = -2$

5. $f(x) = \dfrac{3}{2x - 1}$, $c = 0$

6. $f(x) = \dfrac{3}{2x - 1}$, $c = 2$

7. $f(x) = \dfrac{1}{2x - 5}$, $c = -3$

8. $f(x) = \dfrac{1}{2x - 5}$, $c = 0$

9. $f(x) = \dfrac{3}{x + 2}$, $c = 0$

10. $f(x) = \dfrac{4}{3x + 2}$, $c = 2$

11. $f(x) = \dfrac{3x}{x^2 + x - 2}$, $c = 0$

12. $f(x) = \dfrac{4x - 7}{2x^2 + 3x - 2}$, $c = 0$

13. $f(x) = \dfrac{2}{1 - x^2}$, $c = 0$

14. $f(x) = \dfrac{4}{4 + x^2}$, $c = 0$

In Exercises 15–22, use the power series

$$\frac{1}{1 + x} = \sum_{n=0}^{\infty} (-1)^n x^n$$

to determine a power series representation, centered at 0, for the given function. List the interval of convergence with your solution.

15. $f(x) = -\dfrac{1}{(x + 1)^2}$

16. $f(x) = \dfrac{2}{(x + 1)^3}$

17. $f(x) = \ln (x + 1)$

18. $f(x) = \ln (x^2 + 1)$

19. $f(x) = \dfrac{1}{4x^2 + 1}$

20. $f(x) = \arctan 2x$

21. $f(x) = \ln \sqrt{\dfrac{1 + x}{1 - x}}$

22. $f(x) = \dfrac{1}{x^2 - x + 1} = \dfrac{x + 1}{x^3 + 1}$

23. Complete the following table to demonstrate the inequalities

$$x - \frac{x^2}{2} \le \ln (x + 1) \le x - \frac{x^2}{2} + \frac{x^3}{3}.$$

x	0.0	0.2	0.4	0.6	0.8	1.0
$x - \dfrac{x^2}{2}$						
$\ln (x + 1)$						
$x - \dfrac{x^2}{2} + \dfrac{x^3}{3}$						

24. Complete a table for the same values of x as in Exercise 23 to demonstrate the inequalities

$$x - \frac{x^2}{2} + \frac{x^3}{3} - \frac{x^4}{4} \le \ln (x + 1)$$

$$\le x - \frac{x^2}{2} + \frac{x^3}{3} - \frac{x^4}{4} + \frac{x^5}{5}.$$

In Exercises 25–28, use the series for $f(x) = \arctan x$ to approximate the given value, using $R_N \le 0.001$.

25. $\arctan \dfrac{1}{4}$

26. $\displaystyle\int_0^{3/4} \arctan x^2 \, dx$

27. $\displaystyle\int_0^{1/2} \dfrac{\arctan x^2}{x} \, dx$

28. $\displaystyle\int_0^{1/2} x^2 \arctan x \, dx$

In Exercises 29 and 30, (a) verify the given equation. [Hint: Use trigonometric identities for the tangent of the sum of two angles and the tangent of a double angle.] (b) Use the equation and the series for the arctangent to approximate π to two-decimal-place accuracy.

29. $2 \arctan \dfrac{1}{2} - \arctan \dfrac{1}{7} = \dfrac{\pi}{4}$

30. $2 \arctan \dfrac{2}{3} - \arctan \dfrac{7}{17} = \dfrac{\pi}{4}$

In Exercises 31 and 32, use a computer or graphics calculator to verify the inequality graphically on the interval $0 \leq x \leq 1$.

31. $x - \dfrac{x^2}{2} \leq \ln(x+1) \leq x - \dfrac{x^2}{2} + \dfrac{x^3}{3}$

32. $x - \dfrac{x^2}{2} + \dfrac{x^3}{3} - \dfrac{x^4}{4} \leq \ln(x+1)$

$$\leq x - \dfrac{x^2}{2} + \dfrac{x^3}{3} - \dfrac{x^4}{4} + \dfrac{x^5}{5}$$

33. Use long division to develop the power series for $1/(1+x)$.

9.10 Taylor and Maclaurin Series

Taylor series and Maclaurin series ▪ Binomial series

Colin Maclaurin

In Section 9.9 we derived power series for several functions using geometric series with term-by-term differentiation or integration. In this section we develop a *general* procedure for deriving the power series for a function that has derivatives of all orders.

The development of power series to represent functions is due to the combined work of many seventeenth and eighteenth century mathematicians. Gregory, Newton, John and James Bernoulli, Leibniz, Euler, Lagrange, Wallis, and Fourier all contributed to this work. However, the two names that are most commonly associated with power series are Brook Taylor (1685–1731) and Colin Maclaurin (1698–1746).

We begin with a theorem that gives us the form that *every* (convergent) power series must take.

THEOREM 9.23
THE FORM OF A CONVERGENT POWER SERIES

If f is represented by a power series $f(x) = \sum a_n(x - c)^n$ for all x in an open interval I containing c, then $a_n = f^{(n)}(c)/n!$ and

$$f(x) = f(c) + f'(c)(x - c) + \frac{f''(c)}{2!}(x - c)^2 + \cdots + \frac{f^{(n)}(c)}{n!}(x - c)^n + \cdots.$$

PROOF Suppose the power series $\sum a_n(x - c)^n$ has a radius of convergence R. Then, by Theorem 9.21, we know that the nth derivative of f exists for $|x - c| < R$, and by successive differentiation we obtain the following.

$$f^{(0)}(x) = a_0 + a_1(x - c) + a_2(x - c)^2 + a_3(x - c)^3 + a_4(x - c)^4 + \cdots$$
$$f^{(1)}(x) = a_1 + 2a_2(x - c) + 3a_3(x - c)^2 + 4a_4(x - c)^3 + \cdots$$
$$f^{(2)}(x) = 2a_2 + 3!a_3(x - c) + 4 \cdot 3a_4(x - c)^2 + \cdots$$
$$f^{(3)}(x) = 3!a_3 + 4!a_4(x - c) + \cdots$$

$$\vdots$$

$$f^{(n)}(x) = n!a_n + (n + 1)!a_{n+1}(x - c) + \cdots$$

Now, evaluating each of these derivatives at $x = c$ yields

$$f^{(0)}(c) = 0!a_0, \qquad f^{(1)}(c) = 1!a_1, \qquad f^{(2)}(c) = 2!a_2, \qquad f^{(3)}(c) = 3!a_3$$

and, in general, $f^{(n)}(c) = n!a_n$. By solving for a_n we find that the coefficients of the power series representation of $f(x)$ are

$$a_n = \frac{f^{(n)}(c)}{n!}.$$

Notice that the coefficients of the power series in Theorem 9.23 are precisely the coefficients of the Taylor polynomials for $f(x)$ at c as defined in Section 9.7. For this reason, the series is called the **Taylor series** for $f(x)$ at c.

DEFINITION OF TAYLOR AND MACLAURIN SERIES

If a function f has derivatives of all orders at $x = c$, then the series

$$\sum_{n=0}^{\infty} \frac{f^{(n)}(c)}{n!}(x - c)^n = f(c) + f'(c)(x - c) + \cdots + \frac{f^{(n)}(c)}{n!}(x - c)^n + \cdots$$

is called the **Taylor series for $f(x)$ at c.** Moreover, if $c = 0$, then this series is also called the **Maclaurin series for f.**

If we know the pattern for the coefficients of the Taylor polynomials for a function, then we can extend the pattern easily to form the corresponding Taylor series. For instance, in Example 4 of Section 9.7, we found the fourth Taylor polynomial for $\ln x$, centered at 1, to be

$$P_4(x) = (x - 1) - \frac{1}{2}(x - 1)^2 + \frac{1}{3}(x - 1)^3 - \frac{1}{4}(x - 1)^4.$$

Following this pattern, we obtain the Taylor series for $\ln x$ centered at 1

$$(x - 1) - \frac{1}{2}(x - 1)^2 + \cdots + \frac{(-1)^{n+1}}{n}(x - 1)^n + \cdots.$$

EXAMPLE 1   Forming a power series

Use the function $f(x) = \sin x$ to form the Maclaurin series

$$\sum_{n=0}^{\infty} \frac{f^{(n)}(0)}{n!} x^n = f(0) + f'(0)x + \frac{f''(0)}{2!}x^2 + \frac{f^{(3)}(0)}{3!}x^3 + \frac{f^{(4)}(0)}{4!}x^4 + \cdots$$

and determine the interval of convergence.

SOLUTION

Successive differentiation of $f(x)$ yields

$$f(x) = \sin x \qquad f(0) = 0$$
$$f'(x) = \cos x \qquad f'(0) = 1$$
$$f''(x) = -\sin x \qquad f''(0) = 0$$
$$f^{(3)}(x) = -\cos x \qquad f^{(3)}(0) = -1$$
$$f^{(4)}(x) = \sin x \qquad f^{(4)}(0) = 0$$
$$f^{(5)}(x) = \cos x \qquad f^{(5)}(0) = 1$$

and so on. The pattern repeats after the third derivative. Hence, the power series is as follows.

$$\sum_{n=0}^{\infty} \frac{f^{(n)}(0)}{n!} x^n = f(0) + f'(0)x + \frac{f''(0)}{2!}x^2 + \frac{f^{(3)}(0)}{3!}x^3 + \frac{f^{(4)}(0)}{4!}x^4 + \cdots$$

$$\sum_{n=0}^{\infty} \frac{(-1)^n x^{2n+1}}{(2n+1)!} = x - \frac{x^3}{3!} + \frac{x^5}{5!} - \frac{x^7}{7!} + \cdots$$

By the Ratio Test, we can conclude that this series converges for all x. ▭

Notice that we did not conclude that the power series in Example 1 converges to $\sin x$ for all x. We simply concluded that the power series converges to some function, but we are not sure what function it is. This is a subtle, but important, point in dealing with Taylor or Maclaurin series. To persuade yourself that the series

$$f(c) + f'(c)(x - c) + \frac{f''(c)}{2!}(x - c)^2 + \cdots + \frac{f^{(n)}(c)}{n!}(x - c)^n + \cdots$$

might converge to a function other than f, remember that the derivatives are being evaluated at a single point. It can easily happen that another function will agree with the values of $f^{(n)}(x)$ when $x = c$ and disagree at other x-values. For instance, if we formed the power series (centered at 0) for the function

$$f(x) = \begin{cases} -1, & x < -\dfrac{\pi}{2} \\ \sin x, & |x| \leq \dfrac{\pi}{2} \\ 1, & x > \dfrac{\pi}{2} \end{cases} \qquad \text{(See Figure 9.15.)}$$

we would obtain the same series we obtained in Example 1. We know that the series converges for all x, and yet it obviously cannot converge to both $f(x)$ and $\sin x$ for all x.

FIGURE 9.15

REMARK Don't confuse this observation with the result given in Theorem 9.23. That theorem says that *if a power series converges to $f(x)$*, then the series must be a Taylor series. The theorem does not say that every series formed with the Taylor coefficients $a_n = f^{(n)}(c)/n!$ will converge to $f(x)$.

To determine whether a power series formed with Taylor coefficients actually converges to $f(x)$, it is helpful to consider the series as the sum of the nth Taylor polynomial for $f(x)$ and a remainder term.

$$\underbrace{f(c) + \cdots + \frac{f^{(n)}(c)}{n!}(x - c)^n}_{P_n(x)} + \underbrace{\frac{f^{(n+1)}(c)}{(n + 1)!}(x - c)^{n+1} + \cdots}_{R_n(x)}$$

Recall that Taylor's Theorem, given in Section 9.7, tells us that the remainder for a Taylor polynomial is given by

$$R_n(x) = \frac{f^{(n+1)}(z)}{(n + 1)!}(x - c)^{n+1}$$

where z lies between x and c. The problem we now want to tackle is that of finding those values of x for which the Taylor polynomials $P_n(x)$ actually converge to $f(x)$ as $n \to \infty$. In other words, we want to find the values of x for which

$$f(x) = \lim_{n \to \infty} P_n(x)$$

$$= \lim_{n \to \infty} \sum_{k=0}^{n} \frac{f^{(k)}(c)}{k!}(x - c)^k$$

$$= \sum_{k=0}^{\infty} \frac{f^{(k)}(c)}{k!}(x - c)^k.$$

The conditions that guarantee this convergence are given in Theorem 9.24.

THEOREM 9.24
CONVERGENCE OF
TAYLOR SERIES

If a function f has derivatives of all orders in an interval I centered at c, then the equality

$$f(x) = \sum_{n=0}^{\infty} \frac{f^{(n)}(c)}{n!}(x - c)^n$$

holds if and only if

$$\lim_{n \to \infty} R_n(x) = 0$$

for every x in I.

PROOF For a Taylor series, the nth partial sum coincides with the nth Taylor polynomial. That is, $S_n(x) = P_n(x)$. Moreover, since

$$P_n(x) = f(x) - R_n(x)$$

it follows that

$$\lim_{n \to \infty} S_n(x) = \lim_{n \to \infty} P_n(x) = \lim_{n \to \infty} [f(x) - R_n(x)] = f(x) - \lim_{n \to \infty} R_n(x).$$

Hence, for a given x, the Taylor series (the sequence of partial sums) converges to $f(x)$ if and only if $R_n(x) \to 0$ as $n \to \infty$.

REMARK Stated another way, this theorem says that a power series formed with the Taylor coefficients $a_n = f^{(n)}(c)/n!$ converges to the function from which it was derived at precisely those values for which the remainder approaches zero as $n \to \infty$.

In Example 2, we take another look at the series formed in Example 1. We derived that series from the sine function and we also concluded that the series converges to some function on the entire real line. We now show that the series actually converges to $\sin x$.

EXAMPLE 2 A convergent Maclaurin series

Show that the Maclaurin series for $f(x) = \sin x$ converges to $\sin x$ for all x.

SOLUTION

Using the result in Example 1, we need to show that

$$\sin x = x - \frac{x^3}{3!} + \frac{x^5}{5!} - \frac{x^7}{7!} + \cdots + \frac{(-1)^n x^{2n+1}}{(2n+1)!} + \cdots$$

is true for all x. Since

$$f^{(n+1)}(x) = \pm \sin x \quad \text{or} \quad f^{(n+1)}(x) = \pm \cos x$$

we know that $|f^{(n+1)}(z)| \le 1$ for every real number z. Therefore, for any fixed x, we can apply Taylor's Theorem (Theorem 9.19) to conclude that

$$0 \le |R_n(x)| \le \left| \frac{f^{(n+1)}(z)}{(n+1)!} x^{n+1} \right| \le \frac{|x|^{n+1}}{(n+1)!}.$$

Now, from our discussion in Section 9.1 regarding the relative rate of convergence of exponential and factorial sequences, it follows that for a fixed x

$$\lim_{n \to \infty} \frac{|x|^{n+1}}{(n+1)!} = 0.$$

Finally, by the Squeeze Theorem, it follows that for all x, $R_n(x) \to 0$ as $n \to \infty$. Hence, by Theorem 9.24, the Maclaurin series for $\sin x$ converges to $\sin x$ for all x.

Figure 9.16 visually illustrates the convergence of the Maclaurin series for $\sin x$ by comparing the graphs of the Maclaurin polynomials $P_1(x)$, $P_3(x)$, $P_5(x)$, and $P_7(x)$ with the graph of the sine function. Notice that as the degree of the polynomial increases, its graph more closely resembles that of the sine function.

We summarize the steps for finding a Taylor series for $f(x)$ at c as follows.

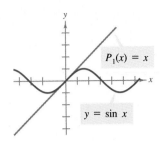

$P_1(x) = x$

$y = \sin x$

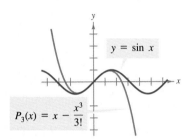

$y = \sin x$

$P_3(x) = x - \dfrac{x^3}{3!}$

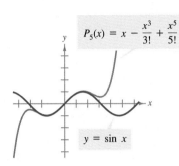

$P_5(x) = x - \dfrac{x^3}{3!} + \dfrac{x^5}{5!}$

$y = \sin x$

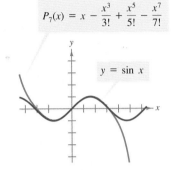

$P_7(x) = x - \dfrac{x^3}{3!} + \dfrac{x^5}{5!} - \dfrac{x^7}{7!}$

$y = \sin x$

FIGURE 9.16

GUIDELINES FOR FINDING A TAYLOR SERIES

1. Differentiate $f(x)$ several times and evaluate each derivative at c.
   $$f(c), f'(c), f''(c), f'''(c), \ldots, f^{(n)}(c), \ldots$$
   Try to recognize a pattern for these numbers.
2. Use the sequence developed in the first step to form the Taylor coefficients $a_n = f^{(n)}(c)/n!$, and determine the interval of convergence for the resulting power series
   $$f(c) + f'(c)(x - c) + \frac{f''(c)}{2!}(x - c)^2 + \cdots + \frac{f^{(n)}(c)}{n!}(x - c)^n + \cdots .$$
3. Within this interval of convergence, determine whether the series converges to $f(x)$.

The direct determination of Taylor or Maclaurin coefficients using successive differentiation can be difficult, and the next example illustrates a shortcut for finding the coefficients indirectly—using the coefficients of a known Taylor or Maclaurin series.

EXAMPLE 3   Maclaurin series for a composite function

Find the Maclaurin series for $f(x) = \sin x^2$.

SOLUTION

To find the coefficients for this Maclaurin series *directly* we must calculate successive derivatives of $f(x) = \sin x^2$. By calculating just the first two,

$$f'(x) = 2x \cos x^2 \qquad \text{and} \qquad f''(x) = -4x^2 \sin x^2 + 2 \cos x^2$$

we recognize this to be cumbersome. Fortunately there is an alternative. Suppose we first consider the Maclaurin series for $\sin x$ found in Example 1.

$$g(x) = \sin x$$

$$= x - \frac{x^3}{3!} + \frac{x^5}{5!} - \frac{x^7}{7!} + \cdots$$

Now, since $\sin x^2 = g(x^2)$, we can apply Theorem 9.22 and substitute x^2 for x in the series for $\sin x$ to obtain

$$\sin x^2 = g(x^2)$$

$$= x^2 - \frac{x^6}{3!} + \frac{x^{10}}{5!} - \frac{x^{14}}{7!} + \cdots .$$

Be sure to understand the point illustrated in Example 3. Since direct computation of Taylor or Maclaurin coefficients can be tedious, the most practical way to find a Taylor or Maclaurin series is to develop power series for a *basic list* of elementary functions. From this list, we can determine power series for other functions by the operations of addition, subtraction, multiplication, division, differentiation, integration, or composition with known power series. Before presenting the basic list for elementary functions, we develop one additional formula—for a function of the form $f(x) = (1 + x)^k$. This produces the **binomial series.**

EXAMPLE 4   Binomial series

Find the Maclaurin series for $f(x) = (1 + x)^k$ and determine its radius of convergence.

SOLUTION

By successive differentiation, we have

$$f(x) = (1 + x)^k \qquad\qquad f(0) = 1$$
$$f'(x) = k(1 + x)^{k-1} \qquad\qquad f'(0) = k$$
$$f''(x) = k(k - 1)(1 + x)^{k-2} \qquad\qquad f''(0) = k(k - 1)$$
$$f'''(x) = k(k - 1)(k - 2)(1 + x)^{k-3} \qquad f'''(0) = k(k - 1)(k - 2)$$
$$\vdots \qquad\qquad\qquad\qquad \vdots$$
$$f^{(n)}(x) = k \cdots (k - n + 1)(1 + x)^{k-n} \quad f^{(n)}(0) = k(k - 1) \cdots (k - n + 1)$$

which produces the series

$$1 + kx + \frac{k(k - 1)x^2}{2} + \cdots + \frac{k(k - 1) \cdots (k - n + 1)x^n}{n!} + \cdots .$$

By the Ratio Test, we have $a_{n+1}/a_n \to 1$ as $n \to \infty$. This implies that the radius of convergence is $R = 1$, and thus the series converges to *some* function in the interval $(-1, 1)$.

Note that in Example 4 we showed that the Taylor series for $(1 + x)^k$ converges to *some* function in the interval $(-1, 1)$. However, we did not show that the series actually converges to $(1 + x)^k$. To do this, we could show that the remainder $R_n(x)$ converges to 0, as we did in Example 2. Another way to show that the Taylor series for f converges to $f(x)$ is to use the series to form a differential equation. Then, the verification of convergence to $f(x)$ can be obtained by solving the differential equation. For example, if we let y be the binomial series obtained in Example 4,

$$y = 1 + kx + \frac{k(k-1)x^2}{2} + \cdots + \frac{k(k-1) \cdots (k-n+1)x^n}{n!} + \cdots$$

then we can derive the differential equation

$$y' + xy' = ky. \qquad \text{Initial condition } y = 1 \text{ if } x = 0$$

In the following list, we provide the power series for several elementary functions with the corresponding intervals of convergence.

POWER SERIES FOR ELEMENTARY FUNCTIONS

Interval of Convergence

$$\frac{1}{x} = 1 - (x-1) + (x-1)^2 - (x-1)^3 + (x-1)^4 - \cdots + (-1)^n(x-1)^n + \cdots, \qquad 0 < x < 2$$

$$\frac{1}{1+x} = 1 - x + x^2 - x^3 + x^4 - x^5 + \cdots + (-1)^n x^n + \cdots, \qquad -1 < x < 1$$

$$\ln x = (x-1) - \frac{(x-1)^2}{2} + \frac{(x-1)^3}{3} - \frac{(x-1)^4}{4} + \cdots + \frac{(-1)^{n-1}(x-1)^n}{n} + \cdots, \qquad 0 < x \leq 2$$

$$e^x = 1 + x + \frac{x^2}{2!} + \frac{x^3}{3!} + \frac{x^4}{4!} + \frac{x^5}{5!} + \cdots + \frac{x^n}{n!} + \cdots, \qquad -\infty < x < \infty$$

$$\sin x = x - \frac{x^3}{3!} + \frac{x^5}{5!} - \frac{x^7}{7!} + \frac{x^9}{9!} - \cdots + \frac{(-1)^n x^{2n+1}}{(2n+1)!} + \cdots, \qquad -\infty < x < \infty$$

$$\cos x = 1 - \frac{x^2}{2!} + \frac{x^4}{4!} - \frac{x^6}{6!} + \frac{x^8}{8!} - \cdots + \frac{(-1)^n x^{2n}}{(2n)!} + \cdots, \qquad -\infty < x < \infty$$

$$\arctan x = x - \frac{x^3}{3} + \frac{x^5}{5} - \frac{x^7}{7} + \frac{x^9}{9} - \cdots + \frac{(-1)^n x^{2n+1}}{2n+1} + \cdots, \qquad -1 \leq x \leq 1$$

$$\arcsin x = x + \frac{x^3}{2 \cdot 3} + \frac{1 \cdot 3 x^5}{2 \cdot 4 \cdot 5} + \frac{1 \cdot 3 \cdot 5 x^7}{2 \cdot 4 \cdot 6 \cdot 7} + \cdots + \frac{(2n)! x^{2n+1}}{(2^n n!)^2 (2n+1)} + \cdots, \qquad -1 \leq x \leq 1$$

$$(1+x)^k = 1 + kx + \frac{k(k-1)x^2}{2!} + \frac{k(k-1)(k-2)x^3}{3!} + \frac{k(k-1)(k-2)(k-3)x^4}{4!} + \cdots, \qquad -1 < x < 1^*$$

*The convergence at $x = \pm 1$ depends on the value k.

REMARK The binomial series is valid for noninteger values of k. Moreover, if k happens to be a positive integer, then the binomial series reduces to a simple binomial expansion.

In the next three examples we show how to use the basic list of power series.

EXAMPLE 5 Deriving a new power series from a given series

Find the power series for $f(x) = \cos \sqrt{x}$.

SOLUTION

Using the power series

$$\cos x = 1 - \frac{x^2}{2!} + \frac{x^4}{4!} - \frac{x^6}{6!} + \frac{x^8}{8!} - \cdots$$

we replace x by $\sqrt{x}$ to obtain the series

$$\cos \sqrt{x} = 1 - \frac{x}{2!} + \frac{x^2}{4!} - \frac{x^3}{6!} + \frac{x^4}{8!} - \cdots$$

which converges for all x in the domain of $\cos \sqrt{x}$, that is, for $x \geq 0$. ▭

EXAMPLE 6 Using the binomial series

Find the power series for $g(x) = \sqrt[3]{1 + x}$.

SOLUTION

Using the binomial series

$$(1 + x)^k = 1 + kx + \frac{k(k-1)x^2}{2!} + \frac{k(k-1)(k-2)x^3}{3!} + \cdots$$

we let $k = \frac{1}{3}$ and write

$$(1 + x)^{1/3} = 1 + \frac{x}{3} - \frac{2x^2}{3^2 2!} + \frac{2 \cdot 5 x^3}{3^3 3!} - \frac{2 \cdot 5 \cdot 8 x^4}{3^4 4!} + \cdots$$

which converges for $-1 \leq x \leq 1$. ▭

EXAMPLE 7 A power series for $\sin^2 x$

Find the power series for $f(x) = \sin^2 x$.

SOLUTION

Instead of squaring the power series for $\sin x$, we write

$$\sin^2 x = \frac{1 - \cos 2x}{2} = \frac{1}{2} - \frac{\cos 2x}{2}.$$

Now, using the series for $\cos x$, we proceed as follows.

$$\cos x = 1 - \frac{x^2}{2!} + \frac{x^4}{4!} - \frac{x^6}{6!} + \frac{x^8}{8!} - \cdots$$

$$\cos 2x = 1 - \frac{2^2}{2!}x^2 + \frac{2^4}{4!}x^4 - \frac{2^6}{6!}x^6 + \frac{2^8}{8!}x^8 - \cdots$$

$$-\frac{1}{2}\cos 2x = -\frac{1}{2} + \frac{2}{2!}x^2 - \frac{2^3}{4!}x^4 + \frac{2^5}{6!}x^6 - \frac{2^7}{8!}x^8 + \cdots$$

$$\sin^2 x = \frac{1}{2} - \frac{1}{2}\cos 2x = \frac{1}{2} - \frac{1}{2} + \frac{2}{2!}x^2 - \frac{2^3}{4!}x^4 + \frac{2^5}{6!}x^6 - \frac{2^7}{8!}x^8 + \cdots$$

$$= \frac{2}{2!}x^2 - \frac{2^3}{4!}x^4 + \frac{2^5}{6!}x^6 - \frac{2^7}{8!}x^8 + \cdots.$$

This series converges for $-\infty < x < \infty$. ▭

As mentioned in the previous section, power series can be used to obtain tables of values of transcendental functions. They are also useful for estimating the value of definite integrals for which antiderivatives cannot be found. The next example demonstrates this use.

EXAMPLE 8 Power series approximation of a definite integral

Use a power series to approximate

$$\int_0^1 e^{-x^2}\,dx$$

with an error of less than 0.01.

SOLUTION

Replacing x with $-x^2$ in the series for e^x, we have

$$e^{-x^2} = 1 - x^2 + \frac{x^4}{2!} - \frac{x^6}{3!} + \frac{x^8}{4!} - \cdots$$

$$\int_0^1 e^{-x^2}\,dx = \left[x - \frac{x^3}{3} + \frac{x^5}{5 \cdot 2!} - \frac{x^7}{7 \cdot 3!} + \frac{x^9}{9 \cdot 4!} - \cdots \right]_0^1$$

$$= 1 - \frac{1}{3} + \frac{1}{10} - \frac{1}{42} + \frac{1}{216} - \cdots.$$

Summing the first *four* terms, we have

$$\int_0^1 e^{-x^2}\, dx \approx 0.74$$

which, by the Alternating Series Test, has an error of less than $\frac{1}{216} \approx 0.005$.

___

EXERCISES for Section 9.10

In Exercises 1–10, find the Taylor series (centered at *c*) for the given function.

1. $f(x) = e^{2x}$, $c = 0$ **2.** $f(x) = e^{-2x}$, $c = 0$

3. $f(x) = \cos x$, $c = \dfrac{\pi}{4}$ **4.** $f(x) = \sin x$, $c = \dfrac{\pi}{4}$

5. $f(x) = \ln x$, $c = 1$ **6.** $f(x) = e^x$, $c = 1$
7. $f(x) = \sin 2x$, $c = 0$
8. $f(x) = \tan x$, $c = 0$ (first three nonzero terms)
9. $f(x) = \sec x$, $c = 0$ (first three nonzero terms)
10. $f(x) = \ln(x^2 + 1)$, $c = 0$

In Exercises 11–16, use the binomial series to find the Maclaurin series for the given function.

11. $f(x) = \dfrac{1}{(1 + x)^2}$ **12.** $f(x) = \dfrac{1}{\sqrt{1 - x}}$

13. $f(x) = \dfrac{1}{\sqrt{4 + x^2}}$ **14.** $f(x) = \sqrt{1 + x}$

15. $f(x) = \sqrt{1 + x^2}$ **16.** $f(x) = \sqrt{1 + x^3}$

In Exercises 17–30, find the Maclaurin series for the given function from a power series in the table of power series for elementary functions.

17. $f(x) = e^{x^2/2}$ **18.** $g(x) = e^{-3x}$
19. $g(x) = \sin 2x$ **20.** $h(x) = x \cos x$
21. $f(x) = \cos \sqrt{x}$ **22.** $f(x) = \sin x^2$

23. $g(x) = \dfrac{\sin x}{x}$ **24.** $f(x) = \dfrac{\arcsin x}{x}$

25. $f(x) = \dfrac{1}{2}(e^x - e^{-x}) = \sinh x$

26. $f(x) = \dfrac{1}{2}(e^x + e^{-x}) = \cosh x$

27. $g(x) = \dfrac{1}{2i}(e^{ix} - e^{-ix}) = \sin x$

   [Hint: $i^2 = -1$]

28. $g(x) = \dfrac{1}{2}(e^{ix} + e^{-ix}) = \cos x$

   [Hint: $i^2 = -1$]
29. $f(x) = \cos^2 x$

   $\left[\text{Hint: } \cos^2 x = \dfrac{1}{2}(1 + \cos 2x)\right]$

30. $f(x) = \sinh^{-1} x = \ln(x + \sqrt{x^2 + 1})$

   $\left[\text{Hint: Integrate the series for } \dfrac{1}{\sqrt{x^2 + 1}}.\right]$

In Exercises 31–38, use power series to approximate the given integral with an error of less than 0.0001.

31. $\displaystyle\int_0^{\pi/2} \dfrac{\sin x}{x}\, dx$ **32.** $\displaystyle\int_0^1 \cos x^2\, dx$

33. $\displaystyle\int_0^{\pi/2} \sqrt{x} \cos x\, dx$ **34.** $\displaystyle\int_0^{1/2} \dfrac{\ln(x + 1)}{x}\, dx$

35. $\displaystyle\int_0^{1/2} \dfrac{\arctan x}{x}\, dx$ **36.** $\displaystyle\int_1^2 e^{-x^2}\, dx$

37. $\displaystyle\int_{0.1}^{0.3} \sqrt{1 + x^3}\, dx$ **38.** $\displaystyle\int_{0.5}^1 \cos \sqrt{x}\, dx$

In Exercises 39 and 40, find a Maclaurin series for $f(x)$ defined by the given integral.

39. $\displaystyle\int_0^x (e^{-t^2} - 1)\, dt$ **40.** $\displaystyle\int_0^x \sqrt{1 + t^3}\, dt$

In Exercises 41 and 42, use a computer to approximate the given probability with an error of less than 0.0001 for the standard normal probability density function

$$P(a < x < b) = \dfrac{1}{\sqrt{2\pi}} \int_a^b e^{-x^2/2}\, dx.$$

41. $P(0 < x < 1)$ **42.** $P(1 < x < 2)$

43. Show that

$$\sum_{n=0}^{\infty} (-1)^n \left[\dfrac{1}{(2n + 1)!}\right] = \sin 1.$$

44. Show that

$$\sum_{n=1}^{\infty} (-1)^{n-1}\left(\dfrac{1}{n!}\right) = \dfrac{e - 1}{e}.$$

45. As discussed in the Chapter 9 Application, a projectile fired from the ground follows the trajectory given by

$$y = \left(\tan \theta - \frac{g}{kv_0 \cos \theta} \right) x - \frac{g}{k^2} \ln \left(1 - \frac{kx}{v_0 \cos \theta} \right)$$

where v_0 is the initial speed, θ is the angle of projection, g is the acceleration due to gravity, and k is the drag factor caused by the air resistance. Using the power series representation

$$\ln (1 + x) = x - \frac{x^2}{2} + \frac{x^3}{3} - \frac{x^4}{4} + \cdots,$$

$$-1 < x < 1$$

verify that the trajectory can be rewritten as

$$y = (\tan \theta)x + \frac{gx^2}{2v_0^2 \cos^2 \theta} + \frac{kgx^3}{3v_0^3 \cos^3 \theta}$$

$$+ \frac{k^2 gx^4}{4v_0^4 \cos^4 \theta} + \cdots.$$

46. Use the result of Exercise 45 to determine the series for the path of a projectile projected from ground level at an angle of $\theta = 60°$ with an initial speed of $v_0 = 64$ feet per second and a drag factor of $k = \frac{1}{16}$.

47. Consider the function f defined by

$$f(x) = \begin{cases} e^{-1/x^2}, & x \neq 0 \\ 0, & x = 0 \end{cases}.$$

(a) Sketch a graph of the function.

(b) Use the alternate form of the definition of the derivative (Section 3.1) and L'Hôpital's Rule to show that $f'(0) = 0$. (By continuing this process, it can be shown that $f^{(n)}(0) = 0$ for $n > 1$.)

(c) Using the result of part (b), find the Maclaurin series for f. Does the series converge to f?

REVIEW EXERCISES for Chapter 9

In Exercises 1 and 2, find the general term of the sequence.

1. $1, \dfrac{1}{2}, \dfrac{1}{6}, \dfrac{1}{24}, \dfrac{1}{120}, \ldots$

2. $\dfrac{1}{2}, \dfrac{2}{5}, \dfrac{3}{10}, \dfrac{4}{17}, \ldots$

In Exercises 3–10, determine the convergence or divergence of the sequence with the given general term.

3. $a_n = \dfrac{n + 1}{n^2}$

4. $a_n = \dfrac{1}{\sqrt{n}}$

5. $a_n = \dfrac{n^3}{n^2 + 1}$

6. $a_n = \dfrac{n}{\ln n}$

7. $a_n = \sqrt{n + 1} - \sqrt{n}$

8. $a_n = \left(1 + \dfrac{1}{2n} \right)^n$

9. $a_n = \dfrac{\sin \sqrt{n}}{\sqrt{n}}$

10. $a_n = (b^n + c^n)^{1/n}$
(b and c are positive real numbers)

In Exercises 11–14, find the first five terms of the sequence of partial sums for the given series.

11. $\displaystyle\sum_{n=0}^{\infty} \left(\dfrac{3}{2} \right)^n$

12. $\displaystyle\sum_{n=1}^{\infty} \dfrac{(-1)^{n+1}}{2n}$

13. $\displaystyle\sum_{n=1}^{\infty} \dfrac{(-1)^{n+1}}{(2n)!}$

14. $\displaystyle\sum_{n=1}^{\infty} \dfrac{1}{n(n + 1)}$

In Exercises 15–18, find the sum of the given series.

15. $\displaystyle\sum_{n=0}^{\infty} \left(\dfrac{2}{3} \right)^n$

16. $\displaystyle\sum_{n=0}^{\infty} \dfrac{2^{n+2}}{3^n}$

17. $\displaystyle\sum_{n=0}^{\infty} \left(\dfrac{1}{2^n} - \dfrac{1}{3^n} \right)$

18. $\displaystyle\sum_{n=0}^{\infty} \left[\left(\dfrac{2}{3} \right)^n - \dfrac{1}{(n + 1)(n + 2)} \right]$

In Exercises 19 and 20, express the repeating decimal as the ratio of two integers.

19. $0.090909 \ldots$

20. $0.923076923076 \ldots$

In Exercises 21–32, determine the convergence or divergence of the given series.

21. $\displaystyle\sum_{n=1}^{\infty} \dfrac{2^n}{n^3}$

22. $\displaystyle\sum_{n=1}^{\infty} \dfrac{1}{\sqrt{n^2 + 2n}}$

23. $\displaystyle\sum_{n=1}^{\infty} \dfrac{1}{\sqrt{n^3 + 2n}}$

24. $\displaystyle\sum_{n=1}^{\infty} \dfrac{n + 1}{n(n + 2)}$

25. $\displaystyle\sum_{n=1}^{\infty} \dfrac{1}{(n^3 + 2n)^{1/3}}$

26. $\displaystyle\sum_{n=1}^{\infty} \dfrac{n!}{e^n}$

27. $\displaystyle\sum_{n=1}^{\infty} \dfrac{(-1)^n n}{\ln n}$

28. $\displaystyle\sum_{n=1}^{\infty} \dfrac{(-1)^n \sqrt{n}}{n + 1}$

29. $\displaystyle\sum_{n=1}^{\infty} \dfrac{(-1)^n 1 \cdot 3 \cdot 5 \cdots (2n - 1)}{2 \cdot 4 \cdot 6 \cdots (2n)(2n + 1)}$

30. $\displaystyle\sum_{n=1}^{\infty} \frac{1 \cdot 3 \cdot 5 \cdots (2n - 1)}{2 \cdot 5 \cdot 8 \cdots (3n - 1)}$

31. $\displaystyle\sum_{n=1}^{\infty} \left(\frac{1}{n^2} - \frac{1}{n} \right)$ **32.** $\displaystyle\sum_{n=1}^{\infty} \left(\frac{1}{n^2} - \frac{1}{2^n} \right)$

In Exercises 33–36, find the interval of convergence of the power series.

33. $\displaystyle\sum_{n=0}^{\infty} \frac{(-1)^n (x - 2)^n}{(n + 1)^2}$ **34.** $\displaystyle\sum_{n=0}^{\infty} (2x)^n$

35. $\displaystyle\sum_{n=0}^{\infty} n!(x - 2)^n$ **36.** $\displaystyle\sum_{n=0}^{\infty} \frac{(x - 2)^n}{2^n}$

In Exercises 37–42, find the power series for $f(x)$ centered at c.

37. $f(x) = \sin x, \ c = \dfrac{3\pi}{4}$ **38.** $f(x) = \sqrt{x}, \ c = 4$

39. $f(x) = 3^x, \ c = 0$

40. $f(x) = \csc x, \ c = \dfrac{\pi}{2}$ (first three terms)

41. $f(x) = \dfrac{1}{x}, \ c = -1$ **42.** $f(x) = \cos x, \ c = -\dfrac{\pi}{4}$

In Exercises 43–46, find the series representation of the function defined by the given integral.

43. $\displaystyle\int_0^x \frac{\sin t}{t} \, dt$ **44.** $\displaystyle\int_0^x \cos \frac{\sqrt{t}}{2} \, dt$

45. $\displaystyle\int_0^x \frac{\ln (t + 1)}{t} \, dt$ **46.** $\displaystyle\int_0^x \frac{e^t - 1}{t} \, dt$

In Exercises 47–50, use a Taylor polynomial to approximate the given value with an error of less than 0.001.

47. $\sin 95°$ **48.** $\cos (0.75)$

49. $\ln (1.75)$ **50.** $e^{-0.25}$

In Exercises 51 and 52, show that the function defined by the series is a solution to the specified equation.

51. $y = \displaystyle\sum_{n=0}^{\infty} (-1)^n \frac{x^{2n}}{4^n (n!)^2}$ Bessel function

    $x^2 y'' + xy' + x^2 y = 0$

52. $y = \displaystyle\sum_{n=0}^{\infty} \frac{(-3)^n x^{2n}}{2^n n!}$

    $y'' + 3xy' + 3y = 0$

53. Sketch the graphs of $f(x) = e^x$ and its fifth Taylor polynomial on the same axes.

54. A ball is dropped from a height of 8 feet. Each time it drops h feet, it rebounds $0.7h$ feet. Find the total distance traveled by the ball.

Chapter 10 Application

A communications satellite receiving dish, because of the distance involved, may receive signal strength from a satellite on the order of only a few billionths of a watt. To increase the strength of the received signal, receiving terminal designs use the reflective property of a parabola.

Communications Receiving Station

To form a model for a satellite-signal receiving dish, imagine that the parabola given by

$$x^2 = 4py$$

is revolved about the y-axis. Each satellite signal that enters the receiving dish, parallel to the y-axis, is reflected to the *focus* of the parabola, p units up from its vertex. For example, the parabola given by

$$x^2 = 20y$$

has its vertex at the origin, and since $4p = 20$, the focus is at $(0, 5)$.

Larger dishes receive stronger signals, but they are more expensive than smaller dishes. Part of the cost of constructing a receiving dish is related to its surface area. To find the surface area, we use the formula for the area of a surface of revolution. For example, if a segment of the parabola $x^2 = 20y$ is revolved about the y-axis to form a dish of radius r, the surface area is

$$2\pi \int_0^r x\sqrt{1 + \left(\frac{x}{10}\right)^2}\, dx = \frac{\pi}{15}[(100 + r^2)^{3/2} - 1000].$$

This means that the surface area is roughly proportional to the cube of the radius of the dish.

Satellite signals

Parabolic reflector

Chapter Overview

This chapter covers **conic sections**—classic curves that were discovered by the ancient Greeks. We will study conic sections from two points of view: as the intersection of a plane and a double-napped cone, and as the graph of a second-degree equation of the form

$$Ax^2 + Bxy + Cy^2 + Dx + Ey + F = 0.$$

There are four basic types of conics. The first type, circles, was discussed in Chapter 1. The other three basic conics are discussed in Sections 10.1, 10.2, and 10.3.

Conic	Standard Form
Circle:	$(x - h)^2 + (y - k)^2 = r^2$
Parabola:	$(x - h)^2 = 4p(y - k)$
Ellipse:	$\dfrac{(x - h)^2}{a^2} + \dfrac{(y - k)^2}{b^2} = 1$
Hyperbola:	$\dfrac{(x - h)^2}{a^2} - \dfrac{(y - k)^2}{b^2} = 1$

Conics whose axes are horizontal or vertical have simpler equations than do conics whose axes are not horizontal or vertical, and in the first three sections, we restrict our discussion to the simpler type. Then, in Section 10.4, we look at a technique for rotating the coordinate axes so that they are parallel to the axes of a given conic.

Conic sections occur in many applications, from planetary orbits to bridge construction. Throughout the chapter we will describe several of these.

See Exercise 56, Section 10.1

Conic Sections

10.1 Parabolas

Conic sections ▪ Parabolas ▪ Standard equation of a parabola ▪ Applications

Hypatia

The Greeks discovered conic sections sometime between 600 and 300 B.C. By the beginning of the Alexandrian period, enough was known about conics for Apollonius (262–190 B.C.) to produce an eight-volume work on the subject. Later, toward the end of the Alexandrian period, Hypatia (370–415 A.D.) wrote a textbook entitled *On the Conics of Apollonius*. Her death marked the end of major mathematical discoveries in Europe for several hundred years.

The early Greeks were largely concerned with the geometrical properties of conics. It was not until 1900 years later, in the early seventeenth century, that the broader applicability of conics became apparent, and conics then played a prominent role in the development of calculus.

Each **conic section** (or simply **conic**) can be described as the intersection of a plane and a double-napped cone. Notice from Figure 10.1 that in the formation of the four basic conics, the intersecting plane does not pass through the vertex of the cone. When the plane does pass through the vertex, we call the resulting figure a **degenerate conic,** as shown in Figure 10.2.

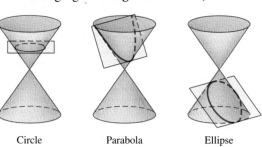

Circle   Parabola   Ellipse   Hyperbola

FIGURE 10.1   Conic Sections

Point   Line   Two intersecting lines

FIGURE 10.2   Degenerate Conics

627

There are several ways to study conics. We could begin as the Greeks did by defining the conics in terms of the intersections of planes and cones, or we could define them algebraically in terms of the general second-degree equation

$$Ax^2 + Bxy + Cy^2 + Dx + Ey + F = 0$$ General second-degree equation

a procedure to be discussed in Section 10.4. However, a third approach, in which each of the conics is defined as a **locus** (collection) of points satisfying a certain geometric property, suits our needs best. For example, in Section 1.2, we defined a circle as the collection of all points (x, y) that are equidistant from a fixed point (h, k). This locus definition easily produced the standard equation of a circle,

$$(x - h)^2 + (y - k)^2 = r^2.$$ Standard equation of a circle

Parabolas

In this and the following two sections, we give similar definitions to the other three types of conics. We will also identify practical geometric properties used in the construction of objects such as bridges, searchlights, telescopes, and radar detectors. Later, in multivariable calculus, we will often use conics in examples.

DEFINITION OF A PARABOLA

A **parabola** is the set of all points (x, y) that are equidistant from a fixed line (**directrix**) and a fixed point (**focus**) not on the line.

FIGURE 10.3

The midpoint between the focus and the directrix is called the **vertex,** and the line passing through the focus and the vertex is called the **axis** of the parabola. Note in Figure 10.3 that a parabola is symmetric with respect to its axis.

Using the definition of a parabola, we derive the following theorem, which gives the **standard form** of the equation of a parabola whose directrix is parallel to the x-axis or to the y-axis.

THEOREM 10.1 STANDARD EQUATION OF A PARABOLA

The **standard form** of the equation of a parabola with vertex (h, k) and directrix $y = k - p$ is

$$(x - h)^2 = 4p(y - k).$$ Vertical axis

For directrix $x = h - p$, the equation is

$$(y - k)^2 = 4p(x - h).$$ Horizontal axis

The focus lies on the axis p units (*directed distance*) from the vertex.

PROOF We prove only the case for which the directrix is parallel to the x-axis and the focus lies above the vertex, as shown in Figure 10.4(a). If (x, y) is any point on the parabola, then by definition (x, y) is equidistant from the focus $(h, k + p)$ and the directrix $y = k - p$, and we have the following.

$$\sqrt{(x - h)^2 + [y - (k + p)]^2} = |y - (k - p)|$$
$$(x - h)^2 + [y - (k + p)]^2 = [y - (k - p)]^2$$
$$(x - h)^2 + y^2 - 2y(k + p) + (k + p)^2 = y^2 - 2y(k - p) + (k - p)^2$$
$$(x - h)^2 - 2py + 2pk = 2py - 2pk$$
$$(x - h)^2 = 4p(y - k)$$

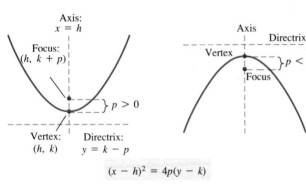

$(x - h)^2 = 4p(y - k)$

(a) Vertical axis: $p > 0$

(b) Vertical axis: $p < 0$

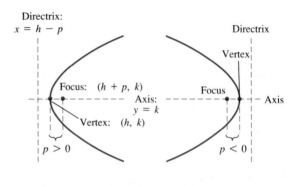

$(y - k)^2 = 4p(x - h)$

(c) Horizontal axis: $p > 0$ (d) Horizontal axis: $p < 0$

Parabolic Orientations

FIGURE 10.4

EXAMPLE 1 Finding the standard equation for a parabola

Find the standard form of the equation of the parabola with vertex $(2, 1)$ and focus $(2, 4)$.

SOLUTION

Since the axis of the parabola is vertical, we consider the equation

$$(x - h)^2 = 4p(y - k)$$

where $h = 2$, $k = 1$, and $p = 3$. Thus the standard form is

$$(x - 2)^2 = 12(y - 1).$$

The graph of this parabola is shown in Figure 10.5. ▭

REMARK By expanding the standard equation in Example 1, we obtain the more common quadratic form $y = \frac{1}{12}(x^2 - 4x + 16)$.

$(x - 2)^2 = 12(y - 1)$

FIGURE 10.5

EXAMPLE 2   Finding the vertex and focus

Find the focus of the parabola given by $y = -\frac{1}{2}x^2 - x + \frac{1}{2}$.

SOLUTION

To find the focus, we convert to standard form by completing the square.

$$y = \frac{1}{2}(1 - 2x - x^2)$$

$$2y = 1 - (x^2 + 2x + \quad)$$

$$2y = 2 - (x^2 + 2x + 1)$$

$$(x + 1)^2 = -2(y - 1) \qquad \text{Standard form}$$

Comparing this equation to $(x - h)^2 = 4p(y - k)$, we conclude that

$$h = -1, \qquad k = 1, \qquad \text{and} \qquad p = -\frac{1}{2}.$$

Since p is negative, the parabola opens downward, as shown in Figure 10.6. Therefore, the focus of the parabola is

$$(h, k + p) = \left(-1, \frac{1}{2}\right). \qquad \text{Focus}$$

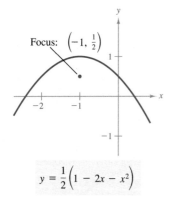

Focus: $\left(-1, \frac{1}{2}\right)$

$y = \frac{1}{2}\left(1 - 2x - x^2\right)$

FIGURE 10.6

If the vertex of a parabola is at the origin, then the standard form

$$(x - h)^2 = 4p(y - k) \qquad \text{Vertex at } (h, k)$$

simplifies to

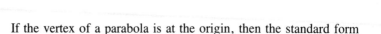

$$x^2 = 4py \qquad \text{Vertex at origin}$$

as demonstrated in the next example.

EXAMPLE 3   Vertex at the origin

Find the standard equation of the parabola with vertex at the origin and focus at (2, 0).

SOLUTION

The axis of the parabola is horizontal, passing through (0, 0) and (2, 0), as shown in Figure 10.7. Thus, we consider the standard form

$$y^2 = 4px$$

where $h = k = 0$ and $p = 2$. Therefore, the equation is

$$y^2 = 8x.$$

$y^2 = 8x$

Focus: (2, 0)

FIGURE 10.7

A line segment that passes through the focus of a parabola and has endpoints on the parabola is called a **focal chord.** The specific focal chord perpendicular to the axis of the parabola is called the **latus rectum.** In the next example we determine the length of the latus rectum and the length of the corresponding intercepted arc.

EXAMPLE 4   Focal chord length and arc length

Find the length of the latus rectum of the parabola given by $x^2 = 4py$. Then find the length of the parabolic arc intercepted by the latus rectum.

SOLUTION

Since the latus rectum passes through the focus $(0, p)$ and is perpendicular to the y-axis, the coordinates of its endpoints are $(-x, p)$ and (x, p). Substituting p for y in the equation of the parabola, we have

$$x^2 = 4p(p) = 4p^2 \quad \Longrightarrow \quad x = \pm 2p.$$

Thus, the endpoints of the latus rectum are $(-2p, p)$ and $(2p, p)$, and we conclude that its length is $4p$, as shown in Figure 10.8. In contrast, the length of the intercepted arc is given by the following.

$$
\begin{aligned}
s &= \int_{-2p}^{2p} \sqrt{1 + (y')^2}\, dx = 2 \int_{0}^{2p} \sqrt{1 + \left(\frac{x}{2p}\right)^2}\, dx \\
&= \frac{1}{p} \int_{0}^{2p} \sqrt{4p^2 + x^2}\, dx \\
&= \frac{1}{2p} \left[x\sqrt{4p^2 + x^2} + 4p^2 \ln \left| x + \sqrt{4p^2 + x^2} \right| \right]_{0}^{2p} \\
&= \frac{1}{2p} [2p\sqrt{8p^2} + 4p^2 \ln (2p + \sqrt{8p^2}) - 4p^2 \ln (2p)] \\
&= 2p[\sqrt{2} + \ln (1 + \sqrt{2})] \approx 4.59p
\end{aligned}
$$

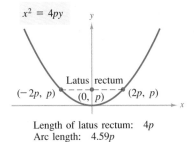

$x^2 = 4py$

$(-2p, p)$ Latus rectum $(2p, p)$

$(0, p)$

Length of latus rectum: $4p$
Arc length: $4.59p$

Latus Rectum and Intercepted Arc

FIGURE 10.8

Applications

One widely used property of a parabola is its reflective property. In physics, a surface is called **reflective** if the tangent line at any point on the surface makes equal angles with an incoming ray and the resulting outgoing ray. The angle corresponding to the incoming ray is called the **angle of incidence,** and the angle corresponding to the outgoing ray is called the **angle of reflection.** One example of a reflective surface is a flat mirror. Another type of reflective surface is that formed by revolving a parabola about its axis. A special property of parabolic reflectors is that they allow us to direct all incoming rays parallel to the axis through the focus of the parabola—this is the principle behind the design of the parabolic mirrors used in reflecting telescopes. Conversely, all light rays emanating from the focus of a parabolic reflector used in a flashlight are parallel, as shown in Figure 10.9.

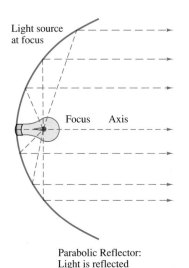

Light source at focus

Focus   Axis

Parabolic Reflector: Light is reflected in parallel rays.

FIGURE 10.9

THEOREM 10.2 REFLECTIVE PROPERTY OF A PARABOLA	The tangent line to a parabola at the point P makes equal angles with the following two lines: 1. the line passing through P and the focus and 2. the line passing through P parallel to the axis of the parabola.

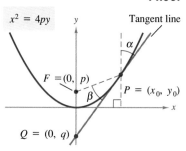

FIGURE 10.10

PROOF We consider the point $P = (x_0, y_0)$ on the parabola given by $x^2 = 4py$, as shown in Figure 10.10. Using differentiation, we can determine the tangent line at P to have a slope of $x_0/2p$, which implies that the equation of the tangent line is

$$y - y_0 = \frac{x_0}{2p}(x - x_0). \qquad \text{Tangent line at } P$$

If P lies at the vertex of the parabola, then the other two lines referred to in the theorem coincide and the theorem is trivially valid. Thus, we assume P is a point other than the vertex. This implies that the tangent line intersects the y-axis at a point $Q = (0, q)$, different from P. Using Figure 10.10, we can show that the angles α and β are equal by showing that the triangle with vertices at F, P, and Q is isosceles. Using the equation of the tangent line at $x = 0$, we see that

$$q = y_0 - \frac{x_0{}^2}{2p} = \frac{x_0{}^2}{4p} - \frac{x_0{}^2}{2p} = -\frac{x_0{}^2}{4p}$$

which implies that the length of $\overline{FQ}$ is

$$p - q = p + \frac{x_0{}^2}{4p}. \qquad \text{Length of } \overline{FQ}$$

Moreover, using the distance formula, we find the length of $\overline{FP}$ to be

$$\sqrt{x_0{}^2 + (y_0 - p)^2} = \sqrt{x_0{}^2 + \left(\frac{x_0{}^2}{4p} - p\right)^2} = \sqrt{\left(\frac{x_0{}^2}{4p} + p\right)^2}$$

$$= \frac{x_0{}^2}{4p} + p. \qquad \text{Length of } \overline{FP}$$

Therefore, ΔFQP is isosceles, and we conclude that $\alpha = \beta$.

EXERCISES for Section 10.1

In Exercises 1–6, match the equation with the correct graph. [The graphs are labeled (a)–(f).]

1. $y^2 = 4x$
2. $x^2 = -2y$
3. $x^2 = 8y$
4. $y^2 = -12x$
5. $(y - 1)^2 = 4(x - 2)$
6. $(x + 3)^2 = -2(y - 2)$

(a)

(b)

(c)

(d)

(e)

(f)

In Exercises 7–26, find the vertex, focus, and directrix of the parabola, and sketch its graph.

7. $y = 4x^2$
8. $y = 2x^2$
9. $y^2 = -6x$
10. $y^2 = 3x$
11. $x^2 + 8y = 0$
12. $x + y^2 = 0$
13. $(x - 1)^2 + 8(y + 2) = 0$
14. $(x + 3) + (y - 2)^2 = 0$
15. $\left(y + \dfrac{1}{2}\right)^2 = 2(x - 5)$
16. $\left(x + \dfrac{1}{2}\right)^2 - 4(y - 3) = 0$
17. $y = \dfrac{1}{4}(x^2 - 2x + 5)$
18. $y = -\dfrac{1}{6}(x^2 + 4x - 2)$
19. $4x - y^2 - 2y - 33 = 0$
20. $y^2 + x + y = 0$
21. $y^2 + 6y + 8x + 25 = 0$
22. $x^2 - 2x + 8y + 9 = 0$
23. $y^2 - 4y - 4x = 0$
24. $y^2 - 4x - 4 = 0$
25. $x^2 + 4x + 4y - 4 = 0$
26. $y^2 + 4y + 8x - 12 = 0$

In Exercises 27–40, find an equation of the specified parabola.

27. Vertex: $(0, 0)$
  Focus: $\left(0, -\dfrac{3}{2}\right)$
28. Vertex: $(0, 0)$
  Focus: $(2, 0)$
29. Vertex: $(3, 2)$
  Focus: $(1, 2)$
30. Vertex: $(-1, 2)$
  Focus: $(-1, 0)$
31. Vertex: $(0, -4)$
  Directrix: $y = 2$
32. Vertex: $(-2, 1)$
  Directrix: $x = 1$
33. Focus: $(0, 0)$
  Directrix: $y = 4$
34. Focus: $(2, 2)$
  Directrix: $x = -2$

35. Axis parallel to y-axis, graph passes through $(0, 3)$, $(3, 4)$, and $(4, 11)$
36. Axis parallel to x-axis, graph passes through $(4, -2)$, $(0, 0)$, and $(3, -3)$

37.

38.

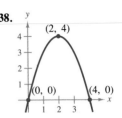

39. Directrix $y = 1$, length of latus rectum is 8, opens downward
40. Directrix $y = -2$, endpoints of latus rectum $(0, 2)$ and $(8, 2)$

41. The filament of a flashlight bulb is $\dfrac{3}{8}$ inch from the vertex of the parabolic reflector and is located at the focus. Find an equation of a cross section of the reflector. (Assume that it is directed toward the right and the vertex is at the origin.)
42. The receiver in a parabolic television dish antenna is 3 feet from the vertex and is located at the focus. Find an equation of a cross section of the reflector. (Assume that the dish is directed upward and the vertex is at the origin.)

In Exercises 43 and 44, find an equation of the tangent line to the parabola at the specified point.

Parabola	Point
43. $x^2 = 2y$	$(4, 8)$
44. $y^2 = -8x$	$(-2, 4)$

In Exercises 45 and 46, find the coordinates of the centroid of the region bounded by the graphs of the given equations.

45. $y^2 = 4x$, $x = 1$
46. $x^2 = \dfrac{1}{2}y$, $y = 8$

47. Find the point on the graph of $x^2 = 8y$ that is closest to the focus of the parabola.
48. Find equations of the tangent lines to $y^2 = 2x$ that pass through the point $(-4, 1)$.

In Exercises 49–52, find the arc length of the parabola over the given interval.

Function	Interval
49. $x^2 + 8y = 0$	$0 \le x \le 4$
50. $x + y^2 = 0$	$0 \le y \le 2$
51. $4x - y^2 = 0$	$0 \le y \le 4$
52. $x^2 - 2y = 0$	$0 \le x \le 1$

53. A cable of a parabolic suspension bridge is suspended between two towers that are 400 feet apart and 50 feet above the roadway (see figure). The cable touches the roadway midway between the towers. Find the length of the cable.

FIGURE FOR 53

54. An earth satellite in a circular orbit 100 miles high has a speed of approximately 17,470 miles per hour. If this speed is multiplied by $\sqrt{2}$, then the satellite will have the minimum velocity necessary to escape the earth's gravitational force and will follow a parabolic path with the center of the earth as the focus (see figure). Find an equation for the resulting parabolic path. (Assume that the radius of the earth is 4000 miles.)

55. Water is flowing from a horizontal pipe 48 feet above the ground with a horizontal velocity of 10 feet per second. The falling stream of water forms a parabola whose vertex, (0, 48), is at the end of the pipe (see figure). Where does the water hit the ground?

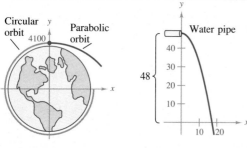

FIGURE FOR 54   FIGURE FOR 55

56. As discussed in the Chapter 10 Application, a satellite-signal receiving dish is formed by revolving the parabola given by the graph of $x^2 = 20y$ about the y-axis. If the radius of the dish is r feet, verify that the surface area of the dish is given by

$$2\pi \int_0^r x\sqrt{1 + \left(\frac{x}{10}\right)^2}\, dx = \frac{\pi}{15}[(100 + r^2)^{3/2} - 1000].$$

57. Find the surface area of the satellite dish in Exercise 56 if its radius is (a) $r = 6$ feet and (b) $r = 10$ feet.

In Exercises 58 and 59, use a computer or graphics calculator to sketch the graph of the parabola and find any x-intercepts.

58. $x^2 - 6.2x - 2.5y + 7.61 = 0$
59. $y^2 - 4y + 4.1x - 4.2 = 0$

60. Prove that the area enclosed by a parabola and a line parallel to its directrix is two-thirds the area of the circumscribed rectangle.
61. Find an equation of the tangent line to the parabola $y = ax^2$ at $x = x_0$. Prove that the x-intercept of this tangent line is $(x_0/2, 0)$.
62. Prove that any two distinct tangent lines to a parabola intersect.
63. Demonstrate the result of Exercise 62 by finding the point of intersection of the tangent lines to the parabola $x^2 - 4x - 4y = 0$ at the points (0, 0) and (6, 3).
64. Prove that if any two tangent lines to a parabola intersect at right angles, then their point of intersection must lie on the directrix.
65. Demonstrate the result of Exercise 64 by showing that the tangent lines to the parabola $x^2 - 4x - 4y + 8 = 0$ at the points $(-2, 5)$ and $\left(3, \frac{5}{4}\right)$ intersect at right angles and that the point of intersection lies on the directrix.
66. Prove that two tangent lines to a parabola intersect at right angles if and only if the focus of the parabola lies on the line segment connecting the two points of tangency.

10.2 Ellipses

Ellipses ■ Standard equation of an ellipse ■ Eccentricity

Nicholas Copernicus

More than a thousand years after the close of the Alexandrian period of Greek mathematics, Western civilization finally began a Renaissance of mathematical and scientific discovery. One of the principal figures in this rebirth was the Polish astronomer Nicholas Copernicus (1473–1543). In his work *On the Revolutions of the Heavenly Spheres*, Copernicus claimed that all of the planets, including the earth, revolved about the sun in circular orbits. Although some of Copernicus's claims were invalid, the controversy set off by his heliocentric theory motivated astronomers to search for a mathematical model to explain the observed movements of the sun and planets. The first to find the correct model was the German astronomer Johannes Kepler (1571–1630). Kepler discovered that the planets move about the sun in elliptical orbits, with the sun not as the center but as a focal point of the orbit.

The use of ellipses to explain the movement of the planets is only one of many practical and aesthetic uses. As in the previous section, we begin our study of this second type of conic by defining it as a locus of points. Now, however, we use *two* focal points rather than one.

DEFINITION OF AN ELLIPSE
An **ellipse** is the set of all points (x, y) the sum of whose distances from two distinct fixed points (**foci**) is constant. (See Figure 10.11.)

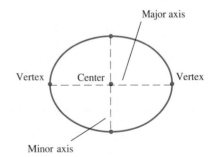

FIGURE 10.11 $d_1 + d_2 =$ constant

FIGURE 10.12

The line through the foci intersects the ellipse at two points called the **vertices.** The chord joining the vertices is called the **major axis,** and its midpoint is called the **center** of the ellipse. The chord perpendicular to the major axis at the center is called the **minor axis** of the ellipse.

REMARK   You can visualize the definition of an ellipse by imagining two thumbtacks placed at the foci, as shown in Figure 10.12. If the ends of a fixed length of string are fastened to the thumbtacks and the string is drawn taut with a pencil, the path traced by the pencil will be an ellipse.

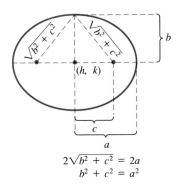

$$2\sqrt{b^2 + c^2} = 2a$$
$$b^2 + c^2 = a^2$$

FIGURE 10.13

To derive the standard form of the equation of an ellipse, consider the ellipse in Figure 10.13 with the following points.

center: (h, k) vertices: $(h \pm a, k)$ foci: $(h \pm c, k)$

The sum of the distances from any point on the ellipse to the two foci is constant. At a vertex, this constant sum is

$$(a + c) + (a - c) = 2a \qquad \text{Length of major axis}$$

or the length of the major axis. Now, for *any* point (x, y) on the ellipse, the sum of the distances between (x, y) and the two foci must also be $2a$:

$$\sqrt{[x - (h - c)]^2 + (y - k)^2} + \sqrt{[x - (h + c)]^2 + (y - k)^2} = 2a$$

which, after expanding and regrouping, reduces to

$$(a^2 - c^2)(x - h)^2 + a^2(y - k)^2 = a^2(a^2 - c^2).$$

Finally, in Figure 10.13 we can see that $b^2 = a^2 - c^2$, which implies that the equation of the ellipse is

$$b^2(x - h)^2 + a^2(y - k)^2 = a^2 b^2$$
$$\frac{(x - h)^2}{a^2} + \frac{(y - k)^2}{b^2} = 1.$$

Choosing a vertical major axis would have produced a similar equation. Both results are summarized in the following theorem.

**THEOREM 10.3
STANDARD EQUATION
OF AN ELLIPSE**

The standard form of the equation of an ellipse, with center (h, k) and major and minor axes of lengths $2a$ and $2b$, where $a > b$, is

$$\frac{(x - h)^2}{a^2} + \frac{(y - k)^2}{b^2} = 1 \qquad \text{Major axis is horizontal}$$

$$\frac{(x - h)^2}{b^2} + \frac{(y - k)^2}{a^2} = 1. \qquad \text{Major axis is vertical}$$

The foci lie on the major axis, c units from the center, with $c^2 = a^2 - b^2$.

Figure 10.14 shows an ellipse's vertical and horizontal orientations.

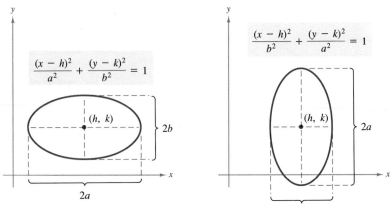

FIGURE 10.14   Standard Equations for Ellipses

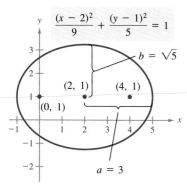

$$\frac{(x - 2)^2}{9} + \frac{(y - 1)^2}{5} = 1$$

$b = \sqrt{5}$

(2, 1)   (4, 1)

(0, 1)

$a = 3$

FIGURE 10.15

EXAMPLE 1   Finding the standard equation for an ellipse

Find the standard form of the equation of the ellipse having foci at $(0, 1)$ and $(4, 1)$, with a major axis of length 6, as shown in Figure 10.15.

SOLUTION

Since the foci occur at $(0, 1)$ and $(4, 1)$, the center of the ellipse is $(2, 1)$. This implies that the distance from the center to one of the foci is $c = 2$, and since $2a = 6$ we know that $a = 3$. Now, using $c^2 = a^2 - b^2$, we have

$$b = \sqrt{a^2 - c^2} = \sqrt{9 - 4} = \sqrt{5}.$$

Since the major axis is parallel to the x-axis, the standard equation is

$$\frac{(x - 2)^2}{9} + \frac{(y - 1)^2}{5} = 1.$$

EXAMPLE 2   Completing the square to find the standard equation

Find the center, vertices, and foci of the ellipse given by

$$4x^2 + y^2 - 8x + 4y - 8 = 0.$$

SOLUTION

By completing the square, we can write the given equation in standard form.

$$4x^2 + y^2 - 8x + 4y - 8 = 0$$
$$4(x^2 - 2x + 1) + (y^2 + 4y + 4) = 8 + 4 + 4$$
$$4(x - 1)^2 + (y + 2)^2 = 16$$
$$\frac{(x - 1)^2}{4} + \frac{(y + 2)^2}{16} = 1$$

Thus, the major axis is parallel to the y-axis, where $h = 1$, $k = -2$, $a = 4$, $b = 2$, and $c = \sqrt{16 - 4} = 2\sqrt{3}$. Therefore, we obtain the following.

center: $(1, -2)$	vertices: $(1, -6)$	foci: $(1, -2 - 2\sqrt{3})$
	$(1, 2)$	$(1, -2 + 2\sqrt{3})$

The graph of the ellipse is shown in Figure 10.16.

y

Vertex

Focus

x

-2   2   4

Center

Focus

-6

Vertex

$4x^2 + y^2 - 8x + 4y - 8 = 0$

FIGURE 10.16

REMARK   If the constant term $F = -8$ in the equation in Example 2 had been greater than or equal to 8, we would have obtained one of the following degenerate cases.

1. Single point, $(1, -2)$: $\dfrac{(x - 1)^2}{4} + \dfrac{(y + 2)^2}{16} = 0$ $F = 8$

2. No solution points: $\dfrac{(x - 1)^2}{4} + \dfrac{(y + 2)^2}{16} < 0$ $F > 8$

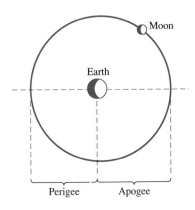

FIGURE 10.17

EXAMPLE 3 An application involving an elliptical orbit

The moon orbits the earth in an elliptical path with the center of the earth at one focus, as shown in Figure 10.17. The major and minor axes of the orbit have lengths of 768,806 kilometers and 767,746 kilometers, respectively. Find the greatest and least distances (the apogee and perigee) from the earth's center to the moon's center.

SOLUTION

Since $2a = 768,806$ and $2b = 767,746$, we have $a = 384,403$, $b = 383,873$, and

$$c = \sqrt{a^2 - b^2} \approx 20,179.$$

Therefore, the greatest distance between the center of the earth and the center of the moon is

$$a + c \approx 404,582 \text{ km}$$

and the least distance is

$$a - c \approx 364,224 \text{ km}.$$

EXAMPLE 4 Finding the area of an ellipse

Find the area of an ellipse whose major and minor axes have lengths of $2a$ and $2b$, respectively.

SOLUTION

For simplicity, we choose an ellipse centered at the origin

$$\frac{x^2}{a^2} + \frac{y^2}{b^2} = 1.$$

Then, using symmetry, we can find the area of the entire region lying within the ellipse by finding the area of the region in the first quadrant and multiplying by 4, as indicated in Figure 10.18. In the first quadrant, we have

$$y = \frac{b}{a}\sqrt{a^2 - x^2}$$

which implies that the entire area is

$$A = 4 \int_0^a \frac{b}{a}\sqrt{a^2 - x^2} \, dx.$$

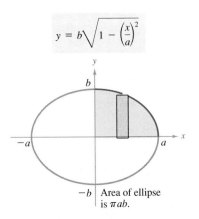

FIGURE 10.18

Using the trigonometric substitution $x = a \sin \theta$, we have

$$A = \frac{4b}{a} \int_0^{\pi/2} a^2 \cos^2 \theta \, d\theta$$

$$= 4ab \int_0^{\pi/2} \frac{1 + \cos 2\theta}{2} \, d\theta$$

$$= 2ab \left[\theta + \frac{\sin 2\theta}{2} \right]_0^{\pi/2}$$

$$= 2ab \left(\frac{\pi}{2} \right) = \pi ab.$$

REMARK Note that if $a = b$, then the formula for the area of an ellipse reduces to the area of a circle.

In the previous section we looked at a reflective property of parabolas. Ellipses have a similar reflective property. You are asked to prove the following theorem in Exercise 60.

THEOREM 10.4 **REFLECTIVE PROPERTY** **OF AN ELLIPSE**	The tangent line to an ellipse at the point P makes equal angles with the lines through P and the foci.

Eccentricity

One of the reasons that astronomers had difficulty in detecting that the orbits of the planets are ellipses is that the foci of the planetary orbits are relatively close to the center of the sun, making the orbits nearly circular. To measure the ovalness of an ellipse, we use the concept of **eccentricity.**

DEFINITION OF ECCENTRICITY **OF AN ELLIPSE**	The **eccentricity** e of an ellipse is given by the ratio $$e = \frac{c}{a}.$$

To see how this ratio is used to describe the shape of an ellipse, note that since the foci of an ellipse are located along the major axis between the vertices and the center, it follows that

$$0 < c < a.$$

(a) $\dfrac{c}{a}$ is small.

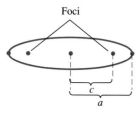

(b) $\dfrac{c}{a}$ is close to 1.

FIGURE 10.19

For an ellipse that is nearly circular, the foci are close to the center and the ratio c/a is small, as shown in Figure 10.19. For an elongated ellipse, the foci are close to the vertices and the ratio is close to 1. Note that $0 < e < 1$ for every ellipse.

The orbit of the moon has an eccentricity of $e = 0.0549$, and the eccentricities of the nine planetary orbits are as follows.

Mercury:	$e = 0.2056$	Saturn:	$e = 0.0543$
Venus:	$e = 0.0068$	Uranus:	$e = 0.0460$
Earth:	$e = 0.0167$	Neptune:	$e = 0.0082$
Mars:	$e = 0.0934$	Pluto:	$e = 0.2481$
Jupiter:	$e = 0.0484$		

In Example 4, we found that the formula for the *area* of an ellipse is $A = \pi a b$. However, it is not so simple to find the *circumference* of an ellipse. The next example shows how to use eccentricity to set up an "elliptic integral" for the circumference of an ellipse.

EXAMPLE 5   Finding the circumference of an ellipse

Show that the circumference of the ellipse given by $(x^2/a^2) + (y^2/b^2) = 1$ is

$$4a \int_0^{\pi/2} \sqrt{1 - e^2 \sin^2 \theta}\, d\theta. \qquad e = c/a$$

SOLUTION

Since the given ellipse is symmetric with respect to both the x-axis and the y-axis, we know that its circumference C is four times the arc length of $y = (b/a)\sqrt{a^2 - x^2}$ in the first quadrant. Thus

$$C = 4 \int_0^a \sqrt{1 + (y')^2}\, dx = 4 \int_0^a \sqrt{1 + \frac{b^2 x^2}{a^2(a^2 - x^2)}}\, dx.$$

Using the trigonometric substitution $x = a \sin \theta$, we obtain

$$C = 4 \int_0^{\pi/2} \sqrt{1 + \frac{b^2 \sin^2 \theta}{a^2 \cos^2 \theta}}(a \cos \theta)\, d\theta$$

$$= 4 \int_0^{\pi/2} \sqrt{a^2 \cos^2 \theta + b^2 \sin^2 \theta}\, d\theta$$

$$= 4 \int_0^{\pi/2} \sqrt{a^2(1 - \sin^2 \theta) + b^2 \sin^2 \theta}\, d\theta$$

$$= 4 \int_0^{\pi/2} \sqrt{a^2 - (a^2 - b^2) \sin^2 \theta}\, d\theta.$$

Since $e^2 = c^2/a^2 = (a^2 - b^2)/a^2$, we can rewrite this integral as

$$C = 4a \int_0^{\pi/2} \sqrt{1 - e^2 \sin^2 \theta}\, d\theta.$$

A great deal of time has been devoted to the study of elliptic integrals. Such integrals generally do not have elementary antiderivatives. To find the circumference of an ellipse, we usually resort to an approximation technique, as illustrated in the following example.

EXAMPLE 6   Approximating the value of an elliptic integral

Use the elliptic integral in Example 5 to approximate the circumference of the ellipse

$$\frac{x^2}{25} + \frac{y^2}{16} = 1.$$

SOLUTION

Since $e^2 = c^2/a^2 = (a^2 - b^2)/a^2 = 9/25$, we have

$$C = (4)(5) \int_0^{\pi/2} \sqrt{1 - \frac{9 \sin^2 \theta}{25}} \, d\theta.$$

Applying Simpson's Rule with $n = 4$ produces

$$C \approx 20\left(\frac{\pi}{6}\right)\left(\frac{1}{4}\right)[1 + 4(0.9733) + 2(0.9055) + 4(0.8323) + 0.8] \approx 28.36.$$

EXERCISES for Section 10.2

In Exercises 1–6, match the equation with the correct graph. [The graphs are labeled (a)–(f).]

1. $\dfrac{x^2}{1} + \dfrac{y^2}{9} = 1$ **2.** $\dfrac{x^2}{9} + \dfrac{y^2}{1} = 1$

3. $\dfrac{x^2}{9} + \dfrac{y^2}{4} = 1$ **4.** $\dfrac{x^2}{9} + \dfrac{y^2}{9} = 1$

5. $\dfrac{(x-2)^2}{16} + \dfrac{(y+1)^2}{4} = 1$

6. $\dfrac{(x+2)^2}{4} + \dfrac{(y+2)^2}{25} = 1$

(c)

(d)

(a)

(b)

(e)

(f)
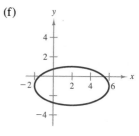

In Exercises 7–26, find the center, foci, vertices, and eccentricity of the ellipse, and sketch its graph.

7. $\dfrac{x^2}{25} + \dfrac{y^2}{16} = 1$

8. $\dfrac{x^2}{144} + \dfrac{y^2}{169} = 1$

9. $\dfrac{x^2}{16} + \dfrac{y^2}{25} = 1$

10. $\dfrac{x^2}{169} + \dfrac{y^2}{144} = 1$

11. $\dfrac{x^2}{9} + \dfrac{y^2}{5} = 1$

12. $\dfrac{x^2}{28} + \dfrac{y^2}{64} = 1$

13. $x^2 + 4y^2 = 4$

14. $5x^2 + 3y^2 = 15$

15. $3x^2 + 2y^2 = 6$

16. $5x^2 + 7y^2 = 70$

17. $4x^2 + y^2 = 1$

18. $16x^2 + 25y^2 = 1$

19. $\dfrac{(x-1)^2}{9} + \dfrac{(y-5)^2}{25} = 1$

20. $(x+2)^2 + 4(y+4)^2 = 1$

21. $9x^2 + 4y^2 + 36x - 24y + 36 = 0$

22. $9x^2 + 4y^2 - 36x + 8y + 31 = 0$

23. $16x^2 + 25y^2 - 32x + 50y + 31 = 0$

24. $9x^2 + 25y^2 - 36x - 50y + 61 = 0$

25. $12x^2 + 20y^2 - 12x + 40y - 37 = 0$

26. $36x^2 + 9y^2 + 48x - 36y + 43 = 0$

In Exercises 27–36, find an equation for the specified ellipse.

27. Center: (0, 0)
Focus: (2, 0)
Vertex: (3, 0)

28. Center: (0, 0)
Vertex: (2, 0)
Minor axis length: 3

29. Vertices: (± 5, 0)
Eccentricity: $\frac{3}{5}$

30. Vertices: (0, ± 8)
Eccentricity: $\frac{1}{2}$

31. Vertices: (3, 1), (3, 9)
Minor axis length: 6

32. Vertices: (0, 2), (4, 2)
Minor axis length: 2

33. Foci: (0, ± 5)
Major axis length: 14

34. Foci: (± 2, 0)
Major axis length: 8

35. Center: (0, 0)
Major axis: horizontal
Points on ellipse:
   (3, 1), (4, 0)

36. Center: (1, 2)
Major axis: vertical
Points on ellipse:
   (1, 6), (3, 2)

In Exercises 37 and 38, use the following definition. The **latera recta** (plural of **latus rectum**) of an ellipse are the chords passing through the foci and perpendicular to the major axis.

37. Show that the length of a latus rectum of an ellipse is $2b^2/a$.

38. With the result of Exercise 37, sketch the graph of each of the following, making use of the endpoints of the latera recta.

(a) $\dfrac{x^2}{4} + \dfrac{y^2}{1} = 1$

(b) $5x^2 + 3y^2 = 15$

In Exercises 39 and 40, determine the points at which dy/dx is zero or undefined to locate the endpoints of the major and minor axes of the ellipse.

39. $16x^2 + 9y^2 + 96x + 36y + 36 = 0$

40. $9x^2 + 4y^2 + 36x - 24y + 36 = 0$

In Exercises 41 and 42, consider a particle traveling clockwise on the given elliptical path. The particle leaves the orbit at the indicated point and travels in a straight line tangent to the ellipse. At what point will the particle cross the y-axis?

41. $\dfrac{x^2}{100} + \dfrac{y^2}{25} = 1$, $(-8, 3)$

42. $\dfrac{x^2}{16} + \dfrac{y^2}{25} = 1$, $\left(3, \dfrac{5\sqrt{7}}{4}\right)$

In Exercises 43–46, use Simpson's Rule with $n = 8$ to approximate the elliptic integral representing the circumference of the given ellipse.

43. $\dfrac{x^2}{9} + \dfrac{y^2}{16} = 1$

44. $\dfrac{x^2}{9} + \dfrac{y^2}{1} = 1$

45. $\dfrac{x^2}{3} + \dfrac{y^2}{2} = 1$

46. $\dfrac{(x-3)^2}{4} + \dfrac{(y+1)^2}{3} = 1$

In Exercises 47 and 48, find (a) the area of the region bounded by the given ellipse, (b) the volume and surface area of the solid generated by revolving the region about its major axis (prolate spheroid), and (c) the volume and surface area of the solid generated by revolving the region about its minor axis (oblate spheroid).

47. $\dfrac{x^2}{4} + \dfrac{y^2}{1} = 1$

48. $\dfrac{x^2}{16} + \dfrac{y^2}{9} = 1$

49. Find the dimensions of the rectangle of maximum area that can be inscribed in the ellipse

$$\dfrac{x^2}{a^2} + \dfrac{y^2}{b^2} = 1.$$

(Assume that the sides of the rectangle are parallel to the coordinate axes.)

50. A solid has an elliptical base given by

$$\dfrac{x^2}{25} + \dfrac{y^2}{16} = 1.$$

Cross sections perpendicular to the major axis are isosceles triangles of height 6. Find the volume of the solid.

51. Show that the tangent line to the ellipse

$$\frac{x^2}{a^2} + \frac{y^2}{b^2} = 1$$

at the point (x_0, y_0) is given by

$$\frac{x_0}{a^2}x + \frac{y_0}{b^2}y = 1.$$

52. The equation of an ellipse with its center at the origin can be written as

$$\frac{x^2}{a^2} + \frac{y^2}{a^2(1 - e^2)} = 1.$$

Show that as $e \to 0$, with a remaining fixed, the ellipse approaches a circle.

53. Show that the eccentricity of the ellipse

$$\frac{x^2}{a^2} + \frac{y^2}{b^2} = 1$$

is identical to the eccentricity of

$$\frac{(tx)^2}{a^2} + \frac{(ty)^2}{b^2} = 1$$

for any real t. Give a geometrical explanation of this result.

54. A line segment 9 inches long moves so that one endpoint is always on the y-axis and the other always on the x-axis. Find the equation of the curve traced by a point on the line segment 6 inches from the endpoint that is on the y-axis.

55. A fireplace arch is to be constructed in the shape of a semiellipse (see figure). The opening is to be 2 feet high at the center and 5 feet wide along the base. To sketch the outline of the fireplace, the contractor uses a 5-foot string tied to two thumbtacks. Where should the thumbtacks be placed?

FIGURE FOR 55

56. The length of half of the major axis of the earth's orbit is 14,957,000 kilometers, and the eccentricity is 0.0167. Find the least and greatest distances of the earth from the sun.

57. If the distances to the apogee and the perigee of an elliptical orbit of an earth satellite are measured from the center of the earth, show that the eccentricity of the orbit is given by

$$e = \frac{A - P}{A + P}$$

where A and P are the apogee and perigee distances, respectively.

58. The first artificial satellite to orbit the earth was Sputnik I (launched by Russia in 1957). Its highest point above the earth's surface was 583 miles, and its lowest point was 132 miles. Find the eccentricity of its orbit.

59. Probably the most famous of all comets, **Halley's Comet,** has an elliptical orbit with the sun at the focus. Its maximum distance from the sun is approximately 35.34 au (astronomical unit $\approx$ 92.956 $\times$ 10^6 miles), and its minimum distance (perihelion) is approximately 0.59 au. Find the eccentricity of the orbit.

60. Prove that the tangent line to an ellipse at a point P makes equal angles with the lines through P and the foci (see figure). [Hint: (a) Find the slope of the tangent line at P, (b) find the slopes of the lines through P and each focus, and (c) use the formula for the tangent of the angle between two lines.]

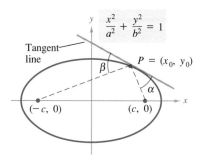

Reflective property of an ellipse: $\alpha = \beta$

FIGURE FOR 60

In Exercises 61 and 62, use a computer or graphics calculator to sketch the graph of the ellipse.

61. $x^2 + 2y^2 - 3x + 4y + 0.25 = 0$

62. $2x^2 + y^2 + 4.8x - 6.4y + 3.12 = 0$

10.3 Hyperbolas

Hyperbola ▪ Standard equation of a hyperbola ▪ Asymptotes ▪ Eccentricity ▪ Applications

The definition of a hyperbola is similar to that of an ellipse. For an ellipse the *sum* of the distances between the foci and a point on the ellipse is fixed, whereas for a hyperbola the *difference* of these distances is fixed.

DEFINITION OF A HYPERBOLA	A **hyperbola** is the set of all points (x, y) the difference of whose distances from two distinct fixed points (foci) is constant. (See Figure 10.20.)

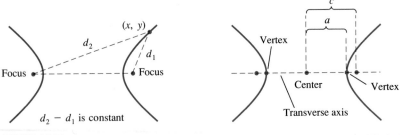

FIGURE 10.20

The line through the two foci intersects a hyperbola at two points called the **vertices.** The line segment connecting the vertices is called the **transverse axis,** and the midpoint of the transverse axis is called the **center** of the hyperbola. One distinguishing feature of hyperbolas is that their graphs have two separate *branches*.

The development of the standard form of the equation of a hyperbola is similar to that for an ellipse, and we list the following theorem without proof.

THEOREM 10.5 STANDARD EQUATION OF A HYPERBOLA	The standard form of the equation of a hyperbola with center at (h, k) is $$\frac{(x - h)^2}{a^2} - \frac{(y - k)^2}{b^2} = 1 \qquad \text{Transverse axis is } \textit{horizontal}$$ $$\frac{(y - k)^2}{a^2} - \frac{(x - h)^2}{b^2} = 1. \qquad \text{Transverse axis is } \textit{vertical}$$ The vertices are a units from the center, and the foci are c units from the center. Moreover, $b^2 = c^2 - a^2$.

Figure 10.21 shows both the horizontal and vertical orientations for a hyperbola.

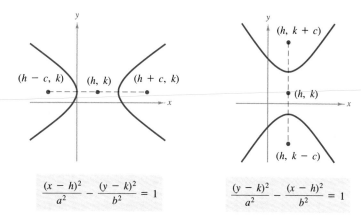

$$\frac{(x - h)^2}{a^2} - \frac{(y - k)^2}{b^2} = 1 \qquad \frac{(y - k)^2}{a^2} - \frac{(x - h)^2}{b^2} = 1$$

FIGURE 10.21

Standard Equations for Hyperbolas

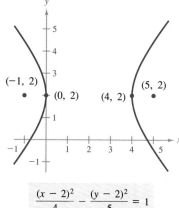

$$\frac{(x - 2)^2}{4} - \frac{(y - 2)^2}{5} = 1$$

FIGURE 10.22

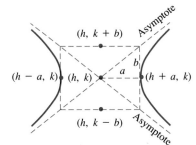

FIGURE 10.23

EXAMPLE 1 Finding the standard equation for a hyperbola

Find the standard form of the equation of the hyperbola with foci at $(-1, 2)$ and $(5, 2)$ and vertices at $(0, 2)$ and $(4, 2)$.

SOLUTION

By the Midpoint Formula, the center of the hyperbola occurs at the point $(2, 2)$. Furthermore, $c = 3$ and $a = 2$, and it follows that

$$b^2 = 3^2 - 2^2 = 9 - 4 = 5.$$

Thus, the equation of the hyperbola is

$$\frac{(x - 2)^2}{4} - \frac{(y - 2)^2}{5} = 1.$$

Figure 10.22 shows the graph of the hyperbola. ▭

An important aid in sketching the graph of a hyperbola is the determination of its **asymptotes,** as shown in Figure 10.23. Each hyperbola has two asymptotes that intersect at the center of the hyperbola. The asymptotes pass through the vertices of a rectangle of dimension $2a$ by $2b$, with its center at (h, k). The line segment of length $2b$ joining $(h, k + b)$ and $(h, k - b)$ is referred to as the **conjugate axis** of the hyperbola. The following theorem identifies the equation for the asymptotes.

THEOREM 10.6
ASYMPTOTES OF A HYPERBOLA

For a *horizontal* transverse axis, the equations of the asymptotes are

$$y = k + \frac{b}{a}(x - h) \qquad \text{and} \qquad y = k - \frac{b}{a}(x - h).$$

For a *vertical* transverse axis, the equations of the asymptotes are

$$y = k + \frac{a}{b}(x - h) \qquad \text{and} \qquad y = k - \frac{a}{b}(x - h).$$

PROOF   To simplify the algebra, we assume that the center of the hyperbola is $(h, k) = (0, 0)$. Solving for y in the equation

$$\frac{x^2}{a^2} - \frac{y^2}{b^2} = 1$$

produces

$$y = \pm \frac{b}{a}\sqrt{x^2 - a^2}.$$

If $y = (b/a)x$ is to be an asymptote of the hyperbola, then the difference of the y-values of a point on the hyperbola and a point on the asymptote

$$\frac{b}{a}\sqrt{x^2 - a^2} - \frac{b}{a}x$$

must approach zero as $x \to \infty$. To prove this, observe that

$$\lim_{x \to \infty} \left[\frac{b}{a}\sqrt{x^2 - a^2} - \frac{b}{a}x \right] = \lim_{x \to \infty} \left[\frac{b}{a}(\sqrt{x^2 - a^2} - x) \right]$$

$$= \lim_{x \to \infty} \left[\frac{b}{a}\left(\frac{x^2 - a^2 - x^2}{\sqrt{x^2 - a^2} + x} \right) \right]$$

$$= \lim_{x \to \infty} \left[\frac{-ab}{\sqrt{x^2 - a^2} + x} \right]$$

$$= 0.$$

The asymptotic behavior of the other three portions of the hyperbola can be established similarly.

REMARK   In Figure 10.23 we can see that the asymptotes coincide with the diagonals of the rectangle with dimensions $2a$ and $2b$, centered at (h, k). This provides us with a quick means of sketching the asymptotes, which in turn aids in sketching the hyperbola.

EXAMPLE 2   Using asymptotes to sketch a hyperbola

Sketch the graph of the hyperbola whose equation is

$$4x^2 - y^2 = 16.$$

SOLUTION

Rewriting this equation in standard form produces

$$\frac{4x^2}{16} - \frac{y^2}{16} = \frac{16}{16}$$

$$\frac{x^2}{2^2} - \frac{y^2}{4^2} = 1.$$

From this, we conclude that the transverse axis is horizontal and the vertices occur at $(-2, 0)$ and $(2, 0)$. Moreover, the ends of the conjugate axis occur at $(0, -4)$ and $(0, 4)$, and we are able to sketch the rectangle shown in Figure 10.24. Finally, by drawing the asymptotes through the corners of this rectangle, we complete the sketch shown in Figure 10.25.

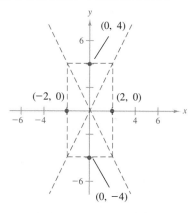

FIGURE 10.24

$4x^2 - y^2 = 16$

FIGURE 10.25

EXAMPLE 3   Finding the asymptotes of a hyperbola

Sketch the hyperbola given by

$$4x^2 - 3y^2 + 8x + 16 = 0$$

and find the equations of its asymptotes.

SOLUTION

Rewriting this equation in standard form produces

$$4x^2 - 3y^2 + 8x + 16 = 0$$
$$4(x^2 + 2x) - 3y^2 = -16$$
$$-4(x^2 + 2x + 1) + 3y^2 = 16 - 4$$
$$-4(x + 1)^2 + 3y^2 = 12$$
$$\frac{y^2}{4} - \frac{(x + 1)^2}{3} = 1.$$

From this equation we conclude that the hyperbola is centered at $(-1, 0)$ and has vertices at $(-1, 2)$ and $(-1, -2)$, and the ends of the conjugate axis occur at $(-1 - \sqrt{3}, 0)$ and $(-1 + \sqrt{3}, 0)$. To sketch the graph of the hyperbola, we draw a rectangle through these four points. The asymptotes are the lines passing through the corners of the rectangle, as shown in Figure 10.26. Finally, using $a = 2$ and $b = \sqrt{3}$, we apply Theorem 10.6 to conclude that the equations of the asymptotes are

$$y = \frac{2}{\sqrt{3}}(x + 1) \qquad \text{and} \qquad y = -\frac{2}{\sqrt{3}}(x + 1).$$

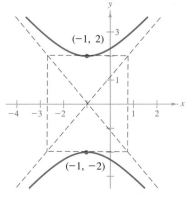

$4x^2 - 3y^2 + 8x + 16 = 0$

FIGURE 10.26

REMARK   If the constant term in the equation in Example 3 had been $F = -4$ instead of 16, then we would have obtained the following degenerate case.

$$\text{Two intersecting lines:} \quad \frac{y^2}{4} - \frac{(x+1)^2}{3} = 0$$

EXAMPLE 4   Using asymptotes to find the standard equation

Find the standard form of the equation of the hyperbola having vertices at $(3, -5)$ and $(3, 1)$ and asymptotes $y = 2x - 8$ and $y = -2x + 4$, as shown in Figure 10.27.

SOLUTION

By the Midpoint Formula the center of the hyperbola is at $(3, -2)$. Furthermore, the hyperbola has a vertical transverse axis with $a = 3$. By Theorem 10.6 the asymptotes have equations whose slopes are

$$m_1 = \frac{a}{b} \quad \text{and} \quad m_2 = -\frac{a}{b}.$$

From the given equations of the asymptotes, we know that

$$\frac{a}{b} = 2 \quad \text{and} \quad -\frac{a}{b} = -2$$

and since $a = 3$, we conclude that $b = \frac{3}{2}$. Therefore, the standard equation is

$$\frac{(y+2)^2}{9} - \frac{(x-3)^2}{9/4} = 1.$$

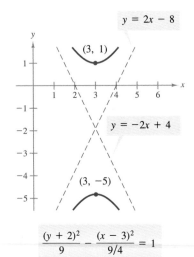

$$\frac{(y+2)^2}{9} - \frac{(x-3)^2}{9/4} = 1$$

FIGURE 10.27

As with ellipses, the **eccentricity** of a hyperbola is $e = c/a$. Because $c > a$, it follows that $e > 1$. If the eccentricity is large, then the branches of the hyperbola are nearly flat. If the eccentricity is close to 1, then the branches of the hyperbola are more pointed, as shown in Figure 10.28.

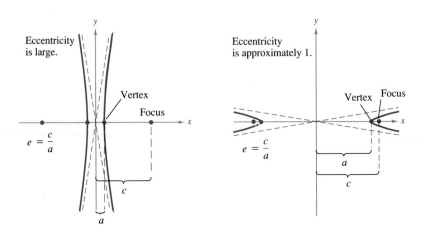

FIGURE 10.28

DEFINITION OF ECCENTRICITY OF A HYPERBOLA	The eccentricity e of a hyperbola is given by the ratio $$e = \frac{c}{a}.$$

Applications

The following application was developed during World War II. It shows how the properties of hyperbolas can be used in radar or other detection systems.

EXAMPLE 5   An application involving hyperbolas

Two microphones, 1 mile apart, record an explosion. Microphone A received the sound 2 seconds before microphone B. Where did the explosion come from?

SOLUTION

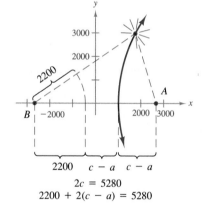

$$2c = 5280$$
$$2200 + 2(c - a) = 5280$$

FIGURE 10.29

Assuming that sound travels at 1100 feet per second, we know that the explosion took place 2200 feet farther from B than from A, as shown in Figure 10.29. The locus of all points that are 2200 feet closer to A than to B is one branch of the hyperbola $(x^2/a^2) - (y^2/b^2) = 1$, where

$$c = \frac{5280}{2} = 2640 \quad \text{and} \quad a = \frac{2200}{2} = 1100.$$

Thus, $b^2 = c^2 - a^2 = 5{,}759{,}600$, and we conclude that the explosion occurred somewhere on the right branch of the hyperbola given by

$$\frac{x^2}{1{,}210{,}000} - \frac{y^2}{5{,}759{,}600} = 1.$$

In Example 5 we were able to determine only the hyperbola on which the explosion occurred, but not the exact location of the explosion. If, however, we had received the sound at a third position, C, then two other hyperbolas would be determined. The exact location of the explosion would be the point at which these three hyperbolas intersect.

Another interesting application of conics involves the orbits of comets in our solar system. Of the 610 comets identified prior to 1970, 245 have elliptical orbits, 295 have parabolic orbits, and 70 have hyperbolic orbits. The first woman to be credited with detecting a new comet was the English astronomer Caroline Herschel (1750–1848). During her long life, Caroline Herschel discovered a total of eight new comets.

Caroline Herschel

The center of the sun is a focus of each orbit, and each orbit has a vertex at the point at which the comet is closest to the sun, as shown in Figure 10.30. Undoubtedly, many comets with parabolic or hyperbolic orbits have not been identified. We only get to see such comets *once*. Only comets with elliptical orbits such as Halley's Comet remain in our solar system.

The type of orbit for a comet can be determined as follows.

1. Ellipse: $v < \sqrt{2GM/p}$
2. Parabola: $v = \sqrt{2GM/p}$
3. Hyperbola: $v > \sqrt{2GM/p}$

In these three formulas, p is the distance between the vertex and the focus of the comet's orbit (in meters), v is the velocity of the comet at the vertex (in meters per second), $M \approx 1.991 \times 10^{30}$ kilograms is the mass of the sun, and $G \approx 6.67 \times 10^{-11}$ cubic meters per kilogram-second squared is the gravitational constant.

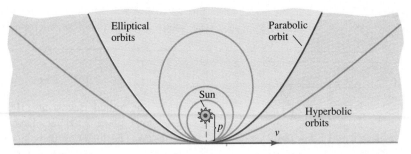

FIGURE 10.30

EXERCISES for Section 10.3

In Exercises 1–6, match the equation with the correct graph. [The graphs are labeled (a)–(f).]

1. $\dfrac{x^2}{9} - \dfrac{y^2}{4} = 1$ **2.** $\dfrac{y^2}{9} - \dfrac{x^2}{4} = 1$

3. $\dfrac{y^2}{1} - \dfrac{x^2}{16} = 1$ **4.** $\dfrac{y^2}{16} - \dfrac{x^2}{1} = 1$

5. $\dfrac{(x - 2)^2}{9} - \dfrac{y^2}{4} = 1$

6. $\dfrac{(x + 1)^2}{16} - \dfrac{(y - 3)^2}{9} = 1$

(c)

(d)

(a)

(b)

(e)

(f)

In Exercises 7–26, find the center, vertices, and foci of the hyperbola and sketch its graph, using asymptotes as an aid.

7. $x^2 - y^2 = 1$

8. $\dfrac{x^2}{9} - \dfrac{y^2}{16} = 1$

9. $y^2 - \dfrac{x^2}{4} = 1$

10. $\dfrac{y^2}{9} - x^2 = 1$

11. $\dfrac{y^2}{25} - \dfrac{x^2}{144} = 1$

12. $\dfrac{x^2}{36} - \dfrac{y^2}{4} = 1$

13. $2x^2 - 3y^2 = 6$

14. $3y^2 = 5x^2 + 15$

15. $5y^2 = 4x^2 + 20$

16. $7x^2 - 3y^2 = 21$

17. $\dfrac{(x-1)^2}{4} - (y+2)^2 = 1$

18. $\dfrac{(x+1)^2}{144} - \dfrac{(y-4)^2}{25} = 1$

19. $(y+6)^2 - (x-2)^2 = 1$

20. $4(y-1)^2 - 9(x+3)^2 = 1$

21. $9x^2 - y^2 - 36x - 6y + 18 = 0$

22. $x^2 - 9y^2 + 36y - 72 = 0$

23. $9y^2 - x^2 + 2x + 54y + 62 = 0$

24. $16y^2 - x^2 + 2x + 64y + 63 = 0$

25. $x^2 - 9y^2 + 2x - 54y - 80 = 0$

26. $9x^2 - y^2 + 54x + 10y + 55 = 0$

In Exercises 27–36, find an equation for the hyperbola.

27. Center: (0, 0)
Vertex: (0, 2)
Focus: (0, 4)

28. Center: (0, 0)
Vertex: (3, 0)
Focus: (5, 0)

29. Vertices: (±1, 0)
Asymptotes: $y = \pm3x$

30. Vertices: (0, ±3)
Asymptotes: $y = \pm3x$

31. Vertices: (0, 2), (6, 2)
Asymptotes: $y = \frac{2}{3}x$
$y = 4 - \frac{2}{3}x$

32. Vertices: (2, ±3)
Foci: (2, ±5)

33. Vertices: (2, ±3)
Point on graph: (0, 5)

34. Focus: (10, 0)
Asymptotes: $y = \pm\frac{3}{4}x$

35. For any point on the hyperbola, the difference of its distances from the points (2, 2) and (10, 2) is 6.

36. For any point on the hyperbola, the difference of its distances from the points (-3, 0) and (-3, 3) is 2.

In Exercises 37 and 38, find equations for (a) the tangent lines and (b) the normal lines to the hyperbola for the given values of x.

37. $\dfrac{x^2}{9} - y^2 = 1$, $x = 6$

38. $\dfrac{y^2}{4} - \dfrac{x^2}{2} = 1$, $x = 4$

In Exercises 39 and 40, the region bounded by the graphs of the given equations revolves about the x-axis. Find the volume and surface area of the resulting solid.

39. $x^2 - y^2 = 1$, $y = 0$, $x = 2$

40. $\dfrac{x^2}{16} - \dfrac{y^2}{9} = 1$, $y = 0$, $x = 5$

41. LORAN (long distance radio navigation) for aircraft and ships uses synchronized pulses transmitted by widely separated transmitting stations. These pulses travel at the speed of light (186,000 miles per second). The difference in the times of arrival of these pulses at an aircraft or ship is constant on a hyperbola having the two transmitting stations as foci. Assume that two stations, 300 miles apart, are positioned on the rectangular coordinate system at points with coordinates (-150, 0) and (150, 0) and that a ship is traveling on a path with coordinates (x, 75) (see figure). Find the x-coordinate of the position of the ship if the time difference between the pulses from the transmitting stations is 1000 microseconds (0.001 second).

FIGURE FOR 41

42. A rifle, positioned at point ($-c$, 0), is fired at a target positioned at point (c, 0). A person hears the sound of the rifle and the sound of the bullet hitting the target at the same time. Show that the person is positioned on one branch of the hyperbola given by

$$\frac{x^2}{c^2 v_s^2 / v_m^2} - \frac{y^2}{c^2(v_m^2 - v_s^2)/v_m^2} = 1$$

where v_m is the muzzle velocity of the rifle and the speed of sound is $v_s = 1100$ feet per second.

43. Show that the equation of the tangent line to

$$\frac{x^2}{a^2} - \frac{y^2}{b^2} = 1$$

at the point (x_0, y_0) is

$$\frac{x_0}{a^2}x - \frac{y_0}{b^2}y = 1.$$

44. Show that the ellipse

$$\frac{x^2}{a^2} + \frac{2y^2}{b^2} = 1$$

and the hyperbola

$$\frac{x^2}{a^2 - b^2} - \frac{2y^2}{b^2} = 1$$

intersect at right angles.

In Exercises 45 and 46, use the following general form of the equation of the conics.

$$Ax^2 + Cy^2 + Dx + Ey + F = 0$$

45. Prove that the graph of the equation is one of the following (except in degenerate cases).

Conic	Condition
(a) Circle	$A = C$
(b) Parabola	$A = 0$ or $C = 0$ (but not both)
(c) Ellipse	$AC > 0$
(d) Hyperbola	$AC < 0$

46. If $AC < 0$, determine the relationships among A, C, D, E, and F so that the graph of the equation is
(a) a hyperbola with transverse axis parallel to the x-axis,
(b) a hyperbola with transverse axis parallel to the y-axis,
(c) two intersecting lines.

In Exercises 47–56, use Exercise 45 to classify the graph of each equation as a circle, a parabola, an ellipse, or a hyperbola.

47. $x^2 + 4y^2 - 6x + 16y + 21 = 0$
48. $4x^2 - y^2 - 4x - 3 = 0$
49. $y^2 - 4y - 4x = 0$
50. $25x^2 - 10x - 200y - 119 = 0$
51. $4x^2 + 4y^2 - 16y + 15 = 0$
52. $y^2 - 4y = x + 5$
53. $9x^2 + 9y^2 - 36x + 6y + 34 = 0$
54. $2x(x - y) = y(3 - y - 2x)$
55. $3(x - 1)^2 = 6 + 2(y + 1)^2$
56. $9(x + 3)^2 = 36 - 4(y - 2)^2$

In Exercises 57 and 58, use a computer or graphics calculator to sketch the graph of the hyperbola.

57. $3x^2 - 2y^2 - 6x - 12y - 27 = 0$
58. $3y^2 - x^2 + 6x - 12y = 0$

10.4 Rotation and the General Second-Degree Equation

General second-degree equations ▪ Rotation of axes ▪ Invariants under rotation ▪ Discriminant

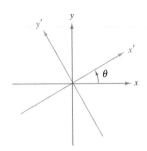

Rotated: x'-axis
          y'-axis

FIGURE 10.31

In previous sections we have shown that the equation of a conic with axes parallel to one of the coordinate axes has a standard form that can be written in the general form

$$Ax^2 + Cy^2 + Dx + Ey + F = 0. \qquad \text{Horizontal or vertical axes}$$

In this section we study the equations of conics whose axes are rotated so that they are not parallel to either the x-axis or the y-axis. The general equation for such conics contains an xy term.

$$Ax^2 + Bxy + Cy^2 + Dx + Ey + F = 0 \qquad \text{Equation in } xy\text{-plane}$$

To eliminate this xy term, we use a procedure called **rotation of axes.** We want to rotate the x- and y-axes until they are parallel to the axes of the conic. (We denote the rotated axes as the x'-axis and the y'-axis, as shown in Figure 10.31.) After the rotation has been accomplished, the equation of the conic in the new $x'y'$-plane will have the form

$$A'(x')^2 + C'(y')^2 + D'x' + E'y' + F' = 0. \qquad \text{Equation in } x'y'\text{-plane}$$

Since this equation has no $x'y'$ term, we can obtain a standard form by completing the square.

The following theorem identifies how much to rotate the axes to eliminate an xy term, and also the equations for determining the new coefficients A', C', D', E', and F'.

THEOREM 10.7
ROTATION OF AXES TO
ELIMINATE AN *XY* TERM

$Ax^2 + Bxy + Cy^2 + Dx + Ey + F = 0$ can be rewritten as

$$A'(x')^2 + C'(y')^2 + D'x' + E'y' + F' = 0$$

by rotating the coordinate axes through an angle θ, where

$$\cot 2\theta = \frac{A - C}{B}.$$

The coefficients of the new equation are obtained by making the substitutions

$$x = x' \cos \theta - y' \sin \theta$$
$$y = x' \sin \theta + y' \cos \theta.$$

PROOF

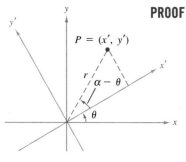

Rotated: $x' = r \cos (\alpha - \theta)$
$y' = r \sin (\alpha - \theta)$

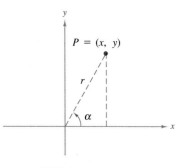

Original: $x = r \cos \alpha$
$y = r \sin \alpha$

FIGURE 10.32

To discover how the coordinates in the xy-system are related to the coordinates in the $x'y'$-system, we choose a point $P = (x, y)$ in the original system and attempt to find its coordinates (x', y') in the rotated system. In either system the distance r between the point P and the origin is the same, thus the equations for x, y, x', and y' are those given in Figure 10.32. Using the formulas for the sine and cosine of the difference of two angles, we obtain

$$x' = r \cos (\alpha - \theta) = r(\cos \alpha \cos \theta + \sin \alpha \sin \theta)$$
$$= r \cos \alpha \cos \theta + r \sin \alpha \sin \theta = x \cos \theta + y \sin \theta$$
$$y' = r \sin (\alpha - \theta) = r(\sin \alpha \cos \theta - \cos \alpha \sin \theta)$$
$$= r \sin \alpha \cos \theta - r \cos \alpha \sin \theta = y \cos \theta - x \sin \theta.$$

Solving this system for x and y yields

$$x = x' \cos \theta - y' \sin \theta \quad \text{and} \quad y = x' \sin \theta + y' \cos \theta.$$

Finally, by substituting these values for x and y into the original equation and collecting terms, we obtain the following.

$$A' = A \cos^2 \theta + B \cos \theta \sin \theta + C \sin^2 \theta$$
$$C' = A \sin^2 \theta - B \cos \theta \sin \theta + C \cos^2 \theta$$
$$D' = D \cos \theta + E \sin \theta$$
$$E' = -D \sin \theta + E \cos \theta$$
$$F' = F$$

Now, in order to eliminate the $x'y'$ term, we must select θ so that $B' = 0$ as follows.

$$B' = 2(C - A) \sin \theta \cos \theta + B(\cos^2 \theta - \sin^2 \theta)$$
$$= (C - A) \sin 2\theta + B \cos 2\theta$$
$$= B(\sin 2\theta)\left(\frac{C - A}{B} + \cot 2\theta\right) = 0, \quad \sin 2\theta \neq 0$$

If $B = 0$, no rotation is necessary since the xy term is not present in the original equation. If $B \neq 0$, then the only way to make $B' = 0$ is to let

$$\cot 2\theta = \frac{A - C}{B}, \quad B \neq 0.$$

Thus, we have established the desired results.

EXAMPLE 1   Rotation of a hyperbola

Write the equation $xy - 1 = 0$ in standard form.

SOLUTION

Since $A = 0$, $B = 1$, and $C = 0$, we have (for $0 < \theta < \pi/2$)

$$\cot 2\theta = \frac{A - C}{B} = 0 \ \Longrightarrow\ 2\theta = \frac{\pi}{2} \ \Longrightarrow\ \theta = \frac{\pi}{4}.$$

Therefore, the equation in the $x'y'$-system is given by making the following substitutions.

$$x = x' \cos \frac{\pi}{4} - y' \sin \frac{\pi}{4} = x'\left(\frac{\sqrt{2}}{2}\right) - y'\left(\frac{\sqrt{2}}{2}\right) = \frac{x' - y'}{\sqrt{2}}$$

$$y = x' \sin \frac{\pi}{4} + y' \cos \frac{\pi}{4} = x'\left(\frac{\sqrt{2}}{2}\right) + y'\left(\frac{\sqrt{2}}{2}\right) = \frac{x' + y'}{\sqrt{2}}$$

Substituting these expressions into the equation $xy - 1 = 0$ produces

$$\left(\frac{x' - y'}{\sqrt{2}}\right)\left(\frac{x' + y'}{\sqrt{2}}\right) - 1 = 0$$

$$\frac{(x')^2 - (y')^2}{2} - 1 = 0$$

$$\frac{(x')^2}{(\sqrt{2})^2} - \frac{(y')^2}{(\sqrt{2})^2} = 1. \qquad \text{Standard form}$$

This is the equation of a hyperbola centered at the origin with vertices at $(\pm\sqrt{2}, 0)$ in the $x'y'$-system as shown in Figure 10.33. To find the coordinates of the vertices in the xy-system, we substituted the $x'y'$-coordinates into the equations

$$x = \frac{x' - y'}{\sqrt{2}} \qquad \text{and} \qquad y = \frac{x' + y'}{\sqrt{2}}.$$

From these substitutions, we find the vertices to be $(1, 1)$ and $(-1, -1)$ in the xy-system. Note also that the asymptotes of the hyperbola have equations $y' = \pm x'$, which correspond to the original x- and y-axes.

$$\frac{(x')^2}{(\sqrt{2})^2} - \frac{(y')^2}{(\sqrt{2})^2} = 1$$
$$xy - 1 = 0$$

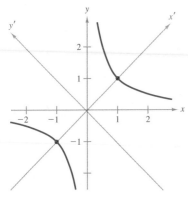

Vertices:
$(\sqrt{2}, 0)$, $(-\sqrt{2}, 0)$ in $x'y'$-system
$(1, 1)$, $(-1, -1)$ in xy-system

FIGURE 10.33

Remember that the substitutions

$$x = x' \cos \theta - y' \sin \theta \qquad \text{and} \qquad y = x' \sin \theta + y' \cos \theta$$

were developed to eliminate the $x'y'$ term in the rotated system. You can use this as a check on your work. In other words, if your final equation contains an $x'y'$ term, you know that you made a mistake.

EXAMPLE 2 Rotation of an ellipse

Sketch the graph of $7x^2 - 6\sqrt{3}xy + 13y^2 - 16 = 0$.

SOLUTION

Since $A = 7$, $B = -6\sqrt{3}$, and $C = 13$, we have (for $0 < \theta < \pi/2$)

$$\cot 2\theta = \frac{A - C}{B} = \frac{7 - 13}{-6\sqrt{3}} = \frac{1}{\sqrt{3}} \quad \Longrightarrow \quad \theta = \frac{\pi}{6}.$$

Therefore, the equation in the $x'y'$-system is given by making the following substitutions.

$$x = x' \cos \frac{\pi}{6} - y' \sin \frac{\pi}{6} = x'\left(\frac{\sqrt{3}}{2}\right) - y'\left(\frac{1}{2}\right) = \frac{\sqrt{3}x' - y'}{2}$$

$$y = x' \sin \frac{\pi}{6} + y' \cos \frac{\pi}{6} = x'\left(\frac{1}{2}\right) + y'\left(\frac{\sqrt{3}}{2}\right) = \frac{x' + \sqrt{3}y'}{2}$$

Substituting these expressions into the original equation produces

$$7\left(\frac{\sqrt{3}x' - y'}{2}\right)^2 - 6\sqrt{3}\left(\frac{\sqrt{3}x' - y'}{2}\right)\left(\frac{x' + \sqrt{3}y'}{2}\right)$$
$$+ 13\left(\frac{x' + \sqrt{3}y'}{2}\right)^2 = 16$$

which simplifies to

$$4(x')^2 + 16(y')^2 = 16$$
$$\frac{(x')^2}{4} + \frac{(y')^2}{1} = 1. \qquad \text{Standard form}$$

This is the equation of an ellipse centered at the origin with vertices at $(\pm 2, 0)$ in the $x'y'$-system, as shown in Figure 10.34. ▭

In constructing Examples 1 and 2 we carefully chose the equations so that θ would be one of the common angles 30°, 45°, and so forth. Of course, many second-degree equations do not yield such common solutions to the equation $\cot 2\theta = (A - C)/B$. Example 3 illustrates such a case.

EXAMPLE 3 Rotation of a parabola

Sketch the graph of $x^2 - 4xy + 4y^2 + 5\sqrt{5}y + 1 = 0$.

SOLUTION

Since $A = 1$, $B = -4$, and $C = 4$, we have

$$\cot 2\theta = \frac{A - C}{B} = \frac{1 - 4}{-4} = \frac{3}{4}.$$

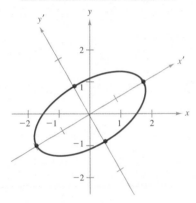

$7x^2 - 6\sqrt{3}xy + 13y^2 - 16 = 0$
$\frac{(x')^2}{(2)^2} + \frac{(y')^2}{(1)^2} = 1$

Vertices:

$(\pm 2, 0)$, $(0, \pm 1)$ in $x'y'$-system

$(\pm\sqrt{3}, \pm 1)$, $\left(\pm\frac{1}{2}, \mp\frac{\sqrt{3}}{2}\right)$ in xy-system

FIGURE 10.34

Using the trigonometric identity $\cot 2\theta = (\cot^2 \theta - 1)/(2 \cot \theta)$ produces

$$\cot 2\theta = \frac{3}{4} = \frac{\cot^2 \theta - 1}{2 \cot \theta}$$

from which we obtain the equation

$$6 \cot \theta = 4 \cot^2 \theta - 4 \quad \Longrightarrow \quad 4 \cot^2 \theta - 6 \cot \theta - 4 = 0$$
$$(2 \cot \theta - 4)(2 \cot \theta + 1) = 0.$$

Considering $0 < \theta < \pi/2$, we have $2 \cot \theta = 4$. Thus,

$$\cot \theta = 2 \quad \Longrightarrow \quad \theta \approx 26.6°.$$

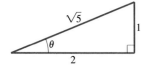

FIGURE 10.35

From the triangle in Figure 10.35 we obtain $\sin \theta = 1/\sqrt{5}$ and $\cos \theta = 2/\sqrt{5}$. Consequently, we obtain the following.

$$x = x' \cos \theta - y' \sin \theta = x'\left(\frac{2}{\sqrt{5}}\right) - y'\left(\frac{1}{\sqrt{5}}\right) = \frac{2x' - y'}{\sqrt{5}}$$

$$y = x' \sin \theta + y' \cos \theta = x'\left(\frac{1}{\sqrt{5}}\right) + y'\left(\frac{2}{\sqrt{5}}\right) = \frac{x' + 2y'}{\sqrt{5}}$$

Substituting these expressions into the original equation, we have

$$\left(\frac{2x' - y'}{\sqrt{5}}\right)^2 - 4\left(\frac{2x' - y'}{\sqrt{5}}\right)\left(\frac{x' + 2y'}{\sqrt{5}}\right) + 4\left(\frac{x' + 2y'}{\sqrt{5}}\right)^2$$
$$+ 5\sqrt{5}\left(\frac{x' + 2y'}{\sqrt{5}}\right) + 1 = 0$$

which simplifies to

$$5(y')^2 + 5x' + 10y' + 1 = 0.$$

By completing the square, we can obtain the standard form

$$5(y' + 1)^2 = -5x' + 4$$
$$(y' + 1)^2 = (-1)\left(x' - \frac{4}{5}\right). \qquad \text{Standard form}$$

The graph of the equation is a parabola with its vertex at $\left(\frac{4}{5}, -1\right)$ and its axis parallel to the x'-axis in the $x'y'$-system, as shown in Figure 10.36.

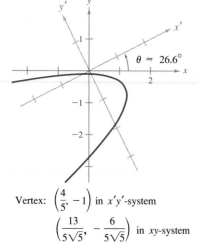

$$x^2 - 4xy + 4y^2 + 5\sqrt{5}y + 1 = 0$$
$$(y' + 1)^2 = 4\left(-\frac{1}{4}\right)\left(x' - \frac{4}{5}\right)$$

Vertex: $\left(\frac{4}{5}, -1\right)$ in $x'y'$-system

$\left(\frac{13}{5\sqrt{5}}, -\frac{6}{5\sqrt{5}}\right)$ in xy-system

FIGURE 10.36

Invariants under rotation

In Theorem 10.7 note that the constant term $F' = F$ is the same in both equations. We say that it is **invariant under rotation.** Theorem 10.8 lists some other rotation invariants. We leave the proof of this theorem as an exercise (see Exercise 32).

THEOREM 10.8 **ROTATION INVARIANTS**	The rotation of coordinate axes through an angle θ that transforms the equation $Ax^2 + Bxy + Cy^2 + Dx + Ey + F = 0$ into the form $A'(x')^2 + C'(y')^2 + D'x' + E'y' + F' = 0$ has the following rotation invariants. 1. $F = F'$ 2. $A + C = A' + C'$ 3. $B^2 - 4AC = (B')^2 - 4A'C'$

We can use the results of this theorem to classify the graph of a second-degree equation *with* an xy term in much the same way we did for second-degree equations *without* an xy term (see Exercise 45 in Section 10.3). Note that since $B' = 0$, the invariant $B^2 - 4AC$ reduces to

$$B^2 - 4AC = -4A'C'. \qquad \text{Discriminant}$$

We call this quantity the **discriminant** of the equation

$$Ax^2 + Bxy + Cy^2 + Dx + Ey + F = 0.$$

Now, because the sign of $A'C'$ determines the type of graph for the equation

$$A'(x')^2 + C'(y')^2 + D'x' + E'y' + F' = 0$$

we can conclude that the sign of $B^2 - 4AC$ must determine the type of graph for the original equation. This result is stated in Theorem 10.9.

THEOREM 10.9
CLASSIFICATION OF CONICS
BY THE DISCRIMINANT

The graph of the equation $Ax^2 + Bxy + Cy^2 + Dx + Ey + F = 0$ is, except in degenerate cases, determined by its discriminant as follows.

1. Ellipse or circle: $B^2 - 4AC < 0$
2. Parabola: $B^2 - 4AC = 0$
3. Hyperbola: $B^2 - 4AC > 0$

EXAMPLE 4   Using the discriminant

Classify the graph of each of the following equations.
(a) $4xy - 9 = 0$
(b) $2x^2 - 3xy + 2y^2 - 2x = 0$
(c) $x^2 - 6xy + 9y^2 - 2y + 1 = 0$
(d) $3x^2 + 8xy + 4y^2 - 7 = 0$

SOLUTION

(a) The graph is a hyperbola because

$$B^2 - 4AC = 16 - 0 > 0.$$

(b) The graph is a circle or an ellipse because

$$B^2 - 4AC = 9 - 16 < 0.$$

(c) The graph is a parabola because

$$B^2 - 4AC = 36 - 36 = 0.$$

(d) The graph is a hyperbola because

$$B^2 - 4AC = 64 - 48 > 0.$$

EXERCISES for Section 10.4

In Exercises 1–16, rotate the axes to eliminate the xy term. Sketch the graph of the resulting equation, showing both sets of axes.

1. $xy + 1 = 0$
2. $xy - 4 = 0$
3. $9x^2 + 24xy + 16y^2 + 90x - 130y = 0$
4. $9x^2 + 24xy + 16y^2 + 80x - 60y = 0$
5. $x^2 - 10xy + y^2 + 1 = 0$
6. $xy + x - 2y + 3 = 0$
7. $xy - 2y - 4x = 0$
8. $2x^2 - 3xy - 2y^2 + 10 = 0$
9. $5x^2 - 2xy + 5y^2 - 12 = 0$
10. $13x^2 + 6\sqrt{3}xy + 7y^2 - 16 = 0$
11. $3x^2 - 2\sqrt{3}xy + y^2 + 2x + 2\sqrt{3}y = 0$
12. $16x^2 - 24xy + 9y^2 - 60x - 80y + 100 = 0$
13. $17x^2 + 32xy - 7y^2 = 75$
14. $40x^2 + 36xy + 25y^2 = 52$
15. $32x^2 + 50xy + 7y^2 = 52$
16. $4x^2 - 12xy + 9y^2 + (4\sqrt{13} - 12)x - (6\sqrt{13} + 8)y = 91$

In Exercises 17–24, use the discriminant to determine if the graph of the equation is a parabola, ellipse, or hyperbola.

17. $16x^2 - 24xy + 9y^2 - 30x - 40y = 0$
18. $x^2 - 4xy - 2y^2 - 6 = 0$
19. $13x^2 - 8xy + 7y^2 - 45 = 0$
20. $2x^2 + 4xy + 5y^2 + 3x - 4y - 20 = 0$
21. $x^2 - 6xy - 5y^2 + 4x - 22 = 0$
22. $36x^2 - 60xy + 25y^2 + 9y = 0$
23. $x^2 + 4xy + 4y^2 - 5x - y - 3 = 0$
24. $x^2 + xy + 4y^2 + x + y - 4 = 0$

In Exercises 25–28, sketch the graph (if possible) of the degenerate conic.

25. $y^2 - 4x^2 = 0$
26. $x^2 + y^2 - 2x + 6y + 10 = 0$
27. $5x^2 - 2xy + 5y^2 = 0$
28. $x^2 - 10xy + y^2 = 0$

In Exercises 29 and 30, use a computer or graphics calculator to sketch the graph of the conic.

29. $x^2 + xy + y^2 = 10$
30. $x^2 - 4xy + 2y^2 = 6$

31. Show that the equation $x^2 + y^2 = r^2$ is invariant under rotation of axes.
32. Prove Theorem 10.8.
33. Consider $Ax^2 + Bxy + Cy^2 = 1$, where $B^2 - 4AC < 0$. Show that the area of this ellipse is

$$\pi ab = \frac{2\pi}{\sqrt{4AC - B^2}}$$

where $2a$ and $2b$ are the lengths of the major and minor axes, respectively.

REVIEW EXERCISES for Chapter 10

In Exercises 1–10, match the equation with the correct graph. [The graphs are labeled (a)–(j).]

1. $4x^2 + y^2 = 4$
2. $x^2 = 4y$
3. $4x^2 - y^2 = 4$
4. $y^2 = -4x$
5. $x^2 + 4y^2 = 4$
6. $y^2 - 4x^2 = 4$
7. $x^2 = -6y$
8. $x^2 + 5y^2 = 10$
9. $x^2 - 5y^2 = -5$
10. $y^2 - 8x = 0$

(c)

(d)

(a)

(b)

(e)

(f)

(g)

(h)

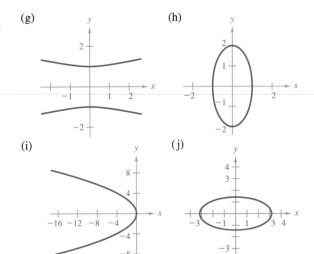

(i)

(j)

In Exercises 11–22, analyze each equation and sketch its graph.

11. $16x^2 + 16y^2 - 16x + 24y - 3 = 0$
12. $y^2 - 12y - 8x + 20 = 0$
13. $3x^2 - 2y^2 + 24x + 12y + 24 = 0$
14. $4x^2 + y^2 - 16x + 15 = 0$
15. $3x^2 + 2y^2 - 12x + 12y + 29 = 0$
16. $4x^2 - 4y^2 - 4x + 8y - 11 = 0$
17. $x^2 - 6x + 2y + 9 = 0$
18. $x^2 + y^2 - 2x - 4y + 5 = 0$
19. $x^2 + y^2 + 2xy + 2\sqrt{2}x - 2\sqrt{2}y + 2 = 0$
20. $9x^2 + 4xy + 6y^2 - 20 = 0$
21. $4x^2 + 9y^2 - 8x + 9y + 4 = 0$
22. $9x^2 + 4y^2 - 36x + 8y + 31 = 0$

In Exercises 23–32, find an equation of the specified conic.

23. Hyperbola; vertices $(0, \pm 1)$, foci $(0, \pm 3)$
24. Hyperbola; vertices $(\pm 2, 2)$, foci $(\pm 4, 2)$
25. Ellipse; foci $(0, 0)$ and $(4, 0)$, sum of the distances from a point on the ellipse to the foci is 10
26. Parabola; vertex $(4, 2)$, focus $(4, 0)$
27. Parabola; vertex $(0, 0)$, focus $(1, 1)$
28. Ellipse; vertices $(2, 0)$ and $(2, 4)$, foci $(2, 1)$ and $(2, 3)$
29. Ellipse; points on graph $(1, 2)$ and $(2, 0)$, center at $(0, 0)$
30. Hyperbola; foci $(\pm 4, 0)$, asymptotes $y = \pm 2x$
31. Hyperbola; foci $(\pm 4, 0)$, absolute value of the difference of the distances from a point on the hyperbola to the foci is 4
32. Parabola; vertex $(0, 2)$, point on graph $(-1, 0)$

33. Find the equation of the line tangent to the parabola $y = x^2 - 2x + 2$ and perpendicular to the line $y = x - 2$.

34. Find the equations of the lines tangent to the ellipse $x^2 + 4y^2 - 4x - 8y - 24 = 0$ and parallel to the line $x - 2y - 12 = 0$.

35. Find a so that the hyperbola given by

$$\frac{x^2}{a^2} - \frac{y^2}{4} = 1$$

is tangent to the line $2x - y - 4 = 0$.

36. A large parabolic antenna is described as the surface formed by revolving the parabola $y = x^2/200$ on the interval $[0, 100]$ about the y-axis. The receiving and transmitting equipment is positioned at the focus.
(a) Find the focus.
(b) Find the antenna's surface area.

In Exercises 37–39, consider the region bounded by the ellipse

$$\frac{x^2}{a^2} + \frac{y^2}{b^2} = 1, \text{ eccentricity } e = c/a.$$

37. Show that the area of the region is πab.
38. Show that the solid (oblate spheroid) generated by revolving the region about the minor axis of the ellipse has a volume of $V = 4\pi a^2 b/3$ and a surface area of

$$S = 2\pi a^2 + \pi\left(\frac{b^2}{e}\right) \ln\left(\frac{1 + e}{1 - e}\right).$$

39. Show that the solid (prolate spheroid) generated by revolving the region about the major axis of the ellipse has a volume of $V = 4\pi ab^2/3$ and a surface area of

$$S = 2\pi b^2 + 2\pi\left(\frac{ab}{e}\right) \arcsin e.$$

40. Use the results of Exercises 38 and 39 to find the volume and surface areas of the prolate and oblate spheroids

$$\frac{x^2}{9} + \frac{y^2}{4} = 1.$$

41. Approximate, to two decimal places, the perimeter of the ellipse

$$\frac{x^2}{9} + \frac{y^2}{4} = 1.$$

In Exercises 42–45, consider a firetruck with a water tank 16 feet long whose vertical cross sections are ellipses described by the equation

$$\frac{x^2}{16} + \frac{y^2}{9} = 1.$$

42. Find the volume of the tank.
43. Find the depth of the water in the tank if it is $\frac{3}{4}$ full (by volume) and the truck is on level ground.
44. Find the force on the end of the tank, full of water.
45. Approximate, to two decimal places, the surface area of the tank.

Chapter 11 Application

Saturn is the sixth planet from the sun. It is the second largest planet in our solar system, with a volume 815 times that of the earth—only Jupiter is larger. The Voyager I and II flights to Saturn determined that a Saturnian day is short— only 10 hours, 39 minutes, and 24 seconds.

The Orbit of Saturn

The planets in our solar system travel in elliptical orbits with the sun as a focus. In polar coordinates, the equation for the orbit of Saturn is

$$r = \frac{1.4228}{1 - 0.0543 \cos \theta}$$

where r is measured in billions of kilometers. The eccentricity of Saturn's orbit is $e = 0.0543$.

The point on the orbit that is farthest from the sun is called the *aphelion* and occurs when $\theta = 0$. This point is

$$r = \frac{1.4228}{1 - 0.0543 \cos 0} \approx 1.5045 \times 10^9 \text{ kilometers}$$

from the sun.

The point on the orbit that is closest to the sun is called the *perihelion* and occurs when $\theta = \pi$. This point is

$$r = \frac{1.4228}{1 - 0.0543 \cos \pi} \approx 1.3495 \times 10^9 \text{ kilometers}$$

from the sun.

Chapter Overview

The chapter is divided into two parts—**parametric equations** (Sections 11.1 and 11.2) and the **polar coordinate system** (Sections 11.3–11.6).

In Section 11.1, we show how to use a parameter to represent a curve in the plane. For instance, the set of equations

$$x = t \quad \text{and} \quad y = t^2$$

uses the parameter t to represent the curve whose rectangular equation is $y = x^2$. Parametric representation of a curve is especially useful for studying motion in the plane. In Section 11.2, we discuss the calculus of parametric equations—the parametric form of the derivative, arc length in parametric form, and the area of a surface of revolution.

In Sections 11.3 and 11.4, we introduce the polar coordinate system and discuss techniques for sketching the graph of a polar equation. Then, in Section 11.5, we look at the problem of finding the area of a region bounded by a polar graph, and finding the arc length of a polar graph. Finally, in Section 11.6, we look at polar equations for conics and Kepler's Laws, which can be used to describe the orbits of the planets about the sun.

$$r = \frac{1.4228}{1 - 0.0543 \cos \theta}$$

Elliptical Orbit of Saturn

See Exercise 53, Section 11.6.

Plane Curves, Parametric Equations, and Polar Coordinates

11.1 Plane Curves and Parametric Equations

Plane curves ▪ Parametric equations ▪ Curve sketching ▪ Eliminating the parameter ▪ Finding parametric equations ▪ The brachistochrone problem

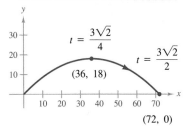

Rectangular equation:
$$y = -\frac{x^2}{72} + x$$
Parametric equations:
$$x = 24\sqrt{2}t, \quad y = -16t^2 + 24\sqrt{2}t$$

Two variables for position
One variable for time

FIGURE 11.1

Until now, we have been representing a graph by a single equation involving the *two* variables x and y. In this section we look at situations in which it is useful to introduce a *third* variable to represent a curve in the plane.

To see the usefulness of this procedure, consider the path followed by an object that is propelled into the air at an angle of 45°. If the initial velocity of the object is 48 feet per second, it follows the parabolic path given by

$$y = -\frac{x^2}{72} + x \qquad \text{Rectangular equation}$$

as shown in Figure 11.1. However, this equation does not tell the whole story. Although it does tell us *where* the object has been, it doesn't tell us *when* the object was at a given point (x, y). To determine this time, we introduce a third variable t, called a **parameter.** By writing both x and y as functions of t, we obtain the **parametric equations**

$$x = 24\sqrt{2}t \qquad \text{and} \qquad y = -16t^2 + 24\sqrt{2}t. \qquad \text{Parametric equations}$$

Now, from this set of equations, we can determine that at time $t = 0$, the object is at the point $(0, 0)$. Similarly, at time $t = 1$, the object is at the point $(24\sqrt{2}, 24\sqrt{2} - 16)$, and so on.*

For this particular motion problem, x and y are continuous functions of t, and we call the resulting path a **plane curve.**

DEFINITION OF A PLANE CURVE

If f and g are continuous functions of t on an interval I, then the set of ordered pairs $(f(t), g(t))$ is called a **plane curve** C. The equations $x = f(t)$ and $y = g(t)$ are called **parametric equations** for C, and t is called the **parameter.**

---

*We will discuss a method for determining this particular set of parametric equations—the equations of motion—later in the text.

The set of points $(x, y) = (f(t), g(t))$ in the plane is called the **graph of the curve** C. For simplicity, we will not distinguish between a curve and its graph. When sketching a curve represented by a pair of parametric equations, we still plot points in the xy-plane. Each set of coordinates (x, y) is determined from a value chosen for the parameter t. By plotting the resulting points in the order of *increasing* values of t, we trace the curve out in a specific direction. This is called the **orientation** of the curve.

EXAMPLE 1   Sketching a curve

Sketch the curve described by the parametric equations

$$x = t^2 - 4 \quad \text{and} \quad y = \frac{t}{2}, \quad -2 \le t \le 3.$$

SOLUTION

For values of t on the given interval, the parametric equations yield the points (x, y) shown in Table 11.1.

TABLE 11.1

t	-2	-1	0	1	2	3
x	0	-3	-4	-3	0	5
y	-1	$-\frac{1}{2}$	0	$\frac{1}{2}$	1	$\frac{3}{2}$

FIGURE 11.2

By plotting these points in the order of increasing t and using the continuity of f and g, we obtain the curve C shown in Figure 11.2. Note that the arrows on the curve indicate its orientation as t increases from -2 to 3. ▭

REMARK   By the vertical line test, we can see that the graph shown in Figure 11.2 does not define y as a function of x. This points out one benefit of parametric equations—they can be used to represent graphs that are more general than graphs of functions.

It often happens that two different sets of parametric equations have the same graph. For example, the set of parametric equations

$$x = 4t^2 - 4 \quad \text{and} \quad y = t, \quad -1 \le t \le \frac{3}{2}$$

FIGURE 11.3

has the same graph as the set given in Example 1. However, comparing the values of t in Figures 11.2 and 11.3, we see that this second graph is traced out more *rapidly* (considering t as time) than the first graph. Thus, in applications, different parametric representations can be used to represent various *speeds* at which objects travel along a given path.

Eliminating the parameter

In Example 1 we used a simple point-plotting method to sketch the given curve. This tedious process can sometimes be simplified by finding a rectangular equation (in x and y) that has the same graph. We call this process **eliminating the parameter.**

| Parametric equations | → | Solve for t in one equation | → | Substitute into second equation | → | Rectangular equation |

$$x = t^2 - 4 \qquad\qquad t = 2y \qquad\qquad x = (2y)^2 - 4 \qquad\qquad x = 4y^2 - 4$$
$$y = t/2$$

Now, we recognize that the equation $x = 4y^2 - 4$ represents a parabola with a horizontal axis and vertex at $(-4, 0)$.

The range of x and y implied by the parametric equations may be altered by the change to rectangular form. In such instances the domain of the rectangular equation must be adjusted so that its graph matches the graph of the parametric equations. Such a situation is demonstrated in the next example.

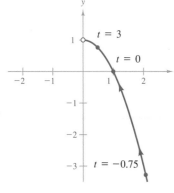

$t = 3$
$t = 0$
$t = -0.75$

Parametric equations:
$$x = \frac{1}{\sqrt{t+1}}, \; y = \frac{t}{t+1}$$

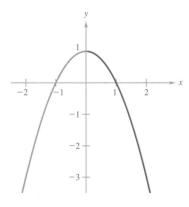

Rectangular equation:
$$y = 1 - x^2$$

FIGURE 11.4

EXAMPLE 2   Adjusting the domain after eliminating the parameter

Sketch the curve represented by the equations

$$x = \frac{1}{\sqrt{t+1}} \qquad \text{and} \qquad y = \frac{t}{t+1}$$

by eliminating the parameter and adjusting the domain of the resulting rectangular equation.

SOLUTION

Solving for t in the equation for x, we have

$$x = \frac{1}{\sqrt{t+1}} \quad \Longrightarrow \quad x^2 = \frac{1}{t+1}$$

which implies that

$$t = \frac{1 - x^2}{x^2}.$$

Now, substituting into the equation for y, we obtain

$$y = \frac{t}{t+1} = \frac{(1 - x^2)/x^2}{[(1 - x^2)/x^2] + 1} = 1 - x^2.$$

The rectangular equation, $y = 1 - x^2$, is defined for all values of x, but from the parametric equation for x we see that the curve is defined only when $-1 < t$. This implies that we should restrict the domain of x to positive values, as shown in Figure 11.4.

It is not necessary for the parameter in a set of parametric equations to represent time. Our next example uses an *angle* as the parameter. In this example we use *trigonometric identities* to eliminate the parameter.

EXAMPLE 3   Using a trigonometric identity to eliminate a parameter

Sketch the curve represented by

$$x = 3 \cos \theta \quad \text{and} \quad y = 4 \sin \theta, \quad 0 \le \theta \le 2\pi$$

by eliminating the parameter and finding the corresponding rectangular equation.

SOLUTION

We begin by solving for $\cos \theta$ and $\sin \theta$ in the given equations.

$$\cos \theta = \frac{x}{3} \quad \text{and} \quad \sin \theta = \frac{y}{4} \qquad \text{Solve for } \sin \theta \text{ and } \cos \theta$$

Now we make use of the identity $\sin^2 \theta + \cos^2 \theta = 1$ to form an equation involving only x and y.

$$\cos^2 \theta + \sin^2 \theta = 1 \qquad \text{Trigonometric identity}$$

$$\cos^2 \theta + \sin^2 \theta = \left(\frac{x}{3}\right)^2 + \left(\frac{y}{4}\right)^2 = 1 \qquad \text{Substitute}$$

$$\frac{x^2}{9} + \frac{y^2}{16} = 1 \qquad \text{Rectangular equation}$$

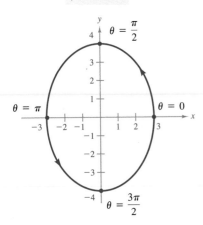

FIGURE 11.5

From this rectangular equation we see that the graph is an ellipse centered at $(0, 0)$, with vertices at $(0, 4)$ and $(0, -4)$ and minor axis of length $2b = 6$, as shown in Figure 11.5. Note that the elliptic curve is traced out *counterclockwise* as θ varies from 0 to 2π.

In Examples 2 and 3 it is important to realize that eliminating the parameter is primarily an *aid to curve sketching*. If the parametric equations represent the path of a moving object, the graph alone is not sufficient to describe the object's motion. We still need the parametric equations to tell us the *position*, *direction*, and *speed* at a given time.

Finding parametric equations for a graph

We have been looking at techniques for sketching the graph represented by a set of parametric equations. We now look at the reverse problem. How can we determine a set of parametric equations for a given graph or a given physical description? From the discussion following Example 1, we know that such a representation is not unique. This is demonstrated further in the following example, in which we find two different parametric representations for a given graph.

EXAMPLE 4   Finding parametric equations for a given graph

Find a set of parametric equations to represent the graph of $y = 1 - x^2$, using the following parameters.
(a) $t = x$ (b) the slope $m = dy/dx$ at the point (x, y)

SOLUTION

(a) Letting $x = t$, we obtain the parametric equations

$$x = t \quad \text{and} \quad y = 1 - x^2 = 1 - t^2.$$

(b) To express x and y in terms of the parameter m, we proceed as follows.

$$m = \frac{dy}{dx} = -2x \qquad\qquad \text{Differentiate } y = 1 - x^2$$

$$x = -\frac{m}{2} \qquad\qquad \text{Solve for } x$$

$$y = 1 - x^2 = 1 - \left(-\frac{m}{2}\right)^2 = 1 - \frac{m^2}{4} \qquad \text{Substitute}$$

Thus, the parametric equations are

$$x = -\frac{m}{2} \quad \text{and} \quad y = 1 - \frac{m^2}{4}.$$

In Figure 11.6, note that the resulting curve has a right-to-left orientation as determined by the direction of increasing values of slope m. For part (a) the curve would have the opposite orientation.

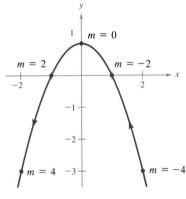

$x = -\dfrac{m}{2}, \; y = 1 - \dfrac{m^2}{4}$

Graph of $y = 1 - x^2$

FIGURE 11.6

EXAMPLE 5   Parametric equations for a cycloid

Determine the curve traced by a point P on the circumference of a circle of radius a rolling along a straight line in a plane. Such a curve is called a **cycloid.**

SOLUTION

We let the parameter θ be the measure of the circle's rotation, and we let the point $P = (x, y)$ begin at the origin. When $\theta = 0$, P is at the origin; when $\theta = \pi$, P is at a maximum point $(\pi a, 2a)$; and when $\theta = 2\pi$, P is back on the x-axis at $(2\pi a, 0)$. From Figure 11.7 we see that $\angle APC = 180° - \theta$. Hence,

$$\sin \theta = \sin (180° - \theta) = \sin (\angle APC) = \frac{AC}{a} = \frac{BD}{a}$$

$$\cos \theta = -\cos (180° - \theta) = -\cos (\angle APC) = \frac{AP}{-a}$$

which implies that

$$AP = -a \cos \theta \quad \text{and} \quad BD = a \sin \theta.$$

Now, since the circle rolls along the *x*-axis, we know that $OD = \overset{\frown}{PD} = a\theta$. Furthermore, since $BA = DC = a$, we have

$$x = OD - BD = a\theta - a \sin \theta$$
$$y = BA + AP = a - a \cos \theta.$$

Therefore, the parametric equations are

$$x = a(\theta - \sin \theta) \qquad \text{and} \qquad y = a(1 - \cos \theta).$$

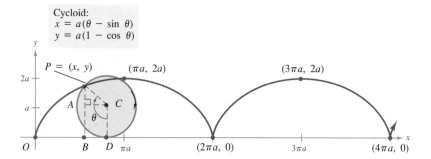

Cycloid:
$x = a(\theta - \sin \theta)$
$y = a(1 - \cos \theta)$

FIGURE 11.7

The cycloid in Figure 11.7 has sharp corners at the values $x = 2n\pi a$. Between these points, the cycloid is called **smooth.**

DEFINITION OF A SMOOTH CURVE

A curve *C* represented by $x = f(t)$ and $y = g(t)$ on an interval *I* is called **smooth** if f' and g' are continuous on *I* and not simultaneously zero, except possibly at the endpoints of *I*. The curve *C* is called **piecewise smooth** if it is smooth on each subinterval of some partition of *I*.

The brachistochrone problem

The type of curve described in Example 5 is related to one of the most famous pairs of problems in the history of calculus. The first problem began with Galileo's discovery that the time required to complete a full swing of a given pendulum is *approximately* the same, whether it makes a large movement at high speeds or a small movement at lower speeds. Late in his life, Galileo (1564–1642) realized that he could use this principle to construct a clock. However, he was not able to conquer the mechanics of actual construction. Christian Huygens (1629–1695) was the first to design and construct a working model. In his work with pendulums, Huygens realized that a pendulum does not take exactly the same time to complete swings of varying lengths. (This doesn't affect a pendulum clock, since the length of the circular arc is kept constant by giving the pendulum a slight boost each time it passes its lowest point.) But, in studying the problem, Huygens discovered that a ball rolling back and forth on an inverted cycloid does complete each cycle in exactly the same time.

The second problem, posed by John Bernoulli in 1696, is called the **brachistochrone problem**—in Greek *brachys* means short and *chronos* means time. The problem was to determine the path down which a particle will slide from point *A* to point *B* in the *shortest time*. Several mathematicians

James Bernoulli

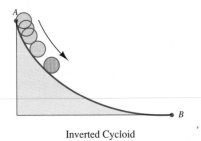

Inverted Cycloid

FIGURE 11.8

took up the challenge, and the following year the problem was solved by Newton, Leibniz, L'Hôpital, John Bernoulli, and James Bernoulli. As it turns out, the solution is not a straight line from A to B, but an inverted cycloid passing through the points A and B, as shown in Figure 11.8. The amazing part of the solution is that a particle starting at rest at *any* other point C of the cycloid between A and B will take exactly the same time to reach B.

James Bernoulli (1654–1705), also called Jacques, was the older brother of John. He was one of several accomplished mathematicians of the Swiss Bernoulli family. James's mathematical accomplishments have given him a prominent place in the early development of calculus.

EXERCISES for Section 11.1

In Exercises 1–30, sketch the curve represented by the parametric equations (indicate the orientation of the curve), and find the corresponding rectangular equation by eliminating the parameter.

1. $x = 3t - 1$
$y = 2t + 1$

2. $x = 3 - 2t$
$y = 2 + 3t$

3. $x = \sqrt{t}$
$y = 1 - t$

4. $x = \sqrt[3]{t}$
$y = 1 - t$

5. $x = t + 1$
$y = t^2$

6. $x = t + 1$
$y = t^3$

7. $x = t^3$
$y = \dfrac{t^2}{2}$

8. $x = 1 + \dfrac{1}{t}$
$y = t - 1$

9. $x = t - 1$
$y = \dfrac{t}{t - 1}$

10. $x = t^2 + t$
$y = t^2 - t$

11. $x = 2t$
$y = |t - 2|$

12. $x = |t - 1|$
$y = t + 2$

13. $x = \sec \theta$
$y = \cos \theta$

14. $x = \tan^2 \theta$
$y = \sec^2 \theta$

15. $x = 3 \cos \theta$
$y = 3 \sin \theta$

16. $x = \cos \theta$
$y = 3 \sin \theta$

17. $x = 4 \sin 2\theta$
$y = 2 \cos 2\theta$

18. $x = \cos \theta$
$y = 2 \sin 2\theta$

19. $x = \cos \theta$
$y = 2 \sin^2 \theta$

20. $x = 4 \cos^2 \theta$
$y = 2 \sin \theta$

21. $x = 4 + 2 \cos \theta$
$y = -1 + \sin \theta$

22. $x = 4 + 2 \cos \theta$
$y = -1 + 2 \sin \theta$

23. $x = 4 + 2 \cos \theta$
$y = -1 + 4 \sin \theta$

24. $x = \sec \theta$
$y = \tan \theta$

25. $x = 4 \sec \theta$
$y = 3 \tan \theta$

26. $x = \cos^3 \theta$
$y = \sin^3 \theta$

27. $x = t^3$
$y = 3 \ln t$

28. $x = e^{2t}$
$y = e^t$

29. $x = e^{-t}$
$y = e^{3t}$

30. $x = \ln 2t$
$y = t^2$

In Exercises 31 and 32, determine how the plane curves differ from each other.

31. (a) $x = t$
$y = 2t + 1$
(c) $x = e^{-t}$
$y = 2e^{-t} + 1$

(b) $x = \cos \theta$
$y = 2 \cos \theta + 1$
(d) $x = e^t$
$y = 2e^t + 1$

32. (a) $x = 2 \cos \theta$
$y = 2 \sin \theta$
(c) $x = \sqrt{t}$
$y = \sqrt{4 - t}$

(b) $x = \dfrac{\sqrt{4t^2 - 1}}{|t|}$
$y = \dfrac{1}{t}$
(d) $x = -\sqrt{4 - e^{2t}}$
$y = e^t$

In Exercises 33–38, sketch the curve represented by the parametric equations.

33. Cycloid: $x = 2(\theta - \sin \theta)$
$y = 2(1 - \cos \theta)$

34. Cycloid: $x = \theta + \sin \theta$
$y = 1 - \cos \theta$

35. Witch of Agnesi: $x = 2 \cot \theta$
$y = 2 \sin^2 \theta$

36. Curtate cycloid: $x = 2\theta - \sin \theta$
$y = 2 - \cos \theta$

37. Prolate cycloid: $x = \theta - \dfrac{3}{2} \sin \theta$
$y = 1 - \dfrac{3}{2} \cos \theta$

38. Folium of Descartes: $x = \dfrac{3t}{1 + t^3}$
$y = \dfrac{3t^2}{1 + t^3}$

In Exercises 39–44, eliminate the parameter and obtain the standard form of the rectangular equation of the curve.

39. Line through (x_1, y_1) and (x_2, y_2):
$$x = x_1 + t(x_2 - x_1)$$
$$y = y_1 + t(y_2 - y_1)$$
40. Circle: $x = r \cos \theta$
$$y = r \sin \theta$$
41. Circle: $x = h + r \cos \theta$
$$y = k + r \sin \theta$$
42. Ellipse: $x = h + a \cos \theta$
$$y = k + b \sin \theta$$
43. Hyperbola: $x = h + a \sec \theta$
$$y = k + b \tan \theta$$
44. Hyperbola: $x = h + a\sqrt{t + 1}$
$$y = k + b\sqrt{t}$$

In Exercises 45–52, find a set of parametric equations for the given graph.

45. Line: passes through $(0, 0)$ and $(5, -2)$
46. Line: passes through $(1, 4)$ and $(5, -2)$
47. Circle: center at $(2, 1)$ and radius of 4
48. Circle: center at $(-3, 1)$ and radius of 3
49. Ellipse: vertices at $(\pm 5, 0)$ and foci at $(\pm 4, 0)$
50. Ellipse: vertices at $(4, 7)$ and $(4, -3)$ and foci at $(4, 5)$ and $(4, -1)$
51. Hyperbola: vertices at $(\pm 4, 0)$ and foci at $(\pm 5, 0)$
52. Hyperbola: vertices at $(0, \pm 1)$ and foci at $(0, \pm 2)$

In Exercises 53 and 54, find two different sets of parametric equations for the given rectangular equation.

53. $y = x^3$ **54.** $y = x^2$

55. A wheel of radius a rolls along a straight line without slipping. The curve traced by a point P that is b units from the center ($b < a$) is called a **curtate cycloid** (see figure). Use the angle θ shown in the figure to find a set of parametric equations for the curve.

FIGURE FOR 55

56. A circle of radius 1 rolls around the outside of a circle of radius 2 without slipping. The curve traced by a point on the circumference of the smaller circle is called an **epicycloid** (see figure). Use the angle θ shown in the figure to find a set of parametric equations for the curve.

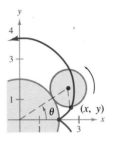

FIGURE FOR 56

In Exercises 57 and 58, sketch the **Lissajous curve**. A Lissajous curve can be represented by

$$x = a \cos nt \quad \text{and} \quad y = b \sin mt$$

where n/m is a rational number.

57. $x = 2 \cos \dfrac{t}{2}$ **58.** $x = \cos 2t$
$\quad\ y = \sin t$ $\qquad\quad y = \sin t$

In Exercises 59–62, use a computer or graphics calculator to sketch the curve described by the parametric equations over the specified interval.

Equations	Interval
59. $x = 4\sqrt{t} - 1$	$0 \le t \le 4$
$\quad\ y = 4t^2 - t^3$	
60. $x = 3 \cos^3 \theta$	$0 \le \theta \le 2\pi$
$\quad\ y = 3 \sin^3 \theta$	
61. $x = 2\theta - 4 \sin \theta$	$-\pi \le \theta \le 3\pi$
$\quad\ y = 2 - 4 \cos \theta$	
62. $x = 3(\theta + \sin \theta)$	$-\pi \le \theta \le 3\pi$
$\quad\ y = 3(1 - \cos \theta)$	

11.2 Parametric Equations and Calculus

Slope and tangent lines ▪ Arc length ▪ Area of a surface of revolution

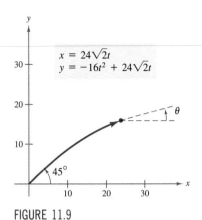

FIGURE 11.9

Now that we can represent a curve in the plane by a set of parametric equations, it is natural to ask how to use calculus to study curves. To begin, let's take another look at the projectile represented by the parametric equations

$$x = 24\sqrt{2}\,t \qquad \text{and} \qquad y = -16t^2 + 24\sqrt{2}\,t$$

as shown in Figure 11.9. From the previous section we know that these parametric equations enable us to locate the position of the projectile at a given time. We also know that the object is initially projected at an angle of 45°. But, how can we find the angle θ representing the object's direction at some other time t? The following theorem answers this question by giving us a formula for the slope of the tangent line as a function of t.

**THEOREM 11.1
PARAMETRIC FORM
OF THE DERIVATIVE**

If a smooth curve C is given by the equations $x = f(t)$ and $y = g(t)$, then the slope of C at (x, y) is

$$\frac{dy}{dx} = \frac{dy/dt}{dx/dt}, \qquad \frac{dx}{dt} \neq 0.$$

PROOF

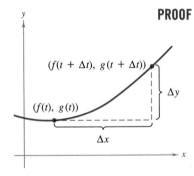

FIGURE 11.10

In Figure 11.10 consider $\Delta t > 0$ and let

$$\Delta y = g(t + \Delta t) - g(t)$$

and

$$\Delta x = f(t + \Delta t) - f(t).$$

Since

$$\frac{dy}{dx} = \lim_{\Delta x \to 0} \frac{\Delta y}{\Delta x}$$

and $\Delta x \to 0$ as $\Delta t \to 0$, we can write

$$\frac{dy}{dx} = \lim_{\Delta t \to 0} \frac{g(t + \Delta t) - g(t)}{f(t + \Delta t) - f(t)}.$$

Dividing both the numerator and denominator by Δt, we can use the differentiability of f and g to conclude that

$$\frac{dy}{dx} = \lim_{\Delta t \to 0} \frac{[g(t + \Delta t) - g(t)]/\Delta t}{[f(t + \Delta t) - f(t)]/\Delta t}$$

$$= \frac{\displaystyle\lim_{\Delta t \to 0} \frac{g(t + \Delta t) - g(t)}{\Delta t}}{\displaystyle\lim_{\Delta t \to 0} \frac{f(t + \Delta t) - f(t)}{\Delta t}} = \frac{g'(t)}{f'(t)} = \frac{dy/dt}{dx/dt}.$$

Since dy/dx is a function of t, we can use Theorem 11.1 repeatedly to find *higher-order* derivatives. For instance,

$$\frac{d^2y}{dx^2} = \frac{d}{dx}\left[\frac{dy}{dx}\right] = \frac{\dfrac{d}{dt}\left[\dfrac{dy}{dx}\right]}{dx/dt} \qquad \text{Second derivative}$$

$$\frac{d^3y}{dx^3} = \frac{d}{dx}\left[\frac{d^2y}{dx^2}\right] = \frac{\dfrac{d}{dt}\left[\dfrac{d^2y}{dx^2}\right]}{dx/dt}. \qquad \text{Third derivative}$$

EXAMPLE 1 Finding slope and concavity

For the curve given by

$$x = \sqrt{t} \qquad \text{and} \qquad y = \frac{1}{4}(t^2 - 4)$$

find the slope and concavity at the point $(2, 3)$.

SOLUTION

Since

$$\frac{dy}{dx} = \frac{dy/dt}{dx/dt} = \frac{(1/2)t}{(1/2)t^{-1/2}} = t^{3/2}$$

we have

$$\frac{d^2y}{dx^2} = \frac{\dfrac{d}{dt}\left[\dfrac{dy}{dx}\right]}{dx/dt} = \frac{(3/2)t^{1/2}}{(1/2)t^{-1/2}} = 3t.$$

At $(x, y) = (2, 3)$, it follows that $t = 4$, and the slope is

$$\frac{dy}{dx} = (4)^{3/2} = 8.$$

Moreover, when $t = 4$, the second derivative is

$$\frac{d^2y}{dx^2} = 3(4) = 12 > 0$$

and we conclude that the graph is concave upward at $(2, 3)$, as shown in Figure 11.11.

$$x = \sqrt{t}, \quad y = \frac{1}{4}(t^2 - 4)$$

FIGURE 11.11

Since the parametric equations $x = f(t)$ and $y = g(t)$ need not define y as a function of x, it follows that a curve defined parametrically can loop and cross itself in the plane. At such points the curve may have more than one tangent line, as shown in the next example.

EXAMPLE 2   A curve with two tangent lines at a point

The **prolate cycloid** given by

$$x = 2t - \pi \sin t \qquad \text{and} \qquad y = 2 - \pi \cos t$$

crosses itself at the point $(0, 2)$, as shown in Figure 11.12. Find the equations of both tangent lines at this point.

SOLUTION

Since $x = 0$ and $y = 2$ when $t = \pm\pi/2$, and since

$$\frac{dy}{dx} = \frac{dy/dt}{dx/dt} = \frac{\pi \sin t}{2 - \pi \cos t}$$

we have $dy/dx = -\pi/2$ when $t = -\pi/2$ and $dy/dx = \pi/2$ when $t = \pi/2$. Therefore, the two tangent lines at $(0, 2)$ are

$$y - 2 = -\left(\frac{\pi}{2}\right)x \qquad \text{Tangent line when } t = -\frac{\pi}{2}$$

$$y - 2 = \left(\frac{\pi}{2}\right)x. \qquad \text{Tangent line when } t = \frac{\pi}{2}$$

If $dy/dt = 0$ and $dx/dt \neq 0$ when $t = t_0$, then the curve represented by $x = f(t)$ and $y = g(t)$ has a *horizontal* tangent at $(f(t_0), g(t_0))$. For instance, in Example 2, the given curve has a horizontal tangent at the point $(0, 2 - \pi)$ (when $t = 0$). Similarly, if $dx/dt = 0$ and $dy/dt \neq 0$ when $t = t_0$, then the curve represented by $x = f(t)$ and $y = g(t)$ has a *vertical* tangent at $(f(t_0), g(t_0))$.

Arc length

We have some idea how parametric equations can be used to describe the path of a particle moving in the plane. We now develop a formula for determining the *distance* traveled by the particle along its path.

Recall from Section 7.4 that the formula for the arc length of a curve C given by $y = h(x)$ over the interval $[x_0, x_1]$ is

$$s = \int_{x_0}^{x_1} \sqrt{1 + [h'(x)]^2} \, dx = \int_{x_0}^{x_1} \sqrt{1 + \left(\frac{dy}{dx}\right)^2} \, dx.$$

If C is represented by the parametric equations $x = f(t)$ and $y = g(t)$, $a \leq t \leq b$, and if $dx/dt = f'(t) > 0$, then we can write

$$\begin{aligned} s &= \int_{x_0}^{x_1} \sqrt{1 + \left(\frac{dy}{dx}\right)^2} \, dx = \int_{x_0}^{x_1} \sqrt{1 + \left(\frac{dy/dt}{dx/dt}\right)^2} \, dx \\ &= \int_a^b \sqrt{\frac{(dx/dt)^2 + (dy/dt)^2}{(dx/dt)^2}} \frac{dx}{dt} \, dt \\ &= \int_a^b \sqrt{\left(\frac{dx}{dt}\right)^2 + \left(\frac{dy}{dt}\right)^2} \, dt = \int_a^b \sqrt{[f'(t)]^2 + [g'(t)]^2} \, dt. \end{aligned}$$

$x = 2t - \pi \sin t$
$y = 2 - \pi \cos t$

FIGURE 11.12

This formula for arc length can be shown to be valid for *any* smooth curve *C* that does not intersect itself.

THEOREM 11.2
ARC LENGTH IN
PARAMETRIC FORM

If a smooth curve *C* is given by $x = f(t)$ and $y = g(t)$ such that *C* does not intersect itself on the interval $a \le t \le b$ (except possibly at the endpoints), then the arc length of *C* over the interval is given by

$$s = \int_a^b \sqrt{\left(\frac{dx}{dt}\right)^2 + \left(\frac{dy}{dt}\right)^2}\, dt = \int_a^b \sqrt{[f'(t)]^2 + [g'(t)]^2}\, dt.$$

In the previous section we saw that if a circle rolls along a line, then a point on its circumference will trace a path called a cycloid. If the circle rolls around the circumference of another circle, then the path of the point is called an **epicycloid.** In the next example, we find the arc length of an epicycloid.

EXAMPLE 3 Finding arc length

A circle of radius 1 rolls around the circumference of a larger circle of radius 4, as shown in Figure 11.13. The epicycloid traced by a point on the circumference of the smaller circle is given by

$$x = 5\cos t - \cos 5t \qquad \text{and} \qquad y = 5\sin t - \sin 5t.$$

Find the distance traveled by the point in one complete trip about the larger circle.

SOLUTION

Before applying Theorem 11.2, we note from Figure 11.13 that the curve has sharp points when $t = 0$ and $t = \pi/2$. Between these two points, dx/dt and dy/dt are not simultaneously zero. Thus, the portion of the curve generated from $t = 0$ to $t = \pi/2$ is smooth. To find the total distance traveled by the point, we find the arc length of that portion lying in the first quadrant and multiply by 4.

$$s = 4\int_0^{\pi/2} \sqrt{\left(\frac{dx}{dt}\right)^2 + \left(\frac{dy}{dt}\right)^2}\, dt$$

$$= 4\int_0^{\pi/2} \sqrt{(-5\sin t + 5\sin 5t)^2 + (5\cos t - 5\cos 5t)^2}\, dt$$

$$= 20\int_0^{\pi/2} \sqrt{2 - 2\sin t \sin 5t - 2\cos t \cos 5t}\, dt$$

$$= 20\int_0^{\pi/2} \sqrt{2 - 2\cos 4t}\, dt = 20\int_0^{\pi/2} \sqrt{4\sin^2 2t}\, dt$$

$$= 40\int_0^{\pi/2} \sin 2t\, dt = -20\Big[\cos 2t\Big]_0^{\pi/2} = 40$$

For the epicycloid shown in Figure 11.13, an arc length of 40 seems about right because the circumference of a circle of radius 6 is $2\pi r = 12\pi \approx 37.7$.

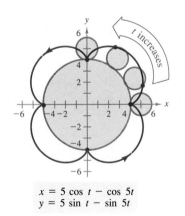

$x = 5\cos t - \cos 5t$
$y = 5\sin t - \sin 5t$

FIGURE 11.13

0.5 in.

0.001 in.

2 in.

$x = r \cos \theta$
$y = r \sin \theta$

(x, y)

r

θ

FIGURE 11.14

EXAMPLE 4 An application involving arc length

A recording tape 0.001 inch thick is wound around a reel whose inner radius is 0.5 inch and outer radius is 2 inches, as shown in Figure 11.14. How much tape is required to fill the reel?

SOLUTION

To create a model for this problem, we assume that as the tape is wound around the reel its distance r from the center increases linearly at the rate of 0.001 inch per revolution, or

$$r = (0.001)\frac{\theta}{2\pi} = \frac{\theta}{2000\pi}, \quad 1000\pi \le \theta \le 4000\pi$$

where θ is measured in radians. Now, we can determine the coordinates of the point (x, y) corresponding to a given radius to be

$$x = r \cos \theta$$

and

$$y = r \sin \theta.$$

Substituting for r, we obtain the parametric equations

$$x = \left(\frac{\theta}{2000\pi}\right) \cos \theta$$

and

$$y = \left(\frac{\theta}{2000\pi}\right) \sin \theta.$$

Now, we use the arc length formula to determine the total length of the tape to be

$$s = \int_{1000\pi}^{4000\pi} \sqrt{\left(\frac{dx}{d\theta}\right)^2 + \left(\frac{dy}{d\theta}\right)^2} \, d\theta$$

$$= \frac{1}{2000\pi} \int_{1000\pi}^{4000\pi} \sqrt{(-\theta \sin \theta + \cos \theta)^2 + (\theta \cos \theta + \sin \theta)^2} \, d\theta$$

$$= \frac{1}{2000\pi} \int_{1000\pi}^{4000\pi} \sqrt{\theta^2 + 1} \, d\theta \qquad \text{Integration tables, Formula 26}$$

$$= \frac{1}{2000\pi}\left(\frac{1}{2}\right)\left[\theta\sqrt{\theta^2 + 1} + \ln\left|\theta + \sqrt{\theta^2 + 1}\right| \right]_{1000\pi}^{4000\pi}$$

$$\approx 11{,}781 \text{ in.} \approx 982 \text{ ft.}$$

REMARK The graph of $r = a\theta$ is called the **Spiral of Archimedes.**

Area of a surface of revolution

As we did for arc length, we can use the formula for the area of a surface of revolution in rectangular form to develop a formula for surface area in parametric form.

THEOREM 11.3 AREA OF A SURFACE OF REVOLUTION	If a smooth curve C given by $x = f(t)$ and $y = g(t)$ does not cross itself on an interval $a \le t \le b$, then the area S of the surface of revolution formed by revolving C about the coordinate axes is given by the following.

$$1. \quad S = 2\pi \int_a^b g(t) \sqrt{\left(\frac{dx}{dt}\right)^2 + \left(\frac{dy}{dt}\right)^2}\, dt \qquad \text{Revolution about } x\text{-axis: } g(t) \ge 0$$

$$2. \quad S = 2\pi \int_a^b f(t) \sqrt{\left(\frac{dx}{dt}\right)^2 + \left(\frac{dy}{dt}\right)^2}\, dt \qquad \text{Revolution about } y\text{-axis: } f(t) \ge 0$$

EXAMPLE 5   Finding the area of a surface of revolution

Let C be the arc of the circle $x^2 + y^2 = 9$ from $(3, 0)$ to $(3/2, 3\sqrt{3}/2)$, as shown in Figure 11.15. Find the area of the surface formed by revolving C about the x-axis.

SOLUTION

We can represent C parametrically by the equations

$$x = 3 \cos t \qquad \text{and} \qquad y = 3 \sin t, \quad 0 \le t \le \pi/3.$$

(Note that we determine the interval for t by observing that $t = 0$ when $x = 3$ and $t = \pi/3$ when $x = 3/2$.) On this interval, C is smooth and y is nonnegative, and we can apply Theorem 11.3 to obtain a surface area of

$$S = 2\pi \int_0^{\pi/3} (3 \sin t) \sqrt{(-3 \sin t)^2 + (3 \cos t)^2}\, dt$$

$$= 6\pi \int_0^{\pi/3} \sin t \sqrt{9(\sin^2 t + \cos^2 t)}\, dt$$

$$= 6\pi \int_0^{\pi/3} 3 \sin t\, dt$$

$$= -18\pi \left[\cos t \right]_0^{\pi/3}$$

$$= -18\pi \left[\frac{1}{2} - 1 \right]$$

$$= 9\pi.$$

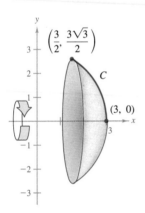

Surface of Revolution

FIGURE 11.15

EXERCISES for Section 11.2

In Exercises 1–10, find dy/dx and d^2y/dx^2, and evaluate each at the specified value of the parameter.

Parametric equations	Point
1. $x = 2t,\ y = 3t - 1$	$t = 3$
2. $x = \sqrt{t},\ y = 3t - 1$	$t = 1$
3. $x = t + 1,\ y = t^2 + 3t$	$t = -1$
4. $x = t^2 + 3t,\ y = t + 1$	$t = 0$
5. $x = 2\cos\theta,\ y = 2\sin\theta$	$\theta = \dfrac{\pi}{4}$
6. $x = \cos\theta,\ y = 3\sin\theta$	$\theta = 0$
7. $x = 2 + \sec\theta,\ y = 1 + 2\tan\theta$	$\theta = \dfrac{\pi}{6}$
8. $x = \sqrt{t},\ y = \sqrt{t} - 1$	$t = 2$
9. $x = \cos^3\theta,\ y = \sin^3\theta$	$\theta = \dfrac{\pi}{4}$
10. $x = \theta - \sin\theta,\ y = 1 - \cos\theta$	$\theta = \pi$

In Exercises 11–16, find an equation of the tangent line to the curve at the specified value of the parameter.

Parametric equations	Point
11. $x = 2t,\ y = t^2 - 1$	$t = 2$
12. $x = t - 1,\ y = \dfrac{1}{t} + 1$	$t = 1$
13. $x = t^2 - t + 2,\ y = t^3 - 3t$	$t = -1$
14. $x = 4\cos\theta,\ y = 3\sin\theta$	$\theta = \dfrac{3\pi}{4}$
15. $x = 2\cot\theta,\ y = 2\sin^2\theta$	$\theta = \dfrac{\pi}{4}$
16. $x = 2 - 3\cos\theta,\ y = 3 + 2\sin\theta$	$\theta = \dfrac{5\pi}{3}$

In Exercises 17–28, find all points (if any) of horizontal and vertical tangency.

17. $x = 1 - t$
$y = t^2$

18. $x = t + 1$
$y = t^2 + 3t$

19. $x = 1 - t$
$y = t^3 - 3t$

20. $x = t^2 - t + 2$
$y = t^3 - 3t$

21. $x = 3\cos\theta$
$y = 3\sin\theta$

22. $x = \cos\theta$
$y = 2\sin 2\theta$

23. $x = 4 + 2\cos\theta$
$y = -1 + \sin\theta$

24. $x = 4\cos^2\theta$
$y = 2\sin\theta$

25. $x = \sec\theta$
$y = \tan\theta$

26. $x = \cos^2\theta$
$y = \cos\theta$

27. $x = 2\theta$
$y = 2(1 - \cos\theta)$

28. Involute of a circle:
$x = \cos\theta + \theta\sin\theta$
$y = \sin\theta - \theta\cos\theta$

In Exercises 29–34, find the arc length of the given curve.

Parametric equations	Interval
29. $x = e^{-t}\cos t,\ y = e^{-t}\sin t$	$0 \le t \le \dfrac{\pi}{2}$
30. $x = t^2,\ y = 4t^3 - 1$	$-1 \le t \le 1$
31. $x = t^2,\ y = 2t$	$0 \le t \le 2$
32. $x = \arcsin t,\ y = \ln\sqrt{1 - t^2}$	$0 \le t \le \dfrac{1}{2}$
33. $x = \sqrt{t},\ y = 3t - 1$	$0 \le t \le 1$
34. $x = t,\ y = \dfrac{t^5}{10} + \dfrac{1}{6t^3}$	$1 \le t \le 2$

In Exercises 35–38, find the arc length of the given curve.

35. Perimeter of a hypocycloid:
$x = a\cos^3\theta,\ y = a\sin^3\theta,\ 0 \le \theta \le 2\pi$

36. Circumference of a circle:
$x = a\cos\theta,\ y = a\sin\theta,\ 0 \le \theta \le 2\pi$

37. Length of one arch of a cycloid:
$x = a(\theta - \sin\theta),\ y = a(1 - \cos\theta),\ 0 \le \theta \le 2\pi$

38. Circumference of an ellipse:
$x = 3\cos\theta,\ y = 4\sin\theta,\ 0 \le \theta < 2\pi$
(Use Simpson's Rule with $n = 6$.)

In Exercises 39–44, find the area of the surface generated by revolving the curve about the given axis.

39. $x = t,\ y = 2t,\ 0 \le t \le 4$
   (a) x-axis   (b) y-axis

40. $x = t,\ y = 4 - 2t,\ 0 \le t \le 2$
   (a) x-axis   (b) y-axis

41. $x = 4\cos\theta,\ y = 4\sin\theta,\ 0 \le \theta \le \dfrac{\pi}{2}$
   y-axis

42. $x = t^3,\ y = t + 2,\ 1 \le t \le 2$
   y-axis

43. $x = a\cos^3\theta,\ y = a\sin^3\theta,\ 0 \le \theta \le \pi$
   x-axis

44. $x = a\cos\theta,\ y = b\sin\theta,\ 0 \le \theta \le 2\pi$
   (a) x-axis   (b) y-axis

45. A portion of a sphere is removed by a circular cone with its vertex at the center of the sphere. Find the surface area removed from the sphere if the vertex of the cone forms an angle 2θ.

46. Use integration by substitution to show that if y is a continuous function of x on the interval $a \le x \le b$, where $x = f(t)$ and $y = g(t)$, then

$$\int_a^b y \, dx = \int_{t_1}^{t_2} g(t) f'(t) \, dt,$$

where $f(t_1) = a$, $f(t_2) = b$, and both g and f' are continuous on $[t_1, t_2]$.

In Exercises 47–50, find the area of the indicated region. (Use the result of Exercise 46.)

47. $x = \cos^3 \theta$
$y = \sin^3 \theta$

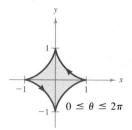

$0 \le \theta \le 2\pi$

48. $x = a \cos \theta$
$y = b \sin \theta$

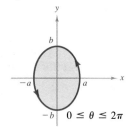

$0 \le \theta \le 2\pi$

49. $x = 2 \sin^2 \theta$
$y = 2 \sin^2 \theta \tan \theta$

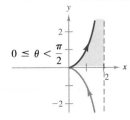

$0 \le \theta < \dfrac{\pi}{2}$

50. $x = 2 \cot \theta$
$y = 2 \sin^2 \theta$

$0 < \theta < \pi$

In Exercises 51 and 52, find the centroid of the region bounded by the graph of the parametric equations and the coordinate axes. (Use the result of Exercise 46.)

51. $x = \sqrt{t}$
$y = 4 - t$

52. $x = \sqrt{4 - t}$
$y = \sqrt{t}$

In Exercises 53 and 54, find the volume of the solid formed by revolving the region bounded by the graphs of the given equations about the x-axis. (Use the result of Exercise 46.)

53. $x = 3 \cos \theta$
$y = 3 \sin \theta$

54. $x = \cos \theta$
$y = 3 \sin \theta$

55. Given the parametric equations

$$x = \frac{4t}{1 + t^3} \quad \text{and} \quad y = \frac{4t^2}{1 + t^3}$$

use a computer or graphics calculator to do the following.
(a) Sketch the curve described by the parametric equations.
(b) Find the points of horizontal tangency to the curve.
(c) Use Simpson's Rule with $n = 10$ to approximate the arc length of the closed loop. [Hint: Because of symmetry we need only integrate for $0 \le t \le 1$.]

11.3 Polar Coordinates and Polar Graphs

Polar coordinates ▪ Coordinate conversion ▪ Graphs of polar equations ▪ Symmetry

So far, we have been representing plane curves as collections of points (x, y) in the rectangular coordinate system, where x and y represent the directed distances from the coordinate axes to the point (x, y). The corresponding equations for these curves have been in either rectangular or parametric form. In this section we introduce a third system called the **polar coordinate system.**

To form the polar coordinate system in the plane, we fix a point O, called the **pole** (or **origin**), and construct from O an initial ray called the **polar axis,** as shown in Figure 11.16. Then each point P in the plane can be assigned **polar coordinates** (r, θ), as follows.

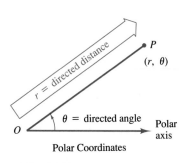

Polar Coordinates

FIGURE 11.16

$r = $ *directed distance* from O to P

$\theta = $ *directed angle*, counterclockwise from the polar axis to segment $\overline{OP}$

(a)

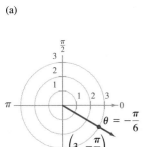

(b)

(c)

FIGURE 11.17

In the polar coordinate system, it is convenient to locate points with respect to a grid of concentric circles intersected by **radial lines** through the pole. This procedure is shown in the following example.

EXAMPLE 1 Plotting points in the polar coordinate system

(a) The point $(r, \theta) = (2, \pi/3)$ lies at the intersection of a circle of radius $r = 2$ and a ray that is the terminal side of the angle $\theta = \pi/3$, as shown in Figure 11.17(a).
(b) The point $(r, \theta) = (3, -\pi/6)$ lies in the fourth quadrant, 3 units from the pole. Note that we measure negative angles *clockwise*, as shown in Figure 11.17(b).
(c) The point $(r, \theta) = (3, 11\pi/6)$ coincides with the point $(3, -\pi/6)$, as shown in Figure 11.17(c). ▭

With rectangular coordinates, each point (x, y) has a unique representation. This is not true with polar coordinates. For instance, the coordinates (r, θ) and $(r, 2\pi + \theta)$ represent the same point, as illustrated in Example 1. Another way to obtain multiple representations of a point is to use negative values for r. Since r is a *directed distance*, the coordinates (r, θ) and $(-r, \theta + \pi)$ represent the same point. In general, the point (r, θ) can be written as

$$(r, \theta) = (r, \theta + 2n\pi) \qquad \text{or} \qquad (r, \theta) = (-r, \theta + (2n + 1)\pi)$$

where n is any integer. Moreover, the pole is represented by $(0, \theta)$, where θ is any angle.

EXAMPLE 2 Multiple representations of points

Plot the point $(3, -3\pi/4)$ and find three additional polar coordinate representations of this point, using $-2\pi < \theta < 2\pi$.

SOLUTION

The point is shown in Figure 11.18. Three other representations are as follows.

$$\left(3, -\frac{3\pi}{4} + 2\pi\right) = \left(3, \frac{5\pi}{4}\right) \qquad \text{Add } 2\pi \text{ to } \theta$$

$$\left(-3, -\frac{3\pi}{4} - \pi\right) = \left(-3, -\frac{7\pi}{4}\right) \qquad \text{Replace } r \text{ by } -r \text{ and subtract } \pi \text{ from } \theta$$

$$\left(-3, -\frac{3\pi}{4} + \pi\right) = \left(-3, \frac{\pi}{4}\right) \qquad \text{Replace } r \text{ by } -r \text{ and add } \pi \text{ to } \theta \quad ▭$$

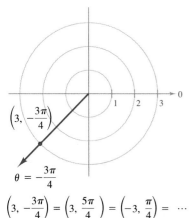

$$\left(3, -\frac{3\pi}{4}\right) = \left(3, \frac{5\pi}{4}\right) = \left(-3, \frac{\pi}{4}\right) = \cdots$$

FIGURE 11.18

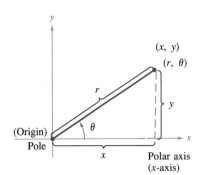

FIGURE 11.19

Relating Polar
and Rectangular Coordinates

To establish the relationship between polar and rectangular coordinates, we let the polar axis coincide with the positive x-axis and the pole with the origin, as shown in Figure 11.19. Since (x, y) lies on a circle of radius r, it follows that $r^2 = x^2 + y^2$. Moreover, for $r > 0$, the definition of the trigonometric functions implies that

$$\tan \theta = \frac{y}{x}, \qquad \cos \theta = \frac{x}{r}, \qquad \text{and} \qquad \sin \theta = \frac{y}{r}.$$

If $r < 0$, we can show that the same relationships hold. For example, consider the point (r, θ), where $r < 0$. Then, since $(-r, \theta + \pi)$ represents the same point and $-r > 0$, we have $-\sin \theta = \sin (\theta + \pi) = -y/r$, which implies that $\sin \theta = y/r$.

These relationships allow us to convert *coordinates* or *equations* from one system to the other, as indicated in the following theorem.

THEOREM 11.4
COORDINATE CONVERSION

The polar coordinates (r, θ) are related to the rectangular coordinates (x, y) as follows.

$$x = r \cos \theta \qquad \text{and} \qquad y = r \sin \theta$$

$$\tan \theta = \frac{y}{x} \qquad \text{and} \qquad r^2 = x^2 + y^2$$

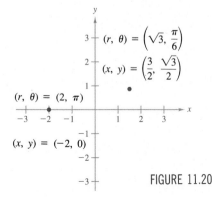

FIGURE 11.20

EXAMPLE 3   Polar-to-rectangular conversion

(a) For the point $(r, \theta) = (2, \pi)$, we have

$$x = r \cos \theta = 2 \cos \pi = -2 \qquad \text{and} \qquad y = r \sin \theta = 2 \sin \pi = 0.$$

Thus, the rectangular coordinates are $(x, y) = (-2, 0)$, as shown in Figure 11.20.

(b) For the point $(r, \theta) = (\sqrt{3}, \pi/6)$, we have

$$x = \sqrt{3} \cos \frac{\pi}{6} = \sqrt{3}\left(\frac{\sqrt{3}}{2}\right) = \frac{3}{2}$$

and

$$y = \sqrt{3} \sin \frac{\pi}{6} = \sqrt{3}\left(\frac{1}{2}\right) = \frac{\sqrt{3}}{2}.$$

Thus, the rectangular coordinates are $(x, y) = (3/2, \sqrt{3}/2)$, as shown in Figure 11.20. ◻

EXAMPLE 4 Rectangular-to-polar conversion

(a) For the second quadrant point $(x, y) = (-1, 1)$, we have

$$\tan \theta = \frac{y}{x} = -1 \quad \Longrightarrow \quad \theta = \frac{3\pi}{4}.$$

Since our choice of θ lies in the same quadrant as (x, y) we use positive r.

$$r = \sqrt{x^2 + y^2} = \sqrt{(-1)^2 + (1)^2} = \sqrt{2}$$

Thus, *one* set of polar coordinates is $(r, \theta) = (\sqrt{2}, 3\pi/4)$, as shown in Figure 11.21.

(b) Since the point $(x, y) = (0, 2)$ lies on the positive y-axis, we choose $\theta = \pi/2$ and $r = 2$, and one set of polar coordinates is $(r, \theta) = (2, \pi/2)$, as shown in Figure 11.21. ◻

$(r, \theta) = \left(\sqrt{2}, \dfrac{3\pi}{4}\right)$ $(r, \theta) = \left(2, \dfrac{\pi}{2}\right)$

$(x, y) = (-1, 1)$ $(x, y) = (0, 2)$

FIGURE 11.21

By comparing Examples 3 and 4, we see that point conversion from the polar to the rectangular system is straightforward, whereas point conversion from the rectangular to the polar system is more involved. For equations, the opposite is true. To convert a rectangular equation to polar form, we simply replace x by $r \cos \theta$ and y by $r \sin \theta$. For instance, the rectangular equation $y = x^2$ has the polar form

$$r \sin \theta = (r \cos \theta)^2.$$

On the other hand, to convert a polar equation to rectangular form can require considerable ingenuity.

Graphs of polar equations

Curve sketching in polar coordinates is similar to that for rectangular coordinates. We rely heavily on point-by-point plotting, aided by intercepts and symmetry. Converting to rectangular coordinates and then making the sketch is also an option, as demonstrated in the next example.

EXAMPLE 5 Graphing polar equations

Describe the graphs of the following polar equations and find the corresponding rectangular equations.

(a) $r = 2$ (b) $\theta = \dfrac{\pi}{3}$ (c) $r = \sec \theta$

SOLUTION

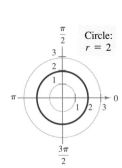

(a) The graph of the polar equation $r = 2$ consists of all points that are 2 units from the pole. In other words, this graph is a circle centered at the origin and having a radius of 2, as shown in Figure 11.22(a). We can confirm this by converting to rectangular coordinates, using the relationship $r^2 = x^2 + y^2$.

$$r = 2 \quad \Longrightarrow \quad r^2 = 2^2 \quad \Longrightarrow \quad x^2 + y^2 = 2^2$$

Polar equation $\qquad\qquad\qquad$ Rectangular equation

(b) The graph of the polar equation $\theta = \pi/3$ consists of all points on the line that makes an angle of $\pi/3$ with the positive x-axis, as shown in Figure 11.22(b). To convert to rectangular form, we make use of the relationship $\tan \theta = y/x$.

$$\theta = \frac{\pi}{3} \quad \Longrightarrow \quad \tan \theta = \sqrt{3} \quad \Longrightarrow \quad y = \sqrt{3}x$$

Polar equation $\qquad\qquad\qquad$ Rectangular equation

(c) The graph of the polar equation $r = \sec \theta$ is not evident by simple inspection, so we convert to rectangular form by using the relationship $r \cos \theta = x$.

$$r = \sec \theta \quad \Longrightarrow \quad r \cos \theta = 1 \quad \Longrightarrow \quad x = 1$$

Polar equation $\qquad\qquad\qquad$ Rectangular equation

Now we see that the graph is a vertical line, as shown in Figure 11.22(c).

FIGURE 11.22

Curve sketching by converting to rectangular form is not always convenient, and in the next example we demonstrate a straightforward point-plotting technique.

EXAMPLE 6 Graphing a polar equation by point-plotting

Sketch the graph of the polar equation

$$r = 4 \sin \theta.$$

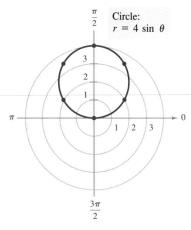

FIGURE 11.23

SOLUTION

Since the sine function is periodic, we can get a full range of r values by considering values of θ in the interval $0 \le \theta \le 2\pi$, as shown in Table 11.2.

TABLE 11.2

θ	0	$\frac{\pi}{6}$	$\frac{\pi}{3}$	$\frac{\pi}{2}$	$\frac{2\pi}{3}$	$\frac{5\pi}{6}$	π	$\frac{7\pi}{6}$	$\frac{3\pi}{2}$	$\frac{11\pi}{6}$	2π
r	0	2	$2\sqrt{3}$	4	$2\sqrt{3}$	2	0	-2	-4	-2	0

By plotting these points as shown in Figure 11.23, it appears that the graph is a circle of radius 2 whose center is at the point $(x, y) = (0, 2)$. (In Exercise 33 you are asked to confirm this by converting to rectangular form.)

Symmetry

When making a table of points, it is often helpful to use at least one θ-value from each of the four quadrants. However, in Figure 11.23 we see that no points are plotted in Quadrants III and IV because r is negative for these values of θ. Thus, on the interval from $\theta = 0$ to $\theta = 2\pi$ the curve is traced twice. Moreover, we also see that this curve is *symmetric with respect to the line $\theta = \pi/2$*. Had we known about this symmetry and retracing ahead of time we could have plotted fewer points.

Symmetry with respect to the line $\theta = \pi/2$ is one of three important types of symmetry to consider in polar curve sketching. (See Figure 11.24.)

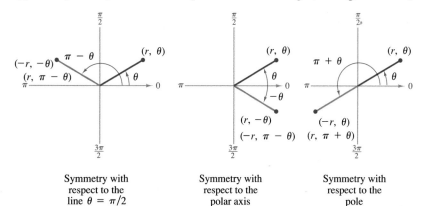

Symmetry with respect to the line $\theta = \pi/2$

Symmetry with respect to the polar axis

Symmetry with respect to the pole

FIGURE 11.24

The following theorem lists a test for each of the three types.

**THEOREM 11.5
TESTS FOR SYMMETRY IN
POLAR COORDINATES**

The graph of a polar equation is symmetric with respect to the following if the indicated substitution produces an equivalent equation.

1. *The line $\theta = \pi/2$:* Replace (r, θ) by $(r, \pi - \theta)$ or $(-r, -\theta)$.
2. *The polar axis:* Replace (r, θ) by $(r, -\theta)$ or $(-r, \pi - \theta)$.
3. *The pole:* Replace (r, θ) by $(r, \pi + \theta)$ or $(-r, \theta)$.

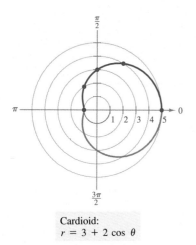

Cardioid:
$r = 3 + 2 \cos \theta$

FIGURE 11.25

EXAMPLE 7 Using symmetry to sketch the graph of a polar equation

Use symmetry to sketch the graph of $r = 3 + 2 \cos \theta$.

SOLUTION

Since the cosine is an even function, we replace (r, θ) by $(r, -\theta)$.

$$r = 3 + 2 \cos (-\theta) \quad \Longrightarrow \quad r = 3 + 2 \cos \theta$$

Since the substitution produces an equivalent equation, the curve is symmetric with respect to the polar axis. This means that we need only use θ-values from the first two quadrants, as shown in Table 11.3.

TABLE 11.3

θ	0	$\dfrac{\pi}{3}$	$\dfrac{\pi}{2}$	$\dfrac{2\pi}{3}$	π
r	5	4	3	2	1

Plotting these points and using polar axis symmetry produces the graph shown in Figure 11.25. ▭

The equations discussed in Examples 6 and 7 are of the following form.

$$r = 4 \sin \theta = f(\sin \theta) \qquad \text{r is a function of $\sin \theta$}$$
$$r = 3 + 2 \cos \theta = g(\cos \theta) \qquad \text{r is a function of $\cos \theta$}$$

From Figures 11.23 and 11.25 we see that the graphs of these equations are symmetric with respect to the line $\theta = \pi/2$ and the polar axis, respectively. This observation can be generalized to form the following *quick test for symmetry*.

1. The graph of $r = f(\sin \theta)$ is symmetric with respect to the line $\theta = \pi/2$.
2. The graph of $r = g(\cos \theta)$ is symmetric with respect to the polar axis.

When applying Theorem 11.5 note that we are using only two of the many possible polar representations of the point (r, θ). Consequently, the conditions of Theorem 11.5 are *sufficient* to guarantee symmetry, but they are not necessary (that is, a graph can have symmetry without satisfying the conditions of the theorem). For instance, Figure 11.26 shows the graph of $r = \theta + 2\pi$ to be symmetric with respect to the line $\theta = \pi/2$. Yet the test fails to indicate symmetry because neither of the following replacements yields an equivalent equation.

$r = \theta + 2\pi, \ -4\pi \le \theta \le 0$
Spiral of Archimedes

FIGURE 11.26

Original equation	Replacement	New equation
$r = \theta + 2\pi$	(r, θ) by $(-r, -\theta)$	$-r = -\theta + 2\pi$
$r = \theta + 2\pi$	(r, θ) by $(r, \pi - \theta)$	$r = -\theta + 3\pi$

EXERCISES for Section 11.3

In Exercises 1–8, the polar coordinates of a point are given. Plot the point and find the corresponding rectangular coordinates.

1. $\left(4, \dfrac{3\pi}{6}\right)$ 2. $\left(4, \dfrac{3\pi}{2}\right)$

3. $\left(-1, \dfrac{5\pi}{4}\right)$ 4. $(0, -\pi)$

5. $\left(4, -\dfrac{\pi}{3}\right)$ 6. $\left(-1, -\dfrac{3\pi}{4}\right)$

7. $(\sqrt{2}, 2.36)$ 8. $(-3, -1.57)$

In Exercises 9–16, the rectangular coordinates of a point are given. Find two sets of polar coordinates for the point, using $0 \le \theta < 2\pi$.

9. $(1, 1)$ 10. $(0, -5)$
11. $(-3, 4)$ 12. $(3, -1)$
13. $(-\sqrt{3}, -\sqrt{3})$ 14. $(-2, 0)$
15. $(4, 6)$ 16. $(5, 12)$

In Exercises 17–32, convert the given rectangular equation to polar form.

17. $x^2 + y^2 = 9$ 18. $x^2 + y^2 = a^2$
19. $x^2 + y^2 - 2ax = 0$ 20. $x^2 + y^2 - 2ay = 0$
21. $y = 4$ 22. $y = b$
23. $x = 10$ 24. $x = a$
25. $3x - y + 2 = 0$ 26. $4x + 7y - 2 = 0$
27. $xy = 4$ 28. $y = x$
29. $y^2 = 9x$ 30. $y^2 - 8x - 16 = 0$
31. $x^2 - 4ay - 4a^2 = 0$
32. $(x^2 + y^2)^2 - 9(x^2 - y^2) = 0$

In Exercises 33–44, convert the given polar equation to rectangular form.

33. $r = 4 \sin \theta$ 34. $r = 4 \cos \theta$
35. $\theta = \dfrac{\pi}{6}$ 36. $r = 4$
37. $r = 2 \csc \theta$ 38. $r^2 = \sin 2\theta$
39. $r = 1 - 2 \sin \theta$ 40. $r = \dfrac{1}{1 - \cos \theta}$
41. $r = \dfrac{6}{2 - 3 \sin \theta}$ 42. $r = 2 \sin 3\theta$
43. $r^2 = 4 \sin \theta$ 44. $r = \dfrac{6}{2 \cos \theta - 3 \sin \theta}$

In Exercises 45–54, sketch the graph of the polar equation. Indicate any symmetry possessed by the graph.

45. $r = 5$ 46. $r = -2$
47. $r = \theta$ 48. $r = \dfrac{\theta}{\pi}$
49. $r = \sin \theta$ 50. $r = 3 \cos \theta$
51. $r = 2 \sec \theta$ 52. $r = 3 \csc \theta$
53. $r = 4(2 + \sin \theta)$ 54. $r = 2 + \cos \theta$

55. Convert the equation

$$r = 2(h \cos \theta + k \sin \theta)$$

to rectangular form and verify that it is the equation of a circle. Find the radius and the rectangular coordinates of the center of the circle.

56. Verify that the distance between the two points (r_1, θ_1) and (r_2, θ_2) in polar coordinates is given by

$$d = \sqrt{r_1^2 + r_2^2 - 2r_1 r_2 \cos(\theta_1 - \theta_2)}.$$

11.4 Tangent Lines and Curve Sketching in Polar Coordinates

Relative extrema of r ▪ Tangent lines to polar graphs ▪ Special polar graphs

We have seen that polar curve sketching can be simplified by considering the periodic nature and symmetry of the graph. We begin this section by using calculus to develop two additional curve sketching aids.

If r is a differentiable function of θ, then we can determine the **relative extrema of r** using the procedures discussed in Chapter 4. However, since the distance between a point (r, θ) and the pole is $|r|$, it is possible that both the relative maxima and relative minima of r yield points whose distance from the pole is a relative maximum. We illustrate this with an example.

$r = 1 + 2 \cos \theta$

Set $dr/d\theta = 0$ to find the relative extrema of r.

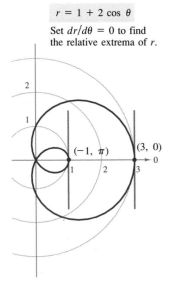

Polar Coordinates: $\dfrac{dr}{d\theta} \neq$ slope

FIGURE 11.27

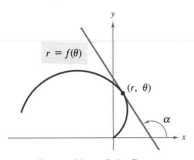

Tangent Line to Polar Curve

FIGURE 11.28

EXAMPLE 1   Relative extrema of r

Find the relative extrema of the function given by $r = 1 + 2 \cos \theta$ and locate the corresponding points on the graph of the function.

SOLUTION

By setting the derivative equal to zero, we have

$$\frac{dr}{d\theta} = -2 \sin \theta = 0$$

and we conclude that on the interval $0 \le \theta < 2\pi$ there are two critical numbers

$$\theta = 0 \qquad \text{and} \qquad \theta = \pi.$$

The corresponding points $(3, 0)$ and $(-1, \pi)$ are shown in Figure 11.27. The graph shown in this figure is called a **limaçon.**   ▭

Slope and tangent lines

Be sure you see that $dr/d\theta$ does not represent the slope of a polar graph. To find the slope of a tangent line to a polar graph, consider a differentiable function given by $r = f(\theta)$. If α is the angle from the polar axis to the tangent line at the point (r, θ), then $dy/dx = \tan \alpha$, as shown in Figure 11.28. To convert to polar form, we use the parametric equations

$$x = r \cos \theta = f(\theta) \cos \theta \qquad \text{and} \qquad y = r \sin \theta = f(\theta) \sin \theta.$$

Using the parametric form of dy/dx given in Theorem 11.1, we have

$$\frac{dy}{dx} = \frac{dy/d\theta}{dx/d\theta} = \frac{f(\theta) \cos \theta + f'(\theta) \sin \theta}{-f(\theta) \sin \theta + f'(\theta) \cos \theta}$$

which establishes the following theorem.

**THEOREM 11.6
SLOPE IN POLAR FORM**

If f is a differentiable function of θ, then the *slope* of the tangent line to the graph of $r = f(\theta)$ at the point (r, θ) is

$$\frac{dy}{dx} = \frac{dy/d\theta}{dx/d\theta} = \frac{f(\theta) \cos \theta + f'(\theta) \sin \theta}{-f(\theta) \sin \theta + f'(\theta) \cos \theta}$$

provided $dx/d\theta \neq 0$ at (r, θ).

From Theorem 11.6, we can make the following observations.

1. Solutions to $dy/d\theta = 0$ yield horizontal tangents, provided $dx/d\theta \neq 0$.
2. Solutions to $dx/d\theta = 0$ yield vertical tangents, provided $dy/d\theta \neq 0$.

If $dy/d\theta$ and $dx/d\theta$ are *simultaneously* zero, then no conclusions can be drawn about tangent lines.

EXAMPLE 2 Finding horizontal and vertical tangents to polar curves

Find the horizontal and vertical tangents to the graph of $r = 2(1 - \cos \theta)$.

SOLUTION

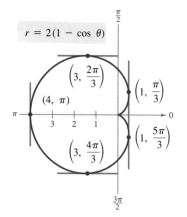

$r = 2(1 - \cos \theta)$

FIGURE 11.29

Using $y = r \sin \theta$, we differentiate and set $dy/d\theta$ equal to zero.

$$y = r \sin \theta = 2(1 - \cos \theta) \sin \theta$$

$$\frac{dy}{d\theta} = 2[(1 - \cos \theta)(\cos \theta) + \sin \theta(\sin \theta)]$$

$$= 2(1 + \cos \theta - 2 \cos^2 \theta) = -2(2 \cos \theta + 1)(\cos \theta - 1) = 0$$

Thus, $\cos \theta = -\frac{1}{2}$ and $\cos \theta = 1$, and we conclude that $dy/d\theta = 0$ when $\theta = 2\pi/3$, $4\pi/3$, and 0. Similarly, using $x = r \cos \theta$, we have

$$x = r \cos \theta = 2 \cos \theta - 2 \cos^2 \theta$$

$$\frac{dx}{d\theta} = -2 \sin \theta + 4 \cos \theta \sin \theta$$

$$= 2 \sin \theta(2 \cos \theta - 1) = 0.$$

Thus, $\sin \theta = 0$ or $\cos \theta = \frac{1}{2}$, and we conclude that $dx/d\theta = 0$ when $\theta = 0$, π, $\pi/3$, or $5\pi/3$. Finally, we eliminate $\theta = 0$ because both derivatives are zero there, and conclude that the graph, shown in Figure 11.29, has horizontal and vertical tangents at the following points.

$$\left(3, \frac{2\pi}{3}\right), \left(3, \frac{4\pi}{3}\right) \qquad \text{Horizontal tangents}$$

$$\left(1, \frac{\pi}{3}\right), \left(1, \frac{5\pi}{3}\right), (4, \pi) \qquad \text{Vertical tangents}$$

The graph is called a **cardioid**. ▭

Theorem 11.6 has an important consequence. Suppose that the graph of $r = f(\theta)$ passes through the pole when $\theta = \alpha$ and $f'(\alpha) \neq 0$. Then the formula for dy/dx simplifies as follows.

$$\frac{dy}{dx} = \frac{f'(\alpha) \sin \alpha + f(\alpha) \cos \alpha}{f'(\alpha) \cos \alpha - f(\alpha) \sin \alpha} = \frac{f'(\alpha) \sin \alpha + 0}{f'(\alpha) \cos \alpha - 0} = \frac{\sin \alpha}{\cos \alpha} = \tan \alpha$$

This means that the line $\theta = \alpha$ is tangent to the graph at the pole, $(0, \alpha)$. We summarize this result in the following theorem.

THEOREM 11.7 **TANGENT LINES AT THE POLE**	If $f(\alpha) = 0$ and $f'(\alpha) \neq 0$, then the line $\theta = \alpha$ is tangent at the pole to the graph of $r = f(\theta)$.

Theorem 11.7 is useful because it tells us that the zeros of $r = f(\theta)$ can be used to find the tangent lines at the pole. Note that since a polar curve can cross the pole more than once, it can have more than one tangent line at the pole. Study the next example carefully to see how the various aids to polar curve sketching are combined.

EXAMPLE 3 Sketching a polar graph

Sketch the graph of $r = 2 \cos 3\theta$.

SOLUTION

The graph is symmetric with respect to the polar axis because

$$\cos 3\theta = \cos(-3\theta).$$

Setting $dr/d\theta$ equal to zero, we have

$$\frac{dr}{d\theta} = -6 \sin 3\theta = 0.$$

Thus, in the interval $0 \leq \theta \leq \pi$, the extrema of r occur when $\theta = 0$, $\pi/3$, and $2\pi/3$. Moreover, by setting r equal to zero, we have

$$2 \cos 3\theta = 0$$

and in the interval $0 \leq \theta \leq \pi$ the tangent lines to the pole occur when $\theta = \pi/6$, $\pi/2$, and $5\pi/6$. Finally, by plotting a few points, we obtain the graph shown in Figure 11.30. It is interesting to note in this figure that both the upper and lower halves of the graph are traced as θ increases from 0 to π. Were we to continue plotting points as θ increases from π to 2π, the graph would be traced a second time. The graph shown in this figure is called a **rose curve.**

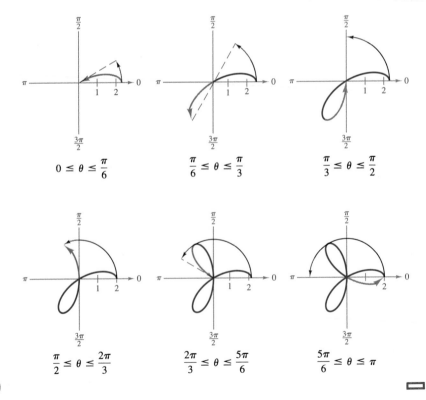

$$0 \leq \theta \leq \frac{\pi}{6} \qquad \frac{\pi}{6} \leq \theta \leq \frac{\pi}{3} \qquad \frac{\pi}{3} \leq \theta \leq \frac{\pi}{2}$$

$$\frac{\pi}{2} \leq \theta \leq \frac{2\pi}{3} \qquad \frac{2\pi}{3} \leq \theta \leq \frac{5\pi}{6} \qquad \frac{5\pi}{6} \leq \theta \leq \pi$$

FIGURE 11.30

Special polar graphs

Several important types of graphs have equations that are simpler in polar form than in rectangular form. For example, the polar equation of a circle having a radius of a and centered at the origin is simply $r = a$. Later in the text you will come to appreciate this benefit. For now, we summarize some other types of graphs that have simpler equations in polar form. (Conics are considered in Section 11.6.)

Limaçons

$r = a \pm b \cos \theta$
$r = a \pm b \sin \theta$
$(0 < a, \; 0 < b)$

Rose Curves

n petals if n is odd
$2n$ petals if n is even
$(n \geq 2)$

Circles and Lemniscates

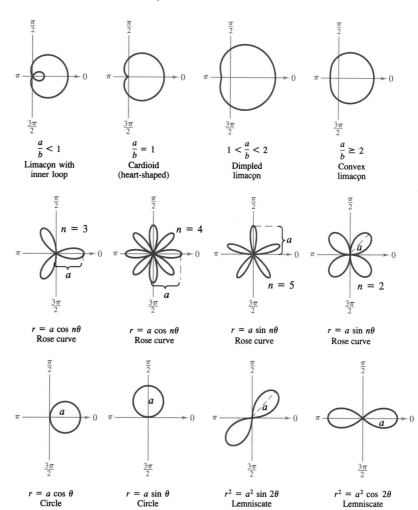

$\dfrac{a}{b} < 1$	$\dfrac{a}{b} = 1$	$1 < \dfrac{a}{b} < 2$	$\dfrac{a}{b} \geq 2$
Limaçon with inner loop	Cardioid (heart-shaped)	Dimpled limaçon	Convex limaçon

$r = a \cos n\theta$ Rose curve	$r = a \cos n\theta$ Rose curve	$r = a \sin n\theta$ Rose curve	$r = a \sin n\theta$ Rose curve

$r = a \cos \theta$ Circle	$r = a \sin \theta$ Circle	$r^2 = a^2 \sin 2\theta$ Lemniscate	$r^2 = a^2 \cos 2\theta$ Lemniscate

EXAMPLE 4 Sketching a polar graph

Sketch the graph of $r = 3 \cos 2\theta$.

SOLUTION

Type of curve: Rose curve with $2n = 4$ petals
Symmetry: With respect to the polar axis and the line $\theta = \pi/2$
Extrema of r: $(3, 0)$, $(-3, \pi/2)$, $(3, \pi)$, and $(-3, 3\pi/2)$
Tangents at pole: $r = 0$ when $\theta = \pi/4, \; 3\pi/4$

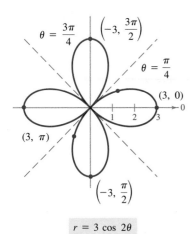

$r = 3 \cos 2\theta$

FIGURE 11.31

By using this information and the additional points shown in Table 11.4, we obtain the graph shown in Figure 11.31.

TABLE 11.4

θ	0	$\dfrac{\pi}{6}$	$\dfrac{\pi}{4}$	$\dfrac{\pi}{3}$
r	3	$\dfrac{3}{2}$	0	$-\dfrac{3}{2}$

EXAMPLE 5   Sketching a polar graph

Sketch the graph of $r^2 = 9 \sin 2\theta$.

SOLUTION

Type of curve: Lemniscate
Symmetry: With respect to the pole
Extrema of r: $r = \pm 3$ when $\theta = \pi/4$
Tangents at pole: $r = 0$ when $\theta = 0, \pi/2$

If $\sin 2\theta < 0$, then this equation has no solution points. Thus, we restrict the values of θ to those for which $\sin 2\theta \geq 0$. That is,

$$0 \leq \theta \leq \frac{\pi}{2}$$

or

$$\pi \leq \theta \leq \frac{3\pi}{2}.$$

Moreover, using symmetry, we need only consider the first of these two intervals. In Table 11.5 we find a few additional points, and the graph is shown in Figure 11.32.

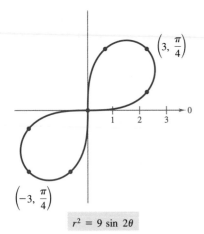

$r^2 = 9 \sin 2\theta$

FIGURE 11.32

TABLE 11.5

θ	0	$\dfrac{\pi}{12}$	$\dfrac{\pi}{4}$	$\dfrac{5\pi}{12}$	$\dfrac{\pi}{2}$
$r = \pm 3\sqrt{\sin 2\theta}$	0	$\pm\dfrac{3}{\sqrt{2}}$	± 3	$\pm\dfrac{3}{\sqrt{2}}$	0

EXERCISES for Section 11.4

In Exercises 1–6, find the relative extrema of r.

1. $r = 5 \cos 3\theta$

2. $r = \cos \theta + \sqrt{3} \sin \theta$

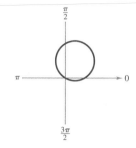

3. $r = 5 + 3 \sin \theta$

4. $r^2 = 9 \sin 2\theta$

5. $r = \dfrac{6}{1 - \cos \theta}$

6. $r = 2 \sin \theta \cos^2 \theta$

In Exercises 7–16, find dy/dx and the slope of the graph of the polar curve at the given value of θ.

Polar equation	Point
7. $r = 3(1 - \cos \theta)$	$\theta = \dfrac{\pi}{2}$
8. $r = 2(1 - \sin \theta)$	$\theta = \dfrac{\pi}{6}$
9. $r = 2 + 3 \sin \theta$	$\theta = \dfrac{\pi}{2}$
10. $r = 3 - 2 \cos \theta$	$\theta = 0$
11. $r = \theta$	$\theta = \pi$
12. $r = 3 \cos \theta$	$\theta = \dfrac{\pi}{4}$
13. $r = 3 \sin \theta$	$\theta = \dfrac{\pi}{3}$
14. $r = 4$	$\theta = \dfrac{\pi}{4}$
15. $r = 2 \sec \theta$	$\theta = \dfrac{\pi}{4}$
16. $r = \dfrac{6}{2 \sin \theta - 3 \cos \theta}$	$\theta = \pi$

In Exercises 17 and 18, find the points of horizontal and vertical tangency (if any) to the polar curve.

17. $r = 1 + \sin \theta$

18. $r = a \sin \theta$

In Exercises 19 and 20, find the points of horizontal tangency (if any) to the polar curve.

19. $r = 2 \csc \theta + 3$

20. $r = a \sin \theta \cos^2 \theta$

In Exercises 21–54, identify and sketch the graph of the given equation. In each case, find the tangents at the pole.

21. $r = 4$

22. $r = -2$

23. $r = 3 \sin \theta$

24. $r = 3 \cos \theta$

25. $r = 3(1 - \cos \theta)$

26. $r = 2(1 - \sin \theta)$

27. $r = 4(1 + \sin \theta)$

28. $r = 1 + \cos \theta$

29. $r = 2 + 3 \sin \theta$

30. $r = 4 + 5 \cos \theta$

31. $r = 3 - 4 \cos \theta$

32. $r = 2(1 - 2 \sin \theta)$

33. $r = 3 - 2 \cos \theta$

34. $r = 5 - 4 \sin \theta$

35. $r = 2 + \sin \theta$

36. $r = 4 + 3 \cos \theta$

37. $r = 2 \cos 3\theta$

38. $r = -\sin 5\theta$

39. $r = 3 \sin 2\theta$

40. $r = 3 \cos 2\theta$

41. $r = 2 \sec \theta$

42. $r = 3 \csc \theta$

43. $r = \dfrac{3}{\sin \theta - 2 \cos \theta}$

44. $r = \dfrac{6}{2 \sin \theta - 3 \cos \theta}$

45. $r^2 = 4 \cos 2\theta$

46. $r^2 = 4 \sin \theta$

47. $r^2 = 4 \sin 2\theta$

48. $r^2 = \cos 3\theta$

49. $r = \sin \theta \cos^2 \theta$

50. $r^2 = \dfrac{1}{\theta}$

51. $r = 2\theta$

52. $r = \dfrac{1}{\theta}$

53. $r = 2 \cos \left(\dfrac{3\theta}{2}\right)$

54. $r = 3 \sin \left(\dfrac{5\theta}{2}\right)$

In Exercises 55–58, sketch the graph of the given equation and show that the indicated line is an asymptote to the graph.

Polar equation	Asymptote
55. $r = 2 - \sec \theta$	$x = -1$
56. $r = 2 + \csc \theta$	$y = 1$
57. $r = \dfrac{2}{\theta}$	$y = 2$
58. $r = 2 \cos 2\theta \sec \theta$	$x = -2$

59. Verify that if the curve whose polar equation is $r = f(\theta)$ is rotated about the pole through an angle ϕ, then an equation for the rotated curve is $r = f(\theta - \phi)$.

60. If the polar form of an equation for a curve is $r = f(\sin \theta)$, show that the form becomes

(a) $r = f(-\cos \theta)$ if the curve is rotated counterclockwise $\pi/2$ radians about the pole.

(b) $r = f(-\sin \theta)$ if the curve is rotated counterclockwise π radians about the pole.

(c) $r = f(\cos \theta)$ if the curve is rotated counterclockwise $3\pi/2$ radians about the pole.

In Exercises 61–64, use the results of Exercises 59 and 60.

61. Write an equation for the limaçon

$$r = 2 - \sin \theta$$

after it has been rotated by the given amount.

(a) $\dfrac{\pi}{4}$ (b) $\dfrac{\pi}{2}$ (c) π (d) $\dfrac{3\pi}{2}$

62. Write an equation for the rose curve

$$r = 2 \sin 2\theta$$

after it has been rotated by the given amount.

(a) $\dfrac{\pi}{6}$ (b) $\dfrac{\pi}{2}$ (c) $\dfrac{2\pi}{3}$ (d) π

63. Sketch the graphs of the equations.

(a) $r = 1 - \sin \theta$ (b) $r = 1 - \sin\left(\theta - \dfrac{\pi}{4}\right)$

64. Sketch the graphs of the equations.

(a) $r = 3 \sec \theta$

(b) $r = 3 \sec\left(\theta - \dfrac{\pi}{4}\right)$

(c) $r = 3 \sec\left(\theta + \dfrac{\pi}{3}\right)$

In Exercises 65–68, use a computer or graphics calculator to sketch the graph of the polar equation and find all points of horizontal tangency.

65. $r = 4 \sin \theta \cos^2 \theta$

66. $r = 3 \cos 2\theta \sec \theta$

67. $r = 2 \csc \theta + 5$

68. $r = 2 \cos (3\theta - 2)$

11.5 Area and Arc Length in Polar Coordinates

Area of a polar region ▪ Points of intersection of polar graphs ▪ Arc length in polar form ▪ Area of a surface of revolution

Area of Sector: $\dfrac{1}{2}\theta r^2$

FIGURE 11.33

The development of a formula for the area of a polar region parallels that for the area of regions in the rectangular coordinate system, but uses *sectors* of a circle instead of rectangles as the basic element of area. In Figure 11.33 note that the area of a circular sector of radius r is given by

$$\text{area of circular sector} = \frac{1}{2}\theta r^2$$

provided θ is measured in radians.

Consider the function given by $r = f(\theta)$, where f is continuous and nonnegative in the interval $[\alpha, \beta]$. The region bounded by the graph of f and the radial lines $\theta = \alpha$ and $\theta = \beta$ is shown in Figure 11.34. To find the area

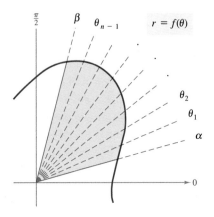

Partition $[\alpha, \beta]$ into n equal subintervals.

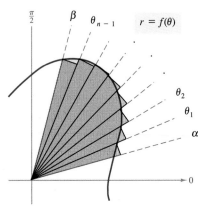

Approximate area by n sectors.

FIGURE 11.34

of this region, we partition the interval $[\alpha, \beta]$ into n equal subintervals,

$$\alpha = \theta_0 < \theta_1 < \theta_2 < \cdots < \theta_{n-1} < \theta_n = \beta.$$

Then we approximate the area of the region by the sum of the areas of the n sectors.

$$\text{radius of } i\text{th sector} = f(\theta_i)$$

$$\text{central angle of } i\text{th sector} = \frac{\beta - \alpha}{n} = \Delta\theta$$

$$A \approx \sum_{i=1}^{n} \left(\frac{1}{2}\right) \Delta\theta f(\theta_i)^2$$

Taking the limit as $n \to \infty$, we have

$$A = \lim_{n \to \infty} \frac{1}{2} \sum_{i=1}^{n} f(\theta_i)^2 \, \Delta\theta = \frac{1}{2} \int_{\alpha}^{\beta} [f(\theta)]^2 \, d\theta$$

which leads to the following theorem.

THEOREM 11.8
AREA IN POLAR COORDINATES

If f is continuous and nonnegative on the interval $[\alpha, \beta]$, then the area of the region bounded by the graph of $r = f(\theta)$ between the radial lines $\theta = \alpha$ and $\theta = \beta$ is given by

$$A = \frac{1}{2} \int_{\alpha}^{\beta} [f(\theta)]^2 \, d\theta = \frac{1}{2} \int_{\alpha}^{\beta} r^2 \, d\theta.$$

REMARK   We can use the same formula to find the area of a region bounded by the graph of a continuous *nonpositive* function. However, the formula is not necessarily valid if f takes on both positive *and* negative values in the interval $[\alpha, \beta]$.

Sometimes the most difficult part of finding the area of a polar region is determining the limits of integration. A good sketch of the region helps.

EXAMPLE 1   Finding the area of a polar region

Find the area of *one petal* of the rose curve given by

$$r = 3 \cos 3\theta.$$

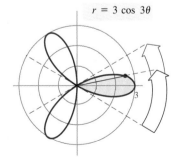
$r = 3 \cos 3\theta$

FIGURE 11.35

SOLUTION

In Figure 11.35 we see that the right petal is traced as θ increases from $-\pi/6$ to $\pi/6$. Thus, the area is

$$A = \frac{1}{2} \int_{\alpha}^{\beta} r^2 \, d\theta = \frac{1}{2} \int_{-\pi/6}^{\pi/6} (3 \cos 3\theta)^2 \, d\theta$$

$$= \frac{9}{2} \int_{-\pi/6}^{\pi/6} \frac{1 + \cos 6\theta}{2} \, d\theta$$

$$= \frac{9}{4} \left[\theta + \frac{\sin 6\theta}{6} \right]_{-\pi/6}^{\pi/6} = \frac{9}{4} \left[\frac{\pi}{6} + \frac{\pi}{6} \right] = \frac{3\pi}{4}.$$

Suppose you were asked to find the area of the region lying inside all three petals of the rose curve in Example 1. You could not simply integrate between 0 and 2π. In doing this you would obtain $9\pi/2$, which is twice the area of the three petals—the duplication occurs because the rose curve is traced *twice* as θ increases from 0 to 2π. To avoid this type of error, be sure that the graph of $r = f(\theta)$ does not retrace itself between your chosen limits of integration.

EXAMPLE 2   Finding the area bounded by a single curve

Find the area of the region lying between the inner and outer loops of the limaçon $r = 1 - 2 \sin \theta$.

SOLUTION

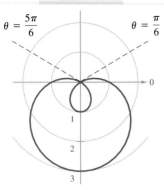

FIGURE 11.36

In Figure 11.36 we see that the inner loop is traced as θ increases from $\pi/6$ to $5\pi/6$. Thus, the area of the region lying inside the *inner loop* is given by

$$
\begin{aligned}
A_1 &= \frac{1}{2} \int_{\pi/6}^{5\pi/6} (1 - 2 \sin \theta)^2 \, d\theta \\
&= \frac{1}{2} \int_{\pi/6}^{5\pi/6} (1 - 4 \sin \theta + 4 \sin^2 \theta) \, d\theta \\
&= \frac{1}{2} \int_{\pi/6}^{5\pi/6} \left[1 - 4 \sin \theta + 4\left(\frac{1 - \cos 2\theta}{2}\right) \right] d\theta \\
&= \frac{1}{2} \int_{\pi/6}^{5\pi/6} (3 - 4 \sin \theta - 2 \cos 2\theta) \, d\theta \\
&= \frac{1}{2} \left[3\theta + 4 \cos \theta - \sin 2\theta \right]_{\pi/6}^{5\pi/6} \\
&= \frac{1}{2} [2\pi - 3\sqrt{3}] = \pi - \frac{3\sqrt{3}}{2}.
\end{aligned}
$$

In a similar way, we can integrate from $5\pi/6$ to $13\pi/6$ to find that the area of the region lying inside the *outer loop* is $A_2 = 2\pi + (3\sqrt{3}/2)$. Finally, to find the area of the region lying between the two loops, we subtract the smaller area from the larger to obtain

$$
\begin{aligned}
A &= A_2 - A_1 \\
&= \left(2\pi + \frac{3\sqrt{3}}{2} \right) - \left(\pi - \frac{3\sqrt{3}}{2} \right) \\
&= \pi + 3\sqrt{3} \approx 8.34.
\end{aligned}
$$

Points of intersection

Because a point may be represented in different ways in polar coordinates, care must be taken in determining the points of intersection of two polar graphs. For example, consider the points of intersection of the graphs of $r = 1 - 2 \cos \theta$ and $r = 1$, as shown in Figure 11.37. If, as with rectangular equations, we attempted to find the points of intersection by solving the two equations simultaneously, we would obtain the following.

$$1 = 1 - 2 \cos \theta \implies \cos \theta = 0 \implies \theta = \frac{\pi}{2}, \frac{3\pi}{2}$$

The corresponding points of intersection are $(1, \pi/2)$ and $(1, 3\pi/2)$. However, from Figure 11.37 we see that there is a *third* point of intersection that did not show up when we solved the two polar equations simultaneously. (This is one reason we stress sketching a graph when finding the area of a polar region.) The reason we did not find the third point is that it does not occur with the same coordinates in the two graphs. On the graph of $r = 1$, the point occurs with the coordinates $(1, \pi)$, but on the graph of $r = 1 - 2 \cos \theta$, the point occurs with coordinates $(-1, 0)$.

We can compare the problem of finding points of intersection of two polar graphs with that of finding collision points of two satellites in intersecting orbits about the earth, as shown in Figure 11.38. The satellites will not collide as long as they reach the points of intersection at different times (θ-values). A collision will occur only at the points of intersection that are "simultaneous points"—those reached at the same time (θ-value).

REMARK   Because the pole can be represented by $(0, \theta)$, where θ is *any* angle, you should check separately for the pole when hunting for points of intersection.

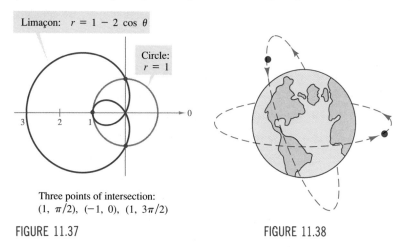

Limaçon: $r = 1 - 2 \cos \theta$

Circle: $r = 1$

Three points of intersection:
$(1, \pi/2), (-1, 0), (1, 3\pi/2)$

FIGURE 11.37   FIGURE 11.38

EXAMPLE 3   Finding the area of a region between two curves

Find the area of the region common to the two regions bounded by the circle $r = -6 \cos \theta$ and the cardioid $r = 2 - 2 \cos \theta$.

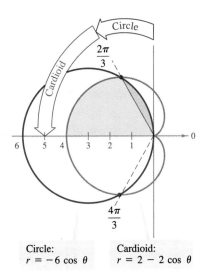

Circle:
$r = -6 \cos \theta$

Cardioid:
$r = 2 - 2 \cos \theta$

FIGURE 11.39

SOLUTION

By symmetry we can work with the upper half-plane, so we begin by dividing the region into two parts, as shown in Figure 11.39. The gray shaded region lies between the circle and the radial line $\theta = 2\pi/3$. Since the circle has coordinates $(0, \pi/2)$ at the pole, we integrate between $\pi/2$ and $2\pi/3$ to obtain the area of this region. The colored region lies between the radial lines $\theta = 2\pi/3$ and $\theta = \pi$ and the cardioid. Thus, we can find the area of this second region by integrating between $2\pi/3$ and π. The sum of these two integrals gives the area of the common region lying *above* the polar axis.

$$
\frac{A}{2} = \underbrace{\frac{1}{2} \int_{\pi/2}^{2\pi/3} (-6 \cos \theta)^2 \, d\theta}_{\substack{\text{Region between circle} \\ \text{and radial line } \theta = 2\pi/3}} + \underbrace{\frac{1}{2} \int_{2\pi/3}^{\pi} (2 - 2 \cos \theta)^2 \, d\theta}_{\substack{\text{Region between cardioid and} \\ \text{radial lines } \theta = 2\pi/3 \text{ and } \theta = \pi}}
$$

$$
= 18 \int_{\pi/2}^{2\pi/3} \cos^2 \theta \, d\theta + \frac{1}{2} \int_{2\pi/3}^{\pi} (4 - 8 \cos \theta + 4 \cos^2 \theta) \, d\theta
$$

$$
= 9 \int_{\pi/2}^{2\pi/3} (1 + \cos 2\theta) \, d\theta + \int_{2\pi/3}^{\pi} (3 - 4 \cos \theta + \cos 2\theta) \, d\theta
$$

$$
= 9 \left[\theta + \frac{\sin 2\theta}{2} \right]_{\pi/2}^{2\pi/3} + \left[3\theta - 4 \sin \theta + \frac{\sin 2\theta}{2} \right]_{2\pi/3}^{\pi}
$$

$$
= 9 \left(\frac{2\pi}{3} - \frac{\sqrt{3}}{4} - \frac{\pi}{2} \right) + \left(3\pi - 2\pi + 2\sqrt{3} + \frac{\sqrt{3}}{4} \right)
$$

$$
= \frac{5\pi}{2}
$$

Finally, multiplying by 2, we conclude that the total area is 5π. ▭

Arc length

The formula for the length of a polar arc can be obtained from the arc length formula for a curve described by parametric equations.

THEOREM 11.9 ARC LENGTH OF A POLAR CURVE	Let f be a function whose derivative is continuous in an interval $\alpha \le \theta \le \beta$. The length of the graph of $r = f(\theta)$ from $\theta = \alpha$ to $\theta = \beta$ is $$ s = \int_{\alpha}^{\beta} \sqrt{[f(\theta)]^2 + [f'(\theta)]^2} \, d\theta. $$

PROOF   Considering the parametric form

$$ x = r \cos \theta = f(\theta) \cos \theta \qquad \text{and} \qquad y = r \sin \theta = f(\theta) \sin \theta $$

we have

$$ \frac{dx}{d\theta} = f'(\theta) \cos \theta - f(\theta) \sin \theta \qquad \text{and} \qquad \frac{dy}{d\theta} = f'(\theta) \sin \theta + f(\theta) \cos \theta $$

and it follows that

$$\left(\frac{dx}{d\theta}\right)^2 + \left(\frac{dy}{d\theta}\right)^2 = [f(\theta)]^2 + [f'(\theta)]^2.$$

Consequently, we can apply Theorem 11.2, using θ in place of t, to obtain the desired result.

EXAMPLE 4   Finding the length of a polar curve

Find the length of the arc from $\theta = 0$ to $\theta = 2\pi$ for the cardioid

$$r = f(\theta) = 2 - 2 \cos \theta$$

as shown in Figure 11.40.

SOLUTION

Since $f'(\theta) = 2 \sin \theta$, we have

$$s = \int_\alpha^\beta \sqrt{[f(\theta)]^2 + [f'(\theta)]^2}\, d\theta$$

$$= \int_0^{2\pi} \sqrt{(2 - 2\cos\theta)^2 + (2\sin\theta)^2}\, d\theta$$

$$= 2\sqrt{2} \int_0^{2\pi} \sqrt{1 - \cos\theta}\, d\theta$$

$$= 2\sqrt{2} \int_0^{2\pi} \sqrt{2 \sin^2 \frac{\theta}{2}}\, d\theta$$

$$= 4 \int_0^{2\pi} \sin \frac{\theta}{2}\, d\theta$$

$$= -8 \cos \frac{\theta}{2} \bigg]_0^{2\pi}$$

$$= 8 + 8$$

$$= 16.$$

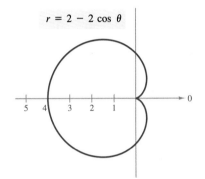

$r = 2 - 2 \cos \theta$

FIGURE 11.40

REMARK   Using Figure 11.40 we can determine the reasonableness of this answer by comparing it to the circumference of a circle. For example, a circle of radius $\frac{5}{2}$ has a circumference of $5\pi \approx 15.7$.

Area of a surface of revolution

The polar coordinate version of the formulas for the area of a surface of revolution can be obtained from the parametric versions given in Theorem 11.3, using the equations $x = r \cos \theta$ and $y = r \sin \theta$.

THEOREM 11.10
AREA OF A SURFACE
OF REVOLUTION

Let f be a function whose derivative is continuous in an interval $\alpha \leq \theta \leq \beta$. The area of the surface formed by revolving the graph of $r = f(\theta)$ from $\theta = \alpha$ to $\theta = \beta$ about the indicated line is as follows.

1. $S = 2\pi \displaystyle\int_{\alpha}^{\beta} f(\theta) \sin \theta \sqrt{[f(\theta)]^2 + [f'(\theta)]^2} \, d\theta$ About polar axis

2. $S = 2\pi \displaystyle\int_{\alpha}^{\beta} f(\theta) \cos \theta \sqrt{[f(\theta)]^2 + [f'(\theta)]^2} \, d\theta$ About line $\theta = \dfrac{\pi}{2}$

REMARK When using Theorem 11.9 or 11.10, check to see that the graph of $r = f(\theta)$ is traced only once on the interval $\alpha \leq \theta \leq \beta$. For example, the circle given by $r = \sin \theta$ is traced once on the interval $0 \leq \theta \leq \pi$.

EXAMPLE 5   Finding the area of a surface of revolution

Find the area of the surface formed by revolving the circle given by

$$r = f(\theta) = \cos \theta$$

about the line $\theta = \pi/2$, as shown in Figure 11.41.

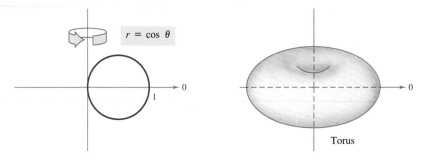

FIGURE 11.41

SOLUTION

We use the second formula given in Theorem 11.10 with $f'(\theta) = -\sin \theta$. Moreover, since the circle is traced once as θ increases from 0 to π, we have the following integral.

$$S = 2\pi \int_{\alpha}^{\beta} f(\theta) \cos \theta \sqrt{[f(\theta)^2 + [f'(\theta)]^2} \, d\theta$$

$$= 2\pi \int_{0}^{\pi} \cos \theta \, (\cos \theta) \sqrt{\cos^2 \theta + \sin^2 \theta} \, d\theta$$

$$= 2\pi \int_{0}^{\pi} \cos^2 \theta \, d\theta = \pi \int_{0}^{\pi} (1 + \cos 2\theta) \, d\theta$$

$$= \pi \left[\theta + \frac{\sin 2\theta}{2} \right]_{0}^{\pi} = \pi^2$$

EXERCISES for Section 11.5

In Exercises 1–12, find the points of intersection of the graphs of the given equations.

1. $r = 1 + \cos \theta$
$r = 1 - \cos \theta$

2. $r = 3(1 + \sin \theta)$
$r = 3(1 - \sin \theta)$

3. $r = 1 + \cos \theta$
$r = 1 - \sin \theta$

4. $r = 2 - 3 \cos \theta$
$r = \cos \theta$

5. $r = 4 - 5 \sin \theta$
$r = 3 \sin \theta$

6. $r \overset{\underline{?}}{=} 1 + \cos \theta$
$r = 3 \cos \theta$

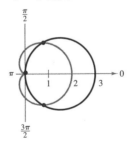

7. $r = \dfrac{\theta}{2}, \quad r = 2$

8. $\theta = \dfrac{\pi}{4}, \quad r = 2$

9. $r = 4 \sin 2\theta, \quad r = 2$

10. $r = 3 + \sin \theta, \quad r = 2 \csc \theta$

11. $r = 2 + 3 \cos \theta, \quad r = \dfrac{\sec \theta}{2}$

12. $r = 3(1 - \cos \theta), \quad r = \dfrac{6}{1 - \cos \theta}$

In Exercises 13–32, find the area of the given region.

13. One petal of $r = 2 \cos 3\theta$
14. One petal of $r = 4 \sin 2\theta$
15. One petal of $r = \cos 2\theta$
16. One petal of $r = \cos 5\theta$
17. Interior of $r = 1 - \sin \theta$
18. Interior of $r = 1 - \sin \theta$ (above the polar axis)
19. Inner loop of $r = 1 + 2 \cos \theta$
20. Inner loop of $r = 3 + 4 \sin \theta$
21. Between the loops of $r = 1 + 2 \cos \theta$
22. Between the loops of $r = 2(1 + 2 \sin \theta)$
23. Common interior of $r = 4 \sin 2\theta$ and $r = 2$
24. Common interior of $r = 3(1 + \sin \theta)$ and
$r = 3(1 - \sin \theta)$
25. Common interior of $r = 3 - 2 \sin \theta$ and
$r = -3 + 2 \sin \theta$
26. Common interior of $r = 3 - 2 \sin \theta$ and
$r = 3 - 2 \cos \theta$
27. Common interior of $r = 4 \sin \theta$ and $r = 2$
28. Inside $r = 3 \sin \theta$ and outside $r = 2 - \sin \theta$
29. Inside $r = a(1 + \cos \theta)$ and outside $r = a \cos \theta$
30. Inside $r = 2a \cos \theta$ and outside $r = a$
31. Common interior of $r = a(1 + \cos \theta)$ and
$r = a \sin \theta$
32. Region bounded by the graphs of

$$r = \frac{ab}{a \sin \theta + b \cos \theta}, \quad \theta = 0, \quad \text{and} \quad \theta = \frac{\pi}{2}$$

In Exercises 33–40, find the length of the given graph over the indicated interval.

33. $r = a, \quad 0 \leq \theta \leq 2\pi$
34. $r = a \cos \theta, \quad 0 \leq \theta \leq 2\pi$
35. $r = 1 + \sin \theta, \quad 0 \leq \theta \leq 2\pi$
36. $r = 5(1 + \cos \theta), \quad 0 \leq \theta \leq 2\pi$

37. $r = 2\theta, \quad 0 \leq \theta \leq \dfrac{\pi}{2}$

38. $r = \sec \theta, \quad 0 \leq \theta \leq \dfrac{\pi}{3}$

39. $r = \dfrac{1}{\theta}, \quad \pi \leq \theta \leq 2\pi$

40. $r = e^\theta, \quad 0 \leq \theta \leq \pi$

In Exercises 41–46, find the area of the surface formed by revolving the indicated curve about the given line.

41. $r = 2 \cos \theta, \quad 0 \leq \theta \leq \dfrac{\pi}{2}$

revolved about the polar axis

42. $r = a \cos \theta, \quad 0 \le \theta \le \dfrac{\pi}{2}$

revolved about the line $\theta = \dfrac{\pi}{2}$

43. $r = e^{a\theta}, \quad 0 \le \theta \le \dfrac{\pi}{2}$

revolved about the line $\theta = \dfrac{\pi}{2}$

44. $r = a(1 + \cos \theta), \quad 0 \le \theta \le \pi$
revolved about the polar axis

45. $r = 4 \cos 2\theta, \quad 0 \le \theta \le \dfrac{\pi}{4}$

revolved about the polar axis
[Hint: Use Simpson's Rule with $n = 4$.]

46. $r = \theta, \quad 0 \le \theta \le \pi$
revolved about the polar axis
[Hint: Use Simpson's Rule with $n = 4$.]

47. Find the surface area of the torus generated by revolving the circle given by $r = a$ about the line $r = b \sec \theta$ where $b > a > 0$. [Hint: Derive the integral that yields the surface area for revolving the graph of $r = f(\theta)$ about the line $r = b \sec \theta$.]

48. The graph of the polar equation $r = a\theta$ is called **Archimedes' spiral.** Let A_n be the area between turns $n - 1$ and n of the spiral for $n \ge 2$ (see figure).

(a) Complete the following table.

n	2	3	4	5	6	7
A_n						

[Hint: Note that

$$A_n = \frac{1}{2} \int_{(n-1)(2\pi)}^{n(2\pi)} (a\theta)^2 \, d\theta - \frac{1}{2} \int_{(n-2)(2\pi)}^{(n-1)(2\pi)} (a\theta)^2 \, d\theta.]$$

(b) Verify that $A_n = (n - 1)A_2$ for $n \ge 3$.

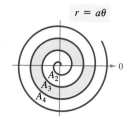

$r = a\theta$

FIGURE FOR 48

In Exercises 49 and 50, use a computer or graphics calculator to sketch the graph of the equation. Use Simpson's Rule with $n = 12$ to find the arc length of the curve and the area of the region enclosed.

49. $r = \sin (3 \cos \theta)$ **50.** $r = 2 \sin (2 \cos \theta)$

11.6 Polar Equations for Conics and Kepler's Laws

Polar equations for conics ▪ Kepler's Laws

In Chapter 10 we saw that the rectangular equations of ellipses and hyperbolas take simple forms when the origin lies at their *center*. As it happens, there are many important applications of conics in which it is more convenient to use one of the *foci* as the reference point (the origin) for the coordinate system. For example, the sun lies at a focus of the earth's orbit. Similarly, the light source of a parabolic reflector lies at its focus. In this section we will see that polar equations of conics take simple forms if one of the foci lies at the pole.

The following theorem uses the concept of *eccentricity*, as defined in Chapter 10, to classify the three basic types of conics. A proof of this theorem is given in Appendix A.

THEOREM 11.11
CLASSIFICATION OF CONICS
BY ECCENTRICITY

The locus of a point in the plane whose distance from a fixed point (*focus*) has a constant ratio to its distance from a fixed line (*directrix*) is a conic. The constant ratio e is the *eccentricity* of the conic.

1. The conic is an ellipse if $0 < e < 1$.
2. The conic is a parabola if $e = 1$.
3. The conic is a hyperbola if $e > 1$.

Parabola: $e = 1$
$PF = PQ$

Ellipse: $0 < e < 1$
$\dfrac{PF}{PQ} < 1$

Hyperbola: $e > 1$
$\dfrac{PF}{PQ} = \dfrac{P'F}{P'Q'} > 1$

FIGURE 11.42

In Figure 11.42 note that for each type of conic, the pole corresponds to the fixed point (focus) given in the definition. The benefit of this location is seen in the proof of the following theorem.

THEOREM 11.12 POLAR EQUATIONS FOR CONICS	The graph of a polar equation of the form

$$r = \frac{ed}{1 \pm e \cos \theta} \qquad \text{or} \qquad r = \frac{ed}{1 \pm e \sin \theta}$$

is a conic, where $e > 0$ is the eccentricity and $|d|$ is the distance between the focus at the pole and its corresponding directrix.

PROOF We give a proof for $r = ed/(1 + e \cos \theta)$ with $d > 0$. The proofs of the other cases are similar. In Figure 11.43 consider a vertical directrix, d units to the right of the focus $F = (0, 0)$. If $P = (r, \theta)$ is a point on the graph of $r = ed/(1 + e \cos \theta)$, then the distance between P and the directrix is

$$PQ = |d - x| = |d - r \cos \theta|$$

$$= \left| d - \left(\frac{ed}{1 + e \cos \theta} \right) \cos \theta \right|$$

$$= \left| d \left(1 - \frac{e \cos \theta}{1 + e \cos \theta} \right) \right|$$

$$= \left| \frac{d}{1 + e \cos \theta} \right| = \left| \frac{r}{e} \right|.$$

Moreover, since the distance between P and the pole is simply $PF = |r|$, the ratio of PF to PQ is

$$\frac{PF}{PQ} = \frac{|r|}{|r/e|} = |e| = e$$

and, by Theorem 11.11, the graph of the equation must be a conic.

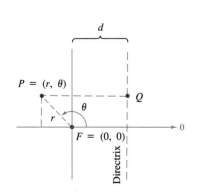

FIGURE 11.43

The four types of equations indicated in Theorem 11.12 can be classified as follows, where $d > 0$.

POLAR EQUATIONS FOR CONICS

1. Horizontal directrix above the pole: $r = \dfrac{ed}{1 + e \sin \theta}$

2. Horizontal directrix below the pole: $r = \dfrac{ed}{1 - e \sin \theta}$

3. Vertical directrix to the right of the pole: $r = \dfrac{ed}{1 + e \cos \theta}$

4. Vertical directrix to the left of the pole: $r = \dfrac{ed}{1 - e \cos \theta}$

Figure 11.44 illustrates these four possibilities for a parabola.

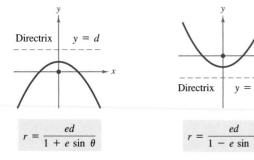

$$r = \frac{ed}{1 + e \sin \theta}$$

(1)

$$r = \frac{ed}{1 - e \sin \theta}$$

(2)

$$r = \frac{ed}{1 + e \cos \theta}$$

(3)

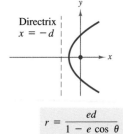

$$r = \frac{ed}{1 - e \cos \theta}$$

(4)

FIGURE 11.44

EXAMPLE 1 Determining a conic from its equation

Sketch the graph of the conic given by

$$r = \frac{15}{3 - 2 \cos \theta}.$$

SOLUTION

To determine the type of conic, we rewrite the equation as

$$r = \frac{15}{3 - 2 \cos \theta} = \frac{5}{1 - (2/3) \cos \theta}.$$

From this form we conclude that the graph is an ellipse with $e = \frac{2}{3}$. We sketch the upper half of the ellipse by plotting points from $\theta = 0$ to $\theta = \pi$, as shown in Figure 11.45. Then, using symmetry with respect to the polar axis, we sketch the lower half.

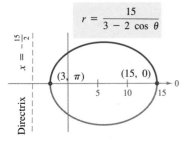

FIGURE 11.45

For the ellipse in Figure 11.45, the major axis is horizontal and the vertices lie at (15, 0) and (3, π). Thus, the length of the *major* axis is $2a = 18$. To find the length of the *minor* axis, we use the equations $e = c/a$ and $b^2 = a^2 - c^2$ and conclude that

$$b^2 = a^2 - c^2 = a^2 - (ea)^2 = a^2(1 - e^2).$$ Ellipse

Since $e = \frac{2}{3}$, we have $b^2 = 9^2\left[1 - \left(\frac{2}{3}\right)^2\right] = 45$, which implies that $b = \sqrt{45} = 3\sqrt{5}$. Thus, the length of the minor axis is $2b = 6\sqrt{5}$. A similar analysis for hyperbolas yields

$$b^2 = c^2 - a^2 = (ea)^2 - a^2 = a^2(e^2 - 1).$$ Hyperbola

EXAMPLE 2   Sketching a conic from its polar equation

Sketch the graph of the polar equation

$$r = \frac{32}{3 + 5 \sin \theta}.$$

SOLUTION

Dividing the numerator and denominator by 3 produces

$$r = \frac{32/3}{1 + (5/3) \sin \theta}.$$

Since $e = \frac{5}{3} > 1$, the graph is a hyperbola. The transverse axis of the hyperbola lies on the line $\theta = \pi/2$, and the vertices occur at

$$\left(4, \frac{\pi}{2}\right) \quad \text{and} \quad \left(-16, \frac{3\pi}{2}\right).$$

Since the length of the transverse axis is 12, we see that $a = 6$. To find b, we write

$$b^2 = a^2(e^2 - 1) = 6^2\left[\left(\frac{5}{3}\right)^2 - 1\right] = 64.$$

Therefore, $b = 8$. Finally, we use a and b to determine the asymptotes of the hyperbola and obtain the sketch shown in Figure 11.46.   ▭

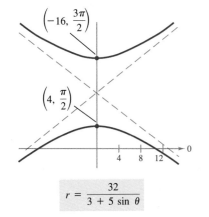

$$r = \frac{32}{3 + 5 \sin \theta}$$

FIGURE 11.46

EXAMPLE 3   Finding the polar equation for a conic

Find a polar equation for the parabola whose focus is the pole and directrix is the line $y = 3$.

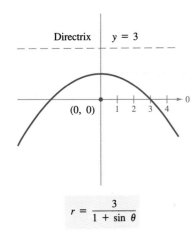

Directrix | $y = 3$

(0, 0)   1   2   3   4

$$r = \frac{3}{1 + \sin \theta}$$

FIGURE 11.47

Johannes Kepler

SOLUTION

From Figure 11.47 we see that the directrix is horizontal, and we choose an equation of the form

$$r = \frac{ed}{1 + e \sin \theta}.$$

Furthermore, since the eccentricity of a parabola is $e = 1$ and the distance between the pole and the directrix is $d = 3$, we have the equation

$$r = \frac{3}{1 + \sin \theta}.$$

Kepler's Laws

Kepler's Laws, named after the German astronomer Johannes Kepler (1571–1630), can be used to describe the orbits of the planets about the sun.

1. Each planet moves in an elliptical orbit with the sun as a focus.
2. The ray from the sun to the planet sweeps out equal areas of the ellipse in equal times.
3. The square of the period is proportional to the cube of the mean distance between the planet and the sun.*

Although Kepler derived these laws empirically, they were later validated by Newton. In fact, Newton was able to show that each law can be deduced from a set of universal laws of motion and gravitation that govern the movement of all heavenly bodies, including comets and satellites. This is illustrated in the next example, involving the comet named after the English mathematician and physicist Edmund Halley (1656–1742).

EXAMPLE 4   An application

Halley's comet has an elliptical orbit with an eccentricity of $e \approx 0.97$. The length of the major axis of the orbit is approximately 36.18 astronomical units. (An astronomical unit is defined to be the mean distance between the earth and the sun, 93 million miles.) Find a polar equation for the orbit. How close does Halley's comet come to the sun?

*If the earth is used as a reference with a period of 1 year and a distance of 1 astronomical unit, the proportionality constant is 1. For example, since Mars has a mean distance to the sun of $D = 1.523$ AU, its period P is given by $D^3 = P^2$. Thus, the period for Mars is $P = 1.88$ years.

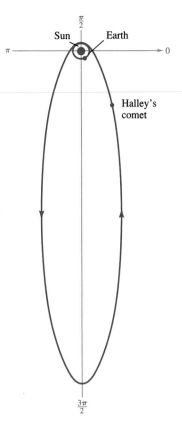

FIGURE 11.48

SOLUTION

Using a vertical axis, we choose an equation of the form $r = ed/(1 + e \sin \theta)$. Since the vertices of the ellipse occur when $\theta = \pi/2$ and $\theta = 3\pi/2$, we can determine the length of the major axis to be the sum of the r-values of the vertices, as shown in Figure 11.48. That is,

$$2a = \frac{0.97d}{1 + 0.97} + \frac{0.97d}{1 - 0.97} \approx 32.83d \approx 36.18.$$

Thus, $d \approx 1.102$ and $ed \approx (0.97)(1.102) \approx 1.069$. Using this value in the equation produces

$$r = \frac{1.069}{1 + 0.97 \sin \theta}$$

where r is measured in astronomical units. To find the closest point to the sun (the focus), we write $c = ea \approx (0.97)(18.09) \approx 17.55$. Since c is the distance between the focus and the center, the closest point is

$$a - c \approx 18.09 - 17.55$$
$$\approx 0.54 \text{ AU}$$
$$\approx 50{,}000{,}000 \text{ mi.}$$

Kepler's Second Law states that as a planet moves about the sun, a ray from the sun to the planet sweeps out equal areas in equal times. This law can also be applied to comets or asteroids with elliptical orbits. For example, Figure 11.49 shows the orbit of the asteroid Apollo about the sun. Applying Kepler's Second Law to this asteroid, we know that the closer it is to the sun, the greater its velocity, since a short ray must be moving quickly to sweep out as much area as a long ray.

FIGURE 11.49

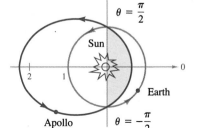

FIGURE 11.50

EXAMPLE 5 An application to elliptical orbits

The asteroid Apollo has a period of 478 earth days, and its orbit is approximated by the ellipse

$$r = \frac{1}{1 + (5/9) \cos \theta} = \frac{9}{9 + 5 \cos \theta}$$

where r is measured in astronomical units. How long does it take Apollo to move from the position given by $\theta = -\pi/2$ to $\theta = \pi/2$, as shown in Figure 11.50?

SOLUTION

We begin by finding the area swept out as θ increases from $-\pi/2$ to $\pi/2$.

$$A = \frac{1}{2}\int_\alpha^\beta r^2\, d\theta = \frac{1}{2}\int_{-\pi/2}^{\pi/2}\left(\frac{9}{9 + 5\cos\theta}\right)^2 d\theta$$

Using the substitution $u = \tan(\theta/2)$ as discussed in Section 8.6, we can obtain

$$A = \frac{81}{112}\left[\frac{-5\sin\theta}{9 + 5\cos\theta} + \frac{18}{\sqrt{56}}\arctan\frac{\sqrt{56}\tan(\theta/2)}{14}\right]_{-\pi/2}^{\pi/2} \approx 0.90429.$$

Now, since the major axis of the ellipse has length $2a = 81/28$ and the eccentricity is $e = 5/9$, we can determine that $b = a\sqrt{1 - e^2} = 9/\sqrt{56}$. Thus, the area of the ellipse is

$$\text{area of ellipse} = \pi ab = \pi\left(\frac{81}{56}\right)\left(\frac{9}{\sqrt{56}}\right) \approx 5.46507.$$

Since the time required to complete the orbit is 478 days, we can apply Kepler's Second Law to conclude that the time t required to move from the position $\theta = -\pi/2$ to $\theta = \pi/2$ is given by

$$\frac{t}{478} = \frac{\text{area of elliptical segment}}{\text{area of ellipse}} \approx \frac{0.90429}{5.46507}$$

which implies that

$$t = 478\left(\frac{0.90429}{5.46507}\right) \approx 79 \text{ days.}$$

EXERCISES for Section 11.6

In Exercises 1–16, sketch the graph of the equation and identify the curve.

1. $r = \dfrac{2}{1 - \cos\theta}$

2. $r = \dfrac{4}{1 + \sin\theta}$

3. $r = \dfrac{5}{1 + \sin\theta}$

4. $r = \dfrac{6}{1 + \cos\theta}$

5. $r = \dfrac{2}{2 - \cos\theta}$

6. $r = \dfrac{3}{3 + 2\sin\theta}$

7. $r(2 + \sin\theta) = 4$

8. $r(3 - 2\cos\theta) = 6$

9. $r = \dfrac{-1}{1 - \sin\theta}$

10. $r = \dfrac{-3}{2 + 4\sin\theta}$

11. $r = \dfrac{5}{-1 + 2\cos\theta}$

12. $r = \dfrac{3}{-4 + 2\cos\theta}$

13. $r = \dfrac{3}{2 - 6\cos\theta}$

14. $r = \dfrac{4}{1 - 2\cos\theta}$

15. $r = \dfrac{3}{2 + 6\sin\theta}$

16. $r = \dfrac{4}{1 + 2\cos\theta}$

In Exercises 17–28, find a polar equation for the specified conic. In each case consider the focus to be at the pole. (For convenience, the equation for the directrix is given in rectangular form.)

Conic	Eccentricity	Directrix
17. Ellipse	$e = \dfrac{1}{2}$	$y = 1$
18. Ellipse	$e = \dfrac{3}{4}$	$y = -2$
19. Parabola	$e = 1$	$x = -1$
20. Parabola	$e = 1$	$y = 1$
21. Hyperbola	$e = 2$	$x = 1$
22. Hyperbola	$e = \dfrac{3}{2}$	$x = -1$

23. Parabola: vertex at $\left(1, -\dfrac{\pi}{2}\right)$

24. Parabola: directrix is $x = 8$

25. Ellipse: vertices at $\left(6, \dfrac{\pi}{2}\right)$ and $\left(2, \dfrac{3\pi}{2}\right)$

26. Ellipse: vertex at $(2, 0)$, directrix is $x = 6$

27. Hyperbola: vertex at $\left(2, \dfrac{3\pi}{2}\right)$, directrix is $y = -3$

28. Hyperbola: vertices at $(2, 0)$ and $(6, 0)$

29. Show that the polar equation for the ellipse

$$\frac{x^2}{a^2} + \frac{y^2}{b^2} = 1$$

is

$$r^2 = \frac{b^2}{1 - e^2 \cos^2 \theta}.$$

30. Show that the polar equation for the hyperbola

$$\frac{x^2}{a^2} - \frac{y^2}{b^2} = 1$$

is

$$r^2 = \frac{-b^2}{1 - e^2 \cos^2 \theta}.$$

In Exercises 31–34, use the results of Exercises 29 and 30 to find the polar equation of the given conic.

31. Ellipse: focus at $(4, 0)$, vertices at $(5, 0)$, $(5, \pi)$
32. Hyperbola: focus at $(5, 0)$, vertices at $(4, 0)$, $(4, \pi)$
33. $\dfrac{x^2}{9} - \dfrac{y^2}{16} = 1$ **34.** $\dfrac{x^2}{4} + y^2 = 1$

In Exercises 35–38, sketch the graph of the rotated conic.

35. $r = \dfrac{2}{1 - \cos (\theta - \pi/4)}$ (See Exercise 1)

36. $r = \dfrac{4}{1 + \sin (\theta - \pi/3)}$ (See Exercise 2)

37. $r = \dfrac{4}{2 + \sin (\theta + \pi/6)}$ (See Exercise 7)

38. $r = \dfrac{4}{1 + 2 \cos (\theta + 2\pi/3)}$ (See Exercise 16)

In Exercises 39 and 40, use Simpson's Rule (with $n = 6$) to find the area of the region bounded by the graph of the given polar equation.

39. $r = \dfrac{3}{2 - \cos \theta}$ **40.** $r = \dfrac{2}{3 - 2 \sin \theta}$

41. Show that the graphs of the equations

$$r = \frac{ed}{1 + \sin \theta} \quad \text{and} \quad r = \frac{ed}{1 - \sin \theta}$$

intersect at right angles.

42. Prove that the tangent of the angle ψ $(0 \leq \psi \leq \pi/2)$ between the radial line and the tangent line at the point (r, θ) on the graph of $r = f(\theta)$ (see figure) is given by

$$\tan \psi = \frac{r}{dr/d\theta}.$$

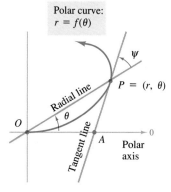

Polar curve:
$r = f(\theta)$

Angle Between Radial
and Tangent Line   FIGURE FOR 42

In Exercises 43–48, use the result of Exercise 42 to find the angle ψ between the radial and tangent lines to the graph at the given point.

Polar equation	Point
43. $r = 2(1 - \cos \theta)$	$\theta = \pi$
44. $r = 3(1 - \cos \theta)$	$\theta = \dfrac{3\pi}{4}$
45. $r = 2 \cos 3\theta$	$\theta = \dfrac{\pi}{6}$
46. $r = 4 \sin 2\theta$	$\theta = \dfrac{\pi}{6}$
47. $r = \dfrac{6}{1 - \cos \theta}$	$\theta = \dfrac{2\pi}{3}$
48. $r = 5$	$\theta = \dfrac{\pi}{6}$

49. The planets travel in elliptical orbits with the sun as a focus. If this focus is at the pole, then the major axis lies on the polar axis, and the length of the major axis is $2a$ (see figure). Show that the polar equation of the orbit is given by

$$r = \frac{(1 - e^2)a}{1 - e \cos \theta}$$

where e is the eccentricity.

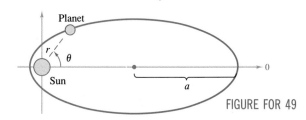

FIGURE FOR 49

50. Use the result of Exercise 49 to show that the minimum distance (*perihelion distance*) from the sun to the planet is $r = a(1 - e)$ and the maximum distance (*aphelion distance*) is $r = a(1 + e)$.

In Exercises 51–53, use the results of Exercises 49 and 50 to find the polar equation of the orbit of the planet and the perihelion and aphelion distances.

51. Earth: $a = 92.957 \times 10^6$ miles
$e = 0.0167$
52. Pluto: $a = 3.666 \times 10^9$ miles
$e = 0.2481$

53. Saturn: $a = 1.427 \times 10^9$ kilometers
$e = 0.0543$

54. Use a computer or graphics calculator to sketch the graph of

$$r = \frac{2e}{1 + e \cos \theta}$$

on the same set of coordinate axes for each of the following eccentricities.
(a) $e = 1$ (b) $e = 0.5$
(c) $e = 0.7$ (d) $e = 1.5$

REVIEW EXERCISES for Chapter 11

In Exercises 1–10, (a) find dy/dx and all points of horizontal tangency, (b) eliminate the parameter where possible, and (c) sketch the curve represented by the parametric equations.

1. $x = 1 + 4t, y = 2 - 3t$
2. $x = e^t, y = e^{-t}$
3. $x = 3 + 2 \cos \theta, y = 2 + 5 \sin \theta$
4. $x = t^2 - 3t + 2, y = t^3 - 3t^2 + 2$
5. $x = \frac{1}{t}, y = 2t + 3$
6. $x = 2t - 1, y = \frac{1}{t^2 - 2t}$
7. $x = \frac{1}{2t + 1}, y = \frac{2t(t + 1)}{2t + 1}$
8. $x = \cot \theta, y = \sin 2\theta$
9. $x = \cos^3 \theta, y = 4 \sin^3 \theta$
10. $x = 2\theta - \sin \theta, y = 2 - \cos \theta$

In Exercises 11 and 12, find a parametric representation of the conic.

11. Ellipse: center at $(-3, 4)$, horizontal major axis of length 8, and minor axis of length 6
12. Hyperbola: vertices at $(0, \pm 4)$ and foci at $(0, \pm 5)$

13. Eliminate the parameter from

$$x = a(\theta - \sin \theta)$$

and

$$y = a(1 - \cos \theta)$$

to show that the rectangular equation of a cycloid is

$$x = a \arccos\left(\frac{a - y}{a}\right) \pm \sqrt{2ay - y^2}.$$

14. The **involute of a circle** is described by the endpoint P of a string that is held taut as it is unwound from a spool that does not turn (see figure). Show that a parametric representation of the involute is given by

$$x = r(\cos \theta + \theta \sin \theta)$$

and

$$y = r(\sin \theta - \theta \cos \theta).$$

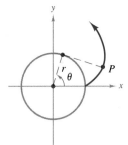

FIGURE FOR 14

In Exercises 15 and 16, find the length of the curve represented by the parametric equations over the given interval.

15. $x = r(\cos \theta + \theta \sin \theta), y = r(\sin \theta - \theta \cos \theta),$
$0 \le \theta \le \pi$
16. $x = 6 \cos \theta, y = 6 \sin \theta, 0 \le \theta \le \pi$

In Exercises 17–32, sketch the graph of the equation.

17. $r = -2(1 + \cos \theta)$ **18.** $r = 3 - 4 \cos \theta$
19. $r = 4 - 3 \cos \theta$ **20.** $r = \cos 5\theta$
21. $r = -3 \cos 2\theta$ **22.** $r^2 = \cos 2\theta$
23. $r = 4$ **24.** $r = 2\theta$

25. $r = -\sec \theta$

26. $r = 3 \csc \theta$

27. $r^2 = 4 \sin^2 2\theta$

28. $r = 2 \sin \theta \cos^2 \theta$

29. $r = \dfrac{2}{1 - \sin \theta}$

30. $r = \dfrac{4}{5 - 3 \cos \theta}$

31. $r = 4 \cos 2\theta \sec \theta$

32. $r = 4(\sec \theta - \cos \theta)$

In Exercises 33–38, convert the given polar equation to rectangular form.

33. $r = 3 \cos \theta$

34. $r = 4 \sec \left(\theta - \dfrac{\pi}{3} \right)$

35. $r = -2(1 + \cos \theta)$

36. $r = 1 + \tan \theta$

37. $r = 4 \cos 2\theta \sec \theta$

38. $\theta = \dfrac{3\pi}{4}$

In Exercises 39–42, convert the given rectangular equation to polar form.

39. $(x^2 + y^2)^2 = ax^2 y$

40. $x^2 + y^2 - 4x = 0$

41. $x^2 + y^2 = a^2 \left(\arctan \dfrac{y}{x} \right)^2$

42. $(x^2 + y^2) \left(\arctan \dfrac{y}{x} \right)^2 = a^2$

In Exercises 43–48, find a polar equation for the given graph.

43. Parabola: focus at the pole, vertex at $(2, \pi)$

44. Ellipse: a focus at the pole, vertices at $(5, 0)$ and $(1, \pi)$

45. Hyperbola: a focus at the pole, vertices at $(1, 0)$ and $(7, 0)$

46. Circle: center $(0, 5)$, passing through the origin

47. Line: intercepts at $(3, 0)$ and $(0, 4)$

48. Line: through the origin, slope $\sqrt{3}$

In Exercises 49 and 50, (a) find the tangents at the pole, (b) find all points of horizontal and vertical tangency, and (c) sketch the graph of the equation.

49. $r = 1 - 2 \cos \theta$

50. $r^2 = 4 \sin 2\theta$

In Exercises 51 and 52, show that the graphs of the given polar equations are orthogonal at the points of intersection.

51. $r = 1 + \cos \theta$, $r = 1 - \cos \theta$

52. $r = a \sin \theta$, $r = a \cos \theta$

In Exercises 53–60, find the area of the given region.

53. Interior of $r = 2 + \cos \theta$

54. Interior of $r = 5(1 - \sin \theta)$

55. Interior of $r = \sin \theta \cos^2 \theta$

56. Interior of $r = 4 \sin 3\theta$

57. Interior of $r^2 = a^2 \sin 2\theta$

58. Common interior of $r = a$ and $r^2 = 2a^2 \sin 2\theta$

59. Common interior of $r = 4 \cos \theta$ and $r = 2$

60. Region bounded by the polar axis and $r = e^\theta$ for $0 \le \theta \le \pi$

In Exercises 61 and 62, find the perimeter of the curve represented by the polar equation.

61. $r = a(1 - \cos \theta)$

62. $r = a \cos 2\theta$

63. Find the angle between the circle $r = 3 \sin \theta$ and the limaçon $r = 4 - 5 \sin \theta$ at the point of intersection $(3/2, \pi/6)$.

APPENDIX A

Proofs of Selected Theorems

THEOREM 2.3 **PROPERTIES OF LIMITS** **(Properties 3, 4, and 5)** **(page 78)**	Let b and c be real numbers, n a positive integer, and let f and g be functions whose limits exist as $x \to c$. Then the following properties are true:

3. $\lim\limits_{x \to c} f(x)g(x) = \left[\lim\limits_{x \to c} f(x)\right]\left[\lim\limits_{x \to c} g(x)\right]$

4. $\lim\limits_{x \to c} \dfrac{f(x)}{g(x)} = \dfrac{\lim\limits_{x \to c} f(x)}{\lim\limits_{x \to c} g(x)}$, provided $\lim\limits_{x \to c} g(x) \neq 0$

5. $\lim\limits_{x \to c} [f(x)]^n = \left[\lim\limits_{x \to c} f(x)\right]^n$

PROOF Let

$$\lim_{x \to c} f(x) = L \qquad \text{and} \qquad \lim_{x \to c} g(x) = K.$$

To prove Property 3, we write

$$f(x)g(x) = [f(x) - L][g(x) - K] + [Lg(x) + Kf(x)] - LK.$$

Since the limit of $f(x)$ is L, and the limit of $g(x)$ is K, we have

$$\lim_{x \to c} [f(x) - L] = 0 \qquad \text{and} \qquad \lim_{x \to c} [g(x) - K] = 0.$$

Let $0 < \varepsilon < 1$. Then there exists $\delta > 0$ such that if $0 < |x - c| < \delta$, then

$$|f(x) - L - 0| < \varepsilon \qquad \text{and} \qquad |g(x) - K - 0| < \varepsilon,$$

which implies that

$$|[f(x) - L][g(x) - K] - 0| = |f(x) - L|\,|g(x) - K| < \varepsilon\varepsilon < \varepsilon.$$

Hence,

$$\lim_{x \to c} [f(x) - L][g(x) - K] = 0.$$

Furthermore, by Property 1 of Theorem 2.3, we have

$$\lim_{x \to c} Lg(x) = LK \qquad \text{and} \qquad \lim_{x \to c} Kf(x) = KL.$$

Finally, by Property 2 of Theorem 2.3, we obtain

$$\lim_{x \to c} f(x)g(x) = \lim_{x \to c} [f(x) - L][g(x) - K] + \lim_{x \to c} Lg(x) + \lim_{x \to c} Kf(x) - \lim_{x \to c} LK$$

$$= 0 + LK + KL - LK$$

$$= LK.$$

To prove Property 4, we use Property 3 and the limit of a reciprocal function and write

$$\lim_{x \to c} \frac{f(x)}{g(x)} = \lim_{x \to c} \left[f(x) \frac{1}{g(x)} \right] = \left[\lim_{x \to c} f(x) \right]\left[\lim_{x \to c} \frac{1}{g(x)} \right] = \frac{L}{K}.$$

Finally, the proof of Property 5 can be obtained by a straightforward application of mathematical induction coupled with Property 3.

THEOREM 2.6
LIMIT OF A FUNCTION INVOLVING
A RADICAL   (page 80)

If $c > 0$ and n is *any* positive integer, or if $c \le 0$ and n is an *odd* positive integer, then

$$\lim_{x \to c} \sqrt[n]{x} = \sqrt[n]{c}.$$

PROOF Consider the case for which $c > 0$ and n is any positive integer. For a given $\varepsilon > 0$, we need to find $\delta > 0$ such that

$$|\sqrt[n]{x} - \sqrt[n]{c}| < \varepsilon \qquad \text{whenever} \qquad 0 < |x - c| < \delta,$$

which is the same as saying

$$-\varepsilon < \sqrt[n]{x} - \sqrt[n]{c} < \varepsilon \qquad \text{whenever} \qquad -\delta < x - c < \delta.$$

We assume $\varepsilon < \sqrt[n]{c}$, which implies that $0 < \sqrt[n]{c} - \varepsilon < \sqrt[n]{c}$. Now, we let δ be the smaller of the two numbers.

$$c - \left(\sqrt[n]{c} - \varepsilon \right)^n \qquad \text{and} \qquad \left(\sqrt[n]{c} + \varepsilon \right)^n - c$$

Then we have

$$-\delta < x - c \qquad < \delta$$

$$-[c - \left(\sqrt[n]{c} - \varepsilon \right)^n] < x - c \qquad < \left(\sqrt[n]{c} + \varepsilon \right)^n - c$$

$$\left(\sqrt[n]{c} - \varepsilon \right)^n - c < x - c \qquad < \left(\sqrt[n]{c} + \varepsilon \right)^n - c$$

$$\left(\sqrt[n]{c} - \varepsilon \right)^n < x \qquad < \left(\sqrt[n]{c} + \varepsilon \right)^n$$

$$\sqrt[n]{c} - \varepsilon < \sqrt[n]{x} \qquad < \sqrt[n]{c} + \varepsilon$$

$$-\varepsilon < \sqrt[n]{x} - \sqrt[n]{c} < \varepsilon$$

THEOREM 2.7
LIMIT OF A COMPOSITE FUNCTION
(page 80)

If f and g are functions such that

$$\lim_{x \to c} g(x) = L \qquad \text{and} \qquad \lim_{x \to L} f(x) = f(L)$$

then

$$\lim_{x \to c} f(g(x)) = f(L).$$

PROOF

For a given $\varepsilon > 0$, we must find $\delta > 0$ such that

$$|f(g(x)) - f(L)| < \varepsilon \qquad \text{whenever} \qquad 0 < |x - c| < \delta.$$

Since the limit of $f(x)$ as $x \to L$ is $f(L)$, we know there exists $\delta_1 > 0$ such that

$$|f(u) - f(L)| < \varepsilon \qquad \text{whenever} \qquad 0 < |u - L| < \delta_1.$$

Moreover, since the limit of $g(x)$ as $x \to c$ is L, we know there exists $\delta > 0$ such that

$$|g(x) - L| < \delta_1 \qquad \text{whenever} \qquad 0 < |x - c| < \delta.$$

Finally, letting $u = g(x)$, we have

$$|f(g(x)) - f(L)| < \varepsilon \qquad \text{whenever} \qquad 0 < |x - c| < \delta.$$

THEOREM 2.15
VERTICAL ASYMPTOTES
(page 103)

Let f and g be continuous on an open interval containing c. If $f(c) \neq 0$, $g(c) = 0$, and there exists an open interval containing c such that $g(x) \neq 0$ for all $x \neq c$ in the interval, then the graph of the function given by

$$F(x) = \frac{f(x)}{g(x)}$$

has a vertical asymptote at $x = c$.

PROOF

We consider the case for which $f(c) > 0$, and there exists $b > c$ such that $c < x < b$ implies $g(x) > 0$. Then for $M > 0$, we choose δ_1 such that

$$0 < x - c < \delta_1 \qquad \text{implies that} \qquad \frac{f(c)}{2} < f(x) < \frac{3f(c)}{2}$$

and δ_2 such that

$$0 < x - c < \delta_2 \qquad \text{implies that} \qquad 0 < g(x) < \frac{f(c)}{2M}.$$

Now we let δ be the smaller of δ_1 and δ_2. Then, it follows that

$$0 < x - c < \delta \qquad \text{implies that} \qquad \frac{f(x)}{g(x)} > \frac{f(c)}{2} \left(\frac{2M}{f(c)} \right) = M.$$

Therefore, it follows that

$$\lim_{x \to c^+} \frac{f(x)}{g(x)} = \infty$$

and the line $x = c$ is a vertical asymptote of the graph of F.

THEOREM 3.11 **THE CHAIN RULE**	**(page 148)**	If $y = f(u)$ is a differentiable function of u, and $u = g(x)$ is a differentiable function of x, then $y = f(g(x))$ is a differentiable function of x, and $$\frac{dy}{dx} = \frac{dy}{du} \cdot \frac{du}{dx}.$$

PROOF In Section 3.5, we let $F(x) = f(g(x))$ and used the alternate form of the derivative to show that $F'(c) = f'(g(c))g'(c)$, provided $g(x) \neq g(c)$ for values of x other than c. We now consider a more general proof. We begin by considering the derivative of f.

$$f'(x) = \lim_{\Delta x \to 0} \frac{f(x + \Delta x) - f(x)}{\Delta x} = \lim_{\Delta x \to 0} \frac{\Delta y}{\Delta x}$$

For a fixed value of x, we define a function η such that

$$\eta(\Delta x) = \begin{cases} 0, & \Delta x = 0 \\ \dfrac{\Delta y}{\Delta x} - f'(x), & \Delta x \neq 0. \end{cases}$$

Since the limit of $\eta(\Delta x)$ as $\Delta x \to 0$ doesn't depend upon the value of $\eta(0)$, we have

$$\lim_{\Delta x \to 0} \eta(\Delta x) = \lim_{\Delta x \to 0} \left[\frac{\Delta y}{\Delta x} - f'(x) \right] = 0$$

and we can conclude that η is continuous at 0. Moreover, since $\Delta y = 0$ when $\Delta x = 0$, the equation

$$\Delta y = \Delta x \eta(\Delta x) + \Delta x f'(x)$$

is valid whether Δx is a zero or not. Now, by letting $\Delta u = g(x + \Delta x) - g(x)$, we use the continuity of g to conclude that

$$\lim_{\Delta x \to 0} \Delta u = \lim_{\Delta x \to 0} [g(x + \Delta x) - g(x)] = 0,$$

which implies that

$$\lim_{\Delta x \to 0} \eta(\Delta u) = 0.$$

Finally,

$$\Delta y = \Delta u \eta(\Delta u) + \Delta u f'(u) \quad \Longrightarrow \quad \frac{\Delta y}{\Delta x} = \frac{\Delta u}{\Delta x} \eta(\Delta u) + \frac{\Delta u}{\Delta x} f'(u), \quad \Delta x \neq 0$$

and taking the limit as $\Delta x \to 0$, we have

$$\frac{dy}{dx} = \frac{du}{dx}\left[\lim_{\Delta x \to 0} \eta(\Delta u)\right] + \frac{du}{dx}f'(u) = \frac{dy}{dx}(0) + \frac{du}{dx}f'(u)$$

$$= \frac{du}{dx}f'(u) = \frac{du}{dx}\cdot\frac{dy}{du}.$$

EQUIVALENCE OF CONCAVITY DEFINITIONS (page 199)

Let f be a function that is differentiable on some open interval containing c.

The graph of f is concave *upward* at $f(c)$ if and only if the graph of f lies *above* the tangent line to the graph of f at $(c, f(c))$ on some open interval containing c.

The graph of f is concave *downward* at $f(c)$ if and only if the graph of f lies *below* the tangent line to the graph of f at $(c, f(c))$ on some open interval containing c.

PROOF

There are many parts to this proof, but they are all similar. We give only one part here, and ask you to supply the remaining parts. Assume f is concave upward at c, then by definition, there exists an interval (a, b) containing c such that f' is increasing on (a, b). The equation of the tangent line to the graph of f at c is given by

$$g(x) = f(c) + f'(c)(x - c).$$

If x is in the open interval (c, b), then the directed distance from the point $(x, f(x))$ (on the graph of f) to the point $(x, g(x))$ (on the tangent line) is given by

$$d = f(x) - [f(c) + f'(c)(x - c)]$$
$$= f(x) - f(c) - f'(c)(x - c).$$

Moreover, by the Mean Value Theorem there exists a number z in (c, x) such that

$$f'(z) = \frac{[f(x) - f(c)]}{(x - c)}.$$

Thus, we have

$$d = f(x) - f(c) - f'(c)(x - c)$$
$$= f'(z)(x - c) - f'(c)(x - c)$$
$$= [f'(z) - f'(c)](x - c).$$

The second factor $(x - c)$ is positive because $c < x$. Moreover, since f' is increasing, it follows that the first factor $[f'(z) - f'(c)]$ is also positive. Therefore, $d > 0$ and we conclude that the graph of f lies above the tangent line. If x is in the open interval (a, c), a similar argument can be given.

THEOREM 4.10
LIMITS AT INFINITY
(page 208)

If r is a positive rational number, and c is any real number, then

$$\lim_{x \to \infty} \left(\frac{c}{x^r} \right) = 0.$$

Furthermore, if x^r is defined when $x < 0$, then

$$\lim_{x \to -\infty} \left(\frac{c}{x^r} \right) = 0.$$

PROOF   We begin by proving that

$$\lim_{x \to \infty} \frac{1}{x} = 0.$$

For $\varepsilon > 0$, let $M = 1/\varepsilon$. Then, for $x > M$, we have

$$x > M = \frac{1}{\varepsilon} \implies \frac{1}{x} < \varepsilon \implies \left| \frac{1}{x} - 0 \right| < \varepsilon.$$

Therefore, by the definition of a limit at infinity, we conclude that the limit of $1/x$ as $x \to \infty$ is 0. Now, using this result, and letting $r = m/n$, we can write the following.

$$\lim_{x \to \infty} \left(\frac{c}{x^r} \right) = \lim_{x \to \infty} \left(\frac{c}{x^{m/n}} \right)$$

$$= c \left[\lim_{x \to \infty} \left(\frac{1}{\sqrt[n]{x}} \right)^m \right]$$

$$= c \left[\lim_{x \to \infty} \sqrt[n]{\frac{1}{x}} \right]^m$$

$$= c \left[\sqrt[n]{\lim_{x \to \infty} \frac{1}{x}} \right]^m$$

$$= c \left[\sqrt[n]{0} \right]^m$$

$$= 0$$

The proof of the second part of the theorem is similar.

THEOREM 5.4
SUMMATION FORMULAS
(page 270)

1. $\displaystyle\sum_{i=1}^{n} c = cn$

2. $\displaystyle\sum_{i=1}^{n} i = \frac{n(n+1)}{2}$

3. $\displaystyle\sum_{i=1}^{n} i^2 = \frac{n(n+1)(2n+1)}{6}$

4. $\displaystyle\sum_{i=1}^{n} i^3 = \frac{n^2(n+1)^2}{4}$

PROOF   The proof of part 1 is straightforward. By adding c to itself n times, we obtain a sum of nc.

To prove part 2, we write the sum in increasing and decreasing order and add corresponding terms as follows.

$$\sum_{i=1}^{n} i = 1 + 2 + 3 + \cdots + (n-1) + n$$

$$\downarrow \quad \downarrow \quad \downarrow \quad\quad \downarrow \quad \downarrow$$

$$\sum_{i=1}^{n} i = n + (n-1) + (n-2) + \cdots + 2 + 1$$

$$\downarrow \quad \downarrow \quad \downarrow \quad\quad \downarrow \quad \downarrow$$

$$2\sum_{i=1}^{n} i = (n+1) + (n+1) + (n+1) + \cdots + (n+1) + (n+1)$$

n terms

Therefore,

$$\sum_{i=1}^{n} i = \frac{n(n+1)}{2}.$$

To prove part 3, we use mathematical induction. First, if $n = 1$, the result is true because

$$\sum_{i=1}^{1} i^2 = 1^2 = 1 = \frac{1(1+1)(2+1)}{6}.$$

Now, assuming the result is true for $n = k$, we show that it is true for $n = k + 1$, as follows.

$$\sum_{i=1}^{k+1} i^2 = \sum_{i=1}^{k} i^2 + (k+1)^2$$

$$= \frac{k(k+1)(2k+1)}{6} + (k+1)^2$$

$$= \frac{k+1}{6}(2k^2 + k + 6k + 6)$$

$$= \frac{k+1}{6}[(2k+3)(k+2)]$$

$$= \frac{(k+1)(k+2)[2(k+1)+1]}{6}$$

Part 4 can be proved using a similar argument with mathematical induction.

THEOREM 5.10
PRESERVATION OF INEQUALITY
(page 289)

1. If f is integrable and nonnegative on the closed interval $[a, b]$, then

$$0 \le \int_a^b f(x)dx.$$

2. If f and g are integrable on the closed interval $[a, b]$, and $f(x) \le g(x)$ for every x in $[a, b]$, then

$$\int_a^b f(x)dx \le \int_a^b g(x)dx.$$

PROOF   To prove part 1, we suppose, on the contrary, that

$$\int_a^b f(x)dx = I < 0.$$

Then, let $a = x_0 < x_1 < x_2 < \cdots < x_n = b$ be a partition of $[a, b]$, and let

$$R = \sum_{i=1}^n f(c_i)\Delta x_i$$

be a Riemann sum. Since $f(x) \geq 0$, it follows that $R \geq 0$. Now, for $\|\Delta\|$ sufficiently small, we have $|R - I| < -I/2$, which implies that

$$\sum_{i=1}^n f(c_i)\Delta x_i = R < I - \frac{I}{2} < 0,$$

which is not possible. From this contradiction, we can conclude that

$$0 \leq \int_a^b f(x)dx.$$

To prove the second part of the theorem, we note that $f(x) \leq g(x)$ implies that $g(x) - f(x) \geq 0$. Hence, we can apply the result of part 1 to conclude that

$$0 \leq \int_a^b (g(x) - f(x))dx$$

$$0 \leq \int_a^b g(x)dx - \int_a^b f(x)dx$$

$$\int_a^b f(x)dx \leq \int_a^b g(x)dx.$$

PROPERTIES OF THE NATURAL LOGARITHMIC FUNCTION (page 327)

$$\lim_{x \to 0^+} \ln x = -\infty \quad \text{and} \quad \lim_{x \to \infty} \ln x = \infty$$

PROOF   To begin, we show that $\ln 2 > \frac{1}{2}$. From the Mean Value Theorem of Integrals, we can write

$$\ln 2 = \int_1^2 \frac{1}{x} dx = (2 - 1)\frac{1}{c} = \frac{1}{c}$$

where c is in $[1, 2]$. This implies that

$$1 \leq c \quad \leq 2$$

$$1 \geq \frac{1}{c} \quad \geq \frac{1}{2}$$

$$1 \geq \ln 2 \geq \frac{1}{2}.$$

Now, let N be any positive (large) number. Since $\ln x$ is increasing, it follows that if $x > 2^{2N}$, then

$$\ln x > \ln 2^{2N} = 2N \ln 2.$$

However, since $\ln 2 > \frac{1}{2}$, it follows that

$$\ln x > 2N \ln 2 \geq 2N \left(\frac{1}{2}\right) = N.$$

This verifies the second limit. To verify the first limit, we let $z = 1/x$. Then, $z \to \infty$ as $x \to 0^+$, and we can write

$$\lim_{x \to 0^+} \ln x = \lim_{x \to 0^+} \left(-\ln \frac{1}{x}\right) = \lim_{z \to \infty} (-\ln z) = -\lim_{z \to \infty} \ln z = -\infty.$$

**THEOREM 6.9
CONTINUITY AND
DIFFERENTIABILITY OF INVERSE
FUNCTIONS   (page 346)**

1. If f is continuous on its domain, then f^{-1} is continuous on its domain.
2. If f is increasing on its domain, then f^{-1} is increasing on its domain.
3. If f is decreasing on its domain, then f^{-1} is decreasing on its domain.
4. If f is differentiable at c and $f'(c) \neq 0$, then f^{-1} is differentiable at $f(c)$.

PROOF

To prove part 1, assume that a is in an open interval in the domain of f^{-1}, and let $f^{-1}(a) = c$. (A similar proof using one-sided limits can be given for the endpoints.) For $\varepsilon > 0$, let $\delta_1 = a - f(c - \varepsilon)$ and $\delta_2 = f(c + \varepsilon) - a$. If y is in the interval

$$(a - \delta_1, a + \delta_2) = (f(c - \varepsilon), \ f(c + \varepsilon))$$

then by the one-to-one nature of f and f^{-1}, it follows that $f^{-1}(y)$ is in the interval $(c - \varepsilon, c + \varepsilon)$. By taking δ to be the minimum of δ_1 and δ_2, it follows that y is in $(a - \delta, a + \delta)$, and since $f^{-1}(y)$ is in $(c - \varepsilon, c + \varepsilon)$, we can conclude that

$$\lim_{y \to a} f^{-1}(y) = f^{-1}(a),$$

which implies that f^{-1} is continuous at a.

To prove part 2, let y_1 and y_2 be in the domain of f^{-1}, with $y_1 < y_2$. Then, there exist x_1 and x_2 in the domain of f such that

$$f(x_1) = y_1 < y_2 = f(x_2).$$

Since f is increasing, $f(x_1) < f(x_2)$ holds precisely when $x_1 < x_2$. Therefore,

$$f^{-1}(y_1) = x_1 < x_2 = f^{-1}(y_2),$$

which implies that f^{-1} is increasing. (Part 3 can be proved in a similar way.)

Finally, to prove part 4, we consider the limit

$$(f^{-1})'(a) = \lim_{y \to a} \frac{f^{-1}(y) - f^{-1}(a)}{y - a}$$

where a is in the domain of f^{-1} and $f^{-1}(a) = c$. Since f is differentiable, f is continuous, and so is f^{-1}. Thus, $y \to a$ implies that $x \to c$, and we have

$$(f^{-1})'(a) = \lim_{x \to c} \frac{x - c}{f(x) - f(c)}$$

$$= \lim_{x \to c} \frac{1}{\dfrac{f(x) - f(c)}{x - c}}$$

$$= \frac{1}{\displaystyle\lim_{x \to c} \frac{f(x) - f(c)}{x - c}}$$

$$= \frac{1}{f'(c)}.$$

Hence, $(f^{-1})'(a)$ exists, and f^{-1} is differentiable at $f(c)$.

THE EXTENDED MEAN VALUE THEOREM   (page 520)

If f and g are differentiable on an open interval (a, b) and continuous on $[a, b]$ such that $g(x) \neq 0$ for any x in (a, b), then there exists a point c in (a, b) such that

$$\frac{f'(c)}{g'(c)} = \frac{f(b) - f(a)}{g(b) - g(a)}.$$

PROOF We can assume that $g(a) \neq g(b)$, since otherwise by Rolle's Theorem it would follow that $g'(x) = 0$ for some x in (a, b). Now, we define $h(x)$ to be

$$h(x) = f(x) - \left(\frac{f(b) - f(a)}{g(b) - g(a)} \right) g(x).$$

Then

$$h(a) = f(a) - \left(\frac{f(b) - f(a)}{g(b) - g(a)} \right) g(a) = \frac{f(a)g(b) - f(b)g(a)}{g(b) - g(a)}$$

and

$$h(b) = f(b) - \left(\frac{f(b) - f(a)}{g(b) - g(a)} \right) g(b) = \frac{f(a)g(b) - f(b)g(a)}{g(b) - g(a)}$$

and by Rolle's Theorem there exists a point c in (a, b) such that

$$h'(c) = f'(c) - \frac{f(b) - f(a)}{g(b) - g(a)} g'(c) = 0,$$

which implies that

$$\frac{f'(c)}{g'(c)} = \frac{f(b) - f(a)}{g(b) - g(a)}.$$

We can use the Extended Mean Value Theorem to prove L'Hôpital's Rule. Of the several different cases of this rule, we illustrate the proof of only one case and leave the remaining cases for you to prove.

**THEOREM 8.3
L'HÔPITAL'S RULE
(page 520)**

Let f and g be functions that are differentiable on an open interval (a, b) containing c, except possibly at c itself. If the limit of $f(x)/g(x)$ as x approaches c produces the indeterminate form $0/0$ or ∞/∞, then

$$\lim_{x \to c} \frac{f(x)}{g(x)} = \lim_{x \to c} \frac{f'(x)}{g'(x)},$$

provided the limit on the right exists (or is infinite).

PROOF   We consider the case for which

$$\lim_{x \to a^+} f'(x) = 0 \qquad \text{and} \qquad \lim_{x \to a^+} g'(x) = 0.$$

Moreover, we assume that

$$\lim_{x \to a^+} \frac{f'(x)}{g'(x)} = L.$$

Now, we choose b such that both f and g are differentiable in the interval (a, b) and define the following two new functions:

$$F(x) = \begin{cases} f(x), & x \neq a \\ 0, & x = a \end{cases} \qquad \text{and} \qquad G(x) = \begin{cases} g(x), & x \neq a \\ 0, & x = a. \end{cases}$$

It follows that F and G are continuous on $[a, b]$, and we can apply the Extended Mean Value Theorem to conclude that there exists a number c in (a, b) such that

$$\frac{F'(c)}{G'(c)} = \frac{F(b) - F(a)}{G(b) - G(a)} = \frac{F(b) - 0}{G(b) - 0} = \frac{F(b)}{G(b)}.$$

Finally, by letting b approach a, we also force c to approach a, and it follows that

$$\lim_{b \to a^+} \frac{f(b)}{g(b)} = \lim_{b \to a^+} \frac{F(b)}{G(b)} = \lim_{c \to a^+} \frac{F'(c)}{G'(c)} = \lim_{c \to a^+} \frac{f'(c)}{g'(c)}$$

and we conclude that

$$\lim_{x \to a^+} \frac{f(x)}{g(x)} = \lim_{x \to a^+} \frac{f'(x)}{g'(x)}.$$

**THEOREM 11.11
CLASSIFICATION OF CONICS BY
ECCENTRICITY   (page 698)**

The locus of a point in the plane whose distance from a fixed point (*focus*) has a constant ratio to its distance from a fixed line (*directrix*) is a conic. The constant ratio e is the *eccentricity* of the conic.

1. The conic is an ellipse if $0 < e < 1$.
2. The conic is a parabola if $e = 1$.
3. The conic is a hyperbola if $e > 1$.

PROOF   If $e = 1$, then by definition, the conic must be a parabola. If $e \neq 1$, then we consider the focus F to lie at the origin and the directrix $x = d$ to lie to the

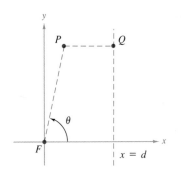

FIGURE A.1

right of the origin, as shown in Figure A.1. For the point $P = (r, \theta) = (x, y)$, we have $|PF| = r$ and $|PQ| = d - r \cos \theta$. Given that $e = |PF|/|PQ|$, it follows that

$$|PF| = |PQ|e \quad \Longrightarrow \quad r = e(d - r \cos \theta).$$

By converting to rectangular coordinates and squaring both sides, we obtain

$$x^2 + y^2 = e^2(d - x)^2 = e^2(d^2 - 2dx + x^2).$$

Completing the square produces

$$\left(x + \frac{e^2 d}{1 - e^2}\right)^2 + \frac{y^2}{1 - e^2} = \frac{e^2 d^2}{(1 - e^2)^2}.$$

If $e < 1$, then this equation represents an ellipse. If $e > 1$, then $1 - e^2 < 0$, and the equation represents a hyperbola.

APPENDIX B

Basic Differentiation Rules for Elementary Functions

1. $\dfrac{d}{dx}[cu] = cu'$

2. $\dfrac{d}{dx}[u \pm v] = u' \pm v'$

3. $\dfrac{d}{dx}[uv] = uv' + vu'$

4. $\dfrac{d}{dx}\left[\dfrac{u}{v}\right] = \dfrac{vu' - uv'}{v^2}$

5. $\dfrac{d}{dx}[c] = 0$

6. $\dfrac{d}{dx}[u^n] = nu^{n-1}u'$

7. $\dfrac{d}{dx}[x] = 1$

8. $\dfrac{d}{dx}[|u|] = \dfrac{u}{|u|}(u'),\ u \neq 0$

9. $\dfrac{d}{dx}[\ln u] = \dfrac{u'}{u}$

10. $\dfrac{d}{dx}[e^u] = e^u u'$

11. $\dfrac{d}{dx}[\sin u] = (\cos u)u'$

12. $\dfrac{d}{dx}[\cos u] = -(\sin u)u'$

13. $\dfrac{d}{dx}[\tan u] = (\sec^2 u)u'$

14. $\dfrac{d}{dx}[\cot u] = -(\csc^2 u)u'$

15. $\dfrac{d}{dx}[\sec u] = (\sec u \tan u)u'$

16. $\dfrac{d}{dx}[\csc u] = -(\csc u \cot u)u'$

17. $\dfrac{d}{dx}[\arcsin u] = \dfrac{u'}{\sqrt{1 - u^2}}$

18. $\dfrac{d}{dx}[\arccos u] = \dfrac{-u'}{\sqrt{1 - u^2}}$

19. $\dfrac{d}{dx}[\arctan u] = \dfrac{u'}{1 + u^2}$

20. $\dfrac{d}{dx}[\text{arccot } u] = \dfrac{-u'}{1 + u^2}$

21. $\dfrac{d}{dx}[\text{arcsec } u] = \dfrac{u'}{|u|\sqrt{u^2 - 1}}$

22. $\dfrac{d}{dx}[\text{arccsc } u] = \dfrac{-u'}{|u|\sqrt{u^2 - 1}}$

APPENDIX C

Integration Tables

Forms Involving u^n

1. $\displaystyle\int u^n\, du = \frac{u^{n+1}}{n+1} + C, \; n \neq -1$

2. $\displaystyle\int \frac{1}{u}\, du = \ln|u| + C$

Forms Involving $a + bu$

3. $\displaystyle\int \frac{u}{a+bu}\, du = \frac{1}{b^2}(bu - a\ln|a+bu|) + C$

4. $\displaystyle\int \frac{u}{(a+bu)^2}\, du = \frac{1}{b^2}\left(\frac{a}{a+bu} + \ln|a+bu|\right) + C$

5. $\displaystyle\int \frac{u}{(a+bu)^n}\, du = \frac{1}{b^2}\left[\frac{-1}{(n-2)(a+bu)^{n-2}} + \frac{a}{(n-1)(a+bu)^{n-1}}\right] + C, \; n \neq 1, 2$

6. $\displaystyle\int \frac{u^2}{a+bu}\, du = \frac{1}{b^3}\left(-\frac{bu}{2}(2a-bu) + a^2\ln|a+bu|\right) + C$

7. $\displaystyle\int \frac{u^2}{(a+bu)^2}\, du = \frac{1}{b^3}\left(bu - \frac{a^2}{a+bu} - 2a\ln|a+bu|\right) + C$

8. $\displaystyle\int \frac{u^2}{(a+bu)^3}\, du = \frac{1}{b^3}\left[\frac{2a}{a+bu} - \frac{a^2}{2(a+bu)^2} + \ln|a+bu|\right] + C$

9. $\displaystyle\int \frac{u^2}{(a+bu)^n}\, du = \frac{1}{b^3}\left[\frac{-1}{(n-3)(a+bu)^{n-3}} + \frac{2a}{(n-2)(a+bu)^{n-2}} - \frac{a^2}{(n-1)(a+bu)^{n-1}}\right] + C,$
$n \neq 1, 2, 3$

10. $\displaystyle\int \frac{1}{u(a+bu)}\, du = \frac{1}{a}\ln\left|\frac{u}{a+bu}\right| + C$

11. $\displaystyle\int \frac{1}{u(a+bu)^2}\, du = \frac{1}{a}\left(\frac{1}{a+bu} + \frac{1}{a}\ln\left|\frac{u}{a+bu}\right|\right) + C$

12. $\displaystyle\int \frac{1}{u^2(a+bu)}\, du = -\frac{1}{a}\left(\frac{1}{u} + \frac{b}{a}\ln\left|\frac{u}{a+bu}\right|\right) + C$

13. $\int \dfrac{1}{u^2(a + bu)^2}\, du = -\dfrac{1}{a^2}\left[\dfrac{a + 2bu}{u(a + bu)} + \dfrac{2b}{a}\ln\left|\dfrac{u}{a + bu}\right|\right] + C$

Forms Involving $a + bu + cu^2,\ b^2 \neq 4ac$

14. $\int \dfrac{1}{a + bu + cu^2}\, du = \begin{cases} \dfrac{2}{\sqrt{4ac - b^2}}\arctan\dfrac{2cu + b}{\sqrt{4ac - b^2}} + C, & b^2 < 4ac \\[4mm] \dfrac{1}{\sqrt{b^2 - 4ac}}\ln\left|\dfrac{2cu + b - \sqrt{b^2 - 4ac}}{2cu + b + \sqrt{b^2 - 4ac}}\right| + C, & b^2 > 4ac \end{cases}$

15. $\int \dfrac{u}{a + bu + cu^2}\, du = \dfrac{1}{2c}\left(\ln|a + bu + cu^2| - b\int \dfrac{1}{a + bu + cu^2}\, du\right)$

Forms Involving $\sqrt{a + bu}$

16. $\int u^n\sqrt{a + bu}\, du = \dfrac{2}{b(2n + 3)}\left[u^n(a + bu)^{3/2} - na\int u^{n-1}\sqrt{a + bu}\, du\right]$

17. $\int \dfrac{1}{u\sqrt{a + bu}}\, du = \begin{cases} \dfrac{1}{\sqrt{a}}\ln\left|\dfrac{\sqrt{a + bu} - \sqrt{a}}{\sqrt{a + bu} + \sqrt{a}}\right| + C, & 0 < a \\[4mm] \dfrac{2}{\sqrt{-a}}\arctan\sqrt{\dfrac{a + bu}{-a}} + C, & a < 0 \end{cases}$

18. $\int \dfrac{1}{u^n\sqrt{a + bu}}\, du = \dfrac{-1}{a(n - 1)}\left[\dfrac{\sqrt{a + bu}}{u^{n-1}} + \dfrac{(2n - 3)b}{2}\int \dfrac{1}{u^{n-1}\sqrt{a + bu}}\, du\right],\ n \neq 1$

19. $\int \dfrac{\sqrt{a + bu}}{u}\, du = 2\sqrt{a + bu} + a\int \dfrac{1}{u\sqrt{a + bu}}\, du$

20. $\int \dfrac{\sqrt{a + bu}}{u^n}\, du = \dfrac{-1}{a(n - 1)}\left[\dfrac{(a + bu)^{3/2}}{u^{n-1}} + \dfrac{(2n - 5)b}{2}\int \dfrac{\sqrt{a + bu}}{u^{n-1}}\, du\right],\ n \neq 1$

21. $\int \dfrac{u}{\sqrt{a + bu}}\, du = \dfrac{-2(2a - bu)}{3b^2}\sqrt{a + bu} + C$

22. $\int \dfrac{u^n}{\sqrt{a + bu}}\, du = \dfrac{2}{(2n + 1)b}\left(u^n\sqrt{a + bu} - na\int \dfrac{u^{n-1}}{\sqrt{a + bu}}\, du\right)$

Forms Involving $a^2 \pm u^2,\ 0 < a$

23. $\int \dfrac{1}{a^2 + u^2}\, du = \dfrac{1}{a}\arctan\dfrac{u}{a} + C$

24. $\int \dfrac{1}{u^2 - a^2}\, du = -\int \dfrac{1}{a^2 - u^2}\, du = \dfrac{1}{2a}\ln\left|\dfrac{u - a}{u + a}\right| + C$

25. $\int \dfrac{1}{(a^2 \pm u^2)^n}\, du = \dfrac{1}{2a^2(n - 1)}\left[\dfrac{u}{(a^2 \pm u^2)^{n-1}} + (2n - 3)\int \dfrac{1}{(a^2 \pm u^2)^{n-1}}\, du\right],\ n \neq 1$

Forms Involving $\sqrt{u^2 \pm a^2},\ 0 < a$

26. $\int \sqrt{u^2 \pm a^2}\, du = \dfrac{1}{2}\left(u\sqrt{u^2 \pm a^2} \pm a^2\ln\left|u + \sqrt{u^2 \pm a^2}\right|\right) + C$

27. $\int u^2\sqrt{u^2 \pm a^2}\, du = \dfrac{1}{8}\Big[u(2u^2 \pm a^2)\sqrt{u^2 \pm a^2} - a^4 \ln|u + \sqrt{u^2 \pm a^2}|\Big] + C$

28. $\int \dfrac{\sqrt{u^2 + a^2}}{u}\, du = \sqrt{u^2 + a^2} - a \ln\left|\dfrac{a + \sqrt{u^2 + a^2}}{u}\right| + C$

29. $\int \dfrac{\sqrt{u^2 - a^2}}{u}\, du = \sqrt{u^2 - a^2} - a \operatorname{arcsec} \dfrac{|u|}{a} + C$

30. $\int \dfrac{\sqrt{u^2 \pm a^2}}{u^2}\, du = \dfrac{-\sqrt{u^2 \pm a^2}}{u} + \ln|u + \sqrt{u^2 \pm a^2}| + C$

31. $\int \dfrac{1}{\sqrt{u^2 \pm a^2}}\, du = \ln|u + \sqrt{u^2 \pm a^2}| + C$

32. $\int \dfrac{1}{u\sqrt{u^2 + a^2}}\, du = \dfrac{-1}{a} \ln\left|\dfrac{a + \sqrt{u^2 + a^2}}{u}\right| + C$

33. $\int \dfrac{1}{u\sqrt{u^2 - a^2}}\, du = \dfrac{1}{a} \operatorname{arcsec} \dfrac{|u|}{a} + C$

34. $\int \dfrac{u^2}{\sqrt{u^2 \pm a^2}}\, du = \dfrac{1}{2}\left(u\sqrt{u^2 \pm a^2} \mp a^2 \ln|u + \sqrt{u^2 \pm a^2}|\right) + C$

35. $\int \dfrac{1}{u^2\sqrt{u^2 \pm a^2}}\, du = \mp\dfrac{\sqrt{u^2 \pm a^2}}{a^2 u} + C$

36. $\int \dfrac{1}{(u^2 \pm a^2)^{3/2}}\, du = \dfrac{\pm u}{a^2\sqrt{u^2 \pm a^2}} + C$

Forms Involving $\sqrt{a^2 - u^2}$, $0 < a$

37. $\int \sqrt{a^2 - u^2}\, du = \dfrac{1}{2}\left(u\sqrt{a^2 - u^2} + a^2 \arcsin \dfrac{u}{a}\right) + C$

38. $\int u^2\sqrt{a^2 - u^2}\, du = \dfrac{1}{8}\left[u(2u^2 - a^2)\sqrt{a^2 - u^2} + a^4 \arcsin \dfrac{u}{a}\right] + C$

39. $\int \dfrac{\sqrt{a^2 - u^2}}{u}\, du = \sqrt{a^2 - u^2} - a \ln\left|\dfrac{a + \sqrt{a^2 - u^2}}{u}\right| + C$

40. $\int \dfrac{\sqrt{a^2 - u^2}}{u^2}\, du = \dfrac{-\sqrt{a^2 - u^2}}{u} - \arcsin \dfrac{u}{a} + C$

41. $\int \dfrac{1}{\sqrt{a^2 - u^2}}\, du = \arcsin \dfrac{u}{a} + C$

42. $\int \dfrac{1}{u\sqrt{a^2 - u^2}}\, du = \dfrac{-1}{a} \ln\left|\dfrac{a + \sqrt{a^2 - u^2}}{u}\right| + C$

43. $\int \dfrac{u^2}{\sqrt{a^2 - u^2}}\, du = \dfrac{1}{2}\left(-u\sqrt{a^2 - u^2} + a^2 \arcsin \dfrac{u}{a}\right) + C$

44. $\int \dfrac{1}{u^2\sqrt{a^2 - u^2}}\, du = \dfrac{-\sqrt{a^2 - u^2}}{a^2 u} + C$

45. $\int \dfrac{1}{(a^2 - u^2)^{3/2}}\, du = \dfrac{u}{a^2\sqrt{a^2 - u^2}} + C$

Forms Involving sin *u* or cos *u*

46. $\displaystyle\int \sin u \; du = -\cos u + C$

47. $\displaystyle\int \cos u \; du = \sin u + C$

48. $\displaystyle\int \sin^2 u \; du = \frac{1}{2}(u - \sin u \cos u) + C$

49. $\displaystyle\int \cos^2 u \; du = \frac{1}{2}(u + \sin u \cos u) + C$

50. $\displaystyle\int \sin^n u \; du = -\frac{\sin^{n-1} u \cos u}{n} + \frac{n-1}{n}\int \sin^{n-2} u \; du$

51. $\displaystyle\int \cos^n u \; du = \frac{\cos^{n-1} u \sin u}{n} + \frac{n-1}{n}\int \cos^{n-2} u \; du$

52. $\displaystyle\int u \sin u \; du = \sin u - u \cos u + C$

53. $\displaystyle\int u \cos u \; du = \cos u + u \sin u + C$

54. $\displaystyle\int u^n \sin u \; du = -u^n \cos u + n \int u^{n-1} \cos u \; du$

55. $\displaystyle\int u^n \cos u \; du = u^n \sin u - n \int u^{n-1} \sin u \; du$

56. $\displaystyle\int \frac{1}{1 \pm \sin u} \; du = \tan u \mp \sec u + C$

57. $\displaystyle\int \frac{1}{1 \pm \cos u} \; du = -\cot u \pm \csc u + C$

58. $\displaystyle\int \frac{1}{\sin u \cos u} \; du = \ln |\tan u| + C$

Forms Involving tan *u*, cot *u*, sec *u*, csc *u*

59. $\displaystyle\int \tan u \; du = -\ln |\cos u| + C$

60. $\displaystyle\int \cot u \; du = \ln |\sin u| + C$

61. $\displaystyle\int \sec u \; du = \ln |\sec u + \tan u| + C$

62. $\displaystyle\int \csc u \; du = \ln |\csc u - \cot u| + C$

63. $\displaystyle\int \tan^2 u \; du = -u + \tan u + C$

64. $\displaystyle\int \cot^2 u \; du = -u - \cot u + C$

65. $\displaystyle\int \sec^2 u \; du = \tan u + C$

66. $\displaystyle\int \csc^2 u \; du = -\cot u + C$

67. $\displaystyle\int \tan^n u \; du = \frac{\tan^{n-1} u}{n-1} - \int \tan^{n-2} u \; du, \; n \neq 1$

68. $\displaystyle\int \cot^n u \; du = -\frac{\cot^{n-1} u}{n-1} - \int (\cot^{n-2} u) \; du, \; n \neq 1$

69. $\displaystyle\int \sec^n u \; du = \frac{\sec^{n-2} u \tan u}{n-1} + \frac{n-2}{n-1} \int \sec^{n-2} u \; du, \; n \neq 1$

70. $\displaystyle\int \csc^n u \; du = -\frac{\csc^{n-2} u \cot u}{n-1} + \frac{n-2}{n-1} \int \csc^{n-2} u \; du, \; n \neq 1$

71. $\displaystyle\int \frac{1}{1 \pm \tan u} \; du = \frac{1}{2}(u \pm \ln |\cos u \pm \sin u|) + C$

72. $\displaystyle\int \frac{1}{1 \pm \cot u} \; du = \frac{1}{2}(u \mp \ln |\sin u \pm \cos u|) + C$

73. $\displaystyle\int \frac{1}{1 \pm \sec u} \; du = u + \cot u \mp \csc u + C$

74. $\displaystyle\int \frac{1}{1 \pm \csc u} \; du = u - \tan u \pm \sec u + C$

Forms Involving Inverse Trigonometric Functions

75. $\displaystyle\int \arcsin u \; du = u \arcsin u + \sqrt{1 - u^2} + C$

76. $\displaystyle\int \arccos u \; du = u \arccos u - \sqrt{1 - u^2} + C$

77. $\displaystyle\int \arctan u \; du = u \arctan u - \ln \sqrt{1 + u^2} + C$

78. $\displaystyle\int \text{arccot } u \; du = u \text{ arccot } u + \ln \sqrt{1 + u^2} + C$

79. $\displaystyle\int \text{arcsec } u \; du = u \text{ arcsec } u - \ln |u + \sqrt{u^2 - 1}| + C$

80. $\displaystyle\int \text{arccsc } u \; du = u \text{ arccsc } u + \ln |u + \sqrt{u^2 - 1}| + C$

Forms Involving e^u

81. $\displaystyle\int e^u \; du = e^u + C$

82. $\displaystyle\int u e^u \; du = (u - 1)e^u + C$

83. $\displaystyle\int u^n e^u \; du = u^n e^u - n \int u^{n-1} e^u \; du$

84. $\displaystyle\int \frac{1}{1 + e^u} \; du = u - \ln(1 + e^u) + C$

85. $\displaystyle\int e^{au} \sin bu \; du = \frac{e^{au}}{a^2 + b^2}(a \sin bu - b \cos bu) + C$

86. $\displaystyle\int e^{au} \cos bu \; du = \frac{e^{au}}{a^2 + b^2}(a \cos bu + b \sin bu) + C$

Forms Involving ln u

87. $\displaystyle\int \ln u \; du = u(-1 + \ln u) + C$

88. $\displaystyle\int u \ln u \; du = \frac{u^2}{4}(-1 + 2 \ln u) + C$

89. $\displaystyle\int u^n \ln u \; du = \frac{u^{n+1}}{(n+1)^2}[-1 + (n+1) \ln u] + C, \; n \neq -1$

90. $\displaystyle\int (\ln u)^2 \; du = u[2 - 2 \ln u + (\ln u)^2] + C$

91. $\displaystyle\int (\ln u)^n \; du = u(\ln u)^n - n \int (\ln u)^{n-1} \; du$

Answers to Odd-Numbered Exercises

Chapter 1

Section 1.1

1. Rational   **3.** Irrational   **5.** Rational
7. Rational   **9.** Rational   **11.** $\frac{4}{11}$ **13.** $\frac{11}{37}$
15. (a) True   **(b)** False   **(c)** True
     (d) False   **(e)** False   **(f)** False
17.

Interval Notation	Set Notation	Graph
$[-2, 0)$	$\{x: -2 \le x < 0\}$	
$(-\infty, -4]$	$\{x: x \le -4\}$	
$[3,.11/2]$	$\{x: 3 \le x \le 11/2\}$	
$(-1, 7)$	$\{x: -1 < x < 7\}$	

19. $x \ge 12$

21. $x < -\frac{1}{2}$

23. $x \ge \frac{1}{2}$

25. $-\frac{1}{2} < x < \frac{7}{2}$

27. $x < -4$

29. $x > 6$

31. $-1 < x < 1$

33. $x \ge 13, \quad x \le -7$

35. $a - b < x < a + b$

37. $-3 < x < 2$

39. $0 < x < 3$

41. $-3 \le x \le 1$

43. $-3 \le x \le 2$

45. $4, -4, 4$

47. (a) $-51, 51, 51$ **(b)** $51, -51, 51$
49. 1 **51. (a)** 14 **(b)** 10
53. $|x| \le 2$ **55.** $|x - 2| > 2$
57. (a) $|x - 12| \le 10$ **(b)** $|x - 12| \ge 10$
59. $r > 12.5\%$ **61.** $m < 24{,}062.5$ miles
63. $x \le 41$ or $x \ge 59$
65. (a) $\frac{355}{113} > \pi$ **(b)** $\frac{22}{7} > \pi$

Section 1.2

1. $d = 2\sqrt{5}$

3. $d = 2\sqrt{10}$

5. $d = \sqrt{8 - 2\sqrt{3}}$

7. Right triangle,
$d_1 = \sqrt{45}, d_2 = \sqrt{5},$
$d_3 = \sqrt{50}$
$(d_1)^2 + (d_2)^2 = (d_3)^2$

9. Rhombus; length of each side is $\sqrt{5}$

45. $\left(x - \frac{1}{2}\right)^2 + \left(y - \frac{1}{2}\right)^2 = 2$

47. $\left(x + \frac{1}{2}\right)^2 + \left(y + \frac{5}{4}\right)^2 = \frac{9}{4}$

11. Collinear, since
$d_1 + d_2 = d_3$

13. Not collinear, since
$d_1 + d_2 > d_3$

49. $x^2 + y^2 = 26{,}000^2$
51. $(x - 1)^2 + (y + 1)^2 = 25$
53.   **55.**

15. $x = \pm 3$ **17.** $y = \pm\sqrt{55}$
19. $3x - 2y - 1 = 0$
21. $\left(\dfrac{3x_1 + x_2}{4}, \dfrac{3y_1 + y_2}{4}\right), \left(\dfrac{x_1 + x_2}{2}, \dfrac{y_1 + y_2}{2}\right),$
$\left(\dfrac{x_1 + 3x_2}{4}, \dfrac{y_1 + 3y_2}{4}\right)$

23. (a) $\left(x + \frac{5}{2}\right)^2 - \frac{25}{4}$ **(b)** $(x + 4)^2 - 9$ **25.** d
26. c   **27.** b   **28.** e   **29.** a   **30.** f
31. $x^2 + y^2 - 9 = 0$
33. $x^2 + y^2 - 4x + 2y - 11 = 0$
35. $x^2 + y^2 + 2x - 4y = 0$
37. $x^2 + y^2 - 6x - 4y + 3 = 0$
39. $x^2 + y^2 - 6x - 8y = 0$
41. $(x - 1)^2 + (y + 3)^2 = 4$

Section 1.3

1. c   **2.** d   **3.** b   **4.** f   **5.** a   **6.** e
7. $(0, -3), \left(\frac{3}{2}, 0\right)$ **9.** $(0, -2), (-2, 0), (1, 0)$
11. $(0, 0), (-3, 0), (3, 0)$ **13.** $\left(0, \frac{1}{2}\right), (1, 0)$
15. $(0, 0)$ **17.** Symmetric with respect to the y-axis
19. Symmetric with respect to the y-axis
21. Symmetric with respect to the x-axis
23. Symmetric with respect to the origin
25. Symmetric with respect to the origin
27. $y = x$ **29.** $y = x + 3$

43. $(x - 1)^2 + (y + 3)^2 = 0$

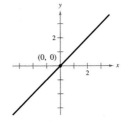

31. $y = -3x + 2$

33. $y = \frac{1}{2}x - 4$

65. (a)

67. (a)

35. $y = 1 - x^2$

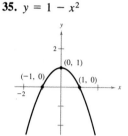

37. $y = -2x^2 + x + 1$

(b) 1995: 471.4

69. $(-1.2164, 0), (0, 2)$

(b) 1995: 1.9%

39. $y = x^3 + 2$

41. $x^2 + 4y^2 = 4$

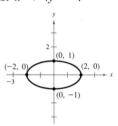

Section 1.4

1. $m = 1$ **3.** $m = 0$ **5.** $m = -3$
7. $m = 3$ **9.** $m = 0$

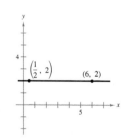

43. $y = (x + 2)^2$

45. $y = \dfrac{1}{x}$

11. $m = -\frac{2}{3}$ **13.** $(0, 1), (1, 1), (3, 1)$

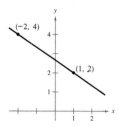

47. $(1, 1)$ **49.** $(5, 2)$ **51.** $(-1, -2), (2, 1)$
53. $(-1, -1), (0, 0), (1, 1)$
55. $(-1, -5), (0, -1), (2, 1)$ **57.** $x \approx 192$ units
59. $(1, 2)$ is not on the graph
  $(1, -1)$ is on the graph
  $(4, 5)$ is on the graph
61. $\left(1, \frac{1}{3}\right)$ is on the graph
  $\left(2, \frac{1}{2}\right)$ is on the graph
  $(-1, -2)$ is not on the graph
63. (a) $k = 4$ **(b)** $k = -\frac{1}{8}$
  (c) $k =$ any real number   **(d)** $k = 1$

15. $(0, 10), (2, 4), (3, 1)$ **17.** $m = -\frac{1}{5}, (0, 4)$
19. m is undefined, no y-intercept

21. $2x - y - 3 = 0$

23. $3x + y = 0$

25. $y + 2 = 0$

27. $3x - 4y + 12 = 0$

29. $2x - 3y = 0$

31. $4x - y + 2 = 0$

33. $x - 3 = 0$

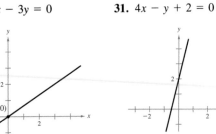

35. $3x + 2y - 6 = 0$

37. $12x + 3y + 2 = 0$ **39.** $x + y - 3 = 0$

41. (a) $2x - y - 3 = 0$ **(b)** $x + 2y - 4 = 0$

43. (a) $40x + 24y - 53 = 0$ **(b)** $24x - 40y + 9 = 0$

45. (a) $x - 2 = 0$ **(b)** $y - 5 = 0$

47.

49.

51.

53. $2x - y = 0$ **55.** Not collinear, since $m_1 \neq m_2$

57. $\left(0, \dfrac{-a^2 + b^2 + c^2}{2c}\right)$ **59.** $\left(b, \dfrac{a^2 - b^2}{c}\right)$

61. $5F - 9C - 160 = 0$ **63.** $C = 0.25x + 95$

65. $y = 875 - 175t$ where $0 \leq t \leq 5$

67. (a) $x = \frac{1}{15}(1130 - p)$ **(b)** 45 units

     (c) 49 units

69. 2   **71.** $\dfrac{5\sqrt{2}}{2}$ **73.** $2\sqrt{2}$

Section 1.5

1. (a) -3 **(b)** -9 **(c)** $2b - 3$ **(d)** $2x - 5$

3. (a) 1   **(b)** 3   **(c)** $\sqrt{c + 3}$

     (d) $\sqrt{x + \Delta x + 3}$

5. (a) 1   **(b)** -1 **(c)** 1   **(d)** $\dfrac{|x - 1|}{x - 1}$

7. $3 + \Delta x$ **9.** $3x^2 + 3x\Delta x + (\Delta x)^2$

11. $\dfrac{-1}{\sqrt{x - 1}(1 + \sqrt{x - 1})}$

13. $f(x) = 4 - x$

     Domain: $(-\infty, \infty)$

     Range: $(-\infty, \infty)$

15. $f(x) = 4 - x^2$

     Domain: $(-\infty, \infty)$

     Range: $(-\infty, 4]$

17. $h(x) = \sqrt{x - 1}$

     Domain: $[1, \infty)$

     Range: $[0, \infty)$

19. $f(x) = \sqrt{9 - x^2}$
Domain: $[-3, 3]$
Range: $[0, 3]$

21. $f(x) = |x - 2|$
Domain: $(-\infty, \infty)$
Range: $[0, \infty)$

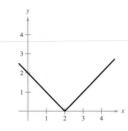

23. y is a function of x
27. y is a function of x
31. y is a function of x
35. y is not a function of x
37. **(a)** $y = \sqrt{x} + 2$

25. y is not a function of x
29. y is not a function of x
33. y is a function of x

(b) $y = -\sqrt{x}$

(c) $y = \sqrt{x - 2}$ **(d)** $y = \sqrt{x + 3}$

(e) $y = \sqrt{x - 4}$ **(f)** $y = 2\sqrt{x}$

39. **(a)** $y = (x + 3)^2$ **(b)** $y = x^2 + 3$
41. **(a)** 0   **(b)** 0   **(c)** -1 **(d)** $\sqrt{15}$
      (e) $\sqrt{x^2 - 1}$ **(f)** $x - 1$
43. $(f \circ g)(x) = x$
Domain: $[0, \infty)$
$(g \circ f)(x) = |x|$
Domain: $(-\infty, \infty)$

45. $(f \circ g)(x) = \dfrac{x + 1}{x}$
Domain: $(-\infty, 0), (0, \infty)$
$(g \circ f)(x) = \dfrac{1}{x + 1}$
Domain: $(-\infty, -1), (-1, \infty)$

47. $x = \pm 3$ **49.** $\frac{10}{7}$ **51.** Even   **53.** Odd
59. $R(x) = 4 - \dfrac{x^2}{2}$, $r(x) = 2$
61. $h(x) = x^2$, $p(x) = x$
63. $A = xy = x\left(\dfrac{100 - 2x}{2}\right) = x(50 - x)$
65. $V = x(12 - 2x)^2$ **67.** $V = 108x^2 - 4x^3$
69. $T = \dfrac{\sqrt{x^2 + 4}}{2} + \dfrac{\sqrt{x^2 - 6x + 10}}{4}$
71. **(a)**

      (b) $x = 0, \pm 3$
      (c) $[-3, 3]$

Section 1.6

1. **(a)** $396°, -324°$ **(b)** $240°, -480°$
3. **(a)** $\dfrac{19\pi}{9}, -\dfrac{17\pi}{9}$ **(b)** $\dfrac{10\pi}{3}, -\dfrac{2\pi}{3}$
5. **(a)** $\dfrac{\pi}{6}$ **(b)** $\dfrac{5\pi}{6}$ **(c)** $\dfrac{7\pi}{4}$ **(d)** $\dfrac{2\pi}{3}$
7. **(a)** $270°$ **(b)** $210°$ **(c)** $-105°$ **(d)** $20°$
9.

r	8 ft	15 in.	85 cm	24 in.	$\dfrac{12,963}{\pi}$ mi
s	12 ft	24 in.	63.75π cm	96 in.	8642 mi
θ	1.5	1.6	$\dfrac{3\pi}{4}$	4	$\dfrac{2\pi}{3}$

11. **(a)** $\dfrac{5\pi}{12}$ **(b)** 7.8125π in.
13. **(a)** $\sin\theta = \frac{4}{5}$, $\csc\theta = \frac{5}{4}$
      $\cos\theta = \frac{3}{5}$, $\sec\theta = \frac{5}{3}$
      $\tan\theta = \frac{4}{3}$, $\cot\theta = \frac{3}{4}$
      (b) $\sin\theta = -\frac{15}{17}$, $\csc\theta = -\frac{17}{15}$
      $\cos\theta = \frac{8}{17}$, $\sec\theta = \frac{17}{8}$
      $\tan\theta = -\frac{15}{8}$, $\cot\theta = -\frac{8}{15}$
15. Quadrant III   **17.** 2   **19.** $\frac{4}{3}$ **21.** $\frac{17}{15}$

23. (a) $\sin 60° = \dfrac{\sqrt{3}}{2}$ **(b)** $\sin \dfrac{2\pi}{3} = \dfrac{\sqrt{3}}{2}$

$\cos 60° = \dfrac{1}{2}$ $\cos \dfrac{2\pi}{3} = -\dfrac{1}{2}$

$\tan 60° = \sqrt{3}$ $\tan \dfrac{2\pi}{3} = -\sqrt{3}$

(c) $\sin \dfrac{\pi}{4} = \dfrac{\sqrt{2}}{2}$ **(d)** $\sin \dfrac{5\pi}{4} = -\dfrac{\sqrt{2}}{2}$

$\cos \dfrac{\pi}{4} = \dfrac{\sqrt{2}}{2}$ $\cos \dfrac{5\pi}{4} = -\dfrac{\sqrt{2}}{2}$

$\tan \dfrac{\pi}{4} = 1$ $\tan \dfrac{5\pi}{4} = 1$

25. (a) $\sin 225° = -\dfrac{\sqrt{2}}{2}$ **(b)** $\sin(-225°) = \dfrac{\sqrt{2}}{2}$

$\cos 225° = -\dfrac{\sqrt{2}}{2}$ $\cos(-225°) = -\dfrac{\sqrt{2}}{2}$

$\tan 225° = 1$ $\tan(-225°) = -1$

(c) $\sin 300° = -\dfrac{\sqrt{3}}{2}$ **(d)** $\sin 330° = -\dfrac{1}{2}$

$\cos 300° = \dfrac{1}{2}$ $\cos 330° = \dfrac{\sqrt{3}}{2}$

$\tan 300° = -\sqrt{3}$ $\tan 330° = -\dfrac{\sqrt{3}}{3}$

27. (a) 0.1736
(b) 5.759

29. (a) 0.3640
(b) 0.3640

31. (a) $\theta = \dfrac{\pi}{4}, \dfrac{7\pi}{4}$ **33. (a)** $\theta = \dfrac{\pi}{4}, \dfrac{5\pi}{4}$

(b) $\theta = \dfrac{3\pi}{4}, \dfrac{5\pi}{4}$ **(b)** $\theta = \dfrac{5\pi}{6}, \dfrac{11\pi}{6}$

35. $\theta = \dfrac{\pi}{4}, \dfrac{3\pi}{4}, \dfrac{5\pi}{4}, \dfrac{7\pi}{4}$ **37.** $\theta = 0, \dfrac{\pi}{4}, \pi, \dfrac{5\pi}{4}$

39. $\theta = \dfrac{\pi}{3}, \dfrac{5\pi}{3}$ **41.** $\theta = 0, \dfrac{\pi}{2}, \pi$

43. $y = \dfrac{100\sqrt{3}}{3}$ $r = \dfrac{200\sqrt{3}}{3}$

45. $x = \dfrac{25\sqrt{3}}{3}$ $r = \dfrac{50\sqrt{3}}{3}$

47. $20 \sin 75° \approx 19.32$ ft
49. $150 \cot 4° \approx 2145.1$ ft
51. (a) Period: π
Amplitude: 2
(b) Period: 2
Amplitude: $\frac{1}{2}$

53. Period: $\dfrac{\pi}{5}$
Amplitude: 2

55. Period: $\dfrac{1}{2}$ **57.** Period: $\dfrac{\pi}{2}$
Amplitude: 3

59. Period: $\dfrac{2\pi}{5}$

61.

63.

65.

67.

69.

71.

73.

75. (a) 6 seconds
(b) 10 cycles/min
(c)

77. (a) $\frac{1}{440}$ **(b)** 440
(c)

79. (a)

(b)

81.

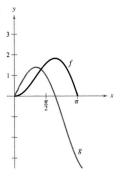

The zero of $g(x)$ corresponds to the maximum of $f(x)$.

Review Exercises for Chapter 1

1. $-1 \le x \le 5$

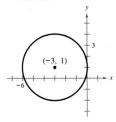

3. $x < -5, \quad x > -1$

5. $\frac{27}{16}$ **7.** $(-1, 3), (3, 2), (1, 1)$

9. Center: $(-3, 1)$ **11.** Point: $(-3, 1)$
Radius: 3

13. $c = -21$ **15.** $x^2 + y^2 - 2x - 4y - 4 = 0$
   (a) on the circle
   (b) inside the circle
   (c) outside the circle
   (d) inside the circle

17.

19.

21. The points are not collinear.

23.

25.

27. (a) $7x - 16y + 78 = 0$ **29.** $(-4, 5)$ **31.** $(4, 1)$
   (b) $5x - 3y + 22 = 0$
   (c) $y + 2x = 0$
   (d) $x + 2 = 0$
33. $v = 850a + 300,000$ **35.** $s = 6x^2$
   Domain: $\{a: a \ge 0\}$ Domain: $\{x: x \ge 0\}$
37. $d = 45t$ **39.** $P(x) = 500x - x^2$
   Domain: $\{t: t \ge 0\}$
41. Function   **43.** Not a function

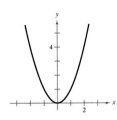

45. Function   **47. (a)** $-x^2 + 2x + 2$
   (b) $-x^2 - 2x$
   (c) $-2x^3 - x^2 + 2x + 1$
   (d) $\dfrac{1 - x^2}{2x + 1}$
   (e) $-4x^2 - 4x$
   (f) $3 - 2x^2$

49. (a)

(b) Speed of the plane is 560 mi/hr

51. $C = 0.30x + 150$

53. (a) 1.732   **(b)** 1.732   **(c)** -0.6052
(d) -1.001

55. $\dfrac{7\pi}{6}$ rad $= 210°$,   **57.** $\dfrac{5\pi}{6}$ rad $= 150°$,

$\dfrac{11\pi}{6}$ rad $= 330°$ $\dfrac{7\pi}{6}$ rad $= 210°$

59. $\theta = 60°$, $c = 10$ **61.** $\theta = 30°$; $a = 4\sqrt{3}$

63. $\theta = 40°$ **65.** 15 feet

67.   **69.**

71.   **73.**

75.

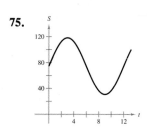

Chapter 2

Section 2.1

1.

x	1.9	1.99	1.999
$f(x)$	0.3448	0.3344	0.3334

x	2.001	2.01	2.1
$f(x)$	0.3332	0.3322	0.3226

$\displaystyle\lim_{x\to 2} \frac{x-2}{x^2-x-2} \approx 0.3333 \left(\text{Actual limit is } \frac{1}{3}.\right)$

3.

x	-0.1	-0.01	-0.001
$f(x)$	0.2911	0.2889	0.2887

x	0.001	0.01	0.1
$f(x)$	0.2887	0.2884	0.2863

$\displaystyle\lim_{x\to 0} \frac{\sqrt{x+3}-\sqrt{3}}{x} \approx 0.2887$

$\left(\text{Actual limit is } \dfrac{1}{2\sqrt{3}}.\right)$

5.

x	2.9	2.99	2.999
$f(x)$	-0.0641	-0.0627	-0.0625

x	3.001	3.01	3.1
$f(x)$	-0.0625	-0.0623	-0.0610

$\displaystyle\lim_{x\to 3} \frac{[1/(x+1)]-(1/4)}{x-3} \approx -0.0625$

$\left(\text{Actual limit is } -\dfrac{1}{16}.\right)$

7.

x	0.70	0.75	0.784
$f(x)$	1.8468	1.9324	1.9972

x	0.786	0.80	0.85
$f(x)$	2.0012	2.0298	2.1413

$\displaystyle\lim_{x\to \pi/4} \frac{\tan x - 1}{x-(\pi/4)} \approx 2.0000 \text{ (Actual limit is 2.)}$

9.

x	-0.1	-0.01	-0.001
$f(x)$	0.9983	0.99998	1.0000

x	0.001	0.01	0.1
$f(x)$	1.0000	0.99998	0.9983

$\lim\limits_{x \to 0} \dfrac{\sin x}{x} \approx 1.0000$ (Actual limit is 1.)

11. 1   **13.** 2   **15.** Limit does not exist.
17. Limit does not exist.   **19.** Limit does not exist.
21. $L = 8$. Let $\delta = \dfrac{0.01}{3} \approx 0.0033$.   **23.** $L = 1$.

   Assume $1 < x < 3$ and let $\delta = \dfrac{0.01}{5} = 0.002$.

25. Limit is 5. Let $\delta = \varepsilon$.
27. Limit is 3. Any δ will work.
29. Limit is 0. Let $\delta = \varepsilon^3$.

31. Limit is 2. Let $\delta = \dfrac{\varepsilon}{3}$ for $0 < x < 2$.

33. No. The value of $f(x)$ at $x = 2$ has no bearing on the
behavior of f as x approaches 2.

35. $\frac{1}{6}$

23.

x	-0.1	-0.01	-0.001
$f(x)$	0.358	0.354	0.354

x	0.001	0.01	0.1
$f(x)$	0.354	0.353	0.349

$\lim\limits_{x \to 0} \dfrac{\sqrt{x + 2} - \sqrt{2}}{x} \approx 0.354$ $\left(\text{Actual limit is } \dfrac{1}{2\sqrt{2}}.\right)$

25.

x	-0.1	-0.01	-0.001
$f(x)$	-0.263	-0.251	-0.250

x	0.001	0.01	0.1
$f(x)$	-0.250	-0.249	-0.238

$\lim\limits_{x \to 0} \dfrac{[1/(2 + x)] - (1/2)}{x} \approx -0.250$

$\left(\text{Actual limit is } -\dfrac{1}{4}.\right)$

27. $\frac{1}{5}$ **29.** 0   **31.** 0   **33.** 1   **35.** 1   **37.** $\frac{2}{3}$
39. 0   **41.** 4   **43.** 0   **45.** -16 ft/sec
51. $\frac{5}{2}$

53.

x	± 0.1	± 0.01	± 0.001
$\dfrac{\sin x}{x}$	0.9983	1.0000	1.0000

Section 2.2

1. (a) 1   **(b)** 3   **3. (a)** $\dfrac{\sqrt{3}}{2}$ **(b)** 1   **5.** 4

7. -1 **9.** -4 **11.** 2   **13.** 1   **15.** $\frac{1}{2}$
17. -2 **19.** 1   **21.** -1 **23.** 1   **25.** $\frac{1}{2}$
27. -1 **29. (a)** 15   **(b)** 5   **(c)** 6   **(d)** $\frac{2}{3}$
31. (a) 64   **(b)** 2   **(c)** 12   **(d)** 8

33. $f(x) = -\dfrac{1}{x}$, $g(x) = \dfrac{1}{x}$

Section 2.3

1. (a) 1   **(b)** 3   **3. (a)** 2   **(b)** 0   **5.** -2
7. $\frac{1}{6}$ **9.** 12   **11.** 2   **13.** $2x - 2$ **15.** $\frac{1}{10}$
17. $\dfrac{3}{2}$ **19.** $\dfrac{\sqrt{3}}{6}$ **21.** $-\dfrac{1}{4}$

Section 2.4

1. (a) 1   **(b)** 1   **(c)** 1
3. (a) 0   **(b)** 0   **(c)** 0
5. (a) 3   **(b)** -3 **(c)** Limit does not exist.
7. $\frac{1}{10}$ **9.** Limit does not exist.   **11.** 2

13. $-\dfrac{1}{x^2}$ **15.** Limit does not exist.

17. Limit does not exist.   **19.** 2

21. Limit does not exist.

23. Continuous for all real x

25. Discontinuous at $x = -1$

27. Discontinuous at $x = 1$

29. Continuous for all real x

31. Continuous for all real x

33. Nonremovable discontinuity at $x = 1$

35. Continuous for all real x

37. Removable discontinuity at $x = -2$;
Nonremovable discontinuity at $x = 5$

39. Continuous for all real x

41. Nonremovable discontinuity at $x = 2$

43. Nonremovable discontinuity at $x = -2$

45. Continuous for all real x

47. Nonremovable discontinuities at integer multiples
of $\pi/2$

49. Nonremovable discontinuities at each integer

51. Continuous for all real x

53. Continuous for all real x

55. Removable discontinuity at $x = 4$

57. Nonremovable discontinuity at each integer

59. $(-\infty, -6), (-6, 6), (6, \infty)$ **61.** $(-\infty, \infty)$

63. $\ldots, (-2\pi, 0), (0, 2\pi), (2\pi, 4\pi), \ldots$

67. (a) $(0.6, 0.7)$ **(b)** $(0.68, 0.69)$

69. $f(3) = 11$ **71.** $f(2) = 4$ **73.** $a = 2$

75. Discontinuous at every positive integer

77. Discontinuous at every positive even integer
(Must replenish inventory every two months)

79. Discontinuous at $x = 3$

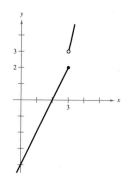

Section 2.5

1. $\displaystyle\lim_{x \to -2^+} \dfrac{1}{(x + 2)^2} = \infty$ $\displaystyle\lim_{x \to -2^-} \dfrac{1}{(x + 2)^2} = \infty$

3. $\displaystyle\lim_{x \to -2^+} \tan \dfrac{\pi x}{4} = -\infty$ $\displaystyle\lim_{x \to -2^-} \tan \dfrac{\pi x}{4} = \infty$

5. $\displaystyle\lim_{x \to -3^+} \dfrac{1}{x^2 - 9} = -\infty$ $\displaystyle\lim_{x \to -3^-} \dfrac{1}{x^2 - 9} = \infty$

7. $\displaystyle\lim_{x \to -3^+} \dfrac{x^2}{x^2 - 9} = -\infty$ $\displaystyle\lim_{x \to -3^-} \dfrac{x^2}{x^2 - 9} = \infty$

9. $x = 0$ **11.** $x = 2, x = -1$ **13.** $x = \pm 1$

15. $x = 0$ **17.** $x = -2, x = 1$

19. Removable discontinuity at $x = -1$

21. Vertical asymptote at $x = -1$ **23.** $-\infty$

25. ∞ **27.** $-\infty$ **29.** ∞ **31.** $\frac{1}{2}$ **33.** ∞

35. (a) $\frac{7}{12}$ ft/sec   **(b)** $\frac{3}{2}$ ft/sec   **(c)** ∞

37. (a) \$176 million   **(b)** \$528 million
(c) \$1584 million   **(d)** ∞

39.   **41.** $-\infty$

Review Exercises for Chapter 2

1. 7 **3.** 77 **5.** $\frac{10}{3}$ **7.** $-\frac{1}{4}$ **9.** -1
11. 3 **13.** $-\infty$ **15.** $-\infty$ **17.** $\frac{1}{3}$ **19.** 0
21. $\frac{4}{5}$ **23.** ∞ **25.** $\frac{\sqrt{3}}{2}$

27.

x	1.1	1.01	1.001	1.0001
$f(x)$	0.5680	0.5764	0.5773	0.5773

$$\lim_{x \to 1^+} \frac{\sqrt{2x+1} - \sqrt{3}}{x - 1} \approx 0.577$$

29. $\frac{1}{\sqrt{3}}$ **31.** False **33.** False **35.** True
37. Nonremovable discontinuity at each integer
39. Removable discontinuity at $x = 1$
41. Nonremovable discontinuity at $x = 2$
43. Nonremovable discontinuity at $x = -1$
45. Nonremovable discontinuity at each even integer
47. $c = -\frac{1}{2}$
49. Nonremovable discontinuity every 6 months

51. $x = 0$ **53.** $x = 10$
55. (a) 2 **(b)** $f(x) = \begin{cases} \dfrac{\tan 2x}{x}, & x \neq 0 \\ 2, & x = 0 \end{cases}$

Chapter 3

Section 3.1

1.

3. (a) $m = 0$
(b) $m = -3$

5. $f'(x) = 0$ **7.** $f'(x) = -5$
9. $f'(x) = 4x + 1$ **11.** $f'(x) = -\dfrac{1}{(x-1)^2}$
13. $f'(t) = 3t^2 - 12$
15. $f'(x) = 2x$ **17.** $f'(x) = 3x^2$
Tangent line: Tangent line:
$y = 4x - 3$ $y = 12x - 16$

19. $f'(x) = \dfrac{1}{2\sqrt{x+1}}$
Tangent line: $4y = x + 5$

21. $f(x) = x^2 - 1$
$$\begin{aligned} f'(2) &= \lim_{x \to 2} \frac{f(x) - f(2)}{x - 2} \\ &= \lim_{x \to 2} \frac{(x^2 - 1) - 3}{x - 2} \\ &= \lim_{x \to 2} (x + 2) = 4 \end{aligned}$$
23. $f(x) = x^3 + 2x^2 + 1$
$$\begin{aligned} f'(-2) &= \lim_{x \to -2} \frac{f(x) - f(-2)}{x + 2} \\ &= \lim_{x \to -2} \frac{(x^3 + 2x^2 + 1) - 1}{x + 2} \\ &= \lim_{x \to -2} x^2 = 4 \end{aligned}$$
25. $f(x) = (x - 1)^{2/3}$
$$\begin{aligned} f'(1) &= \lim_{x \to 1} \frac{f(x) - f(1)}{x - 1} \\ &= \lim_{x \to 1} \frac{(x - 1)^{2/3} - 0}{x - 1} \\ &= \lim_{x \to 1} \frac{1}{(x - 1)^{1/3}} \end{aligned}$$
Limit does not exist.
f is not differentiable at $x = 1$.

27. $(-\infty, -3), (-3, \infty)$ **29.** $(-\infty, -1), (-1, \infty)$
31. $(-\infty, 3), (3, \infty)$ **33.** $(1, \infty)$
35. $(-\infty, 0), (0, \infty)$
37. f is not differentiable at $x = 1$.   **39.** $f'(1) = 0$
41. $y = 3x - 2$ **43.** $y = 2x + 1$
    $y = 3x + 2$ $y = -2x + 9$

45. (a) 3   **(b)** -3 **47.** True   **49.** True

51.

x	-2	-1.5	-1	-0.5
$f(x)$	-2	-0.84375	-0.25	-0.03125
$f'(x)$	3	1.6875	0.75	0.1875

x	0	0.5	1	1.5	2
$f(x)$	0	0.03125	0.25	0.84375	2
$f'(x)$	0	0.1875	0.75	1.6875	3

Section 3.2

1. Average rate: 2
Instantaneous rates:
$f'(1) = 2$
$f'(2) = 2$

3. Average rate: $-\frac{1}{4}$
Instantaneous rates:
$f'(0) = -1$
$f'(3) = -\frac{1}{16}$

5. Average rate: 4.1
Instantaneous rates:
$f'(2) = 4$
$f'(2.1) = 4.2$

7. (a) -48 ft/sec
(b) $s'(1) = -32$ ft/sec
    $s'(2) = -64$ ft/sec
(c) $t = \dfrac{15\sqrt{6}}{4} \approx 9.2$ sec
(d) -293.9 ft/sec

9. $v(5) = 224$ ft/sec, $v(10) = 64$ ft/sec

11. $v\left(\dfrac{5\sqrt{6}}{2}\right) = -80\sqrt{6}$ ft/sec ≈ -195.96 ft/sec

13. 740 ft

15.

t	0	1	2	3	4
$s(t)$	0	57.75	99	123.75	132
$v(t)$	66	49.5	33	16.5	0
$a(t)$	-16.5	-16.5	-16.5	-16.5	-16.5

17. [0, 1], 57.75 ft/sec   **19.** $2x$ **21.** $\dfrac{2}{x^2}$
    [1, 2], 41.25 ft/sec
    [2, 3], 24.75 ft/sec
    [3, 4], 8.25 ft/sec
23. 0   **25.** $-\$1.91, -\$1.93$
27. (a) 83.67   **(b)** 40.79
29. $\dfrac{dF}{dr} = -2K\dfrac{m_1 m_2}{r^3}$ **31.** $t \approx 3.48$ sec
    Force decreases as $v(3.48) = -84.4$ ft/sec
    the distance between
    the particles increases.

Section 3.3

1. (a) $\frac{1}{2}$ **(b)** $\frac{3}{2}$ **(c)** 2   **(d)** 3   **3.** 0
5. 1   **7.** $2x$ **9.** $-4t + 3$ **11.** $3t^2 - 2$
13. $2x + \dfrac{1}{2}\sin x$ **15.** $-\dfrac{1}{x^2} - 3\cos x$
17. -1 **19.** 0   **21.** 4   **23.** 3
25. $2x + \dfrac{4}{x^2}$ **27.** $3x^2 - 3 + \dfrac{8}{x^5}$ **29.** $\dfrac{x^3 - 8}{x^3}$
31. $3x^2 + 1$ **33.** $\dfrac{4}{5x^{1/5}}$ **35.** $\dfrac{1}{3x^{2/3}} + \dfrac{1}{5x^{4/5}}$
37. $\dfrac{2}{\sqrt{x}} - 3\sin x$

Function	Rewrite	Derivative	Simplify
39. $y = \dfrac{1}{3x^3}$	$y = \dfrac{1}{3}x^{-3}$	$y' = -x^{-4}$	$y' = -\dfrac{1}{x^4}$
41. $y = \dfrac{1}{(3x)^3}$	$y = \dfrac{1}{27}x^{-3}$	$y' = -\dfrac{1}{9}x^{-4}$	$y' = -\dfrac{1}{9x^4}$
43. $y = \dfrac{\sqrt{x}}{x}$	$y = x^{-1/2}$	$y' = -\dfrac{1}{2}x^{-3/2}$	$y' = -\dfrac{1}{2x^{3/2}}$

45. $2x + y - 2 = 0$
47. $(0, 2), \left(\sqrt{\frac{3}{2}}, -\frac{1}{4}\right), \left(-\sqrt{\frac{3}{2}}, -\frac{1}{4}\right)$
49. No horizontal tangents   **51.** (π, π)

53. $y = 2x - 1$ ‎ ‎ ‎ ‎ $y = 4x - 4$

55. 8
57. (a) 10,000 ‎ ‎ **(b)** 4,000 ‎ ‎ **(c)** 0 ‎ ‎ **(d)** $-8,000$
59. (a) $\frac{4}{9}$ ‎ ‎ **(b)** $\frac{1}{3}$ ‎ ‎ **(c)** 0 ‎ ‎ **(d)** $-\frac{5}{9}$
61. -44.1 ft/sec ‎ ‎ ‎ **63.** -5.4 ft/sec^2
‎ ‎ ‎ ‎ ‎ ‎ ‎ ‎ ‎ ‎ ‎ ‎ ‎ ‎ Approximately $\frac{1}{6}$ the accelera-
‎ ‎ ‎ ‎ ‎ ‎ ‎ ‎ ‎ ‎ ‎ ‎ ‎ ‎ tion due to gravity on earth.
65. True ‎ ‎ ‎ ‎ **67.** (0.1104, 0.1353)
‎ ‎ ‎ ‎ ‎ ‎ ‎ ‎ ‎ ‎ ‎ ‎ ‎ ‎ (1.8408, -10.4862)

Section 3.4

1. $f'(x) = 2x^2,$ ‎ $f'(0) = 0$
3. $f'(x) = 5[x^{-2}(1) + (x + 3)(-2x^{-3})]$
‎ ‎ ‎ ‎ $= -\dfrac{5(x + 6)}{x^3}$
‎ ‎ ‎ ‎ $f'(1) = -35$
5. $f'(x) = (x^3 - 3x)(4x + 3) + (2x^2 + 3x + 5)(3x^2 - 3)$
‎ ‎ ‎ ‎ $= 10x^4 + 12x^3 - 3x^2 - 18x - 15$
‎ ‎ ‎ ‎ $f'(0) = -15$
7. ‎ $f'(x) = (x^5 - 3x)\left(-\dfrac{2}{x^3}\right) + \left(\dfrac{1}{x^2}\right)(5x^4 - 3)$
‎ ‎ ‎ ‎ ‎ $= 3x^2 + \dfrac{3}{x^2}$
‎ ‎ ‎ ‎ $f'(-1) = 6$
9. $\dfrac{(2x - 3)(3) - (3x - 2)(2)}{(2x - 3)^2} = -\dfrac{5}{(2x - 3)^2}$
11. $\dfrac{(x^2 - 1)(-2 - 2x) - (3 - 2x - x^2)(2x)}{(x^2 - 1)^2} = \dfrac{2}{(x + 1)^2}$
13. $\dfrac{\sqrt{x}(1) - (x + 1)[1/(2\sqrt{x})]}{x} = \dfrac{x - 1}{2x^{3/2}}$
15. $\dfrac{(t^2 + 2t + 2)(1) - (t + 1)(2t + 2)}{(t^2 + 2t + 2)^2} = \dfrac{-t^2 - 2t}{(t^2 + 2t + 2)^2}$
17. $6s^2(s^3 - 2)$

19. $\left[\dfrac{x + 1}{x + 2}\right](2) + (2x - 5)\left[\dfrac{(x + 2)(1) - (x + 1)(1)}{(x + 2)^2}\right]$
‎ ‎ ‎ ‎ $= \dfrac{2x^2 + 8x - 1}{(x + 2)^2}$
21. $15x^4 - 48x^3 - 33x^2 - 32x - 20$
23. $\dfrac{(c^2 + x^2)(-2x) - (c^2 - x^2)(2x)}{(c^2 + x^2)^2} = -\dfrac{4xc^2}{(c^2 + x^2)^2}$

	Function	Rewrite	Derivative	Simplify
25.	$y = \dfrac{x^2 + 2x}{x}$	$y = x + 2$	$y' = 1$	$y' = 1$
27.	$y = \dfrac{7}{3x^3}$	$y = \dfrac{7}{3}x^{-3}$	$y' = -7x^{-4}$	$y' = -\dfrac{7}{x^4}$
29.	$y = \dfrac{3x^2 - 5}{7}$	$y = \dfrac{1}{7}(3x^2 - 5)$	$y' = \dfrac{1}{7}(6x)$	$y' = \dfrac{6x}{7}$

31. $t(t \cos t + 2 \sin t)$ ‎ ‎ ‎ ‎ **33.** $-\dfrac{t \sin t + \cos t}{t^2}$
35. $f'(x) = -1 + \sec^2 x = \tan^2 x$
37. $-5x \csc x \cot x + 5 \csc x = 5 \csc x \, (1 - x \cot x)$
39. $\csc x \cot x - \cos x = \cos x \cot^2 x$
41. $x^2 \cos x + 2x \sin x - 2x \sin x + 2 \cos x$
‎ ‎ ‎ ‎ ‎ ‎ ‎ ‎ ‎ ‎ $= x^2 \cos x + 2 \cos x$
43. $-\sin^2 x + \cos^2 x = \cos 2x$
45. $x(x \sec^2 x + 2 \tan x)$
47. $\dfrac{1 - \sin \theta + \theta \cos \theta}{(1 - \sin \theta)^2}$ ‎ ‎ ‎ ‎ **49.** $\dfrac{\sec t(t \tan t - 1)}{t^2}$
51. $\dfrac{3}{\sqrt{x}}$ ‎ ‎ ‎ **53.** $\dfrac{2}{(x - 1)^3}$ ‎ ‎ ‎ **55.** $-3 \sin x$
57. $y = -x + 4$ ‎ ‎ ‎ ‎ **59.** $y = -x - 2$
61. $4x - 2y - \pi + 2 = 0$ ‎ ‎ ‎ **63.** (0, 0), (2, 4)
65. (a) -0.48 ‎ ‎ ‎ ‎ **(b)** 0.12 ‎ ‎ ‎ ‎ **(c)** 0.0149
67. 31.55 ‎ ‎ ‎ **69.** 0
71. $a(2) = 4.069$

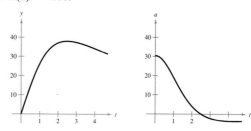

Section 3.5

$y = f(g(x))$	$u = g(x)$	$y = f(u)$
1. $y = (6x - 5)^4$	$u = 6x - 5$	$y = u^4$
3. $y = \sqrt{x^2 - 1}$	$u = x^2 - 1$	$y = \sqrt{u}$
5. $y = \csc^3 x$	$u = \csc x$	$y = u^3$

7. $6(2x - 7)^2$
9. $12(9x - 4)^3(9) = 108(9x - 4)^3$

11. $-\dfrac{1}{(x-2)^2}$ **13.** $-2(t-3)^{-3}(1) = -\dfrac{2}{(t-3)^3}$

15. $-3(x^3-4)^{-2}(3x^2) = -\dfrac{9x^2}{(x^3-4)^2}$

17. $x^2[4(x-2)^3(1)] + (x-2)^4(2x) = 2x(x-2)^3(3x-2)$

19. $\dfrac{1}{2}(1-t)^{-1/2}(-1) = -\dfrac{1}{2\sqrt{1-t}}$

21. $\dfrac{1}{2}(t^2+2t-1)^{-1/2}(2t+2) = \dfrac{t+1}{\sqrt{t^2+2t-1}}$

23. $\dfrac{1}{3}(9x^2+4)^{-2/3}(18x) = \dfrac{6x}{(9x^2+4)^{2/3}}$

25. $(4-x^2)^{-1/2}(-2x) = -\dfrac{2x}{\sqrt{4-x^2}}$

27. $\dfrac{2}{3}(9-x^2)^{-1/3}(-2x) = -\dfrac{4x}{3(9-x^2)^{1/3}}$

29. $-\dfrac{1}{2(x+2)^{3/2}}$

31. $x\left(\dfrac{1}{2}\right)(1-x^2)^{-1/2}(-2x) + (1-x^2)^{1/2}(1) = \dfrac{1-2x^2}{\sqrt{1-x^2}}$

33. $\dfrac{(x^2+1)^{1/2}(1) - x(1/2)(x^2+1)^{-1/2}(2x)}{x^2+1} = \dfrac{1}{(x^2+1)^{3/2}}$

35. $\dfrac{(x^2+1)(1/2)(x^{-1/2}) - (\sqrt{x}+1)(2x)}{(x^2+1)^2} = \dfrac{1-3x^2-4x^{3/2}}{2\sqrt{x}(x^2+1)^2}$

37. $\dfrac{(t-1)(3) - (3t+2)}{(t-1)^2} = -\dfrac{5}{(t-1)^2}$

39. $3t^2\left(-\dfrac{1}{2}\right)(t^2+2t-1)^{-3/2}(2t+2) + (t^2+2t-1)^{-1/2}(6t)$

$= \dfrac{3t(t^2+3t-2)}{(t^2+2t-1)^{3/2}}$

41. $\dfrac{1}{2}\left(\dfrac{x+1}{x}\right)^{-1/2}\left(\dfrac{x-(x+1)}{x^2}\right) = -\dfrac{1}{2x^{3/2}\sqrt{x+1}}$

43. $-\dfrac{2}{3}\left[(2-t)\dfrac{1}{2\sqrt{1+t}} + (-1)\sqrt{1+t}\right] = \dfrac{t}{\sqrt{1+t}}$

45. $9x - 5y - 2 = 0$ **47.** $2x - y - 2\pi = 0$

49. $12(5x^2-1)(x^2-1)$

51. $2[\cos x^2 - 2x^2 \sin x^2]$ **53. (a)** 1 **(b)** 2

55. $-3 \sin 3x$ **57.** $12 \sec^2 4x$ **59.** $\dfrac{1}{2}\sin 4x$

61. $2x \cos (2x)^2$ **63.** $6 \sec^3 2x \tan 2x$

65. $-\pi \csc^2\left(\pi x - \dfrac{\pi}{2}\right) = -\pi \sec^2 (\pi x)$

67. $\dfrac{-x \sin x - \cos x - 1}{x^2}$

71. $(3x^2+1)\left(\dfrac{x^3+x}{|x^3+x|}\right)$

73. (a) 1.461 **(b)** -1.016

75. 0.2 rad, 1.45 rad/sec   **77.** 0.04224

Section 3.6

1. $-\dfrac{x}{y}, \ -\dfrac{3\sqrt{7}}{7}$ **3.** $-\dfrac{y}{x}, \ -\dfrac{1}{4}$ **5.** $-\sqrt{\dfrac{y}{x}}, \ -\dfrac{5}{4}$

7. $\dfrac{y-3x^2}{2y-x}, \ \dfrac{1}{2}$ **9.** $\dfrac{18x}{(x^2+9)^2 y}$, undefined

11. $\dfrac{1-3x^2y^3}{3x^3y^2-1}, \ -1$ **13.** $-\sqrt[3]{\dfrac{y}{x}}, \ -\dfrac{1}{2}$

15. $\dfrac{4xy-3x^2-3y^2}{2x(3y-x)}, \ -\dfrac{15}{28}$

17. $\dfrac{\cos x}{2 \sin 2y}$, undefined

19. $\dfrac{\cos x - \tan y - 1}{x \sec^2 y}, \ -\dfrac{1}{2\pi}$

21. $-\dfrac{x^2}{x^2+1}, \ 0$ **23.** $\dfrac{y \cos (xy)}{1 - x \cos (xy)}, \ 0$

25. $-\dfrac{1}{2}$ **27.** 0

29. $y' = -\dfrac{x}{y}$ **31.** $y' = -\dfrac{9x}{16y}$

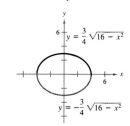

33. $\dfrac{10}{x^3}$ **35.** $-\dfrac{16}{y^3}$ **37.** $\dfrac{3x}{4y}$

39. (a) At $(4, 3)$:
  Tangent line: $4x + 3y - 25 = 0$
  Normal line: $3x - 4y = 0$
  (b) At $(-3, 4)$:
  Tangent line: $3x - 4y + 25 = 0$
  Normal line: $4x + 3y = 0$

41. Horizontal tangents: $(-4, 0), (-4, 10)$
  Vertical tangents: $(0, 5), (-8, 5)$

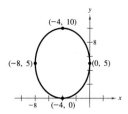

43. At $(1, 2)$:
  Slope of ellipse: -1
  Slope of parabola: 1
  At $(1, -2)$:
  Slope of ellipse: 1
  Slope of parabola: -1

45. At $(0, 0)$:
Slope of line: -1
Slope of sine curve: 1

47. $(x - 1 - 2\sqrt{2})^2 + (y - 2 + 2\sqrt{2})^2 = 16$
$(x - 1 + 2\sqrt{2})^2 + (y - 2 - 2\sqrt{2})^2 = 16$

49. $-\dfrac{\pi}{2} < y < \dfrac{\pi}{2}, \dfrac{dy}{dx} = \dfrac{1}{1 + x^2}$

51. $x + 2y - 6 = 0$

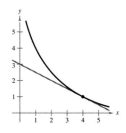

Section 3.7

1. (a) $\frac{3}{4}$ (b) 20 **3.** (a) $-\frac{5}{8}$ (b) $\frac{3}{2}$
5. (a) 24π in.^{2}/min (b) 96π in.^{2}/min
7. If dr/dt is constant, dA/dt is proportional to r.

9. (a) $\dfrac{5}{\pi}$ ft/min (b) $\dfrac{5}{4\pi}$ ft/min

11. $\dfrac{8}{405\pi}$ ft/min

13. (a) 9 cm^3/sec (b) 900 cm^3/sec
15. (a) 0 cm/min (b) 12 cm/min
17. (a) $\frac{8}{25}$ cm/min (b) 0 cm/min
(c) $-\frac{8}{25}$ cm/min (d) -0.0039 cm/min
19. (a) 24.6% (b) $\frac{1}{64}$ ft/min
21. (a) $-\frac{7}{12}$ ft/sec (b) $-\frac{3}{2}$ ft/sec (c) $-\frac{48}{7}$ ft/sec
23. (a) 21.96 ft^2/sec (b) $\frac{1}{12}$ rad/sec
25. (a) -750 mi/hr (b) 20 minutes
27. $-\dfrac{28}{\sqrt{10}} \approx -8.85$ ft/sec

29. (a) $\frac{25}{3}$ ft/sec (b) $\frac{10}{3}$ ft/sec

33. $v^{0.3}\left(1.3p\dfrac{dv}{dt} + v\dfrac{dp}{dt}\right) = 0$ **35.** $\dfrac{1}{20}$ rad/sec

37. (a) $\frac{1}{2}$ rad/min (b) $\frac{3}{2}$ rad/min
(c) 1.87 rad/min
39. (a) 0 ft/sec (b) 10π ft/sec (c) $10\sqrt{3}\pi$ ft/sec

Review Exercises for Chapter 3

1. $3x(x - 2)$ **3.** $\dfrac{x + 1}{2x^{3/2}}$ **5.** $-\dfrac{4}{3t^3}$

7. $\dfrac{3x^2}{2\sqrt{x^3 + 1}}$ **9.** $2(6x^3 - 9x^2 + 16x - 7)$

11. $s(s^2 - 1)^{3/2}(8s^3 - 3s + 25)$ **13.** $\dfrac{2x(2 - x)}{(x - 1)^2}$

15. $-\dfrac{x^2 + 1}{(x^2 - 1)^2}$ **17.** $32x - 128x^3$

19. $\dfrac{6x}{(4 - 3x^2)^2}$ **21.** $\dfrac{x + 2}{(x + 1)^{3/2}}$ **23.** $\dfrac{5}{6(t + 1)^{1/6}}$

25. $-\csc 2x \cot 2x$ **27.** $\tan^2 x$
29. $\frac{1}{2}(1 - \cos 2x) = \sin^2 x$ **31.** $-9 \sin (3x + 1)$

33. $-\dfrac{\sec^2 \sqrt{1 - x}}{2\sqrt{1 - x}}$ **35.** $\dfrac{x \cos x - 2 \sin x}{x^3}$

37. $-x \sec^2 x - \tan x$ **39.** $\cos 4x + 1$

41. $\dfrac{9}{(x^2 + 9)^{3/2}}$ **43.** $\dfrac{2(t + 2)}{(1 - t)^4}$

45. $\dfrac{2(6x^3 - 15x^2 - 18x + 5)}{(x^2 + 1)^3}$ **47.** $2 \csc^2 x \cot x$

49. $\dfrac{-x^2 \cos x + 2x \sin x + 2 \cos x}{x^3}$

51. $-\dfrac{2x + 3y}{3(x + y^2)}$ **53.** $\dfrac{2y\sqrt{x} - y\sqrt{y}}{2x\sqrt{y} - x\sqrt{x}}$ **55.** $\dfrac{x}{y}$

57. $\dfrac{y \sin x + \sin y}{\cos x - x \cos y}$

59. Tangent line: $3x - y + 7 = 0$
Normal line: $x + 3y - 1 = 0$
61. Tangent line: $x + 2y - 10 = 0$
Normal line: $2x - y = 0$
63. Tangent line: $2x - 3y - 3 = 0$
Normal line: $3x + 2y - 11 = 0$

65. (a) $(0, -1), \left(-2, \dfrac{7}{3}\right)$ (b) $(-3, 2), \left(1, -\dfrac{2}{3}\right)$
(c) $\left(-1 + \sqrt{2}, \dfrac{2[1 - 2\sqrt{2}]}{3}\right), \left(-1 - \sqrt{2}, \dfrac{2[1 + 2\sqrt{2}]}{3}\right)$

67. $f'(x) = -\dfrac{2}{x^3}$ **69.** $f'(x) = \dfrac{1}{2\sqrt{x + 2}}$

71. $y'' + y = -(2 \sin x + 3 \cos x)$
$+ (2 \sin x + 3 \cos x) = 0$

73. $v(t) = 1 - \dfrac{1}{(t + 1)^2}$, $a(t) = \dfrac{2}{(t + 1)^3}$

75. (a) -18.667 (b) -7.284 (c) -3.240
(d) -0.747 **77.** 56 ft/sec
79. (a) (b) 50
(c) $x = 25$

(d) $y' = 1 - 0.04x$

x	0	10	25	30	50
y'	1	0.6	0	-0.2	-1

(e) $y'(25) = 0$

83. (a) $2\sqrt{2}$ units/sec   **(b)** 4 units/sec

(c) 8 units/sec   **85.** $\frac{2}{25}$ ft/min

89. $f(x) = 4 - |x - 2|$

(a) Yes

(b) No, sharp turn in graph

Chapter 4

Section 4.1

1. $f'(0) = 0$ **3.** $f'(4) = 0$

5. $f'(-2)$ is undefined

7. Minimum: (2, 2)

Maximum: (−1, 8)

9. Minimum: (0, 0) and (3, 0)

Maximum: $\left(\frac{3}{2}, \frac{9}{4}\right)$

11. Minimum: (−1, −4) and (2, −4)

Maximum: (0, 0) and (3, 0)

13. Minimum: (0, 0)   **15.** Minimum: (1, 1)

Maximum: (−1, 5)   Maximum: (4, 4)

17. Minimum: (1, −1)   **19.** Minimum: $\left(\frac{1}{6}, \frac{\sqrt{3}}{2}\right)$

Maximum: $\left(0, -\frac{1}{2}\right)$ Maximum: (0, 1)

21. Continuous on $\left[0, \dfrac{\pi}{4}\right]$ **23. (a)** Yes

Not continuous on $[0, \pi]$ **(b)** No

25. (a) No   **27. (a)** Minimum: (0, −3)

(b) Yes   Maximum: (2, 1)

(b) Minimum: (0, −3)

(c) Maximum: (2, 1)

(d) No extrema

29. Maximum: $\left|f''(0)\right| = 2$

31. Maximum: $\left|f''\left(\sqrt[3]{-10 + \sqrt{108}}\right)\right| \approx 1.47$

33. Maximum: $\left|f^{(4)}\left(\frac{1}{2}\right)\right| = 360$

35. Maximum: $\left|f^{(4)}(0)\right| = \frac{56}{81}$

37. Maximum: $P(12) = 72$

39. The part of the lawn farthest from the sprinkler

41. 0.4398

Section 4.2

1. $f(0) = f(2) = 0$

f is not differentiable on (0, 2)

3. $f'(1) = 0$ **5.** [1, 2], $f'\left(\dfrac{6 - \sqrt{3}}{3}\right) = 0$

[2, 3], $f'\left(\dfrac{6 + \sqrt{3}}{3}\right) = 0$

7. Not differentiable at $x = 0$

9. Not differentiable at $x = 0$

11. $f'(-2 + \sqrt{5}) = 0$ **13.** $[0, \pi]$, $f'\left(\dfrac{\pi}{2}\right) = 2$

$[\pi, 2\pi]$, $f'\left(\dfrac{3\pi}{2}\right) = 0$

15. $f'\left(\dfrac{\pi}{4}\right) = 0$ **17.** $f'(-0.5756) = 0$

19. Not continuous on $[0, \pi]$ **21.** $f'\left(-\dfrac{1}{2}\right) = -1$

23. $f'\left(\dfrac{8}{27}\right) = 1$ **25.** $f'\left(\dfrac{-2 + \sqrt{6}}{2}\right) = \dfrac{2}{3}$

27. $f'\left(\dfrac{\sqrt{3}}{3}\right) = 1$ **29.** $f'\left(-\dfrac{\pi}{2}\right) = f'\left(\dfrac{\pi}{2}\right) = 1$

31. (a) $f(1) = f(2) = 64$

(b) Velocity = 0 for some t in [1, 2]

33. (a) −48 ft/sec   **(b)** $t = \frac{3}{2}$ sec

35. $f(x)$ is not continuous on [2, 6]

43. (a) $x - 4y + 3 = 0$ **(b)** $c = 4$, $x - 4y + 4 = 0$

Section 4.3

1. Increasing on (3, ∞)

Decreasing on (−∞, 3)

3. Increasing on (−∞, −2) and (2, ∞)

Decreasing on (−2, 2)

5. Increasing on $(-\infty, 0)$
Decreasing on $(0, \infty)$
7. Critical number: $x = 1$
Increasing on $(-\infty, 1)$
Decreasing on $(1, \infty)$
Relative maximum: $(1, 5)$
9. Critical number: $x = 3$
Increasing on $(3, \infty)$
Decreasing on $(-\infty, 3)$
Relative minimum: $(3, -9)$
11. Critical numbers: $x = -2, 1$
Increasing on $(-\infty, -2)$ and $(1, \infty)$
Decreasing on $(-2, 1)$
Relative maximum: $(-2, 20)$
Relative minimum: $(1, -7)$
13. Critical numbers: $x = 0, 4$
Increasing on $(-\infty, 0)$ and $(4, \infty)$
Decreasing on $(0, 4)$
Relative maximum: $(0, 15)$
Relative minimum: $(4, -17)$
15. Critical number: $x = 1$
Increasing on $(1, \infty)$
Decreasing on $(-\infty, 1)$
Relative minimum: $(1, 0)$
17. Critical number: $x = 0$
Increasing on $(-\infty, \infty)$
No relative extrema
19. Critical numbers: $x = -1, 1$
Discontinuity: $x = 0$
Increasing on $(-\infty, -1)$ and $(1, \infty)$
Decreasing on $(-1, 0)$ and $(0, 1)$
Relative maximum: $(-1, -2)$
Relative minimum: $(1, 2)$
21. Critical number: $x = 0$
Discontinuities: $x = -3, 3$
Increasing on $(-\infty, -3)$ and $(-3, 0)$
Decreasing on $(0, 3)$ and $(3, \infty)$
Relative maximum: $(0, 0)$
23. Critical numbers: $x = -1, 1$
Increasing on $(-\infty, -1)$ and $(1, \infty)$
Decreasing on $(-1, 1)$
Relative maximum: $\left(-1, \frac{4}{5}\right)$
Relative minimum: $\left(1, -\frac{4}{5}\right)$
25. Critical numbers: $x = -3, 1$
Discontinuity: $x = -1$
Increasing on $(-\infty, -3)$ and $(1, \infty)$
Decreasing on $(-3, -1)$ and $(-1, 1)$
Relative maximum: $(-3, -8)$
Relative minimum: $(1, 0)$
27. Critical numbers: $x = \pi/6, 5\pi/6$
Increasing on $(0, \pi/6), (5\pi/6, 2\pi)$
Decreasing on $(\pi/6, 5\pi/6)$
Relative maximum: $(\pi/6, [\pi + 6\sqrt{3}]/12)$
Relative minimum: $(5\pi/6, [5\pi - 6\sqrt{3}]/12)$

29. Critical numbers: $x = \pi/2, 7\pi/6, 3\pi/2, 11\pi/6$
Increasing on $(0, \pi/2), (7\pi/6, 3\pi/2), (11\pi/6, 2\pi)$
Decreasing on $(\pi/2, 7\pi/6), (3\pi/2, 11\pi/6)$
Relative maxima: $(\pi/2, 2), (3\pi/2, 0)$
Relative minima: $(7\pi/6, -1/4), (11\pi/6, -1/4)$
31. (a) Not monotonic   **33. (a)** Not monotonic
   (b) Strictly monotonic   **(b)** Not monotonic
   (c) Strictly monotonic   **(c)** Strictly monotonic
35. Moving upward when $0 < t < 3$
Moving downward when $3 < t < 6$
Maximum height: $s(3) = 144$ ft
37. $r = \dfrac{2R}{3}$
39. Increasing when $0 < t < 84.3388$ minutes
Decreasing when $84.3388 < t < 120$ minutes
41. Increasing when $6.02 < t < 14$ days
Decreasing when $0 < t < 6.02$ days
43. $T = 10°$ **45.** May 31
47. $f(x) = -\frac{1}{2}x^3 + \frac{3}{2}x^2$ **49.** $g'(0) < 0$
51. $g'(-6) < 0$ **53.** $g'(0) > 0$
55. (a)

   (b) Critical numbers: $x = \pm\dfrac{3\sqrt{2}}{2}$
   (c) $f' > (0)$ on $\left(-\dfrac{3\sqrt{2}}{2}, \dfrac{3\sqrt{2}}{2}\right)$
       $f' < (0)$ on $\left(-3, -\dfrac{3\sqrt{2}}{2}\right), \left(\dfrac{3\sqrt{2}}{2}, 3\right)$

57. (a)

   (b) Critical numbers: $x = 2.2889, 5.0870$
   (c) $f' > 0$ on $(0, 2.889), (5.0870, 2\pi)$
       $f' < 0$ on $(2.2889, 5.0870)$

Section 4.4

1. Concave upward: $(-\infty, \infty)$
3. Concave upward: $(-\infty, 1)$
Concave downward: $(1, \infty)$
5. Concave upward: $(-\infty, -1)$, $(1, \infty)$
Concave downward: $(-1, 1)$
7. Relative maximum: $(3, 9)$
9. Relative minimum: $(5, 0)$
11. Relative maximum: $(0, 3)$
Relative minimum: $(2, -1)$
13. Relative minimum: $(3, -25)$
15. Relative minimum: $(0, -3)$
17. Relative maximum: $(-2, -4)$
Relative minimum: $(2, 4)$
19. No relative extrema since f is non-increasing
21. Relative maximum: $(-2, 16)$
Relative minimum: $(2, -16)$
Point of inflection: $(0, 0)$

23. Point of inflection: $(2, 0)$

25. Relative minima: $(\pm 2, -4)$
Relative maximum: $(0, 0)$
Points of inflection: $\left(\pm \dfrac{2}{\sqrt{3}}, -\dfrac{20}{9}\right)$

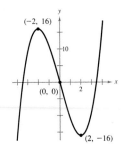

27. Relative minimum: $(1, -27)$
Points of inflection: $(2, -16)$, $(4, 0)$

29. Relative minima: $(-1, 2)$, $(1, 2)$

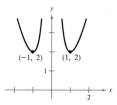

31. Relative minimum: $(-2, -2)$

33. Point of inflection: $(0, 0)$

35. Point of inflection: $(2, 0)$

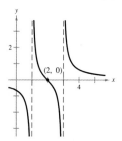

37. Relative minimum: $(3\pi, -1)$
Relative maximum: $(\pi, 1)$
Point of inflection: $(2\pi, 0)$

39. Relative minima: $\left(\frac{\pi}{2}, 1\right), \left(\frac{5\pi}{2}, 1\right)$

Relative maxima: $\left(\frac{3\pi}{2}, -1\right), \left(\frac{7\pi}{2}, -1\right)$

41. Relative minimum: $\left(\frac{5\pi}{3}, -2.598\right)$

Relative maximum: $\left(\frac{\pi}{3}, 2.598\right)$

Points of inflection:
$(0, 0), (\pi, 0), (1.823, 1.452), (4.46, -1.452)$

43. (a)

(b)

45.

47.

49. $f(x) = (x - c)^n$ has a point of inflection at $(c, 0)$ if n is odd and $n \geq 3$.

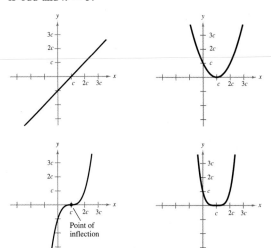

51. $f(x) = \frac{1}{2}x^3 - 6x^2 + \frac{45}{2}x - 24$

53. Relative extrema: $(2, 32), (6, 0)$

Point of inflection: $(4, 16)$

55. $x = \left(\frac{15 - \sqrt{33}}{16}\right) L \approx 0.578 L$

57. $x = 100$ units

59. $\theta = \frac{\pi}{2} + 2n\pi, \theta = \frac{3\pi}{2} + 2n\pi$

61. (a)

(b) Critical numbers: $x = -1, 1$

Relative minimum: $\left(-1, -\frac{2}{3}\right)$

Relative maximum: $(1, 2)$

(c) Points of inflection:
$(-1.8794, -0.5863), (0.3473, 0.8982),$
$(1.5321, 1.6881)$

63. (a)

(b) Critical numbers: $x = \frac{\pi}{6}, \frac{\pi}{2}, \frac{5\pi}{6}$

Relative maximum: $\left(\frac{\pi}{2}, 1.53333\right)$

(c) Points of inflection:
$(0, 0)$, $(0.5236, 0.2667)$, $(1.1731, 0.9638)$,
$(1.9685, 0.9637)$, $(2.6180, 0.5563)$, $(\pi, 0)$

Section 4.5

1. h **2.** c **3.** e **4.** a **5.** d **6.** g
7. b **8.** f **9.** $\frac{2}{3}$ **11.** 0
13. Limit does not exist.
15. Limit does not exist. **17.** 5 **19.** -1
21. 2 **23.** 1 **25.** 0 **27.** 0 **29.** 1
31. 0 **33.** $-\frac{1}{2}$

35.

37.

39.

41.

43.

45.

47.

x	1	10	10^2	10^3
$f(x)$	2.000	0.348	0.101	0.032

x	10^4	10^5	10^6
$f(x)$	0.010	0.0032	0.001

$$\lim_{x \to \infty} \frac{x + 1}{x\sqrt{x}} = 0$$

49.

x	1	10	10^2	10^3
$f(x)$	0.479	0.500	0.500	0.500

x	10^4	10^5	10^6
$f(x)$	0.500	0.500	0.500

$$\lim_{x \to \infty} x \sin \frac{1}{2x} = \frac{1}{2}$$

51. 0.5 **53.** 100%
55. Horizontal asymptotes: $y = \pm\frac{3}{2}$

Section 4.6

1.

3.

5.

7.

9.

11.

13.

(2, 0)

(0, −16)

−25

(−1, −27)

15.

6

(−1, 4)

(0, 0)

2

(1, −4)

33.

3

(0, 0)

4

35.

3

(−2, 0)

3

17.

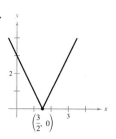

2

3

$\left(\frac{3}{2}, 0\right)$

19.

1

$\left(-1, \frac{1}{4}\right)$ $\left(1, \frac{1}{4}\right)$

(0, 0)

3

37.

3

(1, 2)

−3

(−1, −2)

39.

$\left(2\sqrt{3}, \frac{3\sqrt{3}}{2}\right)$

2

(0, 0)

4

$\left(-2\sqrt{3}, -\frac{3\sqrt{3}}{2}\right)$

21.

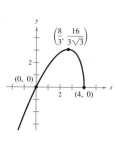

$\left(\frac{8}{3}, \frac{16}{3\sqrt{3}}\right)$

2

(0, 0)

2 (4, 0)

23.

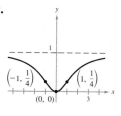

4

(1, 1)

(0, 0) 3

41. $a < 0$ and $b^2 < 3ac$ **43.** $a < 0$ and $b^2 = 3ac$

45. $a < 0$ and $b^2 > 3ac$ **47.**

3

−2

f

25.

1

(0, 0)

3

27.

2 $\left(\frac{\pi}{6}, \frac{4}{9}\right)$

$\left(\frac{\pi}{2}, \frac{19}{18}\right)$

$\left(\frac{5\pi}{6}, \frac{4}{9}\right)$

2π

(π, 0)

$\left(\frac{7\pi}{6}, -\frac{4}{9}\right)$

$\left(\frac{3\pi}{2}, -\frac{19}{18}\right)$

$\left(\frac{11\pi}{6}, -\frac{4}{9}\right)$

49.

f

2

2

51.

f

4

2

29.

2

$\left(\frac{\pi}{4}, \frac{\pi}{2} - 1\right)$

$-\frac{\pi}{2}$

$\frac{\pi}{2}$

$\left(-\frac{\pi}{4}, 1 - \frac{\pi}{2}\right)$

31.

$\left(\frac{7}{3}, 0\right)$

4

−2

$\left(0, -\frac{7}{2}\right)$

Section 4.7

1. 55 and 55   **3.** $\sqrt{192}$ and $\sqrt{192}$

5. 1 and 1   **7.** $l = w = 25$ ft

9. $l = w = 8$ ft   **11.** $\left(\frac{7}{2}, \sqrt{\frac{7}{2}}\right)$

13. $x = \frac{Q}{2}$ **15.** $600m \times 300m$

17. $V = 128$ when $x = 2$

19. $\frac{5 - \sqrt{7}}{6}$ ft $\times \frac{1 + \sqrt{7}}{3}$ ft $\times \frac{4 + \sqrt{7}}{3}$ ft

21. Rectangular portion: $\dfrac{16}{\pi + 4}$ ft $\times$ $\dfrac{32}{\pi + 4}$ ft

23. $x = 3$, $y = \frac{3}{2}$ **25.** $(0, 0)$, $(4, 0)$, $(0, 6)$

27. Height: $\dfrac{5\sqrt{2}}{2}$; width: $5\sqrt{2}$

29. Bases: r and $2r$; altitude: $\dfrac{\sqrt{3}r}{2}$

31. $r \approx 1.51$, $h = 2r \approx 3.02$

33. 18 in. $\times$ 18 in. $\times$ 36 in.   **35.** $\dfrac{32\pi r^3}{81}$

37. $r = \sqrt[3]{\dfrac{9}{\pi}} \approx 1.42$ in.

39. Side of square: $\dfrac{10\sqrt{3}}{9 + 4\sqrt{3}}$

Side of triangle: $\dfrac{30}{9 + 4\sqrt{3}}$

41. (a) $\dfrac{10\pi}{\pi + 3 + 2\sqrt{2}} \approx 3.5$ ft   (b) 10 ft

43. $w = 8\sqrt{3}$, $h = 8\sqrt{6}$

45. 1 mile from the nearest point on the coast

49. $\theta = \arctan k$, $F = \dfrac{kW}{\sqrt{k^2 + 1}}$

51. $\theta = \dfrac{2\pi}{3}(3 - \sqrt{6}) \approx 66°$

53. 14.1421×7.071

Section 4.8

1.

n	x_n	$f(x_n)$	$f'(x_n)$	$\dfrac{f(x_n)}{f'(x_n)}$	$x_n - \dfrac{f(x_n)}{f'(x_n)}$
1	1.700	-0.110	3.400	-0.032	1.732

3.

n	x_n	$f(x_n)$	$f'(x_n)$	$\dfrac{f(x_n)}{f'(x_n)}$	$x_n - \dfrac{f(x_n)}{f'(x_n)}$
1	3	0.141	-0.990	-0.143	3.143

5. 0.682   **7.** 1.146   **9.** 3.317   **11.** -0.489

13. 0.569   **15.** 4.493   **17.** $f'(x_1) = 0$

19. $1 = x_1 = x_3 = \ldots$ **21.** $x_{i+1} = \dfrac{x_i^2 + a}{2x_i}$
$0 = x_2 = x_4 = \ldots$

23. 2.646   **25.** 1.565   **29.** 3.141   **31.** 0.860

33.

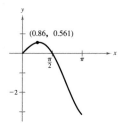

35. $(1.939, 0.240)$ **37.** $x \approx 1.563$ mi
39. $x \approx 11.803$

Section 4.9

1. $6x\,dx$ **3.** $-\dfrac{3}{(2x - 1)^2}\,dx$ **5.** $\dfrac{1 - 2x^2}{\sqrt{1 - x^2}}\,dx$

7. $\dfrac{2\sec^2 x(x^2 \tan x + \tan x - x)}{(x^2 + 1)^2}\,dx$

9. $-\pi \sin\left(\dfrac{6\pi x - 1}{2}\right)\,dx$

11.

$dx = \Delta x$	dy	Δy	$\Delta y - dy$	$\dfrac{dy}{\Delta y}$
1.000	4.000	5.000	1.000	0.800
0.500	2.000	2.250	0.250	0.889
0.100	0.400	0.410	0.010	0.976
0.010	0.040	0.040	0.000	1.000
0.001	0.004	0.004	0.000	1.000

13.

$dx = \Delta x$	dy	Δy	$\Delta y - dy$	$\dfrac{dy}{\Delta y}$
1.000	80.000	211.000	131.000	0.379
0.500	40.000	65.656	25.656	0.609
0.100	8.000	8.841	0.841	0.905
0.010	0.800	0.808	0.008	0.990
0.001	0.080	0.080	0.000	1.000

15.

$dx = \Delta x$	dy	Δy	$\Delta y - dy$	$\dfrac{dy}{\Delta y}$
1.000	-0.4161	-0.7682	-0.3520	0.5417
0.500	-0.2081	-0.3108	-0.1028	0.6694
0.100	-0.0416	-0.0461	-0.0045	0.9029
0.010	-0.0042	-0.0042	0.000	0.9892
0.001	-0.0004	-0.0004	0.000	0.9988

17. (a) $dA = 2x\Delta x$, $\Delta A = 2x\Delta x + (\Delta x)^2$
(b)

(c) $\Delta A - dA = (\Delta x)^2$
19. $\pm 7\pi$ in.^{2} **21. (a)** $\frac{2}{3}\%$ **(b)** 1.25%
23. (a) $\pm 2.88\pi$ in.^{3} **(b)** $\pm 0.96\pi$ in.^{2}
(c) 1%, $\frac{2}{3}\%$
25. (a) $\frac{1}{4}\%$ **(b)** 216 sec $= 3.6$ min
29. (a) 0.87% **(b)** 2.16% **31.** $4{,}961$ ft

Section 4.10

1. 4500   **3.** 300   **5.** 200   **7.** 200
9. $x = 30$ **11.** $x = 1500$ **13.** 20   **15.** 200
$\quad p = 60 \qquad\qquad p = 35$
17. Line should run from power station to point across
the river $3/(2\sqrt{7})$ mile downstream.
19. 8% **21.** $x = 3$
23. (a) $p = \$68$ **(b)** $\overline{C} \approx \$50.88$
25. $x \approx 40$ units   **27.** $\$30{,}000$
29. (a) $t = \frac{1}{2}$ (January 15)
$\qquad Q\!\left(\frac{1}{2}\right) = 7.2$ quads/month
(b) $t = \frac{13}{2}$ (July 15)
$\qquad Q\!\left(\frac{13}{2}\right) = 5.2$ quads/month
31. (a) $t = \frac{1}{2}$ (January 15)
$\qquad Q\!\left(\frac{1}{2}\right) = 7.2$ quads/month
(b) $t = \frac{9}{2}$ (May 15), $t = \frac{17}{2}$ (September 15)
$\qquad Q\!\left(\frac{9}{2}\right) = 5.7$ quads/month
33. $\eta = -\frac{17}{3}$, elastic   **35.** $\eta = -\frac{1}{2}$, inelastic

Review Exercises for Chapter 4

1.

3.

5.

7.

9.

11.

13.

15.

17.

19.

21.

23.

17. $\frac{2}{15}x^{1/2}(3x^2 + 5x + 15) + C$

19. $x^3 + \frac{1}{2}x^2 - 2x + C$ **21.** $t - \frac{2}{t} + C$

23. $\frac{2}{7}y^{7/2} + C$ **25.** $x + C$

27. $-2\cos x + 3\sin x + C$ **29.** $t + \csc t + C$

31. $\tan \theta + \cos \theta + C$ **33.** $\tan y + C$

35. $y = x^2 - x + 1$ **37.** $y = \sin x + 4$

39. $f(x) = x^2 + x + 4$

41. $f(x) = -4x^{1/2} + 3x = -4\sqrt{x} + 3x$

43. $s(t) = -16t^2 + 1600$, $t = 10$ sec

45. $v_0 \approx 187.617$ ft/sec

47. (a) $\dfrac{1 + \sqrt{17}}{2} \approx 2.562$ sec

   (b) $-16\sqrt{17} \approx -65.970$ ft/sec

49. (a) $\frac{154}{39} \approx 3.95$ ft/sec^2 **(b)** $\frac{1859}{3} \approx 619.67$ ft

51. (a) 300 ft   **(b)** 60 ft/sec $\approx$ 41 mi/hr

57. $R = x\left(100 - \frac{5}{2}x\right)$, $p = 100 - \frac{5}{2}x$

25.

27. Maximum: (1, 3)
   Minimum: (1, 1)

29. $f'\left(\dfrac{-2 + \sqrt{85}}{3}\right) = -\dfrac{1}{17}$ **31.** $f'\left(\dfrac{2744}{729}\right) = \dfrac{3}{7}$

33. $f'(0) = 1$ **35.** No, f is discontinuous at $x = 0$

37. $c = \dfrac{x_1 + x_2}{2}$ **39.** $t \approx 4.92 \approx 4{:}55$ P.M.

   $d \approx 64$ mi

41. \$48   **43.** (0, 0), (5, 0), (0, 10)   **45.** 120

49. 14.05 ft   **51.** $3(3^{2/3} + 2^{2/3})^{3/2} \approx 21.07$ ft

53. (a) $y = \dfrac{1}{4}$ in.   **55.** $x = \sqrt{\dfrac{2Qs}{r}}$

   $v = 4$ in./sec

   (c) Period: $\dfrac{\pi}{6}$

   Frequency: $\dfrac{6}{\pi}$

57. -0.347 **59.** -0.453 **61. (a)** 2%   **(b)** 3%

Chapter 5

Section 5.1

Given	Rewrite	Integrate	Simplify
1. $\displaystyle\int \sqrt[3]{x}\, dx$	$\displaystyle\int x^{1/3}\, dx$	$\dfrac{x^{4/3}}{4/3} + C$	$\dfrac{3}{4}x^{4/3} + C$
3. $\displaystyle\int \dfrac{1}{x\sqrt{x}}\, dx$	$\displaystyle\int x^{-3/2}\, dx$	$\dfrac{x^{-1/2}}{-1/2} + C$	$-\dfrac{2}{\sqrt{x}} + C$
5. $\displaystyle\int \dfrac{1}{2x^3}\, dx$	$\dfrac{1}{2}\displaystyle\int x^{-3}\, dx$	$\dfrac{1}{2}\left(\dfrac{x^{-2}}{-2}\right) + C$	$-\dfrac{1}{4x^2} + C$

7. $\frac{1}{4}x^4 + 2x + C$ **9.** $\frac{2}{5}x^{5/2} + x^2 + x + C$

11. $\frac{3}{5}x^{5/3} + C$ **13.** $-\dfrac{1}{2x^2} + C$ **15.** $-\dfrac{1}{4x} + C$

Section 5.2

1. 35   **3.** $\dfrac{158}{85}$ **5.** $4c$ **7.** 238   **9.** $\displaystyle\sum_{i=1}^{9} \dfrac{1}{3i}$

11. $\displaystyle\sum_{j=1}^{8}\left[2\left(\dfrac{j}{8}\right) + 3\right]$ **13.** $\dfrac{1}{6}\displaystyle\sum_{k=1}^{6}\left[\left(\dfrac{k}{6}\right)^2 + 2\right]$

15. $\dfrac{2}{n}\displaystyle\sum_{i=1}^{n}\left[\left(\dfrac{2i}{n}\right)^3 - \left(\dfrac{2i}{n}\right)\right]$ **17.** $\dfrac{3}{n}\displaystyle\sum_{i=1}^{n}\left[2\left(1 + \dfrac{3i}{n}\right)^2\right]$

19. 420   **21.** 2470   **23.** $\dfrac{1015}{n^3}$

25. $\frac{8}{3}$ **27.** $\frac{81}{4}$ **29.** 9

31. $\displaystyle\lim_{n\to\infty} \dfrac{1}{6}\left[\dfrac{2n^3 - 3n^2 + n}{n^3}\right] = \dfrac{1}{3}$

33. $\displaystyle\lim_{n\to\infty}\left[8\left(\dfrac{n^2 + n}{n^2}\right)\right] = 8$

35. $\displaystyle\lim_{n\to\infty} 2\left[\dfrac{10n^4 + 13n^3 + 4n^2}{n^4}\right] = 20$

37. $S \approx 0.768$ **39.** $S \approx 0.746$ **41.** $S \approx 0.859$
   $s \approx 0.518$ $s \approx 0.646$ $s \approx 0.659$

43.

n	5	10	50	100
$s(n)$	1.6	1.8	1.96	1.98
$S(n)$	2.4	2.2	2.04	2.02

45. $A = 2$

47. $A = \frac{7}{3}$

49. $A = \frac{52}{3}$

51. $A = \frac{3}{4}$

53. $A = \frac{2}{3}$

55. $A = 6$

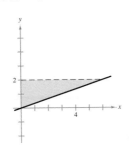

57. $\frac{69}{8}$ **59.** 0.345
61. $N(5) \approx 167$, $N(10) \approx 250$, $N(25) \approx 400$
$$\lim_{t \to \infty} \frac{10(5 + 3t)}{1 + 0.04t} = 750$$
63.

n	4	8	12	16	20
$s(n)$	4.1463	4.7650	4.9623	5.0587	5.1156
$S(n)$	6.1463	5.7650	5.6290	5.5587	5.5156

Section 5.3

1. $\int_0^5 3 \, dx$ **3.** $\int_{-4}^4 (4 - |x|) \, dx$

5. $\int_{-2}^2 (4 - x^2) \, dx$ **7.** $\int_0^2 y^3 \, dy$ **9.** $\int_0^\pi \sin x \, dx$

11. $A = 12$

Rectangle

13. $A = 8$

Triangle

15. $A = 14$

Trapezoid

17. $A = 1$

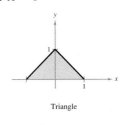

Triangle

19. $A = \dfrac{9\pi}{2}$

Semicircle

21. (a) 13 (b) -10 (c) 0 (d) 30
23. (a) 8 (b) -12 (c) -4 (d) 30
25. 36 **27.** 0 **29.** $\dfrac{10}{3}$ **31.** $\dfrac{4\sqrt{2}}{3}$
33. $\int_{-1}^5 (3x + 10) \, dx$ **35.** $\int_0^3 \sqrt{x^2 + 4} \, dx$
39. Not integrable because there are an infinite number of discontinuities.

Section 5.4

1. 1 **3.** $-\frac{5}{2}$ **5.** $-\frac{10}{3}$ **7.** $\frac{1}{3}$ **9.** $\frac{1}{2}$
11. 36 **13.** -4 **15.** $\frac{2}{3}$ **17.** $-\frac{1}{18}$
19. $-\frac{27}{20}$ **21.** 1 **23.** 4 **25.** $\pi + 2$
27. $\dfrac{2\sqrt{3}}{3}$ **29.** 0 **31.** $\dfrac{1}{6}$ **33.** $\dfrac{8}{5}$ **35.** 6
37. 1 **39.** 10 **41.** 6 **43.** $\sqrt[3]{2}$
45. $\pm\arccos \dfrac{\sqrt{\pi}}{2} \approx \pm 0.4817$

47. Average $= \dfrac{8}{3}$

$x = \pm\dfrac{2\sqrt{3}}{3} \approx \pm 1.155$

49. Average $= \dfrac{2}{\pi}$

$x \approx 0.690, \ x \approx 2.451$

51. $\frac{1}{2}x^2 + 2x$ **53.** $\frac{3}{4}x^{4/3} - 12$ **55.** $\tan x - 1$
57. $x^2 - 2x + 5$ **59.** $\sqrt{x^4 + 1}$ **61.** $x \cos x$
63. 0.5318 liter **65.** (a) $64.4°$ (b) $68.6°$

Section 5.5

$\int f(g(x))g'(x)\,dx$	$u = g(x)$	$du = g'(x)\,dx$
1. $\displaystyle\int (5x^2 + 1)^2(10x)\,dx$	$5x^2 + 1$	$10x\,dx$
3. $\displaystyle\int \dfrac{x}{\sqrt{x^2 + 1}}\,dx$	$x^2 + 1$	$2x\,dx$
5. $\displaystyle\int \tan^2 x \sec^2 x\,dx$	$\tan x$	$\sec^2 x\,dx$

7. $\dfrac{(1 + 2x)^5}{5} + C$ **9.** $\dfrac{2}{3}(9 - x^2)^{3/2} + C$

11. $\dfrac{(x^3 - 1)^5}{15} + C$ **13.** $-\dfrac{15}{8}(1 - x^2)^{4/3} + C$

15. $-\dfrac{1}{3(1 + x^3)} + C$ **17.** $-4\sqrt{16 - x^2} + C$

19. $-\dfrac{1}{2(x^2 + 2x - 3)} + C$ **21.** $-\dfrac{1}{4}\left(1 + \dfrac{1}{t}\right)^4 + C$

23. $\sqrt{2x} + C$ **25.** $\frac{2}{5}\sqrt{x}(x^2 + 5x + 35) + C$
27. $\frac{1}{4}t^4 - t^2 + C$ **29.** $\frac{2}{5}y^{3/2}(15 - y) + C$
31. $-\frac{1}{2}\cos 2x + C$ **33.** $\frac{1}{2}\sin x^2 + C$
35. $\frac{1}{4}\sin^2 2x + C$ or $-\frac{1}{4}\cos^2 2x + C_1$
37. $\frac{1}{5}\tan^5 x + C$ **39.** $\frac{1}{2}\tan^2 x + C$ or $\frac{1}{2}\sec^2 x + C_1$
41. $-\cot x - x + C$
43. $2\left[\frac{1}{5}(x + 2)^{5/2} - \frac{2}{3}(x + 2)^{3/2}\right] + C$

$\qquad = \frac{2}{15}(x + 2)^{3/2}(3x - 4) + C$

45. $-2\left[\frac{1}{3}(1 - x)^{3/2} - \frac{2}{5}(1 - x)^{5/2} + \frac{1}{7}(1 - x)^{7/2}\right] + C$

$\qquad = -\frac{2}{105}(1 - x)^{3/2}(15x^2 + 12x + 8) + C$

47. $\dfrac{1}{4}\left[\dfrac{1}{5}(2x - 1)^{5/2} + \dfrac{2}{3}(2x - 1)^{3/2} - 3(2x - 1)^{1/2}\right] + C$

$\qquad = \dfrac{\sqrt{2x - 1}}{15}(3x^2 + 2x - 13) + C$

49. $-x - 1 - 2\sqrt{x + 1} + C$
$\quad$ or $-(x + 2\sqrt{x + 1}) + C_1$
51. $\frac{1}{2}\left[\frac{1}{3}(2x + 1)^{3/2} - (2x - 1)\right] + C$

$\qquad = \frac{1}{3}\sqrt{2x + 1}(x - 1) + C$

53. 0 **55.** 2 **57.** $\frac{1}{2}$ **59.** $\frac{4}{15}$ **61.** $\frac{144}{5}$
63. $\dfrac{1209}{28}$ **65.** $\dfrac{7088}{105}$ **67.** $\dfrac{3\sqrt{3}}{4}$ **69.** $2(\sqrt{3} - 1)$
71. $\frac{1}{6}(2x - 1)^3 + C_1 = \frac{4}{3}x^3 - 2x^2 + x - \frac{1}{6} + C_1$
$\quad$ or $\frac{4}{3}x^3 - 2x^2 + x + C_2$
$\quad$ Answers differ by a constant: $C_2 = C_1 - \frac{1}{6}$
73. 4 **75.** (a) $\frac{8}{3}$ (b) $\frac{16}{3}$ (c) $-\frac{8}{3}$ (d) 8
77. $f(x) = -\frac{1}{3}[(1 - x^2)^{3/2} - 8]$
79. (a) $C(x) = \frac{3}{2}(12x + 1)^{2/3} + 56.35$
$\quad$ (b)

81. (a) 102.352 thousand units
$\quad$ (b) 102.352 thousand units
$\quad$ (c) 74.5 thousand units

Section 5.6

	Exact	Trapezoidal	Simpson's
1.	2.6667	2.7500	2.6667
3.	4.0000	4.2500	4.0000
5.	4.0000	4.0625	4.0000
7.	12.6667	12.6640	12.6667
9.	0.1667	0.1676	0.1667

	Trapezoidal	Simpson's
11.	0.957	0.978
13.	3.41	3.22
15.	0.342	0.372
17.	0.089	0.089
19.	0.194	0.186
21.	0.500	0.000
23.	0.258	0.199
25.	$n = 366$	$n = 26$
27.	$n = 130$	$n = 12$

29. 0.70 **31.** 3.14159 **33.** 89,250 ft²
35. (a) 9,920 ft² (b) $10,413\frac{1}{3}$ ft²

37.	n	*Trapezoidal Rule*	*Simpson's Rule*
	4	15.6055	15.4845
	8	15.5010	15.4662
	12	15.4814	15.4657
	16	15.4745	15.4657
	20	15.4713	15.4657

39. 2.467

Review Exercises for Chapter 5

1. $x^{2/3} + C$ **3.** $\frac{2}{3}x^3 + \frac{1}{2}x^2 - x + C$

5. $\frac{2\sqrt{x}}{15}(15 + 10x + 3x^2) + C$

7. $\frac{2}{3}\sqrt{x^3 + 3} + C$ **9.** $\frac{1}{7}x^7 + \frac{3}{5}x^5 + x^3 + x + C$

11. $\frac{1}{4}\sin^4 x + C$ **13.** $\frac{\tan^{n+1} x}{n + 1} + C, n \neq -1$

15. $\frac{1}{2}\sec 2x + C$ **17.** $\frac{1}{2}x^2 - \frac{1}{x} + C$

19. (a) $\sum_{i=1}^{10} (2i - 1)$ **(b)** $\sum_{i=1}^{n} i^3$ **(c)** $\sum_{i=1}^{10} (4i + 2)$

21. 16   **23.** 0   **25.** 2   **27.** $\frac{422}{5}$

29. $\frac{28\pi}{15}$ **31.** $y = 2 - x^2$ **33.** 240 ft/sec

35. (a) 3 sec   **(b)** 144 ft   **(c)** $\frac{3}{2}$ sec   **(d)** 108 ft

37. (a) $S = \frac{5mb^2}{8}, s = \frac{3mb^2}{8}$

(b) $S(n) = \frac{mb^2(n + 1)}{2n}$

$s(n) = \frac{mb^2(n - 1)}{2n}$

(c) $\frac{1}{2}mb^2$ **(d)** $\frac{1}{2}mb^2$

39. 6   **41.** $\frac{10}{3}$

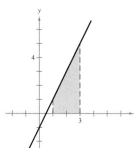

43. $\frac{1}{4}$ **45.** Average $= \frac{2}{5}, x = \frac{29}{4}$

47. Average $= 2, x = 2$ **49.** 0.254

51. 1.6234 liters

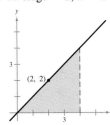

53. (a) $C \approx \$9.17$

(b) $C \approx \$3.14$ Savings $\approx \$6.03$

55. (a) $\frac{20,650}{M}$ **(b)** $\frac{43,150}{M}$

57. (a) $0.025 = 2.5\%$ **(b)** $0.736 = 73.6\%$

Chapter 6

Section 6.1

1.

3.

5.

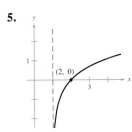

7. (a) 1.7917

(b) -0.4055

(c) 4.3944

(d) 0.5493

9. $\ln 2 - \ln 3$ **11.** $\ln x + \ln y - \ln z$

13. $\frac{3}{2}\ln 2$ **15.** $3[\ln(x + 1) + \ln(x - 1) - 3\ln x]$

17. $\ln z + 2\ln(z - 1)$ **19.** $\ln \frac{x - 2}{x + 2}$

21. $\ln \sqrt[3]{\frac{x(x + 3)^2}{x^2 - 1}}$ **23.** $\ln \frac{9}{\sqrt{x^2 + 1}}$ **25.** 3

27. 2   **29.** $\frac{2}{x}$ **31.** $\frac{2(x^3 - 1)}{x(x^3 - 4)}$ **33.** $\frac{4(\ln x)^3}{x}$

35. $\frac{2x^2 - 1}{x(x^2 - 1)}$ **37.** $\frac{1 - x^2}{x(x^2 + 1)}$ **39.** $\frac{1 - 2\ln x}{x^3}$

41. $\frac{2}{x \ln x^2} = \frac{1}{x \ln x}$ **43.** $\frac{1}{1 - x^2}$

45. $\frac{-4}{x(x^2 + 4)}$ **47.** $\frac{\sqrt{x^2 + 1}}{x^2}$ **49.** $\cot x$

51. $-\tan x + \frac{\sin x}{\cos x - 1}$

53. $\frac{3\cos x}{(\sin x - 1)(\sin x + 2)}$ **55.** $\frac{2x^2 - 1}{\sqrt{x^2 - 1}}$

57. $\frac{3x^3 - 15x^2 + 8x}{2(x - 1)^3\sqrt{3x - 2}}$

59. $\frac{(2x^2 + 2x - 1)\sqrt{x - 1}}{(x + 1)^{3/2}}$

61. $xy'' + y' = x\left(\dfrac{-2}{x^2}\right) + \dfrac{2}{x} = 0$ **63.** $\dfrac{2xy}{3 - 2y^2}$

65. $5x - y - 2 = 0$

67. Relative minimum: **69.** Relative minimum:
$\left(1, \frac{1}{2}\right)$ $(e^{-1}, -e^{-1})$

71. Relative minimum: (e, e)
Point of inflection: $\left(e^2, \dfrac{e^2}{2}\right)$

73. 0.567   **75.** $c = e - 1$ **79.** 1.16, 1.08, 1.05
81.   **83.** 1.099

Section 6.2

1. $\ln |x + 1| + C$ **3.** $-\frac{1}{2} \ln |3 - 2x| + C$

5. $\ln \sqrt{x^2 + 1} + C$ **7.** $\dfrac{x^2}{2} - 4 \ln |x| + C$

9. $\frac{1}{4}$ **11.** $\frac{7}{3}$ **13.** $-\ln 3$ **15.** $2\sqrt{x + 1} + C$

17. $\frac{1}{3} \ln |x^3 + 3x^2 + 9x| + C$

19. $3 \ln |1 + x^{1/3}| + C$

21. $2[\sqrt{x} - \ln(1 + \sqrt{x})] + C$

23. $x + 6\sqrt{x} + 18 \ln |\sqrt{x} - 3| + C$

25. $-\frac{2}{3} \ln |1 - x\sqrt{x}| + C$

27. $\ln |x - 1| + \dfrac{1}{2(x - 1)^2} + C$

29. $-\frac{1}{2} \ln |\csc 2x + \cot 2x| + C$

31. $-\sin(1 - x) + C$ **33.** $\ln |\sec x - 1| + C$

35. $\ln |1 + \sin t| + C$

37. $-\ln |\csc x + \cot x| + \cos x + C$
39. $\ln |\theta - \sin \theta| + C$
45. $\frac{15}{2} + 8 \ln 2 \approx 13.045$ square units

47. $P(t) = 1000[12 \ln |1 + 0.25t| + 1]$ **49.** $\dfrac{\ln 5}{4}$
$P(3) \approx 7715$

Section 6.3

1.

3.

5.

7.

9. $f^{-1}(x) = \dfrac{x + 3}{2}$ **11.** $f^{-1}(x) = x^{1/5}$

13. $f^{-1}(x) = x^2, x \geq 0$

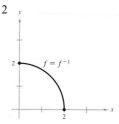

15. $f^{-1}(x) = \sqrt{4 - x^2}, 0 \leq x \leq 2$

17. $f^{-1}(x) = x^3 + 1$ **19.** $f^{-1}(x) = x^{3/2}$, $x \geq 0$

21. $f^{-1}(x) = \dfrac{\sqrt{7x}}{\sqrt{1 - x^2}}$, $-1 < x < 1$

23.

x	1	2	3	4
$f^{-1}(x)$	0	1	2	4

25. 32   **27.** 600   **29.** Inverse exists
31. Inverse does not exist   **33.** Inverse exists
35. Inverse exists   **37.** Inverse exists
39. Inverse does not exist   **41.** Inverse exists
43. $f'(x) = 2(x - 4) > 0$ on $(4, \infty)$

45. $f'(x) = -\dfrac{8}{x^3} < 0$ on $(0, \infty)$

47. $f'(x) = -\sin x < 0$ on $(0, \pi)$
49. $f'\left(\frac{1}{2}\right) = \frac{3}{4}$, $(f^{-1})'\left(\frac{1}{8}\right) = \frac{4}{3}$
51. $f'(5) = \frac{1}{2}$, $(f^{-1})'(1) = 2$

53. Not continuous at $x = \dfrac{(2n - 1)\pi}{2}$

55. $f^{-1}(x) = \dfrac{-8x}{1 + \sqrt{1 + 16x^2}}$

Section 6.4

1. (a) $\ln 1 = 0$ **(b)** $\ln 7.389 \ldots = 2$
3. (a) $e^{0.6931\ldots} = 2$ **(b)** $e^{2.128\ldots} = 8.4$
5. (a) $x = 4$ **(b)** $x = \frac{3}{2}$
7. (a) $x = e^2$ **(b)** $x = \ln 4$
9.   **11.**

13.   **15.**

17. (a) $\frac{271,801}{99,990} = 2.7182818281 \ldots$ **19. (a)** 3
      $e = 2.7182818284 \ldots$ **(b)** -3
   (b) $\frac{299}{110} = 2.718181818 \ldots$
      $e = 2.7182818284 \ldots$

21. $2e^{2x}$ **23.** $2(x - 1)e^{-2x+x^2}$ **25.** $\dfrac{e^{\sqrt{x}}}{2\sqrt{x}}$

27. $3(e^{-x} + e^x)^2(e^x - e^{-x})$ **29.** $2x$ **31.** $\dfrac{2e^{2x}}{1 + e^{2x}}$

33. $\dfrac{e^{2x} - 1}{e^{2x} + 1} = \dfrac{e^x - e^{-x}}{e^x + e^{-x}}$ **35.** $x^2 e^x$

37. $e^{-x}\left(\dfrac{1}{x} - \ln x\right)$ **39.** $2e^x \cos x$

41. $2e^x \tan e^x \sec^2 e^x$ **43.** $\dfrac{10 - e^y}{xe^y + 3}$

47. Relative maximum:
   $(0, 1/\sqrt{2\pi})$
   Points of inflection:
   $(\pm 1, 1/\sqrt{2\pi e})$

49. Relative minimum: $(0, 1)$

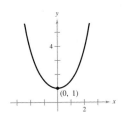

51. Relative minimum: $(0, 0)$
Relative maximum: $(2, 4e^{-2})$
Points of inflection: $(2 \pm \sqrt{2}, (6 \pm 4\sqrt{2})e^{-(2\pm\sqrt{2})})$

53. $y = x + 1$ **55.** $A = \sqrt{2}e^{-1/2}$

57. $\dfrac{e^2 - 1}{2e^2}$ **59.** $\dfrac{e}{3}(e^2 - 1)$

61. $x - \ln(e^x + 1) + C_1$ or $-\ln(1 + e^{-x}) + C_2$

63. $\dfrac{1}{2a}e^{ax^2} + C$ **65.** $\dfrac{e}{3}(e^2 - 1)$

67. $-\dfrac{2}{3}(1 - e^x)^{3/2} + C$ **69.** $\ln |e^x - e^{-x}| + C$

71. $-\dfrac{5}{2}e^{-2x} + e^{-x} + C$ **73.** $\dfrac{1}{\pi}e^{\sin \pi x} + C$

75. $\ln |\cos e^{-x}| + C$ **77.** $e^5 - 1 \approx 147.41$

79. $1 - e^{-1} \approx 0.632$ **83.** $f(x) = \frac{1}{2}(e^x + e^{-x})$

85. (a)

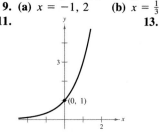

(b) $y = 1$
(c) Limit does not exist

Section 6.5

1. (a) $\log_2 8 = 3$ **(b)** $\log_3 (1/3) = -1$
3. (a) $10^{-2} = 0.01$ **(b)** $(1/2)^{-3} = 8$
5. (a) $x = 3$ **(b)** $x = -1$
7. (a) $x = \frac{1}{3}$ **(b)** $x = \frac{1}{16}$
9. (a) $x = -1, 2$ **(b)** $x = \frac{1}{3}$
11.

13.

15.

17. $(\ln 4)4^x$
19. $(\ln 5)5^{x-2}$

21. $x2^x(x \ln 2 + 2)$ **23.** $\dfrac{1}{x(\ln 3)}$

25. $\dfrac{x - 2}{(\ln 2)x(x - 1)}$ **27.** $\dfrac{x}{(\ln 5)(x^2 - 1)}$

29. $2(1 - \ln x)x^{(2/x)-2}$

31. $(x - 2)^{x+1}\left[\dfrac{x + 1}{x - 2} + \ln (x - 2)\right]$ **33.** $\dfrac{3^x}{\ln 3} + C$

35. $\dfrac{7}{\ln 4}$ **37.** $-\dfrac{1}{2(\ln 5)}5^{-x^2} + C$ **39.** $\dfrac{26}{\ln 3}$

41. Double: 9.24 years; Triple: 14.65 years
43.

x	1	10^{-1}	10^{-2}	10^{-4}	10^{-6}
$(1 + x)^{1/x}$	2.0000	2.5937	2.7048	2.7181	2.7183

45. (a) \$2061.03   **(b)** \$2088.15   **(c)** \$2102.35
(d) \$2112.06   **(e)** \$2116.84   **(f)** \$2117.00
47. (a) 6.7 million ft^3
(b) $V'(20) = 0.073$ million ft^3/yr
$V'(60) = 0.040$ million ft^3/yr
49. (a) $0.83 = 83\%$
(b) $P'(3) = 0.038$, $P'(10) = 0.017$
51. (a) \$332.01   **(b)** \$604.96   **(c)** \$1,102.32

Section 6.6

1. $y = \frac{1}{2}e^{0.4605t}$ **3.** $y = 0.6687e^{0.4024t}$
5. Time to double: 5.78 years
Amount after 10 years: \$3,320.12
Amount after 25 years: \$20,085.54
7. Annual rate: 8.94%
Amount after 10 years: \$1,833.67
Amount after 25 years: \$7,009.86
9. Annual rate: 9.50%
Time to double: 7.30 years
Amount after 25 years: \$5,375.51
11. (a) $N = 30(1 - e^{-0.0502t})$ **(b)** 36 days
13. (a) $S = 30e^{-1.7918/t}$ **(b)** 20,965 units
(c)

15. 900   **17.** 6015
19. Amount after 1,000 years: 6.52 grams
Amount after 10,000 years: 0.14 gram
21. Initial quantity: 6.70 grams
Amount after 1,000 years: 5.94 grams
23. Initial quantity: 2.16 grams
Amount after 10,000 years: 1.63 grams
25. 95.81%   **27.** 15,683 years   **29.** 22.35°
31. 11.75°   **33.** 527.06 mm Hg   **35.** $583,275.41

31.   **33.**

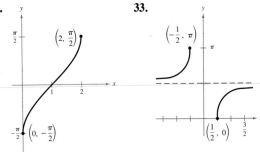

Section 6.7

1. $\dfrac{\pi}{6}$ **3.** $\dfrac{\pi}{3}$ **5.** $\dfrac{\pi}{6}$ **7.** $\dfrac{\pi}{4}$

9. $\arccos\left(\dfrac{1}{1.269}\right) \approx 0.663$

11. (a) $\dfrac{1}{2}$ **(b)** $\cos 2\theta = 1 - 2\sin^2\theta = \dfrac{1}{2}$

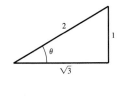

13. (a) $\dfrac{3}{5}$ **(b)** $\dfrac{5}{3}$

15. (a) $\cot\left[\arcsin\left(-\dfrac{1}{2}\right)\right] = -\cot\left[\arcsin\left(\dfrac{1}{2}\right)\right] = -\sqrt{3}$

(b) $-\dfrac{13}{5}$

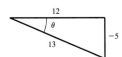

17. x **19.** $\sqrt{1 - 4x^2}$ **21.** $\dfrac{\sqrt{x^2 - 1}}{x}$

23. $\dfrac{\sqrt{x^2 - 9}}{3}$ **25.** $\dfrac{\sqrt{x^2 + 2}}{x}$

27. $\arcsin\left(\dfrac{9}{\sqrt{x^2 + 81}}\right)$

35. $x = \dfrac{1}{3}\left[\sin\left(\dfrac{1}{2}\right) + \pi\right] \approx 1.207$ **37.** $x = \dfrac{1}{3}$

39. $\dfrac{2}{\sqrt{1 - 4x^2}}$ **41.** $\dfrac{2}{\sqrt{2x - x^2}}$ **43.** $-\dfrac{3}{\sqrt{4 - x^2}}$

45. $\dfrac{5}{1 + 25x^2}$ **47.** $\dfrac{1}{|x|\sqrt{x^2 - 1}}$ **49.** 0

51. $-\dfrac{t}{\sqrt{1 - t^2}}$ **53.** $\dfrac{1}{2 + 3t^2}$ **55.** $\dfrac{1}{1 - x^4}$

57. $\arcsin x$ **59.** (0, 0)

61. Relative maximum: $(1.272, -0.606)$
Relative minimum: $(-1.272, 3.747)$

63. (0.7862, 0.6662)   **65.** $\dfrac{\sqrt{3}}{40}$ rad/sec

Section 6.8

1. $\dfrac{\pi}{18}$ **3.** $\dfrac{\pi}{6}$ **5.** $\text{arcsec } 2x + C$

7. $\dfrac{1}{2}x^2 - \dfrac{1}{2}\ln(x^2 + 1) + C$ **9.** $\arcsin(x + 1) + C$

11. $\dfrac{1}{2}\arcsin t^2 + C$ **13.** $\dfrac{1}{2}(\arctan x)^2 + C$

15. $\dfrac{\pi^2}{32} \approx 0.308$ **17.** $\dfrac{\sqrt{3} - 2}{2} \approx -0.134$

19. $\arcsin e^x + C$ **21.** $\dfrac{1}{3}\arctan\left(\dfrac{x - 3}{3}\right) + C$

23. $2\arctan\sqrt{x} + C$ **25.** $\dfrac{\pi}{4}$ **27.** $\dfrac{\pi}{2}$

29. $\ln|x^2 + 6x + 13| - 3\arctan\left(\dfrac{x + 3}{2}\right) + C$

31. $\arcsin\left(\dfrac{x + 2}{2}\right) + C$

33. $-\sqrt{-x^2 - 4x} + C$ **35.** $4 - 2\sqrt{3} + \dfrac{\pi}{6} \approx 1.059$

37. $\dfrac{1}{2}\arctan(x^2 + 1) + C$
39. $\dfrac{1}{4}\arcsin(4x - 2) + C$
41. $2(\sqrt{x - 1} - \arctan\sqrt{x - 1}) + C$
43. $2\sqrt{e^t - 3} - 2\sqrt{3}\arctan\left(\dfrac{\sqrt{e^t - 3}}{\sqrt{3}}\right) + C$

45. $\dfrac{\pi}{4}$ **47.** $\dfrac{\pi}{8}$

49. $\sqrt{\dfrac{32}{k}}\tan\left[\arctan\left(500\sqrt{\dfrac{k}{32}}\right) - \sqrt{32k}\,t\right]$

51. (b) 3.14159   **53. (a)** arcsin $x + C$
  (b) $-\sqrt{1 - x^2} + C$.
  (c) Does not fit Basic Formulas
55. (a) $\frac{2}{3}(x - 1)^{3/2} + C$
  (b) $\frac{2}{15}(x - 1)^{3/2}(3x + 2) + C$
  (c) $\frac{2}{3}\sqrt{x - 1}(x + 2) + C$

Section 6.9

1. (a) 10.018   **3. (a)** $\frac{4}{3}$ **5. (a)** 1.317
  (b) -0.964 **(b)** $\frac{13}{12}$ **(b)** 0.962
7. $\left(\dfrac{e^x - e^{-x}}{e^x + e^{-x}}\right)^2 + \left(\dfrac{2}{e^x + e^{-x}}\right)^2 = 1$
9. $\left(\dfrac{e^x - e^{-x}}{2}\right)\left(\dfrac{e^y + e^{-y}}{2}\right) + \left(\dfrac{e^x + e^{-x}}{2}\right)\left(\dfrac{e^y - e^{-y}}{2}\right)$
  $= \dfrac{e^{(x+y)} - e^{-(x+y)}}{2} = \sinh(x + y)$
11. $\left(\dfrac{e^x - e^{-x}}{2}\right)\left[3 + 4\left(\dfrac{e^x - e^{-x}}{2}\right)^2\right]$
  $= \dfrac{e^{3x} - e^{-3x}}{2} = \sinh(3x)$
13. $-2x \cosh(1 - x^2)$ **15.** $\coth x$
17. $\operatorname{csch} x$ **19.** $\sinh^2 x$ **21.** $\operatorname{sech} x$
23. $\dfrac{y}{x}[\cosh x + x(\sinh x) \ln x]$ **25.** $-2e^{-2x}$
27. $\dfrac{3}{\sqrt{9x^2 - 1}}$ **29.** $|\sec x|$ **31.** 2 sec 2x
33. $2 \sinh^{-1}(2x)$ **35.** $-\frac{1}{2}\cosh(1 - 2x) + C$
37. $\frac{1}{3}\cosh^3(x - 1) + C$ **39.** $\ln|\sinh x| + C$
41. $-\coth\dfrac{x^2}{2} + C$ **43.** $\operatorname{csch}\dfrac{1}{x} + C$
45. $\dfrac{1}{5}\ln 3$ **47.** $\dfrac{\pi}{4}$ **49.** $\dfrac{1}{2}\arctan x^2 + C$
51. $-\operatorname{csch}^{-1}(e^x) + C = -\ln\left(\dfrac{1 + \sqrt{1 + e^{2x}}}{e^x}\right) + C$
53. $2 \sinh^{-1}\sqrt{x} + C = 2 \ln(\sqrt{x} + \sqrt{1 + x}) + C$
55. $-\ln\left[\dfrac{1 + \sqrt{(x - 1)^2 + 1}}{|x - 1|}\right] + C$
57. $\dfrac{1}{2\sqrt{6}}\ln\left|\dfrac{\sqrt{2}(x + 1) + \sqrt{3}}{\sqrt{2}(x + 1) - \sqrt{3}}\right| + C$
59. $\dfrac{1}{4}\arcsin\left(\dfrac{4x - 1}{9}\right) + C$
61. $-\dfrac{x^2}{2} - 4x - \dfrac{10}{3}\ln\left|\dfrac{x - 5}{x + 1}\right| + C$
63. Point of inflection: (0, 0)

65. $y''' - y' = a \cosh x - a \cosh x = 0$
67. $-\dfrac{\sqrt{a^2 - x^2}}{x}$ **69.** $\dfrac{52}{31}$ kg   **71.** $2a \sinh\dfrac{b}{a}$

Review Exercises for Chapter 6

1. $f^{-1}(x) = 2x + 6$ **3.** $f^{-1}(x) = x^2 - 1$, $x \geq 0$

5. $f^{-1}(x) = \sqrt{x + 5}$

7. $f^{-1}(x) = \sqrt{\dfrac{x}{2} + 4}$, $x \geq 4$ **9.** $x = 3$
11. $x = 5 \pm \sqrt{7}$ **13.** $\dfrac{1}{2x}$ **15.** $\dfrac{1 + 2 \ln x}{2\sqrt{\ln x}}$
17. $\dfrac{-y}{x(2y + \ln x)}$ **19.** $\dfrac{x}{(a + bx)^2}$ **21.** $\dfrac{1}{x(a + bx)}$
23. $-2x$ **25.** $xe^x(x + 2)$ **27.** $\dfrac{e^{2x} - e^{-2x}}{\sqrt{e^{2x} + e^{-2x}}}$
29. $3^{x-1} \ln 3$ **31.** $\dfrac{y - ye^x - e^y}{e^x + xe^y - x}$
33. $-\dfrac{2x \sin x^2 + e^y}{xe^y} = -2x \tan x^2 - \dfrac{1}{x}$
35. $(1 - x^2)^{-3/2}$ **37.** $\dfrac{x}{|x|\sqrt{x^2 - 1}} + \operatorname{arcsec} x$
39. $(\arcsin x)^2$ **41. (a)** axa^{-1} **(b)** $(\ln a)a^x$
  (c) $x^x(1 + \ln x)$ **(d)** 0
43. $\frac{1}{7}\ln|7x - 2| + C$ **45.** $-\ln|1 + \cos x| + C$
47. $\dfrac{x^2}{2} + 3 \ln|x| + C$ **49.** $3 + \ln 4$
51. $\ln(2 + \sqrt{3})$ **53.** $-\frac{1}{6}e^{-3x^2} + C$
55. $\dfrac{e^{4x} - 3e^{2x} - 3}{3e^x} + C$ **57.** $\ln|e^x - 1| + C$
59. $-e^{-x^2/2} + C$ **61.** $-\frac{1}{2}\ln(1 + e^{-2x}) + C$
63. $\frac{1}{2}\arctan(e^{2x}) + C$ **65.** $\frac{1}{2}\arcsin x^2 + C$

67. $\dfrac{1}{2}\ln(16 + x^2) + C$ **69.** $\dfrac{1}{4}\left(\arctan\dfrac{x}{2}\right)^2 + C$

71. $\ln 2$ **73.** $\frac{1}{2}(1 - e^{-16}) \approx 0.500$

75. 1.346

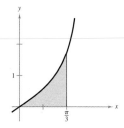

77. (a) $525.64   **(b)** $824.36   **(c)** $74,206.58

79. $3,499.38   **81. (a)** 27.73 yrs   **(b)** 43.94 yrs

83. (a) -24.26% **(b)** -14.72%

85. (a) 0.3935   **(b)** 0.7769

 (c) 0.2492   **(d)** 0.9502

87. (a) 0.60   **(b)** 0.85

91. $v(t) = \dfrac{1}{k}[32 + (kv_0 - 32)e^{kt}]$

93. $s(t) = \dfrac{32}{k}t - \dfrac{32}{k^2}(1 - e^{kt}) + s_0$

Chapter 7

Section 7.1

1. $A = 36$ **3.** $A = 9$ **5.** $A = \frac{3}{2}$

7. $A = \frac{32}{3}$ **9.** $A = \frac{9}{2}$

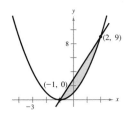

11. $A = 1$ **13.** $A = \frac{500}{27}$

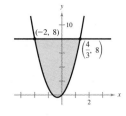

15. $A = 2$ **17.** $A = \frac{3}{2}$

19. $A = \frac{64}{3}$ **21.** $A = \frac{9}{2}$

23. $A = 6$

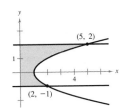

25. $A = \dfrac{7 + \sqrt{13}}{3} - \ln\dfrac{3 + \sqrt{13}}{2} \approx 2.34$

27. $A = \dfrac{\pi}{2} - \dfrac{1}{3} \approx 1.237$

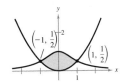

29. $A = 2(1 - \ln 2) \approx 0.614$

31. $A = 4$

33. $A = \dfrac{1}{2}\left(1 - \dfrac{1}{e}\right) \approx 0.316$

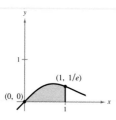

35. $A = 3 \ln 10 \approx 6.908$ **37.** $A = 8$

39. $A = \frac{1}{2}ac$

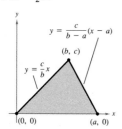

41. $b = 9\left(1 - \dfrac{1}{\sqrt[3]{4}}\right) \approx 3.330$

43. $x^4 - 2x^2 + 1 \le 1 - x^2$ on $[-1, 1]$

$$A = \int_{-1}^{1} [(1 - x^2) - (x^4 - 2x^2 + 1)]\, dx = \dfrac{4}{15}$$

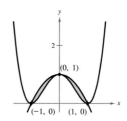

45. $A = \displaystyle\int_{-2}^{1} [x^3 - (3x - 2)]\, dx = \dfrac{27}{4}$

47. 1.760   **49.** $\frac{1}{6}$

51. \$1.625 billion   **53.** Consumer surplus $= 1600$
Producer surplus $= 400$

55. Consumer surplus $= 50,000$ **57.** 4.773
Producer surplus $= 25,497$

Section 7.2

1. $\dfrac{\pi}{3}$ **3.** $\dfrac{16\pi}{3}$ **5.** $\dfrac{15\pi}{2}$ **7.** $\dfrac{2\pi}{35}$ **9.** 8π

11. $\dfrac{\pi}{4}$

13. (a) 8π **(b)** $\dfrac{128\pi}{5}$ **(c)** $\dfrac{256\pi}{15}$ **(d)** $\dfrac{192\pi}{5}$

15. (a) $\dfrac{32\pi}{3}$ **(b)** $\dfrac{64\pi}{3}$ **17.** 18π

19. $\pi\left(-\frac{3}{4} + 8 \ln 4\right) \approx 32.49$ **21.** $\dfrac{208\pi}{3}$

23. $\dfrac{384\pi}{5}$ **25.** $\pi \ln 4$ **27.** $\dfrac{3\pi}{4}$

29. $\dfrac{\pi}{2}\left(1 - \dfrac{1}{e^2}\right) \approx 1.358$ **31.** π **33.** $\dfrac{512\pi}{15}$

35. 60π **37.** 18π **41.** $\pi r^2 h\left(1 - \dfrac{h}{H} + \dfrac{h^2}{3H^2}\right)$

45. $\dfrac{\pi}{30}$ **47.** 1.959

49. (a) $\dfrac{128}{3}$ **(b)** $\dfrac{32\sqrt{3}}{3}$ **(c)** $\dfrac{16\pi}{3}$ **(d)** $\dfrac{32}{3}$

51. (a) $\dfrac{1}{10}$ **(b)** $\dfrac{\pi}{80}$ **(c)** $\dfrac{\sqrt{3}}{40}$ **(d)** $\dfrac{3}{80}$ **(e)** $\dfrac{\pi}{20}$

53. $\dfrac{2r^3}{3}$ **55.** 49.023

Section 7.3

1. $\dfrac{16\pi}{3}$ **3.** $\dfrac{8\pi}{3}$ **5.** $\dfrac{128\pi}{5}$ **7.** 8π **9.** $\dfrac{16\pi}{3}$

11. 16π **13.** 64π **15.** $\dfrac{8\pi}{3}$ **17.** $\dfrac{192\pi}{5}$

19. $\sqrt{2\pi}\left(1 - \dfrac{1}{\sqrt{e}}\right) \approx 0.986$

21. (a) $\dfrac{128\pi}{7}$ **(b)** $\dfrac{64\pi}{5}$ **(c)** $\dfrac{96\pi}{5}$ **(d)** $\dfrac{320\pi}{7}$

23. (a) $\dfrac{\pi a^3}{15}$ **(b)** $\dfrac{\pi a^3}{15}$ **(c)** $\dfrac{4\pi a^3}{15}$

25. Diameter $= 2\sqrt{4 - 2\sqrt{3}} \approx 1.464$

29. 2.97 (Shell), 3.01 (Disc)

31. 122,313 ft^3 **33.** $4\pi^2$ **35.** 186.055

Section 7.4

1. 13   **3.** $\frac{2}{3}(\sqrt{8} - 1) \approx 1.219$ **5.** $\frac{33}{16}$

7. $\frac{779}{240}$ **9.** $s = \displaystyle\int_{-2}^{1} \sqrt{2 + 4x + 4x^2}\, dx$

11. $s = \displaystyle\int_{0}^{2} \sqrt{1 + 4x^2}\, dx$ **13.** $s = \displaystyle\int_{0}^{2} \dfrac{\sqrt{e^{2y} + 1}}{e^y}\, dy$

15. 2.1517   **17.** 3.829   **19.** $\frac{2}{3}$

21. $3 \arcsin \frac{2}{3} \approx 2.1892$

23. $\dfrac{\pi}{9}(82\sqrt{82} - 1) \approx 258.85$ **25.** $\dfrac{47\pi}{16}$

27. $\dfrac{\pi}{27}(145\sqrt{145} - 10\sqrt{10}) \approx 199.48$

29. $S = 2\pi \displaystyle\int_{0}^{\pi} \sin x \sqrt{1 + \cos^2 x}\, dx$

33. $6\pi(3 - \sqrt{5}) \approx 14.40$

35. Surface area $= \dfrac{\pi}{27}$ ft$^2 \approx 16.8$ in.^{2} **39.** 50.86

Amount of glass $= \dfrac{\pi}{27}\left(\dfrac{0.015}{12}\right)$

≈ 0.00015 ft^3

≈ 0.25 in.^{3}

Section 7.5

1. 1000 ft · lb   **3.** 300 ft · lb

5. 30.625 in · lb ≈ 2.55 ft · lb

7. 360 in · lb $= 30$ ft · lb

9. 180 in · lb $= 15$ ft · lb

11. (a) 2496 ft · lb   **(b)** 9984 ft · lb

13. 431,308.8π ft · lb   **15.** 20,217.6π ft · lb

17. 2995.2π ft · lb   **19.** 2457π ft · lb

21. (a) 761.905 mi · ton $\approx 8,046,000,000$ ft · lb

     (b) 1,454.545 mi · ton $\approx 15,360,000,000$ ft · lb

23. (a) 2.93×10^4 mi · tons $\approx 3.10 \times 10^{11}$ ft · lb

     (b) 3.38×10^4 mi · tons $\approx 3.57 \times 10^{11}$ ft · lb

25. $\dfrac{3k}{4}$ **27.** 337.5 ft · lb   **29.** 300 ft · lb

31. 168.75 ft · lb   **33.** 7987.5 ft · lb

35. $2000 \ln \frac{3}{2} \approx 810.93$ ft · lb   **37.** 3250 lb

Section 7.6

1. 936 lb   **3.** 748.8 lb   **5.** 1123.2 lb

7. 748.8 lb   **9.** 1064.96 lb   **11.** 748.8 lb

13. 15,163.2 lb   **15.** 2814 lb   **17.** 3376.8 lb

19. 94.5 lb   **23.** 960 lb

25. 9,984 lb (on the wall at the shallow end)

     39,936 lb (on the wall at the deep end)

     46,592 lb (on the side walls)

27. 6483 lb

Section 7.7

1. $\bar{x} = -\frac{6}{7}$ **3.** $\bar{x} = 12$ **5.** $\bar{x} = 17$

7. $(\bar{x}, \bar{y}) = \left(\frac{10}{9}, -\frac{1}{9}\right)$ **9.** $(\bar{x}, \bar{y}) = \left(-\frac{7}{8}, -\frac{7}{16}\right)$

11. $(\bar{x}, \bar{y}) = \left(\dfrac{4 + 3\pi}{4 + \pi}, 0\right)$ **13.** $(\bar{x}, \bar{y}) = \left(0, \dfrac{135}{34}\right)$

15. $(\bar{x}, \bar{y}) = \left(\dfrac{2 + 3\pi}{2 + \pi}, 0\right)$

17. $M_x = 4\rho,\ M_y = \dfrac{64\rho}{5},\ (\bar{x}, \bar{y}) = \left(\dfrac{12}{5}, \dfrac{3}{4}\right)$

19. $M_x = \dfrac{\rho}{12},\ M_y = \dfrac{\rho}{15},\ (\bar{x}, \bar{y}) = \left(\dfrac{2}{5}, \dfrac{1}{2}\right)$

21. $M_x = \dfrac{99\rho}{5},\ M_y = \dfrac{27\rho}{4},\ (\bar{x}, \bar{y}) = \left(\dfrac{3}{2}, \dfrac{22}{5}\right)$

23. $M_x = 0,\ M_y = \dfrac{256\rho}{15},\ (\bar{x}\ \bar{y}) = \left(\dfrac{8}{5}, 0\right)$

25. $M_x = \dfrac{27\rho}{4},\ M_y = -\dfrac{27\rho}{10},\ (\bar{x}, \bar{y}) = \left(-\dfrac{3}{5}, \dfrac{3}{2}\right)$

27. $M_x = \dfrac{192\rho}{7},\ M_y = 96\rho,\ (\bar{x}, \bar{y}) = \left(5, \dfrac{10}{7}\right)$

29. $(\bar{x}, \bar{y}) = \left(0, \dfrac{4}{3\pi}\right)$ **31.** $(\bar{x}, \bar{y}) = \left(\dfrac{1}{2}, \dfrac{2}{5}\right)$

33. $(\bar{x}, \bar{y}) = \left(\dfrac{12}{7}, \dfrac{26}{7}\right)$ **35.** $(\bar{x}, \bar{y}) = \left(\dfrac{b}{3}, \dfrac{c}{3}\right)$

37. $(\bar{x}, \bar{y}) = (3.0, 127.6)$ **39.** $\bar{y} = 16.5$ ft

41. $V = 160\pi^2 \approx 1579.14$

43. $V = \dfrac{128\pi}{3}$ **45.** $(\bar{x}, \bar{y}) = \left(0, \dfrac{2r}{\pi}\right)$

Review Exercises for Chapter 7

1. $A = \dfrac{4}{5}$ **3.** $A = \dfrac{\pi}{2}$

5. $A = \frac{4}{3}$ **7.** $A = \frac{1}{2}$

9. $A = \frac{512}{3}$ **11.** $A = \frac{14}{3}$

13. $A = \frac{1}{6}$ **15.** $A = e^2 + 1$

17. $A = 2\sqrt{2}$

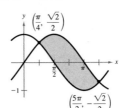

19. (a) $\dfrac{64\pi}{3}$

(b) $\dfrac{128\pi}{3}$

(c) $\dfrac{64\pi}{3}$

(d) $\dfrac{160\pi}{3}$

21. (a) 64π **(b)** 48π **23.** $\dfrac{\pi^2}{4}$ **25. (a)** $\dfrac{512\pi}{15}$

(b) 64π

27. 50 in · lb ≈ 4.167 ft · lb

29. 104,000π ft · lb ≈ 163.4 ft · ton   **31.** 250 ft · lb

33. 72,800 lb (on side walls)

62,400 lb (on wall at deep end)

15,600 lb (on wall at shallow end)

35. 4992π lb   **37.** $(\bar{x}, \bar{y}) = \left(\dfrac{a}{5}, \dfrac{a}{5}\right)$

39. $(\bar{x}, \bar{y}) = \left(0, \dfrac{2a^2}{5}\right)$

41. $s = \displaystyle\int_0^{\sqrt{3}} \dfrac{4}{\sqrt{4-x^2}}\,dx = \dfrac{4\pi}{3}$ **43.** 15π

45. $\dfrac{4}{15}$ **47.** $\dfrac{32\pi}{105}$ **49.** $\dfrac{8}{15}(1 + 6\sqrt{3}) \approx 6.076$

Chapter 8

Section 8.1

1. $\frac{1}{15}(3x-2)^5 + C$ **3.** $-\frac{1}{5}(-2x+5)^{5/2} + C$

5. $\dfrac{1}{2}v^2 - \dfrac{1}{6(3v-1)^2} + C$

7. $-\frac{1}{3}\ln|-t^3 + 9t + 1| + C$

9. $\frac{1}{2}x^2 + x + \ln|x - 1| + C$

11. $\dfrac{1}{3}\ln\left|\dfrac{3x-1}{3x+1}\right| + C$ **13.** $-\dfrac{1}{2}\cos t^2 + C$

15. $e^{\sin x} + C$ **17.** $-e^{-t} + 2t + e^t + C$

19. $\frac{1}{3}\sec 3x + C$ **21.** $2\ln(1 + e^x) + C$

23. $-\cot x - \csc x + C$

25. $\ln(t^2 + 4) - \dfrac{1}{2}\arctan\dfrac{t}{2} + C$

27. $3\arctan t + C$ **29.** $\dfrac{1}{2}\operatorname{arcsec}\dfrac{|x|}{2} + C$

31. $-\dfrac{1}{2}\arcsin(2t-1) + C$ **33.** $-\dfrac{1}{\pi}\csc(\pi x) + C$

35. $\dfrac{1}{2}\arcsin t^2 + C$ **37.** $\dfrac{1}{2}\arctan\dfrac{\tan x}{2} + C$

39. $\dfrac{1}{2}\ln\left|\cos\dfrac{2}{t}\right| + C$

41. $\dfrac{x}{15}(12x^4 + 20x^2 + 15) + C$

43. $\dfrac{1}{2}x^2 + 3x + 3\ln|x| - \dfrac{1}{x} + C$

45. $\dfrac{1}{2}e^{2x} + 2e^x + x + C$ **47.** $3\arcsin\dfrac{x-3}{3} + C$

49. $\dfrac{1}{4}\arctan\dfrac{2x+1}{8} + C$ **51.** $\dfrac{1}{2}(1 - e^{-1}) \approx 0.316$

53. $\dfrac{1}{2}$ **55.** 4 **57.** $\dfrac{1}{2}\ln 2$

59. $\dfrac{1}{2}\left(\text{arcsec } 4 - \dfrac{\pi}{3}\right) \approx 0.135$ **61.** $\dfrac{4}{3}$

63. $a = \dfrac{1}{2}$ **65.** $b = \sqrt{\ln\left(\dfrac{3\pi}{3\pi-4}\right)} \approx 0.743$

67. $\dfrac{2}{\text{arcsin}(4/5)} \approx 2.157$ **69.** 1.0320

59. $\dfrac{e^{2x}}{13}(2\cos 3x + 3\sin 3x) + C$

61. $1 - \dfrac{5}{e^4} \approx 0.908$ **63.** $\dfrac{\pi}{1+\pi^2}\left(\dfrac{1}{e} + 1\right) \approx 0.395$

65. (a) 1 (b) $\pi(e - 2) \approx 2.257$

(c) $\dfrac{(e^2+1)\pi}{2} \approx 13.177$

(d) $\left(\dfrac{e^2+1}{4}, \dfrac{e-2}{2}\right) \approx (2.097, 0.359)$

67. (a) $3.2(\ln 2) - 0.2 \approx 2.018$

(b) $12.8(\ln 4) - 7.2(\ln 3) - 1.8 \approx 8.035$

69. \$771,721.44 **71.** $\dfrac{8h}{(n\pi)^2}\sin\left(\dfrac{n\pi}{2}\right)$

73. 125.5 mi

75. For any integrable function, **77.** 5.739
$\int f(x)\,dx = C + \int f(x)\,dx$,
but this does not imply that $C = 0$.

Section 8.2

1. $\dfrac{e^{2x}}{4}(2x - 1) + C$ **3.** $\dfrac{1}{2}e^{x^2} + C$

5. $-\dfrac{1}{4e^{2x}}(2x + 1) + C$

7. $e^x(x^3 - 3x^2 + 6x - 6) + C$

9. $\dfrac{x^4}{16}(4\ln x - 1) + C$

11. $\dfrac{1}{4}[2(t^2 - 1)\ln|t + 1| - t^2 + 2t] + C$

13. $x(\ln x)^2 - 2x\ln x + 2x + C$

15. $\dfrac{(\ln x)^3}{3} + C$ **17.** $\dfrac{e^{2x}}{4(2x+1)} + C$

19. $(x - 1)^2 e^x + C$ **21.** $\dfrac{2(x-1)^{3/2}}{15}(3x + 2) + C$

23. $\dfrac{2}{3}x^2(2 + 3x)^{1/2} - \dfrac{8}{27}x(2 + 3x)^{3/2} + \dfrac{16}{405}(2 + 3x)^{5/2}$
$= \dfrac{2}{405}(27x^2 - 24x + 32)\sqrt{2 + 3x} + C$

25. $x\sin x + \cos x + C$

27. $x\tan x + \ln|\cos x| + C$

29. $x\arcsin 2x + \dfrac{1}{2}\sqrt{1 - 4x^2} + C$

31. $x\arctan x - \dfrac{1}{2}\ln(1 + x^2) + C$

33. $\dfrac{1}{5}e^{2x}(2\sin x - \cos x) + C$

35. $-\dfrac{\pi}{2}$ **37.** $\dfrac{e[\sin(1) - \cos(1)] + 1}{2} \approx 0.909$

39. $\dfrac{\pi}{2} - 1$ **41.** $\dfrac{e^{2x}}{4}(2x^2 - 2x + 1) + C$

43. $(3x^2 - 6)\sin x - (x^3 - 6x)\cos x + C$

45. $x\tan x + \ln|\cos x| + C$

47. $\dfrac{2}{5}(2x - 3)^{3/2}(x + 1) + C$

49. $\dfrac{1}{3}\sqrt{4 + x^2}(x^2 - 8) + C$

57. $\dfrac{x^4}{16}(4\ln x - 1) + C$

Section 8.3

1. $-\dfrac{1}{4}\cos^4 x + C$ **3.** $\dfrac{1}{12}\sin^6 2x + C$

5. $-\dfrac{1}{3}\cos^3 x + \dfrac{2}{5}\cos^5 x - \dfrac{1}{7}\cos^7 x + C$

7. $\dfrac{1}{12}(6x + \sin 6x) + C$

9. $\dfrac{1}{32\pi}(12\pi x - 8\sin 2\pi x + \sin 4\pi x) + C$

11. $\dfrac{1}{32}(4x - \sin 4x) + C$

13. $\dfrac{1}{8}(2x^2 - 2x\sin 2x - \cos 2x) + C$

15. $\dfrac{1}{3}\ln|\sec 3x + \tan 3x| + C$

17. $\dfrac{1}{15}\tan 5x(3 + \tan^2 5x) + C$

19. $\dfrac{1}{2\pi}(\sec \pi x\tan \pi x + \ln|\sec \pi x + \tan \pi x|) + C$

21. $-\dfrac{\tan^2(1-x)}{2} - \ln|\cos(1 - x)| + C$

23. $\tan^4\left(\dfrac{x}{4}\right) - 2\tan^2\left(\dfrac{x}{4}\right) - 4\ln\left|\cos\dfrac{x}{4}\right| + C$

25. $\dfrac{1}{2}\tan^2 x + C$ **27.** $\dfrac{\tan^3 x}{3} + C$

29. $\dfrac{1}{5\pi}\sec^5 \pi x + C$ **31.** $\dfrac{\sec^6 4x}{24} + C$

33. $\dfrac{1}{3}\sec^3 x + C$ **35.** $\dfrac{1}{9}\sec^3 3x - \dfrac{1}{3}\sec 3x + C$

37. $\dfrac{1}{4}[\ln|\csc^2(2x)| - \cot^2(2x)] + C$

39. $-\cot\theta - \dfrac{1}{3}\cot^3\theta + C$

41. $\ln|\csc t - \cot t| + \cos t + C$

43. $-\dfrac{1}{10}(\cos 5x + 5\cos x) + C$

45. $\dfrac{1}{8}(2\sin 2\theta - \sin 4\theta) + C$

47. $\ln|\csc x - \cot x| + \cos x + C$

49. $t - 2\tan t + C$ **51.** π **53.** $\dfrac{1}{2}(1 - \ln 2)$

55. $\ln 2$ **57.** $\dfrac{3\sqrt{2}}{10}$ **59.** $\dfrac{4}{3}$

61. $\dfrac{\tan^6 3x}{18} + \dfrac{\tan^4 3x}{12} + C_1 = \dfrac{\sec^6 3x}{18} - \dfrac{\sec^4 3x}{12} + C_2$

63. $\dfrac{1}{2}$ **65.** $2\pi\left(1 - \dfrac{\pi}{4}\right) \approx 1.348$

67. (a) $\dfrac{\pi^2}{2}$ **(b)** $(\bar{x}, \bar{y}) = \left(\dfrac{\pi}{2}, \dfrac{\pi}{8}\right)$

79. $-\dfrac{1}{15}\cos x(3\sin^4 x + 4\sin^2 x + 8) + C$

81. $\dfrac{1}{6}\cos^5 x \sin x + \dfrac{5}{24}\cos^3 x \sin x$
$\qquad\qquad + \dfrac{5}{16}\cos x \sin x + \dfrac{5}{16}x + C$

83. $\dfrac{1}{8\pi}(-2\cos^3 \pi x \sin \pi x + \cos \pi x \sin \pi x + \pi x) + C$

87. 8.586

Section 8.4

1. $\dfrac{x}{25\sqrt{25 - x^2}} + C$

3. $5\ln\left|\dfrac{5 - \sqrt{25 - x^2}}{x}\right| + \sqrt{25 - x^2} + C$

5. $\ln|x + \sqrt{x^2 - 4}| + C$

7. $\dfrac{1}{15}(x^2 - 4)^{3/2}(3x^2 + 8) + C$ **9.** $\dfrac{1}{3}(1 + x^2)^{3/2} + C$

11. $\dfrac{1}{2}\left(\arctan x + \dfrac{x}{1 + x^2}\right) + C$ **13.** $\sqrt{x^2 + 9} + C$

15. 2π **17.** $\ln|x + \sqrt{x^2 - 9}| + C$

19. $\sqrt{3} - \dfrac{\pi}{3} \approx 0.685$ **21.** $-\dfrac{(1 - x^2)^{3/2}}{3x^3} + C$

23. $-\dfrac{1}{3}\ln\left|\dfrac{\sqrt{4x^2 + 9} + 3}{2x}\right| + C$

25. $-\dfrac{1}{\sqrt{x^2 + 3}} + C$ **27.** $\dfrac{1}{3}(1 + e^{2x})^{3/2} + C$

29. $\dfrac{1}{3}(x^2 + 2x + 2)^{3/2} + C$

31. $\sqrt{x^2 + 4x + 8}$
$\qquad - 2\ln|\sqrt{x^2 + 4x + 8} + (x + 2)| + C$

33. $\dfrac{1}{2}(x - 15)\sqrt{x^2 + 10x + 9}$
$\qquad + 33\ln|\sqrt{x^2 + 10x + 9} + (x + 5)| + C$

35. $\dfrac{1}{2}(\arcsin e^x + e^x\sqrt{1 - e^{2x}}) + C$

37. $\dfrac{1}{4}\left(\dfrac{x}{x^2 + 2} + \dfrac{1}{\sqrt{2}}\arctan\dfrac{x}{\sqrt{2}}\right) + C$

39. $\dfrac{1}{6}(3x\sqrt{9x^2 + 4} + 4\ln|3x + \sqrt{9x^2 + 4}|) + C$

41. $\dfrac{1}{2}(x\sqrt{x^2 - 1} + \ln|x + \sqrt{x^2 - 1}|) + C$

43. $x\,\text{arcsec}\,2x - \dfrac{1}{2}\ln|2x + \sqrt{4x^2 - 1}| + C$

45. $9(2 - \sqrt{2})$ **47.** $1 - \dfrac{\sqrt{3}\pi}{6} \approx 0.093$

49. 187.2π lb   **51.** πr^2 **53.** $6\pi^2$

55. $\ln\left[\dfrac{5(\sqrt{2} + 1)}{\sqrt{26} + 1}\right] + \sqrt{26} - \sqrt{2} \approx 4.367$

57. $100\sqrt{2} + 50\ln\left(\dfrac{\sqrt{2} + 1}{\sqrt{2} - 1}\right) \approx 229.559$

59. $\dfrac{\pi}{32}[102\sqrt{2} - \ln(3 + 2\sqrt{2})] \approx 13.989$

61. 121.3 lb   **63.** $(\bar{x}, \bar{y}) \approx (5.30, 1.43)$

Section 8.5

1. $\dfrac{1}{2}\ln\left|\dfrac{x - 1}{x + 1}\right| + C$ **3.** $\ln\left|\dfrac{x - 1}{x + 2}\right| + C$

5. $\dfrac{3}{2}\ln|2x - 1| - 2\ln|x + 1| + C$

7. $5\ln|x - 2| - \ln|x + 2| - 3\ln|x| + C$

9. $x^2 + \dfrac{3}{2}\ln|x - 4| - \dfrac{1}{2}\ln|x + 2| + C$

11. $\dfrac{1}{x} + \ln|x^4 + x^3| + C$

13. $\dfrac{1}{2}x^2 + 3x + 6\ln|x - 1| - \dfrac{4}{x - 1} - \dfrac{1}{2(x - 1)^2} + C$

15. $3\ln|x - 3| - \dfrac{9}{x - 3} + C$

17. $\ln\left|\dfrac{x^2 + 1}{x}\right| + C$

19. $\dfrac{1}{6}\left[\ln\left|\dfrac{x - 2}{x + 2}\right| + \sqrt{2}\arctan\left(\dfrac{x}{\sqrt{2}}\right)\right] + C$

21. $\dfrac{1}{16}\ln\left|\dfrac{4x^2 - 1}{4x^2 + 1}\right| + C$

23. $\dfrac{\sqrt{2}}{2}\arctan\dfrac{x}{\sqrt{2}} - \dfrac{1}{2(x^2 + 2)} + C$

25. $\ln|x + 1| + \sqrt{2}\arctan\left(\dfrac{x - 1}{\sqrt{2}}\right) + C$

27. $4\ln|x - 2| + \ln(x^2 + 4) + \dfrac{1}{2}\arctan\dfrac{x}{2} + C$

29. $\ln|x - 2| + \dfrac{1}{2}\ln|x^2 + x + 1|$
$\qquad\qquad - \sqrt{3}\arctan\left(\dfrac{2x + 1}{\sqrt{3}}\right) + C$

31. $\ln 2$ **33.** $\dfrac{1}{2}\ln\left(\dfrac{8}{5}\right) - \dfrac{\pi}{4} + \arctan(2) \approx 0.557$

35. $\ln 2 + \arctan 2 - \arctan 3$

37. $\ln\left|\dfrac{\cos x}{\cos x - 1}\right| + C$ **39.** $\ln\left|\dfrac{-1 + \sin x}{2 + \sin x}\right| + C$

41. $\dfrac{1}{5}\ln\left|\dfrac{e^x - 1}{e^x + 4}\right| + C$ **47.** $6 - \dfrac{7}{4}\ln 7 \approx 2.595$

49. $2\pi\left[\arctan 3 - \dfrac{3}{10}\right] \approx 5.963$

51. $x = \dfrac{n[e^{(n+1)kt} - 1]}{n + e^{(n+1)kt}}$

53. (a) As $t \to \infty$, $x(t) \to y_0 = z_0$ **55.** 14.666
(b) $y_0 > z_0$: As $t \to \infty$,
$\qquad\qquad x(t) \to z_0$
$\qquad y_0 = z_0$: As $t \to \infty$,
$\qquad\qquad x(t) \to y_0 = z_0$

Section 8.6

1. $-\dfrac{1}{2}x(2 - x) + \ln|1 + x| + C$

3. $-\dfrac{\sqrt{1 - x^2}}{x} + C$ **5.** $\dfrac{1}{9}x^3(-1 + 3\ln x) + C$

7. $\dfrac{\sqrt{x^2 - 4}}{4x} + C$ **9.** $\dfrac{1}{2}e^{x^2} + C$

11. $\dfrac{2}{9}\left(\ln|1 - 3x| + \dfrac{1}{1 - 3x}\right) + C$

13. $e^x \arccos(e^x) - \sqrt{1 - e^{2x}} + C$

15. $\frac{1}{16}x^4(4 \ln|x| - 1) + C$

17. $\frac{1}{27}\left(3x - \frac{25}{3x - 5} + 10 \ln|3x - 5|\right) + C$

19. $\frac{1}{2}(x^2 + \cot x^2 + \csc x^2) + C$

21. $\arctan(\sin x) + C$ **23.** $x - \frac{1}{2}\ln(1 + e^{2x}) + C$

25. $\frac{\sqrt{2}}{2}\arctan\left(\frac{1 + \sin \theta}{\sqrt{2}}\right) + C$

27. $-\frac{\sqrt{2 + 9x^2}}{2x} + C$

29. $\frac{1}{2}(e^x\sqrt{e^{2x} + 1} + \ln|e^x + \sqrt{e^{2x} + 1}|) + C$

31. $\frac{1}{16}(6x - 3\sin 2x \cos 2x - 2\sin^3 2x \cos 2x) + C$

33. $-2(\cot \sqrt{x} + \csc \sqrt{x}) + C$

35. $(t^4 - 12t^2 + 24)\sin t + (4t^3 - 24t)\cos t + C$

37. $\frac{1}{2}[(x^2 + 1)\operatorname{arcsec}(x^2 + 1)$
$\qquad\qquad - \ln|(x^2 + 1) + \sqrt{x^4 + 2x^2}|] + C$

39. $\frac{1}{4}(2 \ln|x| - 3 \ln|3 + 2 \ln|x||) + C$

41. $\sqrt{2 - 2x - x^2}$
$\qquad\qquad - \sqrt{3} \ln\left|\frac{\sqrt{3} + \sqrt{2 - 2x - x^2}}{x + 1}\right| + C$

43. $\frac{1}{2}\arctan(x^2 - 3) + C$

45. $\frac{1}{2}\ln|(x^2 - 3) + \sqrt{x^4 - 6x^2 + 5}| + C$

47. $-\frac{1}{3}\sqrt{4 - x^2}(x^2 + 8) + C$ **49.** $-\frac{2\sqrt{1 - x}}{\sqrt{x}} + C$

51. $\frac{2}{1 + e^x} - \frac{1}{2(1 + e^x)^2} + \ln|1 + e^x| + C$

59. $\frac{1}{\sqrt{5}}\ln\left|\frac{2\tan(\theta/2) - 3 - \sqrt{5}}{2\tan(\theta/2) - 3 + \sqrt{5}}\right| + C$

61. $\ln 2$ **63.** $\frac{1}{2}\ln(3 - 2\cos \theta) + C$

65. $2\sin \sqrt{\theta} + C$ **67.** $-\csc \theta + C$

69. 1919.145 ft · lb

Section 8.7

1. 3   **3.** 0   **5.** 2   **7.** $n = 1$: 0   **9.** 0
$\qquad\qquad\qquad\qquad n = 2$: $\frac{1}{2}$
$\qquad\qquad\qquad\qquad n \geq 3$: ∞

11. $\frac{3}{2}$ **13.** ∞ **15.** 0   **17.** $-\frac{3}{2}$ **19.** 1

21. 0   **23.** 1   **25.** 1   **27.** $\frac{2}{3}$ **29.** 1

31. 1   **33.** 1   **35.** 0   **37.** 0   **39.** 0

41.

x	10	10^2	10^4
$\dfrac{(\ln x)^4}{x}$	2.811	4.498	0.720

x	10^6	10^8	10^{10}
$\dfrac{(\ln x)^4}{x}$	0.036	0.001	0.000

45. Horizontal asymptote: $y = 1$
Relative maximum: $(e, e^{1/e})$

47. Horizontal asymptote: $y = 0$
Relative maximum: $\left(1, \dfrac{2}{e}\right)$

49. Limit is not of the form $0/0$ or ∞/∞.

51. Limit is not of the form $0/0$ or ∞/∞.   **53.** 1

55. $v = 32t + v_0$ **57.** $\frac{1}{2}$

59. $\frac{5}{2}$ **61.** $\lim\limits_{k \to 0^+} \dfrac{x^k - 1}{k} = \ln x$

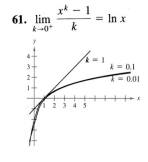

Section 8.8

1. 4   **3.** 6   **5.** 1   **7.** Diverges   **9.** 2

11. 1   **13.** $\frac{1}{2}$ **15.** π **17.** $\frac{\pi}{4}$

19. Diverges   **21.** Diverges   **23.** 6

25. $-\frac{1}{4}$ **27.** Diverges   **29.** $\ln(2 + \sqrt{3})$

31. 0   **33.** $p > 1$ **37.** Diverges

39. Converges   **41.** Converges   **43.** Diverges

45. Converges   **47. (a)** 1   **(b)** $\dfrac{\pi}{3}$ **(c)** Diverges

49. 6

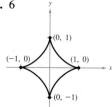

51. (a) $\Gamma(1) = 1$
   (b) $\Gamma(2) = 1$
   (c) $\Gamma(3) = 2$

53. (b) 51.05%
   (c) 7

55. (a) \$743,997.58
   (b) \$795,584.54
   (c) \$858,333.33

57. $k \dfrac{\sqrt{a^2 + 1} - 1}{a^2 \sqrt{a^2 + 1}}$

Review Exercises for Chapter 8

1. $x + \frac{9}{8} \ln |x - 3| - \frac{25}{8} \ln |x + 5| + C$

3. $\tan \theta + \sec \theta + C$

5. $\dfrac{e^{2x}}{13}(2 \sin 3x - 3 \cos 3x) + C$

7. $-\dfrac{1}{x}(1 + \ln 2x) + C$

9. $\dfrac{1}{2}\left(4 \arcsin \dfrac{x}{2} + x\sqrt{4 - x^2}\right) + C$

11. $\dfrac{3\sqrt{4 - x^2}}{x} + C$

13. $\dfrac{2}{3}\left[\tan^3\left(\dfrac{x}{2}\right) + 3 \tan\left(\dfrac{x}{2}\right)\right] + C$

15. $\dfrac{3}{2} \ln \left|\dfrac{x - 3}{x + 3}\right| + C$

17. $\frac{1}{4}[6 \ln |x - 1| - \ln (x^2 + 1) + 6 \arctan x] + C$

19. $\dfrac{3}{2} \ln (x^2 + 1) - \dfrac{1}{2(x^2 + 1)} + C$

21. $16 \arcsin \dfrac{x}{4} + C$ **23.** $\dfrac{1}{2} \arctan \dfrac{e^x}{2} + C$

25. $\dfrac{1}{2} \ln |x^2 + 4x + 8| - \arctan\left(\dfrac{x + 2}{2}\right) + C$

27. $\frac{1}{8}(\sin 2\theta - 2\theta \cos 2\theta) + C$

29. $\frac{1}{2}(2\theta - \cos 2\theta) + C$

31. $\frac{4}{3}[x^{3/4} - 3x^{1/4} + 3 \arctan (x^{1/4})] + C$

33. $2\sqrt{1 - \cos x} + C$

35. $x \ln |x^2 + x| - 2x + \ln |x + 1| + C$

37. $\sin x \ln (\sin x) - \sin x + C$

39. $x - \dfrac{1}{2(x^2 + 1)} + C$ **41.** $-\dfrac{\sqrt{4 + x^2}}{4x} + C$

43. $\frac{1}{3}\sqrt{4 + x^2}(x^2 - 8) + C$

45. (a) $e - 1 \approx 1.72$ **(b)** 1
   (c) $\frac{1}{2}(e - 1) \approx 0.86$ **(d)** 1.46 (Simpson's Rule)

47. 0   **49.** ∞ **51.** 1   **53.** $1000e^{0.09} \approx 1094.17$

55. $\dfrac{\pi}{3}$ **57.** 3.82   **59.** $(\bar{x}, \bar{y}) = \left(0, \dfrac{4}{3\pi}\right)$

61. $0.015846 < \displaystyle\int_2^\infty \dfrac{1}{x^5 - 1}\, dx < 0.015851$

Chapter 9

Section 9.1

1. 2, 4, 8, 16, 32   **3.** $-\frac{1}{2}, \frac{1}{4}, -\frac{1}{8}, \frac{1}{16}, -\frac{1}{32}$

5. $3, \frac{9}{2}, \frac{27}{6}, \frac{81}{24}, \frac{243}{120}$ **7.** $-1, -\frac{1}{4}, \frac{1}{9}, \frac{1}{16}, -\frac{1}{25}$

9. $3n - 2$ **11.** $n^2 - 2$ **13.** $\dfrac{n + 1}{n + 2}$

15. $\dfrac{(-1)^{n-1}}{2^{n-2}}$ **17.** $\dfrac{n + 1}{n}$ **19.** $\dfrac{n}{(n + 1)(n + 2)}$

21. $\dfrac{2^n n!}{(2n)!}$ **23.** $(-1)^{n(n-1)/2}$ **25.** Converges to 1

27. Diverges   **29.** Converges to $\frac{3}{2}$ **31.** Diverges

33. Converges to 0   **35.** Diverges

37. Converges to 0   **39.** Diverges

41. Converges to 3   **43.** Converges to e^k

45. Converges to 0   **47.** Monotonic

49. Not monotonic   **51.** Not monotonic

53. Monotonic   **55.** Not monotonic   **57.** 5

59. $\frac{1}{3}$

61. (a) No
   (b)

Month	1	2	3
Amount	\$9,086.25	\$9,173.33	\$9,261.24

Month	4	5
Amount	\$9,349.99	\$9,340.60

Month	6	7	8
Amount	\$9,530.06	\$9,621.39	\$9,713.59

Month	9	10
Amount	\$9,806.68	\$9,900.66

63. (a) $\$2,500,000,000(0.8)^n$
   (b)

Year	1	2
Budget	\$2,000,000,000	\$1,600,000,000

Year	3	4
Budget	\$1,280,000,000	\$1,024,000,000

   (c) Converges to 0

65. $S_6 = 240$, $S_7 = 440$, $S_8 = 810$,
$S_9 = 1490$, $S_{10} = 2740$

67. 1, 1.4142, 1.4422, 1.4142, 1.3797, 1.3480
Converges to 1

71. **(a)** 1, 1, 2, 3, 5, 8, 13, 21, 34, 55, 89, 144
 (b) 1, 2, 1.5, 1.6667, 1.6, 1.6250, 1.6154, 1.6190,
 1.6176, 1.6182
 (d) $\rho = \dfrac{1 + \sqrt{5}}{2} \approx 1.6180$

Section 9.2

1. 1, 1.25, 1.361, 1.424, 1.464
3. 3, −1.5, 5.25, −4.875, 10.3125
5. 3, 4.5, 5.25, 5.625, 5.8125
7. $\lim\limits_{n\to\infty} a_n = 1 \neq 0$ **9.** $\lim\limits_{n\to\infty} a_n = 3 \neq 0$
11. $\lim\limits_{n\to\infty} a_n = 1 \neq 0$ **13.** Geometric series: $r = \frac{3}{2} > 1$
15. Geometric series: $r = 1.055 > 1$
17. $\lim\limits_{n\to\infty} a_n = \frac{1}{2} \neq 0$ **19.** Geometric series: $r = \frac{3}{4} < 1$
21. Geometric series: $r = 0.9 < 1$
23. Telescoping series: $a_n = \dfrac{1}{n} - \dfrac{1}{n+1}$, converges to 1
25. 2 **27.** $\frac{2}{3}$ **29.** $\frac{10}{9}$ **31.** $\frac{9}{4}$ **33.** $5\left(\frac{5}{8}\right)^3 = \frac{625}{512}$
35. $\frac{3}{4}$ **37.** 3 **39.** $\frac{1}{2}$ **41.** $\sum\limits_{n=0}^{\infty} \frac{3}{5}(0.1)^n = \frac{2}{3}$
43. $\sum\limits_{n=0}^{\infty} \frac{3}{40}(0.01)^n = \frac{5}{66}$ **45.** Diverges
47. Converges **49.** Diverges **51.** Diverges
53. Diverges **55.** Diverges **57.** $80{,}000(1 - 0.9^n)$
59. 152.42 ft **61.** $\frac{1}{3}$ **63.** \$11,616.95
65. \$235,821.51
71. $\sum\limits_{n=0}^{\infty} 1$, $\sum\limits_{n=0}^{\infty} (-1)$ (Answer is not unique.)

Section 9.3

1. Diverges **3.** Converges **5.** Converges
7. Diverges **9.** Diverges **11.** Converges
13. Diverges **15.** Diverges **17.** Converges
19. Converges **21.** Diverges **23.** Converges
25. Converges **27.** Diverges **29.** Diverges
31. Converges **33.** $p > 1$ **35.** 31
37. $R_6 \approx 0.0015$, $S_6 \approx 1.0811$
39. $R_{10} \approx 0.0997$, $S_{10} \approx 0.9818$
41. $R_4 \approx 5.6 \times 10^{-8}$, $S_4 \approx 0.4049$
43. $N \geq 7$ **45.** $N \geq 2$

Section 9.4

1. Converges **3.** Diverges **5.** Converges
7. Diverges **9.** Converges **11.** Converges
13. Diverges **15.** Diverges **17.** Converges
19. Diverges **21.** Converges **23.** Diverges
25. Diverges **27.** Diverges; p-Series Test
29. Converges; Comparison Test with $\sum\limits_{n=1}^{\infty} \left(\frac{1}{3}\right)^n$
31. Diverges; nth-Term Test
33. Converges; Integral Test **37.** Diverges
39. Converges

Section 9.5

1. Converges **3.** Converges **5.** Diverges
7. Converges **9.** Diverges **11.** Converges
13. Diverges **15.** Converges **17.** Converges
19. Converges **21.** Converges
23. Converges absolutely
25. Converges conditionally
27. Converges conditionally
29. Converges absolutely **31.** Converges absolutely
33. Converges conditionally
35. Converges absolutely
37. Converges conditionally **39.** 0.947
41. 0.368 **43.** 0.842 **45.** 0.406
47. At least 499 **51.** **(a)** $-1 < x < 1$
 (b) $-1 \leq x < 1$

Section 9.6

1. Diverges **3.** Converges **5.** Converges
7. Diverges **9.** Converges **11.** Diverges
13. Converges **15.** Converges **17.** Diverges
19. Converges **21.** Converges **23.** Converges
25. Diverges **27.** Converges **29.** Converges
31. Converges; Alternating Series Test
33. Converges; p-Series Test
35. Diverges; nth-Term Test
37. Diverges; Ratio Test
39. Converges; Limit Comparison Test with $b_n = \dfrac{1}{2^n}$
41. Converges; Alternating Series Test
43. Converges; Comparison Test with $b_n = \dfrac{1}{2^n}$
45. Converges; Ratio Test
47. Converges; Ratio Test **49.** Converges; Ratio Test

Section 9.7

1. $1 - x + \frac{1}{2}x^2 - \frac{1}{6}x^3$

3. $1 + 2x + 2x^2 + \frac{4}{3}x^3 + \frac{2}{3}x^4$

5. $x - \frac{1}{6}x^3 + \frac{1}{120}x^5$ **7.** $x + x^2 + \frac{1}{2}x^3 + \frac{1}{6}x^4$

9. $1 - x + x^2 - x^3 + x^4$ **11.** $x + \frac{1}{3}x^3$

13. $2 - 3x^3 + x^4$

15. $1 - (x - 1) + (x - 1)^2 - (x - 1)^3 + (x - 1)^4$

17. $(x - 1) - \frac{1}{2}(x - 1)^2 + \frac{1}{3}(x - 1)^3 - \frac{1}{4}(x - 1)^4$

19.

x	$\sin x$	$P_1(x)$	$P_3(x)$	$P_5(x)$
0	0	0	0	0
0.25	0.2474	0.25	0.2474	0.2474
0.50	0.4794	0.50	0.4792	0.4794
0.75	0.6816	0.75	0.6797	0.6817
1.00	0.8415	1.00	0.8333	0.8417

21. 0.6042   **23.** -6.7954

25. For e^x: $P_4(x) = 1 + x + \frac{1}{2}x^2 + \frac{1}{6}x^3 + \frac{1}{24}x^4$
For xe^x: $Q_4(x) = x + x^2 + \frac{1}{2}x^3 + \frac{1}{6}x^4$
$Q_4(x) = xP_4(x) - \frac{1}{24}x^5$

27. $-0.3936 < x < 0$ **29.** 0.0038

31. (a) $P_3(x) = x + \frac{1}{6}x^3$

(b)

x	-1	-0.75	-0.50	-0.25
$f(x)$	-1.571	-0.848	-0.524	-0.253
$P_3(x)$	-1.167	-0.820	-0.521	-0.253

x	0	0.25	0.50	0.75	1
$f(x)$	0	0.253	0.524	0.848	1.571
$P_3(x)$	0	0.253	0.521	0.820	1.167

(c)

Section 9.8

1. $R = 1$ **3.** $R = \frac{1}{2}$ **5.** $R = \infty$

7. $(-2, 2)$ **9.** $(-1, 1]$ **11.** $(-\infty, \infty)$

13. $x = 0$ **15.** $(-4, 4)$ **17.** $(0, 10]$

19. $(0, 2]$ **21.** $(0, 2c)$ **23.** $\left(-\frac{1}{2}, \frac{1}{2}\right)$

25. $(-\infty, \infty)$ **27.** $(-1, 1)$ **29.** $x = 3$

31. (a) $(-2, 2)$ **33. (a)** $(0, 2]$
    (b) $(-2, 2)$ **(b)** $(0, 2)$
    (c) $(-2, 2)$ **(c)** $(0, 2)$
    (d) $[-2, 2)$ **(d)** $[0, 2]$

35. (a) For $f(x)$: $(-\infty, \infty)$ **37. (a)** $(-\infty, \infty)$
        For $g(x)$: $(-\infty, \infty)$ **(d)** $f(x) = e^x$
    (d) $f(x) = \sin x$
        $g(x) = \cos x$

Section 9.9

1. $\displaystyle\sum_{n=0}^{\infty} \frac{x^n}{2^{n+1}}$, $(-2, 2)$ **3.** $\displaystyle\sum_{n=0}^{\infty} \frac{(x - 5)^n}{(-3)^{n+1}}$, $(2, 8)$

5. $\displaystyle-3\sum_{n=0}^{\infty} (2x)^n$, $\left(-\frac{1}{2}, \frac{1}{2}\right)$

7. $\displaystyle-\frac{1}{11}\sum_{n=0}^{\infty} \left[\frac{2}{11}(x + 3)\right]^n$, $\left(-\frac{17}{2}, \frac{5}{2}\right)$

9. $\displaystyle\frac{3}{2}\sum_{n=0}^{\infty} \left(\frac{x}{-2}\right)^n$, $(-2, 2)$

11. $\displaystyle\sum_{n=0}^{\infty} \left[\frac{1}{(-2)^n} - 1\right]x^n$, $(-1, 1)$

13. $\displaystyle2\sum_{n=0}^{\infty} x^{2n}$, $(-1, 1)$

15. $\displaystyle\sum_{n=1}^{\infty} n(-1)^n x^{n-1}$, $(-1, 1)$

17. $\displaystyle\sum_{n=0}^{\infty} \frac{(-1)^n x^{n+1}}{n + 1}$, $(-1, 1]$

19. $\displaystyle\sum_{n=0}^{\infty} (-1)^n (2x)^{2n}$, $\left(-\frac{1}{2}, \frac{1}{2}\right)$

21. $\displaystyle\sum_{n=0}^{\infty} \frac{x^{2n+1}}{2n + 1}$, $(-1, 1)$

33.

(figure)

23.

x	$x - \dfrac{x^2}{2}$	$\ln(x+1)$	$x - \dfrac{x^2}{2} + \dfrac{x^3}{3}$
0.0	0.000	0.000	0.000
0.2	0.180	0.182	0.183
0.4	0.320	0.336	0.341
0.6	0.420	0.470	0.492
0.8	0.480	0.588	0.651
1.0	0.500	0.693	0.833

25. 0.245   **27.** 0.125   **29.** 3.14

31.

33. $\displaystyle\sum_{n=0}^{\infty} (-1)^n x^n$

Section 9.10

1. $\displaystyle\sum_{n=0}^{\infty} \frac{(2x)^n}{n!}$ **3.** $\dfrac{\sqrt{2}}{2} \displaystyle\sum_{n=0}^{\infty} \frac{(-1)^{n(n+1)/2}}{n!}\left(x - \frac{\pi}{4}\right)^n$

5. $\displaystyle\sum_{n=0}^{\infty} \frac{(-1)^n(x-1)^{n+1}}{n+1}$ **7.** $\displaystyle\sum_{n=0}^{\infty} \frac{(-1)^n(2x)^{2n+1}}{(2n+1)!}$

9. $1 + \dfrac{x^2}{2!} + \dfrac{5x^4}{4!} + \cdots$

11. $\displaystyle\sum_{n=0}^{\infty} (-1)^n(n+1)x^n$

13. $\dfrac{1}{2}\left[1 + \displaystyle\sum_{n=1}^{\infty} \frac{(-1)^n 1 \cdot 3 \cdot 5 \cdots (2n-1)x^{2n}}{2^{3n}n!}\right]$

15. $1 + \dfrac{x^2}{2} + \displaystyle\sum_{n=2}^{\infty} \frac{(-1)^{n+1} 1 \cdot 3 \cdot 5 \cdots (2n-3)x^{2n}}{2^n n!}$

17. $1 + \dfrac{x^2}{2} + \dfrac{x^4}{2^2 2!} + \dfrac{x^6}{2^3 3!} + \cdots$

19. $\displaystyle\sum_{n=0}^{\infty} \frac{(-1)^n(2x)^{2n+1}}{(2n+1)!}$ **21.** $\displaystyle\sum_{n=0}^{\infty} \frac{(-1)^n x^n}{(2n)!}$

23. $\displaystyle\sum_{n=0}^{\infty} \frac{(-1)^n x^{2n}}{(2n+1)!}$ **25.** $\displaystyle\sum_{n=0}^{\infty} \frac{x^{2n+1}}{(2n+1)!}$

27. $\displaystyle\sum_{n=0}^{\infty} \frac{(-1)^n x^{2n+1}}{(2n+1)!}$

29. $\dfrac{1}{2}\left[1 + \displaystyle\sum_{n=0}^{\infty} \frac{(-1)^n(2x)^{2n}}{(2n)!}\right]$ **31.** 1.3708

33. 0.7040   **35.** 0.4872   **37.** 0.2010

39. $\displaystyle\sum_{n=0}^{\infty} \frac{(-1)^{n+1} x^{2n+3}}{(2n+3)(n+1)!}$ **41.** 0.3413

47. (a)

(c) $\displaystyle\sum_{n=0}^{\infty} 0\, x^n = 0 \neq f(x)$

Review Exercises for Chapter 9

1. $a_n = \dfrac{1}{n!}$ **3.** Converges to 0   **5.** Diverges

7. Converges to 0   **9.** Converges to 0

11. 1, 2.5, 4.75, 8.125, 13.1875

13. 0.5, 0.45833, 0.45972, 0.45970, 0.45970

15. 3   **17.** $\frac{1}{2}$ **19.** $\frac{1}{11}$ **21.** Diverges

23. Converges   **25.** Diverges   **27.** Diverges

29. Converges   **31.** Diverges   **33.** [1, 3]

35. Converges only at $x = 2$

37. $\dfrac{\sqrt{2}}{2} \displaystyle\sum_{n=0}^{\infty} \frac{(-1)^{n(n+1)/2}}{n!}\left(x - \frac{3\pi}{4}\right)^n$

39. $\displaystyle\sum_{n=0}^{\infty} \frac{(x \ln 3)^n}{n!}$ **41.** $-\displaystyle\sum_{n=0}^{\infty} (x+1)^n$

43. $\displaystyle\sum_{n=0}^{\infty} \frac{(-1)^n x^{2n+1}}{(2n+1)(2n+1)!}$ **45.** $\displaystyle\sum_{n=0}^{\infty} \frac{(-1)^n x^{n+1}}{(n+1)^2}$

47. 0.996   **49.** 0.560

53.

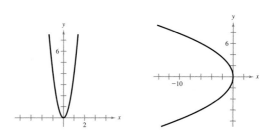

Chapter 10

Section 10.1

1. e   **2.** f   **3.** a   **4.** c   **5.** d   **6.** b

7. Vertex: (0, 0)   **9.** Vertex: (0, 0)
Focus: $\left(0, \frac{1}{16}\right)$ Focus: $\left(-\frac{3}{2}, 0\right)$
Directrix: $y = -\frac{1}{16}$ Directrix: $x = \frac{3}{2}$

11. Vertex: $(0, 0)$
Focus: $(0, -2)$
Directrix: $y = 2$

13. Vertex: $(1, -2)$
Focus: $(1, -4)$
Directrix: $y = 0$

15. Vertex: $\left(5, -\frac{1}{2}\right)$
Focus: $\left(\frac{11}{2}, -\frac{1}{2}\right)$
Directrix: $x = \frac{9}{2}$

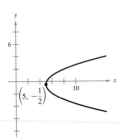

17. Vertex: $(1, 1)$
Focus: $(1, 2)$
Directrix: $y = 0$

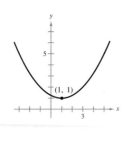

19. Vertex: $(8, -1)$
Focus: $(9, -1)$
Directrix: $x = 7$

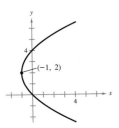

21. Vertex: $(-2, -3)$
Focus: $(-4, -3)$
Directrix: $x = 0$

23. Vertex: $(-1, 2)$
Focus: $(0, 2)$
Directrix: $x = -2$

25. Vertex: $(-2, 2)$
Focus: $(-2, 1)$
Directrix: $y = 3$

27. $x^2 + 6y = 0$ **29.** $y^2 - 4y + 8x - 20 = 0$
31. $x^2 + 24y + 96 = 0$ **33.** $x^2 = -16y$
35. $5x^2 - 14x - 3y + 9 = 0$
37. $x^2 + y - 4 = 0$ **39.** $(x - h)^2 = -8(y + 1)$
41. $3x - 2y^2 = 0$ **43.** $4x - y - 8 = 0$
45. $\left(\frac{3}{5}, 0\right)$ **47.** $(0, 0)$
49. $2[\sqrt{2} + \ln(\sqrt{2} + 1)]$
51. $2\sqrt{5} + \ln(\sqrt{5} + 2)$
53. $100\left[\sqrt{5} + 4\ln\left(\dfrac{1 + \sqrt{5}}{2}\right)\right] \approx 416.1$ ft
55. $10\sqrt{3}$ ft from end of pipe
57. (a) 122.7 ft^2
(b) 382.9 ft^2
59. $(1.02439, 0)$ **61.** $y = 2ax_0x - ax_0^2$

63. Tangent lines: $x + y = 0$, $2x - y - 9 = 0$
Point of intersection: $(3, -3)$
65. Tangent lines: $2x + y - 1 = 0$, $2x - 4y - 1 = 0$
Point of intersection (on the directrix): $\left(\frac{1}{2}, 0\right)$

Section 10.2

1. e   **2.** a   **3.** c   **4.** b   **5.** f   **6.** d
7. Center: $(0, 0)$
Foci: $(\pm 3, 0)$
Vertices: $(\pm 5, 0)$
$e = \frac{3}{5}$

9. Center: $(0, 0)$
Foci: $(0, \pm 3)$
Vertices: $(0, \pm 5)$
$e = \frac{3}{5}$

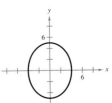

11. Center: $(0, 0)$
Foci: $(\pm 2, 0)$
Vertices: $(\pm 3, 0)$
$e = \frac{2}{3}$

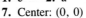

13. Center: $(0, 0)$
Foci: $(\pm\sqrt{3}, 0)$
Vertices: $(\pm 2, 0)$
$e = \dfrac{\sqrt{3}}{2}$

15. Center: $(0, 0)$
Foci: $(0, \pm 1)$
Vertices: $(0, \pm\sqrt{3})$
$e = \dfrac{\sqrt{3}}{3}$

17. Center: $(0, 0)$
Foci: $\left(0, \pm\dfrac{\sqrt{3}}{2}\right)$
Vertices: $(0, \pm 1)$
$e = \dfrac{\sqrt{3}}{2}$

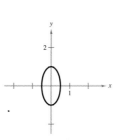

19. Center: $(1, 5)$
Foci: $(1, 9)$, $(1, 1)$
Vertices: $(1, 10)$, $(1, 0)$
$e = \frac{4}{5}$

21. Center: $(-2, 3)$
Foci: $(-2, 3 \pm \sqrt{5})$
Vertices: $(-2, 6)$, $(-2, 0)$
$e = \dfrac{\sqrt{5}}{3}$

23. Center: $(1, -1)$
Foci: $\left(1 \pm \dfrac{3\sqrt{10}}{20}, -1\right)$
Vertices: $\left(1 \pm \dfrac{\sqrt{10}}{4}, -1\right)$
$e = \frac{3}{5}$

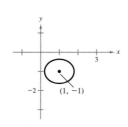

25. Center: $\left(\dfrac{1}{2}, -1\right)$
Foci: $\left(\dfrac{1}{2} \pm \sqrt{2}, -1\right)$
Vertices: $\left(\dfrac{1}{2} \pm \sqrt{5}, -1\right)$
$e = \dfrac{\sqrt{10}}{5}$

27. $\dfrac{x^2}{9} + \dfrac{y^2}{5} = 1$ **29.** $\dfrac{x^2}{25} + \dfrac{y^2}{16} = 1$

31. $\dfrac{(x - 3)^2}{9} + \dfrac{(y - 5)^2}{16} = 1$ **33.** $\dfrac{x^2}{24} + \dfrac{y^2}{49} = 1$

35. $\dfrac{x^2}{16} + \dfrac{7y^2}{16} = 1$

39. Minor axis: $(-6, -2)$, $(0, -2)$
Major axis: $(-3, -6)$, $(-3, 2)$

41. $\left(0, \frac{25}{3}\right)$ **43.** 22.10   **45.** 9.91

47. **(a)** 2π

   (b) Volume $= \dfrac{8\pi}{3}$

   Surface area $= \dfrac{2\pi(9 + 4\sqrt{3}\pi)}{9} \approx 21.48$

   (c) Volume $= \dfrac{16\pi}{3}$

   Surface area $= \dfrac{4\pi[6 + \sqrt{3}\ln(2 + \sqrt{3})]}{3} \approx 34.69$

49. $\sqrt{2}a \times \sqrt{2}b$ **55.** 1.5 ft from center

59. $e \approx 0.96716$

61.

Section 10.3

1. e   **2.** a   **3.** f   **4.** c   **5.** d   **6.** b

7. Center: $(0, 0)$
Vertices: $(\pm 1, 0)$
Foci: $(\pm\sqrt{2}, 0)$

9. Center: $(0, 0)$
Vertices: $(0, \pm 1)$
Foci: $(0, \pm\sqrt{5})$

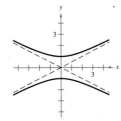

11. Center: $(0, 0)$
Vertices: $(0, \pm 5)$
Foci: $(0, \pm 13)$

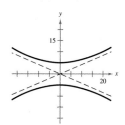

13. Center: $(0, 0)$
Vertices: $(\pm\sqrt{3}, 0)$
Foci: $(\pm\sqrt{5}, 0)$

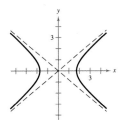

21. Center: $(2, -3)$
Vertices: $(1, -3), (3, -3)$
Foci: $(2 \pm \sqrt{10}, -3)$

15. Center: $(0, 0)$
Vertices: $(0, \pm 2)$
Foci: $(0, \pm 3)$

23. Center: $(1, -3)$
Vertices: $(1, -3 \pm \sqrt{2})$
Foci: $(1, -3 \pm 2\sqrt{5})$

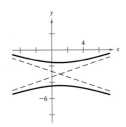

17. Center: $(1, -2)$
Vertices: $(-1, -2), (3, -2)$
Foci: $(1 \pm \sqrt{5}, -2)$

25. Degenerate hyperbola
Graph is two lines intersecting
at $(-1, -3)$.

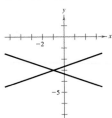

27. $\dfrac{y^2}{4} - \dfrac{x^2}{12} = 1$

29. $\dfrac{x^2}{1} - \dfrac{y^2}{9} = 1$ **31.** $\dfrac{(x - 3)^2}{9} - \dfrac{(y - 2)^2}{4} = 1$

33. $\dfrac{y^2}{9} - \dfrac{(x - 2)^2}{9/4} = 1$

35. $\dfrac{(x - 6)^2}{9} - \dfrac{(y - 2)^2}{7} = 1$

37. **(a)** At $(6, \sqrt{3})$: $2x - 3\sqrt{3}y - 3 = 0$
At $(6, -\sqrt{3})$: $2x + 3\sqrt{3}y - 3 = 0$
(b) At $(6, \sqrt{3})$: $9x + 2\sqrt{3}y - 60 = 0$
At $(6, -\sqrt{3})$: $9x - 2\sqrt{3}y - 60 = 0$

39. Volume $= \dfrac{4\pi}{3}$

Surface area $=$

$$\pi\left[2\sqrt{7} - 1 + \frac{\sqrt{2}}{2}\ln\left(\frac{\sqrt{2} + 1}{2\sqrt{2} + \sqrt{7}}\right)\right] \approx 11.66$$

41. $x \approx 110.3$ mi   **47.** Ellipse   **49.** Parabola
51. Circle   **53.** Circle   **55.** Hyperbola

19. Center: $(2, -6)$
Vertices: $(2, -5), (2, -7)$
Foci: $(2, -6 \pm \sqrt{2})$

57.

Section 10.4

1. $\dfrac{(y')^2}{2} - \dfrac{(x')^2}{2} = 1$

3. $y' = \dfrac{(x')^2}{6} - \dfrac{x'}{3}$

5. $\dfrac{(x')^2}{1/4} - \dfrac{(y')^2}{1/6} = 1$

7. $\dfrac{(x' - 3\sqrt{2})^2}{16} - \dfrac{(y' - \sqrt{2})^2}{16} = 1$

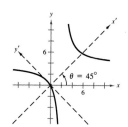

9. $\dfrac{(x')^2}{3} + \dfrac{(y')^2}{2} = 1$

11. $x' = -(y')^2$

13. $\dfrac{(x')^2}{3} - \dfrac{(y')^2}{5} = 1$

15. $\dfrac{(x')^2}{1.096} - \dfrac{(y')^2}{6.153} = 1$

17. Parabola **19.** Ellipse **21.** Hyperbola
23. Parabola **25.** Two lines

27. Point

29.

Review Exercises for Chapter 10

1. h **2.** a **3.** e **4.** i **5.** f **6.** b
7. c **8.** j **9.** g **10.** d
11. Circle
   Center: $\left(\frac{1}{2}, -\frac{3}{4}\right)$
   Radius: 1

13. Hyperbola
   Center: $(-4, 3)$
   Vertices: $(-4 \pm \sqrt{2}, 3)$

15. Ellipse
Center: $(2, -3)$
Vertices: $\left(2, -3 \pm \dfrac{\sqrt{2}}{2}\right)$

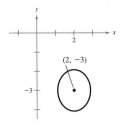

17. Parabola
Vertex: $(3, 0)$

19. Parabola
Vertex: $\left(-\dfrac{\sqrt{2}}{4}, \dfrac{\sqrt{2}}{4}\right)$

21. Ellipse
Center: $\left(1, -\tfrac{1}{2}\right)$
Vertices: $\left(1 \pm \tfrac{3}{4}, -\tfrac{1}{2}\right)$

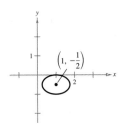

23. $\dfrac{y^2}{1} - \dfrac{x^2}{8} = 1$ **25.** $\dfrac{(x-2)^2}{25} + \dfrac{y^2}{21} = 1$

27. $x^2 - 2xy + y^2 - 8x - 8y = 0$

29. $\dfrac{x^2}{4} + \dfrac{y^2}{16/3} = 1$ **31.** $\dfrac{x^2}{4} - \dfrac{y^2}{12} = 1$

33. $4x + 4y - 7 = 0$ **35.** $a = \sqrt{5}$

41. 15.87   **43.** 4.212 ft   **45.** 428.68 ft^2

Chapter 11

Section 11.1

1. $2x - 3y + 5 = 0$ **3.** $y = 1 - x^2, x \geq 0$

5. $y = (x - 1)^2$ **7.** $y = \tfrac{1}{2}x^{2/3}$

9. $y = \dfrac{x + 1}{x}$ **11.** $y = \dfrac{|x - 4|}{2}$

13. $y = \dfrac{1}{x}, |x| \geq 1$ **15.** $x^2 + y^2 = 9$

 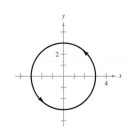

17. $\dfrac{x^2}{16} + \dfrac{y^2}{4} = 1$

19. $y = 2 - 2x^2, \ -1 \le x \le 1$

21. $\dfrac{(x-4)^2}{4} + \dfrac{(y+1)^2}{1} = 1$

23. $\dfrac{(x-4)^2}{4} + \dfrac{(y+1)^2}{16} = 1$

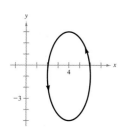

25. $\dfrac{x^2}{16} - \dfrac{y^2}{9} = 1$

27. $y = \ln x$

29. $y = \dfrac{1}{x^3}, \ x > 0$

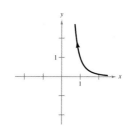

31. Each curve represents a portion of the line $y = 2x + 1$.

Domain	Orientation
(a) $-\infty < x < \infty$	Up
(b) $-1 \le x \le 1$	Oscillates
(c) $0 < x < \infty$	Down
(d) $0 < x < \infty$	Up

33.

35.

37.

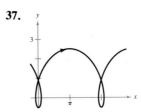

39. $y - y_1 = \dfrac{y_2 - y_1}{x_2 - x_1}(x - x_1)$

41. $(x - h)^2 + (y - k)^2 = r^2$

43. $\dfrac{(x-h)^2}{a^2} - \dfrac{(y-k)^2}{b^2} = 1$

45. $x = 5t$
$y = -2t$
(Solution is not unique.)

47. $x = 2 + 4\cos\theta$
$y = 1 + 4\sin\theta$
(Solution is not unique.)

49. $x = 5\cos\theta$
$y = 3\sin\theta$
(Solution is not unique.)

51. $x = 4\sec\theta$
$y = 3\tan\theta$
(Solution is not unique.)

53. $x = t$
$y = t^3$
$x = \tan t$
$y = \tan^3 t$
(Solution is not unique.)

55. $x = a\theta - b\sin\theta$
$y = a - b\cos\theta$

57.

59.

61.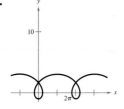

9. At $\theta = \dfrac{\pi}{4}$, $\dfrac{dy}{dx} = -\tan\theta$, $\dfrac{d^2y}{dx^2} = \dfrac{\sec^4\theta\csc\theta}{3}$
$\dfrac{dy}{dx} = -1$, $\dfrac{d^2y}{dx^2} = \dfrac{4\sqrt{2}}{3}$

11. $2x - y - 5 = 0$ **13.** $y = 2$
15. $x + 2y - 4 = 0$ **17.** Horizontal: $(1, 0)$
                     Vertical: None
19. Horizontal: $(0, -2)$, $(2, 2)$
Vertical: None
21. Horizontal: $(0, 3)$, $(0, -3)$
Vertical: $(3, 0)$, $(-3, 0)$
23. Horizontal: $(4, 0)$, $(4, -2)$
Vertical: $(2, -1)$, $(6, -1)$
25. Horizontal: None
Vertical: $(1, 0)$, $(-1, 0)$
27. Horizontal: $(4n\pi, 0)$, $(2[2n - 1]\pi, 4)$
Vertical: None
29. $\sqrt{2}(1 - e^{-\pi/2}) \approx 1.12$
31. $2\sqrt{5} + \ln(2 + \sqrt{5}) \approx 5.916$
33. $\dfrac{1}{12}[\ln(\sqrt{37} + 6) + 6\sqrt{37}] \approx 3.249$ **35.** $6a$
37. $8a$ **39.** **(a)** $32\pi\sqrt{5}$ **(b)** $16\pi\sqrt{5}$
41. 32π **43.** $\dfrac{12\pi a^2}{5}$ **45.** $2\pi r^2(1 - \cos\theta)$
47. $\dfrac{3\pi}{8}$ **49.** $\dfrac{3\pi}{2}$ **51.** $\left(\dfrac{3}{4}, \dfrac{8}{5}\right)$ **53.** 36π
55. **(a)**   **(b)** $(0, 0)$, $(1.6799, 2.1165)$
                      (c) 6.557

Section 11.3

1.

3.

5.

7.

Section 11.2

1. $\dfrac{dy}{dx} = \dfrac{3}{2}$, $\dfrac{d^2y}{dx^2} = 0$

3. At $t = -1$, $\dfrac{dy}{dx} = 2t + 3$, $\dfrac{d^2y}{dx^2} = 2$
$\dfrac{dy}{dx} = 1$, $\dfrac{d^2y}{dx^2} = 2$

5. At $\theta = \dfrac{\pi}{4}$, $\dfrac{dy}{dx} = -\cot\theta$, $\dfrac{d^2y}{dx^2} = -\dfrac{\csc^3\theta}{2}$
$\dfrac{dy}{dx} = -1$, $\dfrac{d^2y}{dx^2} = -\sqrt{2}$

7. At $\theta = \dfrac{\pi}{6}$, $\dfrac{dy}{dx} = 2\csc\theta$, $\dfrac{d^2y}{dx^2} = -2\cot^3\theta$
$\dfrac{dy}{dx} = 4$, $\dfrac{d^2y}{dx^2} = -6\sqrt{3}$

9. $\left(\sqrt{2}, \frac{\pi}{4}\right), \left(-\sqrt{2}, \frac{5\pi}{4}\right)$

11. $(5, 2.214), (-5, 5.356)$

13. $\left(\sqrt{6}, \frac{5\pi}{4}\right), \left(-\sqrt{6}, \frac{\pi}{4}\right)$

15. $(2\sqrt{13}, 0.983), (-2\sqrt{13}, 4.124)$ **17.** $r = 3$

19. $r = 2a \cos \theta$ **21.** $r = 4 \csc \theta$

23. $r = 10 \sec \theta$ **25.** $r = -\dfrac{2}{3 \cos \theta - \sin \theta}$

27. $r^2 = 8 \csc 2\theta$ **29.** $r = 9 \csc^2 \theta \cos \theta$

31. $r = \dfrac{2a}{1 - \sin \theta}, \quad r = \dfrac{-2a}{1 + \sin \theta}$

33. $x^2 + y^2 - 4y = 0$ **35.** $\sqrt{3}x - 3y = 0$

37. $y = 2$ **39.** $(x^2 + y^2 + 2y)^2 = x^2 + y^2$

41. $4x^2 - 5y^2 - 36y - 36 = 0$

43. $(x^2 + y^2)^3 = 16y^2$ **45.** Symmetry: Pole

Polar axis

$\theta = \dfrac{\pi}{2}$

47. Symmetry: $\theta = \dfrac{\pi}{2}$ **49.** Symmetry: $\theta = \dfrac{\pi}{2}$

51. Symmetry: Polar axis   **53.** Symmetry: $\theta = \dfrac{\pi}{2}$

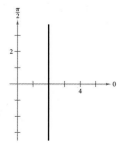

55. $(x - h)^2 + (y - k)^2 = h^2 + k^2$

Center: (h, k)

Radius: $\sqrt{h^2 + k^2}$

Section 11.4

1. Relative maxima:

$(5, 0), \left(5, \dfrac{2\pi}{3}\right), \left(5, \dfrac{4\pi}{3}\right)$

Relative minima:

$\left(0, \dfrac{\pi}{6}\right), \left(0, \dfrac{\pi}{2}\right), \left(0, \dfrac{5\pi}{6}\right), \left(0, \dfrac{7\pi}{6}\right), \left(0, \dfrac{3\pi}{2}\right),$

$\left(0, \dfrac{11\pi}{6}\right)$

3. Relative maximum: $\left(8, \dfrac{\pi}{2}\right)$

Relative minimum: $\left(2, \dfrac{3\pi}{2}\right)$

5. Relative minimum: $(3, \pi)$

7. $\dfrac{(1 + 2 \cos \theta)(1 - \cos \theta)}{\sin \theta(2 \cos \theta - 1)}, -1$

9. $\dfrac{2 \cos \theta(3 \sin \theta + 1)}{3 \cos 2\theta - 2 \sin \theta}, 0$

11. $\dfrac{\sin \theta + \theta \cos \theta}{\cos \theta - \theta \sin \theta}, \pi$ **13.** $\dfrac{2 \sin \theta \cos \theta}{1 - 2 \sin^2 \theta}, -\sqrt{3}$

15. Graph of $r = 2 \sec \theta$ is a vertical line.

Therefore, $\dfrac{dy}{dx}$ is undefined.

17. Horizontal: $\left(2, \dfrac{\pi}{2}\right), \left(\dfrac{1}{2}, \dfrac{7\pi}{6}\right), \left(\dfrac{1}{2}, \dfrac{11\pi}{6}\right)$

Vertical: $\left(\dfrac{3}{2}, \dfrac{\pi}{6}\right), \left(\dfrac{3}{2}, \dfrac{5\pi}{6}\right)$

19. $\left(5, \dfrac{\pi}{2}\right), \left(1, \dfrac{3\pi}{2}\right)$

21.   **23.** $\theta = 0$

25.   **27.**

29. $\theta = \arcsin\left(-\frac{2}{3}\right)$
$\theta = \pi + \arcsin\left(\frac{2}{3}\right)$

31. $\theta = \arccos\left(\frac{3}{4}\right)$
$\theta = 2\pi - \arccos\left(\frac{3}{4}\right)$

49. $\theta = 0, \dfrac{\pi}{2}$

51. $\theta = 0$

33.

35.

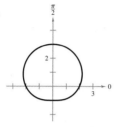

53. $\theta = \dfrac{\pi}{3}, \pi, \dfrac{5\pi}{3}$

55.

37. $\theta = \dfrac{\pi}{6}, \dfrac{\pi}{2}, \dfrac{5\pi}{6}$

39. $\theta = 0, \dfrac{\pi}{2}$

57.

$y = 2$

41.

43.

61. (a) $r = 2 - \sin\left[\theta - \left(\dfrac{\pi}{4}\right)\right]$
$= 2 - \dfrac{\sqrt{2}(\sin\theta - \cos\theta)}{2}$
(b) $r = 2 + \cos\theta$
(c) $r = 2 + \sin\theta$ (d) $r = 2 - \cos\theta$

45. $\theta = \dfrac{\pi}{4}, \dfrac{3\pi}{4}$

47. $\theta = 0, \dfrac{\pi}{2}$

63. (a)

(b)

65. (0, 0), (1.4142, 0.7854), **67.** (7, 1.5708), (3, 4.7124)
(1.4142, 2.3562)

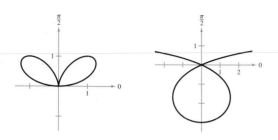

Section 11.5

1. $\left(1, \dfrac{\pi}{2}\right), \left(1, \dfrac{3\pi}{2}\right),$ (0, 0)

3. $\left(\dfrac{2 - \sqrt{2}}{2}, \dfrac{3\pi}{4}\right), \left(\dfrac{2 + \sqrt{2}}{2}, \dfrac{7\pi}{4}\right),$ (0, 0)

5. $\left(\dfrac{3}{2}, \dfrac{\pi}{6}\right), \left(\dfrac{3}{2}, \dfrac{5\pi}{6}\right),$ (0, 0)   **7.** (2, 4), (−2, −4)

9. $\left(2, \dfrac{\pi}{12}\right), \left(2, \dfrac{5\pi}{12}\right), \left(2, \dfrac{7\pi}{12}\right), \left(2, \dfrac{11\pi}{12}\right),$

   $\left(2, \dfrac{13\pi}{12}\right), \left(2, \dfrac{17\pi}{12}\right), \left(2, \dfrac{19\pi}{12}\right), \left(2, \dfrac{23\pi}{12}\right)$

11. (0.581, ±0.535), (2.581, ±1.376)   **13.** $\dfrac{\pi}{3}$

15. $\dfrac{\pi}{8}$ **17.** $\dfrac{3\pi}{2}$ **19.** $\dfrac{2\pi - 3\sqrt{3}}{2}$

21. $\pi + 3\sqrt{3}$ **23.** $\frac{4}{3}(4\pi - 3\sqrt{3})$

25. $11\pi - 24$ **27.** $\frac{2}{3}(4\pi - 3\sqrt{3})$ **29.** $\dfrac{5\pi a^2}{4}$

31. $\dfrac{a^2}{2}(\pi - 2)$ **33.** $2\pi a$ **35.** 8

37. $\dfrac{\pi}{4}\sqrt{\pi^2 + 4} + \ln\left(\dfrac{\pi + \sqrt{\pi^2 + 4}}{2}\right)$

39. $\ln\left(\dfrac{\sqrt{4\pi^2 + 1} + 2\pi}{\sqrt{\pi^2 + 1} + \pi}\right) + \dfrac{2\sqrt{\pi^2 + 1} - \sqrt{4\pi^2 + 1}}{2\pi}$

                                 ≈ 0.7112

41. 4π **43.** $\dfrac{2\pi\sqrt{1 + a^2}}{1 + 4a^2}(e^{\pi a} - 2a)$ **45.** 21.88

47. $4\pi^2 ab$ **49.** $A \approx 0.667,\ s \approx 4.439$

Section 11.6

1. Parabola

3. Parabola

5. Ellipse

7. Ellipse

9. Parabola

11. Hyperbola

13. Hyperbola

15. Hyperbola

17. $r = \dfrac{1}{2 + \sin\theta}$ **19.** $r = -\dfrac{1}{1 - \cos\theta}$

21. $r = \dfrac{2}{1 + 2\cos\theta}$ **23.** $r = \dfrac{2}{1 - \sin\theta}$

25. $r = \dfrac{6}{2 - \sin\theta}$ **27.** $r = \dfrac{6}{1 - 2\sin\theta}$

31. $r^2 = \dfrac{9}{1 - \left(\frac{16}{25}\right)\cos^2\theta}$

33. $r^2 = \dfrac{-16}{1 - \left(\frac{25}{9}\right)\cos^2\theta}$

35.

37.

39. 10.87   **43.** $\dfrac{\pi}{2}$ **45.** 0   **47.** $\dfrac{\pi}{3}$

51. $r = \dfrac{92{,}931{,}075.2223}{1 - 0.0167\cos\theta}$
Perihelion: 91,404,618 mi
Aphelion: 94,509,382 mi

53. $r = \dfrac{1{,}422{,}792{,}504.77}{1 - 0.0543\cos\theta}$
Perihelion: 1,349,513,900 mi
Aphelion: 1,504,486,100 mi

Review Exercises for Chapter 11

1. (a) $\dfrac{dy}{dx} = -\dfrac{3}{4}$
Horizontal tangents: None
(b) $y = \dfrac{-3x + 11}{4}$
(c)

3. (a) $\dfrac{dy}{dx} = -\dfrac{5}{2}\cot\theta$
Horizontal tangents: (3, 7), (3, −3)
(b) $\dfrac{(x - 3)^2}{4} + \dfrac{(y - 2)^2}{25} = 1$
(c)

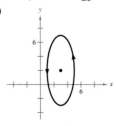

5. (a) $\dfrac{dy}{dx} = -2t^2$
Horizontal tangents: None
(b) $y = 3 + \dfrac{2}{x}$
(c)

7. (a) $\dfrac{dy}{dx} = -(2t^2 + 2t + 1)$
Horizontal tangents: None
(b) $y = \dfrac{1 - x^2}{2x}$
(c)

9. (a) $\dfrac{dy}{dx} = -4\tan\theta$
Horizontal tangents: (±1, 0)
(b) $x^{2/3} + (y/4)^{2/3} = 1$
(c)

11. $x = -3 + 4\cos\theta$
$y = 4 + 3\sin\theta$

15. $\dfrac{\pi^2 r}{2}$

17. Cardioid

19. Limaçon

21. Rose curve

23. Circle

25. Line

27. Rose curve

29. Parabola

31. Strophoid

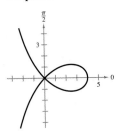

33. $x^2 + y^2 - 3x = 0$

35. $(x^2 + y^2 + 2x)^2 = 4(x^2 + y^2)$

37. $y^2 = x^2\left(\dfrac{4 - x}{4 + x}\right)$ **39.** $r = a \cos^2 \theta \sin \theta$

41. $r^2 = a^2\theta^2$ **43.** $r = \dfrac{4}{1 - \cos \theta}$

45. $r = \dfrac{7}{3 + 4 \cos \theta}$ **47.** $r = \dfrac{12}{4 \cos \theta + 3 \sin \theta}$

49. (a) $\pm\dfrac{\pi}{3}$

   (b) Vertical: $(-1, 0)$, $(3, \pi)$, $\left(\frac{1}{2}, \pm 1.318\right)$
       Horizontal: $(-0.686, \pm 0.568)$, $(2.186, \pm 2.206)$

   (c)

53. $\dfrac{9\pi}{2}$ **55.** $\dfrac{\pi}{32}$ **57.** a^2 **59.** $\dfrac{8\pi - 6\sqrt{3}}{3}$

61. $8a$ **63.** $\arctan\left(\dfrac{2\sqrt{3}}{3}\right)$

Index

D. C. Heath and Company
proudly presents

The
Calculus Multimedia Library

- *Derive*®
- *The Calculus Series*
- *Interactive Calculus*
- *Math Utilities*
- *Hewlett Packard Graphing Calculators*
- *BestGrapher*
- *Casio Graphing Calculators*
- *Kemeny/Kurtz Calculus*
- *TrueBASIC™, inc. Multivariate Calculus: MacFunction*
- *Brøderbund Calculus™*
- *Work-Out!*
- *Computer Activities for Calculus*

Students: Please order through your college bookstore or use coupons provided here for your convenience.

Professors: For information regarding availability of site licensing, networking, and/or quantity discounts, please contact your local D. C. Heath representative.

Nobody said *Calculus was easy...*

VTS Calculus Tape 6
VTS Calculus Tape 5
VTS Calculus Tape 4
VTS Calculus Tape 3
VTS Calculus Tape 2
VTS Calculus Tape 1

VTS Calculus Tape 7
VTS Calculus Tape 8
VTS Calculus Tape 9
VTS Calculus Tape 10
VTS Calculus Tape 11
VTS Calculus Tape 12

*I*ncreasing your success in mastering calculus is now as easy as turning on your VCR, thanks to *The Calculus Series*, a comprehensive and compelling new video program created by Video Tutorial Service (VTS). Keyed to each chapter of *Calculus with Analytic Geometry*, Fourth Edition, these twelve video-tapes (six per semester for the first two semesters) offer the latest in computer and video technology. Featuring expert lecturers, the tapes will enable you to:

- Review calculus thoroughly at your own pace and convenience.
- Build your understanding of challenging textbook material, using problems and step-by-step solutions.
- Master information covered in complicated or missed lectures, simplifying exam preparation.

A Special Offer to Students:

Any individual using *Calculus with Analytic Geometry*, Fourth Edition, by Larson et al. is entitled to substantial discounts off *The Calculus Series* (student version) whose retail price is $49.95 per videotape:

- Purchase Tape I of either or both semesters' six-part package for only $14.95 each—*save 70%!*
- Purchase each six-part package for only $124.95 (a $175.00 savings off the regular price of $299.95) and receive a comprehensive workbook and answer key at no cost.
- Receive a $5.00 rebate for each tape returned in usable condition within a year of its purchase.
- Receive a no-risk 30-day money back guarantee upon placing your order.

Prices subject to change. For more information or rush orders, call **1-800-USA-MATH.**

D. C. Heath and Company
Educational Software
presents

INTERACTIVE CALCULUS

Chris Avery/Frank Soler of The Math Lab

System Requirements: IBM-PC and compatibles, 256K, DOS 2.0 or higher, color monitor, CGA or Hercules graphics card
Apple II (48K plus language card) or IIe (64K)

Package: One disk, User's Manual, and a set of 80 lab assignments keyed to Calculus, Fourth Edition and Calculus, Alternate Edition by Larson, Hostetler and Edwards.

Features:
- Graphing and computational program offers you power and flexibility
- Keyed to the Larson **Calculus** texts and can be used to solve every computer/calculator exercise in the books.
- Features two- and three-dimensional graphing
- Designed for the three-semester course
- 80 lab assignments feature comprehensive topic coverage, worksheet format for solutions, and models of mathematical situations to reinforce concepts
- Easy toggling between graph and menu.

$22.99

Available for IBM (3 1/2" and 5 1/4") and Apple

*Suggested list price; bookstore price may vary slightly.

Students: Order directly through your bookstore or clip the coupon on the last page of this special section.

Professors: For information regarding availability of site licensing, networking, and/or quantity discounts, please contact your local D. C. Heath representative.

D. C. Heath and Company
Educational Software

presents

MATH UTILITIES

CURVES • SURFS • DIFFS

Mark Bridger

System Requirements:
IBM-PC or compatibles, 256K (320K for high resolution output), DOS 2.0 or higher, CGA, EGA, Hercules, monochrome graphics adapter or compatible.

Package:
One disk and User's Guide
The guide explains the features as well as a walk-through "tutorial".

Features:
- Highly praised program (now available as Version 4) consists of three programs - CURVES, for plotting functions of a single variable, SURFS, for plotting mathematical surfaces, and DIFFS, for graphically displaying ordinary first order differential equations.

CURVES

Plots functions of a single variable in either rectangular, polar, or parametric form • Provides a movable magnification window so that user can place box over area of interest and magnify area to fill entire screen • Handles functions which have vertical asymptotes • Draws axes and tick marks

SURFS

Plots mathematical surfaces $z = f(x,y)$ in 3-D with hidden lines removed and coordinate axes displayed where visible. • Rotates a surface around any of the axes through any angle • Scales an image to fit the screen either automatically or manually.

DIFFS

Displays graphically ordinary first-order differential equations and their solutions providing an intuitive understanding of concepts • Permits drawing of many solution curves on the same flow field and allows for quantitative solutions • Visuals allow for explaining differential equations to non-technically trained audience.

MATH UTILITIES (Save! All 3 programs) - $75.00

SURFS - $30.00 • CURVES - $30.00 • DIFFS - $25.00

Available for IBM-PC (3 1/2" and 5 1/4")

Students:
For single copies, order from **Bridge Software** at;

P. O. Box 118
New Town Branch
Boston, MA 02258

Checks or money orders only payable to Bridge Software (U.S. dollars only)
Add $3.00 for Shipping and Handling. Massachusetts orders, please include 5% Sales Tax.

Professors:
If you are using a D. C. Heath mathematics text, please contact your local representative for details on availability of site licensing, networking, and quantity discounts.

IBM is a registered trademark of International Business Machines Corporation.

D. C. Heath and Company
Educational Software
presents

BESTGRAPHER

George Best

System Requirements: IBM-PC or compatibles, 256K, DOS 2.0 or higher, CGA, VGA, or EGA graphics card
Macintosh Plus, SE or II, 512K

Package: IBM-PC (One disk - 3 1/2" or 5 1/4"), User's Guide and Workbook
Macintosh (One 3 1/2" disk), User's Guide and Workbook

The worktext, Using BestGrapher, provides activities and problems that encourage experimentation and discovery about functions, derivatives, areas, and other key topics.

Features:
- Emphasizes concepts so that you spend more time understanding calculus and less time on rote learning. Capabilities include:
  - 2-D and 3-D function grapher • finding zeros over an interval • graphing inequalities • practice with limit concept (IBM-PC version searches for delta for any epsilon supplied) • simulation on calculating volume (rotating an interval of a graph about a horizontal or vertical line and graphically approximating the volume swept out) • graphing parametric equations • graphing polar equations • integration • symbolic differentiation • drawing secants and tangents; drawing line to a function graph • evaluates functions for specified arguments • understands pi and a large library of functions • color graphics and printouts possible with appropriate hardware

$35.99

Available for IBM (3 1/2" and 5 1/4") and Macintosh
*Suggested list price; bookstore price may vary slightly.

Students: Order directly through your bookstore or clip the coupon on the last page of this special section.

Professors: For information regarding availability of site licensing, networking, and/or quantity discounts, please contact your local D. C. Heath representative.

D. C. Heath and Company
Educational Software
presents

WORK-OUT!

Joe Mazur

System Requirements: IBM-PC or compatibles, 256K, DOS 2.0 or higher
CGA card helpful but not required

Package: 3 1/2" version - 2 disks and User's Guide
5 1/4" version - 4 disks and User's Guide

Features:

- Keyed on-screen with problems covering topics from the first twelve chapters of the Fourth and Fourth Alternate editions of **Calculus** by Larson/Hostetler/Edwards.

- Generates problems at four levels of difficulty and adjusts the level to <u>your</u> level of accomplishment.

- Problems include on-line hints and guidance, as well as a handy reference tutorial of helpful facts for problem-solving.

- <u>Exercise</u> your skills with the following topics:
  - Functions and Equations • Limits and Continuity • Differentiation and Rates of Change • Applications of Differentiation • Integration and the Fundamental Theorem • Inverse Functions • Applications of Integration • Integration Techniques • L'Hopital's Rule and Improper Integrals • Infinite Series • Conic Sections • Plane Curves, Parametric Equations, and Polar Coordinates • Vectors and the Geometry of Space

$35.99

Available for IBM (3 1/2" and 5 1/4"), includes User's Guide

*Suggested list price; bookstore price may vary slightly.

Students: Order directly through your bookstore or clip the coupon on the last page of this special section.

Professors: For information regarding availability of site licensing, networking, and/or quantity discounts, please contact your local D. C. Heath representative.

D. C. Heath and Company
Educational Software

Please contact your bookstore manager to place an order or clip this coupon and mail to D. C. Heath.

Interactive Calculus

____	IBM 3 1/2" (CGA)	21839-1	$22.99	_____
____	IBM 3 1/2" (Hercules)	21501-5	$22.99	_____
____	IBM 5 1/4" (CGA)	21838-3	$22.99	_____
____	IBM 5 1/4" (Hercules)	21500-7	$22.99	_____
____	Apple	21502-3	$22.99	_____

BestGrapher

____	IBM 3 1/2"	24650-6	$35.99	_____
____	IBM 5 1/4"	24651-4	$35.99	_____
____	Macintosh	24652-2	$35.99	_____

Work-Out!

____	IBM 3 1/2"	24663-8	$35.99	_____
____	IBM 5 1/4"	24664-6	$35.99	_____

Computer Activities for Calculus

____	IBM 3 1/2"	19602-9	$28.99	_____
____	IBM 5 1/4"	10312-8	$28.99	_____

Subtotal _____

Shipping and Handling _____
(Please add $3.50 per software package ordered.)

Sales Tax _____
(Where applicable, please add your state tax.)

Amount Enclosed _____
(Payable only in U.S. dollars)

Please allow 30 days for shipping and handling. The price shown is a suggested price and may differ from the bookstore price. Orders placed with Heath must include Shipping and Handling and Sales Tax.

Method of Payment ☐ check ☐ Mastercard ☐ Visa

Credit Card # _____ Exp. _____

Signature _____

Name _____ Phone _____

Address _____

City/State/Zip Code _____

Course Name and Number _____ Professor's Name _____

Please mail coupon to:
D. C. Heath and Company
Order Entry
2700 North Richardt Avenue
Indianapolis, IN 46219